Archaea

Archaea

Evolution, Physiology, and Molecular Biology

Edited by Roger A. Garrett and Hans-Peter Klenk

Blackwell Publishing

© 2007 by Blackwell Publishing Ltd
except for editorial material and organization © 2007 by Roger Garrett and
Hans-Peter Klenk

BLACKWELL PUBLISHING
350 Main Street, Malden, MA 02148-5020, USA
9600 Garsington Road, Oxford OX4 2DQ, UK
550 Swanston Street, Carlton, Victoria 3053, Australia

The right of Roger Garrett and Hans-Peter Klenk to be identified as the Authors of the Editorial Material in this Work has been asserted in accordance with the UK Copyright, Designs, and Patents Act 1988.

All rights reserved. No part of this publication may be reproduced, stored in a retrieval system, or transmitted, in any form or by any means, electronic, mechanical, photocopying, recording or otherwise, except as permitted by the UK Copyright, Designs, and Patents Act 1988, without the prior permission of the publisher.

First published 2007 by Blackwell Publishing Ltd

1   2007

Library of Congress Cataloging-in-Publication Data

Archaea : evolution, physiology, and molecular biology / edited by Roger Garrett and
   Hans-Peter Klenk.
      p. ; cm.
   Includes bibliographical references and index.
   ISBN-13: 978-1-4051-4404-9 (hardback : alk. paper)
   ISBN-10: 1-4051-4404-1 (hardback : alk. paper)
   1. Archaebacteria—Genetic aspects. 2. Molecular biology.
I. Garrett, Roger A. (Roger Antony), 1943– .   II. Klenk, Hans-Peter.
   [DNLM: 1. Archaea. 2. Archaea—genetics. 3. Evolution, Molecular.
QW 50 A669 2007]
QR82.A69A73 2007
579.3′21—dc22

2006009421

A catalogue record for this title is available from the British Library.

Set in 10 on 12 pt Meridien
by SNP Best-set Typesetter Ltd, Hong Kong
Printed and bound in Singapore
by Fabulous Printers Pte Ltd

The publisher's policy is to use permanent paper from mills that operate a sustainable forestry policy, and which has been manufactured from pulp processed using acid-free and elementary chlorine-free practices. Furthermore, the publisher ensures that the text paper and cover board used have met acceptable environmental accreditation standards.

For further information on
Blackwell Publishing, visit our website:
www.blackwellpublishing.com

Contents

	List of contributors	vii
	Preface	xi
1	Introductory Chapter **The birth of the Archaea: a personal retrospective** *Carl R. Woese*	1
2	Review Natural history of the archaeal domain *Patrick Forterre, Simonetta Gribaldo and Celine Brochier-Armanet*	17
3	Specialist Article The root of the tree: lateral gene transfer and the nature of the domains *David A. Walsh, Mary Ellen Boudreau, Eric Bapteste and W. Ford Doolittle*	29
4	Review Diversity of uncultivated Archaea: perspectives from microbial ecology and metagenomics *Christa Schleper*	39
5	Specialist Article Nanoarchaeota *Harald Huber and Reinhard Rachel*	51
6	Review Families of DNA viruses infecting hyperthermophilic Crenarchaea *David Prangishvili*	59
7	Review Features of the genomes *Hans-Peter Klenk*	75
8	Specialist Article *Sulfolobus* genomes: mechanisms of rearrangement and change *Kim Brügger, Xu Peng and Roger A. Garrett*	95
9	Specialist Article Plasmids *Georg Lipps*	105
10	Specialist Article Integration mechanisms: possible role in genome evolution *Qunxin She, Haojun Zhu and Xiaoyu Xiang*	113
11	Review Genetics *Moshe Mevarech and Thorsten Allers*	125
12	Specialist Article Genetic properties of *Sulfolobus acidocaldarius* and related Archaea *Dennis W. Grogan*	137
13	Review Chromatin and regulation *John N. Reeve and Kathleen Sandman*	147
14	Review DNA replication and the cell cycle *Victoria L. Marsh and Stephen D. Bell*	159
15	Review DNA repair *Malcolm F. White*	171
16	Review Transcriptional mechanisms *Michael Thomm and Winfried Hausner*	185
17	Specialist Article Transcriptional regulation in Haloarchaea *Felicitas Pfeifer, Torsten Hechler, Sandra Scheuch and Simone Sartorius-Neef*	199

18 **Specialist Article**
Aminoacyl-tRNAs: deciphering and
defining the genetic message 207
*Alexandre Ambrogelly, Juan Carlos Salazar,
Kelly Sheppard, Carla Polycarpo,
Hiroyuki Oshikane, Yuko Nakamura,
Shuya Fukai, Osamu Nureki and Dieter Söll*

19 **Specialist Article**
Translational mechanisms and protein
synthesis 217
Paola Londei

20 **Review**
Expanding world of small noncoding
RNAs in Archaea 229
*Arina Omer, Maria Zago and
Patrick P. Dennis*

21 **Specialist Article**
Transcriptomics, proteomics, and
structural genomics of *Pyrococcus furiosus* 239
*Michael W. W. Adams, Francis E. Jenney Jr,
Chung-Jung Chou, Scott Hamilton-Brehm,
Farris L. Poole II, Keith R. Shockley,
Sabrina Tachdjian and Robert M. Kelly*

22 **Specialist Article**
The glycolytic pathways of Archaea:
evolution by tinkering 247
John van der Oost and Bettina Siebers

23 **Specialist Article**
Metabolism of inorganic sulfur
compounds in Archaea 261
Arnulf Kletzin

24 **Specialist Article**
Methyl-coenzyme M reductase in
methanogens and methanotrophs 275
Rudolf K. Thauer and Seigo Shima

25 **Specialist Article**
Methylation of metal(loid)s by
Methanoarchaea: production of volatile
derivatives with high ecotoxicological
impact and health concern 285
*Klaus Michalke, Jörg Meyer and
Reinhard Hensel*

26 **Review**
Biotechnology 295
*Ksenia Egorova and
Garabed Antranikian*

Wolfram Zillig 323
References 325
Index 377

Contributors

M. W. W. Adams Department of Biochemistry and Molecular Biology, University of Georgia Athens, Georgia 30602-7229, USA

T. Allers Institute of Genetics, University of Nottingham, Queen's Medical Centre, Nottingham NG7 2UH, United Kingdom

A. Ambrogelly Department of Molecular Biophysics and Biochemistry, Yale University, New Haven, Connecticut 06520-8114, USA

G. Antranikian Institute of Technical Microbiology, Hamburg University of Technology, Kasernenstraße 12, 21073 Hamburg, Germany

E. Bapteste Department of Biochemistry and Molecular Biology, Dalhousie University, 5850 College Street, Halifax, Nova Scotia B3H 1X5, Canada

S. D. Bell MRC Cancer Cell Unit, Hutchison MRC Research Centre, Hills Road, Cambridge CB2 2XZ, United Kingdom

M. E. Boudreau Department of Biochemistry and Molecular Biology, Dalhousie University, 5850 College Street, Halifax, Nova Scotia B3H 1X5, Canada

C. Brochier-Armanet EA EGEE (Evolution, Génome, Environnement), Université Aix-Marseille I, Centre Saint-Charles, Case 36, 3 Place Victor Hugo, 13331 Marseille, Cedex 3, France

K. Brügger Danish Archaea Centre, Institute of Molecular Biology, Copenhagen University, Sølvgade 83H, DK-1307 Copenhagen K, Denmark

C.-J. Chou Department of Chemical and Biomolecular Engineering, North Carolina State University, Raleigh, NC 27695-7905, USA

P. P. Dennis Division of Molecular and Cellular Biosciences, National Science Foundation, 4201 Wilson Blvd., Alexandria, Virginia 22230, USA

W. F. Doolittle Department of Biochemistry and Molecular Biology, Dalhousie University, 5850 College Street, Halifax, Nova Scotia B3H 1X5, Canada

K. Egorova Institute of Technical Microbiology, Hamburg University of Technology, Kasernenstraße 12, 21073 Hamburg, Germany

P. Forterre Unité Biologie Moléculaire du Gène chez les Extremophiles, Institut Pasteur, 25 rue du Dr. Roux, 75724 Paris Cedex 15, France

S. Fukai Department of Biological Information, Graduate School of Bioscience and Biotechnology, Tokyo Institute of Technology, Yokohama-shi, Kanagawa 226-8501, Japan

R. A. Garrett Danish Archaea Centre, Institute of Molecular Biology, Copenhagen University, Sølvgade 83H, DK-1307 Copenhagen K, Denmark

S. Gribaldo Unité Biologie Moléculaire du Gène chez les Extremophiles, Institut Pasteur, 25 rue du Dr. Roux, 75724 Paris Cedex 15, France

D. W. Grogan Department of Biological Sciences, 614 Rieveschl Hall, ML 0006, University of Cincinnati, Cincinnati, Ohio 45221-0006, USA

S. Hamilton-Brehm Department of Biochemistry and Molecular Biology, University of Georgia Athens, Georgia 30602-7229, USA

W. Hausner University of Regensburg, Department of Microbiology, Universitätsstraße 31, 93053 Regensburg, Germany

T. Hechler Institute of Microbiology and Genetics, Darmstadt University of Technology, Schnittspahnstrasse 10, 64287 Darmstadt, Germany

R. Hensel Department of Microbiology and Geography, Universität Duisburg-Essen, Campus Essen, Universitätsstraße 5, 45117 Essen, Germany

H. Huber University of Regensburg, Department of Microbiology and Archaea Center, Universitätsstraße 31, 93053 Regensburg, Germany

F. E. Jenney Jr Department of Biochemistry and Molecular Biology, University of Georgia Athens, Georgia 30602-7229, USA

R. M. Kelly Department of Chemical and Biomolecular Engineering, North Carolina State University, Raleigh, NC 27695-7905, USA

H.-P. Klenk e.gene Biotechnologie, Pöckinger Fußweg 7a, 82340 Feldafing, Germany

A. Kletzin Institute of Microbiology and Genetics, Darmstadt University of Technology, Schnittspahnstrasse 10, 64287 Darmstadt, Germany

G. Lipps University of Bayreuth, Biochemistry, 95440 Bayreuth, Germany

P. Londei Department of Medical Biochemistry, Biology and Physics (DIBIFIM), Università degli Studi di Bari (Policlinico), Bari, Italy

V. L. Marsh MRC Cancer Cell Unit, Hutchison MRC Research Centre, Hills Road, Cambridge CB2 2XZ, United Kingdom

J. Meyer Department of Microbiology and Geography, University Duisburg-Essen, Campus Essen, Universitätsstraße 5, 45117 Essen, Germany

M. Mevarech Department of Molecular Microbiology & Biotechnology, Tel-Aviv University, Tel-Aviv 69978, Israel

K. Michalke Department of Microbiology and Geography, University Duisburg-Essen, Campus Essen, Universitätsstraße 5, 45117 Essen, Germany

Y. Nakamura Department of Biological Information, Graduate School of Bioscience and Biotechnology, Tokyo Institute of Technology, Yokohama-shi, Kanagawa 226-8501, Japan

O. Nureki Department of Biological Information, Graduate School of Bioscience and Biotechnology, Tokyo Institute of Technology, Yokohama-shi, Kanagawa 226-8501, Japan

A. Omer Department of Biochemistry and Molecular Biology, University of British Columbia, 2146 Health Sciences Mall, Vancouver, British Columbia V6T 1Z3, Canada

H. Oshikane Department of Biological Information, Graduate School of Bioscience and Biotechnology, Tokyo Institute of Technology, Yokohama-shi, Kanagawa 226-8501, Japan

X. Peng Danish Archaea Centre, Institute of Molecular Biology, Copenhagen University, Sølvgade 83H, DK-1307 Copenhagen K, Denmark

F. Pfeifer Institute of Microbiology and Genetics, Darmstadt University of Technology, Schnittspahnstrasse 10, 64287 Darmstadt, Germany

C. Polycarpo Department of Molecular Biophysics and Biochemistry, Yale University, New Haven, Connecticut 06520-8114, USA

F. L. Poole II Department of Biochemistry and Molecular Biology, University of Georgia Athens, Georgia 30602-7229, USA

D. Prangishvili Unité de Biologie Moléculaire du Gène chez les Extrêmophiles, Institut Pasteur, rue Dr. Roux 25, 75724 Paris Cedex 15, France

R. Rachel University of Regensburg, Department of Microbiology and Archaea Center, Universitätsstraße 31, 93053 Regensburg, Germany

J. N. Reeve Department of Microbiology, Ohio State University, Columbus, Ohio 43210, USA

J. C. Salazar Department of Molecular Biophysics and Biochemistry, Yale University, New Haven, Connecticut 06520-8114, USA

K. Sandman Department of Microbiology, Ohio State University, Columbus, Ohio 43210, USA

S. Sartorius-Neef Institute of Microbiology and Genetics, Darmstadt University of Technology, Schnittspahnstrasse 10, 64287 Darmstadt, Germany

S. Scheuch Institute of Microbiology and Genetics, Darmstadt University of Technology, Schnittspahnstrasse 10, 64287 Darmstadt, Germany

C. Schleper University of Bergen, Department of Biology, Jahnebakken 5, N-5020 Bergen, Norway

K. Sheppard Department of Molecular Biophysics and Biochemistry, Yale University, New Haven, Connecticut 06520-8114, USA

S. Shima Max-Planck-Institute for Terrestrial Microbiology, Karl-von-Frisch-Strasse, 35043 Marburg, Germany

Q. She Danish Archaea Centre, Institute of Molecular Biology and Physiology, Copenhagen University, Sølvgade 83H, DK-1307 Copenhagen K, Denmark

K. R. Shockley Department of Chemical and Biomolecular Engineering, North Carolina State University, Raleigh, NC 27695-7905, USA

B. Siebers Department of Microbiology and Geography, University Duisburg-Essen, Campus Essen, Universitätsstraße 5, 45117 Essen, Germany

D. Söll Departments of Molecular Biophysics and Biochemistry and Chemistry, Yale University, New Haven, Connecticut 06520-8114, USA

S. Tachdjian Department of Chemical and Biomolecular Engineering, North Carolina State University, Raleigh, NC 27695-7905, USA

R. K. Thauer Max-Planck-Institute for Terrestrial Microbiology, Karl-von-Frisch-Strasse, 35043 Marburg, Germany

M. Thomm University of Regensburg, Department of Microbiology, Universitätsstraße 31, 93053 Regensburg, Germany

J. van der Oost Laboratory of Microbiology, Wageningen University, Hesselink van Suchtelenweg 4, 6703 CT Wageningen, The Netherlands

D. A. Walsh Department of Biochemistry and Molecular Biology, Dalhousie University, 5850 College Street, Halifax, Nova Scotia B3H 1X5, Canada

M. F. White Centre for Biomolecular Sciences, University of St Andrews, St Andrews, Fife KY16 9ST, United Kingdom

C. R. Woese Department of Microbiology, University of Illinois, 601 South Goodwin Ave., Urbana, Illinois 61801-3709, USA

X. Xiang Danish Archaea Centre, Institute of Molecular Biology and Physiology, Copenhagen University, Sølvgade 83H, DK-1307 Copenhagen K, Denmark

M. Zago Department of Biochemistry and Molecular Biology, University of British Columbia, 2146 Health Sciences Mall, Vancouver, British Columbia V6T 1Z3, Canada

H. Zhu Danish Archaea Centre, Institute of Molecular Biology and Physiology, Copenhagen University, Sølvgade 83H, DK-1307 Copenhagen K, Denmark

Preface

The inspiration for this book came from an international meeting in Hohenkammer Castle, near Munich, entitled "Archaea – The First Generation" held on the 2–4 June 2005. It followed on from earlier Meetings in the Munich area, the first of which, on "Archaebacteria", was held in June 1981. The earlier meetings monitored the rapid and exciting developments which followed on from the discovery, by Carl Woese and colleagues, of a completely new Domain of life, the Archaea, in 1977 (Photo 1). These new developments rapidly attracted a broad range of microbiologists, molecular biologists, chemists, geologists and paleontologists who created a very stimulating and creative research environment and community.

As described in Chapter 1, many of the early results, and ensuing hypotheses, met with a great deal of skepticism from the well established and conservative, bacterial and eukaryotic research communities. By 1990, however, the experimental evidence in support of a third Domain was overwhelming and has been strengthened regularly since, especially through analyses of whole genome sequences.

The purpose of the book was to provide a useful reference work for introducing younger scientists, and newcomers, to archaeal research, as well as for researchers who are more established in the field. Some of the chapters provide reviews of the more mature research areas, while newly emerging fields are covered by shorter, more specialized, chapters. Given the rapid developments over the past few years, we could not do justice to all of archaeal research in one book. What we have done is to provide a broad coverage of the research emphasising the more important and exciting developments that are taking place.

Photo 1 Carl R. Woese in front of Sanger oligonucleotide patterns from archaeal 16S rRNA taped to a back-lit translucent wall. Source: Dave Graham (Univ. of Texas).

Photos 2 and 3 Wolfram Zillig and Karl O. Stetter on hyperthermophile sampling trips in Naples, Italy (1978) and on Iceland (1980): Source: K.O. Stetter.

One of the purposes of the recent meeting was to celebrate two distinguished pioneers of Archaea research. Wolfram Zillig who spent most of his working life at the Max-Planck-Institute for Biochemistry in Martinsried, Germany, and Karl O. Stetter who worked until recently at the University of Regensburg, Germany. They are shown together in Italy on one of many sampling trips in photos 2 and 3. Both have made a major impact by isolating and characterizing many novel archaeal species and Wolfram Zillig has also pioneered the study of archaeal viruses and plasmids. Their influence extended well beyond their own laboratories to supporting, encouraging and inspiring colleagues, and especially younger scientists, throughout the world. Unfortunately, Wolfram Zillig died of a long-term illness a few days before the Meeting.

1
The birth of the Archaea: a personal retrospective

Carl R. Woese

Let there be light

For me the moment was, I believe, the afternoon of June 11, 1976. I had just taped the film of a primary "Sanger pattern" to a back-lit translucent "wall" in the lab and had begun to "interpret" the pattern in terms of the "secondary cuts" taken from it, the corresponding films of which were lying on a huge light table directly beneath the "primary"; the object being to infer the sequences of all the oligonucleotides (of significant length) in the primary pattern. Except for this eerie lighting arrangement, the room was fairly dark, with the only prominent features being the pattern of back-lit black spots on the "primary" film and the corresponding transluminated lines of black "sub-cut" spots on the "secondary" film lying below.

The spots on the "primary" film represented specific oligonucleotide fragments into which a (radiolabeled) 16S rRNA (ribosomal RNA) had been cut by T1 ribonuclease, then subjected to a two-dimensional paper electrophoretic separation, with the resulting oligonucleotide "spots" detected by means of X-ray film (Uchida et al., 1974). The "isopleth" pattern on the film of the "Sanger pattern" (Sanger et al., 1965) already revealed a great deal about the sequence(s) of the oligonucleotide(s) in the individual spots; for instance, the length of an oligo, the number of uracil residues it contained (a primary determinant of the overall structure of the isopleth pattern), and the C (cytosine) versus A (adenine) contents of the individual spots in each isopleth (Sanger et al., 1965). (Each oligonucleotide had but one G residue, at its 3' end, the cut site of ribonuclease T1 (Sanger et al., 1965)).

My job was to determine the complete sequence of every oligonucleotide of significant length (five or more nucleotides) in the primary pattern, which required the aforementioned "secondary" patterns. These in turn were created by removing little snippets of paper at the appropriate places in the corresponding original electrophoretogram and further digesting the oligonucleotide(s) therein (*in situ*) with one or a few ribonucleases of different cutting specificities than that of T1 RNAse (thereby creating sub-fragments). These enzymatically treated snippets were then individually reinserted (mashed) into a very large sheet of (DEAE cellulose) paper (about 30 of them in a line near the "bottom" of such a sheet). Each large "secondary" sheet was then subject to one-dimensional electrophoresis to resolve the sub-fragments in each of the 30-odd secondary digestions from one another. From the one or several "secondary" cuts taken from a primary spot, the exact sequence of the oligonucleotide(s) in the corresponding primary spot could (almost always) be deduced (Uchida et al., 1974).

"Reading" a Sanger pattern in this fashion was painstaking work, requiring a good fraction of the day to work up a single "primary," something I at the time had been doing for several days a week off and on for a long time. It was routine work, boring, but demanding full concentration. (There were days when I would walk home from work saying to myself: "Woese, you have destroyed your mind again today.") But this day was special: I and biology were in for a surprise. First, however, more background.

Starting down the Yellow Brick Road

I had had an abiding interest in the translation process since the latter part of the 1950s; first with the ribosome and its subunits, then, starting in 1960, with the genetic code – **the** hot topic of molecular biology at the time. The code had come into prominence on the heels of Watson and Crick's two world-shaking 1953 publications. The physicist George Gamow thought he could see "pockets" in the double stranded structure of DNA, pockets of just the right size and spacing to hold and discriminate among amino acids, suggesting the basis for a direct templating mechanism upon which translation could be based (Watson & Crick, 1953; Gamow, 1954).

Then came a thrilling but brief period when a clique of physicists and molecular biologists worked together and competed to see who would be first to derive the "code" from "first principles." The prospect of theoretically solving the genetic code, the "language of life," was so seductive that cameo appearances on the coding stage were made by Feynmann and Teller (no doubt prompted by the charismatic Gamow). The decoders soon split into two camps, however, those who, like Gamow, believed that the basis of the code lay in specific recognition of amino acids by nucleic acids, and those who, like Francis Crick, believed it impossible that nucleic acids could recognize anything except other nucleic acids/nucleotides, which they did through base pairing (F. H. C. Crick, unpublished letter to the RNA Tie Club; see Judson, 1996). When I belatedly entered the area, my intuition sided with Gamow.

However, I differed from the whole lot of them in perceiving the nature of the code as inseparable from the problem of the nature and origin of the decoding mechanism. Thus, translation to me was **the** central biological concern. It represented one of a new class of major evolutionary problems that molecular probings of the cell were bringing to light. Now was the time to start thinking about the evolution of the cell and its macromolecular componentry. How this evolution occurred is almost as much a mystery today as it was four decades ago. But one thing was certain from the start: approaching these sorts of "deep" universal evolutionary problems would require a universal phylogenetic framework within which to work effectively. Since no universal phylogeny then existed – our understanding of evolutionary relationships being effectively confined to plants and animals – this meant taking on the rather large task of determining genealogical relationships for the microbial world, the bacteria and single-celled eukaryotes, which, as it turned out, meant determining the missing 95% or more of the "tree of life." A slight diversion in my research program would be necessary – a diversion that lasted a good two decades!

A method for my madness

In 1965, on his way to developing nucleic acid sequencing technology, Fred Sanger had spun off an "oligonucleotide cataloging" methodology (Sanger et al., 1965). This procedure, applied to ribosomal RNA (the small subunit rRNA, it turned out), was exactly what we needed to determine genealogical relationships across the entire breadth of the phylogenetic spectrum. It was already apparent from DNA–rRNA hybridization studies that the sequence of a ribosomal RNA tended to be highly conserved, probably to the point that recognizable sequence similarity would extend across the full taxonomic spectrum (Yankofsky & Spiegelman, 1962). Ribosomal RNAs are obviously ubiquitous; they occur in the cell in thousands of copies; and they can be radiolabeled and isolated with relative ease. In addition, they are functionally about as constant as one could wish for – they are **not** adaptive characters. And last but not least, rRNAs are integral parts of a complex, integrated molecular aggregate (genetically dispersed within the genome), which would make them as insensitive as can be to the vicissitudes of reticulate evolution (Fox et al., 1977a). Only technological problems seemed to stand in our way: growing the various organisms and doing so in a low phosphate, radioactive medium; tweaking the Sanger method to fit our needs; finding needed help; and so on. Scientists do not want to, or often cannot, create all the things needed in their work. In our case the chief problem was the organisms required for the project. Half of them at least would be too fastidious for anyone but an expert to grow. Striking up collaborations with experts in the culture of particular organisms was essential.

Learning our way around

Coming into the game as a biophysicist/molecular biologist, my knowledge of bacteria and bacteriologists didn't extend far beyond *E. coli*, *Bacillus*, and Louis Pasteur; and I didn't have the foggiest notion of how bacteria were related to one another. It was time to ask real microbiologists for help in choosing the right organisms. Each, of course, had a different opinion (the bacteria they themselves worked with, that is). At that stage, I didn't know that actually there **were** no experts on bacterial relationships (those above the level, say, of genus and occasionally family, that is). And I was completely unaware of the bizarre state that the microbiologists' search for these relationships had gotten itself into.

The best advice I had solicited regarding organism choice came from my colleague in the Microbiology Department at Illinois, Ralph Wolfe. By now I had gotten used to microbiologists suggesting that we work on their favorite bugs, and in this respect, Wolfe was no different. But what he had to say was; his advice was more compelling than any other I had received! I can almost remember his words: he told me the methanogens were united as a group by a unique biochemistry that involved a set of unusual coenzymes. Yet the organisms showed no uniformity in their morphologies, which latter fact had caused taxonomists initially to scatter them throughout the various taxa in the seventh edition of *Bergey's Manual* (Breed et al., 1957). (In the eighth edition, however, they had all been grouped on the basis of their common biochemistry; Murray, 1974.) Finally, here was the kind of phylogenetic challenge I was hoping for. I longed to characterize a methanogen rRNA as soon as possible. But it wasn't possible – at least not yet. Wolfe and I had spoken in early 1974 (if I recall correctly), and the technology needed for growing and radiolabeling the methanogens safely was not at that time in place. Now, back to the main thread.

Epiphany!

By the beginning of 1976 my lab had "cataloged" (generated T1 oligonucleotide lists) for roughly 30 organisms, mainly "prokaryotes" and a smattering of eukaryotes. It had become obvious that the two groups could be readily distinguished from each other on the basis of "oligonucleotide signatures," which were lists of oligonucleotides characteristic of one of the two groups to the exclusion of the other. The two apposing oligonucleotide signatures were remarkably distinct. (In addition, a set of "universal" oligonucleotides existed, those found in all the rRNA catalogs we had so far generated.) In working up a Sanger pattern for an organism, one had only to "read" a small number of oligos into it before being able to smile and say: "Oh, that's a prokaryote," or "that's a euk." There were two spots on the primary films of all prokaryotic rRNAs that easily caught one's eye, for they contained modified nucleotides and, so, were located at places in the Sanger pattern where normally there would be no oligonucleotides. These "odd oligos" allowed one to declare "prokaryote" at first glance: after that, it was just a matter of detailing the rest of the pattern to figure out the relationship of the new prokaryote to ones already cataloged.

By 1976 Wolfe and his student Bill Balch had developed a technique for growing methanogens (in pressurized serum bottles) that was sterile, fast, and (most important from our point of view) safe enough to permit cells to be radiolabeled (Balch & Wolfe, 1976). George Fox, then my post-doc, had known Bill from a course they had taken together at Woods Hole the summer before George arrived at Illinois. Their acquaintance made it easy for George to approach Bill about a collaboration to work on methanogens – which George did on his own initiative in the year the Balch–Wolfe method was being published. It was on that aforementioned day in June 1976 that I began to read the Sanger pattern produced (by my technician Linda Magrum) from George and Bill's first successful methanogen rRNA prep. The formal name of the organism was *Methanobacterium thermoautotrophicum*, a 14-syllable monstrosity that was always shortened to "ΔH," the organism's strain designation (Zeikus & Wolfe, 1972).

From the get-go ΔH's Sanger pattern was strange. First of all the two small "odd" oligos on the primary pattern that screamed out "prokaryote" were absent. Intrigued by this appetizer, but afraid to make too much of it, I quickly jumped into the "G isopleth" (oligonucleotides that lack a uracil residue), hoping to find the first of the prokaryotic signature oligos, which would certainly set things back on the prokaryote track! Imagine my surprise when that "signature" oligo was missing as well. Not only that, but the G-isopleth contained the rather large 3' terminal oligonucleotide of this 16S rRNA, which **did not** belong there! What was going on? This methanogen rRNA was not feeling prokaryotic. The more oligos I sequenced, the less prokaryotic it felt, as signature oligo after prokaryotic signature oligo failed to turn up. However, a number of them were still there, as, surprisingly, were some oligos from the eukaryotic signature, and, thankfully, quite a few of the oligos we had considered universal in distribution. What was this RNA? It was not that of a prokaryote. It was not eukaryotic. Nor was it from Mars (because of the "universals"). Then it dawned on me. Was there something out there other than prokaryotes and eukaryotes – perhaps a distant relative of theirs that no one had realized was there? Why not? But the idea surely wasn't in keeping with conventional wisdom.

I rushed to share my out-of-biology experience with George, a skeptical George Fox to be sure. George was always skeptical. That is what made him a good scientist; and because of that, whatever skepticism he initially evinced quickly dissipated. Yes, he agreed, there probably was something else out there: it wasn't just prokaryotes and eukaryotes all the way down. That was a heady thought, novel enough that we sensed trouble in trying to convince other biologists of the idea. Little did we know how **much** trouble there would be.

A finding like this you do not immediately go out and shout about. You had better have all your ducks in a row and have firmer evidence than would otherwise be needed. We went into fast-forward mode: by the end of the year (1976) we had five additional methanogen catalogs in hand, with more on the way. They would cover all the disparate morphologies associated with the known methanogens. And sure enough, none of the new catalogs was "prokaryotic" (or eukaryotic): and they were all of a kind. The methanogens represented a new highest level taxonomic grouping, which could be defined by a characteristic oligonucleotide signature. (And, tellingly, that signature was no more extensive than were those of the "prokaryotes" or the eukaryotes, implying that within whatever new taxon they represented, the methanogens were quite a highly diverged group.) Darwin had long ago said that there would come a day when there would be "very fairly true genealogical trees of each great kingdom of nature." Perhaps that day was at hand. In any case, there was lots of work still to do.

Build it and they will come

A new "urkingdom" (as we were beginning to call our new highest level phylogenetic group) would be a major evolutionary find. It afforded a rare opportunity to put the theory of evolution to serious predictive test. As I have said, ribosomal RNA is a non-adaptive, universal character. That's what makes its sequence so good for tracing organismal genealogies. It is also what makes it completely uninformative as regards the **phenotypes** of the organisms whose genealogies it traces. Evolution to Darwin was genealogical as descent with variation. In our case there had been a long trail of descent, and, therefore, we should find a comparably huge amount of variation, not only quantitative but qualitative. In other words, there had to be important features characteristic of our new urkingdom that distinguished it sharply from the rest of the living world, and there should be impressive variation among the species in the urkingdom as well. Testing these two main evolutionary predictions drove our work from that point on.

There had already been a promising sign in the biochemistry of the methanogens. We now knew for sure that methanogenesis was indeed confined to a particular phylogenetically defined group of organisms and that the process utilized a set of coenzymes that apparently were found no where else (Balch & Wolfe, 1979). Doubly good. But with their highly specific, restricted distribution these coenzymes were going to be of no use to us in searching for non-methanogenic members of our new urkingdom.

The big question was then: where are the predicted other phenotypes? How could they be found? As it turned out, a couple of the sought-for non-methanogenic archaeal phenotypes lay unknown on our "to-do" list. *Thermoplasma* (which had been described as a mycoplasma growing freely under hot and acid conditions) was among the mycoplasmas tabbed for rRNA characterization in a collaboration with Jack Maniloff (University of Rochester), a collaboration that had begun in earnest in the spring of 1975 (with *Mycoplasma gallisepticum*). And there were the extreme halophiles, whose unusual obligate high salt growth conditions made them obvious candidates for the phylogeny project in any case. I have often mused in recent times about how the history of the Archaea would have played out had their entrance into our world been through one of these other portals. No doubt things would have gone very differently.

Fortunately, the job of hunting for other phenotypes that might belong to the new urkingdom turned out to be easier than I imagined. Unbeknownst to George and me, our collaborator Ralph Wolfe had invited the well known German microbiologist Otto Kandler to Urbana for a visit. (Being a molecular biologist, I had never heard of this Kandler fellow – or almost any other well known microbiologist, for that matter – until the day that Ralph marched him into my office to hear the official word from George and myself about the new urkingdom.) According to his records, Otto (as I would soon come to call him) visited Urbana in January 1977, well before publication of our finding.

The German smiled

Amazingly, Kandler – unlike the others we had tried to convince about what our findings meant – wasn't incredulous or disbelieving or anything. I think he smiled. A "third form of life" was fine with him; he had almost expected it. For some time Kandler had, from his own work and that of others, known that the walls of certain prokaryotes were highly atypical. The walls of the bacteria so far characterized had (almost) all contained peptidoglycan; and possession of peptidoglycan-containing cell walls had come to be considered one (of the few positive) unifying characteristics of prokaryotes (Stanier et al., 1963). There had been no systematic examination of bacterial walls before about 1967, although a fair number of them had been characterized on a hit-or-miss basis. What Kandler had known (which we didn't) was that the walls of at least one methanogen and those of some extreme halophiles did **not** contain peptidoglycan (Kandler & Hippe, 1977).

What did these atypical walls imply? On the basis of the cell wall studies alone (the **lack** of peptidoglycan), one could not reliably infer much, especially infer that the peptidoglycan-less organisms constituted a monophyletic grouping unto themselves. (Bacteria atypical in one way or another were often encountered, and sometimes they lacked some property that microbiologists had come to believe was typical of "prokaryotes." But in all such cases, the idiosyncrasy in question had been passed off as adaptation to some unusual environment. It would require much stronger evidence than one or two out-of-the-blue idiosyncrasies to make a microbiologist question a bacterium's prokaryotic pedigree. And as we were soon to find out, there were some among them who would not question that pedigree even when confronted with **strong** evidence.)

Kandler had fully realized the potential significance of the cell wall studies in the light of our rRNA molecular phylogenetic evidence, and had immediately gone back to Germany intent upon fleshing out the comparative study of cell walls – and to spread the word. But his visit had left us with a critical clue in our hunt for novel archaebacterial phenotypes: if unusual cell walls meant anything, perhaps the extreme halophiles would turn out to be members of our new "far out" group. We were desperate to get our hands on cultures of extreme halophiles.

With the help of Jane Gibson (Cornell) we obtained some halophile cultures from the Woods Hole collection. I was not about to wait for a student or a collaboration to come along to grow the organisms. I donned my acid-eaten lab coat (which had hung on the back of my office door for over a decade) and went back to the bench. I grew the cultures myself, turning them over to one of my students, Kenneth Leuhrsen, for the more exacting extraction and isolation procedures for a radiolabeled 16S rRNA; and Ken's prep would as always be given to our trusty Linda for Sanger pattern production. By late spring of 1977, a year after we'd seen our first methanogen catalog, we were gazing at a 16S rRNA catalog from the first extreme halophile. It didn't disappoint. Here was the first non-methanogenic phenotype to join the group – and a novel phenotype it indeed was. But I'm getting ahead of myself. It is time to see how the public – and the biology establishment – reacted when this strange archaeal chimera was loosed upon their world.

Confrontation and heresy

The press release concerning the "third form of life" was set to coincide with the publication of the first of our two papers in the *Proceedings of the National Academy of Sciences*, November 3, 1977. A telling coincidence was to occur. November 3 just so happened to be the date chosen by the then president of the US National Academy of Sciences, Philip Handler, to release an official statement heralding the dawn of the cloning era (signaled by the recent cloning in bacteria of the gene for the growth hormone somatotropin). At the time, no one could see that this fortuitous coincidence foretold the coming battle over biology's future, and the coincident press releases were tantamount to the first skirmish in the ideological struggle that would pit the forces of what would become the biomedical–industrial complex against those representing evolution resurgent. Our "third form of life," which touched upon one of the deepest chords in human nature (i.e. where we came from), completely wiped the press release announcing the era of "Man-the-medical-miracle" off the front pages of the papers. I was overjoyed at the public's appreciation of our work (Fox et al., 1977b; Woese & Fox, 1977b).

But there were already rumblings from the scientific heights. On the day the front page of *The New York Times* announced our discovery of a "third form of life," my colleague Ralph Wolfe received a telephone call from his friend, the Nobel Laureate Salvador Luria (whom he did not initially name), an upset Salvador Luria. According to Wolfe, Luria told him in no uncertain terms to publicly dissociate himself from this scientific fakery or face the ruination of his career. In a recent recounting of the episode (Wolfe, 2001) Ralph said he was so humiliated he "wanted to crawl under something and hide"; but he managed to tell Luria that supporting evidence for the claim had just been published in the *Proceedings of the National Academy of Sciences* (a fact that had appeared in the *New York Times* account). Luria begrudged that he hadn't known about any publication, but that the journal happened to be there on his desk. Fortunately, Ralph left for a planned family gathering out of town the next day (Wolfe, 2001) and thereby escaped further humiliation.

As you might expect, I saw the episode and its overall significance differently. How could this Luria fellow have the temerity to excoriate his friend and my colleague like that? What pedestal was he standing on? It appears that he had blustered at Ralph something to the effect that: "Everybody knows that all bacteria are prokaryotes; there can't be any such thing as a 'third form of life.'" Irony of ironies! As time and the diligence of a particular scientific historian have shown, the fakery lay not in **our** work, but in the prokaryote concept itself (Sapp, 2005): it is now clear that the "prokaryote" was mere guesswork (more on this below). But in the heyday of the prokaryote, which this was, the true believers were out to pillory us for our heresy: how dare we proclaim that the mighty prokaryotic Emperor wasn't wearing scientific clothes?

Expanding the urkingdom

With the halophiles came another clue – lipids, highly unusual ether-linked, branched chain lipids. In my whole career I had never paid attention to lipids, and here we were with lipids on the brain. The fact that extreme halophiles possessed ether-linked lipids had been published in the mid-1960s (Kates et al., 1965; Kates, 1972), but nothing much had been made of it: these strange lipids seemed to be just one more unexplainable biological idiosyncrasy. Microbiologists and biochemists didn't view things evolutionarily. From a genuinely biological perspective, however, these lipids were significant – **evolutionarily** significant.

It was not simply the extreme halophiles that had the strange lipids, either: two other recently isolated bacteria, *Thermoplasma acidophilum* (Darland et al., 1970) and *Sulfolobus acidocaldarius* (Brock et al., 1972) did as well (Langworthy et al., 1972, 1974). Thomas Brock, whose laboratory had first isolated and characterized both organisms, had noted the lipid coincidence. And he later would say that the unusual lipids the two shared was a clear case of **convergent** evolution. In support of that assertion Brock would invoke the ether-linked lipids of the extreme halophiles: "This hypothesis is strengthened by the fact that *Halobacterium*, another quite different organism, also has lipids similar to those of the two acidophilic thermophiles" (Brock, 1978).

Brock's argument doesn't make sense; at least **today** it doesn't. In its time, however, it was reasonable. Like all biologists of that period, Brock firmly believed all bacteria to be "prokaryotes." If both of the thermophiles were prokaryotes and they were not specifically related (as was believed), then their common idiosyncrasy has to represent convergence, independent adaptations to their respective extreme niches. From this perspective the case for convergent evolution is indeed strengthened by pointing to a third prokaryotic species (the extreme halophiles), from a still different extreme niche, that also had similar strange lipids. Brock's conventional mind set here is a wonderful example of what comes of unquestioned acceptance of the assumption that all bacteria are "prokaryotes" – all idiosyncratic characteristics then become **adaptive** changes.

We didn't know what kind of lipids the methanogens had, and that was critical, for the lipids were a good candidate for a property shared by all of the organisms in the new urkingdom. Determining the lipid type of the methanogens was essential. I frequently discussed with Wolfe whom we could enlist to find the answer. Although Brock wasn't the one to approach, one of his colleagues, Tom Langworthy at the University of South Dakota (an expert in lipid analysis), seemed to me a good bet; and it was decided (I thought) to send cells to Langworthy.

Ralph then threw me a curve ball. He had come up with a different candidate, a young professor at Yale who was about to join our microbiology faculty, who was also an expert lipid chemist, name of John Cronan. Ralph felt (rightly, I had to admit) that it would be more collegial for us to work with Cronan. So Ralph ordered a pellet of methanogen cells sent off to John for lipid analysis. Days went by; weeks went by; it seemed like more; and no word from Yale. Ralph finally agreed to call Cronan up and find out what had happened. It turns out that John's lab had done a quick, but definitive, initial screen of the methanogen for lipid type upon receiving the prep and then dropped the project as uninteresting: there was nothing new there; the methanogen lipids were just like the halophile lipids. John had perceived the whole thing from a strictly biochemical perspective. Hadn't Ralph made clear to him how critical methanogen lipids were to our case?

It was back to plan A. Bill Balch was asked to prepare methanogens again, and these were sent to Tom Langworthy and another lipid biochemist, Thomas Tornabene (a colleague and oft-time collaborator of Langworthy's). And in short order there the answer was: methanogens have ether-linked lipids (Tornabene & Langworthy, 1979). The circle had been closed. The first archaebacterial-universal phenotypic property had been found.

Recently Tom L. recalled to me in touching detail our first contact (he has a photographic kind of memory). It was warm that day in mid-November 1977. He had been eating lunch and reading an article in the latest issue of *Time* magazine about a "third form of life." Since there had been no mention of our unpublished halophile work in the *Time* article, Tom was unaware that we knew about any lipid connection. So Tom had his own "eureka moment," for some of the bacteria he was working with, *Thermoplasma* and *Sulfolobus*, were unusual enough in their properties that they might even represent a **fourth** urkingdom (cf. Brock quote above). He was thinking of contacting *Time* magazine about the possibility when the phone rang. It was Woese calling to ask him for help working with just these organisms. We had decided to grow them ourselves and were seeking cultures from Tom, which he provided on the spot. I was not willing to wait for *Thermoplasma* rRNA to come in under the mycoplasma collaboration; it was at the bottom of Maniloff's list.

The work on *Thermoplasma* went smoothly, and by the beginning of 1978 there existed a catalog defining the second major non-methanogenic archaeal phenotype. I wanted to publish this one in *Nature*, to give wide scientific coverage to the fact that archaebacteria were now a **real** group (comprising at least **two** diverse organismal phenotypes). A manuscript had been prepared accordingly and submitted. In early June of that year a letter from *Nature's* editor arrived, which I opened expectantly. Rejection! "Dear Professor Woese . . . I realize that there is considerable interest in your Archaebacterial group; nevertheless the basic idea has been so well exposed that I am not convinced that the assignation of *T. acidophilum* to the group really demands a place in *Nature*." Perhaps I should have expected it. *Nature* was a mouthpiece for molecular biology, a biology in which the organism's evolutionary history and the organism in its own right don't really count. An evolutionary finding rates only a "so what!" with these people – be it world shaking or not. Fortunately, our work on the halophiles, moving toward publication on a separate track, was accepted (quite rapidly, I might add) – the only drawback being that it had been submitted to a journal having much less coverage than *Nature* had (Magrum et al., 1978). So at least some of the biology community would be aware that the Archaebacteria comprised more than one organismal phenotype. The *Thermoplasma* rRNA catalog was never published in its own right.

The other organism Langworthy had sent us, *Sulfolobus acidocaldarius*, was not being so cooperative. It would turn out to be the first one of what we now call the Crenarchaeota, the second of the two archaeal kingdoms (the first now goes under the name Euryarchaeota) (Woese et al., 1990). We had actually begun the *Sulfolobus* catalog a month before starting in on *Thermoplasma*, but it took seven months in all to bring it to fruition. The primary obstacle here was the hard-to-characterize modified oligonucleotides. Instead of having half a dozen or so modified oligonucleotides (as had the eubacteria and all archaebacteria we'd so far encountered), the 16S rRNA from *Sulfolobus*, the first crenarchaeon, had near an order of magnitude more of them. The high levels of modified nucleotides had proven true of their tRNAs as well (Gupta & Woese, unpublished). (A high degree of modification in rRNA and tRNA, of course, was reminiscent of eukaryotes. Strange coincidence! It was tucked away for future consideration.)

It was clear from the start that *Sulfolobus* was indeed one of the Archaebacteria: it had many of the by-now familiar archaebacterial signature oligonucleotides. Yet its catalog did not fall **within** the grouping defined by the Archaebacteria we had already characterized. Our new urkingdom not only contained several disparate phenotypes, but was deeply divided into two major subgroups within itself. (This latter fact would become more apparent when the less informative oligonucleotide cataloging method was replaced by direct sequencing of whole rRNA molecules – and we started to present the data primarily in the form of phylogenetic trees rather than oligonucleotide signatures (Woese & Olsen, 1986).)

The archaeal family grows

There began to grow – in good measure thanks to the proselytizing and other efforts of Otto Kandler in Germany – an increasingly large coterie of "archaeophiles," each of whom approached the Archaebacteria from their own perspective. Thus by 1980 a fair amount was known about the specific and general characteristics of the Archaebacteria. Notable among the new faces in the movement were Wolfram Zillig and his associate Karl Stetter, then a sort of postdoc. Wolfram, one of the Abteilung Leiters at the Martinsried Max Planck Institute, had built his career around molecular characterizations of the DNA-dependent RNA polymerases. By the late 1970s he (by his own admission) had settled into "the rut of refinement" with bacterial RNA polymerases. Somehow the word had filtered down to him about the "third form"; likely via Kandler. Immediately, Wolfram and Karl turned their attentions to the RNA polymerases of the Archaebacteria, publishing first on the novel RNA polymerase of the halobacteria – in the same year we publicly announced our own findings on their rRNAs (Zillig et al., 1978). This was followed by a mid-1979 description of a similar atypical RNA polymerase from *Sulfolobus acidocaldarius* – whose subunit composition on gross inspection appeared rather eukaryote-like – (Zillig et al., 1979). We were still finishing up the 16S rRNA catalog of *Sulfolobus acidocaldarius* at the time, and hadn't begun to think about publication, which actually occurred several years later in collaboration with Wolfram, a publication now based upon a total of three different (crenarchaeal) extreme thermophiles (Woese et al., 1984).

Wolfram and Karl had figured out the ether-linked lipid connection for themselves (if I am not mistaken). Indeed, it is to these two that we owe any strong emphasis that there was on the crenarchaeal branch of the archaeal tree in the beginning. Wolfram had also deduced that when Tom Brock isolated and characterized his thermoacidophiles, he had probably missed a whole world of anaerobic thermoacidophiles. And here would be the start of his and Karl's colorful (and sometimes dangerous) adventures hunting for hyperthermophilic Archaea around the globe (later carried on by Karl alone). (Given that he frequented the vicinity of boiling sulfurous hot springs so often, Karl came to refer to crenarchaeal hyperthermophiles as "the organisms from hell," even naming one genus *Stygioglobus* and at least one species *infernus* accordingly.)

It was only a matter of time before there would be a formal scientific conference – the field's initiation rite. Thanks to strenuous efforts on the parts of Wolfram and Otto, it happened sooner rather than later. Planning began in 1980, with furious exchanges of letters and phone calls; they also had to obtain financial support (not an easy matter). The meeting got scheduled for 1981 at the Max Planck Institute at Martinsried-bei-München. The Martinsried meeting fulfilled everyone's hopes; it covered the full spectrum of workers and work on the Archaebacteria. Don't think this means that the number of participants was large; there were relatively few of us in those days. A wonderful sharing of knowledge and developing or expanding of scientific relationships occurred at the meeting. Best of all, it gave a feeling of group identity, a camaraderie, to many who attended. The Archaebacteria had arrived!

Starting in 1977 (if I am correct) George and I had begun planning to write a comprehensive publication about all the phylogeny work that had gone on in the phylogeny project – one that would give biology a little surprise. I nicknamed the project "Big Tree." There would be many authors on Big Tree, reflecting the many people who had worked on the project in my lab and all the collaborators who had contributed the radiolabeled rRNA starting materials. The actual writing, however, involved mainly George and myself. That proved difficult enough.

With the hard part over, I submitted Big Tree to *Science* on January 31, 1980 (eschewing *Nature* this time around). With *Science* things went smoothly, and the paper appeared in July of that year: "The phylogeny of prokaryotes," by Fox et al. (1980). It was a heady experience, publishing a phylogenetic tree that (in outline) covered all of life, and said to boot that there were three, not two, primary lineages of organisms on this planet. Before genealogical analysis had been transposed to the molecular level, the best that biologists had been able to do was make trees for animals and/or plants. Here, for the first time, was the skeleton of the full Tree of Life. Monumental! Reviewers' reactions to the paper had been good. The reaction of the **scientific** public this time was good as well.

The dedication unwanted and recognition ungiven

I had dedicated Big Tree to C. B. van Niel (long retired), who was one of the great microbiologists of his era. In 1970 he had won microbiology's highest honor, the Leeuwenhoek Medal – an award given once a decade by the Netherlands Royal Academy of Sciences to "the scientist who has made outstanding contributions to the advancement of microbiology during the preceding ten years." I had thought that van Niel would be pleased and honored to see the final solution to the problem of a natural bacterial classification dedicated to him; for it was a problem that he, Kluyver, and Stanier (among others) had struggled with for a very long time without having the (molecular) technology to cope with it. (A former student and friend of van Niel's, Robert Hungate, had offered to send him the paper personally.)

When dedicating our paper to him I had read only van Niel's early work on natural classification of Bacteria, collaborations, first with Kluyver and then with his student Stanier. Little did I know at the time what van Niel's final judgment on bacterial phylogenetics had been: it was "a waste of time to attempt a natural system of classification for bacteria" (see Stanier et al., 1957). Moreover, in 1962 he and Stanier had even renounced their earlier efforts in bacterial classification as something "neither of us cares any longer to defend." Our paper apparently didn't change van Niel's mind, for I never heard from him. But what I have just said is based upon what I discovered later, from a fairly comprehensive study of van Niel and his cohort's writings on bacterial phylogeny.

Given the nature of our *Science* paper and the reception it received, it should have been nominated for "Paper of the Year" in *Science*, which, indeed, it was. But the "Tree of Life" came in second (as I was told by "sources"). I also found out what may have been the reason: one of the judges had, apparently innocently, praised the work as a contribution "almost as important as the eukaryote–prokaryote dichotomy." "Hmm," I fumed, "how can our work be almost as important as something that is totally wrong-headed, something that has influenced the course of microbiology so adversely! Who is this stupid judge?"

Prokaryote by any other name . . .

The more things progressed, the more prokaryote loomed large, like a current against which no one could swim. I had often reflected on why that day in June 1976 it seemed so incredible to discover a prokaryote that wasn't a prokaryote. The reason was obvious once I thought about it, which I – and it turns out every other biologist – had never felt a need to do previously. **That** is why we encountered such difficulty in convincing others of what we'd discovered. **That** is why the Nobel Laureate Luria had so scornfully rejected the three urkingdom notion. Ralph Wolfe, who frequented microbiology meetings, would tell me at the time about the "talk in the corridors," i.e. the behind-the-scenes, clubby dismissal of the three urkingdom concept. I was itching for them to come after me in print. But none of them would. It was so important to get the "prokaryote" matter out in the open. Something was strange about this "prokaryote"!

Whatever the reason, "prokaryote" had special significance for microbiologists. Yet the reason couldn't be historical usage of the term. "Monera" was the preferred term for bacteria in the first half of the twentieth century; Schizomycetes also being used. And I couldn't find the term "prokaryote" anywhere in the literature before 1962 – and neither (more recently) could the historian Jan Sapp (2005). "Prokaryote" seemed to appear out of nowhere at that point in time (Stanier & van Niel, 1962). So why were microbiologists so wedded to the "prokaryote" that they rejected any suggestion of a third urkingdom without even thinking about it?

By the 1970s a generation of microbiologists and biologists in general had been raised believing that "prokaryote" (term and concept) had originally been the brain child of the protozoologist Edouard Chatton (1938; Stanier & van Niel, 1962; Sapp, 2005). And the recent historical analysis also found no evidence for Chatton's even **presenting** the concept (Sapp, 2005). In any case, how could any biologist working in the early decades of the twentieth century infer anything about the organization of a bacterial cell, much less what features were common among the organizations of **all** bacterial cells: bacteria were a morphologically diverse group, and all that one had to go on as regards their internal organizations was negative evidence: they were not (what we now call) "eukaryotic." None of it made sense.

Although we were unaware of it at the time (or for some time thereafter), we were not faced with an ordinary scientific situation here; it was no simple case of a new, more detailed and factually supported hypothesis displacing an entrenched older one. The strength of that older idea, its dominating, dogmatic hold on microbiology, implied rather more than that. But what? Answering this question has taken me on a somewhat bizarre journey that has lasted over two decades, and only recently shown signs of coming to its end. My understanding has had to pass through three distinct stages, each broader in scope than the previous.

First attempt: apply scientific reason

When we initially realized that the prokaryote concept was the immovable object on the road to microbiology's development, I thought the matter would resolve scientifically: tell biologists what their "prokaryote" really is and they will understand. Accordingly, George and I prepared a paper entitled "The concept of cellular evolution" (Woese & Fox, 1977a). In it we denied the two defining tenets of "prokaryote," but kept the term itself, hoping thereby to ameliorate the rather revolutionary change we were proposing. The prefixes "pro-" and "eu-" had a familiar ring to them; the notion that Bacteria were separate from and older than – some perhaps even ancestral to – eukaryotes was traditional.

Our argument went: yes, the terms "prokaryote" and "eukaryote" do obviously recognize distinctly different kinds of cellular organization, but no, in the case of the prokaryote a common general organization **does not** necessarily denote common ancestry for all! Prokaryote should be looked at only as a **level** of organization, a level distinct from that of the (higher, more complex) eukaryotic cellular organization. Therefore, Archaebacteria and Eubacteria can have the same **general** level of organization, but have arrived at it **independently**. In other words, the prokaryotic level of organization has evolved at least twice!

The general principle here is that biology is a study in emergent levels of increasingly complex organization. This idea, of course, was anathema to twentieth-century reductionist biology; but it had currency in the eighteenth and nineteenth centuries (Burkhardt, 1977). We were simply resurrecting the notion and recasting it in modern scientific terms.

In keeping with this general notion we then postulated a primitive level of organization even simpler than the prokaryotic and eukaryotic, one that had preceded the other two. I named it the "progenote" to signify that the genotype–phenotype link at that early stage had yet to reach the eventual perfection, precision, and sophistication that characterizes modern cells (Woese & Fox, 1977a).

It may have sounded absurd at the time that Archaea and Eubacteria had evolved the same cellular organization independently. But the absurdity actually lies not in our proposal but in the original assertion that **all** bacterial cells were of a kind, had the **same** (prokaryotic) cellular organization – from which monophyly had to follow (Stanier & van Niel, 1962; Stanier et al., 1963). As I have said, there was no way anyone in the early, middle, or even late decades of the twentieth century could have known anything about the nature of the organization of any but the eukaryotic cell (and that was only a superficial description).

Despite what I thought was sound pleading our argument had no takers; I don't think that the vast majority of (micro)biologists even bothered to consider it. We would have to dig deeper to get at the root of the problem.

Second attempt: follow the leader

After this, I essentially went off on my own to try to solve the problem – which meant acquiring the perspective needed to see it in historical terms. Where had "prokaryote" come from in the first place? Microbiologists had traditionally made a strong distinction between Bacteria and the cells of "higher forms" (see above). But none of it was taken as certain or factually supported. It was all just necessary speculation on the road to developing an understanding of bacteria. All that these early speculations really accomplished was to define bacteria negatively: they were not as large as eukaryotic cells; and they lacked this, that, and the other microscopically visible features so characteristic of eukaryotic cells; in addition to their showing no (common) microscopically discernible intracellular structures of their own.

Where, then, did this strongly asserted, definitive, inflexible "prokaryote" come from? Focusing on the "prokaryote's" origin brought me directly to one, and apparently only one, source, R. Y. Stanier and his cohort. As I said above, I couldn't find "prokaryote" in the literature before 1962, when it appeared in the classic paper of Stanier and van Niel (1962; Sapp, 2005). It was also noteworthy that, although the term "prokaryote" was featured in the second edition of *The Microbial World* (Stanier et al., 1963), it was nowhere to be found in the text's first edition (Stanier et al., 1957).

No terminology before 1962. Was it the same for the underlying concept? Yes: 1962 was the first time the commonality of all bacterial cellular organization was asserted (rather than just mused about). There had been the older, questioning, more speculative attitude about the nature of bacteria for some time, but nothing like this, the self-assured, dogmatic "prokaryote" we had encountered in 1977.

It is particularly important to know how "prokaryote" was defined in 1962, for there is a tendency today to adjust the term at will (Judson, 1996), which is counterproductive in that it is tantamount to superficial "surgery" that only further conceals a deeper chronic condition. "Prokaryote" rests on two definitional pillars: (i) the assertion that it represents one of "two different **organizational patterns** of cells" (the other being that of the eukaryote); from which it must follow that (ii) "we can therefore safely infer a **common origin** for [all 'prokaryotes'] in the remote evolutionary past" (Stanier & van Niel, 1962; Stanier et al., 1963; emphasis added). The prokaryote, moreover, was seen to provide "our only hope of more clearly formulating a 'concept of a bacterium'" (Stanier & van Niel, 1962), because **the ultimate scientific goal of biological classification** [i.e. a natural, or phylogenetic classification] **cannot be achieved in the case of bacteria**" (Stanier et al., 1963; original emphasis). Only in 1962–3 does one begin to see such a definitive, strongly worded and inflexible position taken on the nature of bacteria. There was no sound scientific reason for it either, for the "prokaryote" was founded solely upon rhetoric, with **no** intent or attempt to give it a firm foundation subsequently. The whole thing was only scientific "guesswork" (*sensu* Schroedinger, 1954).

What sank matters further into strangeness was the fact that the microbiology community accepted this guesswork (both alien term and out-of-the-blue concept) immediately and overwhelmingly. No criticism, no serious discussion, greeted the prokaryote's debut. An obvious weakness in the authors' presentation had been the lack of adequate comparative evidence upon which to base the assertion of properties common to **all** bacteria – too few properties and too narrow a sampling of bacterial taxa (Murray, 1962), which could have been remedied easily by doing more work, sampling a wider range of taxa. Yet no one did this or apparently saw the need to! It was as though microbiologists simply wanted to make the issue go away: it was time to close the door and move on (Sapp, 2005).

It is revealing to compare the ready and enthusiastic reception of the prokaryote with the irate reaction that the three-urkingdom archaebacterial concept engendered a decade and a half later – even though the three urkingdom concept **had** solid factual support, with more facts obviously to come. I clearly needed to get to the bottom of this, to understand the diametrically opposed responses the microbiology community had given to the two points of view.

What I had now learned was that individuals could not be held responsible for what had happened to microbiology in the middle decades of the twentieth century. They were among the few who had seen the problem, and they had tried to resolve it – unfortunately in the wrong way. Those "responsible" were just the lead birds in a migrating flock: shooting them down does not affect the flock's course. There was a deeper dynamic at work here, one that would be far more difficult to change.

Third attempt: understand the mythology

This last insight proved critical. An intellectual tide was the key to the problem. Molecular biology, in its insistent reductionism, lay at the heart of the issue (Woese, 2004). The intellectual landscape was indeed being restructured! Two factors were going to shape twentieth-century microbiology: the power of molecular biology's vision and the weakness of microbiology's.

Bacteria are **organisms**, not simply bags of biochemistry. Microbiology is *a fortiori* an organismal discipline. Organisms in their fullness must be its study. The organism's parts are surely important, but important in what they contribute to an **overall** understanding; the same being true of the higher-level interactions that structure microbial communities, their ecologies. These, together with an appreciation of microbial variety and long- and short-term evolutionary dynamics, make for a "tetrahedral" synthesis called "organism," quintessential biological organization.

Within the context of his time the great microbiologist Martinus Beijerinck appreciated bacteria in just this full sense. When in 1905 he was asked (on the occasion of being awarded the Leeuwenhoek Medal) to articulate his view of the microbial world and its study, he had replied that the most effective approach to understanding microorganisms was "the study of microbial ecology"; adding that microbes represent the "lowest limits of the organic world, and . . . constantly keep before our minds the profound problem of the origin of life itself" (van Niel, 1949; van Iterson Jr et al., 1983). Here in the making was a rich, genuinely organismal picture of the microbial world.

Unfortunately, Beijerinck's successors did not, nor did they care to, realize his vision. Initially it was not technologically feasible to do so, but doing so in any case would have meant going against the new reductionist tide, which microbiology was conceptually unequipped to do. Therefore, outside influences, not innate tendencies, would shape the discipline.

To Beijerinck's successor A. J. Kluyver (and his "Delft School") the organism in its own right (i.e. as a biological organization) meant nothing: biochemistry was the essence (Kamp et al., 1959). The emphasis would be on uncovering the main biochemical themes among the bacteria (Kluyver & Donker, 1926) – and all of the nuanced variations thereupon (Kluyver, 1931). Yet how many of the organic compounds we know do you think would exist on this planet if some kind of biological organization weren't around to produce them (Kaufmann, 1995). (Most organic compounds we see in nature are like the new elements at the high end of the periodic table in that they require "organismal intervention" to exist at appreciable levels.) After all, bacterial metabolic diversity in the last analysis is not a biochemical problem; it is an evolutionary one, a question of why and how such great biochemical (metabolic) diversity arose in the first place, and why certain biochemical pathways were evolutionarily singled out for (biological) amplification. By the middle of the twentieth century our conception of bacteria had moved from Beijerinck's multifaceted organismal one to the Delft School's disassembled one (see van Niel, 1949).

The reason for roots

Every scientific discipline rests upon an axiomatic foundation, a scientific mythology, which informs it and the world as to what it is and in the process charts its course. Microbiology's mythology, non-organismal and almost non-existent, reflected the Delft School perspective. And that is why in 1962, in the heyday of microbial biochemistry, when all manner of new biochemicals and metabolic pathways were being uncovered, R. Y. Stanier had said (in introducing the "prokaryote" concept): "the abiding intellectual scandal of bacteriology has been the absence of a clear concept of a bacterium... the problem of defining these organisms as a group in terms of their biological organization is clearly still of great importance, and remains unsolved" (Stanier & van Niel, 1962). It was obvious that microbiology was in a state of disarray; it did not understand itself; it had no real conceptual base (mythology) from which to draw support. And this is why the prokaryote concept (or something equivalent) was needed in order for the discipline to have an "organismal" sense of itself.

The "prokaryote" was only scientific guesswork (as said above). But it did seem to give microbiology the badly needed keystone in its conceptual foundation (its mythology). Prokaryote provided an overarching, authoritative, and authenticating framework within which now to work. Yet note that the concept had precluded an evolutionary (phylogenetic) definition of bacteria – in my opinion, the only concept possible.

Missing the train

One needs to recall here that the "prokaryote" concept developed while molecular biology was providing the scientific world its first glimpse of the power that lay in molecular sequencing (Sanger & Tuppy, 1951; Sanger & Thompson, 1953). Zuckerkandl and Pauling (1965) were trumpeting the molecular approach to evolution; and F. H. C. Crick (1958) had said, "Biologists should realize that before long we shall have a subject which might be called 'protein taxonomy' – the study of amino acid sequences of proteins of an organism and the comparison of them between species." He added that "these sequences are the most delicate expression possible of the phenotype of an organism and... vast amounts of evolutionary information may be hidden away within them."

It seems highly unlikely that the vast majority of microbiologists were totally ignorant of the new molecular approach to organismal relationships, especially given the exquisite need the discipline had for evolutionary underpinnings. What is more likely (as discussed) is that a conceptually unsettled microbiology wanted to rid itself of a perspective that was not in keeping with the reductionist tenor of the times (Woese, 2004). The prokaryote accomplished that: it papered over "the problem of defining [bacteria] as a group in terms of their biological organization" (Stanier & van Niel, 1962). But it had cut microbiology away from its organismal, evolutionary roots in the process. (While a considerable amount of nucleic acid hybridization work was subsequently done in microbial taxonomic structure, the work's intent was simply to improve existing bacterial taxonomy. The grand challenge of a natural classification, a universal phylogeny for bacteria (Stanier & van Niel, 1941), had disappeared from the scene.)

There is actually nothing surprising about this strange "prokaryote period" in microbiology's history. The dynamic at work here is encountered in many different fields: "the past needs to be forgotten because it reveals the confusion and lack of cohesion of the present." It is just that in this case the whole thing was so patently unscientific.

Where are we going?

A biology that does not concern itself with evolution is not biology. Contradicting the ill-framed reductionist view of biology is precisely why I had established the program of (molecular) phylogenetic reconstruction in the first place. The program's *raison d'être* was to revive the evolutionary spirit that underlies biology and had been nearly squeezed out of existence by reductionism. Thus, the Archaea were indeed a splendid surprise to me: they were a resonating thunderclap that would awaken the Sleeping Giant of evolution. As I saw it, the discovery of the Archaea had turned over the reductionist rock, and the weakness in that paradigm now lay there for all to see. But things didn't turn out that way.

The evolutionary message inherent in the discovery of the Archaea faded. Consequently, microbiology has yet to resolve its foundational issues. The discipline goes on today rootless as ever, unconcerned with microbial evolution, living in the scientific dream world of the prokaryote. The evidence is everywhere: "prokaryote" remains imprinted across the discipline, structuring its modes of thought, defining its curriculum, its scope, shaping its future. One glance at how the discipline of microbiology responded to the challenge of the Archaebacteria and later to the advent of genomics will show you a discipline lost in its past, blown about by the capricious winds of a society. Microbiology did not meet the challenge of the Archaebacteria, nor has it yet met that of genomics.

Microbial genome sequencing began only in 1995 (Fleischmann et al., 1995). I know from personal experience that had microbiologists (and biologists in general) perceived its significance, microbial genomics could have begun by the mid-1980s (Woese, 1993). Archaea are so unlike Eubacteria, and we didn't (and still don't) know enough about them. Genome sequencing was the only way to bring the "third urkingdom" up to scientific speed (Woese, 1993).

Sadly, when microbiologists finally took a genomic approach, it was only because genomics could be used (and funds obtained for that) to attack a host of practical problems (medical, agricultural, environmental), problems that they had been dealing with for decades. To the extent that microbial genomics affected the intellectual climate in microbiology at all, it did so adversely: each microbial genome rationalized and sequenced for a different reason. No concerted program in, no organismal rationale for, microbial genome sequencing even existed until quite recently.

Moving horizontally

Many microbiologists have developed interests in horizontal gene transfer (HGT) over the past decade. In that HGT provides an important, if not the most important, clue to the dynamic of cellular evolution, there would seem to be hope in this. With a few important exceptions, however, the microbiologist has not seen the fundamental evolutionary significance of HGT: most of the concern centers on health-related matters, and a little also with particulars such as cellular mechanisms involved in HGT. Nevertheless, there are very promising studies involving metagenomic approaches to microbial community structure and dynamics beginning, in which the role of HGT (and viruses) is absolutely central (Rachel et al., 2002a; DeLong, 2005; DeLong & Karl, 2005).

The defining problems of twenty-first-century microbiology (and biology)

The future of microbiology does not lie on the field's current path. If the status quo in microbiology persists, the discipline is on course to become a complete and total service discipline, simply bioengineering. What is needed, what is essential, is that microbiology become far more of a **basic** biological discipline than it has ever been – some sort of modern realization of Beijerinck's holistic vision. Throughout the twentieth century, microbiology's course was charted by the dictates of molecular reductionism and the practical concerns of a biomedically oriented society. Biology itself was the low man on the scientific totem pole; and microbiology the low man on its.

Now things must reverse. Not only must microbiology become the leading discipline, the guide, in biology; but biology itself should become the basal discipline of the sciences. Microbiology's lack of concern with evolution has been the discipline's downfall in the past and could remain its (and biology's) nemesis in the future. Microbiology departments today are products of historical accident, held together in essence by institutional inertia. It is not a good sign that the basic research at microbiology's forefront is increasingly being done outside of the context of formal microbiology departments.

The future of biology lies in understanding the nature of biological organization. Microorganisms are a central concern here, for they are biology's primary window on the problem. Twentieth-century biology (especially microbiology) was structure-oriented, reductionist, and temporal in its view of life. The biology of today must be evolutionary, holistic, and process-oriented.

We meet twenty-first-century biology right now in terms of two grand problems: (i) the evolution of the cell; and (ii) an understanding of the global environment. While these two may seem quite unrelated, the one as fundamental as biology now gets, the other essentially applied (and of pressing concern), this is not so. At base both represent problems in biological organization. And the two will become closely joined when biology comes to study the early stages in the evolution of the cell, when horizontal gene transfer dominated the evolutionary dynamic, leading to an evolution that was essentially communal, not individual (involving distinct lineages) (Woese, 1982, 2002). Only a microbiology that embodies the spirit of biological organization, organism and evolution, will be fit to lead biology in the twenty-first century.

Acknowledgments

This chapter is dedicated to Wolfram Zillig, A founder of the archaeal revolution. My work is supported by grants from the Department of Energy, and the National Aeronautics and Space Administration.

2
Natural history of the archaeal domain

Patrick Forterre, Simonetta Gribaldo and Celine Brochier-Armanet

Introduction

One-third of living organisms on Earth remained unnoticed by human curiosity until the last decades of the twentieth century. However, Archaea have likely been there for more than three billion years, thriving in all terrestrial and marine environments. Archaea exhibit unique features at the molecular level that expand the realm of biochemistry, and in addition they harbor precious relics of our own history. From the very beginning of their appearance in the scientific literature, Archaea have been a goldmine of biological novelties, and many hidden treasures still wait to be deciphered by old and new Archaea lovers. Research on Archaea has produced great scientists and made already great scientists even greater, pushing some of them to cross interdisciplinary boundaries, transforming biochemists and molecular biologists into microbiologists and vice versa.

How did the archaeal domain originate, diversify, and evolve into the myriad of organisms that we observe today? A good knowledge of the evolutionary relationships between archaeal lineages and between Archaea and other organisms is essential to answer these questions and to make sense of the physiological, biochemical, and genomic data that have accumulated at an increasing pace over the past years.

Phylogeny of the Archaea

A brief history of early studies on archaeal evolution

The first glimpses into the natural history of Archaea (formerly Archaebacteria) were rapidly obtained by the same methods that led to their discovery, i.e. comparison of small subunit ribosomal RNA sequences (SSU rRNA) using oligonucleotide catalogues (see Chapter 1 of this volume). In the three years following the establishment of the archaeal concept (Woese & Fox, 1977b), it was already realized that *Thermoplasma* and *Halobacterium* had close evolutionary affinity with methanogens (*Halobacterium* being specifically related to the Methanomicrobiales), whereas *Sulfolobus* remained an outlier. Thus, the division of the archaeal domain into two main branches, now known as Crenarchaeota and Euryarchaeota (Woese et al., 1990), was already established.

The sequencing and phylogenetic analysis of full-length SSU rRNA sequences in the mid-1980s confirmed these two main divisions of Archaea, and the close relationship between the halophilic Archaea (hereafter called Haloarchaeales) and Methanomicrobiales (Woese, 1987). It was also immediately noticed that branches leading to hyperthermophiles grouped at the base of the archaeal SSU rRNA tree using bacterial sequences as the outgroup. Early analyses suggested that *Thermococcus celer* was the earliest emerging euryarchaeal lineage (Achenbach-Richter et al., 1988). Later on, with the availability of novel archaeal SSU rRNA sequences, *T. celer* was displaced from this basal position by another hyperthermophile, the methanogen *Methanopyrus kandleri* (Burggraf et al., 1991). These early observations led the majority of authors to conclude that: (i) Archaea originated from a hyperthermophilic anaerobic ancestor; (ii) methanogenesis was an old phenotype; and (iii) Archaea were indeed very ancient, since they were probably already thriving in the anoxic atmosphere of the primitive Earth, rich in H_2 and CO_2.

In the mid-1990s, the sequencing of environmental SSU rDNA sequences of uncultivated archaea from Yellowstone hot springs led to the proposal of a third archaeal kingdom, the Korarchaeota, that occupied a place even closer to the putative root in the archaeal SSU rRNA tree (Barns et al., 1996). Korarchaeota were recently displaced from this position by the hyperthermophile *Nanoarchaeum equitans*, possibly the first identified parasitic archaeon, and the smallest known cellular organism (Huber et al., 2002).

Although largely settled, the history of Archaea as seen by SSU rRNA sequence comparison was not devoid of problems. For example, these analyses failed to give a clear-cut placement – either monophyletic or not – for the methanogenic lineages (Methanobacteriales, Methanomicrobiales, Methanococcales and Methanopyrales). The basal position of *Methanopyrus kandleri* in the SSU rRNA archaeal tree was also controversial. The presence of atypical (possibly primitive) lipids (Hafenbradl et al., 1993) and some idiosyncrasies (a split reverse gyrase gene (Krah et al., 1996), or else a double-fused histone (Slesarev et al., 1998)) seemed to confirm its basal position in the euryarchaeal branch of the SSU rRNA archaeal tree, away from other methanogens. On the other hand, *Methanopyrus* was a bona fide methanogen, with a classical methanogenic pathway and RNA polymerase (see below).

The inability of SSU rRNA studies to convincingly resolve all these issues led Wolfram Zillig to strongly advocate early on the use of alternative (or complementary) molecular markers such as RNA polymerases. Zillig stressed the usefulness of comparing amino acid sequences and underpinned the large size of RNA polymerase large subunits (Klenk et al., 1994). Analysis of RNA polymerase structures revealed a major division in the archaeal domain, which did not match the euryarchaeal/crenarchaeal one. The large B subunit of archaeal RNA polymerases (the homologue of bacterial β subunit) was found to be split in two subunits – B' and B" – in Haloarchaeales, *Archaeoglobus*, and all methanogens (including *Methanopyrus*). This observation suggested that *Thermoplasma* (with an intact B subunit) diverged from other Archaea before the Haloarchaeales and methanogens. This position was apparently confirmed by an early RNA polymerase phylogeny (Klenk & Zillig, 1994).

In the mid-1990s, the discovery of contradictions between rRNA trees and key protein trees caused a number of scientists to question the validity of deep phylogenetic relationships suggested by rRNA trees. Doubts turned into profound skepticism when the Microsporidia issue unfolded. Microsporidia are eukaryotic parasites lacking mitochondria that had acquired the status of primitive eukaryotes ten years earlier when they appeared to emerge as the first offshoot at the base of the eukaryotic SSU rRNA tree (Vossbrinck et al., 1987). However, later phylogenetic analyses based on markers such as actin, tubulin, and RNA polymerase, as well as the presence of mitochondrial genes, finally revealed that Microsporidia are degenerated fungi. They were misplaced in the SSU rRNA tree due to the fast rate of evolution of their ribosomal apparatus (for a review, see Embley & Hirt, 1998). The shape of the eukaryotic SSU rRNA tree, with a large trunk populated by various protists, and an apical part grouping other protists, animals, plant, and fungi, eventually also turned out to be an artefact of long branch attraction (LBA) (Gribaldo & Philippe, 2002). If the eukaryotic SSU rRNA tree was inaccurate, could it be that the archaeal one was also misleading?

Thus, it was expected that whole genome sequence data would confirm, or not, the main results obtained with the SSU rRNA tree and resolve the remaining questions about the history of the archaeal domain. We will see that these expectations proved to be partly correct when the potentially misleading genomic data were properly analyzed. In the end, apart from a few important exceptions that are discussed below, genomic data have remarkably corroborated most features of the archaeal SSU rRNA tree.

Whole genome based approaches

Genomic sequencing has provided the information needed to produce whole genome trees, either universal trees of life or trees specific to individual domains (Delsuc et al., 2005). Various methods were designed to recover as much information as possible from complete genome sequences, either gene presence/absence (with a correction according to genome sizes), the sharing of orthologous gene clusters, supertrees (combining individual protein trees with different taxonomic sampling into a single one), or even shared metabolic pathways (Wolf et al., 2001, 2002; Clarke et al., 2002; Daubin et al., 2002; Ma & Zeng, 2004; Gophna et al., 2005, and references therein for earlier studies). Most recently, structural genomics has been used to build whole genome trees based on the presence/absence of protein folds (Yang et al., 2005). In all these studies, Archaea turned out to be well separated from Eukarya and Bacteria, thereby confirming the three-domain concept. However, other well established features of the SSU rRNA tree were surprisingly not recovered. In particular, regardless of method, nearly all whole genome trees place Haloarchaeales and Thermoplasmatales either at the base of the Archaea or at the base of Euryarchaeota, or even as sister groups of Crenarchaeota. The late branching of Haloarchaeales and Thermoplasmatales, as in the SSU rRNA tree, was only recovered in the supertrees built by Daubin et al. (2002).

The apparent early emergence of Haloarchaeales and Thermoplasmatales in most whole genome trees is probably due to a bias introduced by horizontal gene transfers (HGT). Indeed, the genome of both *Halobacterium* NRC1 and Thermoplasmatales contains an important number of genes that were recruited from Bacteria (Kennedy et al., 2001), some of which may be shared with Crenarchaeota. The sharing of genes encoding operational proteins by aerobic Crenarchaeota (*Sulfolobus, Aeropyrum*) and Haloarchaeales could also explain why the latter were grouped with Crenarchaeota in a whole genome tree based on shared metabolic pathways (Ma & Zeng, 2004). The genomes of Thermoplasmatales also contain, in addition to a substantial complement of bacterial-related genes, a high proportion of genes shared exclusively with *Sulfolobales* (an archaeal genus inhabiting the same thermoacidophilic environments) (Ruepp et al., 2000; Gophna et al., 2005). As a consequence, in rooted whole genome trees that are not corrected for biases induced by HGT, Haloarchaeales should be attracted towards Bacteria, and Thermoplasmatales towards Crenarchaeota and Bacteria.

Few attempts have been made in the literature to systematically correct whole genome trees for the effect of HGT. Methods used to construct whole genome trees either do not consider this bias, or appear not to handle it correctly (Clarke et al., 2002; Gophna et al., 2005). If HGT is more or less uniform and not prevalent in one direction, the whole genome tree reconstruction approach that appears to be the least affected by HGT is the supertree one. In fact, the biases due to HGT involving particular markers are overwhelmed by the whole phylogenetic signal. For example, *Thermoplasma* and Haloarchaeales showed a late emergence in a universal supertree built from the combination of 730 individual protein trees for 45 species (including nine Archaea) (Daubin et al., 2002). The late emergence of Haloarchaeales and Thermoplasmatales, also observed in the rRNA tree and in trees based on ribosomal proteins (r-protein) and RNA polymerase subunits (see below), is thus most likely correct.

Approaches based on limited gene sets

Faced with the failure of most whole genome trees to solve the problems raised by SSU rRNA phylogenies, several authors have chosen an alternative approach, i.e. the simultaneous analysis (also called supermatrix approach) of protein alignments presenting similar taxonomic samplings (Delsuc et al., 2005). Preliminary individual analyses are usually performed in order to detect and remove from further analysis those markers that show clear evidence of HGT. Remaining datasets are concatenated into a single large dataset that improves the phylogenetic signal and thus the resolution of the resulting trees. Scientists interested in archaeal phylogeny have chosen to focus on the concatenation of informational proteins, since these are thought to be less affected by HGT (Wolf et al., 2001, 2002; Matte-Tailliez et al., 2002; Slesarev et al., 2002; Waters et al., 2003; Brochier et al., 2004, 2005a, b; Baptesté et al., 2005).

Surprisingly, the first concatenated r-protein trees based on 32 r-proteins conserved in Archaea and Bacteria confirmed the odd results of most whole genome trees, with Haloarchaeales, Thermoplasmales, and *Archaeoglobus fulgidus* emerging in that order prior to the divergence between Euryarchaeota and Crenarchaeota (Wolf et al., 2001; Slesarev et al., 2002). In that case, this basal position is most likely due to the attraction of their long branches by that leading to the bacterial outgroup. This emphasizes the importance of building unrooted trees when working within a single domain in order to limit LBA artifacts. In addition, poor sequence conservation among domains usually leads to ambiguous alignments that eliminate numerous informative positions. Following this strategy, unrooted archaeal trees constructed from 53 r-proteins recovered most evolutionary relationships previously seen in SSU rRNA trees, including the late emergence of Haloarchaeales and Thermoplasmatales (Matte-Tailliez et al., 2002; Brochier et al., 2004, 2005a, b; Bapteste et al., 2005). This late emergence was also observed in unrooted trees constructed from the concatenation of RNA polymerase subunits (Brochier et al., 2004, 2005a).

Despite this congruence, trees based on the concatenation of r-proteins and those based on RNA polymerase subunits still exhibited key differences, such as the relative position of Archaeoglobales and Thermoplasmatales within the supergroup also including Haloarchaeales and Methanomicrobiales, or else the relative positions of Methanococcales and Methanobacteriales, which were monophyletic in the transcription tree but not in the translation tree (Brochier et al., 2004). More troublesome, *Methanopyrus kandleri* emerged after Thermococcales in the r-protein trees, whereas it branches at the base of the Euryarchaeota in the RNA polymerase subunits and the SSU rRNA trees. However, the presence of a very long branch suggests the risk of an LBA artifact (Brochier et al., 2004). The incongruence between the two trees in the position of *A. fulgidus*, and in those of Methanobacteriales and Methanococcales, might have been due to either confusion caused by undetected HGT, and/or an insufficient phylogenetic signal in the protein datasets. The latter hypothesis is correct, since, when the number of genomes used for the analysis was increased from 19 to 25, the r-protein and RNA polymerase trees were finally overlapping with only one exception (see below). (See Plate 2.1a, b for updated trees.)

We wish to stress the congruence between the phylogenetic signal harbored by the components of the archaeal ribosome and RNA polymerase. Indeed, since the r-proteins and the SSU rRNA belong to the same macromolecular complex, it could be argued that the convergence between concatenated r-proteins and SSU rRNA trees could be attributed to similar patterns of HGT for these two ribosomal components. In particular, it was recently claimed that "a consistent and extensive pattern of congruence among informational genes in branching patterns at the level of bacterial and archaeal phyla has not been established" (Charlebois & Doolittle, 2004). We think that such congruence is now quite clearly established for Archaea. This highlights the existence of a core of genes that were inherited mainly vertically in Archaea and can thus be confidently used to retrace the natural history of this domain.

The phylogenetic position of *Methanopyrus kandleri*

The fact that Methanopyrales represent the first offshoot of Euryarchaeota in the SSU rRNA tree has for long been taken as a support for an early origin of methanogenesis in Archaea. However, as previously indicated, *M. kandleri* occupies two incongruent positions in previous (Brochier et al., 2005a) and updated r-protein and RNA polymerase trees (Plate 2.1a, b). In the r-protein tree *M. kandleri* groups with *Methanothermobacter thermautotrophicus*, whereas in the RNA polymerase tree *M. kandleri* clusters with *Nanoarchaeum equitans* between Crenarchaeota and Euryarchaeota (Plate 2.1a, b). However, in the RNA polymerase tree *M. kandleri* harbors a relatively long branch (a peculiarity not observed in the r-protein tree; Plate 2.1a), reflecting a higher evolutionary rate of its RNA polymerase subunits (see, for example, the phylogeny of RNA polymerase A1 subunit in Plate 2.2). The fast evolutionary rate of the transcription proteins of *M. kandleri* is exemplified by the presence of a number of insertions higher than that observed in any other archaeal species (27 in *M. kandleri*, whereas the number of insertions specific to other archaeal lineages ranges between one and eight). In addition, in *M. kandleri* the regions surrounding these insertions are frequently flanked by highly divergent regions (Plate 2.2). As a consequence of the fast evolutionary rate of its RNA polymerase, the branch leading to *M. kandleri* may be attracted towards the base of the euryarchaeal branch by the long branch leading to the crenarchaeal outgroup and *N. equitans*.

We thus reckon that the position of *M. kandleri* in the r-protein tree is the correct one. This is in agreement with: (i) a split of the gene coding the RNA polymerase B subunit as seen in other methanogens; and (ii) the presence of pseudomurein, a character shared with Methanobacteriales. Amazingly, the basal positions of *M. kandleri* in the SSU rRNA tree and the RNA polymerase tree are probably based on different artifacts: LBA in the RNA polymerase tree, and G+C content bias in the rRNA tree. Indeed, *M. kandleri* exhibits a very short branch in the rRNA tree, but a very high rRNA G+C content as compared to other methanogens.

The origin of methanogenesis

Phylogenetic analyses indicate that methanogenesis originated only once during archaeal evolution. Indeed, all methanogens (Methanococcales, Methanobacteriales, Methanomicrobiales, Methanosarcinales, and Methanopyrales) share the same set of homologous enzymes and cofactors required for the central pathway of methanogenesis (the hydrogenotrophic pathway) (Bapteste et al., 2005). The grouping of *M. kandleri* with other thermophilic methanogens (Methanobacteriales and Methanococcales) and of Methanomicrobiales and Methanosarcinales in the archaeal tree suggests that methanogenesis originated in Euryarchaeota at least after the divergence of Thermococcales. Methanogens appear interspersed among non-methanogens since methanogenesis enzymes were partly lost in the lineage leading to *Archaeoglobus* (and others re-targeted to other functions) and completely lost in the lineages leading to Haloarchaeales and Thermoplasmatales. Accordingly, Bapteste and colleagues have proposed classifying methanogens based on their position in the archaeal tree: group I includes the Methanopyrales, Methanobacteriales, and Methanococcales, and group II contains the Methanosarcinales and Methanomicrobiales (Bapteste et al., 2005). The division between groups I and II was recovered in a phylogeny based on the concatenation of orthologous enzymes involved in the hydrogenotrophic pathway and in the biosynthesis of coenzymes involved in methanogenesis (Bapteste et al., 2005), as well as in individual trees of these proteins. This indicates that proteins involved in methanogenesis have apparently not been subjected to HGT during archaeal evolution, challenging current thought that successful HGT frequently affects genes coding for operational proteins such as those involved in metabolic pathways. HGT may be more successful when the fixation of the incoming protein is favored by positive selection pressure. In the case of methanogenesis, there was no apparent positive selection pressure for HGT between different methanogens, and the all-in-one transfer of the complete pathway to non-methanogens was probably prevented by the dispersion of these genes in archaeal genomes.

The sudden appearance of the complete methanogenic pathway in the early evolution of Euryarchaea is quite surprising considering the number of proteins involved. Interestingly, three of the archaeal enzymes in the hydrogenotrophic pathway, and most of those required for the biosynthesis of coenzymes required for methanogenesis, have bacterial homologues (Chistoserdova et al., 2004). These are involved in aerobic methanotrophy (the degradation of methane to CO_2) or in formaldehyde detoxification in some α and γ Proteobacteria lineages. These proteins (thereafter called MMF proteins for methanogenesis, methanotrophy, formaldehyde detoxification) have also been identified in the genomes of *Planctomycetales* (Chistoserdova et al., 2004). The presence of MMF proteins in Bacteria can be explained in two ways. Either they were transferred from Archaea to Bacteria (Bapteste et al., 2005), or they were already present in the common ancestor of Archaea and Bacteria (Chistoserdova et al., 2004, and references therein). In the first scenario, aerobic methanotrophy originated as a novel pathway in Bacteria from MMF proteins of archaeal origin, whereas in the second, methanogenesis and anaerobic methanotrophy in Archaea and aerobic methanotrophy in Bacteria originated independently from an ancient formaldehyde detoxification pathway present in their last common ancestor (Chistoserdova et al., 2004). In agreement with the latter hypothesis, formaldehyde is presumed to have been very abundant on early Earth (Chistoserdova et al., 2004).

Several observations support the antiquity of MMF proteins: (i) successful transfer of MMF proteins from Archaea to Bacteria is unlikely, if the dispersion of their genes seen in archaeal genomes is ancestral; (ii) archaeal and bacterial MMF proteins appear well separated in phylogenetic trees (see Fig. 2 in Chistoserdova et al., 2004); and (iii) two MMF proteins are present in Crenarchaeota and one of these is also in the Thermococcales. These do not branch within the methanogens, as would be expected if they were present in the last common ancestor of Archaea and Bacteria rather than being recruited by HGT. If MMF proteins were already present in the last universal common ancestor (LUCA), one has to assume that they were lost independently in most bacterial and archaeal lineages. This might be possible, since the complete loss of all enzymes involved in methanogenesis occurred in Thermoplasmatales and Haloarchaeales. However, if the last archaeal common ancestor (LACA) was a methanogen, then methanogenesis was lost in both Crenarchaeota and Thermococcales.

The origin and evolution of methanogenesis thus appears to be a very important topic that requires new phylogenetic analyses, including more data. It will be especially interesting to analyze MMF proteins that were recently identified from uncultivated Archaea responsible for anaerobic methanotrophy (Hallam et al., 2004). These anaerobic methanotrophic Archaea branch close to Methanosarcinales in the SSU rRNA tree, suggesting that anaerobic methanotrophy originated in Archaea from the methanogenic pathway, although its relationship, if any, with the bacterial aerobic pathway is presently unknown.

The last common archaeal ancestor: a hyperthermophile?

The abundance of hyperthermophiles in the archaeal domain, together with the short branches of their lineages in the SSU rRNA tree, suggested early on that the LACA was a thermophile (Woese, 1987). However, this argument was questioned on the grounds that the high G+C content of hyperthermophilic rRNAs would reduce the available sequence space. This idea is that this might thereby produce short branches in phylogenetic trees, with homologous positions occupied by G or C bases being possibly due to convergence rather than common ancestry (Woese et al., 1991). However, the hyperthermophilic nature of the LACA is now independently supported by the early emergence of both hyperthermophilic Crenarchaeota and Euryarchaeota in r-protein and RNA polymerase subunit trees. One cannot completely exclude that a compositional amino acid bias artificially groups hyperthermophiles at the base of these protein trees as whole proteome analyses have shown that proteins from hyperthermophiles exhibit a clearly identified amino acid bias (Suhre & Claverie, 2003). To confirm (or not) the hyperthermophilic nature of the LACA, it would also be essential to include in the analysis several members of the presently uncultivated Archaea that cluster between Crenarchaeota and Euryarchaeota in the SSU rRNA tree.

The hyperthermophilic LACA hypothesis is also supported by phylogenetic analysis of reverse gyrase, the only protein specific to hyperthermophiles that can be detected by comparative genomic analysis (Forterre et al., 2000; Forterre, 2002). Reverse gyrase is an atypical DNA topoisomerase that adds positive supercoils to circular DNA *in vitro*. This enzyme is formed by the fusion of a classical type I DNA topoisomerase and a large helicase domain. Although the precise role of reverse gyrase *in vivo* is still unclear, this enzyme is certainly essential for life at high temperature. It is systematically present in all hyperthermophiles (Forterre, 2002) and a *Thermococcus kodakarensis* reverse gyrase knock out mutant cannot grow above 90 °C (Atomi et al., 2004). Two lines of evidence suggest that reverse gyrase first originated in Archaea and was later transferred to Bacteria: (i) bacterial reverse gyrases are interspersed within archaeal ones in phylogenetic trees; and (ii) reverse gyrase genes in bacterial genomes are sometimes surrounded by genes of archaeal origin. Interestingly, a reverse gyrase tree containing only archaeal sequences (Plate 2.3) can be superimposed with the "ideal" archaeal tree of Plate 2.4, suggesting that the LACA already contained this enzyme, and thus was a hyperthermophile. An alternative hypothesis would be an origin in the Euryarchaeota or Crenarchaeota and a very early transfer between these domains. In order to reach a definite conclusion it will be essential to obtain additional reverse gyrase sequences from hyperthermophiles branching very deeply in the SSU rRNA tree and to be sure of the rooting of the archaeal tree itself.

Nanoarchaeum equitans: a deep diverging third archaeal phylum?

The discovery of *Nanoarchaeum equitans* is undoubtedly the most fascinating recent finding in archaeal research (Huber et al., 2002, 2003). This tiny hyperthermophile grows and divides at the surface of the crenarchaeal *Ignicoccus* species. It holds the record of the smallest known living cell (with a volume size equal to 1/100 of *E. coli*) and of the smallest known cellular genome (490 Mb) (Waters et al., 2003). Its genome has in fact lost one-third of the core genes present in all other archaeal genomes (Makarova & Koonin, 2005). As a final exciting touch, *N. equitans* turned out to display a very divergent SSU rRNA sequence, with many base changes even in the so-called "highly conserved regions" that are usually employed as primer targets for SSU rDNA PCR (Huber et al., 2002). Huber and coworkers thus proposed a new archaeal phylum, the Nanoarchaeota (Huber et al., 2002). This proposal was rapidly accepted and apparently confirmed by the position of *N. equitans* in an archaeal rooted tree of concatenated r-proteins (Waters et al., 2003). As a consequence, idiosyncrasies observed in the genome of *N. equitans*, such as the presence of split reverse gyrase and tRNA genes, were interpreted as possible ancient traits (Waters et al., 2003; Randau et al., 2005).

N. equitans appears as a separate branch, distinct from those leading to Crenarchaeota and Euryarchaeota, in our recent unrooted trees based on r-protein and RNA polymerase subunits (Plate 2.1a, b; see also Brochier et al., 2005a). Nevertheless, a recent analysis showed that this position may be biased by an LBA artifact due to the high rate of evolution of its informational proteins, possibly triggered by adaptation to its symbiont lifestyle (Brochier et al., 2005a). Indeed, the phylogenetic analysis of a subset of r-proteins indicated, albeit with weak support, the grouping of *Nanoarchaeum* with Thermococcales (Brochier et al., 2005a). Intriguingly, additional crucial informational proteins (elongation factors EF1α and EF2, DNA topoisomerase VI, reverse gyrase, and tyrosyl-tRNA synthetase) seem to support this specific affiliation since they all show the grouping of *N. equitans* with Thermococcales (Moreira et al., 2004; Brochier et al., 2005b). In particular, the phylogenetic position of *N. equitans* in the reverse gyrase phylogenetic tree presently suggests that the encoding of its helicase and topoisomerase domains by two independent genes is not a primitive feature, but is due to a secondary split in the nanoarchaeal lineage (except if the root of the archaeal tree is located in the nanoarchaeal branch).

The branching of several *N. equitans* r-proteins within Crenarchaeota suggests their acquisition by HGT from the *Ignicoccus* host, whereas the branching of the remaining r-proteins within Euryarchaeota (the great majority) suggests that *N. equitans* itself could be a very divergent euryarchaeal lineage. If this is the case, the ribosome of *N. equitans* may exhibit an unusual evolutionary plasticity, being composed of a mixture of r-proteins derived by vertical descent from the unknown free-living euryarchaeote ancestor of Nanoarchaea and by HGT from the *Ignicoccus* host (Brochier et al., 2005b). In our opinion, the most likely interpretation of these data is that *N. equitans* belongs to the Euryarchaeota (as suggested by the majority of individual r-proteins trees), but is attracted towards Crenarchaeota in the whole ribosomal tree by a combination of LBA and HGT. We thus think that Nanoarchaea are bona fide Euryarchaeota that may be distant relatives of modern Thermococcales. Although the long branch of Nanoarchaea indicates a rapid evolution due to profound modifications in its translation apparatus, this divergence and the establishment of Nanoarchaea as symbionts of *Ignicoccus* species are probably ancient, as indicated by the lack of evidence of ongoing reduction in the genome of *N. equitans* (Waters et al., 2003), the wide environmental distribution of Nanoarchaea, and their recent discovery in symbiosis with other Crenarchaeota (K. Stetter, personal communication).

The various steps that surround the discovery of the phylogenetic relationships of *M. kandleri* and *N. equitans* show the power of phylogenomics approaches when they are carefully performed. In general, no outgroup should be used, and placements suspected to be artifactual (displaying long branches or instability) should be investigated further. Plate 2.4 shows what we think is the best possible tree of the archaeal domain that can be presently drawn from the convergence of the ribosomal and the RNA polymerase trees and from the critical analyses of the placements of *M. kandleri* and *N. equitans*. It should not be considered as definitive, since some hidden compositional bias (see below) or LBA artifacts may still affect both ribosomal and RNA polymerase trees in a similar way. The analysis of the genomes of *Ignicoccus*, of a recently isolated Koryarchaeon, and of *Crenarchaeum symbiosum* (all currently under way) should rapidly give new important data, in particular to confirm or not our present interpretation of the phylogenetic position of Nanoarchaea. As additional genomes are completed, the data needed to break remaining long branches and improve the detection of HGT will become available. In the end, the current tree will be either confirmed or subject to further refinement.

The role of horizontal gene transfer in archaeal history

HGT between Archaea and bacteria

The importance of HGT in microbial evolution has been heavily debated in recent years. Whereas some authors have argued that HGT is so frequent that tree building is a futile exercise (Doolittle, 1999), others have argued that HGT is overestimated, and that most puzzling phylogenetic patterns usually attributed to HGT are due to hidden paralogy (Glansdorff, 2000) or weak phylogenetic signal. We have seen that a core of informational genes can be used to infer proper phylogeny in the case of Archaea, and that even operational genes (i.e. those involved in methanogenesis) can be refractory to HGT. However, as indicated by whole genome trees and other global analyses, HGT from Bacteria to Archaea have clearly occurred at a relatively large scale and have greatly influenced the history of the archaeal domain. In a recent analysis, Wiezer and Merkl (2005) concluded that the flux of HGT occurred nearly unidirectionally from Bacteria to mesophilic or moderate thermophilic Archaea, whereas it was approximately balanced between hyperthermophilic Archaea and Bacteria. HGT from mesophilic Bacteria to Archaea have allowed an important expansion of the metabolic repertoire in the archaeal domain. A well documented case is the bacterial origin of several enzymatic steps that allow mesophilic methanogens to use a variety of organic substrates instead of CO_2 for methanogenesis. Other phenotypic traits that were possibly introduced into Archaea by HGT from Bacteria are aerobic respiration, photosynthesis based on bacteriorhodopsin, and ammonium oxidation. Similarly to methanogenesis, these capabilities occurred late in the history of the archaeal domain. The proteins involved in these processes seem to be more widely distributed in Bacteria than in Archaea. Unfortunately, there are presently no updated phylogenetic analyses clarifying these points. However, some proteins involved in aerobic respiration in Haloarchaeales are clearly of bacterial origin (Kennedy et al., 2001). HGT from mesophilic Bacteria to Archaea might have played an important role in the adaptation of some Archaea to cold environments. For instance, several genes branching within mesophilic Bacteria have been detected in genome fragments of uncultivated marine Crenarchaea (Lopez-Garcia et al., 2004). Similarly, as we have previously seen in the case of reverse gyrase, Bacteria probably also recruited some archaeal genes for adaptation to hot environments. A priori, the selection pressure for the fixation of horizontally acquired genes in the course of the adaptation to new environments could have affected all genes, including informational ones. However, with a few exceptions, successful HGT between Archaea and Bacteria has only involved operational proteins. For instance, there is not a single well documented case of acquisition by Archaea of r-proteins or RNA polymerase subunits from the two other domains (Brochier et al., 2004). This clearly indicates that it was much easier for Archaea and Bacteria to adapt their own informational proteins to function at different temperatures than to replace them by proteins from the other domain.

HGT involving informational genes

A few HGT of informational proteins between domains might have had an important impact on archaeal history. For example, a bacterial gene coding for DNA gyrase was acquired by Archaeoglobales, Haloarchaeales, Thermoplasmatales, Methanosarcinales, and Methanobacteriales (Gadelle et al., 2003). At first sight, this distribution would suggest that acquisition of DNA gyrase occurred only once, just before the divergence of the ancestor of all these Archaea from other Euryarchaeota. However, phylogenetic analyses favor instead multiple independent transfers from Bacteria to Archaea (Gadelle et al., 2003). The presence of DNA gyrase has drastically changed the topological state of the intracellular DNA in these Archaea. Whereas all other Archaea (with or without reverse gyrase) have relaxed or slightly positively supercoiled DNA (Charbonnier & Forterre, 1994), those with DNA gyrase have negatively supercoiled DNA, otherwise typical of Bacteria (Sioud et al., 1987b; Lopez-Garcia et al.,

2000). Haloarchaeales have become dependent on DNA gyrase and their growth is inhibited by specific DNA gyrase inhibitors, such as novobiocin (Sioud et al., 1987a, b). Similarly, Thermoplasmatales have lost their original archaeal type II DNA topoisomerase (Topo VI) (Gadelle et al., 2003) and eukaryotic-like histone, typical of Euryarchaeota, which has been replaced by a homologue of the HU type bacterial DNA-binding protein. One can wonder how these organisms may have completely restructured all their protein–DNA interactions after such "informational" HGT.

HGT within Archaea

Although they are more difficult to detect than those between Archaea and Bacteria, transfers between archaeal lineages have likely been more frequent than between domains, as indicated by the fact that a few HGT events were identified for r-proteins and RNA polymerase subunits (Matte-Tailliez et al., 2002; Brochier et al., 2004). Although HGT events are easier to detect between Euryarchaeota and Crenarchaeota, a few cases were identified between different euryarchaeal lineages. For example, in the course of their phylogenetic analysis of archaeal RNA polymerase subunits, Brochier and colleagues noticed that the H subunit of *M. kandleri* RNA polymerase has been replaced by its orthologue from a relative of Thermoplasmatales (Brochier et al., 2004). Diruggiero and coworkers identified the recent transfer of a 16 kb region containing an ABC transporter system for maltose between *Pyrococcus furiosus* and *Thermococcus litoralis* (Diruggiero et al., 2000). Again, a major selection pressure for intra-domain HGT was probably the adaptation to novel environments. It has been suggested in particular that this explains the high number of euryarchaeal-like genes in recently analyzed contigs of marine Crenarchaeota (Lopez-Garcia et al., 2004). If this hypothesis were true, it would suggest that the adaptation of some Crenarchaeota to mesophily occurred after that of Euryarchaeota.

HGT seem to have been especially prevalent between the two thermoacidophilic lineages: Thermoplasmatales and Sulfolobales (Ruepp et al., 2000; Futterer et al., 2004; Ciaramella et al., 2005). It was possible to identify a core of 690 genes common to *Picrophilus oshimae* and *Thermoplasma acidophilum*, two Thermoplasmatales (Euryarchaeota), and *Sulfolobus solfataricus*, a Sulfolobale, as well as 471 genes present in one of the two Thermoplasmatales and *S. solfataricus* (Futterer et al., 2004). For instance, the euryarchaeon *Picrophilus oshimae* shares only 35% of its genes with Pyrococcales (members of the Euryarchaeota), but 58% with the crenarchaeon *S. solfataricus* (Futterer et al., 2004). This suggests that a significant flux of HGT occurred between Thermoplasmatales and Sulfolobales. In this particular case, ecological closeness apparently overrode phylogenetic relatedness (Futterer et al., 2004).

It was initially proposed that HGT from Sulfolobales to Thermoplasmatales was more likely because of the absence of a cell wall in *T. acidophilum* (Ruepp et al., 2000). However, the same trend is also observed in *Picrophilus*, a wall-containing Thermoplasmatale. Moreover, 13% of the proteins shared by *P. oshimae* and *S. solfataricus* are absent in *T. acidophilum*, indicating either recent transfers from Sulfolobales to *Picrophilus* (reviewed in Ciaramella et al., 2005) or recent losses in *T. acidophilum*. Such evidence for important HGT provides a nice explanation of the attraction of Thermoplasmatales towards Crenarchaeota in many whole genome trees. One explanation for the observed HGT trends among thermoacidophilic Archaea is that Thermoplasmatales populated acidic hot springs after Sulfolobales were already established in these particular environments and adapted to their new living style by acquiring genes from Sulfolobales. This fits with the idea that Thermoplasmatales originated from some methanogenic lineages relatively late in the history of Euryarchaeota.

The origin of Archaea and their relationships with Bacteria and Eukarya

The position of Archaea in the universal tree of life

As discussed herein, the three-domains concept based on SSU rRNA trees has been validated by comparative genomics. The vast majority of genes in archaeal genomes have their most closely related homologues in other archaeal genomes, rather than bacterial or eukaryotic ones. The rule is that all archaeal housekeeping proteins (especially informational ones) cluster together in phylogenetic trees, away from bacterial and eukaryal homologues, when these exist. Hence, they correspond to archaeal versions (*sensu* Woese), and the same is true for most informational bacterial and eukaryotic proteins.

Although the three-domains concept is generally accepted, there remains no consensus about the mechanism that led to the formation of the three domains and about their evolutionary relationships. The universal rooted trees that were published between the end of the 1980s and the mid-1990s all converged on what has become the classical model for the evolutionary relationships between domains. In this model, Bacteria derived directly from the LUCA while Archaea and Eukarya shared a common ancestor more recent than the LUCA. According to this model, the similarities between the informational mechanisms of Archaea and Eukaryotes are considered as derived features that appeared in a lineage ancestral to these two domains. However, it should be remembered that archaeal/eukaryal features could also be a priori primitive traits that were lost or modified in Bacteria, if these are a very derived lineage. In a second model, which is increasingly popular, Archaea and Bacteria are both thought to have derived directly from the LUCA and two of their members later fused to produce the eukaryotic lineage (for a review see Lopez-Garcia & Moreira, 1999). These fusion models are designed to explain the chimeric nature of eukaryotes, where many informational proteins are archaeal-like, while membrane lipids and many operational proteins are bacterial-like. In a third model, proposed by Cavalier-Smith, the LUCA was a bacterium and both Archaea and Eukarya derived from a particular lineage of Gram-positive Bacteria (Cavalier-Smith, 2002).

Specific objections can be made against the second and third models. For instance, there is no current molecular evidence that the archaeal-like proteins of eukaryotes are specifically related to a particular archaeal lineage, as fusion models predict. Similarly, the presence of bacterial-like operational proteins in Eukarya, emphasized by chimeric models, can be easily explained by the massive transfer of bacterial genes to the eukaryotic nucleus that occurred at the onset of mitochondrial evolution. Finally, a special affinity between Archaea and Gram-positive Bacteria postulated in the third model has not been observed in any phylogenetic data and not even at the whole genome level. The few cases of archaeal proteins of Gram-positive affinity originally reported in the literature are likely due to HGT (Gribaldo et al., 1999). On a more general ground, Carl Woese has similarly argued strongly against these two models by pointing out that "modern cells are fully evolved entities which are sufficiently complex, integrated and individualized that further major change in their design does not appear possible" (Woese, 2002).

How did the three domains originate?

In order to discuss the relative pertinence of these various models at a more general level, it should be recalled that the three-domain concept was primarily based on the existence of three different "versions" (*sensu* Woese) of rRNA and r-proteins. As recently stated by Woese, "Why canonical patterns exist is a major unanswered question," and this is one of the reasons why several models are still in competition to explain the origin of the domains. To explain these three canonical patterns, it has been suggested that three "dramatic evolutionary events" (Forterre & Philippe, 1999) or "major qualitative evolutionary changes" (Woese, 2000) triggered a drastic modification in the rate of protein evolution (either a reduction or an acceleration) in the lineages leading to modern domains, prior to their diversification. Woese suggested that "the rate of protein evolution was higher in the timeframe between the LUCA and the last common ancestor of each domain, than it is today. As a consequence, subsequent protein evolution occurring at slower rate after the formation of the three domains was unable to erase the signatures of previous divergent evolution that occurred during the fast-track period." The idea that proteins were fast evolving at the time of the LUCA has its root in Woese's conception of the LUCA as a progenote, a primitive organism whose mechanisms for protein synthesis and genome replication were both still error-prone (Woese & Fox, 1977a). Woese and Fox even suggested that the progenote still had an RNA genome and predicted in 1977 that DNA replication proteins should be different (non-homologous) in Bacteria and Eukarya. This prediction was confirmed twenty years later by comparative genomics data (Olsen & Woese, 1996). This led to the suggestion that DNA replication was independently invented twice: once in Bacteria and once in a lineage leading to Archaea and Eukarya (Mushegian & Koonin, 1996). In this model Archaea and Eukarya originated from an ancestor with a DNA genome, whereas Bacteria derived directly from a LUCA with an RNA genome (Leipe et al., 1999). Since DNA genomes can be replicated more faithfully than RNA genomes, the transformation from an RNA LUCA to modern DNA cells would have been immediately followed by a drastic drop in the evolutionary tempo of protein evolution. This could a priori explain the formation of canonical versions very different between Bacteria on one side and Archaea/Eukarya on the other.

A role for viruses?

Recently, it has been suggested that each of the three cellular domains originated from the "fusion" of three different RNA cells with DNA viruses (Forterre, 2005). This model, called the three viruses three domains theory, explains the existence of three domain-specific versions of r-proteins, since these would have come from three separate RNA-cell lineages where fast evolution had produced the three canonical ribosome versions. By chance alone, two of these RNA cell lineages would have been more similar and fused with similar viruses, and these would have given rise to Archaea and Eukarya. The differences between the DNA replication machineries in present-day Archaea and Eukarya would reflect those between the two founder viruses. These differences include the absence of archaeal DNA polymerase D family in Eukarya and the absence of eukaryotic-like DNA topoisomerases IB and IIA in Archaea. Finally, this new theory can also explain the phylogenies of RNA and DNA polymerases where eukaryotic versions appear to stem from different viral lineages (Filee et al., 2002; Raoult et al., 2004), although it is presently unclear whether this reflects true relationships or artifacts of tree reconstruction caused by the high evolutionary rates of viral sequences (Moreira & Lopez-Garcia, 2005).

How can we polarize archaeal traits?

It remains to be determined whether the numerous common characters shared by Archaea and Eukarya informational machineries were already present in the LUCA or appeared after the separation of Bacteria from a lineage common to Archaea and Eukarya. The same question can be asked for the few (but important) characters shared by Archaea and Bacteria, such as conserved order in RNA polymerase and r-protein operons, the similar genomic structure and mode of genome evolution, or else the use of Shine–Dalgarno sequences for the initiation of translation (Londei, 2005). Some of these features could be due to convergent evolution but others could be relics of the LUCA or a common ancestor of Archaea and Bacteria. These questions could be partly solved by rooting the universal tree of life, although it is not clear that this is possible (Bapteste & Brochier, 2004). Initial attempts that rooted the tree in the bacterial branch have been criticized on methodological grounds (Forterre & Philippe, 1999; Lopez et al., 1999; Poole et al., 1999). Several authors have even suggested a root in the eukaryotic branch, based either on the analysis of slowly evolving positions in paralogous proteins that diverged before the LUCA (Brinkmann & Philippe, 1999; Lopez et al., 1999) or on cladistic analyses of rRNA structure or protein folds (Caetano-Anolles & Caetano-Anolles, 2003, 2005). A novel approach to the problem may be required. One possibility might be to try polarizing the evolution of the different molecular machineries by using all structural and mechanistic information available to test the credibility of opposite evolutionary scenarios.

Conclusion and prospects

It is fascinating that, only thirty years after their fortuitous discovery, we are already able to unravel much of the history of Archaea. All available evidence confirms the distinctive nature of the third domain of life. Nevertheless, the origin of the similarities between Archaea and Eukarya remains one of the most fascinating issues in evolution. The analysis of complete genomes has made it possible to highlight a coherent gene complement for the Archaea. The data have also provided insight into the flux of HGT between Archaea and Bacteria and the role of HGT in the evolutionary history of the two prokaryotic domains. Genomic data have also provided the basis for a rather exhaustive reconstruction of the archaeal tree.

A core of suitable genes has provided a phylogenetic signal that converges on a single evolutionary scenario that may be taken as the best current picture of archaeal history. The archaeal ancestor was probably an organism adapted to survive in hyperthermophilic environments, a capacity that was later passed on to a few Bacteria. The future placement of cold-adapted Crenarchaeota and additional species belonging to basal branching lineages in the archaeal tree will provide a significant test of this hypothesis. Methanogenesis also appeared in Archaea and followed a unique and mainly vertical evolutionary history – with an astonishing absence of transfer within Archaea and to Bacteria – and subsequent repeated losses. This, and other evidence, renders open the possibility that, despite the non-basal emergence of present-day methanogens in the archaeal tree, the archaeal ancestor may have been able to perform methanogenesis. This will be highly relevant to discussions on the nature of early Earth biota, when a consensus on the interpretation of fossil evidence will permit us to confidently place these events in geological time.

The assignment of Nanoarchaeota to a third archaeal phylum may be premature, since phylogenetic and genomic evidence points to Nanoarchaea as a fast-evolving euryarchaeal lineage. A more complete knowledge of nanoarchaeal diversity and the sequencing of new genomes as well as that of their hosts will likely allow confirmation of the phylogenetic affiliation of this extremely interesting lineage. Similarly, will Korarchaeota stand the test of time when genomic data will be available? Indeed, it may be that the Crenarchaeota/Euryarchaeota divide will be shown to be a crucial and very ancient point in archaeal history. The evolutionary differences between these two phyla appear far more profound than those observed between any of the 23 currently recognized bacterial phyla. This may be linked to our ability to retrace the divergence between the different archaeal lineages in a more robust way than is possible for the Bacteria. It may be that the first steps in archaeal evolution were not accompanied by the rapid radiation seen in their prokaryotic cousins. Alternatively, this apparent ancient divergence may simply reflect an incomplete sampling of true archaeal diversity. Future environmental studies will provide precious insights into this issue. In any case, whatever the answers ultimately are, the Archaea will remain an essential key to biological research and an invaluable tool to look back into our past.

Acknowledgments

We thank Hans-Peter Klenk, Roger Garrett, and Kim Brügger for kindly providing sequences from *Hyperthermus butylicius*, Yvan Zivanovic and Fabrice Confalonieri for kindly providing sequences from *T. gammatolerans*, and Shiladitya DasSarma and the members of the university of Scranton for the sequences of *H. volcanii* freely available by BLAST on http://halo.umbi.umd.edu/cgi-bin/blast/blast_hvo.pl. We are indebted to three anonymous referees for their accurate revision of the manuscript and a number of valuable comments. We would like to dedicate this chapter to the memory of Wolfram Zillig, who was fascinated by the history of Archaea and always tried to extract as much information as possible from his work either on archaeal RNA polymerases or on archaeal viruses.

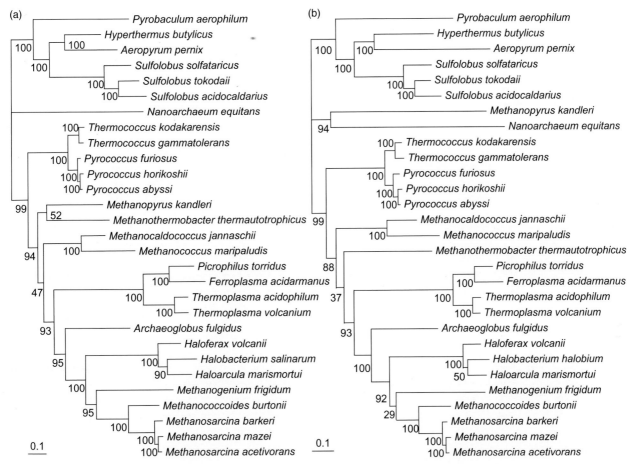

Plate 2.1 Unrooted maximum likelihood trees based on (a) a concatenation of 53 r-proteins (5879 positions) and (b) a concatenation of 12 RNA polymerase subunits and two transcription factors (2629 positions). The trees were calculated by PHYML (JTT model, gamma correction (eight discrete classes), an estimated alpha parameter, and an estimation of the proportion of invariant positions; Guindon & Gascuel, 2003). Numbers at nodes are bootstrap values calculated from 1000 replicates by PHYML. The scale bar represents the percentage of substitutions per site.

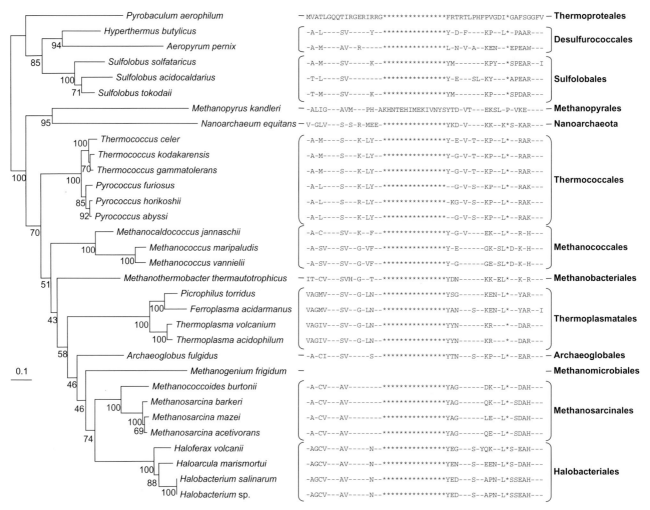

Plate 2.2 Unrooted maximum likelihood tree of RNA polymerase A1 subunit (720 positions). Calculation was made by PHYML (JTT model, gamma correction (eight discrete classes), an estimated alpha parameter, and an estimation of the proportion of invariant positions; Guindon & Gascuel, 2003). Numbers at nodes are bootstrap values calculated from 1000 replicates by PHYML. The scale bar represents the percentage of substitutions per site. An example of insertion specific to *Methanopyrus kandleri* is provided. It corresponds to positions 808–822 of its RNA polymerase A1 subunit.

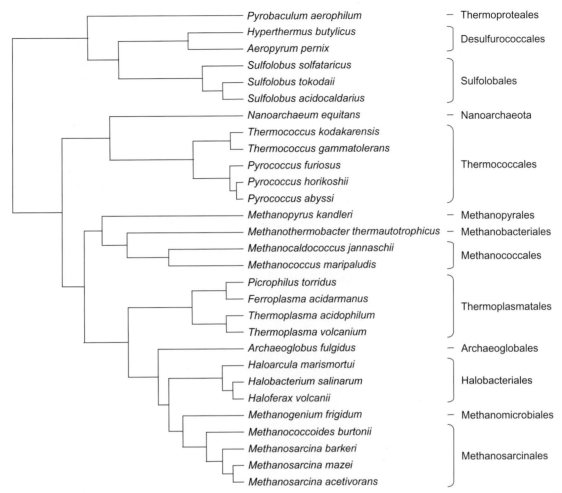

Plate 2.3 Consensual and likely phylogeny of Archaea for which complete genome sequences are available, drawn after the compilation and careful analysis of all genomic studies and data.

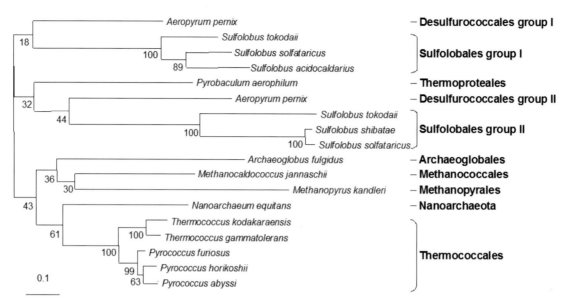

Plate 2.4 Unrooted maximum likelihood tree of reverse gyrase. Calculation was made by PHYML (JTT model, gamma correction (eight discrete classes), an estimated alpha parameter, and an estimation of the proportion of invariant positions; Guindon & Gascuel, 2003). Numbers at nodes are bootstrap values calculated from 1000 replicates by PHYML. The scale bar represents the percentage of substitutions per site.

3

The root of the tree: lateral gene transfer and the nature of the domains

David A. Walsh, Mary Ellen Boudreau, Eric Bapteste and W. Ford Doolittle

Introduction

One problem Darwin claimed to have solved for biologists was how to understand the patterns of similarity and difference between organisms that allowed systematists to classify species hierarchically, in "groups subordinate to groups." On this topic he wrote:

> *The affinities of all the beings of the same class have sometimes been represented by a great tree. I believe this simile largely speaks the truth. The green and budding twigs may represent existing species; and those produced during each former year may represent the long succession of extinct species. At each period of growth all the growing twigs have tried to branch out on all sides, and to overtop and kill the surrounding twigs and branches, in the same manner as species and groups of species have tried to overmaster other species in the great battle for life. The limbs divided into great branches, and these into lesser and lesser branches, were themselves once, when the tree was small, budding twigs; and this connexion of the former and present buds by ramifying branches may well represent the classification of all extinct and living species in groups subordinate to groups. Of the many twigs which flourished when the tree was a mere bush, only two or three, now grown into great branches, yet survive and bear all the other branches; so with the species which lived during long-past geological periods, very few now have living and modified descendants. (Darwin, 1859, p. 171)*

Thus in Darwin's view it was a real branching process, a historical succession of speciation events, that made it possible to represent relationships between living species as a tree, and that must be reconstructed by any classification claiming to be natural. Molecular phylogeneticists have, by and large, taken this "great tree" simile as truth, and considered it their task to recreate that succession of branching speciations as a single, universal, Tree of Life. Seldom have they considered it a hypothesis to be tested – which is precisely what we propose in concluding this chapter.

The division of the living world into three domains (Woese & Fox, 1977b) was the crowning achievement of the first decade of molecular phylogenetic inquiry, and coincidentally a fulfillment of Darwin's conjecture that "of the many twigs which flourished when the tree was a mere bush, only two or three, now grown into great branches, yet survive and bear all the other branches." In Chapter 2 of this volume, Forterre et al. summarize molecular phylogenetic and comparative genomic data that support Woese's three-domain view of life, and, it would seem, justify Darwin's belief in the truth of the tree simile. They present statistically robust trees for Archaea based on concatenations of translational or transcriptional component genes, which are, with one notable exception, fully congruent. They also show how by searching for traits that are shared or not shared between twigs now at the end of the great branches we call Bacteria, Archaea, and Eukarya, we might learn something about the root of the original bush – the last universal common ancestor, or LUCA.

Yet just a few years ago, Woese (2002), musing on the meaning of lateral gene transfer (LGT) for the universal tree, called this triumph of the Darwinian paradigm into question. He wrote:

*although organisms do have a genealogy-defining core of genes whose common history dates back to the root of the universal tree, that core is very small. Our classically motivated notion had been that the genealogy of an organism is reflected in the common history of the **majority** of its genes. What does it mean, then, to speak of an organismal genealogy when nearly all of the genes in the cell – genes that give it its general character – do **not** share a common history? This question again goes beyond the classical Darwinian context.*

Woese was referring to the growing evidence from genome sequencing that an unexpectedly large fraction of many bacterial and archaeal genomes have been introduced by LGT and indeed do have different histories. At the level of genes (to which Darwin of course lacked any access), evolution is not just a process of branching (lineage splitting). Here we ask, in the context of Archaea, how and what we know about the common history of genes and what indeed it might mean to speak of an organismal genealogy, and particularly of LUCA.

How many genes in archaeal genomes have been transferred?

One approach to assessing the impact of LGT on any genome is to look for genes that by their patterns of sequence similarity "don't belong," having closer matches to genes in genomes of seemingly distant taxa than to genes in genomes of close relatives. Many authors have used BLAST-based methods to assess the contribution of LGT from bacteria to the structure of archaeal genomes (for instance, Ng et al., 2000; Koonin et al., 2001; Deppenmeier et al., 2002; Ragan & Charlebois, 2002). Figure 3.1 provides an updated version of such an analysis with currently available archaeal genomes and the "competitive matching" tool developed by R. L. Charlebois (Gophna et al., 2004), which serves specifically to facilitate the detection of LGT. All ORFs of a test genome are reported that match any ORF from a member of one group of genomes (here all bacterial genomes) better than any ORF from a member of a second set of genomes (here all archaeal genomes) by some previously determined difference in normalized BLAST scores. Similarly we estimate the proportion of archaeal genes inherited in a "vertical" manner (by descent) or by transfer from an archaeal genome. From 38 to 70% of the genes in archaeal genomes can be categorized as one or the other in this way. The rest have no acceptable match to any other genome (our expectation value was e^{-10}), or a normalized BLAST score difference of less than 0.05 between groups.

For the majority of archaeal genomes, the genes analysed are more similar to other archaeal genes than to bacterial genes, with the most "archaeal" being the tiny (552 genes) genome of *Nanoarchaeum equitans* (Waters et al., 2003). But a surprising 40–50% of the haloarchaeal and methanogen genes for which we can draw a tentative conclusion have been transferred across the Bacterial/Archaeal domain boundary. Clearly there is a genome size effect here, not unexpected (Fig. 3.2a). One way for genomes to get bigger is to import genes, and if the source can be either bacterial or archaeal, it must follow that larger archaeal genomes will have proportionately more "bacterial genes" – if at their cores there is a conserved cadre of Archaea-defining transcriptional and translational determinants.

Does Fig. 3.1 bespeak a lot of LGT or only a little? There are many reasons why this analysis, and indeed any attempt to ask "how many transferred genes are there in genome X?" is biased, incomplete, and ambiguous from the start. In the direction of overestimating the impact of LGT, we must recognize: (i) that BLAST is a poor way to do phylogeny (Koski & Golding, 2001); (ii) that an archaeal gene can match a bacterial one because the latter is a recent transfer from, rather than to, the archaeal lineage; and (iii) that because there are more bacterial than archaeal genomes to query, genes that are rare and patchily distributed among Bacteria and Archaea are more likely to find bacterial best matches. As Fig. 3.2(b) shows, when we sample random collections of one-tenth of the bacterial genomes (thus roughly as many as all archaeal genomes), the fraction of apparent "bacterial genes" in these Archaea indeed does fall, by about 40%.

In the direction of underestimation: (i) BLAST can as easily produce false negatives as false positives; (ii) transfers from bacteria which are ORFans among sequenced genomes will be missed; and (iii) deep transfers will be missed. This last is a serious and possibly unavoidable cause of underestimation. If tested

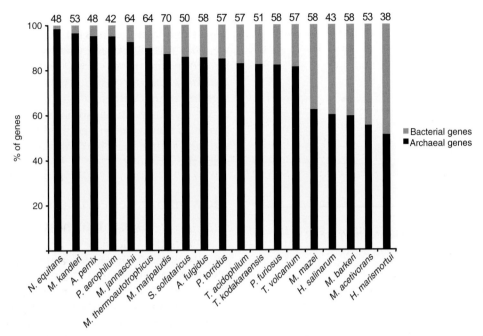

Fig. 3.1 The bacterial gene content of archaeal genomes as reported by "competitive matching" analysis. Competitive matching has been developed specifically to facilitate the detection of LGT (Gophna et al., 2004) where all the ORFs of a genome are reported that match any ORF from a member of one group of genomes (group I) better than any ORF from a member of a second set of genomes (group II) by some previously determined normalized blast score units. In this case, we estimate the number of potential bacterial genes by setting group I to all Bacteria and group II to all Archaea excluding the query genome. Contrastingly, the proportion of archaeal genes, assumed to have been inherited in a vertical manner or by transfer from an archaeal genome, can be estimated by inverting the identity of group I and group II to Archaea (excluding query genome) and Bacteria, respectively. For reasons described in the text, we removed close relatives from the analysis to improve the recovery of potentially transferred genes into the ancestor of established archaeal groups. Bars correspond to the percentage of genes for which we have made a decision on potential origin. Numbers above bars correspond to the percentage of genes in the respective genome for which a decision was made.

archaeal genome X has a closely related genome Y included in the analysis, then transfers from bacteria to their common ancestor that have been retained in both will be missed. In preparing Fig. 3.1, we have partially compensated for this by excluding *Haloarcula marismortui* when assessing transfers into *Halobacterium salinarum* (sp. NRC-1), and vice versa, excluding all other *Sulfobus* genomes in assessing *S. solfataricus*, excluding two of the three Thermoplasmatales (*Picrophilus torridus, Thermoplasma acdiophilum,* and *T. volcanium*) when analyzing any one, and similarly two of the three *Methanosarcina* (*mazei, barkeri,* and *acetivorans*) when examining the third. But this is only a partial correction. With reference to the archaeal tree presented as Plate 2.2 in Chapter 2 of this volume, we would in our analysis of Haloarchaea (for instance) have missed any bacterial trans-fers now resident in a haloarchaeal genome that were deposited there prior to their divergence from Methanosarcinales, or before the joint divergence of halophiles and Methanosarcinales from the Archaeoglobales, or of all these from the Thermoplasmatales, and so forth all the way back to the Crenarchael/Euryarchaeal divergence. All such transferred genes *should* be part of any counting of "bacterial genes" in an archaeal genome, but the more archaeal genomes we exclude, the less discriminating the analysis becomes, because of the patchiness of distribution of many genes, and the vagaries of BLAST scores.

This conundrum underlies any comparative whole genome analysis, we believe, and there is still another enormous difficulty. Although transfers from distant taxa are most likely to be *detected* by

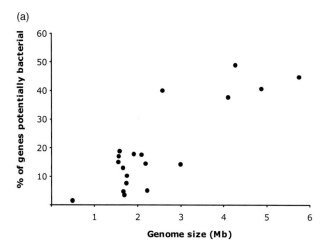

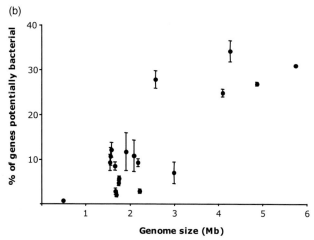

Fig. 3.2 The bacterial gene content of archaeal genomes as described in Figure 1 plotted against genome size. (a) Group I included 213 bacterial genomes. (b) Group I included a subsample of 23 bacterial species. Bacterial genomes were chosen to reflect similar size of archaeal genomes as well as to maximize the phylogenetic and metabolic diversity of Bacteria. Values presented are the means of three unique sets of bacterial genomes. Error bars correspond to one standard deviation of the mean.

integrate with recipient physiology. Only transfers from bacteria were assessed in Fig. 3.1. So whether we take this figure as evidence for a lot of transfer (our attention fixed on Haloarchaea and *Methanosarcina*) or just a little (focusing on *N. equitans* and others with small genomes), we are certain to be seriously underestimating genes in archaeal genomes that have arrived there by LGT – from any source – rather than by continuous vertical descent since the days of LUCA.

How many genes in Archaea have not been transferred?

Another way to assess LGT addresses genes, not the genomes in which they reside, and asks to what extent their phylogenies are congruent. There have been a number of exercises of this sort published in the literature, involving taxa as shallow as strains of *Escherichia coli* and other enteric Bacteria (Daubin et al., 2003) or as deep as all Bacteria, all Archaea or all life (Teichman and Michison, 1999; Creevey et al., 2004; Beiko et al., 2005), and focusing on the core of presumably orthologous genes shared by all the members of such taxa (thousands for strains of *E. coli*, fewer than 100 for all life; Daubin et al., 2003; Charlebois and Doolittle, 2004). Often the result, even at moderate depth, is that many genes shared between the species examined have weak phylogenetic signal, giving different but poorly supported trees. A standard "remedy" is to assess the different genes' collective signal – for instance, by concatenating them and aligning as if a single enormous gene, for phylogenetic treatment. Then each individual gene is tested against the concatentate tree to see if it can reject it by a statistical test such as the SH test of Shimodaira and Hasegawa (1999) or the AU test of Shimodaira (2002). Lerat et al. (2003), for instance, performed such a study of 205 single-copy orthologous genes shared by the genomes of 13 γ-proteobacteria. They found that the tree of the concatenate was rejected at the 5% level by only two of the 205 genes, and concluded that non-rejection by the remaining 203 meant that "single-copy orthologous genes are resistant to horizontal transfer." On re-examination, however, we (Bapteste et al., 2004) could show that more than half of these 203 genes also failed to reject more than half of a collection of 105 trees for the 13 taxa, trees chosen to be

such crude BLAST-based (or indeed any comparative) methods, transfers from more closely related taxa are more likely to occur, for a host of reasons. Recombination (orthologous replacement) is more frequent between more similar sequences, expression systems will be more compatible, agents of exchange (phages and conjugation systems) can be taxon-specific, and imported genes are more likely to biologically reasonable but differing at many nodes from that of the concatenate. Failure of rejection is not exclusive or unambiguous support of a tree. If a single true tree is assumed, then making a tree of a concatenate and assessing the support of individual genes seems a good way to find it. But if the goal is to prove that the individual genes in any core do have the same tree, this is not enough. The fact that more than one tree is not rejected when only one can be true means that wrong trees escape rejection. Hypothesis tests like those of Shimodaira and Hasegawa are designed to give small probabilities that a tree is rejected when it is the true tree. They provide no control over the probability that a wrong tree is not rejected (Bapteste et al., 2005b).

There are fewer genes shared among all Archaea than among all γ-proteobacteria, and their signals are weaker. Mostly these genes are "informational" (involved with transcription or translation), indeed mostly they encode ribosomal proteins. Over the past few years, Forterre and coworkers (Matte-Taillez et al., 2002; Brochier et al., 2004, 2005a) have presented increasingly taxon-rich archaeal trees based on concatenations of such translational/transcriptional genes (see also Chapter 2 in this volume). Although these trees are well resolved and in agreement with much of what is believed about archaeal relationships, they similarly do not **prove** congruence of the genes concatenated – indeed some ribosomal proteins (eight of the 53 tested by Matte-Taillez et al., 2002) revealed themselves to have been involved in transfers. As for the rest, although they do not reject the tree of the concatenate, they do not support it exclusively. Recently we (Bapteste et al., 2005b) have described a new and more sensitive way of testing and displaying congruence (or lack of it) in such datasets of core genes. We make two-dimensional grids, "heat maps" like those now popular in presenting results of microarray-based studies of gene expression under varying conditions. Each square in our grids, corresponding to a particular gene and a particular topology, is colored according to the p-value of the data (the gene alignment) given the toplogy, and then the grid is rearranged (double-clustered) to group genes with similar patterns of support/rejection of topologies along columns, and topologies with similar patterns of support/rejection by genes along rows. The heat map we obtained with the 44-gene (no LGT) dataset of Matte-Taillez et al. (2002) is reproduced as Plate 3.1. We concluded in this case that at least 62 topologies were "supported" (not rejected) by at least some of these genes, while among the best supported topologies, none was supported by all the genes. This is not to say that there is no signal among the genes (some topologies are rejected by all), nor that signals are all conflicting – indeed there appear to be "central tendencies," or clusters of related topologies favored by many genes. But it is to say that even among the 44 pre-screened genes of the archaeal translational core there likely has been some transfer, and that for most of the 44 we simply cannot eliminate that possibility. The existence of a subset of phylogenetically congruent genes is possible but not established. A good term might be "not proven," in the sense defined at www.bbc.co.uk: "a verdict given in Scottish Courts which establishes that the prosecution have not provided enough substantial proof to convict the individual of a crime, but that there remains considerable doubt regarding the offender's innocence."

Speaking of organismal genealogy

Proof would not be required if there were no logical alternatives to the "classical Darwinian context." Darwin lacked a coherent theory of heredity, so that defining the Darwinian context of molecular phylogenetics does involve some conceptual extrapolation. In our extrapolation, we accept Woese's assertion that the "classically motivated notion had been that the genealogy of an organism is reflected in the common history of the majority of its genes." We contend further that the Darwinian context entails the notion that organisms are variously similar (and classifiable into groups subordinate to groups) primarily because of the genes they share through common vertical descent. In this perspective, there are at least three non-classical alternatives:
• A "genealogy-defining core of genes." This is probably the current consensus. It holds that a core of genes, almost exclusively involved with transcription and translation, has escaped transfer since the time of LUCA, probably because their functions are so critical and so highly co-evolved and interdependent that any heterologous substitution would be deleterious (the "complexity hypothesis"; Jain et al., 1999). Because core genes have not been transferred, they reliably track "organismal genealogy"

(Brochier et al., 2002; Matte-Taillez et al., 2002; Lerat et al., 2003). For organisms that multiply by dividing, this genealogy is simply a record of all the events of cell division dating back to LUCA, sometimes thus called the phylogeny of the cell or envelope. The core of shared and potentially congruent genes is very small. Most would admit that the majority of non-ORFan genes in any genome today have at least one LGT in their history. Woese (2002) argues that phylogenetic reconstruction may still be biologically meaningful even for genes outside the core as long as "transfer is basically confined to a natural taxonomic grouping," and he suggests that "tracing organismal genealogies is usefully viewed as the tracing of hierarchically nested gene pools." That is, it is possible that phenotypically significant resemblances between organisms that cause us to classify them together reflect genes shared through LGT within such "natural taxonomic groupings" rather than inherited through common vertical descent (Gogarten et al., 2002). Such situations surely lie outside the classical Darwinian context, as this has been interpreted by molecular phylogeneticists for much of the second half of the twentieth century.

- The rope of life. In his public talks, Gary Olsen (personal communication) has often used a piece of rope as a model for the relationships between genes and organismal genealogies. A rope has continuity even if none of the strands making it up are continuous. Thus all genes in a lineage of genomes might have been introduced by LGT at some time since its origin (one end of the rope), ultimately replacing resident genes serving the same specific function (a step in glycolysis via the Embden–Meyerhoff pathway) or the same general type of function (energy metabolism). In lineage splitting (speciation) events, one imagines the strands themselves splitting, so that such events could be traced by phylogenetic methods, for some distance down in a clade of lineages. Some strands will of course be tougher and thus longer than others (those corresponding to translational or transcriptional protein genes, for instance) and can go deeper. But there is no guarantee, in this model, that gene-based methods will ever take us to the root of such a "tree of ropes."

- The global population model. In its most radical form, this view holds that Bacteria and Archaea comprise a single global species. Gene exchange between its members makes this mega-entity formally analogous to an animal species in terms of the forces and processes driving and constraining within-species cohesion and differentiation of populations. But the mode (LGT versus homologous recombination) and tempo (billions of years, not organismal generation times) of gene exchange are of course very different. Such a model was articulated almost 30 years ago by Sorin Sonea and recalls nineteenth-century perceptions of bacteria as comprising but a single pleiomorphic assemblage (Sonea & Paniset, 1976). It does not require that gene exchange be random with respect to genes, physiology, or ecology, nor does it deny that in the "short term" (which could be hundreds of millions of years) most genes are inherited "vertically." So the patterns of similarity by which bacterial groups were defined in the pre-molecular era and that are often re-created by various whole genome tree methods are expected. Bacterial and archaeal phyla are like "races" or subspecies or breeds of an animal species. We can say something about the recent evolution and genetic coherence of dachshunds and dalmatians and perhaps even speculate about what the canines kept by our Neolithic ancestors looked like, but cannot connect the former to the latter except through the population genetics of dogs as a species. With a population model (as with the rope of life), there need be no expectation that we can trace modern groups back billions of years through the sequences of their genes, individually or collectively. For sure there were prokaryotes fixing nitrogen and carbon, making their livings by photosynthesis and methanogenesis, and enduring various extreme conditions, a billion years ago. But there is no reason to be confident of any simple genealogical (tree-like) connection between such groups and present-day taxa. Nor is there reason to suppose that we can reconstruct from their properties anything like LUCA. Plate 3.2 is an attempt to represent these implications of a thoroughgoing population model.

These models can be imagined to grade into each other, and many might accept the compromise language of Wolf et al. (2002):

> *the concept of the Tree of Life is bound to change in the postgenomic world. It cannot anymore be thought of as a definitive "species tree" (something that does not exist even in reality) but only as a central trend in the rich patchwork of evolutionary history, replete with gene loss and horizontal transfer events.*

Still, the models differ from each other in degree, and in kind (at least the last two) from our extrapolated Darwinian context. One argument in favor of the global population model – an argument that Darwin himself might have entertained – is based on uniformitarian principles, which we take here simply to be the presumption that "natural processes and phenomena have always been and still are due to causes or forces operating continuously and with uniformity" (*Shorter Oxford English Dictionary*, 5th edn). With this as a guide, we might assert that although some genes are clearly less transferrable than others, no gene – including those encoding rRNA and protein components of the transcriptional and translation machinery – is absolutely immune to LGT. Thus there will inevitably come a time in the future when (as in Plate 3.2) no gene in any genome has derived from any gene present in the world today by strictly vertical descent. Microbial taxa will exist, no doubt with many of the capacities of today's microbiota, but packaged differently and with no direct correspondence to divisions of the current edition of *Bergey's Manual*. Molecular phyogenteticists of that remote future time will be unable to reconstruct the microbial world of 2005 using molecular phylogenetics.

How far in the future this point-of-no-looking-back may be we don't know, but it could be mere billions of years. How then can we be confident that we in our era are not well past the point-of-no-looking-back for the origins of life on Earth? Perhaps our situation is that which James Hutton (1727–97), the Scottish founder of geological uniformitarianism, described in a paper to the Royal Society of Edinburgh in 1788:

> We have now got to the end of our reasoning; we have no data further to conclude immediately from that which actually is: But we have got enough; we have the satisfaction to find, that in nature there is wisdom, system, and consistency. For having, in the natural history of this earth, seen a succession of worlds, we may from this conclude that there is a system in nature; in like manner as, from seeing revolutions of the planets, it is concluded, that there is a system by which they are intended to continue those revolutions. But if the succession of worlds is established in the system of nature, it is in vain to look for any thing higher in the origin of the earth. The result, therefore, of our present enquiry is, that we find no vestige of a beginning, no prospect of an end.

The LUCA

Although there have been many attempts to reconstruct the LUCA as if it were a single cell living at some particular time between 3.5 and 4 billion years ago (Penny & Poole, 1999; Forterre, 2001; Chapter 2 in this volume), Woese and Fox maintained in some of their earliest articulations of the three-domain view that the LUCA was a heterogeneous population of primitive entities exchanging genetic determinants quite promiscuously. In 1981 Woese wrote:

> I would contend that the universal ancestor is not some unique entity, but rather, is a universal ancestor **state**, and reflects the nature of and relationships among progenotes, not uniqueness of species. At the progenote stage, many of the processes we associate with extant life, e.g. various biochemical pathways, are still coming in to being. This may even be true of the cell itself, the cell as an entity. In some sense, the world of progenotes may be more a world of semi-autonomous subcellular entities that somehow group to give "loose" (ill-defined) cellular forms. In other words, the cell as defined by the subcellular interactions it comprises may be a less specific, less integrated, more ill-defined, ephemeral entity at the progenote stage than it later would be. At such a stage, it is easy to picture a ready exchange, a flow, among subcellular enitites – be they called genes, plasmids, viruses, selfish DNA, organelles or whatever. Were such genetic mixing to occur, then the progenote stage could give the appearance of being a universal ancestor.

Recently, Woese (2002) described the transition from this primitive state of "genetic communion" to modern cellular life in terms of crossing a "Darwininian threshold," as "a critical point . . . where a more integrated cellular organization emerges, and vertically generated novelty can and does assume greater importance." Bacteria and Archaea crossed this threshold independently (Bacteria first) and so some of the striking differences between them, especially in their informational machinery, represent independently achieved solutions to problems of efficiency and accuracy of form and function not yet solved in the universal ancestral state.

Two ideas, central to Woese's theory from the beginning, distinguish it from the uniformitarian population model sketched above. First, LGT – however important it may be today – was even more the predominant driving force in evolution prior to the divergence of Bacteria and Archaea. This may not be a crucial theoretical distinction: uniformitarians cannot sensibly imagine that microbial evolution has always had its current tempo and mode, only that it has had them as far back as we can know. Second, Bacteria and Archaea diverged from an ancestral state that was primitive, so that differences between them, particularly those involving the more fundamental and less exchangeable machinery of gene expression, can be seen as separately achieved improvements on a primitive condition. This is a crucial distinction, because here we might have a "vestige of a beginning" (at least of modern cellular life), a way to root the universal tree through polarization of traits, and a justification for considering Bacteria and Archaea (at least as defined by those machineries) as the surviving two among the "many twigs which flourished when the tree was a mere bush," rather than just the most recent big winners in the battleground of a global microbial population.

So rather a lot hinges on whether or not we have independent reasons to believe that the differences between bacterial and archaeal informational machineries do represent separately achieved improvements on some more primitive state, and/or

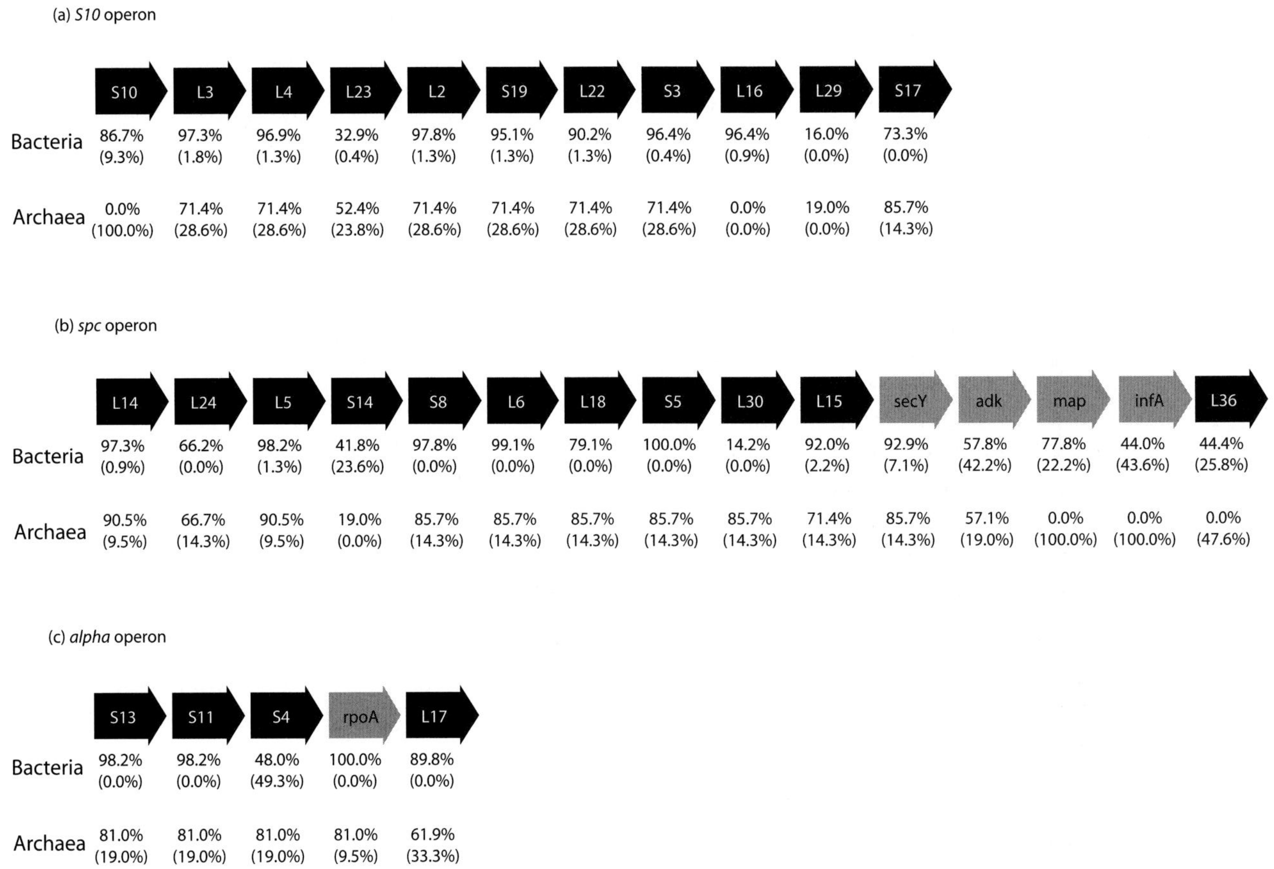

(a) *S10* operon

	S10	L3	L4	L23	L2	S19	L22	S3	L16	L29	S17
Bacteria	86.7% (9.3%)	97.3% (1.8%)	96.9% (1.3%)	32.9% (0.4%)	97.8% (1.3%)	95.1% (1.3%)	90.2% (1.3%)	96.4% (0.4%)	96.4% (0.9%)	16.0% (0.0%)	73.3% (0.0%)
Archaea	0.0% (100.0%)	71.4% (28.6%)	71.4% (28.6%)	52.4% (23.8%)	71.4% (28.6%)	71.4% (28.6%)	71.4% (28.6%)	71.4% (28.6%)	0.0% (0.0%)	19.0% (0.0%)	85.7% (14.3%)

(b) *spc* operon

	L14	L24	L5	S14	S8	L6	L18	S5	L30	L15	secY	adk	map	infA	L36
Bacteria	97.3% (0.9%)	66.2% (0.0%)	98.2% (1.3%)	41.8% (23.6%)	97.8% (0.0%)	99.1% (0.0%)	79.1% (0.0%)	100.0% (0.0%)	14.2% (0.0%)	92.0% (2.2%)	92.9% (7.1%)	57.8% (42.2%)	77.8% (22.2%)	44.0% (43.6%)	44.4% (25.8%)
Archaea	90.5% (9.5%)	66.7% (14.3%)	90.5% (9.5%)	19.0% (0.0%)	85.7% (14.3%)	85.7% (14.3%)	85.7% (14.3%)	85.7% (14.3%)	85.7% (14.3%)	71.4% (14.3%)	85.7% (14.3%)	57.1% (19.0%)	0.0% (100.0%)	0.0% (100.0%)	0.0% (47.6%)

(c) *alpha* operon

	S13	S11	S4	rpoA	L17
Bacteria	98.2% (0.0%)	98.2% (0.0%)	48.0% (49.3%)	100.0% (0.0%)	89.8% (0.0%)
Archaea	81.0% (19.0%)	81.0% (19.0%)	81.0% (19.0%)	81.0% (9.5%)	61.9% (33.3%)

Fig. 3.3 Ribosomal protein clusters in prokaryotic genomes (based on Coenye & Vandamme, 2005, updated by inclusion of 147 additional genomes). Illustrated here is a representation of the frequency and organization of three ribosomal protein gene clusters found in prokaryotes. We performed an all versus all BLAST analysis using the TIGR CMR "genome comparison" utility to determine how often a particular gene is either present in bacterial or archaeal genomes in this cluster (top number) or present elsewhere in the genome (bottom number in parentheses). This analysis includes 246 completely sequenced genomes (225 Bacteria and 21 Archaea), which are found in the TIGR CMR Database. Note: in the *alpha* operon, S11 and S4 are reversed to S4–S11 in all archaeal genomes containing the complete operon; two Archaea (*Nanoarchaeum equitans* and *Pyrobaculum furiosus*) have little to no conserved organization.

whether we can imagine alternative explanations for them. Other chapters in this volume address domain differences in information processing in detail. Much has been made of domain-specific features of replication systems and the possibility that the almost universally believed-in transition from an RNA- to a DNA-protein world occurred independently in two or three domains (Mushegian and Koonin, 1996a; Chapter 2 in this volume).

These are unfortunately the kinds of questions that are easy to argue about and difficult to solve. One alternative to independent emergence from primitivity is Gupta's suggestion that antibiotic pressure forced substantial changes in the ribosomes of Gram-positive bacteria (changes that later spread by LGT), creating Archaea as a sort of derived subclade of Gram positives (Gupta, 1998). Another, following a proposal of Gogarten-Boekels et al. (1995), is that modern life derives from three relatively advanced cellular lineages, the three survivors of a relatively late "nearly ocean-boiling" meteorite impact. Both these scenarios seem to us more in keeping with at least one feature of bacterial and archaeal information processing than divergence from a primitive state, in particular the extreme possibility that the domains arose independently from RNA world antecedents. This feature is synteny in gene clusters encoding ribosomal proteins. Figure 3.3 is an update of a summation of similarities in the S10, *spc* and *alpha* ribosomal protein gene clusters of Bacteria and Archaea recently presented by Coenye and Vandamme (2005). Given that operon structure is in general unstable (Lathe et al., 2000; Tamames, 2001) and that ribosomal protein gene clusters can exhibit different regulatory regimes, we suggest that this remarkable coincidence in gene order holds some secret, yet to be revealed, about the time and nature of the divergence of domains.

Tree as hypothesis

The debate concerning LGT and its meaning for the tree of life is heated, and sometimes inappropriately personal (Kurland, 2000). It is not always clear what this debate is about, what either side would need to be shown before accepting defeat. Tree supporters take the tree as fact, in need only of elaboration, and in this context may be right to consider the objections of critics as obscurantist and ill-motivated. Most dangerously for evolutionary science (because creationists are paying attention, as they do to all disharmony in the discipline), there is no clear "tree of life hypothesis" being tested. We suggest one along the lines implicit in this chapter: that the patterns of similarity and difference which allow for the hierarchical classification of organisms into "groups subordinate to groups" are to be explained primarily by vertical descent, at all taxonomic ranks and at all evolutionary depths back to some single root. We suggest that as a community we need to decide what "primarily" means in this context. But we submit that by any reasonable definition of primarily, the tree of life hypothesis – as it pertains to Bacteria and Archaea – remains not proven.

Acknowledgments

Work of the authors is supported by the Canadian Institutes for Health Research, Genome Atlantic, the Canadian Institute for Advanced Research, and the Canada Research Chair program. This review is dedicated to the memory of Wolfram Zillig.

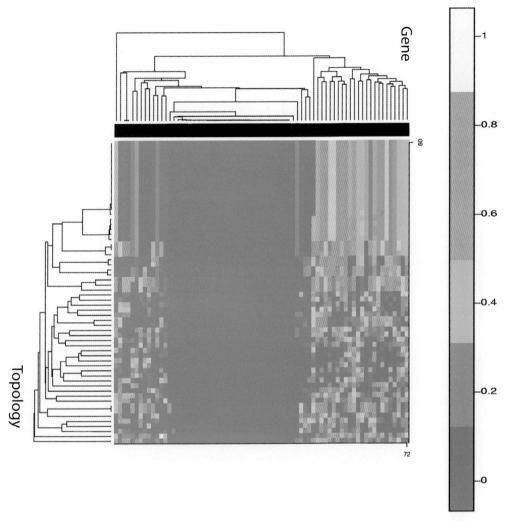

Plate 3.1 Heat map analysis of phylogenetic congruence for archaeal core genes from Bapteste et al. (2005). Reproduced under the Open Access agreement of BioMed Central. The rationale and method of this analysis are discussed in the text and at length in Bapteste et al. (2005). This heat map includes both results with real genes (red) in the 44 gene data set of Matte-Taillez et al. (2002) and manipulated gene data sets (blue), in which up to three "LGT events" have been simulated by assigning the same gene sequence to two taxa. The partial admixture of the two implies that some quite radical LGT events may escape detection although in general such artificial data reject most tested topologies. Real genes with weak phylogenetic signal are unable to distinguish between many topologies and the rows corresponding to such genes have many light-colored squares. The bar at the top shows the colors assigned to p-values for each gene against each topology, ranging from 0 (no support) to 1 (support).

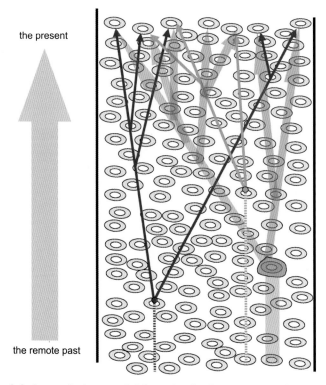

Plate 3.2 Implications of a global population model for microbial genome evolution. Genomes (ovals) comprise a single global population and, over all of evolutionary time, the rate of LGT is such that no two genes have the same phylogeny. Each family of genes (red and green represent two such families) can trace its ancestry to a last common ancestral gene present in some genome in the past but, because of LGT, no two gene families' ancestors will have existed at the same time or in the same genome. The cell envelope, as a token of organismal evolution, will also have a phylogeny (indicated by blue arrows), traceable to a last common ancestral cell, but this particular cell will not have contained the last common ancestral version of any modern genes.

4

Diversity of uncultivated Archaea: perspectives from microbial ecology and metagenomics

Christa Schleper

Introduction

Microorganisms are in many ecosystems, and in the biosphere as a whole, by far the largest component, in terms of both biomass and biological activity (Whitman et al., 1998). By their participation in global biogeochemical cycles they are major players in regulating the ecosphere. Until recently, the role of Archaea in these global cycles, as well as their phylogenetic and physiological diversity, had been largely underestimated. While the distribution of methanogenic Archaea in many different anaerobic habitats woldwide and their role in the global carbon cycle has been well described, all other Archaea were considered as extremophiles with specific adaptations and growth in those habitats on Earth that seem the most inhospitable for all other creatures. With the help of culture-independent molecular techniques, mostly involving the amplification of 16S rRNA genes directly from environmental samples, it has recently been shown that Archaea are not confined to extreme niches. By contrast, they are globally distributed on this planet and occur in significant numbers in soils (Bintrim et al., 1997; Buckley et al., 1998; Sandaa et al., 1999; Jurgens et al., 2000; Ochsenreiter et al., 2003; Simon et al., 2005), the ocean's plankton (DeLong, 1992; Fuhrman et al., 1992), sediments (Boetius et al., 2000; Orphan et al., 2001b), freshwater lakes (MacGregor et al., 1997; Schleper et al., 1997a; Keough et al., 2003), and the deep subsurface (Takai et al., 2001). Even in the well known habitats of Archaea, such as high temperature hydrothermal vent environments (Takai & Horikoshi, 1999; Nercessian et al., 2005) or hypersaline environments (van der Wielen et al., 2005), the microbial diversity of Archaea is much higher than previously assumed. From these studies, one has to conclude that those species of methanogens, halophiles, and thermophiles that have been cultivated in the laboratory and studied in detail represent only a minority of the total archaeal phylo- and phenotypes found on Earth.

The broad distribution and abundance of Archaea in terrestrial and marine habitats implies that they contribute to global energy cycles. Yet many of these organisms have been predicted solely by polymerase chain reaction (PCR) based surveys and no representative has been cultivated in the laboratory. Therefore, their specific metabolisms and cellular features often remain elusive. However, recent advances, particularly in environmental genomic studies (or "metagenomics"), have led to the characterization of some of these organisms in the absence of laboratory cultivation (Schleper et al., 2005).

This review gives an update on the current understanding of the diversity and ecological distribution of Archaea and introduces novel techniques that allow the study of those organisms that are difficult to isolate in pure culture, or that live in synthrophic, interdependent associations. In particular, the review focuses on the novel phenotypes of as yet uncultivated organisms that have been predicted by environmental genomic, or "metagenomic," studies.

While this chapter is intended to serve as a general introduction to the ecology of Archaea for newcomers in the field, it may also be of interest to readers working on specific aspects of archaeal physiology or information processing, who seek new variants of their favorite target in different novel – and perhaps yet uncultivated – groups of non-extremophilic Archaea.

Archaea everywhere?

Around 20 years ago, a new era in microbial ecology began when Norman Pace and colleagues suggested the use of molecular biological tools to study naturally occurring microbial diversity, independent of the cultivation of organisms in the laboratory (Olsen et al., 1986). This approach developed from the pioneering work of Carl Woese, who, based on such phylogenies, discovered the domain Archaea (Woese et al., 1978). It has now become a standard technique in many laboratories to amplify ribosomal RNA genes, mostly small subunit (SSU) or 16S rRNA, directly from environmental samples in order to characterize prokaryotes by inferring phylogenetic relationships. Edward DeLong and Jet Fuhrman were the first to apply this PCR-based approach to study the occurrence of Archaea in the marine plankton and they unexpectedly detected Crenarchaea in high numbers (DeLong, 1992; Fuhrman et al., 1992). Since then Archaea have been continually found in many diverse environments. In August 2005, around 11,000 complete or partial sequences of 16S rRNA genes of mostly uncultivated Archaea were deposited in the public databases and the number is continuously increasing. Plate 4.1 depicts a phylogenetic reconstruction of 1384 mostly full length sequences from this dataset. Thirteen lineages within the kingdom of the Crenarchaea and 32 lineages within the Euryarchaea can be resolved. Three or four tentative novel lineages which are not closely affiliated with Euryarchaea or Crenarchaea based on 16S rRNA phylogeny have been dissected (ancient archaeal group (AAG), pOWA, Korarchaeota, and Nanoarchaeota). Although the phylogenetic placement of Nanoarchaeota (represented by *Nanoarchaeum equitans* in Plate 4.1; Huber et al., 2002a) is currently under debate (Brochier et al., 2005a, b), the Korarchaeota (Barns et al., 1996) and representatives of another "ancient archaeal group" (Takai & Horikoshi, 1999) indicate the presence of more and perhaps deeply branching kingdoms within the Archaea. Of the 49 lineages of Archaea represented in the tree, only about 18 contain one or more cultivated representative (dark triangles in Plate 4.1) or have representative strains in mixed laboratory enrichments (Korarchaeota, Nanoarchaeota, RCI, etc.). All other lineages have been predicted by environmental 16S rRNA gene surveys only (and are labeled with letters and numbers, in most cases referring to the study from which they have been first recovered). This analysis is still even an underestimate, as about 10 or 15 branches would have to be added to the tree if all partial sequences from the databases were included (German Jurgens, personal communication).

Although many of the novel archaeal groups have been predicted indirectly by 16S rRNA gene surveys, a wealth of additional data gives evidence that many (if not all) of these lineages are found as living populations in the environment. In particular, the use of fluorescent *in situ* hybridization (FISH; Amann et al., 1995) allows the visualization of specific phylotypes and direct quantification at the cellular level (see Plate 4.2).

Additionally, the use of biomarkers, particularly archaeal lipids, has been crucial in tracing uncultivated Archaea in the environment and determining their abundance (e.g. Schouten et al., 2000). Furthermore, in conjunction with stable-isotope probing (Wuchter et al., 2003), lipid analysis has been used to trace the activity of Archaea in the environment. With rRNA-based techniques, changes in spatial heterogeneity, abundance, and community structure of soil Archaea have been observed (e.g. Sandaa et al., 1999; Nicol et al., 2003b) and variation in the distribution of planktonic Archaea in the water column has been demonstrated (e.g. Massana et al., 1997). Such analyses clearly indicate the presence of dynamic and active archaeal populations that respond to changing environmental parameters and that contain differently adapted ecotypes.

Many of the novel archaeal groups seem (so far) to be confined to specific geographical locations, or to ecosystems that have similar geochemistry. For example, groups DHVE I/IV and AAG are presently defined only by sequences from a hydrothermal vent environment (Takai & Horikoshi, 1999). Other groups, however, seem to be widely distributed. For example, the two crenarchaeal lineages, which are mostly defined by sequences from marine plankton (group I.1A in Plate 4.1) or soils (group I.1B in Plate 4.1), are both also found in freshwater and deep subsurface samples (Murray et al., 1998; Takai et al., 2001; Ochsenreiter et al., 2003). Sequences of group I.1A have also been found in hydrothermal vents (Takai & Horikoshi, 1999). As the 16S rRNA databases increase even further, it might be possible that those patterns will disappear and we will find that "everything is everywhere." However, the general picture for those habitats that have been studied extensively may not change dramatically. For example, from the considerable number of studies on soil, it is clear that the dominant populations of Archaea in the aerobic layers of most soils are solely representatives of Crenarchaea group I.1b (with fewer sequences, mostly from acidic forest and grassland soils, present in FFS). This finding is in striking contrast to the huge diversity of Bacteria that are found in soils (Gans et al., 2005). However, a considerably larger diversity of Archaea is found in marine sediments (marked by asterisks in Plate 4.1) and an overwhelming diversity is found in hydrothermal vent environments (red dots in Plate 4.1). Sequences of these environments are placed within many lineages of Archaea or define even novel branches (DHVE I, IV, AAG). Could this mean that the origin of Archaea was in hydrothermal vent environments or at least in hot environments, because this is where they diversified the most? Only relatively few organisms would have later adaptively radiated into the more commonplace environments on Earth. This might well be possible. In fact, since the recognition of Archaea as a separate domain, it has been assumed that thermophiles might have arisen first (Stetter, 1992), because Archaea from hot environments are phylogenetically diverse and some of them branch close to the root of the archaeal tree (Barns et al., 1996; Takai & Horikoshi, 1999; Huber et al., 2002a) (see Plate 4.1). Support for this speculation has also been provided by biogeochemical studies and lipid biomarkers (Kuypers et al., 2001).

It is interesting to note that genes involved in the methanogenic pathway are widespread in the Euryarchaea, i.e. in 16 different out of 33 lineages, including all methanogens, Archaeoglobi and the ANME lineages (methanotrophs, see below). This lends support for the hypothesis that the pathway was present in a common ancestor of Euryarchaea, or perhaps even in all Archaea (see also Chapter 2 in this volume). As more of the specific metabolisms of the as yet uncultivated archaeal lineages are identified, it might be possible to dissect more of the earliest chemolithoautotrophic pathways that developed in the Archaea and Bacteria and perhaps even before the split of these two domains.

It is astonishing that most of the newly discovered lineages seem to expand the two major kingdoms – Euryarchaea and Crenarchaea – that were defined as early as 1986 by Carl Woese based on only a few cultured archaeal species (Plate 4.1 and Woese & Olsen, 1986). However, the discovery of a few more distant and deeply branching lineages through molecular surveys (Korarchaeota and AAG in Plate 4.1; Barns et al., 1996; Takai & Horikoshi, 1999) or cultivation (Nanoarchaeota; Huber et al., 2002a) indicate that a greater diversity of Archaea is to be expected and that more kingdoms might be recovered through improved molecular ecological searches, more sophisticated cultivation techniques, and perhaps metagenomic approaches.

Abundant Archaea in the marine plankton

Marine pelagic Archaea were detected more than ten years ago in 16S rRNA based surveys (DeLong, 1992; Fuhrman et al., 1992). The sequences recovered from the picoplankton in Pacific, Atlantic, Mediterranean, and Antarctic waters fell into two groups: one within the Crenarchaea (originally termed group I, now group I.1a in Plate 4.1) and another within the Euryarchaea (group II). Later another less abundant euryarchaeal group III was discovered. From the abundance of 16S rRNA genes in clone libraries and quantitative rRNA hybridization experiments it became apparent that the two archaeal groups thrive in different zones of the water column, with group II predominating at the surface and group I predominating at depth (Massana et al., 1997). Marine planktonic Archaea have also been visualized by FISH using rRNA-specific oligonucleotide probes. The technique has been further extended by DeLong and collaborators, who developed polyribonucleotide probes in order to enhance the sensitivity and applicability of FISH with seawater samples and in particular with Archaea (DeLong et al., 1999). This was successfully applied to identify and quantify archaeal and bacterial cells in seawater even down to several thousand meters depth (down to 4750 m). In an exhaustive one-year-long survey of the north Pacific, marine Crenarchaea were shown to comprise a large fraction of the total marine picoplankton below the euphotic zone (>150 m) with an unusually broad habitat range spanning from mesopelagic to bathypelagic depths and a relative abundance that reaches 39% of total DNA-containing picoplankton in greater depths (Karner et al., 2001). From this study it has been extrapolated that 1.0×10^{28} cells, i.e. approximately 20% of all the picoplankton cells in the world's oceans, appear to be represented by one specific clade, the pelagic Crenarchaea. Beside *Pelagibacter* (SAR11) and *Prochlorococcus*, these Archaea might belong to one of the most abundant groups of microorganisms on this planet (Rappe & Giovannoni, 2003).

Inspired by the rapid advances in genomic techniques applied to cultivated microorganisms, Stein and DeLong used a bacterial artificial chromosome (BAC) derived fosmid vector to prepare a large-insert library from marine plankton of the north-eastern Pacific in order to characterize marine planktonic Archaea beyond their 16S rRNA genes (Stein et al., 1996). This study was the beginning of a novel and now rapidly expanding field of microbial environmental genomics or "metagenomics." A 38.5 kb genomic fragment of an uncultivated mesophilic crenarchaeote was identified within 3552 clones using Archaea-specific 16S rRNA gene probes (FOS-4B7, Group I.1A in Plate 4.1). Further genome fragments of marine Archaea have subsequently been isolated from BAC, fosmid, or cosmid libraries of surface (Beja et al., 2000, 2002) and deep waters (Lopez-Garcia et al., 2004; Moreira et al., 2004) of the Antarctic and the North Pacific. Conservation of gene order around the rRNA operon confirmed the close relationship of the planktonic Crenarchaea, even between strains from different oceanic regions. In contrast, it is noticeable that considerable heterogeneity could be observed in protein encoding regions and intergenic spacers, when genome fragments with otherwise identical or almost identical 16S RNA genes were compared from the same DNA library (Beja et al., 2002). The planktonic archaeal clones shared several genomic features with their hyperthermophilic relatives, including the estimated low G+C content (c.32–36%) as well as the gene repertoire and structure of the rRNA operon. However, some genes that are so far unique to planktonic Crenarchaea have also been identified, such as a putative RNA-binding protein that shares features with the bacterial cold shock family, and a novel Zn finger protein that was found previously only in eukaryotes (Beja et al., 2002).

First hints of the specific metabolism of marine planktonic Crenarchaea stem from stable isotope studies, which suggested that the Archaea incorporate CO_2 and therefore most probably have an autotrophic growth mode (Wuchter et al., 2003). Furthermore, genes encoding proteins that were remotely related to ammonia monooxygenases from bacteria were found in a large-scale environmental genomic survey from the Sargasso Sea (Venter et al., 2004). These genes could be clearly assigned to Crenarchaea with the help of a genomic contig isolated from soil (Schleper et al., 2005), indicating that both soil and marine Crenarchaea could be ammonia oxidizers (see below).

In support of this hypothesis, the first cultivated isolate growing as an autotrophic and aerobic ammonia oxidizer has recently been obtained from an aquarium in Seattle (*Nitrosopumilus*; Könneke et al., 2005). With primers designed against the putative archaeal *amo* genes from the metagenomic studies of soil and the Sargasso Sea, a highly related gene has been amplified from *Nitrosopumilus*. Again this indicates that ammonia oxidation in Archaea might proceed via an ammonia monooxygenase that is distantly related to that of their bacterial counterpart. The isolation of a marine mesophilic crenarchaeote will serve as an important basis for studying the biochemistry and physiology of marine Archaea and for obtaining more strains from other environments. However, it remains to be demonstrated how many of the Crenarchaea within the non-thermophilic terrestrial and marine lineages are indeed ammonia oxidizers. If most of them exhibit this metabolism, then ammonia-oxidizing Archaea should be found in many divergent habitats on Earth, including hot and moderate terrestrial, freshwater, and marine environments.

Cenarchaeum symbiosum: a model marine crenarchaeote

Cenarchaeum symbiosum was first detected by 16S rRNA-based PCR surveys among the complex bacterial community harbored in the tissues of the marine sponge *Axinella mexicana* (Preston et al., 1996). Since it can be "quasi"-cultivated in the laboratory by maintaining it in stable association with its host under controlled conditions (and at temperatures around 40–50 °C below the optimum of its closest cultivated relatives, the hyperthermophilic Crenarchaea), it was the first strain of non-thermophilic

Crenarchaea to be described and named. The *Cenarchaeum/Axinella* association provided a tractable system for the study of non-thermophilic marine Crenarchaea and gave access to relatively large amounts of biomass (and DNA) from this species. Although *C. symbiosum* has not been cultivated or completely physically separated from the host tissues or from the coexisting bacteria, cell fractions that are enriched for the archaeon have served for the construction of large-insert genomic libraries (Schleper et al., 1998), facilitating the isolation of genome fragments and leading to a genome sequencing project of this organism (E. DeLong, personal communication). Many features, including a homologue of a family B DNA polymerase that is encoded on an isolated fosmid, show the close relationship of *C. symbiosum* to cultivated hyperthermophiles (Schleper et al., 1997b). The deep placement of a radA homologue from *Cenarchaeum* in phylogenetic analysis (Sandler et al., 1999) and the discovery of a histone gene (Čuboňová et al., 2005) indicate that the genomes of non-thermophilic Crenarchaea will be very interesting for recalculating and redefining the phylogenetic relationship and particular features of Euryarchaea and Crenarchaea, and the origin of Archaea.

Unexpectedly, the analysis of genomic contigs from *C. symbiosum* revealed the presence of two closely related variants that were found in the majority of sponge individuals analyzed (Schleper et al., 1998), and this observation was confirmed in the analysis of the complete genome (S. Hallam and E. DeLong, personal communication). The two genomic entities have less than 0.7% deviation in the 16S rRNA gene sequence and have an identical gene order, but vary up to 20% in protein encoding regions and even up to 30% within intergenic regions. This study was among the first to reveal genomic heterogeneity at the species level based on environmental genomic studies and indicates a considerable functional diversity and perhaps adaptation to special niches in populations of coexisting closely related prokaryotic strains. This observation was confirmed through the study of closely affiliated archaeal planktonic fosmids (Beja et al., 2002) and was largely extended during the assembly of huge datasets from the shotgun sequencing projects of the Sargasso Sea (Venter et al., 2004) and an acidic mine drainage (Tyson et al., 2004).

The *C. symbiosum* genome has a considerably higher G+C content (>55%; Schleper et al., 1998) than its relatives in marine plankton (approximately 34%), which may reflect adaptation to the symbiotic lifestyle in the metazoan host, rather than a large evolutionary distance. Despite this difference, however, a considerable number of genomic scaffolds with homologous, syntenic regions can be identified using *C. symbiosum* genome fragments to query the environmental database from the Sargasso Sea. In fact, large parts (if not a whole genome) of a marine crenarchaeote can be assembled from the Sargasso Sea dataset, indicating a close relationship between most planktonic Crenarchaea.

Methanotrophic Archaea in marine sediments

In marine sediments deep below the ocean floor, large quantities of methane are stored as solid gas hydrates. It has only been recognized relatively recently that the flux of considerable amounts of this greenhouse gas into the hydrosphere is prevented through the activity of microorganisms. The organisms responsible for the anaerobic oxidation of methane (AOM) all belong to the Euryarchaea. Different lineages affiliated with Methanosarcinales/Methanobacteriales (ANME 1, 2 in Plate 4.1) or with Methanococcoides (ANME 3 in Plate 4.1) have been found in the upper subsurface sediments that lie above the large reservoirs of methane (Boetius et al., 2000; Orphan et al., 2001b). Particularly impressive and massive microbial mats, covering up to 4 m high carbonate buildups, thrive at methane seeps in anoxic waters of the northwestern Black Sea shelf (Plate 4.3, and Michaelis et al., 2002). Strong ^{13}C depletions and *in vitro* incubation experiments indicate that these mats perform anaerobic oxidation of methane, thereby precipitating carbonate structures and producing substantial biomass. *In situ* analyses based on lipid and isotope signatures, combined with 16S rRNA-based surveys of the microbial communities feeding on methane seeps, have suggested that Archaea of the ANME I and II lineages should be methanotrophs, converting methane into CO_2 and reduced by-products in a process that is coupled to sulfate reduction by closely associated bacteria (Hinrichs et al., 1999; Boetius et al., 2000; Orphan et al., 2001b). None of the archaeal organisms that mediate the anaerobic oxidation of methane, nor their sulfate-reducing partners of the *Desulfosarcina/Desulfococcus* branch of the Deltaproteobacteria

(Boetius et al., 2000; Orphan et al., 2001b) have been brought into laboratory culture yet. However, biochemical and metagenomic studies have given crucial insights into the pathways involved in AOM. Genes for methyl coenzyme M reductase (MCR), which performs the terminal step in methanogenesis, were found in association with ANME lineages (Hallam et al., 2003), and a novel nickel compound, a variant of the MCR cofactor F_{430} of methanogenic Archaea associated with an MCR-like protein (Ni-protein I), was characterized in a combined biochemical and environmental genomic study (Krüger et al., 2003). This enzyme might catalyze methane activation in a reverse terminal MCR reaction. The hypothesis, that ANME-Archaea perform a "reverse methanogenesis," was recently further supported by Hallam et al. (2004) through a community genomics approach. The analysis of genome fragments isolated from fractions highly enriched for the ANME I organisms revealed genes encoding factors involved in all but one of the seven steps of the methanogenic pathway. Only *mer*, the gene for methylene-tetrahydromethanopterin reductase, was missing. The absence of this protein, which catalyzes a key reductive step in methanogenesis, might indicate a point in the pathway where methanotrophic Archaea regulate the flux of carbon in the oxidative direction (Hallam et al., 2004). Knittel et al. (2005) have recently investigated to what degree the anaerobic methane oxidizers of the two major groups (ANME I and ANME II), exhibiting the same metabolism, also occupy the same niches. Although the ANME groups were found at all methane environments that have been examined, independent of temperature, pH, or other parameters, differences could be dissected with respect to relative abundance, indicating a specific distribution of various ecotypes in the different microniches.

A number of other archaeal lineages within the Euryarchaea and Crenarchaea have been detected by 16S rRNA surveys in various marine sediments (Plate 4.1), and still await a closer characterization.

Ubiquitous Crenarchaea in soils might be ammonia oxidizers

Soils are the most diverse ecosystems on the planet, with an estimated 12,000–18,000 different dominant microbial species in one small sample or even one million species per gram when rare organisms are included in the estimates (Gans et al., 2005). The number of microorganisms in soil habitats, typically 10^9 cells/g, by far exceeds that in freshwater or marine habitats. Most of them occur in the organically rich surface layers and in association with plant roots, while they are less abundant in the underlying mineral soils. A compilation of 16S rRNA data from studies on Bacteria revealed that at least 20 out of 41 bacterial phyla are represented in the aerobic surface layers of soil (Treusch & Schleper, 2005). In striking contrast to this, the diversity of Archaea in the upper layers of soils seems to be very limited and mostly restricted to Crenarchaea of group I.1B (Plate 4.1) and group FFS (the latter mostly only in acidic forest soils). Albeit less diverse, these particular Crenarchaea can be recovered on all continents from virtually any terrestrial ecosystem. They have been found in sandy ecosystems, pristine forest soil, agricultural fields, contaminated soil, and the rhizosphere (Bintrim et al., 1997; Buckley et al., 1998; Sandaa et al., 1999; Simon et al., 2005; Nicol et al., 2003a; Ochsenreiter et al., 2003) and represent a significant fraction (up to 5%) of the total prokaryotic community (Buckley et al., 1998; Ochsenreiter et al., 2003). They exhibit spatial heterogeneity and changes in abundance and community structure dependent on succession (Fig. 4.1), land-management strategies, heavy-metal contamination, or rhizosphere type (Sandaa et al., 1999; Nicol et al., 2003b, 2005; Sliwinski & Goodman, 2004). Although soil Crenarchaea have been obtained in enrichment cultures from plant roots (Simon et al., 2005), the specific metabolism of these Archaea remained unresolved.

With the help of environmental genomic and postgenomic techniques, it has recently become possible to gain insight into the putative physiology of soil Crenarchaea (and of their marine relatives). Several large genome fragments from Crenarchaea have been isolated from complex soil libraries. They were identified by 16S rRNA or other archaeal-specific core genes (Quaiser et al., 2002; Treusch et al., 2004, 2005). A number of genes on these contigs provided first clues to the energy metabolism of these Archaea. In particular, the identification of two genes encoding proteins related to subunits of ammonia monooxygenases (*amo*AB), the central enzyme of ammonia oxidation in bacterial nitrifiers (Arp et al.,

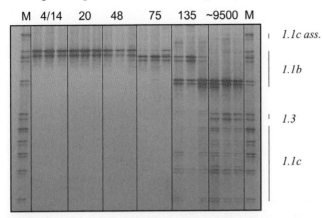

Fig. 4.1 Denaturant gradient gel electrophoresis (DGGE) of 16S rRNA gene fragments to visualize the primary succession of soil Crenarchaea communities associated with soil development across a receding alpine glacier forefield. Receding glaciers provide chronosequences of soil and plant community development from bare substrate to mature acidic grassland soil. Soil substrates were deglaciated for 4 to approximately 9500 years. Lanes labeled M represent a marker lane composed of short PCR products of cloned 16S rRNA gene sequences representative of the most abundant archaeal 16S rRNA gene sequences (1.1b, 1.1c, 1.1c-associated and 1.3 crenarchaeal lineages; see labels to the right of the figure). Sequences from organisms placed within the 1.1b terrestrial lineage were recovered in all samples with successional communities clearly apparent. Sequences from other soil crenarchaeal lineages were only recovered in developed soils indicating a restricted ecological distribution. (Taken from Nicol et al., 2005, with permission of the publisher.)

2002), led to the speculation that non-thermophilic Crenarchaea use ammonia as their primary energy source (Schleper et al., 2005). Proteins from the family of ammonia monooxygenases and particulate methane monooxygenases of methanotrophs have so far only been detected in the γ- and β-Proteobacteria groups. The novel, quite distant homologues from Archaea expand this well known family to a third group. With primers specific to the *amo*A-like gene of soil Crenarchaea, it was possible to demonstrate transcriptional induction of the gene in microcosms from soil that were amended with ammonia (Treusch et al., 2005). This experiment lends further support to the hypothesis that the identified gene encodes an ammonia monooxygenase rather than a particulate methane monooxygenase, which belongs to the same protein family. Interestingly, homologues of the *amo*-like genes from the soil crenarchaeote were found in the dataset from the Sargasso Sea (Plate 4.4a). The cultivation of a chemolithoautotrophic, apparently ammonia-oxidizing, archaeon associated with the marine group I.1a lends further support to the occurrence of this metablism (Könneke et al., 2005; see above). It remains to be shown if soil Crenarchaea are also autotrophs, as has been suggested for the marine relatives. The high conservation of the archaeal *amo*A gene makes it an excellent biomarker to study the distribution and abundance of ammonia-oxidizing Archaea in different habitats (Plate 4.4b). First studies indicate that this group of Archaea is abundant in many different terrestrial and marine environments (S. Leininger and C. Schleper, in preparation).

Interestingly, a homolog of a copper-containing nitrite reductase (NirK), the key enzyme of dissimilatory nitrate reduction, was also identified on the archaeal soil clone (Plate 4.4a), as well as in the Sargasso Sea dataset (Treusch et al., 2005). As this protein is also found in ammonia-oxidizing bacteria and might even play a key role in their metabolism (Schmidt et al., 2004), it lends further support to the hypothesis that the primary energy metabolism of mesophilic terrestrial and marine Crenarchaea is based on ammonia oxidation.

Several but not all archaeal contigs, recovered from soil libraries, show considerable genomic overlap (i.e. syntenic regions of high similarity) with scaffolds from the Sargasso Sea dataset, indicating that some but not all soil groups share metabolic capacities with marine planktonic Crenarchaea (unpublished observation).

Rice Cluster I Archaea are an important group of methanogens

The diversity and abundance of methanogenic Euryarchaea in anoxic soils has been particularly well studied in rice fields. These environments contribute considerably to the global methane emissions (10–25%). After pulse-labeling rice plants with $^{13}CO_2$, Lu and Conrad (2005) demonstrated that most of the stable isotope was ultimately incorporated into the ribosomal RNA of Rice Cluster I Archaea (Euryarchaea RCI in Plate 4.1) in the soil. This archaeal group, of which no representative has been cultivated yet, therefore seems to play a key role in methane production from plant-derived carbon. It is not yet understood how the RCI organisms can live in the rhizosphere, since methanogens are usually confined to anaerobic niches. It has been suggested that either they might be resistant to O_2 toxicity or they might thrive in anaerobic microniches, e.g. in old root segments (Grosskopf et al., 1998). RCI organisms seem to produce methane through CO_2 reduction, rather than acetate cleavage, which is the dominant path of CH_4 production in the bulk soil (Conrad et al., 2002). It will be most interesting to gain more insights into the specific physiology of these important methanogens, as plant-photosynthesized carbon constitutes a major source for methane production in rice soil. A genome of an RCI organism will be available soon, as it has been assembled recently from the DNA library of an enrichment culture (R. Erkel, personal communication).

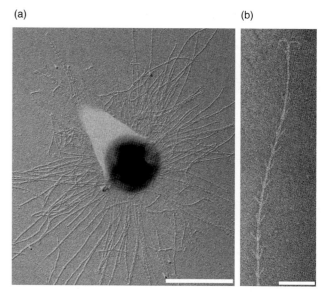

Fig. 4.2 (a) Electron micrograph of platinum-shadowed euryarchaeon SM1 cell isolated from a sulfurous marsh in Bavaria, near Regensburg, exhibiting pili-like appendages. Bar = 1 µm. (b) Electron micrograph of a negatively stained single hamus of SM1. Bar = 100 nm. (Image courtesy of R. Huber, University of Regensburg.)

Archaea in freshwater

From the few studies available, it is apparent that considerably more lineages of Crenarchaea and Euryarchaea are found in freshwater environments than in soil (see MacGregor et al., 1997; Keough et al., 2003; Ochsenreiter et al., 2003). However, so far only the ecology of methanogens has been thoroughly characterized. A glimpse of an interesting novel euryarchaeote from cold sulfurous marsh water at 10°C has recently been obtained. The organism (termed SM1) grows in symbiotic or syntrophic association with bacteria of the genus *Thiotrix* in a macroscopically visible string-of-pearl like structure (Moissl et al., 2003). The assemblage can be cultivated *in situ* (in nature) on polyethylene nets. The specific physiologies of this archaeon remain unclear, but it does not exhibit the typical fluorescence of methanogens and does not contain coenzyme F_{420}. About 100 filamentous cell appendages of 2–3 µm length emanate radially from the surface of each coccoid-shaped SM1 cell (Fig. 4.2a). The ultrastructual examination of these appendages revealed a very unusual and interesting structure never observed before in nature: it resembles a piece of barbed wire with a tripartite barbed grappling hook that strikingly resembles a fishing hook at its distal end (Fig. 4.2b; Moissl et al., 2005). However, this tool was probably developed several hundred million, if not a billion, years before the macroscopic man-made equivalent. The "hamus" might be interesting for nanotechnological applications and represents a wonderful example of the potential for biomimetrics. It also demonstrates nicely how much there is to discover in the unexplored archaeal world, including novel morphological features that can be relevant to the ecology of organisms.

Archaea in association with eukaryotes

Since Archaea have been found in so many commonplace environments, the obvious question arises as to whether they are also typically associated with eukaryotes and perhaps also found as animal or human pathogens. It is well documented that methanogens are found associated with different organisms, including humans. They live, for example, in both endo- and ectosymbiotic associations with many protozoa and use H_2 produced by hydrogenosomes in an anoxic environment, thus acting as an electron sink (Embley & Finlay, 1993–4). Even in protozoa lacking hydrogenosomes, methanogens have been found together with other fermentative bacteria that produce H_2. Through cultivation or molecular PCR-based techniques, methanogens (and in particular Methanobacteriaceae and Methanosarcinaceae) have been found in association with the digestive tract of arthropods (such as termites, millipedes, scarab beetles, and cockroaches) as well as cattle, sheep, and humans (for review see Lange et al., 2005). The nature of the archaeal interactions within the endogenous flora of animals has been best described for methanogens in arthropods, where they are found in different types of symbiosis, including free living organisms that colonize the hindgut lumen or chitinious structures, or intracellular symbionts of protists (Hackstein & Vogels, 1997). In a 16S rRNA survey, Friedrich et al. (2001) detected axial differences in the methanogenic populations in different compartments of the hindgut from a soil-feeding termite, and also found sequences related to Thermoplasmatales and to non-thermophilic Crenarchaea. Beside the detection of crenarchaeal sequences in the holothurian *Oneiruphanta mutabilis* (McInerney et al., 1995), this is the only report so far of Crenarchaea found in the digestive tracts of animals. Other Crenarchaea, affiliated with the marine group I.1b of the planktonic marine Crenarchaea, are found repeatedly in sponges from different oceans (e.g. Webster et al., 2004), with the best described representative being *Cenarchaeum symbiosum* (discussed above).

Multiple niches for Archaea have been described in the human host. Methanogens have been identified in the colonic flora but also in subgingival dental plaque and in the vaginal flora (Eckburg et al., 2003 and references therein). They might be coinhabitants in all strictly anaerobic microenvironments of the human body. It is, however, still unclear if Archaea play a role in human or animal diseases. In principle, their ubiquitous distribution in nature provides ample opportunity for access to susceptible hosts and colonization therein. No principle reasons exist why Archaea could not be pathogens. In fact several general features that we mostly know from genome surveys imply that they might have the prerequisites of pathogens (Eckburg et al., 2003).

Archaeal genomes in large metagenomic datasets

Metagenomic or environmental genomic studies are certainly one of the most promising and expanding technologies to study uncultivated organisms from complex microbial communities. While initially long genome fragments of specific lineages have been targeted in complex large-insert libraries, the shotgun sequencing approach, as pioneered by Craig Venter for sequencing genomes of cultivated organisms, has now also been applied to natural microbial communities. By shotgun sequencing nucleic acids from a very simple community of an acidic biofilm, Tyson et al. (2004) created a dataset of 76 Mb from which a composite genome of roughly 1.8 Mb of a *Ferroplasma* species could be almost completely reconstructed. The genome contained a 16S rRNA with 99% identity to that of *Ferroplasma acidarmanus* (isolated earlier from the same location), but a sequence divergence of 22% on the nucleotide level when comparing the two, largely syntenic, genomes. Most interestingly, the reconstructed genome of *Ferroplasma* type II revealed an extensive degree of nucleotide polymorphism (different in its pattern from that observed with an assembled *Leptospirillum* genome of the same dataset) that was most likely explained by frequent recombination events among closely related *Ferroplasma* type II strains. Beside cryptic prophages and putative mobile genetic (retro)-elements, no evidence was found for the mechanism underlying this frequent gene exchange. However, such genetic exchange has been shown earlier among laboratory strains (Rosenshine et al., 1989; Schleper et al., 1995; Grogan, 1996). Furthermore, frequent events of homologous recombination (Lopez-Garcia et al., 2004), even leading to a degree of linkage equilibrium resembling that of a sexual population (Papke et al., 2004) have recently been suggested for other naturally occuring archaeal assemblages, indicating that extensive genetic exchange could be a more general feature within the Archaea.

The metagenomic approach was taken to yet another scale when Venter and collaborators made a huge random sequencing survey from DNA obtained from filtered surface waters of the Sargasso Sea (Venter et al., 2004). In the roughly one billion basepairs of sequence, representing 1.2 million novel genes, considerable numbers of archaeal sequences can be found, as would be expected from the high abundance of Archaea in marine plankton. Complete and partial 16S rRNA genes and other phylogenetic markers indicated the presence of both Crenarchaea and Euryarchaea. However, since this whole genome shotgun approach was performed on small-insert libraries from an (unexpectedly) complex microbial community, only single sequences or in-silico-assembled scaffolds are obtained and this often impedes a clear assignment of the sequence information to specific lineages. But with more sophisticated annotation tools (that still need to be developed for such large environmental datasets) or simply with the help of large genomic fragments that have been isolated earlier from BAC or fosmid libraries, a considerable number of archaeal genes and scaffolds can be assigned to the euryarchaeotic or crenarchaeotic lineages (own unpublished observation; Venter et al., 2004).

Conclusions and outlook

Research on microorganisms has traditionally been dependent on the pioneering work of microbiologists, who isolated novel strains from environmental samples. In particular, the isolation of "extremophilic" Archaea from remote, pristine, and special places on Earth has been inspiring to this research field (Stetter, 1999). Cultivation of novel organisms will remain crucial in the future too. However, it was through the use of molecular biological tools that Carl Woese in his pioneering studies detected the Archaea and set a basis for our perception of the third domain (Woese et al., 1978). Based on the same technique, i.e. phylogenetic analysis of ribosomal RNA genes, an overwhelming diversity of Archaea has been detected over the past decade with culture-independent approaches. We are only beginning now to get a comprehensive picture about the distribution and diversity of Archaea, and without the use of modern molecular techniques, the ecological distribution and impact of the Archaea would have remained completely obscure. With metagenomics and a number of other techniques (stable isotope probing, FISH), novel tools are at hand that allow us to characterize the physiological potential and activity of microorganisms, even in the absence of their cultivation. Extrapolating from the quickly evolving field of genomics and metagenomics, one can expect that complex communities will be monitored in the future based on "environmental genome tags" (EGTs) in high throughput sequencing efforts. In those studies, the goal might not be to isolate or reconstruct whole genomes, but to get profile characteristics for specific environments (Tringe et al., 2005). This information will in turn be used to study active genes in the environment with meta-transcriptomics and meta-proteomic studies, which have already been used on a smaller scale for less complex environmental samples. Such techniques will help us to get more comprehensive pictures about different naturally occurring microbial communities and how they react to shifting environmental conditions.

In order to transform the huge datasets acquired by such high-throughput technologies into useful scientific knowledge, a number of problems that will require the integration of many different research disciplines will need to be addressed in the near future. But the first metagenomic studies are encouraging (Table 4.1). They have contributed significantly to the identification of two novel metabolisms in the Archaea that play important roles in the ecology of this planet (reverse methanogenesis of deep-sea methanotrophs and ammonia oxidation of non-thermophilic Crenarchaea) and have led to a better understanding of the diversity of microorganisms in natural populations. From the first studies it also becomes evident that genomic analyses of natural microbial communities provide new insights into functional diversity within populations, genome dynamics, and speciation, which cannot be obtained in depth through the analysis of laboratory strains.

It will be exciting to learn more about the ecological impact of Archaea on this planet and their interaction with other life forms. It is also possible that population genomics will lead to the identification of archaeal pathogens or novel, divergent archaeal lineages that have eluded PCR-based molecular surveys.

Cultivation efforts to isolate or enrich for novel microbial species will remain crucial, however, for understanding the physiological diversity of micro-

Table 4.1 Metagenomic studies that include genomic information of uncultivated Archaea.

Environment	Library type	Archaeal group found	Comments	Reference
North Pacific, marine plankton	Fosmids	Marine planktonic Crenarchaea, group I.1A	First large genome fragment from an uncultivated microorganism	Stein et al., 1996; Beja et al., 2002
Californian Pacific coast, marine sponge *Axinella mexicana*	Fosmids	*Cenarchaeum symbiosum* (Crenarchaeote), group I.1A	Archaeal extracellular symbiont, stable association in laboratory aquaria	Schleper et al., 1997b, 1998
Californian coast, surface waters	BACs	Marine planktonic Euryarchaea, group II	60 kb clone identified through 23S rRNA gene	Beja et al., 2000
Antarctic, coastal waters	Fosmids	Planktonic Crenarchaea group I.1A	Comparative analysis of highly related fragments	Beja et al., 2002
Calcerous grassland soil, aerobic layer	Fosmids	Soil Crenarchaea group I.1B	Genes pointing to potential energy metabolism	Quaiser et al., 2002; Treusch et al., 2004, 2005
Eel river basin, Monterey Canyon, microbial mats associated with deep methane seeps	Fosmids	Euryarchaea ANME-1 and -2	Reverse methanogenesis pathway reconstruction	Hallam et al., 2003, 2004
Northwestern Black Sea shelf, microbial mats from methane seeps	Fosmids	Euryarchaea ANME-1	Combination of biochemical and metagenomic approach	Kruger et al., 2003
Antarctic polar front, waters of 500 m depth	Cosmids	Group I.1A Crenarchaea, group II euryarchaeote	Horizontal gene transfer in crenarchaeote	Lopez-Garcia et al., 2004; Moreira et al., 2004
Acid mine drainage biofilm	Plasmids (small inserts)	Ferroplasma	Mosaic genome reconstruction	Tyson et al., 2004
Sargasso Sea surface waters	Plasmids (small inserts)	Planktonic Crenarchaea and Euryarchaea	>1 Gb of environmental DNA sequenced	Venter et al., 2004
Methanogenic consortium	Fosmids	Rice cluster I methanogens	Stably maintained, bacterial/archaeal consortium	Erkel et al., 2005
Michigan soil	Plasmids	Crenarchaea	100 Mbp	Tringe et al., 2005

organisms and identifying gene functions and novel pathways. Considering the complementarity of these approaches, it is disappointing to see that the two disciplines of "classical" organismic microbiology and microbial ecology/metagenomics have mostly remained two independent disciplines so far, instead of merging to form a strong interdisciplinary platform in microbiological science.

The discovery of novel Archaea in non-extreme environments might also open some new perspectives for studying the cellular proteins of Archaea, which are so strikingly similar to their eukaryotic homologues. By obtaining organisms (or at least their genes) with growth requirements that are not extreme with respect to temperature, salt, or oxygen, but instead compatible with those of eukaryotic model organisms, it will become possible to study archaeal proteins in a eukaryotic cell context, by complementing mutants or using eukaryotic *in vitro* systems. Such studies will allow us to address questions about the conservation and evolution of information-processing systems in the archaeal/eukaryotic lineages from a novel perspective.

Acknowledgments

Thanks are due to G. Jurgens and M. Jonuscheit for help with Plate 4.1, to Graeme Nicol and Laila Reigstad for critical reading of the manuscript, and to G. Nicol, K. Knittel, G. Jurgens, K. Seifert, and R. Huber for kindly providing figures for this chapter. Only a subset of studies on archaeal diversity has been cited due to space limitation.

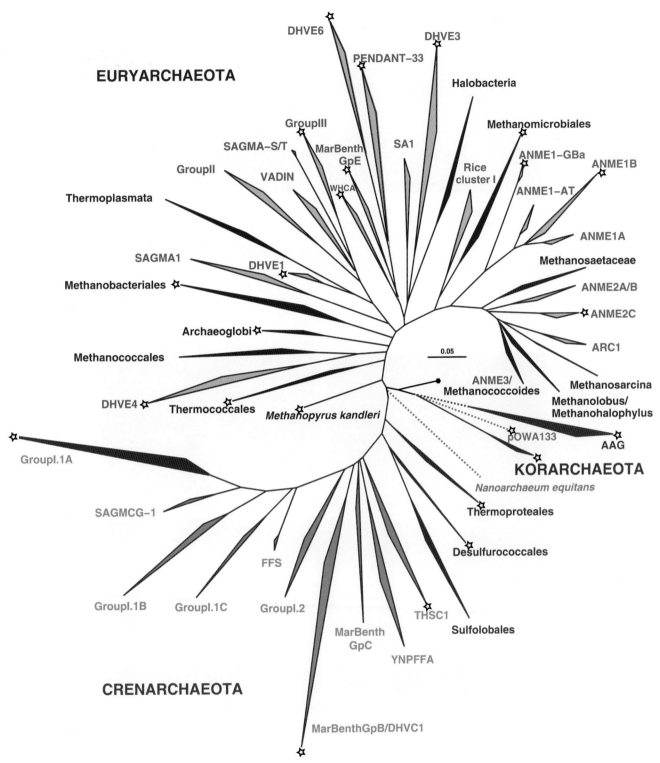

Plate 4.1 The world of Archaea. Phylogenetic tree of 16S rDNA sequences illustrating the major lineages of the domain Archaea. Triangles in light colors represent branches with exclusively uncultivated species, dark triangles show branches in which at least one cultivated species occurs. The size of the triangle is proportional to the number of sequences analysed. In all, 1344 16S rDNA sequences were included in the analysis (mostly full-length). The backbone of the tree was calculated with 55 full-length sequences by using maximum likelihood in combination with filters excluding highly variable positions (ARB software package). Asterisks indicate lineages that contain sequences from hydrothermal vents (see text).

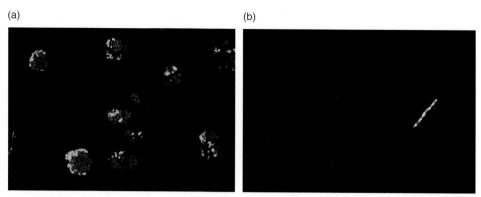

Plate 4.2 Visualization of uncultivated Euryarchaea by fluorescence *in situ* hybridization (FISH). (a) ANMEII Euryarchaea (red) in association with sulfate-reducing Bacteria (green) in sediments above methane hydrates. Image courtesy of T. Lösekann and K. Knittel, Max-Planck-Institute for Marine Microbiology, Bremen, Germany. (b) Diverse Euryarchaea in freshwater plankton (Valkea Kotinen, Southern Finland; see Jurgens et al., 2000). Image courtesy of G. Jurgens, University of Helsinki, Finland.

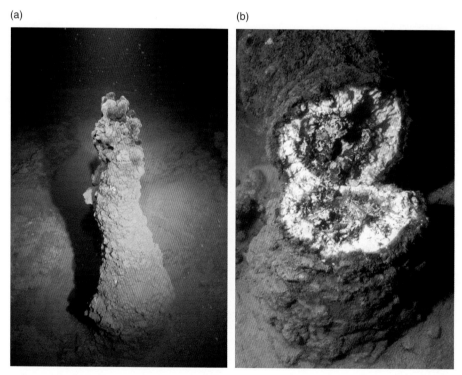

Plate 4.3 Images of microbial reef structures on the bottom of the Black Sea. (a) Chimney-like structure (image courtesy of R. Seifert, Institute of Biogeochemistry and Marine Chemistry, University of Hamburg). (b) Brocken structure of about 1 m height, with microbial mats in pink color inside and grayish outside, with greenish-gray inner part containing porous carbonate. (From Michaelis et al., 2002, with permission from the publisher.)

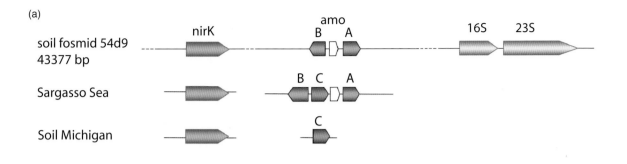

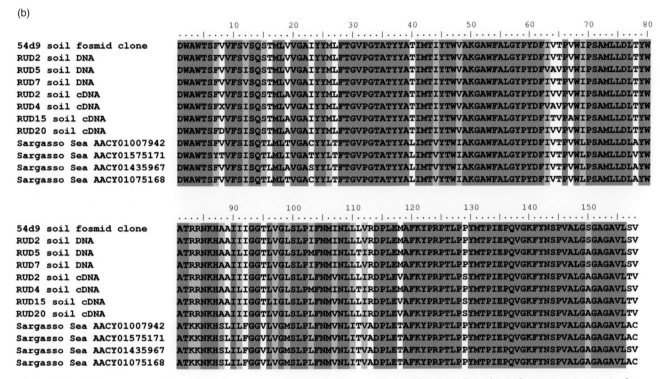

Plate 4.4 Highly conserved genes among terrestrial and marine Crenarchaea coding for subunits A, B, or C of a putative ammonia monooxygenase and a copper-dependent nitrite reductase (nirK). All genes indicate that these non-thermophilic Crenarchaea might be ammonia oxidizers. (a) Terrestrial clone 54d9 recovered from a metagenomic library of a sandy ecosystem (Treusch et al., 2004, 2005) links these genes to the ribosomal RNA operon of Crenarchaea, thereby pointing to their origin. Genes for subunits A, B, and C and nirK were found by similarity screening in the dataset obtained by shotgun sequencing DNA from the Sargasso Sea and amoC and nirK were identified in a shotgun sequencing dataset from a soil in Michigan. (b) The strikingly high conservation of the putative amoA subunit among marine and terrestrial Crenarchaea is depicted, comparing partial sequences from soil DNA, soil cDNA, and the Sargasso Sea dataset.

5
Nanoarchaeota

Harald Huber and Reinhard Rachel

Introduction

Until recently all cultivated Archaea belonged either to the Crenarchaeota or to the Euryarchaeota (Woese et al., 1990). A third archaeal phylum (kingdom), the "Korarchaeota," has been identified by amplification of 16S rDNAs from environmental samples (Barns et al., 1996), but the corresponding organisms still wait for detailed characterization. Archaea show a great variety in terms of morphology and cell size, physiology, adaptations to extreme environments, and genome size. Although they were for a long time regarded as thriving only in extreme environments, it has become more and more evident that Archaea inhabit nearly all natural environments and even anthropogenic biotopes, such as burning heaps or heaters within power plants. Archaea have also been found as partners in a number of symbioses, e.g. *Methanoplanus endosymbiosus* in a marine sapropelic ciliate (van Bruggen et al., 1986), *Cenarchaeum symbiosum* in a marine sponge (Preston et al., 1996), members of the order Methanosarcinales in microbial communities oxidizing methane anaerobically (Michaelis et al., 2002), and *Methanobrevibacter smithii* in the human intestine (Eckburg et al., 2005). In this chapter, we present an overview of the discovery, cultivation, physiology, ultrastructure, and phylogeny of *Nanoarchaeum equitans*. This tiny hyperthermophilic coccus lives in the first host–parasite association described for Archaea and represents in addition the first member of a new kingdom of the Archaea, the Nanoarchaeota (Huber et al., 2002a). *N. equitans* has turned out to be unique in terms of cell and genome size, phylogeny, gene organization, and lifestyle.

Discovery and cultivation

In samples taken at a depth of about 100 m at the Kolbeinsey ridge, north of Iceland, a novel isolate of the genus *Ignicoccus* was found, *Ignicoccus* sp. strain KIN4/I. It grew at temperatures around 90 °C under anoxic conditions. Cultivation was strictly dependent on the presence of molecular hydrogen as electron donor, elemental sulfur as electron acceptor, and, in addition, CO_2 as carbon source. Occasionally, attached to these cells, very tiny cocci were visible, exhibiting a diameter of only 400 nm. They were barely visible in phase contrast microscopy, but could clearly be identified by staining with DAPI, and also in transmission electron micrographs of platinum-shadowed specimens (Plate 5.1a, b, c). The cloning of the *Ignicoccus* sp. strain KIN4/I by single cell isolation using "laser tweezers" was successful, and a stable culture was obtained. In contrast, many attempts failed to grow the tiny cocci alone, after physical separation from the *Ignicoccus*, in defined media with or without a wide range of inorganic or organic compounds. It was, however, possible to establish a stable co-culture of the *Ignicoccus* sp. strain KIN4/I together with the small cocci under laboratory conditions, after isolating them using a "laser-tweezer." This culture grows to an average density of about 2×10^7 cells/ml of each organism, the *Ignicocci* and the small cocci.

In initial attempts to demonstrate that these microorganisms belonged to any of the described phyla within the Archaea, so-called "universal" primers were used in the polymerase chain reaction (PCR) for the amplification of their 16S rDNA genes. These experiments were successful for *Ignicoccus* sp. strain KIN4/I. It is closely related to the described *Ignicoccus* species, *I. islandicus* and *I. pacificus*, with a phylogenetic distance of 4.4 and 3.2, respectively. In contrast, PCR amplification of the 16S rRNA gene failed in the case of the tiny cocci, although many different kinds of primer combinations were employed. In the following, a different approach was successful. The co-culture was screened for all possible rRNA genes present in the co-culture, using Southern blot hybridization, with a *Metallosphaera* 16S rDNA as probe. Two signals were obtained: one, as expected, from the *Ignicoccus* cells and a further one that could be assigned to the small cocci. After the cloning and sequencing of this second 16S RNA gene, its primary and secondary structure showed many typical features for an archaeal 16S rDNA, but an unexpectedly high number of base exchanges in sequence regions, which had been considered to be "highly conserved" among the Archaea. The similarities to the 16S rDNA sequences of species from the archaeal phyla known had been low, with only 0.73–0.81 relative to the Crenarchaeota, 0.69–0.81 relative to the Euryarchaeota, 0.73–0.75 relative to the "Korarchaeota," and 0.59–0.70 relative to bacterial species. This indicated that the tiny cocci, whose proposed name is *Nanoarchaeum equitans* ("the dwarf archaeon, riding the fire sphere"), represented a thus far unknown phylum of Archaea, which we called Nanoarchaeota.

Physiology

The host organism of *N. equitans*, *Ignicoccus* sp. strain KIN4/I, can be grown equally well alone and in co-culture with *N. equitans*. Growth of this *Ignicoccus* species is unchanged or only slightly retarded in the presence of *N. equitans*, and similar cell densities are reached in the presence of the small cocci. Under standard cultivation conditions in closed vessels, the final cell density of *N. equitans* is in the range of the cell density of the host cells (about 2×10^7 cells/ml). About 30–40% of the *Ignicoccus* cells are occupied with *N. equitans* cells (average two to three *N. equitans* cells per occupied *Ignicoccus* cell). Minimal doubling times of about 60 minutes for *Ignicoccus* and 45 minutes for *N. equitans* were determined for such a co-culture. At the end of the stationary growth phase, the majority of the *Ignicoccus* cells are covered with at least ten *N. equitans* cells. During cultivation in fermenters, a significant influence of the gassing rate on the cell concentration of *N. equitans* was detected. By increasing the gas supply from 1 to 20 l/min ($H_2:CO_2=80:20$), the cell density of *N. equitans* could be raised about tenfold (final density about 3×10^8/ml), while the final cell density of *Ignicoccus* remained unchanged. Interestingly, about 70% of the *N. equitans* cells detached from the host cells and were found free in suspension. Further experiments using different gas mixtures demonstrated that this effect is not based on an improved supply of, for example, molecular hydrogen as electron donor, but is the consequence of gas stripping of H_2S, the main metabolic end product of *Ignicoccus*.

Like the other *Ignicoccus* species described so far (*I. islandicus* and *I. pacificus*), *Ignicoccus* sp. strain KIN4/I is a hyperthermophilic archaeon that requires strict anoxic conditions. The only electron donor is molecular hydrogen, which is "freely available" at the site where the original sample was taken, the Kolbeinsey Ridge. As electron acceptor, only elemental sulfur is used. The fixation of CO_2 for the production of biomass is performed by a pathway unknown so far (U. Jahn, H. Huber & G. Fuchs, unpublished), which is, in particular, different to the modified 3-hydroxypropinate cycle recently described for members of the Sulfolobales (Hügler et al., 2003). Accordingly, two main products are formed by *Ignicoccus* cells during growth, H_2S and biomass. In contrast, cells of *N. equitans*, after physical separation from their host *Ignicoccus*, do not grow alone in any artificial medium, either autotrophically or heterotrophically, not even on lysed *Ignicoccus* cells or other cell extracts. All cultivation experiments undertaken so far show that, for thriving of *N. equitans* cells, a direct contact to the surface of *Ignicoccus* sp. strain KIN4/I is mandatory. As a consequence of this dependence, comparable ranges of growth temperature, pH, and salt strength were observed for *N. equitans* and *Ignicoccus* sp. strain KIN4/I.

Genome and genes

With a size of only 490,885 bp *N. equitans* has the smallest archaeal genome and one of the smallest genomes known today. It consists of a single, circular chromosome with an average G+C content of 31.6%. So far, a total of 552 coding DNA sequences (CDS) have been annotated (Waters et al., 2003), exhibiting the highest gene density known, with CDS and stable RNA sequences covering about 95% of the genome. Putative functions could be assigned to about 62% of the annotated genes.

The interdependence between *N. equitans* and its host has been further substantiated by the fact that genes for many metabolic and biosynthetic pathways could not be identified, and are likely to be absent. Specifically, *N. equitans* does not exhibit any genes enabling it to perform a chemolithoautotrophic mode of life, as *Ignicoccus* does. Furthermore, almost all genes coding for enzymes involved in the biosynthesis of lipids, cofactors, amino acids, or nucleotides, as well as genes for glycolysis/gluconeogenesis, the pentose phosphate pathway, the tricarboxylic acid cycle, and carbon assimilation could not be found (Waters et al., 2003). These results suggest that *N. equitans* cells may to a certain degree take up metabolites and cell components from their host (see also below). The conclusion, however, that *N. equitans* is a "parasite" has to be made with some care, as a significant fraction of genes does not show any similarity to genes in the databases, or show similarities to genes coding for "hypothetical proteins." In this respect, two further aspects of the *N. equitans* genome are worth mentioning: (i) according to the annotation achieved, it only has a limited amount of membrane proteins/transporters, compared to other Archaea; (ii) only five or possibly six subunits of the membrane-bound ATPase could be annotated, i.e. it might be either a rudimentary or a reduced version of this type of enzyme. Whether it is able to function as an ATP synthase *in vivo* remains to be shown. The question is still unanswered as to whether *N. equitans* cells are capable of producing sufficient amounts of ATP for metabolic and transport processes, or has to import this molecule. In the meantime, new enzymes have been detected and analyzed for metabolic pathways in various Archaea, like glycolytic enzymes in *Pyrococcus furiosus* and *Sulfolobus solfataricus* (Verhees et al., 2003) and in *Thermoproteus tenax* (Siebers et al., 2004), as well as enzymes used for CO_2 fixation/assimilation in Sulfolobales (Hügler et al., 2003). Therefore, it cannot be excluded that in *N. equitans* novel enzymes, so far completely unknown and undetectable on the gene level, might be present.

In contrast to its paucity of metabolic genes, *N. equitans* harbors a large set of genes coding for enzymes involved in information processing, DNA recombination and repair, and cell cycle (Waters et al., 2003). This is in contrast to many obligate bacterial parasites with small genomes that have lost recombination/repair enzymes (Moran, 1996; Tamas et al., 2002).

A highly remarkable characteristic of the *N. equitans* genome is the high number of split genes whose gene products are encoded by two unlinked CDS that are expressed separately to form subunits of a functional enzyme (e.g. two subunits of alanyl-tRNA synthetase). Similarly, the reverse gyrase is coded by two half genes encoding a helicase and a topoisomerase domain. Very recently a riddle in the annotation of the *N. equitans* genome was solved. Originally, four tRNA genes could not be assigned (Glu, His, Trp, and initiator Met). By the development of a computational approach searching for tRNA signature sequences, nine genes encoding for tRNA halves were detected (Randau et al., 2005a). These halves were demonstrated to mature to full size tRNAs and thus gain acceptor activity. In the meantime biochemical evidence was also presented for the computationally predicted joined tRNAs designated as tRNA (Trp), which could be identified as tRNA (Lys) (Randau et al., 2005b). Therefore, *N. equitans* harbors the complete set of tRNA species: it exhibits 44 tRNAs and can read all 61 sense codons.

Whether the occurrence of split genes (like those for tRNAs) in *N. equitans* is a sign of an ancient genome or is the consequence of a later process of gene reduction is an open question. Nevertheless, the genetic conservation of split genes, along with the paucity of pseudogenes and a minimum of non-coding DNA (below 5%) suggests that the *N. equitans* genome is evolutionarily stable compared with many bacterial parasites (Waters et al., 2003).

Ultrastructure

At first glance, *N. equitans* cells are exceptional due to their extremely small size. Their diameter, as observed in transmission electron micrographs of ultrathin sections, varies from 0.35 to 0.5 μm, with a mean of about 0.4 μm, and is, therefore, the smallest coccoid Archaeum known today. Because the cells are regular cocci, their volume can be estimated to about $0.034\,\mu m^3$ only. For comparison, average cell dimensions and volumina are listed in Table 5.1 for cells of the bacterial genus *Escherichia*, the archaeal genera *Sulfolobus*, *Aeropyrum*, *Thermoproteus*, *Thermofilum*, *Pyrococcus*, *Methanococcus*, *Methanothermobacter*, and, finally, the host *Ignicoccus*. This demonstrates that the volume of an *N. equitans* cell is at least one to two orders of magnitude smaller than that of most archaeal cells (in comparison to its host, *Ignicoccus*, even two to three orders of magnitude).

Electron micrographs of ultrathin sectioned cells, prepared by cryo-immobilization and freeze-substitution, and of freeze-etched cells helped to unravel the structures that might be involved in the cell–cell interaction of *Ignicoccus* and *N. equitans* (Fig. 5.1a, b). The two Archaea have fundamentally different cell surface structures: *Ignicoccus* cells are remarkable in not synthesizing any kind of stabilizing cell wall polymer, like S-layers (Baumeister & Lembcke, 1992; Engelhardt & Peters, 1998) or pseudomurein (Kandler & König, 1993). Instead, they are surrounded by two membranes, a cytoplasmic membrane and an outer membrane, jointly delineating a huge periplasmic space, which is between 20 and 400 nm wide (Huber et al., 2000; Rachel et al., 2002b). Its volume is larger than that of the cytoplasm, and it contains membrane-bounded vesicles, which bleb off the cytoplasmic membrane. In contrast, the cell envelope of *N. equitans* cells is, in its basic architecture, similar to most other archaeal cells, consisting of a cytoplasmic membrane, about 6–8 nm wide, a periplasmic space of constant width, 20 nm and, as outermost sheath, an S-layer. For *N. equitans*, it was shown to consist of protein complexes with p6 symmetry and a center-to-center distance of about 15 nm, as revealed by relief reconstruction (Huber et al., 2002a, 2003b; Schuster and Rachel, in preparation).

On transmission electron micrographs of *N. equitans*, various states of cell division were observed. This is in line with the identification of a gene homologous to *fts*Z (Waters et al., 2003), a protein known to be involved in cell division processes in most Bacteria and Euryarchaeota (which has so far

Table 5.1 Dimensions and volumina of prokaryotic cells.

Genus	Cell dimensions ($\mu m \times \mu m$)	Total volume of cell (μm^3)
Nanoarchaeum	0.4	0.034
Escherichia	0.8 × 3.0	1.508
Sulfolobus, *Aeropyrum*, *Pyrococcus*, *Methanococcus*	1.0	0.524
Thermoproteus, *Methanothermobacter*	0.5 × 5.0	0.982
Thermofilum	0.2 × 10.0	0.314
Ignicoccus: small cell	1.5	1.767
large cell	5.0	65.45

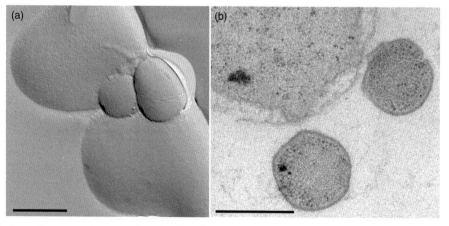

Fig. 5.1 Transmission electron micrographs of cells of *Ignicoccus* sp. strain KIN4/I with cells of *Nanoarchaeum equitans*: (a) freeze etching, showing dividing cells of *N. equitans*; (b) thin section (bar: each 0.5 μm).

not been found in Crenarchaeota), and which is, in its fold and three-dimensional structure, similar to tubulin subunits of eukaryotic cells (van den Ent et al., 2001).

The visualization of the site of interaction between *N. equitans* and *Ignicoccus* is hampered by: (i) the lack of a stabilizing cell wall polymer in the *Ignicoccus* cells; (ii) the low final cell density reached; and (iii) the low diameter of the *N. equitans* cells. Only a small fraction of electron micrographs shows some detail of the corresponding area: the *N. equitans* cell surface appears to be in direct attachment with or at least very close to the outer membrane of *Ignicoccus* cells (Fig. 5.1a, b). At these contact sites, the *Ignicoccus* cytoplasmic membrane is in close vicinity to the outer membrane, suggesting that the cytoplasma and/or the cytoplasmic membrane are also involved in establishing and maintaining the interaction to *N. equitans*. Whether the – so far only rarely detectable – extracellular appendages of both archaeal cells are also involved in the interaction remains to be elucidated.

Lipids

Comparison of the membrane lipids of *N. equitans* and its host *Ignicoccus* revealed that both organisms harbor simple and qualitatively identical lipids, with differences in the relative amounts of certain components (Jahn et al., 2004). As in most Archaea, the total lipid extract contains archaeol and caldarchaeol as the main core lipids. The phytanyl side chains are fully saturated phytane or biphytane, and the glycerol backbone is glycosylated with either mannose (about 95%) or glucose (about 5%). A significant quantitative difference was obtained for the relative amount of biphytane: purified *N. equitans* cells contain only half the amount of *Ignicoccus* sp. strain KIN4/I cells. Patterns of ^{13}C-fractionation of *N. equitans* and *Ignicoccus* isoprenoid chains is almost identical in both organisms (Jahn et al., 2004), suggesting that *N. equitans* does not synthesize its own lipids. Instead, they are imported into the *N. equitans* cells by an as yet unknown mechanism, which, however, must be in some way selective to explain the difference in the relative abundance of, for example, biphytane. In this context it is worth mentioning that the lipid contents of the cytoplasmic and the outer membrane of *Ignicoccus* are slightly, but significantly, different: the latter harbors no caldarchaeol, and therefore no biphytane.

Distribution of the Nanoarchaeota

Marine and terrestrial high temperature environments have been screened for the presence of nanoarchaeal 16S rRNA gene sequences using Nanoarchaeota-specific primers for PCR followed by sequencing (Hohn et al., 2002). From fragments of an abyssal black smoker situated at the East Pacific Rise (9°N; 104°W; Alvin dive no. 3072; sample LPC33), a sequence was obtained that was identical with the original sequence of *N. equitans*. Interestingly, Nanoarchaeota sequences were also found in terrestrial high temperature environments (Hohn et al., 2002). The corresponding samples had been taken from Obsidian Pool at Yellowstone National Park ("OP9"; water and black sediment; $t = 80\,°C$; pH 6.0) and from a little hot waterhole with blackish, vigorously gassed water at Caldeira Uzon, Kamchatka ("CU1"; water and black sediment; $t = 83\,°C$; pH 5.5). These terrestrial sequences revealed about 93% similarity to each other and only about 83% similarity to the (marine) LPC33 sequence and to *N. equitans*. These results demonstrate a great phylogenetic diversity and a worldwide distribution of the Nanoarchaeota within high temperature biotopes, from the deep sea to shallow submarine areas to terrestrial solfataric fields.

Phylogeny

Environmental DNA from different high temperature biotopes was shown to contain 16S rDNA genes closely related to *N. equitans* (see above). All these environmental sequences show the same features as the 16S rDNA of *N. equitans*, with base exchanges at otherwise "conserved" sites, pointing to a significant and close relationship. They form a separate and stable cluster in 16S rRNA based phylogenetic trees (bootstrap support 98–100%), distinct from the other three archaeal phyla. In addition, all approaches yielded essentially the same deep placement of the Nanoarchaeota lineage within the Archaea, irrespective of the algorithms used, e.g. neighbor joining, maximum parsimony, or maximum likelihood (Fig. 5.2). However, the exact

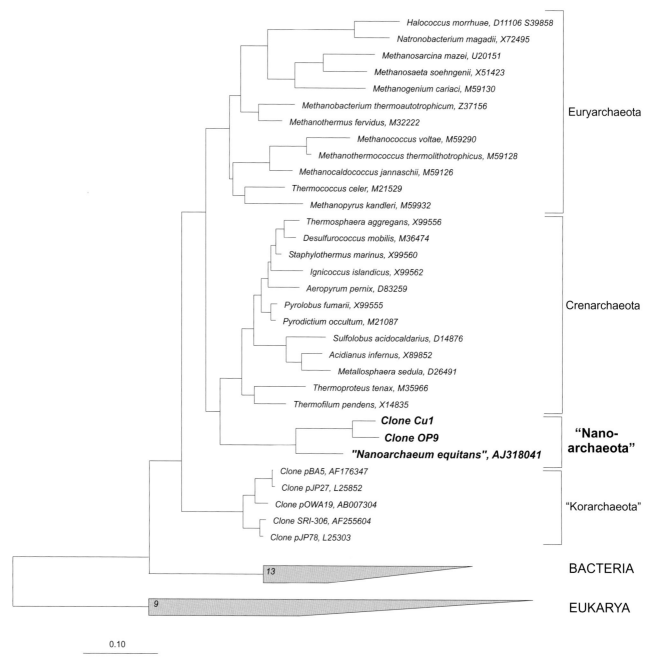

Fig. 5.2 Phylogenetic tree based on 16S rRNA sequence comparisons. The tree was calculated using the maximum likelihood (FastDNAml) program, embedded in the ARB package (Ludwig & Strunk, 2001).

placement of the branching point of the Nanoarchaeota is an open question, because it varies significantly depending on the algorithm and the domain-specific filter sets applied (Huber et al., 2003a). The same situation is true for 23S rRNA data sets. Concatenation and alignment of the amino acid sequences of 35 ribosomal proteins of *N. equitans* revealed with high boot strap support a most deeply branching position within the Archaea, suggesting that the Nanoarchaeota diverged early within the

Archaea and represent an independent lineage (Waters et al., 2003).

This independent phylogenetic position of the Nanoarchaeota was questioned, using other computational approaches of concatenated ribosomal proteins from 25 archaeal genomes (Brochier et al., 2005b). It was suggested that the Nanoarchaeota, instead of being a separate deep branch within the Archaea, represent a highly derived lineage, possibly related to the Thermococcales. However, the same authors have published an updated phylogenetic analysis with an extended dataset (including new taxa) of core proteins involved in translation and transcription and using modified computational analyses (Brochier et al., 2005a). These results confirmed strongly the emergence of *N. equitans* as a separate branch distinct from the Crenarchaeota and Euryarchaeota lineages. Nevertheless, some uncertainty remains concerning the phylogenetic position of *N. equitans*, due to the unknown extent of a possible lateral gene transfer between *N. equitans* and other Archaea or the effects of long branch attractions.

Outlook

The discovery of *N. equitans* and of related environmental 16S rDNA sequences has significantly increased our knowledge of the diversity of Archaea and of hyperthermophiles. Despite the results obtained so far and presented here (based on direct studies on the co-culture), including the analyses of the whole genome sequence of *N. equitans*, several open questions exist as to its physiology, biochemistry, and interaction with the host organism *Ignicoccus* sp. strain KIN4/I. In this respect the outcome of the sequencing of the *Ignicoccus* genome is eagerly awaited. It may give us further clues to the substrates and molecules that might be transferred from one organism to the other and possible transport mechanisms.

The isolation of *N. equitans* and the discovery of further nanoarchaeotal 16S rDNA sequences also demonstrate that environmental diversity studies solely based on PCR-amplified 16S rRNA genes should be interpreted cautiously. So-called "universal" PCR primers will only reveal environmental sequences with 16S rRNA target sequences identical or at least similar to cultivated species. Novel sequences (and organisms) with rRNA sequences too different to be recognized by the primers applied remain undetectable. The discovery of the Nanoarchaeota, however, suggests a much greater diversity of microbial life on Earth than it was formerly expected. Furthermore, it implies the existence of life forms, still unrecognizable by PCR gene amplification, that await discovery and cultivation.

The question remains whether other (or all) representatives of the Nanoarchaeota are as tiny as *N. equitans* and exhibit the same mode of life, strictly depending on a host organism. Phylogenetic staining experiments using a Nanoarchaeota-specific probe, carried out with enrichment cultures of the sample CU1 from Kamchatka, revealed the attachment of tiny cocci to rod-shaped organisms (related to *Pyrobaculum*). However, the isolation of this co-culture has not yet been achieved.

It has been argued (Makarova & Koonin, 2005) that *N. equitans*, with hardly any metabolic genes, an absence of operons, and a noticeable occurrence of split genes, might be a derived organism, with a massively reduced gene set, and not an ancestral archaeon. At present, this interpretation is based purely on comparative genomic predictions. In future, this open discussion calls for experimental feedback, e.g. by analyzing the expression of genes and studying the function of a high number of proteins, especially in comparison with the host organism *Ignicoccus* sp. strain KIN4/I. In addition, the enrichment and isolation of further representatives of the Nanoarchaeota would fundamentally help to elucidate the origin, lifestyle and evolution of these fascinating organisms.

In general, the discovery of *N. equitans* has already stimulated and will further stimulate discussions on the origin of life, minimal cell sizes, gene transfers, and evolution of genes and gene clusters (e.g. Andersson et al., 2005; Brochier et al., 2005a; Randau et al., 2005a, b; Stetter et al., 2005).

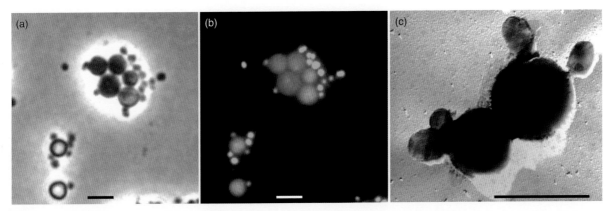

Plate 5.1 (a) Light microscopy (phase contrast), (b) DAPI stain, and (c) transmission electron micrograph (platinum-shadowed) of dividing cells of *Ignicoccus* sp. strain KIN4/I with cells of *Nanoarchaeum equitans* in similar magnification (bar 2 μm).

6
Families of DNA viruses infecting hyperthermophilic Crenarchaea

David Prangishvili

Introduction

By the time of discovery of the domain Archaea, in the late 1970s, there were two known viruses from this domain. Both infected extreme halophiles of the genus *Halobacterium*, had double-stranded (ds) DNA genomes, and resembled morphologically head-tailed bacteriophages from the family *Myoviridae* (Torsvik & Dundas, 1974; Wais et al., 1975). Later, more dsDNA viruses of extreme halophiles and methanogens were discovered. They were also typical head-tailed phages, with non-enveloped virions carrying icosahedral heads and helical tails, contractile or non-contractile, and belonged to the bacteriophage families *Myoviridae* and *Siphoviridae*, respectively (reviewed by Dyall-Smith et al., 2003). The myovirus ΦH of *Halobacterium salinarum* was studied in some detail in the laboratory of Wolfram Zillig. In addition to structural resemblance, it revealed resemblance to bacterial myoviruses in its molecular characteristics. The mode of DNA replication and the type of the lysogeny, conferred to the host cell by a circular plasmid, were similar to those of the bacteriophage P1. The observations suggested common ancestry of archaeal and bacterial head-tailed phages (Zillig et al., 1996). In accord with this distinction, the sequenced genomes of phages of extreme halophiles, such as Chi1, HF1/2, or ΨM encode many proteins homologous to bacteriophage capsid proteins (reviewed by Prangishvili et al., 2006).

Nothing seemed to be unusual in the discovery of similarities between archaeal and bacterial viruses, considering that most cellular characteristics of Archaea resemble those of Bacteria. Moreover, given the existence of an extensive horizontal gene transfer between Bacteria and Archaea (Koonin et al., 1997, 2001; Lawrence & Hendrickson, 2003), it could even be anticipated that archaeal viruses would represent distinct phage varieties. Thus, isolation by Wolfram Zillig and colleagues of several dsDNA viruses of Archaea with unusual morphology appeared rather unexpected. These viruses infect hyperthermophilic Archaea from the kingdom Crenarchaeota, and their morphotypes are so unusual that it was impossible to assign them to the known viral families. For their classification four novel viral families had to be established: *Fuselloviridae*, *Lipothrixviridae*, *Rudiviridae* and *Guttaviridae* (reviewed by Zillig et al., 1994, 1996; Prangishvili et al., 2001). These families have been approved by the International Committee on Taxonomy of Viruses (ICTV).

Later, studies in different laboratories have revealed that viruses with diverse exceptional morphologies are not just an oddity but represent a general picture of viral diversity in geothermally heated hot habitats, where hyperthermophilic Crenarchaea dominate (reviewed by Prangishvili, 2003). Numerous unusual shapes of viral particles associated with hyperthermophilic Crenarchaea were observed in single enrichments from samples from hot terrestrial springs (Rice et al., 2001; Rachel et al., 2002a; Häring et al., 2005a). Several dsDNA viruses isolated from these enrichments revealed virion structure previously never observed in nature. For their classification, three novel viral families have been proposed: *Globuloviridae*, *Ampullaviridae*, and *Bicaudaviridae* (reviewed by Prangishvili & Garrett, 2005). Approval of these families is pending at the ICTV.

Our present knowledge on seven virus families infecting hyperthermophilic Crenarchaea is summarized here. Virion structure and constituents, and virus–host interactions, are described separately for each family. Sequences of dsDNA genomes of all viruses are discussed jointly, in accordance with the idea that crenarchaeal viruses form a distinctive group, unrelated to any other viruses, with a small pool of shared genes and a unique origin, or more likely, multiple origins (Prangishvili & Garrett, 2005; Prangishvili et al., 2006).

Common features of crenarchaeal viruses

Hosts of all the cultured viruses of Crenarchaeota are members of the genera *Sulfolobus, Acidianus, Thermoproteus*, and *Pyrobaculum*. For all of these hyperthermophiles, which grow optimally at 80 °C or above, viral infection occurs most effectively at about 80 °C, such that the viruses can also be considered to be hyperthermophiles. The viruses are also extremely thermostable, e.g. thermal inactivation of unenveloped rod-shaped rudiviruses requires autoclaving at 121 °C for at least 40 minutes. Infectivity of viruses of the anaerobic *Thermoproteus* and *Pyrobaculum* is unaffected by exposure to oxygen.

By contrast to dsDNA viruses of bacteria, which eventually cause death and lysis of host cells, crenarchaeal viruses, except TTV1 and ATV, are nonlytic. Most of them do not vanish from the infected host culture even after prolonged growth and multiple dilutions. This is usually interpreted as a result of continuous replication of a virus in the infected cell, suggesting equilibrium between virus replication and host multiplication. However, this suggestion was never clearly verified. Whatever the mechanism of a stable persistence of viruses in host cells (or host cultures) is, it is obvious that generous virus–host relationships could be beneficial for the virus population to avoid prolonged direct exposure to extremely harsh environmental conditions. In line with this suggestion are reports on extremely low number of virus particles in hot acidic springs compared to any other environments (Breitbart et al., 2004). Under laboratory conditions, replication of most of the nonlytic crenarchaeal viruses cannot be enhanced by treatment of virus-infected cultures with stress factors such as mitomycin C (0.5 μg/ml) or UV radiation.

Family Lipothrixviridae

This family includes the highest number of known species among the crenarchaeal viral families. Virions are filamentous, mostly flexible, and enveloped. Envelopes of all known species contain virus-encoded proteins and, except AFV3, lipids. The genomes are linear dsDNA, 15.9–56 kbp. The structures of virion termini and inner cores are diverse, and due to this diversity, as well as differences in presumable replication strategies, four genera have been established in the family.

Genus *Alphalipothrixvirus*

Species: *Thermoproteus tenax* virus 1, TTV1 (Janekovic et al., 1983; Neumann, 1988; Rettenberger, 1990); genome sequence accession no. X14855 (85% of DNA). Presently does not exist under laboratory collections.

Virion morphology and constituents. Virions of the sole member of the genus, the virus TTV1, are rigid rods, about 400 nm in length and 38 nm in width (Fig. 6.1a). Their structure and composition were studied by TEM and chemical analysis of partially deteriorated particles after treatment with non-ionic detergents and octylglycoside. The envelope was found to have a bilayer structure and to contain proteins, isoprenyl ether lipids, and glycolipids. It encases a helical core, consisting of DNA and at least two highly basic DNA-binding proteins, of 12.9 and 16.3 kDa. Among two other identified structural proteins one, of 18.1 kDa, was found in the envelope. The location of the other structural protein (24.5 kDa) could not be determined.

Host range and virus–host interactions. Host range is limited to *Thermoproteus tenax*. Adsorption and infection proceed via interaction of the terminal protrusions of the virion with pili of the host cell. Virus is temperate and virions are released by lysis. It could be shown that mature virions assemble in the host cell lumen prior to their release. The origin and nature of morphogenetic proteins enabling the assembly of virus envelope remain unclear. The viral DNA exists in cells in linear form, and fragments

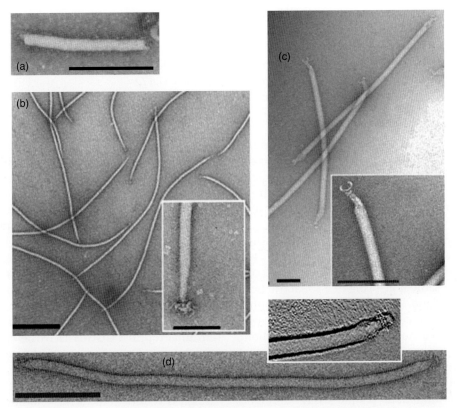

Fig. 6.1 Electron micrographs of virions of representatives of four genera of the family Lipothrixviridae. (a) Alphalipothrixvirus TTV1; (b) betalipothrixvirus SIFV; (c) gammalipothrixvirus AFV1; (d) deltalipothrixvirus AFV2. In insets (b, c) terminal structures magnified from the corresponding electron micrographs; (c) three-dimensional reconstruction of the virion termini. Virions were negative stained with 3% uranyl acetate. Bars: (a) 100 nm; (b) 500 nm, in inset 50 nm; (c) 100 nm; (d) 200 nm. (a, b, Courtesy of Wolfram Zillig; c, modified from Bettstetter et al., 2003; d, modified from Häring et al., 2005c.)

of viral DNA have been detected in the host chromosome.

Genome and its variation. The termini of the 15.9 kb long linear DNA was reported to be masked by hydrophobic ligands, which could not be removed by proteases. Therefore only about 85% of the genome sequence could be determined.

An unknown mechanism of genome variation was observed (Neumann & Zillig, 1990). Viruses isolated from different colony-purified strains of *T. tenax* showed evidence of insertion/deletions in their linear genomes ranging in size from 30 to 102 bp. They were detected within two limited regions of the TTV1 genome, both of which show low complexity nucleotide sequences and probably do not encode proteins. The two regions contain multiple short, imperfect, repeat structures, suggesting that recombination events may occur. The inserted sequences presumably arose either from host chromosome or from other extra-chromosomal elements within the same host.

Genus *Betalipothrixvirus*

Species: *Sulfolobus islandicus* filamentous virus, SIRV (Arnold et al., 2000b); genome sequence accession no. AF440571 (96% of DNA).

Virion morphology and constituents. Virions are flexuous filaments, about 1950 nm in length and 24 nm in width (Fig. 6.1b). The ends of the virion taper and end in mop-like structures in which six tail fibers are attached in a terminal thickening (Fig. 6.1b, inset). The tail fibers appear as angled rods formed by two legs of about equal length.

The virion consists of at least six proteins. Two major proteins, of 17 and 25 kDa, are present in equal amounts and are involved in the formation of the virion core. Three-dimensional reconstruction of the core by image analysis depicted two stacks of doughnut-like subunits, around which, it was suggested, the linear dsDNA was wrapped. The envelope of the virion, encasing the core, is about 4 nm thick. It was shown to contain integral protein component and host-derived phospholipids, differing from those of the host membrane both quantitatively and qualitatively.

Host range and virus–host interactions. Host range is limited to *S. islandicus* strains HVE11/2 (natural host) and HVE10/4 (indicator strain), both isolated from the same hot spring of a solfataric field in Hveragerdi, Iceland. Neither significant amounts of cell debris nor a decrease in cell density was observed in infected host cultures. Upon infection, cultures were inhibited in their growth but eventually surmounted this inhibition and reached the same cell density as the uninfected control culture. Due to growth inhibition, plaque assay could be established on a lawn of *S. islandicus* HVE10/4. The latent period (the time interval between infection and release of the first virus particles) was equal to 6–8 hours. Integration of viral DNA into the host chromosome was not observed, and continuous growth of the infected cell culture for several weeks with multiple successive transfers into fresh medium resulted in virus segregation.

It was suggested that viral receptors are located within the host membrane, and that conformational changes in the virion termini occur at an initial phase of adsorption: the mop-like terminal structures unfold like spiders' legs prior to attachment to the cell membranes.

Mature enveloped virions could be observed in host cells. Apparently, they are assembled and wrapped into their envelopes in host cell lumen prior to the extrusion, rather than in its course.

Genome. The termini of the linear dsDNA genome are modified in unknown manner, precluding exonucleolytic degradation from 3′ and 5′ ends and sequencing this part of the genome. The remainder of the genome (about 97%) has been sequenced and comprises 40,852 bp, including inverted terminal repeats.

Genus *Gammalipothrixvirus*

Species: *Acidianus* filamentous virus 1, AFV1 (Bettstetter et al., 2003); genome sequence accession no. AJ567472.

Virion morphology and constituents. Virions are enveloped flexuous filaments, about 900 nm in length and 24 nm in width (Fig. 6.1c). They consist of five major proteins with molecular masses of 130, 100, 80, 30, and 23 kDa, and modified host lipids. The envelope is 3–4 nm thick and can be removed by treatment with detergents. Both ends carry identical claw-like structures with a diameter of 20 nm.

Genome. The linear dsDNA genome is 20,869 bp long. The nature of the ends of the linear genome is presently unknown. In restriction digests of the DNA terminal fragments are underrepresented and their relative molarities were estimated to be about half of those of the non-terminal fragments. Sequences of the terminal regions are unusual. The short inverted terminal repeat CGGGGGGGGG is followed at either end by an approximately 350 bp region containing numerous short direct repeats of the pentanucleotide TTGTT and close variants thereof. The structural design is reminiscent of the telomeric ends of linear eukaryal chromosomes.

Host range and virus–host interactions. Natural hosts for the viruses are several isolates, all belonging to the genus *Acidianus*, from a hot, acidic spring in Yellowstone National Park, USA: "*Acidianus hospitalis*" strains YS6, YS7, OV5, OV9–13. Indicator strains, in which the virus can be replicated under laboratory conditions, are "*A. hospitalis*" strains YS8, YS9, and W1, isolated from the same spring as the natural hosts. In addition, the virus infects the type strain of the genus *Acidianus*, *Acidianus infernus*.

Viral receptors appear to be located on the pili of host cells rather than within the membrane. Adsorption of virions to the cell surface was never observed. Instead, virions were often observed attached to host pili. The two unusual termini of the virion appear to have a special function in the process of adsorption. As a result of adsorption to the host receptor the claw folds and keeps a virion firmly attached to a pilus.

During a cycle of productive infection with AFV1 neither the formation of any significant amounts of cell debris nor a decrease in cell density were observed. As a result of infection, host growth nearly completely stopped. Later, it slowly recovered, however, with a generation time of 20 hours in contrast to a generation time of 11 hours for uninfected cells. Eventually infected cells reached a density identical to the uninfected control culture. Retardation of host growth could not be used for the development of plaque assays, since none of the hosts is capable of lawn formation. The latent period of virus infection was measured by detection of intracellular viral DNA, and was equal to 4 hours. Infected host cell cultures were not cured of the virus after several successive transfers into fresh medium and continuous growth for at least 4 months. Integration of viral genome into the host chromosome was not observed.

Genus *Deltalipothrixvirus*

Species: *Acidianus* filamentous virus 2, AFV2 (Häring et al., 2005b); genome sequence accession no. AJ854042.

Virion morphology and constituents. Virions are flexuous filaments, about 1100 nm in length and 24 nm in width (Fig. 6.1d). They are enveloped with a 3–4 nm thick envelope which can be removed by treatment with detergents. Both ends of virions present identical structures. Analysis by electron tomography revealed that each terminus constitutes a complex collar with two sets of filaments, resembling a bottle brush with a solid cap 17 nm in diameter on its top.

The virion carries seven major proteins with molecular masses of 65, 50, 45, 40, 35, 26, and 6 kDa. By contrast to other members of the *Lipothrixviridae*, no lipid component could be extracted from virus particles.

Host range and virus–host interactions. The natural host of the virus is an *Acidianus* sp. AciF28 isolated from a hot, acidic spring of a solfataric field of Pozzuoli, Italy, and closely related to *Acidianus brierleyi*. The virus host is devoid of pilus-like structures, and virions attach directly to the cell surface. The terminal structures appear to be involved in adsorption. No conformation changes could be detected in these structures upon adsorption.

On production of AFV2, neither cell debris nor a decrease in cell density was observed in the growing culture of the host strain. Infected cells were not cured of the virus after five successive dilutions in fresh medium, and continuous growth for, at least, 2 months. No circular or integrated form of viral DNA was observed in host cells.

Genome. The genome is linear dsDNA that is 31,787 bp long. The nature of the ends of the linear genome could not be elucidated. No inverted repeats or other symmetrical sequences are present in the terminal regions. The genome contains an unusual 1008 region close to the center and bordered by a short inverted repeat, which carries two large 46 bp direct repeats and multiple imperfect short repeats throughout the region. Its base composition is strongly biased, with one DNA strand containing only guanosines. This region may constitute the origin of replication. Moreover, the central part of the genome also contains a tRNA^{Lys} gene with a 12 bp archaeal intron. Remarkably, a tRNA gene has so far not been detected in other crenarchaeal viral genomes.

Family Rudiviridae

Genus *Rudivirus*

Species: *Sulfolobus islandicus* rod-shaped virus 1, SIRV1 (Prangishvili et al., 1999; Blum et al., 2000; Peng et al., 2000, 2004); genome sequence accession no. AJ414696; *Sulfolobus islandicus* virus rod-shaped virus 2, SIRV2 (Prangishvili et al., 1999; Peng et al., 2000); genome sequence accession no. AJ344259; *Acidianus* rod-shaped virus 1, ARV1 (Vestergaard et al., 2005); genome sequence accession no. AJ875026.

Virion morphology and constituents. Virions are unenveloped rod-shaped particles (Fig. 6.2), with widths of 23 nm and different lengths, proportional to the length of their linear dsDNA genomes: SIRV2, 900 nm long with 35 kb DNA; SIRV1 830 nm long with 32.3 kb DNA, and ARV1, 610 nm long with 24,655 kb DNA. The virions consist of a tube-like superhelix formed by the DNA and a single basic, glycosylated DNA-binding protein of 14.4 kDa. The tube carries plugs, about 50 nm long, at each end. In the plug-like termini of virions three tail fibers are anchored, each about 10 nm in length and about 3 nm in width. It is most likely that the second protein component of the virion (about 124 kDa) is associated with tail fibers. In virions no lipid component could be detected.

Host range and virus–host interactions. Natural hosts for SIRV1 and SIRV2 are *S. islandicus*

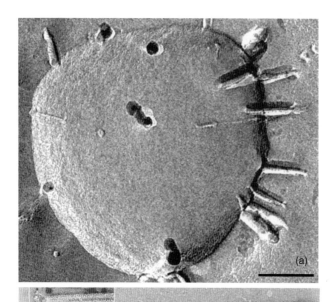

Fig. 6.2 Electron micrographs of a cell of *S. islandicus* and virions of SIRV1, a species in the family Rudiviridae. (a) Extrusion of virions from a SIRV1-infected cell of *S. islandicus*. Platinum shadowing. (b) A single virion with magnified terminal structure in the inset. Negative stain with 3% uranyl acetate. Bars: (a) 500 nm; (b) 200 nm. (a, modified from Vestergaard et al., 2005; b, modified from Prangishvili et al., 1999.)

strains KVEM10H3 and HVE10/2, respectively, isolated from hot acidic springs at two different solfataric fields in Iceland, Kverkfjöll and Hveragerdi. Under laboratory conditions both viruses replicate in *S. islandicus* strains LAL14/1, KVEM10H1, HVE6/3, and HVE10/4, but only SIRV1 replicates in *S. islandicus* KVEM10H1. The natural host for ARV1 is *Acidianus* sp. Acii26, isolated from an acidic, hot spring in Pozzuoli, Italy. Under laboratory conditions the virus also replicates in two other species of *Acidianus*, isolated from the same hot spring as the natural host, *Acidianus* sp. AD1 and *Acidianus pozzuoliensis*. Although it was originally suggested that *Sulfolobus* rudiviruses absorb to the pili of host cells, later investigations demonstrated that viral receptors are present on host cell membranes. The three tail fibers of the virion appear to be involved in adsorption.

Extrusion of virions is not accompanied by cell lysis (Fig. 6.2a) and no decrease of cell density or formation of cell debris is observed during growth of the infected host culture. However, virus infection causes significant retardation of host growth. Therefore plaque assays could be set up on lawns of *S. islandicus*. Latent period is 8 hours for SIRV1 and 6 hours for SIRV2. Infected hosts can be cured of the viruses upon continuous growth.

Genome, its replication and transcription. The three genomes contain very long inverted terminal repeats, comprising up to 7% of the total genome length. The termini of the linear genomes are covalently modified: the two DNA strands are covalently linked at the ends, generating a 4 bp loop that is susceptible to *Bal*31 endonuclease.

Different species show different degrees of genome stability. Sequencing of clone libraries of SIRV2 and ARV1 showed no evidence of sequence heterogeneities even in genomic regions where the clone sequence coverage was 10 to 15-fold. However, the genome of the rudivirus SIRV1 was found to be extremely unstable. The isolated virus invariably contained a population of variants with different but closely related genomes. Upon propagation in a given host strain, one or more genomes dominate in the viral population. However, upon passage in a new host strain the viral population undergoes changes and other variants are selected. Genome sequencing of the variants and genome comparisons demonstrated that they contain a few highly variable genomic regions where deletions/insertions have occurred, as well as gene transpositions (Peng et al., 2004). Comparison of the variant genomes also revealed the presence of small genetic elements, invariably 12 bp in length or multiples thereof, which tend to be concentrated in these variable regions. Although their prevalence in the genome suggests that they were mobile, they show no sequence conservation, nor are they flanked by conserved target sequences for integration or bordered by direct repeats indicative of transposition. However, some of their transcripts could generate "bulge–helix–bulge" splicing motifs, characteristic of archaeal intron splicing junctions. This suggested that they could have been mobilized at the RNA level by splicing followed by reverse splicing and reverse transcription.

The observation of head-to-head and tail-to-tail linked replicative intermediates in cells infected with the virus SIRV1 suggested a self-priming mechanism of replication, similar to that proposed for large eukaryotic dsDNA viruses including poxviruses. Consistent with this proposal are the similarities in structures of linear genomes of these archaeal and eukaryal viruses, including long inverted terminal repeats (ITRs) and covalently closed termini. Moreover, each of these viruses encodes a Holliday junction resolvase, which is likely to resolve Holliday junctions formed during replication.

The large ITRs of the rudiviruses are likely to have a common function in viral genome replication. They all carry repeat sequence motifs, which could act as signals for replication, including the tandem direct repeats TTTTTTTGC located near the genomic termini of the SIRV1 genome. Moreover, all of the rudiviral ITRs contain the internal repeat sequence AATTTAGGAATTTAGGAATTT located 100–150 bp from the genomic termini. Since this is the only highly conserved sequence within all three ITRs, it may be an important signal for DNA replication.

In vivo studies demonstrated a rather simple and ordered transcriptional pattern for both SIRV1 and SIRV2, with very few cases of temporal regulation, consistent with fairly unsophisticated virus–host relationships (Kessler et al., 2004). SIRV promoters, like those of their hosts, carry a TATA-like box and a transcription factor B responsive element. However, many of them exhibit an additional virus-specific consensus element, the trinucleotide GTC located immediately downstream from the TATA-like box. The same pattern is noticeable for about one-third of the ARV1 promoter regions, although there was little evidence of the GTC motif being conserved for homologous genes in the three viral genomes. The results underline the probable importance of both the host transcriptional machinery and virus-specific factors in generating and regulating viral transcripts.

Family Fuselloviridae

Genus *Fusellovirus*

Species: *Sulfolobus* spindle-shaped virus 1, SSV1 (Martin et al., 1984; Palm et al., 1991; Schleper et al., 1992); genome sequence accession no. XO7234; *Sulfolobus* spindle-shaped virus 2, SSV2 (Stedman et al., 2003); genome sequence accession no. AY370762; *Sulfolobus* spindle-shaped virus, Yellowstone 1, SSV-Y1 (Wiedenheft et al., 2004); genome sequence accession no. AY423772; *Sulfolobus* spindle-shaped virus, Kamchatka 1, SSV-K1 (Wiedenheft et al., 2004); genome sequence accession no. AY388628.

Virion morphology and constituents. Virions are enveloped, spindle-shaped, with a short tail attached to one of the two pointed ends and measure about 55–60×80–100 nm (Fig. 6.3). The structure of the virion core, encased by the envelope and apparently responsible for the unusual geometry, has not been studied. Virions carry three proteins, one of which is associated with viral DNA, with two others associated with the envelope. In the literature there is controversial information on lipid composition of SSV1. According to the most reliable data, about 10% of the SSV1 envelope consists of modified host lipids (M. Rettenberger, personal communication). The lipid compositions of other species in the family have not been determined.

Host range and virus–host interactions. The natural host of SSV1 is *Sulfolobus shibatae* strain B12, isolated from a hot spring at Beppu, Japan, and that

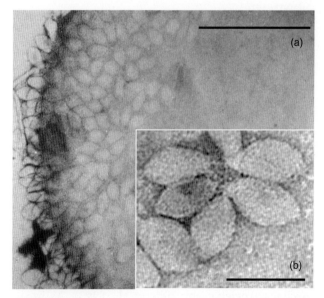

Fig. 6.3 Electron micrographs of a portion of a cell of *S. shibatae* and virions of SSV1, a species in the family Fuselloviridae. (a) Extrusion of virions from a SSV1-infected cell of *S. shibatae*. (b) Virions of SSV1. Samples were negatively stained with 3% uranyl acetate. Bars: (a) 500 nm; (b) 200 nm. (Courtesy of Wolfram Zillig.)

of SSV2 is *S. islandicus* isolate REY15/4 from a hot spring Reykjanes, Iceland. Under laboratory conditions both viruses replicate in *Sulfolobus solfataricus* strains P1 and P2. Natural hosts of SSV-Y1 and SSV-K1 are *Sulfolobus* species from Yellowstone National Park, USA, and Kamchatka, Russia, respectively, both closely related to *S. solfataricus*. In addition to their natural hosts, the two viruses replicate in *S. solfataricus* P2.

The virion tail is involved in adsorption onto the surface of the host cells. Mature virions were observed in host cells (Fig. 6.3a), indicating that coating of the virion by a lipid-containing membrane is apparently not linked to the budding process.

Virus replication and release are not accompanied by cell lysis (Fig. 6.3a). However, virus infection causes significant retardation of host growth and SSV1 and SSV2 plaque assays could be set up on lawns of *S. solfataricus* and *S. islandicus*, respectively. Similar to bacterial temperate phages, replication of SSV1 and SSV2 can be induced, e.g. by UV irradiation, leading to temporary inhibition of host growth. However, in contrast to bacterial examples of lysogeny, the host cells are not lysed and eventually return to the original lysogenic state and normal growth rate.

In infected host cells the viral genomes are found integrated into specific attachments sites on the host chromosomes within tRNA genes (see Chapter 10 in this volume for details). As well as the integrated form, the free circular form of viral genome has been found in host cells.

Genome and its transcription. The viral genome is covalently closed circular (ccc) dsDNA, which in the case of SSV1 was shown to be positively supercoiled, providing one of the first cases of such DNA observed in nature (Nadal et al., 1986). The genome sizes are the smallest among crenarchaeal viruses and are in the range of 14,796 bp for SSV2 to 17,385 bp for SSV-K1.

Transcription of the SSV1 genome was studied following induction of virus replication, and was instrumental in understanding transcription signals in Archaea (reviewed by Zillig et al., 1993). As in the case of the rudiviruses, a rather simple and rarely chronological pattern of transcription was revealed. Most transcripts were constitutional. Examples of transcription regulation following induction by UV irradiation included upregulation of some constitutive transcripts and the appearance of a very short transcript, suggested to be involved in the initiation of DNA replication.

Family Guttaviridae

Genus *Guttavirus*

Species: *Sulfolobus newzealandicus* droplet-shaped virus, SNDV (Arnold et al., 2000a). Presently does not exist in laboratory collections.

Virion morphology and constituents, genome. The virion is droplet-shaped, ranging in size from 110 to 185 nm in length and from 95 to 70 nm in width, with a pointed end densely covered by thin fibers (Fig. 6.4). Its core is covered with a beehive-like structure, the surface of which appears to be either helical or stacked. In the course of purification particles easily deteriorated and even break apart. Looped filaments were observed to be released from broken particles. The virion contains three proteins of 17.5, 13.5 and 13 kDa. The presence of a lipid component has not been investigated. The viral

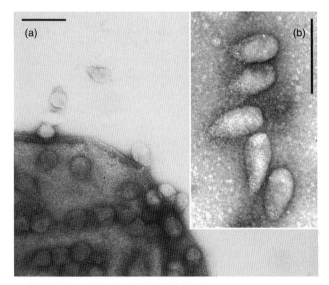

Fig. 6.4 Electron micrographs of a portion of a cell of *S. newzealandicus* and virions of SNDV, a species in the family Guttaviridae. (a) Extrusion of virions from an SNDV-infected cell of *S. newzealandicus*. (b) Virions of SNDV. Samples were negatively stained with 3% uranyl acetate. Bars are 200 nm. (Courtesy of Wolfram Zillig.)

genome is ccc dsDNA, about 20 kb in length, and is extensively modified, most likely by methylation.

Host range and virus–host relationships. The natural host is *Sulfolobus* sp. STH3/1, isolated from a hot spring at Steaming Hill, at Lake Taupo in New Zealand. Under laboratory conditions it also replicated in several strains of *Sulfolobus* isolated from the same environmental sample. Virus release started only in the early stationary growth phase of host cells and this did not result in any detectable cell lysis (Fig. 6.4a). However, infection of cell culture with a large amount of concentrated virus suspension led to an immediate reduction of the optical density, suggesting partial lysis. Upon prolonged incubation of an infected culture the virus was lost.

The virion adsorbs to a receptor on the host cell with its bearded end. Mature virions were observed in host cells prior to their release.

Family Globuloviridae

Genus *Globulovirus*

Species: *Pyrobaculum* spherical virus, PSV (Häring et al., 2004); genome sequence accession no. AJ635162; *Thermoproteus tenax* spherical virus 1, TTSV1 (Ahn et al., 2004, and unpublished); genome sequence accession no. AY722806 (Ahn et al., 2004).

Virion morphology and constituents. The virions are enveloped, spherical, 100 nm in diameter for PSV (Fig. 6.5a). On the virion surface a variable number of spherical protrusions, about 15 nm in diameter, is observed. Most likely these protrusions are involved in adsorption. The envelope encases tightly packed nucleoprotein in a superhelical conformation (Fig. 6.5b). The lipid fraction extracted from PSV virions had a mobility different from host lipids. The major portion of the protein fraction is represented by multimers of 33 kDa protein; in addition, two proteins of 20 and 16 kDa were identified.

Host range and virus–host relationships. The natural host of PSV, *Pyrobaculum* sp., was isolated from Obsidian Pool in Yellowstone National Park, USA. Under laboratory conditions the virus replicates in *Pyrobaculum* sp. D11, isolated from the same environmental sample, and in *T. tenax*. Virus production did not result in cell lysis, nor did it result in any significant increase of the generation time, which is 24 hours. Moreover, host cell membranes in an infected culture remained intact, indicative of a noncytocidal infectious cycle.

Genome. The genome is linear dsDNA, 28,337 bp long for PSV, and 20,993 bp long for TTSV1. Exceptionally for a crenarchaeal dsDNA virus, almost all recognizable genes of PSV are located on one strand. The ends of the PSV genome consist of 190 bp inverted repeats that contain multiple copies of short

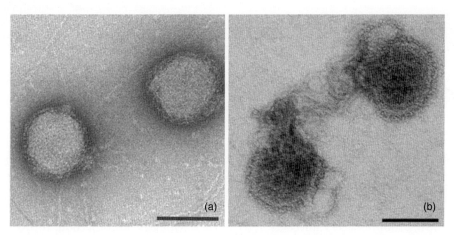

Fig. 6.5 Electron micrographs of virions of PSV, a species in the family Globuloviridae, stained with 3% uranyl acetate. (a) Two intact virions. (b) Two disrupted virions releasing the helical nucleoprotein core. Bars are 100 nm. (Modified from Häring et al., 2004.)

direct repeats. The two DNA strands appear to be covalently linked at their termini.

Family Bicaudaviridae

Genus *Bicaudavirus*

Species: *Acidianus* two-tailed virus, ATV (Häring et al., 2005c); genome sequence accession no. AJ888457.

Virion morphology and constituents. The virion exists in two conformations (Fig. 6.6). When released from the host cell, it is lemon-shaped with an overall length of 250–300 nm and a diameter of 110–120 nm (Fig. 6.6b, c). Specifically at temperatures above 75 °C, an appendage protrudes from each of the pointed ends of the virion and the lemon-shaped virion body shrinks (Fig. 6.6d–h). In fully developed particles it measures 110–180 nm in length and 70–100 nm in width. The two appendages (tails) at each pointed end have a variable length.

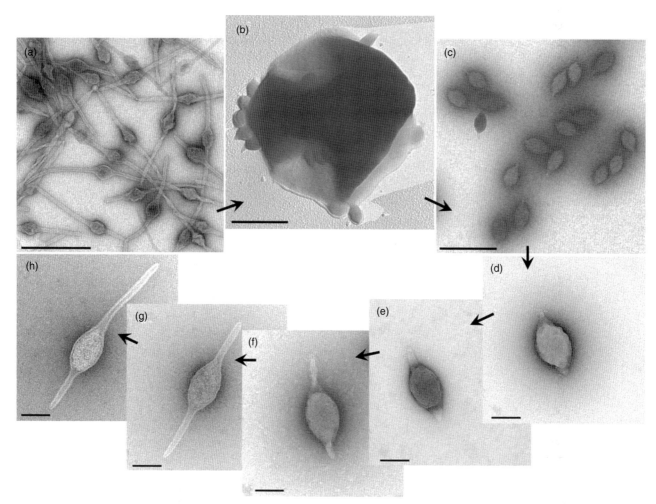

Fig. 6.6 Electron micrographs of a cell of *A. convivator* and different forms of virions of ATV, a species in the family Bicaudaviridae. (a) Two-tailed virions. (b) Extrusion of virions from an ATV-infected cell of "*A. convivator*." (c) Virions in a growing culture of ATV-infected "*A. convivator*," two days post-infection. (d) As for c, but purified by CsCl density gradient. (e–h) As for d, but incubated at 75 °C for 2, 5, 6, and 7 days, respectively. All preparations were negatively stained with 3% uranyl acetate, except for b, which was platinium shadowed. Bars: (a–c) 500 nm; (d–h) 100 nm. (Reproduced from Häring et al., 2005c, with permission.)

The maximum length of the particle reaches about 1 μm. Protrusion of tails does not require the presence of host cells, an exogenous energy source, or cofactors. The only requirement is a temperature in the range of host growth, above 75 °C. The tail has a tube-like structure with a wall about 6 nm thick. It terminates with a narrow channel, 2 nm in width, and an anchor-like structure formed by two furled filaments, each 4 nm wide. Inside the tube resides a filament 2 nm in width. The virion contains at least 11 proteins with molecular masses in the range of 12–90 kDa, as well as modified host lipids.

Adsorption of tail-less particles to host cells or cell-derived membrane vesicles has not been observed. Host cells lack any appendages and viral receptors are apparently located on the cell surface. The anchor-like structure at the terminus of tails appears to be involved in adsorption on host cell surface.

Host range and virus–host relationships. The natural host of ABV is *Acidianus* sp. isolated from a hot spring of a solfataric field in Pozzuoli, Italy. Under laboratory conditions the virus replicates also in *Acidianus convivator*, an isolate from the same environmental sample as the natural host.

ATV is the only known example among crenarchaeal viruses of a typical temperate virus. Infection at 85 °C, the optimal temperature for host growth, results in integration of the viral genome into the host chromosome, leading to a lysogenization of the host cell and a down regulation of lytic genes. The lysogeny can be interrupted by subjecting the host cell to stress conditions, e.g. UV irradiation, treatment with mitomycin C, or freezing–thawing. Induction of virus replication eventually leads to cell lysis. Infection with ATV of cells growing at suboptimal temperatures is immediately followed by virus replication and cell lysis.

Genome. The dsDNA genome is circular and contains 62,730 bp. Unusually for a crenarchaeal viral genome, four IS elements are present. They correspond closely in their transposase sequences to ISC1208, ISC1316, and ISC1476 from *S. solfataricus*.

Family Ampullaviridae

Genus *Ampullavirus*

Species: *Acidianus* bottle-shaped virus, ABV virus (Häring et al., 2005a); genome sequence accession no. X.

Virion morphology and constituents. The virion is enveloped, resembles in its shape a bottle, and has an overall length of 230 nm and a width varying from 75 nm at the broad end to 4 nm at the pointed end (Fig. 6.7). The broad end of the virion exhibits 20 (±2) thin rigid filaments, 20 nm long and 3 nm wide, which appear to be inserted into a disc, or ring, and are interconnected at their bases. The 9 nm thick envelope encases a cone-shaped nucleoprotein core formed by a toroidally supercoiled nucleoprotein filament (Fig. 6.7). The virion contains six major proteins in the size range 15–80 kDa.

Host range and virus–host relationships. The natural host of ABV is *Acidianus* sp., isolated from a hot spring of a solfataric field in Pozzuoli, Italy. Under laboratory conditions the virus also replicates in *A. convivator*, isolated from the same environmental sample. Host lysis was not observed. Virus infection increased the host generation time from about 24 hours to about 48 hours, but the infected culture eventually reached the density of the uninfected cell culture. Release of particles was observed only in the stationary growth phase of the host culture. Which of the two functional structures of the virus, the pointed end or the filaments on the broader end, is involved in adsorption to host cell surface and channeling of viral DNA into host cells is presently unclear.

Genome. The genome is linear dsDNA, 23,794 bp long, and its terminal regions constitute 580 bp long inverted repeats. The genome contains two large non-coding regions, of about 600 and 300 bp (Xu Peng, personal communication).

Unclassified

Species: *Sulfolobus* turreted icosahedral virus, STIV (Rice et al., 2004); genome sequence accession no. AY569307.

Virion morphology and constituents, genome. The virion is icosahedral, built on a pseudo-T = 31 icosahedral lattice. The diameter of the capsid is about 74 nm with a 6.4 nm shell thickness (Fig. 6.8a). The icosahedral asymmetric unit consists of five trimers of the major capsid protein of 37 kDa plus one additional minor capsid protein at the fivefold vertex. A predominant surface feature is a turret-like structure with a central channel. In

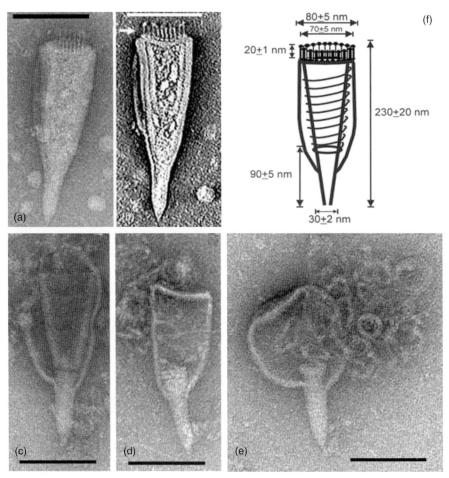

Fig. 6.7 Electron micrographs of virions of ABV, a species in the family Ampullaviridae, stained with 3% uranyl acetate. (a) Intact virion. (b) horizontal slice through the three-dimensional dataset of the three-dimensional reconstruction of the native virion. (c–e) Partially disrupted virions revealing an inner cone-shaped core composed of toroidally supercoiled nucleoprotein filament. (f) A scheme of the structure of an ABV virion. Bars are 100 nm. (Modified from Häring et al., 2004.)

addition to the major capsid protein, the virion carries several minor proteins with estimated masses of 75, 25, 12.5, and 10 kDa. The virion most likely contains an internal lipid envelope.

While virions of all other crenarchaeal viruses have unique morphotypes, the icosahedral capsid of STIV displays a striking resemblance to the capsids of eukaryotic viruses from the families *Adenoviridae* and *Phycodnaviridae* and bacterial viruses from the family *Tectiviridae*. Based on this similarity and the compatibility of the predicted secondary structure of the major capsid protein and the atomic structures of the major capsid proteins of the tectivirus PRD1 and the adenovirus, it has been proposed that the capsids of all these viruses share a common origin. By contrast to linear genomes of *Tectiviridae* and *Adenoviridae*, the genome of STIV is circular dsDNA, 17,663 kb long.

Host range and virus–host relationships. The natural host is a *Sulfolobus solfataricus* isolate from a hot, acidic spring in the Rabbit Creek thermal area in Yellowstone National Park. Under laboratory conditions the virus also propagates in *S. solfataricus* strains P1 and P2. There is no evidence that the virus genome integrates into the host chromosome or induces cell lysis.

Species: *Sulfolobus tengchongensis* spindle-shaped virus, STSV1 (Xiang et al., 2005); genome sequence accession no. AJ783869.

Virion morphology and constituents. The virion has the morphology of a spindle, 230 by 107 nm, and has a tail of variable length, 68 nm on average, at one end (Fig. 6.8b). The virion is enveloped, with a 10–15 nm thick envelope. The virions are nearly entirely composed of a single 15.6 kDa highly basic protein. Among the four minor structural proteins the largest is of about 256 kDa and is most likely involved in adsorption. Furthermore, a lipid component could be identified.

In purified preparations virions occasionally cluster in a rosette-like pattern with tails attached to cell debris, suggesting involvement of tails in viral recognition and attachment to the host cell surface.

Host range and virus–host relationships. The natural host of STV1 was present in an environmental sample collected from hot acidic springs in Tengchong, Yunnan, China, and has not been identified. Under laboratory conditions the virus replicates in *Sulfolobus tengchongensis* strain RT8-4, isolated from a field sample in the same region.

Infection with the virus significantly slowed down host growth, increasing doubling time from 11 to 30 hours. A plaque assay has been developed on host cell lawns. The virus is nonlytic. The eclipse period is about 4 hours. After extrusion from the host cell, virions often remain attached to the cell surface and layers of virus particles are occasionally seen surrounding a single host cell. An infected host culture was able to keep the virus after multiple transfers of cells into a fresh medium. No integration of the viral genome into the host chromosome has been detected.

Genome. The genome is circular dsDNA 75,294 bp long. It appears to be modified in unknown manner by virus-encoded proteins. The genome is highly asymmetric and divides into two equal halves with respect to gene orientation. The biased gene orientation and strand compositional asymmetry, revealed by the cumulative GC skew, suggest that the origin and terminus of virus replication are located in the 1.4 kb long intergenic region with an unusually high AT content, including two sets of tandem repeats and two sets of inverted repeats. Most likely, genome replication proceeds bidirectionally in the θ mode from the proposed origin.

Genomics of crenarchaeal viruses

Genomes of archaeal viruses have been recently analyzed in collaboration with Eugene Koonin and Roger Garrett (Prangishvili & Garrett, 2005; Prangishvili et al., 2006). The present chapter is a summary of this analysis.

The most prominent feature of the genomes of known crenarchaeal viruses is an extremely low number of genes coding for proteins with homologs in the public sequence databases. With the current coverage of viral and cellular genomes a considerable majority of viral genes have no detectable homologs other than in closely related viral species.

Genes with confirmed functions

Putative functions of only a few genes (with homologs in public databases) have been confirmed by their heterologous expression and biochemical

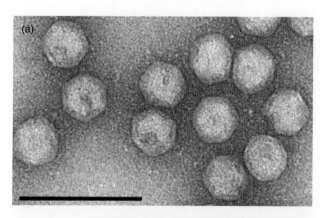

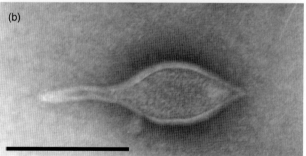

Fig. 6.8 Electron micrographs of virions of unclassified crenarchaeal viruses STIV (a) and STSV1 (b), negatively stained with 3% uranyl acetate. Bars are 200 nm. (Courtesy of Mark Young (a) and Li Huang (b).)

analysis of the recombinant proteins. They encode the dUTPase (Prangishvili et al., 1998) and the Holliday junction resolvase (Birkenbihl et al., 2001) of the rudiviruses SIRV1 and SIRV2 and the integrase/recombinase of the fusellovirus SSV1 (Muskhelishvili et al., 1993). In addition, some structural proteins purified from the virus particles could be sequenced by Edman degradation, which allowed the identification of their genes.

Genes with putative functions unique for individual viruses

A few genes of crenarchaeal viruses have homologs in Archaea and/or Bacteria. These include DNA modification methylases of the STSV1, superfamily 2 helicases of the lipothrixvirus SIFV, ABC-class ATPase of the rudivirus ARV1, and thiol-disulfide isomerase of the globulovirus TTSV1.

Genes with putative functions shared by different viruses

Many crenarchaeal viruses share a repertoire of genes with putative functions that include transcription regulators, ATPases, enzymes of DNA precursor metabolism and RNA modification, and glycosylases.

The most common gene products in crenarchaeal viruses, unexpectedly, turn out to be small proteins containing a ribbon–helix–helix (RHH) domain, which may be considered as a distinct, highly derived version of the classic helix–turn–helix (HTH) domain (Aravind et al., 2005). Typically, the RHH domain proteins are transcription regulators, and were shown to be common in Archaea, being nearly as abundant as typical HTH domains (Aravind and Koonin, 1999; Perez-Rueda et al., 2004). The RHH proteins of crenarchaeal viruses are rather heterogeneous, and apparently have a complex history, probably involving multiple independent acquisitions as well as horizontal gene mobility.

Other putative transcription regulators encompass HTH domains, looped-hinge helix domains, and Zn fingers. Remarkable is the presence of C2H2 Zn finger proteins with moderate similarity to a variety of eukaryotic Zn fingers but no obvious homologs in prokaryotes.

Each of the crenarchaeal viruses encodes at least one of the P-loop ATPases, which are known to be the most abundant protein domain in prokaryotes and the great majority of viruses, and typically are involved in viral replication, transcription, or packaging. Other predicted enzymes with probable functions in DNA replication are RecB-family endonucleases in the *Lipothrixviridae*, *Rudiviridae*, and *Fuselloviridae*, XerC/D-like integrase/recombinase in the Bicaudaviridae and the SSTSV1. As with Holliday junction resolvases in the *Rudiviridae* (Birkenbihl et al., 2001), these enzymes may be involved in intermediate resolution during viral genome replication.

The *Rudiviridae* and STSV1 encode the dUTPase and/or the flavin-dependent thymidylate synthese (ThyX), which are apparently involved in DNA precursor metabolism, a function widely represented in DNA viruses of Bacteria and Eukarya. Another function common to bacterial viruses, RNA modification, is also represented in the *Rudiviridae*: they encode predicted tRNA-ribosyltransferase and S-adenosylmethionine-dependent methyltransferase. The former enzyme is also encoded by STST, and the latter enzyme by the lipothrixvirus SIFV.

All known members of the *Rudiviridae* and *Lipothrixviridae* encode glycosyltransferases that may be involved in modification of virion proteins and/or the host cell wall during viral entry and/or release (Markine-Goriaynoff et al., 2004). In other members of other viral families there might be implicated in this process such enzymes as a membrane-associated acyltransferase (in ATV) and nucleoside-diphosphate-sugar epimerase (in STSV1).

The gene pool shared by crenarchaeal viruses also includes two protein families with unknown functions. One of these, exemplified by the AFV1 protein 03 of 99 amino acids, is found exclusively in the *Rudiviridae* and *Lipothrixviridae* (genera *Beta-* and *Gamma-lipothrixvirus*). Another family of small proteins has a single bacterial representative, the uncharacterized protein YddF of *Bacillus subtilis*, with all other members found in crenarchaeal viruses: in the *Rudiviridae*, *Lipothrixviridae* (genera *Beta-* and *Gamma-lipotrixvirus*), the ATV and STIV.

Orthologous genes

Results for the identification of orthologous genes in the set of genes shared by crenarchaeal viruses are in astounding accord with evolutionary relationships postulated on the basis of viral morphotypes. These results are shown schematically in Plate 6.1. (Because of the uncertainty regarding the very existence of a unique common ancestor virus, orthologs were defined only conditionally, as genes that are more closely related to each other in a given set of viruses than they are to any homologs that may exist outside that set of genomes.) Based on a number of orthologs, different levels of relationships have been distinguished.

A significant number of orthologs (18–44) are observed among viral species assigned to the same genus, which probably derived from relatively recent common ancestors. These are four members of the *Fusellovirus*, three members of the *Rudivirus*, and two members of the *Globuloviridae*.

A smaller number of orthologous genes (five to nine), representing moderately related groups that have probably evolved from a single ancestral virus in a more distant past, were observed for members of different genera in the family *Lipothrixviridae*: the betalipothrixvirus SIFV, gammalipothrix virus AFV1, and deltalipothrixvirus AFV2. A comparable number of orthologous genes were observed between some members of different families, the *Rudiviridae* and *Lipothrixviridae*, suggesting a common ancestry for these two families. Thus, apparently crenarchaeal viruses with linear morphotypes, the rudiviruses and lipothrixviruses, could have evolved, as genetic entities, from the same ancestral linear virus, and can be unified in a superfamily or order. We propose to establish a novel order of crenarchaeal DNA viruses, encompassing the families *Rudiviridae* and *Lipothrixviridae*, and to name it *Ligamenvirales*, from Latin *ligemen*, for string or thread. It is noteworthy that the consistency of orthologous relationships between the rudiviruses and lipothrixviruses is low, mainly formed by non-overlapping sets of orthologs, and the only reliable signature of this superfamily/order of crenarchaeal viruses would be distinct glycosyltransferases and proteins from the AFV1p03 family (see above).

In those cases where the viruses share only one to four orthologous genes, it remains unclear whether a common ancestral virus ever existed. Shared genes could result from horizontal gene transfer between the viruses, or, in some cases, from independent acquisitions from the host. Even less likely appears the existence of common ancestors, when no orthologs are detected among viruses, as is the case with the two members of the *Globuloviridae* and the rest of the crenarchaeal viruses.

Conclusions

Known dsDNA viruses infecting hyperthermophilic Crenarchaea represent a diverse collection of unique morphotypes, not encountered among dsDNA viruses of Bacteria and Eukarya. Based on morphological and genomic characteristics, seven novel families and ten novel genera of dsDNA viruses have been proposed for classification.

Analysis of genome sequences has confirmed a specific nature of crenarchaeal viruses, suggested by structural uniqueness. A considerable majority of their genes have no detectable homologs in other viruses or cellular life forms, implying that dsDNA viruses of Crenarchaea are evolutionarily unrelated to known dsDNA viruses of Bacteria or Eukarya and form a distinctive group in the viral world. Whether this group shares a single origin or is polyphyletic is presently unclear.

Besides the crenarchaeal viruses, in the viral world two other large groups of dsDNA viruses are known that apparently do not share an origin with any other virus. These are: (i) the group of head-tailed bacteriophages (the order Caudovirales), including the families *Myoviridae*, *Siphoviridae*, and *Podoviridae* (Hendrix et al., 1999; Casjens, 2005); and (ii) the monophyletic group of eukaryal nucleoplasmatic large dsDNA viruses, including the families *Poxviridae*, *Phycodnaviridae*, *Iridoviridae*, *Asfarviridae*, and *Mimiviridae* (Iyer et al., 2001; Raoult et al., 2004). The disclosure of the unique group of crenarchaeal viruses provides a new perspective on events related to the origin of DNA viruses and their evolution. It is a challenge for future studies to reveal evolutionary trials of involvement of sets of non-orthologous genes in the establishment of principally similar lifestyles in the three unrelated groups of dsDNA viruses.

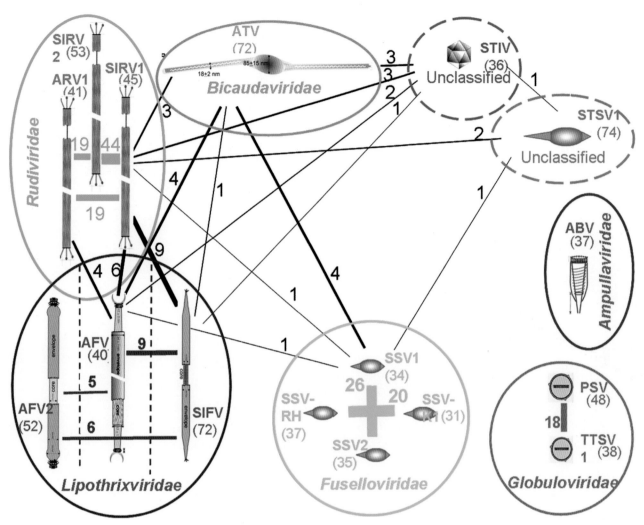

Plate 6.1 Inter-family and intra-family orthologous relationships among genes of crenarchaeal viruses. The numbers over each line indicate the number of inferred orthologous genes between the respective viruses. The thickness of the lines is roughly proportional to the number of orthologs. (Designed by E. Koonin, modified from Prangishvili et al., 2006.)

7
Features of the genomes

Hans-Peter Klenk

Science is an endless search for truth. Any representation of reality we develop can be only partial. There is no finality, sometimes no single best representation. There is only deeper understanding, more revealing and enveloping representation. (Woese, 2004)

Introduction

The term genome was first coined by Hans Winkler (1920) by combining the words **gen**e and chromos**ome**, long before the true nature of genes and DNA was deduced. It represents the complete set of genetic information of an organism, including both the genes and the non-coding sequences located on the chromosome(s) and the extra chromosomal elements. Genomics as a field of scientific investigation analyzes and compares complete genome sequences of organisms. Starting a decade ago with the first completely deciphered bacterial genome, *Haemophilus influenzae* (Fleischmann et al., 1995), genome sequences of more than 300 cellular organisms have been archived in public databases (Table 7.1). The success of these genome analyses has been breathtaking, generating a "genomic revolution" that fundamentally changed biology in two perspectives: it extended our research objectives from single units (e.g. genes or proteins) to complete data sets (e.g. all genes in a genome or all proteins in a proteome), and it spurred the rapid development of high throughput technologies required for large-scale functional analyses: transcript**omics**, prote**omics**, metabol**omics**, structural gen**omics**. Although medical applications played a central role in microbial genomics from the very beginning, it soon became clear that whole genome sequences were also of great interest from organisms with environmental, energy production, and phylogenetic relevance. The latter topics quickly brought the Archaea to the head of the early sequencing pipeline and made *Methano(caldo)coccus jannaschii* the third organism whose genome was sequenced (Bult et al., 1996), even before the first eukaryal genome sequence, that of *Saccharomyces cerevisiae*, was completed (Goffeau et al., 1996).

By the mid-1990s, Archaea were considered to be primarily of evolutionary/phylogenetic interest, with some potential for biotechnologically interesting enzymes from species with an extremophilic lifestyle. Major points of interest that were hoped to be addressed by the analysis of the upcoming archaeal genome sequences included: the phylogenetic unity of the archaeal domain; the correlation of bacterial form and eukaryotic content of archaeal genes (Keeling et al., 1994); a mixed heritage of the Archaea (Koonin et al., 1997; Doolittle & Logston, 1998); and the putative role of the Archaea in the origin of the Eukarya (Forterre, 1997). The publication of the first archaeal genome sequence in 1996 more than doubled the number of known archaeal genes and provided many answers with respect to the molecular basis of origin and diversification of cellular life. It transformed *M. jannaschii* from a rather poorly studied organism (13 Medline records since its initial description in 1983) into a prominent model organism (485 Medline listed papers since the genome publication). As discussed in Chapters 2 and 3 of this volume, comparative and phylogenetic analyses of sequenced archaeal genomes answered many of the evolutionary questions linked to the Archaea, leaving us with even more new questions. This chapter focuses less on evolutionary topics than on the structure and content of archaeal genomes, lessons learned from two dozen sequenced genomes, problems with genome analyses, and future prospects due to even more genomes to come.

Table 7.1 Completed genome sequences.

Year	Total	Archaea	Bacteria	Eukarya
1995	2	0 (0%)	2	0
1996	3	1 (33%)	2	0
1997	7	2 (29%)	4	1
1998	7	1 (14%)	5	1
1999	5	1 (20%)	4	0
2000	18	3 (17%)	13	2
2001	31	3† (10%)	25	3
2002	43	5 (12%)	29	9
2003	51	1 (2%)	48	2
2004	76	3 (4%)	60	13
2005*	77	4 (5%)	64	9
PUB	320	24† (8%)	256	40
CUP	107	6 (6%)	92	9
WIP	1010	29 (3%)	660	320
Total	1437	59 (4%)	1008	369

*To November 10, 2005; †without the proprietary *Pyrolobus fumarii*; PUB, published; CUP, completed but unpublished; WIP, work in progress.

The structure of archaeal genomes

Archaeal genomes can be characterized by several structural parameters that became accessible with the advent of whole genome sequences: genome size and geometry, number of replicons, G+C content, syntheny of genes, operons, origin and terminus of replication, repetitive sequences, and gene structure. Although there is nothing like a typical archaeal genome, the spectrum of variation in these parameters for the completed archaeal genome sequences (Table 7.2) can be compared with the corresponding features in bacterial and eukaryal genomes. Variations in gene (protein) complement and gene conservation, as well as the similarity to homologous bacterial and eukaryal genes, are discussed in a separate section.

Archaeal genome sizes vary across an almost 12-fold range, from 0.49 Mbp of the obligate symbiont *Nanoarchaeum equitans* to 5.75 Mbp of the most metabolically diverse *Methanosarcina acetivorans* (Table 7.2). The size range of bacterial genomes is comparable to that of the archaeal genomes, spanning a 16-fold range from 0.58 Mbp (*Mycoplasma genitalium*) to 9.11 Mbp (*Bradhyrhizobium japonicum*). The more limited size range of the archaeal genomes reflects the significantly smaller sampling size of only 24 genomes as compared to 256 bacterial genomes (Table 7.1). The smallest eukaryal genomes (Protozoa and Fungi, 8–10 Mbp) are in the size range of the largest bacterial genomes, whereas the size of most eukaryal genomes (e.g. mammals, 3000 Mbp) surpasses the archaeal and bacterial ones by several hundred-fold. As in the Bacteria, the genome size of the Archaea is the result of several genetic events, including gene duplication, gene acquisition by lateral gene transfer (LGT), and lineage-specific gene loss. The smallest archaeal and bacterial genomes belong to organisms that are restricted to a stable environment, sometimes associated with a host organism, e.g. *N. equitans*, or *Mycoplasma* and *Buchnera*. The smallest genomes of free living Archaea, Thermoplasmatales (1.57 Mbp) and hydrogenotrophic class I methanogens (1.67 Mbp), surpass the smallest genome of a free living bacterium, *Pelargibacter ubiques* (Giovannoni et al., 2005) by only 20–30%. As in the Bacteria, the species of Archaea with the largest genomes (e.g. *M. acetivorans*) occupy highly complex and variable environments, e.g. marine sediments. It is the possession of a large inventory of genes that enables life in these habitats. The genome size of both Archaea and Bacteria is directly proportional to the number of encoded genes, generally about one gene per kilobase, indicating an almost constant coding density for organisms from both domains (Tables 7.2, 7.3). Deviation from this constant coding density often depends more on variations in annotation procedures than on those in real genetic variation (exception: Methanomicrobiales, with a significantly reduced coding coverage). The constant coding density in Archaea and Bacteria is clearly different from the situation in eukaryal genomes, but supports the idea that archaeal and bacterial genome size (and content) are largely dictated by environmental pressures (Bentley & Parkhill, 2004).

The genome geometry and replicon arithmetic of archaeal and bacterial genomes are very similar, but differ from those of the eukaryal genomes. The main component of archaeal genomes consists always of a single circular chromosome (Table 7.2). Only *Haloarcula marismortui* contains a second, much smaller circular chromosome. All archaeal extrachromosomal elements (ECEs) are circular and are apparently most frequent in extreme halophiles and methanogens, but also present in some species of Crenarchaeota.

Table 7.2 Published archaeal genomes.

Species	Taxonomy	db accession	size [bp]	G+C [%]	CDS [%] cl.	SRSR (copies)	IS-elements fam. (copies)	Lifestyle
Agf E	Archaeoglobales	AE000782	2,178,400	48.5	92.2	3 (150)	3 (16)	anaerobic sulphate-reducer, T_{opt} 83°C
App C	Desulfurococcales	BA000002	1,669,695	56.3	88.8*	4 (101)	1 (2)	aerobic, chemoorganotroph, T_{opt} 95°C
Ham E	Halobacteriales	AY596297	3,131,724	62.4	85.0	n.d.	14 (13)	aerobic, extreme halophile, T_{opt} 37°C
	Chromosome II	AY596298	288,050	57.2			(7)	
	ECE pNG700	AY596296	410,554	59.1			(2)	
	ECEs pNG100-600	AY596290-5	444,314	57.0			10 (18)	
Hbs E	Halobacteriales	AE004437	2,014,239	67.9	85.2	n.d.	† (22)	aerobic, extreme halophile, T_{opt} 37°C
	ECE pNRC200	AE004438	365,425	59.2			12 (40)	
	ECE pNRC100	AF016485	191,346	57.9			† (29)	
Mcj E	Methanococcales	L77117**	1,664,970	31.4	88.0	18 (2–25)	5 (7)	anaerobic, chemoplitoautotroph, T_{opt} 85°C
Mcm E	Methanococcales	BX950229	1,661,137	33.1	88.9	n.d.	none	mesophilic, hydrogenotrophic, T_{opt} 37°C
Mpk E	Methanopyrales	AE009439	1,694,969	61.1	89.1	n.d.	none	hyperthermophilic methanogen, T_{opt} 100°C
Msa E	Methanosarcinales	AE010299	5,751,492	42.7	74.0	none	1 (1)	marine group II methanogen, T_{opt} 37°C
Msb E	Methanosarcinales	CP000099¶	4,837,766	39.2	71.5	n.d.	1 (1)	mesophilic group II methanogen, T_{opt} 35°C
Msm E	Methanosarcinales	AE008384	4,096,345	41.5	75.2	none	(185)	terrestrial group II methanogen, T_{opt} 40°C
Mtt E	Methanobacteriales	AE000666	1,751,377	49.5	90.9	2 (171)	none	thermophilic methanogen, T_{opt} 65°C
Nae	Nanoarchaeota	AE017199	490,885	31.6	93.0	n.d.	n.d.	obligate symbiont, T_{opt} 80°C
Nmp E	Halobacteriales	CR936257	2,595,221	63.4	90.8	n.d.	3 (35)	extreme haloalcaliphile, T_{opt} 80°C, pH_{opt} 8.5
	ECE PL131	CR936258	130,989	57.2	82.3		5 (6)	
	ECE PL23	CR936259	23,486	60.6	83.9		(0)	
Pba C	Thermoproteales	AE009441	2,222,430	51.4	88.8	n.d.	n.d.	hyperthermophilic nitrate-reducer, T_{opt} 100°C
Pca E	Thermococcales	AL096836§	1,765,118	44.7	91.1	n.d.	1 (1)	anaerobic hyperthermophile, T_{opt} 103°C
Pcf E	Thermococcales	AE009950	1,908,256	40.8	91.9	n.d.	3 (24)	anaerobic hyperthermophile, T_{opt} 100°C

Table 7.2 Published archaeal genomes.

Species	Taxonomy	db accession	G+C size [bp]	CDS [%]	SRSR [%] cl.	IS-elements (copies)	fam. (copies)	Lifestyle
Pch E	Thermococcales	BA000001	1,738,505	41.9	90.7	3 (110)	2 (3)	anaerobic, grows with sulphur, T_{opt} 98°C
Plf C	Desulfurococcales	proprietary	1,850,000	n.d.	n.d.	n.d.	n.d.	obligate chemolithoautotroph, T_{opt} 106°C
Ppt E	Thermoplasmatales	AE017261	1,545,895	36.0	91.7	n.d.	none	thermoacidophile, pH_{opt} 0.7, T_{opt} 60°C
Sla C	Sulfolobales	CP000077	2,225,959	36.7	85.1	4 (222)	2 (4)	aerobic thermoacidophile, pH_{opt} 2–3, T_{opt} 75°C
Sls C	Sulfolobales	AE006641	2,992,245	35.8	84.3	7 (420)	25 (201)	aerobic thermoacidophile, pH_{opt} 2–4, T_{opt} 85°C
Slt C	Sulfolobales	BA000023	2,694,756	32.8	83.9	6 (455)	12‡	aerobic thermoacidophile, pH_{opt} 2–3, T_{opt} 80°C
Tck E	Thermococcales	AP006878	2,088,737	52.0	92.1	n.d.	2 (7)	anaerobic extreme thermophile, T_{opt} 85°C
Tpa E	Thermoplasmatales	AL139299	1,564,906	46.0	87.0	1 (46)	4 (4)	thermoacidophile, pH_{opt} 2, T_{opt} 59°C
Tpv E	Thermoplasmatales	BA000011	1,584,804	40.0	86.2	n.d.	24 (27)	thermoacidophile, pH_{opt} 2, T_{opt} 60°C

*CDS from original publication which features 2694 ORFs as compared to 1841 in NCBI database; †the 91 IS-elements in the *Halobacterium* species NRC-1 genome belong altogether to 12 families; ‡see Table 8.1; §also contains a small 3444bp multicopy ECE; ¶also contains a small 36,358bp ECE, pL1; **also contains two ECEs with 16,550bp (small ECE, 28.8% G+C) and 58,407bp (large ECE, 28.2% G+C).

Abbreviations: bp, base pairs; C, crenarchaeon; CDS, total protein coding sequence in percent of genome; cl., number of clusters (copies therein), db, data base; E, Euryarchaeon; ECE, extra chromosomal element; fam., families of IS elements (and number of copies therein); G+C, guanine and cytosine content; IS-element, short mobile insertion element; SRSR, Short Regular Spaced Repeat; T_{opt}, optimal growth temperature; pH_{opt}, optimal pH value.

Species abbreviations: Agf, *Archaeaoglobus fulgidus* (Klenk et al., 1997a); App. *Aeropyrum pernix* (Kawarabayasi et al., 1999); Ham, *Haloarcula marismortui* (Baliga et al., 2004); Hbs, *Halobacterium* species NRC-1 (Ng et al., 2000); Mcj, *Methanocaldococcus jannaschii* (Bult et al., 1996); Mcm, *Methanococcus maripaludis* (Hendrickson et al., 2004); Mpk, *Methanopyrus kandleri* (Slesarev et al., 2002); Msa, *Methanosarcina acetivorans* (Galagan et al., 2002); Msb, *Methanosarcina barkeri* (Copeland et al., Genbank NC0007355, unpublished); Msm, *Methanosarcina mazei* (Deppenmeier et al., 2002); Mtt, *Methanothermobacter thermoautotrophicus* (Smith et al., 1997); Nae, *Nanoarchaeum equitans* (Waters et al. 2003); Nmp, *Natronomonas pharaonis* (Falb et al. 2005); Pba, *Pyrobaculum aerophilum* (Fitz-Gibbon et al. 2002); Pca, *Pyrococcus abyssi* (Lecompte et al. 2002a); Pcf, *Pyrococcus furiosus* (Robb et al. 2001); Pch, *Pyrococcus horikoshii* (Kawarabayasi et al. 1998); Plf, *Pyrolobus fumarii* (proprietary, Diversa press release September 25, 2001); Ppt, *Picrophilus torridus* (Fütterer et al. 2004); Sla, *Sulfolobus acidocaldaricus* (Chen et al. 2005b); Sls, *Sulfolobus solfataricus* (She et al. 2001b); Slt, *Sulfolobus tokodaii* (Kawarabayasi et al. 2001); Tck, *Thermococcus kodakarensis* (Fukui et al. 2005); Tpa, *Thermoplasma acetophilum* (Ruepp et al. 2000); Tpv, *Thermoplasma volcanium* (Kawashima et al. 2000).

Table 7.3 Content of archaeal genomes.

Species	ORFs	COGs [%]	Length av. [bp]	Inteins genes (no.)	Special AAs	ORF fams (members)	rRNAs rD (no.)	tRNAs (introns)	Other stable RNAs
Agf E	2420	84.9	830	none	none	242 (719)	1 (3)	46 (5)	7S, RnaseP, 86 snos
App C	1841	72.6	805	1 (1)	none	185 (532)	1 (4)	47 (14)	none
Ham E	3412†	76.8	780	3 (3)	none	n.d.	3 (10)	50 (?)	7S
Hbs E	2075	78.1	827	2 (2)	none	n.d.	1 (3)	47 (3)	1 unknown, 7S
Mcj E	1729	87.6	847	14 (18)	Sel (9)*	136 (≤16)	2 (5)	37 (2)	7S
Mcm E	1722	88.9	857	none	Sel (9)	n.d.	3 (10)	38 (?)	7S, RNaseP
Mpk E	1687	80.0	895	5 (5)	Sel (8)	n.d.	1 (3)	35 (?)	none reported
Msa E	4540	77.5	936	none	Pyl (1)	539 (2178)	3 (9)	59 (?)	7S
Msb E	3759	81.7	927	none	Pyl (1)	n.d.	3 (9)	62 (?)	1 unknown, 7S
Msm E	3371	81.5	913	none	Pyl (29)	n.d.	3 (10)	58 (?)	1 unknown, 7S
Mtt E	1873	86.3	850	1 (1)	none	111 (409)	2 (6)	39 (3 + 1)	7S, RNaseP
Nae	552	78.4	827	2 (2)	none	n.d.	1 (3)	44 (4)	14 snos
Nmp E	2661	77.1	879	3 (4)	none	n.d.	1 (3)	46 (?)	7S, RNaseP
Pba C	2605	68.1	757	none	none	n.d.	1 (3)	46 (15)	53 snos
Pca E	1784	87.6	911	10 (14)	none	(621)	1 (4)	46 (2)	7S, RnaseP, 45 snos
Pcf E	2116§	84.0	820	9 (10)	none	(845)	1 (4)	46 (2)	7S, RnaseP, 51 snos
Pch E	1955‡	80.9	807	13 (14)	none	(606)	1 (4)	46 (2)	7S, RNaseP
Ppt E	1535	87.8	923	1 (1)	none	n.d.	1 (3)	44 (2)	none
Sla C	2292	81.6	827	none	none	n.d.	1 (3)	48 (19)	7S, 29 snos
Sls C	2977	81.0	848	none	none	52 (2–26)	1 (3)	46 (18)	7S, 6 snos
Slt C	2825	76.1	800	none	none	494 (1471)	1 (3)	46 (24)	none described
Tck E	2306	80.2	833	12 (16)	none	186	1 (4)	46 (2)	7S, RNaseP
Tpa E	1482	89.1	909	1 (1)	none	n.d.	1 (3)	45 (?)	7S
Tpv E	1499	89.0	911	1 (1)	none	n.d.	1 (3)	45 (?)	7S

*Klenk et al., 1997b; †chromosomes I and II only; ‡ORF number from NCBI microbial genomes data base; §estimated from Poole et al., 2005.

General abbreviations: AA, amino acids; av., average; bp, base pairs; C, crenarchaeon; E, euryarchaeon; n.d. and ?, not described; Pyl, pyrrolysine containing genes; rD, rRNA clusters (and genes therein); Sel, selenocysteine containing genes; snos, short non-messenger RNAs.

Species abbreviations as in Table 7.2; ORF numbers taken from updates in various databases; % COGs from NCBI database, November 2005.

Archaeal ECEs (plasmids) are discussed in detail in Chapter 9 of this volume. By far the majority of bacterial genomes also exist as single circular chromosomes. Linear and/or multiple replicons, such as those in several species of *Agrobacterium*, *Borrellia*, *Brucella*, *Deinococcus*, *Ralstonia*, *Streptomyces*, and *Vibrio*, are the exception (Bentley & Parkhill, 2004). Eukaryal genomes consist of few to many linear chromosomes whose lengths often significantly surpass the size of archaeal and bacterial chromosomes.

The average G+C content of archaeal genomes varies between 31.4% (*M. jannaschii*) and 67.9% (*Halobacterium* species NRC-1) (Table 7.2). The C+G range of bacterial genomes is slightly wider, from 26.5% (*Wigglesworthia glossinidia*) to 72.1% (*Streptomyces coelicolor*), again reflecting the larger sampling size of bacterial genomes. A clear correlation between average G+C content and genome size, as described for Bacteria by Bentley and Parkhill (2004), is not obvious for Archaea. Genomes of the *Sulfolobales* feature the lowest average G+C content

(32.8–36.7%), whereas extreme halophiles mark the upper border, ranging from 62.4 to 67.9%. Methanogens contain genomes with G+C contents as low as 33.4% (*M. jannaschii*) and as high as 61.1% (*Methanopyrus kandleri*). The G+C content across archaeal chromosomes is far from uniform and often differs significantly between coding and non-coding regions. Upstream promoter sequences tend to be AT-rich in order to allow a curved, rigid confirmation that unwinds more easily (reviewed by Bentley & Parkhill, 2004), containing AT-rich TATA boxes as part of the archaeal promoter (for details on promoter sequences see Chapter 16 of this volume). Regions around the archaeal termini have a tendency toward higher AT-content, whereas the region(s) around origin(s) of replication have a preference for G and C (archaeal *ori*-regions; see Chapter 14 of this volume). Unequal distribution of bases in leading and lagging strand in DNA replication, commonly referred to as GC skew (G–C/G+C for leading and lagging strand), is used for locating the origin and terminus of replication in both archaeal and bacterial genomes. Whereas bacterial genomes contain only one starting point for DNA replication, it has been demonstrated that some (not all) archaeal genomes contain multiple starting points, but far fewer than eukaryal genomes. Foreign DNA fragments to be integrated in archaeal genomes may arrive with a C+G content significantly different from the average of the host genome. Over time these regions are ameliorated so that the incoming DNA will match the G+C content of the host genome.

Comparison of gene order between genomes informs us about the evolutionary relation between organisms, but can also be used for prediction of gene functions. In this context, **synteny** refers to multi-gene regions where DNA sequences and gene order are conserved between genomes (Bentley & Parkhill, 2004). From genetic maps of bacterial model organisms (*Escherichia coli*, *Bacillus subtilis*) it has long (before the genome sequences) been known that homologous genes are not necessarily located at the same relative position in bacterial genomes, but that only certain gene clusters were syntenic (Tamames, 2001). Figure 3.3 in this volume shows syntenic ribosomal (r-protein) gene clusters in Archaea and Bacteria. It is important to note that the sequences of archaeal r-proteins in these clusters are more similar to their eukaryal homologues than to bacterial homologues, but remain syntenic with the latter ones, whereas syntenic organization of archaeal and eukaryal genes is unknown. There seems to be a positive selection for clustering of physically interacting proteins, but no absolute requirement for juxtaposition of genes in either archaeal or bacterial genomes. Synteny is therefore lost at a much faster rate than sequence similarity (Mushegian & Koonin, 1996b; Rogozin et al., 2004). Nevertheless, synteny remains useful for prediction of gene function (see Chapter 22 in this volume). Only genomes of closely related species maintain a high degree of synteny, whereas genomes of moderately distant species reveal no striking overall synteny at all. Plate 7.1 shows dot plots to represent similarity between three species of Euryarchaeota, two closely related *Pyrococcus* strains (left), and one of the two pyrococci and *M. jannaschii* (right).

Genome comparison by dot blots indicates that large-scale symmetrical inversions centered on the origin of replication are as frequent for the Archaea as for the Bacteria. As in the Bacteria, there are three major mechanisms that promote the genome plasticity in the Cren- and Euryarchaeota: rearrangements linked to the replication terminus (appearing as a broken X pattern in genome dot blots), insertion sequence-mediated recombinations, and DNA integration within tRNA genes. Lecompte et al. (2002a) evaluated all three mechanisms for *Pyrococcus* genomes, and estimated that the level of genome plasticity is at least comparable to the plasticity observed between closely related bacteria. Mechanisms of rearrangement and change in *Sulfolobus* genomes involving autonomous insertion sequences (IS elements), non-autonomous miniature inverted-repeat transposable elements (MITEs), self-transmissible plasmids, prophages, introns, inteins, and short regularly spaced direct repeats (SRSR) are discussed in detail in Chapter 8 in this volume. Possible roles and mechanisms of archaeal integrases and integrated elements, and their role in LGT-based genome evolution, are discussed in Chapter 10. Integrated elements are not restricted to Archaea, but also contribute to the plasticity of bacterial genomes, e.g. pathogenic islands in *E. coli*. Some integrase families even contain archaeal and bacterial members (She et al., 2002). Taking all these mechanisms of genomic flux together, it becomes obvious that Archaea and Bacteria evolve(d) much more quickly than by the classical view of evolution, which describes slow change via the accumulation of point mutations in individual genes, enabling them to rapidly respond to environmental pressures (Bentley & Parkhill, 2004).

The conserved genomic neighborhood in phylogentically distinct organisms is unlikely to occur by coincidence, but suggests relatedness and provides insight to the function of encoded (sometimes functionally unknown) proteins. Groups of adjacent, co-expressed genes that encode functionally linked proteins (operons) represent the principal form of gene co-regulation and co-expression in Archaea and Bacteria (Lawrence, 1999; Rogozin et al., 2004). Interestingly, co-regulation in Archaea and Bacteria may also occur in cases of conserved bidirectionally transcribed gene pairs. When they are conserved in different genomes, the organization of such juxtapositioned genes may indicate shared *cis*-regulatory elements that give rise to transcriptional co-regulation, strongly suggesting a functional relationship between a transcriptional regulator and a non-regulatory protein (Korbel et al., 2004). Some operons have been conserved over long evolutionary time spans and have eventually moved among Archaea and Bacteria via LGT. The "über-operon" concept of Lathe et al. (2000) extended the classical operon concept to gene context conservation of a higher order, demonstrating extensive plasticity of functionally linked genes in Archaea and Bacteria. Until recently, only few operons have been identified in some protists and nematodes. It has recently became clear that many species of Eukarya from protists to chordates encode numerous instances of poly-cistronic transcription units (Blumenthal, 2004). Uncoupling of transcription and translation, and the fusion of adjacent genes during evolution of the eukaryal lineage, may for most genes have removed the need for co-transcription and ribosome binding sequences.

The simple structure of archaeal protein coding genes (ORFs) corresponds to that of bacterial ORFs, without frequent introns that interrupt the coding sequences of many eukaryal genes. Although many archaeal gene products are more similar to their eukaryal homologs than to bacterial homologs (see the next paragraph), the structure of these archaeal genes is still the same as for bacterial genes. Recoding of genetic information by inclusion of selenocysteine as the twenty-first amino acid has long been known from bacterial and eukaryal genes, as well as in several genes encoded in the genomes of *M. jannaschii*, *Methanococcus maripaludis*, and *M. kandleri* (Table 7.3). A twenty-second amino acid, pyrrolysine, has recently been shown to be utilized by three more class II methanogens: *Methanosarcina barkeri*, *M. acetivorans*, and *Methanosarcina mazei* (Table 7.3), as well as by *Methanosarcina thermophila* and *Methanococcoides burtonii*. Hereby the UAG stop codon is utilized in a suppression mechanism with the help of a specialized UAG-tRNA, PylT, and a class II aminoacyl-tRNA synthetase-like enzyme, PylS (see Chapter 18 in this volume). Only one bacterial pyrrolysine-encoding gene has yet been identified, but no eukaryal pyrrolysine-encoding gene. Within Archaea the intrinsic use of **unusual amino acids** is obviously restricted to the methanogens. Only one preliminary report about a third recoding mechanism in Archaea, programmed frameshifting, has been published (reviewed in Cobucci-Ponzano et al., 2005). The proposed process for programmed frameshifting in Archaea corresponds to the respective process known as a niche phenomenon in Bacteria (Cobucci-Ponzano et al., 2005).

Specific archaeal introns in rRNAs and tRNAs were reported long ago (Kjems & Garrett, 1988; Kjems et al., 1989). Cleavage and exon-splicing reactions for these RNAs resemble those found for eukaryal tRNA introns. Group I and group II introns, frequently found in Bacteria and Eukarya, are unknown in Archaea. As in Bacteria, spliceosomal introns in archaeal protein-encoding genes are much less frequent than in Eukarya. In fact, they were unknown until recently when Watanabe et al. (2002) reported the discovery of introns in an archaeal homolog of the eukaryal centromere-binding factor 5 (Cbf5p), a subunit of a small nucleolar ribonucleoprotein in *Aeropyrum pernix*, *Sulfolobus solfataricus*, and *Sulfolobus tokodai*. Introns are removed at the RNA level by a typical archaeal bulge–helix–bulge excision mechanism, suggesting that splicing of pre-mRNAs probably depends on the spicing system elucidated for archaeal pre-tRNAs and rRNAs (Watanabe et al., 2002). *Sulfolobus* introns are described in detail in Chapter 8 in this volume.

Protein splicing is defined as excision of an intervening protein sequence (the intein) from a protein precursor and concomitant ligation of the flanking protein fragments (the exteins) to form a mature extein host protein and the free intein. Protein splicing results in a native peptide bond between the ligated exteins. Extein ligation differentiates protein splicing from other forms of autoproteolysis. Conserved intein motifs differentiate inteins from other types of in-frame sequences present in one homolog and absent in another homolog or from other types of protein rearrangements. Genes encoding intein-containing proteins are relatively more frequent in Archaea (106) than in Bacteria (130) and Eukarya (58), considering that public databases contain much fewer archaeal sequences than sequences from organisms of the other two domains (Intein dataBase; see Table 7.4).

In summary, eukaryal genomes can be clearly distinguished from archaeal and bacterial genomes by structural differences. Archaea and Bacteria, however, differ only in variation of the same features

Table 7.4 Web resources for genome analysis.

National Center for Biotechnology Information	www.ncbi.nlm.nih.gov/
Clusters of Orthologous Groups of proteins	~/COG/
Basic Local Alignment Search Tool	~/BLAST/
Genomic Biology	~/Genomes/
Reference Sequences	~/RefSeq/
European Bioinformatics Institute	www.ebi.ac.uk/
Toolbox	~/Tools/
Protein functional analysis	~/Tools/protein.html
National Institute of Technology and Evaluation	www.bio.nite.go.jp/
Genome Analysis Center	~/ngac/e/project-e.html
Kyoto Encyclopedia of Genes and Genomes	genome.ad.jp/kegg/
Pathways	~/pathway.html
Genes	~/genes.html
DOE Joint Genome Institute	www.jgi.doe.gov/
Programs	~/programs/index.html
Integrated Microbial Genomes system	img.jgi.doe.gov/pub/main.cgi/
Phylogenetic Profiler	
Genomes OnLine Database	www.genomesonline.org/
The Institute for Genome Research	www.tigr.org/
Comprehensive Microbial Research	~/tigr-scripts/CMR/CmrHomePage.cgi
RBSfinder	~/software/
Microbial Database	~/tdb/mdb/mdbinprogress.html
The Wellcome Trust Sanger Institute	www.sanger.ac.uk
Microbial Genomes	~/Projects/Microbes/
Expert Protein Analysis System	www.expasy.org/
ProtParam tool	~/tools/protparam.html
Center for Biological Sequence Analysis	www.cbs.dtu.dk/services/
Genome Atlases	~/GenomeAtlas/
EasyGene, gene finding in prokaryotes	~/EasyGene/
SignalP, prediction of signal peptide cleavage istes	~/SignalP/
TatP, prediction of Twin-arginine signal	~/TatP-1.0/
ArchaeaFun, *ab initio* prediction of enzyme classes	~/ArchaeaFun/
The Intein dataBase and registry	www.neb.com/neb/inteins.html
Biology of Extremophiles Laboratory, Orsay	www.archbac.u-psud.fr/genomics/
Genomics ToolBox	~/GenomicsToolBox.html
MULTALIGN	~/multalin.html
The Sulfolobus database	dac.molbio.ku.dk/dbs/Sulfolobus
ArchaeaWeb, weekly Archaea publication update	www.archaea.unsw.edu.au/

and are not fundamentally different in their gene and genome structure.

The content of archaeal genomes

On average about 87% of archaeal genomes are covered with protein-encoding sequences (CDSs or ORFs), which is equal to the situation known from bacterial genomes and makes CDSs by far the most frequent components of archaeal genomes (Tables 7.2, 7.3). The protein coding fraction of eukaryal genomes is much lower than that of archaeal and bacterial genomes. The range of protein coding coverage in archaeal genomes ranges from 71.5% in *M. barkeri* (Copeland et al., unpublished Genbank accession) to 93% in *N. equitans*, with only the large genomes of the Methanosarcinales coding for proteins on significantly fewer than 85% of their genomes (Table 7.2).

The number of protein coding genes in archaeal genomes ranges from 552 (*N. equitans*) to 4540 (*M. acetivorans*). As stated in the previous section, archaeal genomes harbor on average one protein coding gene per kbp genome length, just like bacterial genomes. The average length of archaeal protein coding genes varies from 757 bp (*Pyrobaculum aerophilum*) to 936 bp (*M. acetivorans*) (Table 7.3). There is no clear correlation between the average length of ORFs and lifestyle or other criteria linked to genome or organism. Variation in annotation strategies (e.g. cutoff values for inclusion or omission of small ORFs without known homologues) can have a significant effect on inferred average ORF lengths. The number of protein families and members therein generally increases with the genome size, because accumulation of paralogous genes is one of the mechanisms leading to increased genome size and metabolic variability (Table 7.3). Unfortunately, strategies applied for inferring protein families are not standardized, so that the numbers given in Table 7.3 can be taken only tentatively. Functionally annotated genes can be categorized in 25 role categories. Plate 7.2 shows the distribution of genes by role category in Crenarchaeota (red) and Euryarchaeota (blue). Whereas the minimal and maximal number of identified genes within individual role categories can vary significantly (depending in part on the quality of annotation), average numbers are about equal for Crenarchaeota and Euryarchaeota, with the largest inter-phylum deviations in cell cycle control (functionally different in the two phyla), cell motility, and lipid transport and metabolism. The absolute number of genes per organism varies least in translation, and nucleotide transport and metabolism, but most in cell motility, defense mechanisms, signal transduction, and biosynthesis of secondary metabolites. Proteins of genome-size independent categories are evenly distributed across all genomes (e.g. information processing), whereas the number of genes involved in metabolism and regulation depend on the genome size. Larger genomes require more complex regulation of gene expression via an increased number of genes encoding regulator proteins. Plate 7.2 also shows a comparison of the relative fraction of genes in archaeal, bacterial, and eukaryal genomes. Archaea and Bacteria, use equal fractions of their coding capacity for many role categories, but Archaea use a significantly higher fraction for translation, energy production, and general function prediction only (reflecting the poor status of knowledge as compared to Bacteria and Eukarya). Bacteria have the highest fraction of genes involved in cell wall/membrane biogenesis (murein sacculus) and the lowest fraction of genes with unknown function, indicating the wealth of information assembled over decades from bacterial model organisms. Eukarya are especially rich in genes for signal transduction and intracellular trafficking (and genes not included in the clusters of orthologous groups (COGs) system). Eukarya use relatively small fractions of their genomes for translation, replication and repair, defense mechanisms, cell wall/membrane biogenesis, cell motility, energy production, and metabolism. The relatively high fraction of archaeal and bacterial genes in category X (not included in COGs) also contains the novel gene pool associated with genomic islands in Archaea and Bacteria (Hsiao et al., 2005).

Makarova and Koonin (2003a, 2005) revealed in a detailed comparative analysis a conserved core of 303 genes represented in all 20 completed archaeal genomes (without *N. equitans*, which is missing nearly all metabolic genes from the core), plus a variable "shell" that is prone to lineage-specific gene loss and LGT. The majority of these Archaea-specific genes have not yet been experimentally characterized. About 80 of these genes belong to the universally conserved core found in genomes of all organisms (mainly information processing). Only 16 of the archaeal core genes are unique to Archaea (with no homologs identified in any bacterial and/or eukaryal genome; again ignoring *N. equitans*). Seven of these Archaea-specific genes belong to information processing or are poorly characterized. Interestingly, only the gene for 23S RNA-specific pseudoouridylate synthase appears to be present in genomes of all Bacteria and Eukarya (through mitochondrial origin), but is absent in archaeal genomes. The conserved core of archaeal genes has been subject to many comparative analyses (Makarova et al., 1999; Graham et al., 2000), mainly for selection of an archaeal genomic signature and for laterally non-transferable and therefore phylogentically valuable marker molecules. As noted by Nesbo et al. (2001) for the Euryarchaeota, the hypothesis of a core of non-transferable genes has not been proven and may even be unprovable (see Chapter 3 in this volume). However, Brochier et al. (2005a) applied phylogenetics analyses to concatenated sets of genes from transcription and translation genes of the archaeal core and could demonstrate that these lead to a coherent signal for archaeal phylogeny, despite the danger of undetected LGT (see Chapter 2 in this volume). Charlebois and Doolittle (2004) proposed to relax the requirement for ubiquity for the assembly of phylogenetically balanced sets of core genes and identified about 150 genes to be members of the archaeal core (as compared to 64 bacterial core genes).

Crenarchaeota and Euryarchaeota were initially divided based on 16S rRNA phylogenies. Comparative genomics strongly supports this division, as a number of genes for key proteins involved in DNA replication, chromosome structure, and replication are exclusively present in either euryarchaeal genomes or crenarchaeal ones: Crenarchaeota lack DNA polymerases of the D family, Eukarya-like histones, replication protein RPA, and cell-division protein FtsZ (reviewed in Brochier et al., 2005b). Euryarchaeal genomes miss genes for r-proteins S30e, S26e, S25e, and L13e. Non-orthologous gene displacements (Koonin et al., 1996) as well as fundamentally different molecular mechanisms might explain some of these differences between the two archaeal phyla. Sixty-one of the 303 COGs shared by all archaeal genomes do have homologs in Eukarya, but not in the bacterial genomes (Makarova & Koonin, 2003a, 2005). Not unexpectedly, all but two of these genes are involved in information processing.

Genes for stable RNAs cover a much smaller fraction of archaeal genomes (only about 1–2%). It is noteworthy that only few of the higher branching Euryarchaea encode two, but never more than three, clusters of rRNAs. In general bacterial genomes encode more genes for rRNAs and eukaryal genomes encode far more rRNA genes. The rapidly growing numbers of small noncoding RNAs in bacterial genomes are described in Chapter 20 in this volume.

In summary, Archaea can be clearly distinguished from Bacteria and Eukarya by their genome content and sequence similarities of genes and encoded proteins. In addition to the features specified above, genes for the synthesis of the structurally unique archaeal membrane glycerolipids, the lack of genes for the synthesis of a murein-containing cell wall, and most strikingly the organization of their information-storage and processing systems (see Chapters 13–19 in this volume) strengthen the unique position of the Archaea as an independent domain of life.

Lessons learned from archaeal genomes

Analysis of two dozen published archaeal genomes has taught us lessons not only about the species from which the individual genomes were derived, but also about common genomic features of the phyla and genera to which they belong, as well as the genetic basis of their phenotypes, e.g. hyperthermophilia, and about Archaea in general. Some of these lessons were essential for our contemporary understanding of the Archaea and the evolution of life on earth. Unexpected knowledge about the enormous size of genetic flux between individual species of Archaea via LGT, as well as between Archaea and Bacteria, arose from comparative genomics and shaped our view of the Archaea as a distinctive domain within the tree of life and our model for molecular evolution (see Chapter 3 in this volume). Comparative genomics also led to a significantly improved and more convincing model for the natural history and internal structure of the Archaea (see Chapter 2 in this volume). Archaeal genomes also contributed significantly to the exciting expansion of our knowledge on how the genetic code is translated not only in the Archaea, but also in Bacteria and Eukarya (see Chapter 18 in this volume).

Lessons from 14 genomes of extreme and hyperthermophiles

More than half of the completed archaeal genome sequences originate from extreme thermophiles (T_{opt} 75–90 °C and hyperthermophiles ($T_{opt} \geq 90$ °C). Due to their small cell size, all cellular components of these extreme and hyperthermophiles must have intrinsic thermostability (Robb, 2004). Comparative analysis of genome sequence derived proteomes revealed that organisms with $T_{opt} > 80$ °C (Table 7.2) possess a significantly larger proportion of charged amino acids (such as Asp, Glu, Lys, and Arg) versus polar (noncharged) amino acids (such as Asn, Gln, Ser, and Tyr) than mesophiles (Cambillau & Claverie, 2000; Suhre & Claverie, 2003). Analysis of the water accessible surfaces of protein structures confirmed this result, indicating the biophysical requirement for the presence of charged residues at the protein surface for further protein stabilization through ion bonds (Cambillau & Claverie, 2000). Disulfide bonds are rarely found in intracellular proteins, but genomic data from hyperthermophilic Archaea contradicted this well established observation from many model organisms. Intracellular proteins of hyperthermophilic Archaea, especially the Crenarchaea *P. aerophilum* and *A. pernix*, are rich in disulfide bonds, implicating disulfide bonding in stabilization of many thermostable proteins (Mallick et al., 2002). In many cases, proteins from hyperthermophiles have smaller subunits than homologous proteins from mesophiles, corresponding to a more compact structure with less internal voids, which are achieved by smaller loops and shorter N- and C-termini (reviewed in Robb, 2004). Genes for short non-messenger RNAs (snos) are much more frequently found in the genomes of hyperthermophiles (Table 7.3). This may be because of an inherent need for methylations of structural RNAs in higher-temperature environments (Omer et al., 2000; Chapter 20 in this volume).

With the exception of *M. kandleri*, and unlike the bacterial hyperthermophiles, the genomes of extreme and hyperthermophiles feature a relatively low G+C content. Stability of the chromosome is instead explained by DNA-binding proteins and positive supercoils introduced into circular chromosomal DNA by reverse gyrase, the only protein whose gene is absolutely specific to genomes of all hyperthermophiles. Bacterial hyperthermophiles most probably received their reverse gyrase by LGT from archaeal hyperthermophiles (Forterre, 2002; Chapter 2 in this volume). Although reverse gyrase is apparently necessary for hyperthermophiles, its presence alone cannot explain the molecular basis of hyperthermophilia. Detailed sequence and genome-context analyses of COGs preferentially present in archaeal and bacterial hyperthermophiles led to functional predictions for several previously uncharacterized protein families, including a putative DNA-repair system that consists of more than 20 COGs, a group of putative molecular chaperones (COGs 2250 and 1895), and a unique transcriptional regulator, COG1318, that might be involved in adaptation to hyperthermal environments (Makarova et al., 2003; modified in Makarova et al., 2006). The extreme codon bias of archaeal hyperthermophiles towards bacterial hyperthermophiles supports the abundant evidence of LGT between thermophiles in general (e.g. observation of 25% archaeal genes in *Thermotoga maritima*), and begs the question of whether LGT in general is pervasive in all organisms sharing the same niche (reviewed in Robb, 2004; see also below on LGT *Thermoplasmatales/Sulfolobales*). In this respective, attempts at phylogenetic reconstruction from individual genes of (hyper-) thermophiles appear to be futile.

Lessons from five crenarchaeotal genomes

With about 1660 protein coding genes (62%) shared by all three *Sulfolobus* genomes, the size of the genomic core of the *Sulfolobales* equals that of the *Pyrococcales* (see below), although *Sulfolobales* are phylogenetically more diverse then *Pyrococcales* (see Plates 2.2, 2.3 in this volume). About half of the shared genes (866 genes) appear to be specific for the *Sulfolobales* (Chen et al., 2005b). The extensive genomic plasticity originally described for genomes of *S. solfataricus* (201 IS elements covering 11% of the genome; She et al., 2001b) and *S. tokodaii* (Kawarabayasi et al., 2001) turned out to be not characteristic for all *Sulfolobales*. *Sulfolobus acidocaldarius* contains no active mobile elements at all and maintained a stable genome organization (Chen et al., 2005b). SRSR elements, however, are much more frequent in the genomes of all three

Sulfolobales than in other archaeal genomes (Table 7.2). Comparative analysis of the three genomes allowed the identification of many short (<120 codons) putative genes conserved between the genomes but previously ignored in annotation (Chen et al., 2005b). The two other Crenarchaea with sequenced genomes, *A. pernix* (*Desulfurococcales*) and *P. aerophilum* (*Thermoproteales*), thrive under significantly higher temperatures than the *Sulfolobales*, with *A. pernix* being phylogenomically closer related to the *Sulfolobales* (see Plate 2.3 in this volume). *A. pernix* ORF0745, a putative gene without known function, is the only predicted gene within the Crenarchaea that might contains an inteine. However, this is only a theoretical prediction without functional proof, and inteins might therefore be restricted to euryarchaeal genes (92 reported inteins; Table 7.3). Disrupted genes for both subunits of adenylylsulfate reductase were identified as a possible reason for *P. aerophilum*'s unusual intolerance to sulfur (Fitz-Gibbon et al., 2002). Repeat-tract instability in *P. aerophilum*, the deepest branching crenarchaeon, was interpreted as a consequence of the lack of a mismatch repair system and a permanent mutator lifestyle of the organism (Fitz-Gibbon et al., 2002). Only later did it become clear that the set of genes involved in crenarchaeal DNA-repair systems differs significantly from that identified in euryarchaeal genomes (see Chapter 15 in this volume). There is only a limited number of genes (21 COGs; Makarova & Koonin, 2003a) identified as signatures of the Crenarchaeota versus the Euryarchaeota. Interestingly, none of the crenarchaeal genomes encodes *ftsZ*, which is present in Euryarchaeota and *Bacteria*, indicating a significant difference in cell division between Cren- and Euryarchaotea. Genes encoding crenarchaeal tRNAs contain on average seven times more introns (14–24) than those encoding euryarchaeal tRNAs (2–5) (Table 7.3).

Lessons from four genomes of *Thermococcales*

The three *Pyrococcus* genomes are significantly polymorphic with respect to pathways for amino acid biosynthesis and uptake and catabolism of several carbohydrates. Genomes of *Pyrococcus abyssi* and *Pyrococcus horikoshii*, which are significantly smaller than that of *Pyrococcus furiosus* (Table 7.2), are lacking the genes for the biosynthesis of histidine, cobalamine, several amino acids, and parts of the TCA cycle, as well as fermentation capacity for polysaccharides, whereas *P. furiosus* encodes the capacity to synthesize all nucleotides as well as vitamins B_6 and B_{12} (reviewed in Robb, 2004). Presumably, *P. furiosus* gained the additional metabolic capacity by gene duplications (more paralogous gene families) and by lateral gene transfer, e.g. the *mal* operon, from *Thermococcus litoralis*, which shares the same habitat. Poole et al. (2005) raised serious doubts about the quality of the annotation of the *P. furiosus* genome (and other genomes) by reporting previously ignored genes to be functional and by revealing dramatic differences in annotation data of this genome in various databases. Two types of long clusters of tandem repeats (LCTRs) as well as eight inteins located at the same insertion site are common to all three *Pyrococcus* genomes, indicating a close relationship between the three species (Lecompte et al., 2001). The newest member of the *Thermococcus* genome family, *Thermococcus kodakaraensis* (Fukui et al., 2005), is significantly larger than and phylogenetically more distant from the *Pyrococcus* genomes, but still shares more than 52% of its genes with all three of them (corresponding to 62% of all pyrococcal genes and almost all genes shared as a core between the three pyrococci), making it a perfect outgroup for comparative genomic analyses. Although no extended segmental synteny could be detected between the genome of *T. kodakaraensis* and the three *Pyrococcus* genomes, there is no indication of recent lateral transfer from other organisms into *T. kodakaraensis* to explain its increased genome size (Fukui et al., 2005).

Lessons from the genome of *N. equitans*

The discovery of *N. equitans* (Huber et al., 2002a) and the analysis of its genome by Waters et al. (2003) belong to the most spectacular publications in Archaea research since the *M. jannaschii* genome in 1996. With only 552 often split protein encoding genes in the 0.49 Mb chromosome, and highly unusual split tRNA genes, this genome supplied several unexpected highlights for archaeal genomics. Huber and Rachel describe in Chapter 5 in this volume most of the outstanding stories about this organism and its genome. What remains to be clarified is the "final" phylogenetic location of *N. equitans*, either within the Euryarchaota as a sister-branch of the *Thermococcales* (Brochier et al., 2005b; Makarova & Koonin, 2005; Plate 2.3 in this volume) or branching off below the Crenarchaeota (Fig. 5.2 in this volume), or even below the separation of Crenarchaeota and Euryarchaeota, representing a new phylum, *Nanoarchaeota*, within the domain Archaea (Waters et al., 2003). Comparative genomics by Makarova and Koonin (2005) strongly supports the first of the three options, because *N. equitans* encodes five of the nine proteins that are shared by the Euryarchaeota and not found in the Crenarchaeota, whereas it lacks all 16 signature crenarchaeal proteins. The availability of the genome sequence of the host strain of *N. equitans*, *Ignicoccus* sp. KIN4/I, is eagerly awaited to finally clarify *N. equitans*'s phylogenetic position within the Archaea and the early history of the domain.

Lessons from seven genomes of the methanogens

Phylogenetically methanogens fall into two separate groups: group I methanogens are specialized in CO_2 reduction and include *Methanococcales* (two genomes), *Methanobacteriales* (one) and *Methanopyrales* (one); group II methanogens are specialized on reduction of acetate and/or methyl compounds and include *Methanosarcinales* (three) and *Methanomicrobiales* (see Plate 2.3 in this volume). Methanogenesis is considered to be the main invention of the Euryarchaeota (for details of the key enzyme of this process, methyl-coenzyme M reductase, see Chapter 24 in this volume). The genomes of all methanogens share the same set of homologous genes encoding enzymes and cofactors required for methanogenesis (40 COGs in Makarova & Koonin, 2003a), indicating that the invention of methanogenesis occurred only once in the Euryarchaeota, probably after divergence from the *Thermococcales* (Bapteste et al., 2005a; Chapter 2 in this volume). *M. jannaschii* was the first and most successful archaeal genome story, because this genome sequence more than doubled the number of known archaeal genes and provided the first opportunity to compare complete genetic complements and pathways among all three domains of life (Bult et al., 1996). Its analysis revealed to the broader public what has been known to *Archaea* researchers from individual pre-genomic studies: the striking sequence similarity of archaeal and eukaryal information-processing systems (transcription, translation, and replication) as compared to the higher similarity of archaeal and bacterial genes encoding components from energy production, cell division, transport, and metabolism (Bult et al., 1996). The presence of 18 inteins in 14 protein coding genes was surprising, because archaeal inteins were previously known only from few thermococcal DNA polymerases. The lack of genes for lysine and cyteine aminoacid-tRNA synthetases spurred the successful search for the corresponding enzymes that are not homologous to their common eukaryal equivalents. Surprisingly, a *Bacteria*-like *ftsZ* gene was identified, which later turned out to be characteristic for the Euryarchaeota only, as opposed to the Crenarchaeota, and which led to the idea that archaeal cell division might occur by a domain-specific mechanism. The closest relative to *M. jannaschii* that has a completely sequenced genome is *M. maripaludis* (Hendrickson et al., 2004). Although the two genomes are almost identical in size, the two methanococci, which differ dramatically in their optimal growth temperature (85 versus 37 °C), show highest pairwise BLASTP scores for only two-thirds of their genes, indicating massive lateral gene transfers (Hendrickson et al., 2004). Equal to *M. jannaschii*, the genetically traceable *M. maripaludis* has the highest number of selenocysteine-containing genes (nine) within the *Archaea*, but lacks any intein. Curiously, the genome encodes for each of the selenocysteine-encoding genes a cysteine-containing paralog. Because both methanococci lack a homolog of Cdc6, as well as discrete transitions in the GC skew, it has been assumed that methanococci may employ multiple origins for chromosome replication (Hendrickson et al., 2004). Analysis of the *Methanothermobacter thermoautotrophicus* (formely *Methanobacterium thermoautotrophicum*) genome confirmed most of the Archaea-specific features described one year earlier for *M. jannaschii*, adding nitrogen fixation, regulatory functions, and interaction with the environment to the more Bacteria-like functions of the Archaea (Smith at al., 1997). *M. thermoautotrophicus* was the first sequenced archaeal genome with many two-component sensor kinase-response regulator systems, as well as Hsp70, DnaK, and DnaJ.

The obscure, deep-branching lineage of *M. kandleri* in pre-genome era 16S rRNA phylogenies was very puzzling. The extreme position has been supported by primitive terpenoid lipids in its membranes, an extremely high intracellular salinity, as well as the possession of several unique enzymes, e.g. a two subunit reverse gyrase (Slesarev et al., 2002). Comparative genomics analyses and phylogenies based on gene content as well as on concatenated r-proteins safely placed *M. kandleri* as the deepest branch within the group I methanogens (Slesarev et al., 2002; Plate 2.3 in this volume). Previous odd placements in various phylogenies, 16S rRNA and DNA-directed RNA polymerase, were explained with fast evolutionary rates in the respective lineages (see Chapter 2 in this volume). *M. kandleri* shares a core of 801 COGs-genes with the other group I methanogens (about 47% of its gene complement), 59 of them to the exclusion of all other genomes (Slesarev et al., 2002). Its genome has almost exactly the same size as those of the other group members (that is, one-third the size of group II methanogen genomes; Table 7.2), and next to the two methanococci *M. kandleri* is the archaeon with the highest number of selenocysteine-containing genes (Table 7.3). *M. kandleri*'s set of methanogenesis genes as well as the organization of some of its operons is clearly shared with the other group I methanogens (Slesarev et al., 2002). The organism is, however, distinct in its paucity of proteins involved in signaling and regulation and an unusually high fraction of negatively charged amino acids in its proteins (pI about 5, the highest within the *Archaea* next to the extreme halophiles).

In addition to their phylogenetic location (Plate 2.3 in this volume) and substrate (acetate) specificity, group II methanogens with completely sequenced genomes also differ from group I methanogens by their significantly larger genome size of 4.90±0.83 Mbp versus 1.69±0.04 Mbp, representing by far the largest archaeal genomes. This may only in part be explained by the lowest coding coverage of all archaeal genomes, 73.6% (compared to 89.2% for group I methanogens; Table 7.2), the longest average ORF size, 925 bp (compared to 804 bp for extreme halophiles; Table 7.3), and unusually long intergenic regions (on average 341 bp in *M. acetivorans* versus 122 bp in *A. fulgidus*). The three species of *Methanosarcinales* whose genomes have been sequenced, *M. acetivorans* (Galagan et al., 2002), *M. mazei* (Deppenmeier et al., 2002), and *M. barkeri* (Copeland et al., Genbank NC0007355, October 2005), are also the only species for which pyrrolysine as the actively used twenty-second amino acid (amber suppression) has been demonstrated (see Chapter 18 in this volume and Table 7.3). Genomes of the *Methanosarcinales* encode a strikingly wide and unanticipated variety of metabolic and cellular capabilities, including three pathways for methanogenesis (nearly 200 genes) and the presence of single subunit CO dehydrogenase, indicating the surprising possibility of non-methanogenic growth (Galagan et al., 2002). Unusually large species-specific gene families also contribute to the extremely large genome size of *M. acetivorans*. In this perspective the second largest multigene family of *M. acetivorans* is especially noteworthy, because it encodes secreted proteins involved in generating the cell envelope as well as an extracellular matrix during the formation of multicellular structures (Galagan et al., 2002). Homologs of this family are also present in *M. mazei*. *Methanosarcinales* contain surprisingly many genes for which homologs are absent in other archaeal genomes, but are frequently found in the *Bacteria*. For *M. mazei* 1043 ORFs (31%) fall into this group (Deppenmeier et al., 2002), including genes for all four GroES/L systems (unique for *Methanosarcinales* within the *Archaea*), indicating massive LGT from Bacteria thriving in the same habitat. With only 376 ORFs (11% in *M. mazei*) the fraction of genus-specific ORFs is relatively small for *Methanosarcinales*, as compared to *Pyrococcales* and *Sulfolobales*, although their phylogenetic depth is much lower (Plate 2.3 in this volume). Unique among methanogens, *M. acetivorans* appears to possess genes for all three types of nitrogenases (molybdenum/iron, vanadium/iron, and iron only), indicating the importance of nitrogen fixation for the organism (Galagan et al., 2002).

Lessons from the genome of *A. fulgidus*

A. fulgidus was one of the very first (archaeal) genomes to be sequenced (Klenk et al., 1997a), with *M. jannaschii* as the only archaeal genome for comparison at the time of annotation. The content of the *A. fulgidus* genome confirmed much of what was first posted for *M. jannaschii*. Archaeoglobales are the only archaeal sulfate-reducers; therefore it was obvious that many of the genes involved in sulfate metabolism were categorized as bacterial-like, received via LGT. Discovery of 57 genes involved in degradation of hydrocarbons and organic acids (previously unknown for *Archaea*) expanded the idea of *A. fulgidus* being the frequent recipient of genes from *Bacteria*. Meanwhile it became clear that the Archaeoglobales share this general feature with their closer phylogenetic neighbors in the upper part of the euryarchaeotal branch of the *Archaea* (Plate 2.3), class II methanogens and extreme halophiles. Comparison of the *A. fulgidus* proteome with the contemporary set of archaeal proteomes using IMG Phylogenetic Profiler (for website see Table 7.4; max. E-value 10^{-5}, minimum 30% identity) revealed that *A. fulgidus* shares more genes with class II methanogens (1031) than with other archaeal genera: *Thermococcales* (740), extreme halophiles (736), *Sulfolobales* (705), class I methanogens (621), *Thermoplasmales* (512), and *N. equitans* (284). All 24 archaeal genomes share 232 genes using these comparative parameters. The 47 genes specifically shared with the methanogens include all those encoding components of methanogenesis from formylmethanofuran to N^5-methyltetrahydromethanopterin, used by *A. fulgidus* for reverse methanogenesis. Results from comparative genomics analysis fit nicely with the phylogenetic location of the hyperthermophilic *A. fulgidus* next to the mesophilic class II methanogens and the extreme halophiles. Like *M. jannaschii*, *A. fulgidus* developed into an archaeal model organism with more than 230 Medline referenced papers in the eight years since the genome publication, as compared to 24 papers in the decade between its discovery and the decoding of the genome.

Lessons from three genomes of *Thermoplasmatales*

Thermoplasmatales and *Sulfolobales* share not only the same hot, acidic habitat, but also a significant fraction of genes (6–11%) not found in genomes of organisms living in different habitats (Fütterer et al., 2004). Due to LGT between phylogenetically only distantly related Cren- and Euryarchaea the proteomes of these organisms are more similar to each other than to phylogentically more closely related organisms from the respective kingdom: ecological closeness overrides phylogenetic relatedness (Fütterer et al., 2004). The set of laterally exchanged genes contains mainly components of the protein degradation system and various transport proteins, but none from information processing systems (Ruepp et al., 2000). Based on the skewed distribution of *Sulfolobus*-like genes in the genome of *Thermoplasma acidophilum*, it has been concluded that only a few large genetic transfers led to this marvelous adaptation to the shared extreme environment (Ruepp et al., 2000). As opposed to *Sulfolobales* and hyperthermophilic bacteria, *Thermoplasmatales* harbor 6–20 times more secondary solute transport systems, indicating the predominant use of proton-driven transport systems as adaptation to the extreme environment (Fütterer et al., 2004). The increased fraction of hydrophobic amino acids on the surface might contribute to acid stability of proteins from *Picrophilus torridus*, which shows an unusually low intracellular pH of 4.6. Putative genes encoding enzymes involved in diether and tetraether lipid biosynthesis, as well as S-layer proteins in *P. torridus*, provide hints for the adaptation of the membrane to the acidic environment, but no deductive reason for acid resistance of the cell wall-less *Thermoplasmatales* (Fütterer et al., 2004). It is still unclear which structure could function as the stator for flaggellar rotation in *Thermoplasmatales* (Ruepp et al., 2000). A domain conserved in many archaeal S-layers might serve as a scaffold for secreted proteases (Ruepp et al., 2000). Although the enhanced living temperature and (for *P. torridus*) the low intracellular pH call for very efficient DNA repair systems, it remains puzzling that *Thermoplasmatales* encode homologs neither for the eukaryal excision repair (NER) genes nor for UvrABC proteins typical for the Bacteria. The complete lack of Eukarya-like genes in the *T. acidophilum* genome ended the debate of *Thermoplasma*-like organisms being a possible ancestor of the eukaryal cytoplasm in the endosymbiosis hypothesis. Computational analysis of the *Thermoplasma volcanium* genome sequence revealed an increasing clustering frequency of purines and pyrimidines, respectively, with rising optimal growth temperature, correlating with a loss of DNA flexibility at higher temperatures (Kawashima et al., 2000).

Lessons from three genomes of extreme halophiles

A general characteristic of genomes of halophilic archaea is a bipartite genome-content organization in the form of one large high G+C content chromosome (62–68%) and multiple (2–8) smaller replicons of lower G+C content. Haloarchaeal origins of replication are found around the maximum, not the minimum, of cumulative GC skew plots (Falb et al., 2005). A relatively large number of IS elements are spread all over the genomes of extreme halophiles, with a higher density of repeat elements in the smaller replicons. It has been suggested that the IS-element rich extra chromosomal elements contribute to the evolution of halophile genomes by facilitating acquisition of new genes (Ng et al., 2000). The predicted proteomes of all extreme halophiles are highly acidic (median pIs between 4.2 and 5), indicating that halophilic archaea use a strategy of high surface negative charge of folded proteins as a means to circumvent the salting-out phenomenon in a hypersaline cytoplasm (Baliga et al., 2004b). Genome comparison suggests a common ancestor for *M. morismortui* and *Halobacterium* species NRC-1, with the latter being reduced in size (Baliga et al., 2004b). The larger genome enables *H. marismortui* to encode an unusually large number of environmental response regulators and metabolic genes, to exploit more diverse environments, and to depend on fewer nutritional requirements than strain NRC-1. Papke et al. (2004) demonstrated recently by multilocus sequence typing that Haloarchaea exchange genetic information promiscuously, exhibiting a degree of linkage equilibrium approaching that of a sexual population. Preliminary analysis of three of the six halophile genomes presently in the sequencing pipeline (Table 7.5; Goo et al., 2004) confirmed the general features described above, with chromosomal G+C contents above 60%, extremely acidic proteomes, and multiple encoded copies of TATA box binding proteins (TBP) and transcription factor IIB (TFB) homologs. The latter indicates complex types of transcriptional regulation through TBP–TFB combinations in response to rapidly changing environmental conditions (Goo et al., 2004). In addition to a complex cell envelope consisting of glycoproteins, which has been predicted from the encoded proteins, the haloalkaliphilic *N. pharaonis* links several of its secreted proteins by lipid anchors to its cell membrane as a protection against alkaline extraction of these proteins from the cell membrane.

There is one lesson from the archaeal genomes that has unfortunately not been learned by all (micro)biologists, not even by all *Archaea* specialists: the genomes of the *Archaea* are not mosaics with a "bacterial" and a "eukaryal" part, and the archaeal genes (proteins) that feature higher sequence similarity to some homologs in the *Bacteria* (or *Eukarya*) are not necessarily bacterial (and probably never eukaryal) in nature. Certainly, some or even many bacterial genes have been laterally transferred from species of *Bacteria* to species of *Archaea*, but in general many of the genes for metabolic enzymes belong to a gene pool that is shared by *Archaea* and *Bacteria*, frequently exchanged through LGT. Higher degrees of sequence similarity between archaeal and eukaryal homologs (as opposed to bacterial homologs), e.g. proteins from information processing categories, indicate that ancient archaeal genes encoding these components might have contributed to the formation of information processing systems of the *Eukarya*, or that the matching eukaryal homolog originated from a previously laterally transferred archaeal gene, but not the other way around. *Archaea* constitute, just like the *Bacteria*, one of the two fundamental natural designs of life (often erroneously joined as "prokaryotes") that differ significantly in the principal components of their information processing systems, and that both contributed to the formation of the third, derived design, the *Eukarya*.

Internet resources and problems in genome annotation

Genomics and widespread use of the Internet arose at about the same time in the mid-1990s. Efficient handling and worldwide availability of the enormous amount of data, unprecedented in biology before the genomic era, would not have been possible without the fast tracks through the web, especially for data comparison with sequence databases. Therefore, microbial genomics was from the very beginning tightly linked to the worldwide web, probably more than any other biological discipline. Table 7.4 lists a selection of uniform resource locators (URLs) for databases and software tools that are useful for archaeal (and bacterial) genomics.

Table 7.5 Genomes in progress.

Species		Size (Mbp)	Status	Institution
Acidianus brierley	*Sulfolobales*	1.8	in progress	UC
Caldivirga maquilingensis	*Thermoproteales*	2.2	in progress	JGI
Cenarchaeum symbiosum	*Cenarchaeales*	3.7	complete	MBARI
Ferroplasma acidarmanus	*Thermoplasmales*	2.0	in progress	JGI
Halobacterium salinarum	*Halobacteriales*	2.0	complete	MPIB
Halobaculum gomorrense	*Halobacteriales*	?	in progress	ISB/UMBI
Halobiforma lacisalsi	*Halobacteriales*	?	in progress	Zhejiang University
Haloferax volcanii DS2	*Halobacteriales*	4.2	in progress	TIGR & IG (ATCC29605)
Halorubrum lacusprofundi	*Halobacteriales*	4.3	in progress	ISB & JGI (ATCC49239)
Haloquadratum walsbyi	*Halobacteriales*	?	in progress	MPIB
Hyperthermus butylicus	*Desulfurococcales*	1.7	complete	UoC/*e*.gene
Ignicoccus species KIN4-1	*Desulfurococcales*	?	complete	Diversa
Methanobrevibacter ruminantium	*Methanobacteriales*	?	in progress	AgResearch
Methanococcoides burtonii	*Methanosarcinales*	2.6	in progress	JGI/UMBI
Methanococcus voltae	*Methanococcales*	?	in progress	IG/MD
Methanocorpusculum labreanum	*Methanomicrobiales*	2.3	in progress	JGI
Methanoculleus marisnigri	*Methanomicrobiales*	2.2	in progress	JGI
Methanogenium frigium	*Methanomicrobiales*	5.0	in progress	UNSW/MD/JCVI/AGRF
Methanosaeta concilii	*Methanosarcinales*	?	in progress	UW/Clamson Univ.
Methanosaeta thermophila	*Methanosarcinales*	3.0	in progress	JGI
Methanosarcina thermophila	*Methanosarcinales*	?	in progress	G_2L
Methanosphaera stadtmanae	*Methanobacteriales*	1.7	complete	G_2L/MPITM
Methanospirillum hungatei	*Methanomicrobiales*	2.8	in progress	JGI
Methanothermococcus thermolitho.	*Methanococcales*	?	in progress	IG/MD
Methanothermus fervidus	*Methanobacteriales*	1.7	in progress	JGI
Natrialba asiatica	*Halobacteriales*	?	in progress	ISB/UMBI
Pyrobaculum arsenaticum	*Thermoproteales*	2.2	in progress	JGI
Pyrobaculum calidifondis	*Thermoproteales*	2.2	in progress	JGI
Pyrobaculum islandicum	*Thermoproteales*	2.2	in progress	JGI
Staphylothermus marinus	*Desulfurococcales*	1.7	in progress	JGI
Sulfolobus islandicus	*Sulfolobales*	?	in progress	UC
Thermococcus gammatolerans	*Thermococcales*	?	in progress	Université Paris Sud
Thermophilum pendens	*Thermoproteales*	2.2	in progress	JGI
Thermoproteus neutrophilus	*Thermoproteales*	2.2	in progress	JGI
Thermoproteus tenax	*Thermoproteales*	1.8	complete	UED/MPIDB/*e*.gene

Abbreviations: AGRF, Australian Genome Research Facility, CNRS, G_2L, Göttingen Genomics Laboratory; IG, Integrated Genomics; ISB, Institute for Systems Biology; JCVI, J. Craig Venter Institute; JGI, Joined Genome Institute; MBARI, Monterey Bay Aquarium Research Institute; MD, Molecular Dynamics; MPIB, Max-Planck-Institute for Biochemistry; MPIE, Max Planck Institute for Developmental Biology; MPITM, Max-Planck-Institute for Terrestrial Microbiology; TIGR, The Institute for Genomic Research; UED, University Essen–Duisburg; UMBI, University of Maryland Biotechnology Institute; UC, University of Copenhagen; UNSW, University of New South Wales; UW, University of Washington.

Institution/institution, collaboration projects; institution & institution, independent sequencing projects.

The enormous growth rate of microbial genomic data, which has doubled every 18 months over the past few years (Table 7.1), will continue at least for several years, even considering only the ongoing genome sequencing projects (Tables 7.1, 7.5), but might even accelerate again due to already foreseeable technological progress in sequencing technology. Data published in printed genome papers describe the actual status at the time of sequence production and annotation, and therefore become outdated much more quickly than conventional publications. Annotation errors cannot be corrected and efficiently transmitted to the scientific community, resulting in propagated errors within the databases. Only routinely updated web resources like those featured in Table 7.4 enable the maintenance of timely and accurate genomic data. The recent update on the *P. furiosus* genome by Poole et al. (2005) provides an excellent overview of the rapid development (*P. furiosus* genome was initially published in 2001 by Robb et al.) and the contemporary difficulties in archaeal genome annotation, starting with such "simple" facts as determining the number of genes encoded in a genome. Most modern publicly available resources, such as the Joint Genome Institute's Integrated Microbial Genomes system (IMG, Table 7.4), the Fellowship for Interpretation of Genomes' SEED system, and CeBiTec's BRIDGE software environment (www.cebitec.uni-bielefeld.de), enable researchers, rather than annotating gene by gene in a given genome, to annotate a specific set of genes across a large number of genomes, facilitating the visualization and exploration of genomes from a functional and evolutionary perspective. However, functional annotation of proteins predicted in genomic sequences is still predominantly based on similarities to homologs deposited in the databases. As a result of the possible strategies for divergent evolution, homologous enzymes frequently do not catalyze the same chemical reaction, and therefore assignment of function from sequence information alone should be viewed with some skepticism (Gerlt & Babbitt, 2000). Another recently realized problem is the discovery that many proteins have multiple functions (**moonlighting proteins**; Jeffery, 1999), e.g. Sso7d, featured in Chapter 13 in this volume. Contemporary protein classification systems and with them the structure of databases are not well prepared for systematic categorization of these enzymes. This holds too for the classification of proteins composed of domains with distinct but different functions (Baliga et al., 2004b).

Considering the problems of contemporary genome databases, e.g. lack of functional annotation in as many as 40% of the predicted genes, lack of functional validation for most predicted genes, and about 5–10% incorrect functional predictions, improvements are desperately needed. An American Academy of Microbiology colloquium on "An experimental approach to genome annotation" (held in Washington, DC, 2004; report available from www.asm.org) recently addressed the critical challenge of microbial genome annotation and proposed a massive annotation initiative to improve the quality status of genome annotation in order to avoid tremendously growing problems in further exploding databases. Accurate and complete annotation of archaeal (and other) genomic data is vital to make full use of genomic data, but there are great deficiencies in currently available annotation sources. A number of early annotation errors have been propagated within the public databases, and the timeliness of incorporation of novel experimental data for previously functionally unknown genes is not satisfying. To address these problems it has been proposed to establish a single central source of annotation information that is regularly updated and is distinct from current archival databases, as a point of reference for future annotators.

Prospects for archaeal genomics

The great days of microbial genomics expired as quickly as they appeared. Premier journals are no longer lining up to publish each microbial genome sequence, but instead turn to a more stringent policy for handling genome manuscripts. Regular reports on newly sequenced genomes, like *Microbiology*'s monthly Genome Update section, get phased out because summarizing the outcomes of genome sequencing studies is no longer of special interest for general (micro)biologists. Within only one decade, genomics lost its glamour and became a "regular" field of life sciences. This does not, however, mean that genome sequencing and analysis is retreating. On the contrary, Table 7.1 indicates more than 1000 genome sequencing projects in progress and more than 100 completely sequenced genomes presently waiting for publication. This is 3.5 times more genomes than are presently stored in public databases (data from GOLD, TIGR, and Sanger Institute databases; see Table 7.4). Table 7.5 lists six completely sequenced but not yet published archaeal genomes, and 30 ongoing genome sequencing projects, whose outcome will soon more than double the archaeal genome data (to more than 140,000 protein coding genes). As impressive as these lists and numbers are, they also indicate a dramatic loss in interest in archaeal genomics, as compared to genomics of *Bacteria* and *Eukarya*. In 1997 one-quarter of all sequenced genomes were of archaeal origin (Table 7.1); now Archaea represent fewer than 8% of the genomes, and barely 3% of the genomes in the contemporary production pipeline. One reason for this dramatically shrinking interest in archaeal genomes might be that a lot of the original interest in these genomes originated from phylogenetics and evolutionary biology, and much of this interest has been satisfied by a fairly good genome coverage of the archaeal part of the "tree of life." Enzymes derived from the genomes of extremophilic Archaea (and Bacteria) had a great impact in biotechnology, calling for many genomes of the extremophiles to be sequenced. The contemporary microbial genome sequencing pipeline is, however, clearly dominated by medically important pathogens in search of novel targets for therapeutic applications, complemented by genomes mostly from organisms involved in global climate change, bioremediation, and energy production.

The interest in archaeal genomics might, however, soon rise again. Over the past decade, large-scale diversity studies have shown that *Archaea* are far more frequent in moderate habitats than previously expected (see Plate 4.1 in this volume). In a number of metagenomic studies (Table 4.1 in this volume) it has been demonstrated that a lot of information about as yet uncultivated *Archaea* can be gained by sequencing genome fragments, which might eventually lead to the cultivation of new organisms with novel physiological properties, e.g. the Korarchaeota (for review see Chapter 4 in this volume). Studying whole communities of *Archaea* (and *Bacteria*) on the genomic level has become feasible over the past few years, and finally put several community sequencing projects in the genomic sequencing pipeline, e.g. at the DEO Joint Genome Institute: sequencing of an obsidian hot spring community from Yellowstone, sequencing of a euryarchaeal community from acid mine drainage, sequencing of root-colonizing Crenarchaea and their community, and sequencing of a Korarchaeota community from which to date there are no representatives in pure culture, with nothing being known about their properties (information from JGI website; see Table 7.4). These community sequencing efforts will also reveal plenty of new viral genomes and thereby enlarge the rapidly growing cornucopia of archaeal viruses described in Chapter 6. No environment will be too extreme or too far off for the pioneers of environmental genomics, like Craig Venter, who now moves on from sequencing the Sargasso Sea (Venter et al., 2004) to sequencing the genomes of all the microbes floating in the air of New York (New York Air metagenomics project, www.venterinstitute.org/). It may sound futuristic, but metagenomics of human wounds, which is about to take off soon, might even lead us to the discovery of species of archaea with impact on human health (see Chapter 25). Only recently, we have learnt from the analysis of the *Nanoarchaeum* genome (see Chapter 5) that there is a good chance for discovering yet unknown species or genera of *Archaea* whose conserved phylogenetic marker genes (16S rRNA) are so different from established standards that the "universal" primers and probes used in biodiversity studies will not match them. Only large scale, unbiased, metagenomics projects, like the Sargasso Sea sequencing and the New York air genomics project, will open the doors into uncharted territory, e.g. new phyla within the *Archaea*.

Over the past decade, since the publication of the first archaeal genome sequence, our knowledge of archaeal genomics (represented by the sequences in public databases) grew at an average annual rate of about 58% (equals 18 months for knowledge duplication). Assuming that this pace might be maintained for the coming years, which appears realistic in view of the growing size of the sequencing pipeline, we can estimate that archaeal sequence databases will hit one million protein coding genes (1000 times the number collected during the pre-genomic era) or 1 billion bp by around 2012. With an approximate 98% decrease in sequencing costs over the past decade, and a similar decrease already announced for sequences to be generated with the new generation of nanotechnology-based genome sequencing systems that are in the process of entering the market, generating this amount of sequence information will not be a budgetary problem. The rate-limiting techniques for the middle-term growth of archaeal genomics knowledge are functional genomics, bioinformatics, and databases to extract knowledge from the flood of sequences. As stated in Chapter 4 in this volume, meta-transcriptomics and meta-proteomics studies should soon be ready to take off to complement the vastly growing amount of genome sequence data for a more comprehensive picture about the naturally occurring archaeal (and bacterial) communities, and how they react to diverse environmental conditions.

As with *Bacteria* and *Eukarya*, there are still plenty of conserved hypothetical and unknown archaeal genes in the databases waiting for functional characterization (about 40% of all cataloged genes). A serious approach to biochemical validation of these gene functions is required to increase the quality of archaeal genome annotation, and to enhance our understanding of archaeal genome sequences. Results from improved annotation procedures that include analysis of domain fusion association, phylogenetic pattern association, operons, and protein families will suggest new ideas for putative functions of many proteins that presently lack any functionally characterized homologs in public databases. The highest priority should therefore be given to conserved hypotheticals found in many different archaeal genomes. Transcriptomic, proteomic, and structural genomic approaches, like the one described in Chapter 21 in this volume for *P. furiosus*, or the systematic approach for characterization of *H. salinarum* by Dieter Oesterhelt (www.biochem.mpg.de/oesterhelt/), will contribute significantly to a new level of understanding of archaeal gene functions in order to solve questions such as the structural basis for life under extreme environmental conditions. As important as the evaluation of novel biochemical functions for currently hypothetical archaeal genes is the validation of already functionally described genes. With the recent reevaluation of some parts of the *P. furiosus* genome (Poole et al., 2005), we can assume that a substantial fraction of the data stored in our databases is erroneous and might therefore seriously endanger functional (expression) and structural analyses that depend on, for example, the correct prediction of ORF-starts.

Understanding of gene function from annotation can only be a first step towards a better comprehension of the complexity of whole archaeal cells. Still, annotation is often not more than the cataloging of protein functions. Many of the initial archaeal (and bacterial) genome papers are striking examples of such approaches to microbial genomics. Much more information about interactions of gene products, about which we know very little at this time, is required for an improved understanding of the whole systems of living archaea. We have to learn how the interaction of the individual components that make up an archaeal cell will result in a functional system that is "more than the sum of its parts," that is alive. Mathematics will join in at a substantial level to integrate genomics, proteomics, and bioinformatics in modeling the interactions between the numerous components of living cells. Systems biologists will be able to understand whole archaeal cells (and communities thereof) by studying the products encoded in the genomes, and how they interact in synergistic networks to accomplish the complex functions of life. Modeling of whole archaeal systems will finally close the circle and guide microbiologists out of the still dominating reductionistic view of organisms, back to Beijerink's holistic vision of the microorganisms, and thereby fulfill the vision of a new biology as described by Carl Woese in Chapter 1 in this volume (see also Woese, 2004).

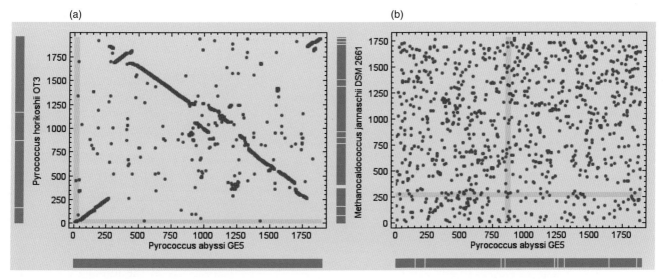

Plate 7.1 Genome organization. Pairwise genome comparisons of protein homology with GenePlot (www.ncbi.nlm.nih.gov/sutils/geneplot). (a) Two closely related *Pyrococcus* species, *P. abyssi* (horizontal) versus *P. horikoshii* (vertical). (b) Two distantly related Euryarchaeota species, *P. abyssi* (horizontal) versus *M. jannaschii* (vertical).

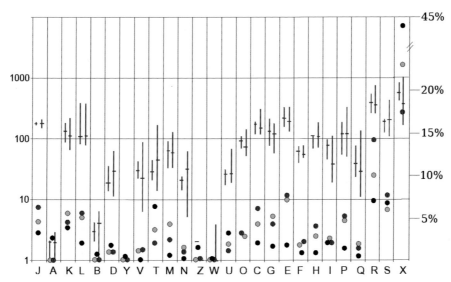

Plate 7.2 Distribution of genes by role category. Left, logarithmic scale and blue and red bars: number of genes categorized within the 25 biological role categories of COGs (Tatusov et al., 2003). Bars indicate minimal, maximal, and average number of genes for five Crenarchaea (red) and 16 Euryarchaea (blue). Species named in Table 7.2, without *M. barkeri* and *N. pharaonis* (not yet analysed in COGs). Abbreviations for role categories: J, translation; A, RNA processing and modification; K, transcription; L, replication, recombination, and repair; B, chromatin structure and dynamics; D, cell cycle control, mitosis, and meiosis; Y, nuclear structure; V, defense mechanisms; T, signal transduction mechanisms; M, cell wall/membrane biogenesis; N, cell motility; Z, cytoskeleton; W, extracellular structures; U, intracellular trafficking and secretion; O, post-translational modification, protein turnover, chaperones; C, energy production and conversion; G, carbohydrate transport and metabolism; E, amino acid transport and metabolism; F, nucleotide transport and metabolism; H, coenzyme transport and metabolism; I, lipid transport and metabolism; P, inorganic ion transport and metabolism; Q, secondary metabolites biosynthesis, transport, and catabolism; R, general functions prediction only; S, function unknown; X, not in COGs. Right, linear scale and blue (Archaea), green (Bacteria), and black (Eukarya) dots: fraction of genes categorized within the role categories of COGs.

8
Sulfolobus genomes: mechanisms of rearrangement and change

Kim Brügger, Xu Peng and Roger A. Garrett

Introduction

Strains of the genus *Sulfolobus* have been isolated globally from acidic solfataric fields and they are facultatively chemolithoautotrophic, growing via oxidation of sulfidic ores, sulfide, S°, or tetrathionate, with the production of sulfuric acid, and organotrophically (Huber & Stetter, 2001). Known strains are strict aerobes and grow optimally at about 80 °C and pH 2–3 and can be cultured on complex organic substrates. *Sulfolobus* strains constitute the most widely studied Crenarchaea because: (i) they are relatively easy to culture; and (ii) many diverse extrachromosomal genetic elements have been characterized for *Sulfolobus*. The latter include novel cryptic and conjugative plasmids as well as viruses exhibiting highly unusual and diverse morphotypes and genomes (reviewed in Zillig et al., 1998; Prangishvili et al., 1998a; Prangishvili & Garrett, 2005).

Thus, much of our current knowledge of crenarchaeal and archaeal mechanisms of DNA folding, DNA replication, DNA repair, integration, and conjugation, as well as RNA transcription, translation, modification, and processing and intron splicing, and properties of the cell cycle, derives from studies on *Sulfolobus* species and many of these earlier developments are considered elsewhere in this book. One of the most important early developments, which had a major impact on the development of the archaeal field, was the demonstration by Zillig and co-workers that the RNA polymerase complex of *Sulfolobus acidocaldarius* and other Archaea is eukaryal in character, not bacterial (summarized in Pühler et al., 1989; Langer et al., 1995).

Studies of these basic cellular processes have developed and expanded rapidly with the availability of chromosomal sequences from *S. acidocaldarius* (Chen et al., 2005b), which was the first hyperthermoacidophile to be isolated and is the type strain of *Sulfolobus* (Brock et al., 1972), the ubiquitous *Sulfolobus solfataricus* P2 (She et al., 2001b), *Sulfolobus tokodaii*, strain 7 (Kawarabayasi et al., 2001) and, recently, *Sulfolobus islandicus* HVE10/4 (K. Brügger, unpublished). In addition, many diverse *Sulfolobus* plasmids and viruses have been sequenced (for an interactive *Sulfolobus* database see http://dac.molbio.ku.dk/dbs/Sulfolobus).

What these sequencing data have revealed is that complex rearrangements can occur in both chromosomes and extrachromosomal elements by a variety of mechanisms. This chapter is limited to describing the properties of these genomes and, in particular, the mechanisms of rearrangement and change that bear on the survival, adaptation, and interaction of *Sulfolobus* strains, and their extrachromosomal elements. Integrative processes in *Sulfolobus* genomes are considered separately (Chapter 10 in this volume).

Chromosome variation

The chromosomal gene pool is well conserved in *Sulfolobus* species. This is illustrated in the circle plot in Plate 8.1, where we show that about three-quarters of the genes of *S. solfataricus*, *S. tokodaii*, and *S. acidocaldarius* are homologs shared with one, or both, of the other two species (Chen et al., 2005b). Moreover, many of these homologous genes are arranged in conserved operon structures. However, the overall gene order is poorly conserved. To illustrate this, we present in Plate 8.2 pairwise alignments of genes in the chromosome of *S. acidocaldarius* with homologous genes in *S. solfataricus* and *S. tokodaii* starting at a conserved region adjacent to one of the conserved replication origins (Brügger et al., 2004). A perfectly conserved gene alignment would yield two crossing diagonal lines. Clearly this is not so. In fact, the almost complete absence of diagonal lines indicates that the gene order shows very little conservation between the two pairs of chromosomes and a similar result is obtained when the chromosomes of *S. solfataricus* and *S. tokodaii* are aligned. We infer that a major contribution to this extensive chromosome shuffling, at least in *S. solfataricus*, *S. tokodaii*, and *S. islandicus*, is the large number of diverse and potentially mobile genetic elements (Brügger et al., 2004).

IS elements

The autonomously mobile insertion sequence (IS) elements invariably encode a transposase that facilitates their transposition and, sometimes, an additional resolvase, which may determine their replicative mechanism. Most of the IS elements carry inverted terminal repeats (ITRs), which provide a target for the IS element-encoded transposase. They have been classified into different families on the basis of the transposase gene/protein sequence and the nature of their ITRs (Brügger et al., 2002). All the sequenced *Sulfolobus* chromosomes, except that of *S. acidocaldarius*, carry representatives from many different IS element families, and some are present as multiple copies of elements that are identical, or almost identical, in sequence. In contrast, the *S. acidocaldarius* chromosome contains only four IS elements, all of which may be inactive (Chen et al., 2005b). A summary of the type and number of these elements in *Sulfolobus* chromosomes is given in Table 8.1.

For *S. solfataricus*, intact IS elements constitute about 10% of the total chromosome. In addition, many fragments of IS elements are present, including fragments of IS elements that are no longer present in an active, intact, form (She et al., 2001b; Brügger et al., 2002; Boult & Grogan, 2005).

MITEs

The nonautonomous miniature inverted-repeat transposable elements (MITEs) of *Sulfolobus* are considered to be mobilized by transposases encoded in IS elements that recognize their ITRs (Redder et al., 2001). These elements fall into two types. Type I MITEs are formed from IS elements as a result of the deletion of part, or all, of the transposase gene, but they retain some of the internal sequence of the original IS element, in addition to the ITR. They can, in principle, be mobilized by a transposase encoded by another copy of a similar IS element, which recognizes the ITR. They occur in nine identical copies in *S. tokodaii* chromosomes (Kawarabayasi et al., 2001) and in multiple copies in *Aeropyrum pernix* and

Table 8.1 Summary of the IS elements in *Sulfolobus* genomes.

Organism	IS elements families (n.a.)	Total no.	Families
S. acidocaldarius	4 (2)	4	IS110, IS605
S. solfataricus	25 (3)	201	IS4, IS5, IS110, IS256, IS605, IS630/Tcl, ISL3
S. tokodaii	12	34	IS4, IS110, IS605
S. islandicus	20 (3)	116	IS1, IS4, IS5, IS110, IS256, IS605, IS630/Tcl, ISL3

n.a., not assigned.

Methanocaldococcus jannaschii (Brügger et al., 2002, 2004). Given that they invariably coexist in the chromosome with the IS element encoding the mobilizing transposase, the type I MITEs have probably arisen within the organism as a result of a deletion prior to spreading.

Until their discovery in *S. solfataricus* (Redder et al., 2001), type II MITEs had only been observed in eukaryal chromosomes and primarily in those of higher eukaryotes. They differ from type I MITEs in that they bear no sequence resemblance to the IS elements apart from their ITRs, and the origin of their intervening sequences is unknown. However, sharing an ITR is considered sufficient for the IS element-encoded transposase to mobilize a MITE. Type II MITEs are common in the chromosomes of *S. solfataricus*, *S. tokodaii*, and *S. islandicus* and a summary of their properties is given in Table 8.2. They fall in the size range 79–335 bp and have been classified into seven families, labeled *Sulfolobus* MITE, SM1 to SM7, one of which, SM3, is divided into two closely related sequence subgroups, SM3A and SM3B. Two of the families, SM1 and SM2, are shared between *S. solfataricus* and *S. tokodaii* and one family, SM3A, is shared between *S. solfataricus* and *S. islandicus*. The IS element with the corresponding transposase is generally present in the chromosome (Table 8.2). Most MITEs lie in intergenic regions or within IS elements. They are rich in stop codons and only three were found in putative protein coding regions that do not disrupt the reading frame (Redder et al., 2001). No MITE-like sequences were detected in the *S. acidocaldarius* genome, which is consistent with its paucity of IS elements (Chen et al., 2005b).

Despite searches for type II MITE-like elements in other archaeal genomes that carry IS elements, no clear examples have been discovered. However, recently, 141 repeat elements were found in the genome of *M. jannaschii* that resemble *Sulfolobus* type II MITEs in size (87–125 bp) and are also predominantly located in intergenic regions and distributed fairly evenly around the genome (Suyama et al., 2005). They fall into three main classes showing about 54–65% sequence identity within each class and one infers, therefore, that they have spread in the chromosome. However, although they exhibit ITRs, no corresponding IS elements are present in the genome that could, via an encoded transposase, facilitate their transposition. They differ from *Sulfolobus* MITEs in that they can generate stable hairpin structures (Suyama et al., 2005).

Mobility and regulation

Several experimental studies have demonstrated the mobility of IS elements in *Sulfolobus* strains and that most spontaneous mutations in *S. solfataricus* involve

Table 8.2 Properties of *Sulfolobus* MITEs.

Organism	MITE	Type	Length (bp)	Copy number	IR (bp)	DR (bp)	Partner IS element	Reference
S. solfataricus	SM1	II	79	40	23	2	ISC1048	Redder et al., 2001
	SM2	II	184	25	16	6	ISC1217	Redder et al., 2001
	SM3A/SM3B	II	131	44	24	9	ISC1058	Redder et al., 2001
	SM4	II	164	34	27	8	ISC1173	Redder et al., 2001
	SM7	II	330	10	13	0	ISC1173	M. E. White & P. Redder unpub.
S. tokodaii	SM1	II	79	1	23	2	ISC1048*	Redder et al., 2001
	SM2	II	184	36	16	6	ISC1217	Redder et al., 2001
	SM5	II	212	7	15	0	ISC774	Brügger et al., 2004
	SM6	II	127	8	12	0–10	ISC794	Brügger et al., 2004
	SMA	I	355	9	4	0	ISC1790	Kawarabayashi et al., 2001
S. islandicus	SM3A	II	131	8	24	9	ISC1058	Boult & Grogan, 2005; K. Brügger, unpublished

*The corresponding IS element is absent from the host.

the transposition of IS elements into genes or into their regulatory regions. Mutational studies have demonstrated that several different IS elements could insert into the contiguous *pyrE/pyrF* genes in the closely related *S. solfataricus* P1 and P2 strains, and in an *S. islandicus* strain (Martusewitsch et al., 2000; Boult & Grogan, 2005; Redder & Garrett, 2006).

The evidence for MITE mobility is more limited. When first discovered in *Sulfolobus* chromosomes, they were considered to be mobile because: (i) they occur in multiple copies with closely similar sequences; (ii) three of the families SM1, SM2, and SM3A occur in multiple copies in more than one *Sulfolobus* species; and (iii) they are often found inserted into IS elements (Redder et al., 2001). However, until recently, experimental evidence for their mobility was lacking, possibly because their small size precluded their being detected in the type of gel electrophresis assay used for following IS element transposition. Recently, using a more sensitive assay, direct evidence was obtained for the transposition into *pyrE/pyrF* genes of an SM3A in an *S. islandicus* strain, and of an SM2 in *S. solfataricus* P2 (Boult & Grogan, 2005; Redder & Garrett, 2006). In both studies, it was concluded that the frequency of MITE transposition was much lower than for IS elements.

One of the major puzzles relating to the high incidence of mobile elements in some *Sulfolobus* chromosomes is that their prevalence is not more destructive for the cell. Both *S. solfataricus* and its close relative *S. islandicus* occur widely in natural environments, in sharp contrast to *S. acidocaldarius*, suggesting that the enhanced capacity for rearrangement and change is a distinct advantage despite the obvious danger of inactivating important genes. Another apparent disadvantage is that *S. solfataricus* must have great difficulty in removing these elements, as evidenced by the high incidence of IS element fragments within its chromosome (Brügger et al., 2002). Indeed, it has been suggested, on the basis of transposition-reversion experiments, that *Sulfolobus* species may lack a mechanism for the precise excision of IS elements (Boult & Grogan, 2005).

One mitigating factor is that mobile elements can inhibit one another and there are many examples of MITEs, and even IS elements, inserting into the transposase genes of IS elements (Brügger et al., 2002). Another more potent regulatory mechanism may be the production of antisense RNAs that target the transposase mRNAs. Several different putative antisense RNAs have been isolated from *S. solfataricus* cells that are specific for different IS elements (Tang et al., 2005). Any one of these can, in principle, block the translation of mRNAs from a whole family of transposases.

Exchange of chromosomal DNA and self-transmissible plasmids

S. acidocaldarius cells are able to undergo exchange of chromosomal DNA. This was first shown for an archaeal 23S rRNA intron from *Desulfurococcus mobilis* that encodes a homing enzyme. It recombined into the single 23S rRNA gene of *S. acidocaldarius* after transformation on a non-replicating phage construct. Furthermore, on culturing the genetically modified cells, and using antibiotic-resistant mutants of *S. acidocaldarius*, it was demonstrated that intercellular transfer occurred, throughout the cell culture, with the intron always recombining at the same site in the 23S rRNA gene (Aagaard et al., 1995). This process did not occur in either *S. solfataricus* or *S. shibatae*. It was also demonstrated that a series of marker genes were able to exchange, in both directions, between chromosomes of different cells and again this phenomenon was only observed in *S. acidocaldarius* (Grogan, 1996; see Chapter 12 in this volume). Subsequent sequencing and analysis of the chromosomal sequence of *S. acidocaldarius* revealed a group of contiguous genes that are homologous, and co-linear, with those found in several conjugative plasmids isolated from different *Sulfolobus* species. It was speculated that these could provide a possible explanation for the intercellular DNA transfer (Garrett et al., 2004; Chen et al., 2005b).

Archaeal self-transmissible plasmids have, to date, only been found in diverse strains of the *Sulfolobus* genus where they occur in about 3% of isolated strains (Prangishvili et al., 1998b). Seven have been sequenced and their genes annotated, pNOB8 from Japan and pING1, pKEF9, pHVE14, pARN3, pARN4, and pSOG from different locations in Iceland. Although very few genes yield significant sequence matches with public sequence databases, all the plasmids share a few conserved regions which encode: (i) the putative major components of the conjugation apparatus; (ii) a putative replication origin; and (iii) an operon of six to nine smaller genes implicated in plasmid replication and its regulation. All the plasmids encode an integrase homolog suggesting they are integrative (Greve et al., 2004) and this has been demonstrated experimentally for pNOB8 (She et al., 2004).

The genome of *S. acidocaldarius* carries most of the conserved region implicated in the conjugative apparatus, and an alignment of the corresponding regions of the *S. acidocaldarius* chromosome and the conjugative plasmid pARN4 is illustrated in Fig. 8.1, where homologous genes are shaded similarly. One ORF shows significant sequence similarity to sections of the bacterial TrbE protein around the ATP-binding Walker A and B and other motifs (She et al., 1998; Stedman et al., 2000) and it may be co-functional in providing energy to facilitate DNA transfer across the cellular membrane, possibly in a double-stranded form. Moreover, ORF600, which carries the 10–12 transmembrane helical motifs, may participate in membrane pore formation (Schleper et al., 1995; Greve et al., 2004).

Introns and inteins

Archaea do not generally carry group I and group II introns, which are common to Bacteria and Eukarya, or the spliceosomal introns found in Eukarya (Lykke-Andersen et al., 1997a). Exceptions are the group II introns found in the large genomes of *Methanosarcina acetovorans* and *Methanosarcina mazei*, which may have been incorporated by lateral gene transfer from bacteria, together with large numbers of other bacterial genes (Rest & Mindell, 2003). However, archaea do carry a class of introns that are ubiquitous in, and exclusive to, archaea (Lykke-Andersen et al., 1997a).

Archaeal introns occur at several different sites in rRNA and tRNA genes, as well as in protein genes. Their excision from transcripts is a prerequisite for producing either a functional RNA (tRNA or mRNA), or a functional site within an RNA (rRNA). The introns are excised by an RNA cleavage enzyme that exists in different forms, generally a homodimer or homotetramer in Euryarchaea, and a heterotetramer in *Sulfolobus* and other Crenarchaea, and it constitutes a simpler form of the more structurally specific eukaryal tRNA splicing enzyme (Kleman-Leyer et al., 1997; Lykke-Andersen et al., 1997b; Tocchini-Valentini et al., 2005).

The archaeal cleavage enzyme appears to have a combined intron splicing and RNA processing function (Kjems et al., 1989; Tang et al., 2002b). It cuts at a "bulge–helix–bulge" structural motif (BHB) symmetrically within two 3-nucleotide bulges. The BHB motif forms at the exon–intron junction, or processing site, and is generally bordered by additional base pairs. Subsequently, the ends of the exons, or processed RNAs, are ligated, by an unknown mechanism, and the larger introns, at least, are circularized, generating a highly ordered structure that is stably maintained within the cell (Kjems & Garrett, 1988; Dalgaard & Garrett, 1992; Lykke-Andersen & Garrett, 1994) and they encode a homing enzyme that can effect site-specific insertion of the DNA intron into a homologous intron-minus site (Aagaard et al., 1995).

In contrast to many rRNA introns, tRNA introns are quite small, lying in the size range 12–104 bp, and there is no direct evidence for their being mobile (Marck & Grosjean, 2003). Thus, no tRNA intron has yet been detected in Archaea that encodes a homing enzyme, considered a prerequisite for insertion at the DNA level. Further doubts about their mobility were raised by a study that demonstrated that some tRNA genes are partitioned in the chromosome of *Nanoarchaeum equitans* and that the two transcripts are linked via G+C-rich extensions that can base-pair and generate a BHB motif. It was implied that this might be an ancient mechanism for tRNA expression

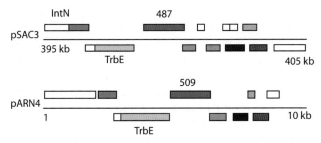

Fig. 8.1 A comparison of the gene content and order of a region of the *S. acidocaldarius* chromosome (corresponding to integrative element pSAC3; see Chapter 10) that carries a closely similar set of genes to a region of the *Sulfolobus* conjugative plasmid pARN4. Homologs are shaded similarly. These genes are considered to encode the proteins that are necessary for forming the conjugative apparatus. ORF509/487 is rich in transmembrane helical motifs and is a candidate for transmembrane pore formation.

and, by extension, that tRNA introns are relics that have not been deleted from whole tRNA genes (Randau et al., 2005a). However, an analysis of the tRNA genes in the *Sulfolobus* chromosomes, with and without introns, indicates a very low level of conservation of the intron-containing tRNAs. The circle plot in Fig. 8.2 reveals that of the 57 intron-containing genes in three *Sulfolobus* chromosomes, most are species-specific, seven are shared between pairs of species, and only four are conserved in all three. This is strongly suggestive that intron insertion into tRNA genes does occur, albeit by an unknown mechanism.

The single known archaeal intron within a protein coding region was detected in a homologous gene of the Cbf5p protein (centromer binding factor 5 in Eukarya) for the crenarchaea *S. solfataricus*, *S. tokodaii*, and *A. pernix* (Watanabe et al., 2002). Each gene exhibited a frame-shift that was removed when the intron was excised from the mRNA, as illustrated in Fig. 8.3(b). This observation raised the question as to whether mRNA introns are more prevalent in Archaea. To address this, all genes of *S. solfataricus*, *S. tokodaii*, and *S. acidocaldarius* were grouped into families based on homology. Then insertions within conserved sequence regions were tested for their capacity to form BHB motifs. For 1856 gene families, about 70 potential insertion sites were found, nine of which could generate stable BHB motifs, suggesting that some inserts could constitute active introns (K. Brügger, unpublished results). Some inserts lie within single gene copies, while others are located in single copies of paralogous genes. An example of the latter is found in the gene of a Rad3-related DNA helicase and the putative BHB motif and protein sequence alignment for the two paralogs of *S. tokodaii* are illustrated in Fig. 8.3(a, b). These indicate that many insertions have occurred within conserved regions of *Sulfolobus* genes and suggest that intron insertion could be one of the mechanisms.

To date, no inteins have been detected in *Sulfolobus* gene products even though they occur, together with their corresponding homing enzyme, in some euryarchaeal polymerases (Belfort et al., 1995).

SRSR clusters

Another possible mechanism for change involves the clusters of short regularly spaced direct repeats (SRSRs, also known as CRISPR). These occur in all except one of the sequenced archaeal chromosomes, in some bacterial chromosomes, and in some *Sulfolobus* conjugative plasmids. The repeat sequences and sizes of the *Sulfolobus* clusters are summarized in Table 8.3. Repeat sequences (24–26 bp) are highly conserved, but not always invariant, within a cluster, and differ between clusters, and they generally exhibit a weak dyad sequence symmetry (Table 8.3). In contrast, the spacers (39–42 bp) show no sequence similarity. Archaeal chromosomes carry multiple clusters, and they can be large, with over 100 repeat units. For example, the 455 repeat units in *S. tokodaii* are distributed between six clusters that constitute

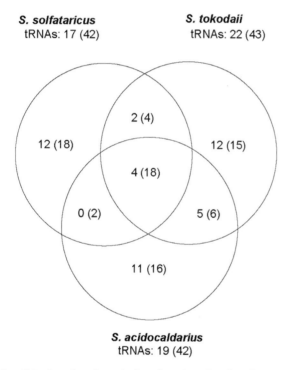

Fig. 8.2 Overlapping circles showing the distributions of similar introns among intron-containing tRNA genes within the three *Sulfolobus* chromosomes. The main number indicates the number of intron-containing tRNA genes, with similar introns, which are (a) exclusive to one species, (b) shared with one or other of the other two species, or (c) shared between all three species. The bracketed number indicates the total number of tRNA genes with or without introns which are specific for, or shared between, the species. Total numbers for each species are given below the species name.

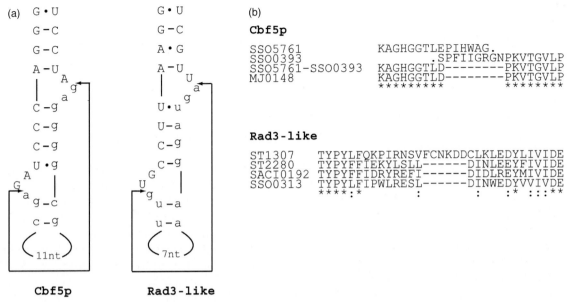

Fig. 8.3 mRNA exon-intron splicing junctions. (a) BHB motif that forms within the transcript of Cbf5p, and an example of a BHB motif that might form in a transcript of a Rad3-like gene. (b) Aligned amino acid sequences across the exon-intron junctions showing the conservation of the exon regions in homologous proteins that lack the intron/insert. Splicing has only been demonstrated experimentally for the Cbf5p transcript (Watanabe et al. 2002). ORF sequences derive from: SSO, *S. solfataricus*; ST, *S. tokodaii*; SAC, *S. acidocaldarius*; and MJ, *M. jannaschii*.

Table 8.3 Properties of the SRSR clusters.

Organism/plasmid	Repeat sequence	Repeats/cluster
S. acidocaldarius	GTTTTAGTTTCTTGTCGTTATTAC	133, 78
	CTTTC**AATCCC**TTWT**GGGATT**CWTC	11, 4
S. solfataricus	CTTTC**AATTCC**TTTT**RGGATT**AATC	103, 48, 46
	CTTTC**AATTCT**ATARK**AGATT**ATC	95, 91, 32, 7
S. tokodaii	CTTTC**AATTCC**TTTTK**GGATT**CATC	73, 64, 47
	CTTTC**AATTCC**ATTAW**GGATT**AKC	120, 104, 47
S. islandicus	CTTTCAATT**CTATAG**TAGATTAGC	>180*
pNOB8	CTTTCAAATT**CTATAG**TAGATTATC	6
pKEF9	CCC**GCACATT**TAGGG**AATTGC**AAC	6

For the repeat sequences, the low level dyad symmetry is shown in bold type when present. K denotes G or T; R denotes A or G; W denotes A or T. The total number of repeats per cluster is also given. *The value given for *S. islandicus* is based on an incomplete genome sequence.

over 1% of the chromosome. A conserved *Sulfolobus* protein has been characterized that specifically binds to, and distorts, a repeat sequence of the *S. solfataricus* chromosome and that of pNOB8 (Peng et al., 2003). It was thought to have a structural role in folding the SRSR clusters but a role in regulating transcription from the clusters is also a possiblity (see below).

Two hypotheses have been proposed for the function of the clusters. One is based on an experiment in which a repeat cluster was inserted into a bacterial–archaeal shuttle vector and transformed into *Haloferax volcanii*. The vector recombined into the host chromosome and produced a decrease in both cellular growth rate and the fraction of viable cells, as well as a difference in the distribution of DNA in daughter cells (Mojica et al., 1995), and it was inferred that the repeat clusters influenced chromosomal segregation. A segregation role is also consistent with the repeat clusters being replicated last in *S. acidocaldarius* and in the euryarchaeon *Pyrococcus* (Zivanovic et al., 2002; Lundgren et al., 2004). There is also circumstantial support for the clusters in pNOB8 facilitating plasmid partitioning (She et al., 1998; Greve et al., 2004).

Another hypothesis has arisen, which addresses the differences in the number and sizes of the clusters, and does not necessarily preclude the former hypothesis. It is based primarily on the observation that some spacer sequences are similar to those found in plasmids and viruses. Thus, spacers in *S. solfataricus* were shown to closely match gene sequences in the conjugative plasmid pNOB8 and the rudivirus SIRV1, while spacers in *S. tokodaii* gave good matches with genes from conjugative plasmids, SIRV1, and the fusellovirus SSV1. The hypothesis proposes that DNA has recombined from the extrachromosomal elements and entered the SRSR cluster as a spacer, thereafter to provide a defense against further invasion of related elements (Mojica et al., 2005). The hypothesis is fueled by two lines of evidence: (i) the observation that a group of genes is often located near to the clusters that were classified as novel repair-recombination enzymes and could be involved in changing the size of the clusters (Makarova et al., 2002, 2006; Jansen et al., 2002); (ii) the clusters produce untranslated RNA transcripts that could act as small antisense RNAs, thereby inhibiting translation of important plasmid- or virus-encoded proteins (Tang et al., 2002b, 2005).

In principle, the SRSR clusters can be multifunctional but many questions remain to be answered. For example, why are there such large variations in the numbers of repeat units per species? What is the function of the spacer regions found in the *Sulfolobus* conjugative plasmids. Why is *S. islandicus* HVE10/4 a good host for SIRV1 despite carrying spacer sequences in its SRSR clusters that match some of the viral genes?

Plasmid variants

Sulfolobus species are hosts for many self-transmissable and cryptic plasmids and some of these can undergo genomic variation. Among the seven characterized conjugative plasmids, pNOB8 and pING1 were found to rearrange in *Sulfolobus* hosts. The former underwent a major deletion (yielding pNOB8-33), resulting from recombination at a large direct repeat, after transferral from its natural host to a different *Sulfolobus* strain (She et al., 1998). The pING plasmid yielded a complex mixture of different variants when plasmid DNA from *S. islandicus* HEN2P2 was propagated in *S. solfataricus* P1 (Prangishvili et al., 1998a; Stedman et al., 2000). Some of these were shown to arise from the uptake of IS elements from the host chromosome, but one of the smaller products, pING2, was produced by recombination at two sites carrying the 16 bp "hairpin" motif, TAAACTGGGGAGTTTA. This motif is present, in 10–16 copies, in each of the sequenced conjugative plasmids (Greve et al., 2004). Interestingly, similar "hairpin" motifs also occur in the large family of cryptic pRN plasmids where they border a variable region (Peng et al., 2000) and it is likely, therefore, that these motifs provide general sites for DNA recombination and rearrangement in *Sulfolobus* plasmids. They are also retained in conjugative and pRN-type plasmids encaptured in *Sulfolobus* chromosomes, where they could also contribute to chromosomal rearrangements (Peng et al., 2000; Greve et al., 2004).

Viral genome variation for *Sulfolobus* and other Crenarchaea

Many archaeal viral genomes have now been sequenced and analyzed. While some appear to be quite stable, undergoing few, if any, changes, even when propagated in a foreign host, others show differing degrees of genomic variability. Two that undergo major changes of a complex, and quite different, nature are the *Sulfolobus* rudivirus SIRV1 and the lipothrixvirus TTV1 from the crenarchaeon *Thermoproteus tenax*, and they are considered separately below.

SIRV1

The genome of the rudivirus SIRV1 was estimated to undergo changes at a much higher rate than previously estimated for a DNA genome (~10^{-3} substitutions per nucleotide per replication cycle) (Prangishvili et al., 1999). A detailed analysis revealed that the isolated virus invariably contained a population of variants with altered genomes. Upon propagation in a given host strain, one or more of the variant genomes was shown to dominate in the viral population, while passage into a new host strain yielded alternative dominant variants (Peng et al., 2004).

Sequence analysis of the variant genomes revealed that they exhibit longer conserved regions interspersed with shorter, highly variable regions that have undergone changes including deletions, duplications, and transposition. The approximate positions of these variable regions, labeled A to F, are superimposed on the larger SIRV2 genome, which shares genomic characteristics with all the SIRV1 variants (Plate 8.3a). One change of exceptional character was the common appearance of 12 bp sequence differences mainly within the variable regions (Plate 8.3a). Some of these were detected during the sequencing step when overlapping clones revealed 12 bp sequence differences at certain positions, as illustrated in Table 8.4 for SIRV1 and the closely related rudivirus SARV1. Several others were revealed by comparing conserved sequence regions of genes common to the SIRV1 variants or the closely related SIRV2. Almost every sequence difference detected was of 12 bp or multiples thereof (Peng et al., 2004). Many of these sequence differences occur within ORFs and they do not interrupt the reading frame. The ratio of clones with extra sequences to those lacking the sequence (Table 8.1) suggests that at least some differences are caused by insertions rather than deletions, and these will have the effect of extending the ORFs, and possibly altering the functions of the gene products (Peng et al., 2004).

Although the 12 bp insertions/deletions are common in the SIRV1 genome, they also appear to be a general property of rudiviruses SIRV2 and SARV1 (Peng et al., 2004; Garrett & Vestergaard, unpublished). However, the mechanism by which they are generated remains unclear. Originally, it was proposed that they constitute mobile introns, supported by the observation that some of their transcripts can generate stable BHB motifs (Peng et al., 2004). However, to date, no experimental evidence has been accrued for transcript splicing. Another possibility is that the changes occur during viral genome replication. Although exceptional, one of the 12 bp elements, underlined in Table 8.4, could be a duplication of the adjacent sequence, and one 24 bp insert in SIRV1 constitutes a perfect 12 bp duplication. Possibly, there is a dynamic interplay between the single 16 kDa DNA-binding protein of the SIRV virus and

Table 8.4   12 bp elements detected in overlapping rudiviral clones.

Genome	Sequence	Clone numbers
SIRV1 pop. I/variant VIII	GTTAAAAATCTGTTTG**TTTAGCAGTTCA**ATTTGTTTCAGCA	10
	GTTAAGAATCGTTTG::::::::::::ATTTGTTTCAGCC	29
SIRV1 pop. I	CTAGATTCTTAGAATG**AACATTCATTAA**ATATAAATTTGTC	7
	CTAGATTTTTAGAATG::::::::::::ATATAAATTTGTC	3
SIRV1 pop. I	CCATCATTAAATTATA**ATACAAATTTCA**AATTACCATCCA	1
	TCTTCTATTAATTATA::::::::::::AATTACCGCCTG	4
SARV1	AGA<u>AAATTAAATTATG</u>**AATTAAATTATG**ATAAGTTAAGATA	5
	AGAAAATTAAATTATG::::::::::::GTAAGTTAAGATA	1
SARV1	ACAACTAGAAATTTTG**AATTATGTTAAG**AAAAACCAGCCAA	8
	ACAACTAGAAATTAAG::::::::::::AAAAACCAGCCAA	8

The 12 bp sequence (bold) and flanking sequences are shown. The underlined sequence represents a perfect direct repeat. The clone library for SIRV1 population I (pop. I) contains a mixture of variant genomes. The SIRV1 virus data derive from Peng et al. (2004) and those of SARV1 are from Garrett and Vestergaard (unpublished).

the DNA polymerase during replication (Prangishvili et al., 1999). Clearly, resolving the mechanism requires further experimentation.

TTV1

Variants of the lipothrixvirus TTV1 were isolated from colonies of *Thermoproteus tenax* and showed marked differences in genomic size (Neumann & Zillig, 1990). Genome sequencing revealed that TTV1 exhibits two sections, I and II, of low complexity sequence, which extend over 250 bp and 438 bp, respectively (Plate 8.3b). Section I consists predominantly of the repeated hexamer CCNACN (where N is any nucleotide) and it forms part of an open reading frame, ORFTPX. TPX is expressed in large amounts within the *Thermoproteus tenax* cell and it may have an important role in combating the cell's defenses. Section II consists of 73 consecutive copies of CCNACN and shows no apparent protein coding capacity.

The viral variants showed evidence of a total of 6 insertion/deletions (30–102 bp) located within these sections (Neumann & Zillig, 1990). Sequence comparison of ORFTPX (section I) in the original TTV1 isolate (WT) and in variant VT3 revealed five insertion/deletions (66 to 84 bp), all of which maintain the reading frame, and show 100% sequence identity to blocks in section II, which are matched by number and color in Plate 8.3(b). This is significant because the sequence identity between differently numbered/colored blocks in sections I and/or II is only 75–81% (Plate 8.3b). Moreover, section II is 100% identical between WT and variant VT3, except that block 8, $(CCNACN)_{17}$, is absent from VT3 (Plate 8.3b).

These observations, together with (i) the finding that blocks inserted into ORFTPX are identical in sequence to blocks in section II, and (ii) the maintenance of a potential pattern of silent codon mutation at positions 3 and 6 of CCNACN in section II, strongly suggest that this region serves as a source of variation for ORFTPX.

Conclusion

A picture has emerged of an extraordinary level of genomic change and gene flux in the *Sulfolobus* genus. This occurs by a variety of different mechanisms, including extensive rearrangements fueled by mobile elements, exchange of genes between chromsomes, plasmids and viruses, the bidirectional and intercellular transfer of chromosomal genes in *S. acidocaldarius*, and many different integrative processes. The complex genomic rearrangements observed in the pING conjugative plasmid, and in the rudivirus SIRV1, may reflect attempts to adapt to foreign hosts, while the changes seen in TTV1 may represent a direct mechanism for weakening host defenses. All these examples illustrate that the genome sequences have provided a platform for studying, in detail, the complex molecular processes involved in genomic rearrangements in Archaea, and their functions, for the first time, and we can expect much progress in understanding the processes underlying genome evolution over the next few years.

Acknowledgments

We thank Peter Redder, Bo Greve, Reidun Lillestøl, and Gisle Vestergaard, and our colleagues involved in genome sequencing projects, for fruitful discussions. A stimulating, productive, and generous collaboration with Wolfram Zillig, over many years, is remembered with pleasure.

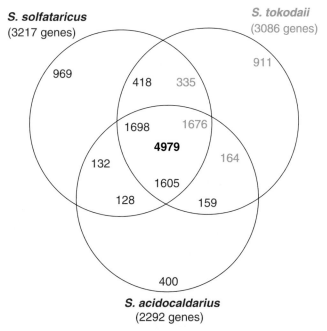

Plate 8.1 Conservation of predicted protein genes. The overlapping circle plot shows the number of genes exclusive to one species, and the number of homologous genes shared between pairs of species or between all three. *S. solfataricus*, red numbers; *S. tokodaii*, green numbers; *S. acidocaldarius*, blue numbers.

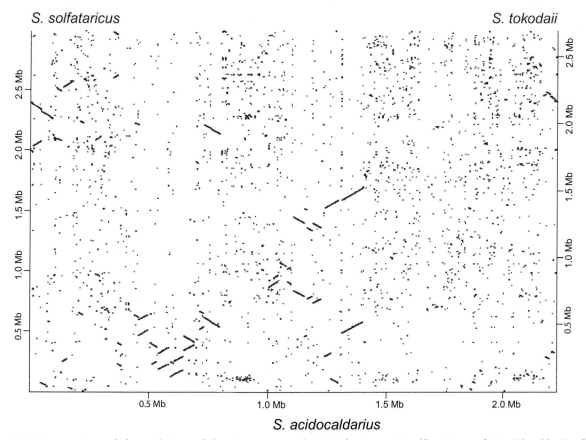

Plate 8.2 Comparison of the ordering of the chromosomal genes between *S. solfataricus* and *S. acidocaldarius* (blue dots), and *S. tokodaii* and *S. acidocaldarius* (red dots). Homologous genes are plotted on the basis of their chromosomal positions such that regions with a conserved gene order yield diagonal lines. Each dot represents a single gene.

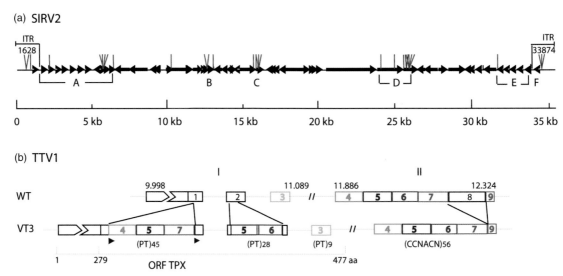

Plate 8.3 A summary of the conserved and variable regions within the genome variants of crenarchaeal viruses. (a) *Sulfolobus* rudivirus SIRV1 where the gene map of the closely related SIRV2 is taken as a reference for the closely related variant genomes and ORFs are represented by arrows. Distributions of both the variable regions (labeled A to F) and the 12 bp elements are illustrated. 12 bp elements occurring only in the SIRV1 variants are denoted by vertical blue lines and those exclusive to SIRV2 are indicated by vertical red lines (Peng et al., 2004). Approximate positions of the ITRs are given. (b) *Thermoproteus tenax* lipothrixvirus TTV1. A scheme is presented for genomic sections I and II from the original isolate (WT) and variant VT3, which exhibit low sequence complexity. Repetitive CCNACN sequences are represented by rectangle blocks and those of identical sequence are assigned the same numbers and colors. Observed insertion/deletions between WT and VT3 are indicated by lines. A 17 bp perfect direct repeat, denoted by arrow heads, occurs at the beginning of block 4, and in the partitioned block 1 in section I of VT3. The numbers of CCNACN repeats and proline-threonine repeats $(PT)_n$ are shown for VT3.

9
Plasmids

Georg Lipps

Introduction

The research on archaeal plasmids is still in its infancy and our knowledge of archaeal plasmids is rudimentary. However, during the past few years significant progress has been made (reviewed by Garrett et al., 2005). Our lack of knowledge is highlighted by the fact that for many archaeal plasmids we do not even know how the plasmid is replicated. For only two plasmids (pGT5 from *Pyrococcus abyssi* and pRN1 from *Sulfolobus islandicus*) have the replication enzymes been characterized biochemically. This review is focused on archaeal plasmids whose genome sequence has been determined. In many cases the plasmidal nucleotide sequence is most if not all of the information we have on a particular plasmid. Clearly archaeal plasmids have not yet entered the "post-genomic" era.

Extrachromosomal elements are instrumental for the development of genetic tools, and early studies aimed to identify plasmids and viruses in order to study various aspects of archaeal genetics and to evaluate their potential to make the newly discovered organisms of the third domain of life genetically tractable. Extrachromosomal elements of the Archaea are not uncommon. Zillig and co-workers screened Sulfolobales in Iceland and discovered four plasmids as well as several viruses in 120 strains (Zillig et al., 1994). Some of the genetic elements discovered at that time are now among the best studied genetic elements, e.g. the plasmid pRN1 and the virus SIRV1 (see also Chapter 6 in this volume). In a further study 11 conjugative plasmids were isolated from about 300 *S. islandicus* strains (Prangishvili et al., 1998a). Plasmids are also abundant in the Euryarchaeota: 40% of 190 screened Thermococcales strains contained plasmids and about 25 different types of plasmids were identified based on restriction analysis (Prieur et al., 2004). Despite their natural abundance the current genome database at the NCBI lists only 38 archaeal plasmids, in contrast to 640 bacterial plasmids.

Several halophilic archaea (*Halobacterium* sp., *Haloferax volcanii*, and *Haloarcula marismortui*) have a complex genome structure with a number of circular double-stranded replicons (Table 9.1). The largest replicon is usually considered as the main chromosome. The medium-sized replicons can be regarded as either minichromosomes or megaplasmids. If a replicon replicates like the main chromosome and contains essential genes then it should be considered as a minichromosome regardless of its size. However, without experimental data a decision is provisional. For example, the mere presence of a plasmid-type replication protein on the replicon is no proof that the replicon is replicated like a plasmid. The fact that a number of archaeal plasmids also exist integrated into the host chromosomes (She et al., 2001a) further complicates the assignment as minichromosomes or megaplasmids. Further criteria for deciding whether the replicon is a minichromosome or a megaplasmid are the copy number and the evolutionary history of the replicon (Ng et al., 1998).

Table 9.1 Haloarchaeal genomes with several replicons.

Haloarcula marismortui	Baliga et al., 2004
chromosome I	3.1 Mbp
chromosome II	290 kbp
pNG100/200/300/400/500/600/700	33–410 kbp
Halobacterium sp. NRC-1	Ng et al., 2000
chromosome	2 Mbp
pNRC100	191 kbp
pNRC200	365 kbp
Halobacterium sp. R1	unpublished
chromosome	2 Mbp
pHS1	148 kbp
pHS2	195 kbp
pHS3	284 kbp
pHS4	41 kbp
Halobacterium sp. GRB	Ebert et al., 1984; St Jean et al., 1994
chromosome	2 Mbp
pGRB305	305 kbp
pGRB90	90 kbp
pGRB37	37 kbp
pGRB1	1.8 kbp
Haloferax volcanii DS2	genome sequencing in progress
chromosome	~2.8 Mbp
pHV4	690 kbp
pHV3	442 kbp
pHV1	86 kbp
pHV2	6 kbp

Archaeal plasmids with rolling circle replication

Rolling circle replication is initiated by an initiator protein, which cuts the plasmid specifically at the double-stranded replication origin (*dso*) liberating a 3′ OH end that is then used as primer by the host replication machinery. The replication of the opposite strand is initiated by a different mechanism at the single-stranded origin (*sso*). Here an RNA primer is synthesized by the host RNA polymerase and then extended by the host DNA polymerase (reviewed in Espinosa et al., 1995; Khan, 1997). The rolling circle plasmid can have a very compact genome, since the only plasmidal protein required for plasmid replication is the initiator protein. The smallest known archaeal plasmid is the plasmid pHSB2 from *Halobacterium salinarum* (1736 bp). This plasmid is similar in size and sequence to the plasmids pHGN1 and pGRB1 (Ebert et al., 1984). These plasmids are probably replicated as rolling circles, since single-stranded replication intermediates have been detected (Akhmanova et al., 1993) and since their putative replication proteins have three motifs typical for rolling circle initiator proteins (Ilyina & Koonin, 1992). Among the archaeal plasmids at least three types of initiator proteins can be identified (Fig. 9.1). The first and largest group contains the initiator proteins from plasmids isolated from halophiles. These proteins are similar and have two catalytic tyrosine residues in motif III. In contrast the initiator proteins from the *Pyrococcus* plasmids pGT5 and pRT1 belong to another protein superfamily and have only one tyrosine residue in motif III. The latter two proteins are only distantly related. In fact the motif I could not be detected within the amino acid sequence of the replication protein from pRT1 (Fig. 9.1).

Amid these initiator proteins only the replication protein from plasmid pGT5 has been studied *in vitro*. The 75 kDa protein (Rep75) has been expressed heterologously in *E. coli*, refolded and purified. The recombinant protein has endonuclease activity on single-stranded oligonucleotides containing the predicted double-stranded origin sequence. The protein nicks the single-stranded substrate between two specific bases and remains covalently bound to the cleaved DNA via the liberated 5′ end, whereas the host DNA polymerase can extend the new 3′ end. The covalently attached DNA can then be ligated back to another strand, the so-called closing activity (Marsin & Forterre, 1998). This activity is required for the termination of a replication round. The active site tyrosine residue of this protein was identified by a point mutational study (Marsin & Forterre, 1999). The protein also exhibits an unexpected nucleotidyl transferase activity, which could have a regulatory function by inactivating the initiator protein (Marsin & Forterre, 1999).

For pGT5 (Lucas et al., 2002), as well as for several of the halophilic plasmids, shuttle vectors, which replicate in the original host as well as in *E. coli*, have been constructed. In particular, the halophilic shuttle vectors (reviewed in Luo & Wasserfallen, 2001; Allers & Mevarech, 2005) have been extensively used as genetic tools for the study of Archaea (see also Chapter 11 in this volume).

```
                                   motif I
pHGN1      84 ER----GLREEWGDLL------HTAMVTLTASTTEEDGGP-------RPLVDHL 120
pHSB2      79 ER----GLRERWGKLL------HTSMVTLAASSTDEDGRP-------RPPLEHL 115
pZMX201   142 ER----GVSEEFGKRL------HTAMLTFTASSRP-NGQP-------IPPVDHL 177
pNB101    127 ER----GI-ERAYQAP------TMVMVTLSASSENAKGGR-------RCPADHM 162
pHK2       71 QRQMGGGERPSGGEALAAWGSPATAMLTFTASSVPNGERLAAGRTHRRAPRRVF 124
pGRB1      79 ER----GLRERWGSLL------HTGMVTLTASSTDDEGRL-------RPPLEHF 115
Cons      163 :*    *               *:*::**:                :       216

pGT5      268 LLEGSLVSYRSGGVETIHHLIPVRHFVLTAPKELSFSIWA---SLKKGDSSLFR 318
pRT1      223 KW-----GDRKTGVENKHPEAEGRQWATWVFR--LSNIHLSEFRADLKLYR--- 266

                                                           motif II
pHGN1     121 RDLLSSWSAVYDALRHTLEDREFEYLAIIEPTTPAGNG---PAGYAHIHLGVFV 171
pHSB2     116 EDLLSSWEAVRRALYRVLDGREWEYLAILEPHE--------SGYVHIHLGVFV 160
pZMX201   178 DELLASWDALTTALDRVLGDRRYARLGILEPH-PGDGV---NNGYLHIHVAVFI 227
pNB101    163 RDIARGWNSARKALHRVLRRFEWEYAKVWEPHQ--------SGYGHMHVAVAV 207
pHK2      125 VRRRARHSAEHDGVPSGLDADEWGYWLQAEPHGMGGDGSGMNACYSHLHVGVYF 178
pGRB1     116 EDLLESWEAVRRALARVLEGREWEYLAILEPHE--------SGYVHIHLGVFV 160
Cons      217        .:    .:    *    .:       **       *:*:*:.* .   270

pGT5      319 AFKDAGAKAIKEFLSYLASKEHISGNLLFGFTINVHVTGDKNPFEPHFHIDAIV 372
pRT1      267 ---------KKKYNYPA---------------IDLSDHPKLEVVHYHVDYSP 294

pHGN1     172 KGP-----VVAEQFQDVLDAHVKNSEGAGREAHRAVVEDDEDEAAVSIRRSA-- 218
pHSB2     161 KGP-----VVAEQFQPVLDAHLDNCPTAGEDAHQ-ILDEDGDEDAVRVRRSS-- 206
pZMX201   228 DGK-----VEQEDFAPVIRSHVNNCEYATEDAHDP-----TSEDTISIRHAGLD 271
pNB101    208 DDP--ADEIEGETFRPVVRSHVENVEPAGSAAHGL--NAVGMGDTVSVNRE--- 254
pHK2      179 DAADLDLEVVGPEFERVIDKHVEECEYASFSAHDY-RNTDYLNDS-----DG-- 224
pGRB1     161 RGP-----VVAEQFQPVLDAHLRNCPTAGEDAQ-VFDENGDEDAVRVRRSS-- 206
Cons      271           :    *  *:  *:    *    **          :        324
pGT5      373 TFICYDKSSTKWFRLNPLLSESDLKKLRDIWKNVLLSYFGELLSEDTKSKDFDV 426
pRT1      295 EAVAKAEREGRE-IIASIIEAANAYDALLPDNEHPVAGYGVANPLIVKAINEED 347

                            motif III
pHGN1     219 -RPDR----------EDGIENLGAYLAAYMAGEYGVEALAMPAHVRAFYAAMW 260
pHSB2     207 -HPSR----------SGGVENLGAYLAAYMAGEYGAEPGEMPAHVRAFYATMW 248
pZMX201   272 RDPKRDHAQFEGIEYDSDVIGELAIYLAEYLGTFGDDDPLEEPEHVQASNTLLW 325
pNB101    255 ------------------VENLGSYISEYI-GIFGEEPTERPVSEQMFYATCW 288
pHK2      225 -CISL----------NAGVENMGSYLAAYMGG-YTEELLDKPVEYLAWGAIYW 265
pGRB1     207 -HPSR----------SGGVENLGAYLAAYMAGEYGSEPSEMPENVRAFYATMW 248
Cons      325                   : ::.*:: *:    :    *    : * 378

pGT5      427 WAGD---NYYSLPLDVPQVFFELKYASRKLFVNFVNYFEQSNFDESSVSDWDF 476
pRT1      348 LKRR---FSEFAPVEALQLDSKERYILMQLMDRRLESEDLKQIAERLGVSVRT 398
```

Fig. 9.1 Alignment of the initiator proteins of rolling circle replication. The sequences of the initiator proteins from halophilic plasmids were aligned and the consensus calculated. The replication proteins from the *Pyrococcus* plasmids pGT5 and pRT1, which belong to another superfamily of initiator proteins (Ilyina & Koonin, 1992), were aligned separately and then added manually to the first alignment. The motifs I to III of the initiator proteins are in bold letters and putative catalytic histidine and tyrosine residues are marked. The motif I of the initiator protein from plasmid pRT1 could not be identified.

Table 9.2 Archaeal plasmids with rolling-circle replication.

Plasmid	Size	Original host	Remarks	Reference
pHGN1	1765 bp	*Halobacterium* strain GN101		Hall & Hackett, 1989
pHSB2	1736 bp	*Halobacterium* strain SB3		Akhmanova et al., 1993
pGRB1	1781 bp	*Halobacterium* strain GRB	Shuttle*	Hackett et al., 1990
pHK2	10,500 bp	*Haloferax volcanii*	Shuttle, only replication region sequenced	Holmes et al., 1994
pML	2158 bp	*Methanohalophilus mahii*		
pZMX201	1668 bp	*Natrinema* sp. CX2021		
pNB101	2538 bp	*Natronobacterium* sp. AS-7091	Shuttle	
pGT5	3444 bp	*Pyrococcus abyssi* GE5	Shuttle	Erauso et al., 1996
pTN1	3367 bp	*Thermococcus* sp.		
pRT1	3373 bp	*Pyrococcus* sp. JT1		Ward et al., 2002

*Shuttle: an *E. coli* shuttle vector has been constructed based on the backbone of the plasmid.

Table 9.3 Crenarchaeal plasmids of the pRN type.

Plasmid	Size	Original host	Remarks	Reference
pRN1	5350 bp	*Sulfolobus islandicus*		Keeling et al., 1996
pRN2	6959 bp	*Sulfolobus islandicus*		Keeling et al., 1998
pHEN7	7830 bp	*Sulfolobus islandicus*		Peng et al., 2000
pSSVx	5705 bp	*Sulfolobus islandicus*	virus-plasmid hybrid	Arnold et al., 1999
pDL10	7598 bp	*Acidianus ambivalens*		Kletzin et al., 1999
pTIK4	13,638 bp	*Sulfolobus neozealandicus*	no prim/pol-domain	Greve et al., 2005
pTAU4	7192 bp	*Sulfolobus neozealandicus*	MCM	Greve et al., 2005
pORA1	9689 bp	*Sulfolobus neozealandicus*	snRNP, no plr/ORF80 homologue	Greve et al., 2005
pIT3	4967 bp	*Sulfolobus solfataricus* strain IT3		Prato et al., 2006

Archaeal plasmids of the pRN type

A number of rather small plasmids of the Crenarchaeota seem to belong to a single but diverse plasmid family (Table 9.3). The first crenarchaeal plasmid sequenced was the plasmid pRN1 (Keeling et al., 1996), which was isolated in 1994 by Zillig and co-workers from a *Sulfolobus* strain from Iceland (Zillig et al., 1994). This plasmid has six open reading frames, but only three of them are conserved within the plasmid family pRN (Peng et al., 2000). The plasmids pRN2, pHEN7 from *S. islandicus*, and pDL10 from *Acidianus ambivalens* also belong to this family. In addition, the plasmid–virus hybrid pSSVx is a fusion between a pRN type plasmid and a virus. This replicon also shares two open reading frames with the *Sulfolobus Fuselloviridae* (SSV1, SSV2, SSV3, SSV-RH). It could be demonstrated that SSV1 and SSV2 act as a helper virus for pSSVx, and that in the presence of these viruses pSSVx can spread through the culture (Arnold et al., 1999).

The three conserved open reading frames of the plasmid pRN1 have been studied by our group. The smallest conserved protein, the protein ORF56, has weak sequence similarity to several eubacterial copy control proteins. In fact the recombinant protein is able to bind double-stranded DNA in a sequence-specific manner. The protein is dimeric and binds as a tetramer to its cognate DNA, which has been shown to be an inverted repeat that is found upstream of its own gene (Lipps et al., 2001b). We propose that binding of ORF56 to its promoter could down-regulate the transcription of *orf56* as well as the downstream gene *orf904*. For that reason ORF56 could be implicated in the copy-control of the plasmid pRN1. The solution structure of ORF56 has recently been determined (G. Lipps, unpublished observation). The protein is a strand–helix–turn–helix DNA-binding protein, which binds DNA over the N-terminal β-sheet formed by the two subunits of the protein. This same fold is adopted by the Arc-repressor and the CopG protein from the plasmid pLS1.

Downstream of all putative *copG* genes a large open reading frame is identified in all these crenarchaeal plasmids. For pRN1 it could be shown that this protein consists at least of two domains (Fig. 9.2). In the N-terminal part of the protein a novel domain has been detected that has primase and DNA polymerase activity (Lipps et al., 2003). The structure of this domain, termed the prim/pol domain, has been solved (Lipps et al., 2004), and together with appropriate point mutants the active site of this domain could be identified. The catalytic core of this domain has the same fold as the archaeal/eukaryal primases, raising the possibility that these proteins might have evolved from a common ancestor capable of replicating small genomes (see also Chapter 2 in this volume).

In the C-terminal half of the protein a helicase domain of superfamily 3 (Gorbalenya et al., 1990) can easily be identified (Lipps et al., 2003). The bioinformatic predictions are in line with the findings that the protein ORF904 from pRN1 has ATPase activity and helicase activity. The ATPase activity is specifically stimulated by double-stranded DNA and the polarity of the helicase activity has been determined to be 3′–5′ (G. Lipps, unpublished observation). The exact role of the replication protein ORF904 during plasmidal replication is not known yet. In principle, the primase, DNA polymerase, and helicase activity of ORF904 would allow ORF904 to carry out three main activities required at a moving replication fork. However, the DNA polymerase activity is quite low and there is no proofreading activity associated with this multifunctional enzyme. It is therefore more plausible that the replication protein is involved in initiating the plasmidal replication by melting the origin of replication and by synthesizing a primer. It is currently not known whether the replication protein ORF904 remains at the replication fork and serves there as a replicative helicase or as a primase. Nevertheless, its conservation as a multidomain protein within the crenarchaeal plasmids is intriguing and is paralleled by well known primase–helicase protein fusions, such as the bacteriophage P4 α protein and the bacteriophage T7 gp4.

The crenarchaeal plasmids isolated from *Sulfolobus* strains and *Acidianus* strains from Iceland, i.e. plasmids pRN1, pRN2, pHEN7, and pDL10, and the virus plasmid-hybrid pSSVx, have very similar replication proteins (Fig. 9.2). In contrast, the replication proteins from the plasmids isolated from Italy and New Zealand are quite different. In some cases (e.g. pIT3 from *Sulfolobus solfataricus*) there is a clear match with the prim/pol domain but not with the helicase domain. However, in all proteins a Walker A motif is detectable in the C-terminal half. In contrast, the replication protein from the New Zealand plasmid pTAU4 is very similar to the MCM (minichromosome maintenance) proteins. These proteins are believed to be the replicative helicases of the Archaea and the eukaryotes. The MCM proteins and the helicase superfamily 3 proteins are structurally and functionally closely related (Iyer et al., 2004a). The pTAU4

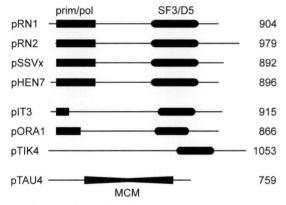

Fig. 9.2 Domain structures of the *Sulfolobus* plasmid replication proteins of the pRN type. The replication proteins can be divided into three groups. The plasmids from *S. islandicus* strains (pRN1, pRN2, pSSVx, and pHEN7) make up the first group. These replication proteins are very similar and have a prim/pol domain as well as a Pox_D5 (helicase superfamily 3) domain. The plasmids pIT3 (*S. solfataricus*) and pORA1 and pTIK4 (*S. neozealandicus*) belong to the second group. These proteins are similar over about 700 amino acids, including the C-terminal helicase domain. The N-terminal prim/pol domain is only weakly conserved and not present in the replication protein from pTIK4. The replication protein pTAU4 is the only member of a third, more distant group. This protein possesses an MCM (mini chromosome maintenance) domain. Proteins (total number of amino acids indicated) and domains are drawn to scale.

replication protein, however, does not contain a prim/pol domain. Except for pTAU4 and pTIK4 it appears that all the plasmids, although they are quite diverse, tend to have a large replication protein with a prim/pol domain and a helicase superfamily 3 domain.

Another characteristic of these crenarchaeal plasmids is that they have a highly conserved short gene that has been termed *plrA* (plasmid regulatory). This gene is not present on the New Zealand plasmid pORA1, but is also highly conserved within the group of conjugative plasmids from *Sulfolobus* (see below). The Plr protein from pRN1 (ORF80) has been studied. This protein is a sequence-specific DNA-binding protein that binds to two neighboring binding sites upstream of its own gene (Lipps et al., 2001a). Most interestingly, the binding sites and the distance between the binding sites are conserved, suggesting that a larger protein–DNA complex could assemble at this specific site of the plasmid. It is possible that this protein marks the origin of replication. However, there are no experimental data that support this proposition. For the time being the physiological function of this highly conserved protein remains completely unknown.

Although these small cryptic plasmids were isolated from diverse geographic locations they all appear to share a common replication machinery, with the involvement of a *plr*-homolog and/or a two-domain protein with a superfamily 3 helicase.

Conjugative plasmids

The crenarchaeote *Sulfolobus* is a host for a number of conjugative plasmids (Table 9.4), and some of them have been studied in more detail (Schleper et al., 1995; She et al., 1998; Stedman et al., 2000). The conjugation involves direct cell–cell contacts without the contribution of pili-like structures (Prangishvili et al., 1998a). It is not known whether single- or double-stranded DNA is transported. From comparison of the genomes of several conjugative plasmids, a cluster of six genes that might be responsible for the conjugational DNA transport has been proposed (Greve et al., 2004). Two of the encoded proteins show sequence similarity to enzymes of type IV transport and contain the conserved domains VirB4 and TraG_VirD4, respectively. It is noteworthy that the minichromosome pNG600 from *H. marismortui* and the megaplasmid pNRC200 from *H. salinarum* also contain these two genes. On these two replicons the respective genes overlap and could be cotranscribed.

The archaeal conjugation apparatus has not been studied yet and it is completely unclear how the conjugational DNA transport is realized, what proteins are involved, and whether there is an origin of transfer. It seems, however, that the archaeal conjugation machinery is quite different from its bacterial counterpart.

The genome of *Sulfolobus acidocaldarius* contains an integrated conjugative plasmid (Garrett et al., 2005), raising the possibility that the conjugational marker exchange observed for this organism (Grogan, 1996; Reilly & Grogan, 2001) is carried out by conjugative proteins supplied by the integrated plasmid.

The replication machinery of these plasmids is currently unknown. A number of conserved relatively short proteins have been suggested to be involved in the plasmid replication. Furthermore, an origin of replication has been proposed. The putative origin region of about 170 bp is highly conserved and contains numerous repeats (Greve et al., 2004).

Table 9.4 Conjugative plasmids.

Plasmid	Size	Original host	Remarks	Reference
pNOB8	41,229 bp	*Sulfolobus* sp. NOB8H2	ParA, ParB	Schleper et al., 1995
pING1	24,554 bp	*Sulfolobus islandicus*		Stedman et al., 2000
pKEF9	28,930 bp	*Sulfolobus islandicus*		Greve et al., 2004
pARN3	26,200 bp	*Sulfolobus islandicus*		Greve et al., 2004
pARN4	26,476 bp	*Sulfolobus islandicus*		Greve et al., 2004
pHVE14	35,422 bp	*Sulfolobus islandicus*		Greve et al., 2004
pTC	20,417 bp	*Sulfolobus tengchongensis*		

Archaeal plasmids with poorly characterized replication proteins

The halophilic Archaea *H. volcanii*, *Halobacterium* sp., and *H. marismortui* host plasmids with *repH* replication proteins (Table 9.5). The precise function of these proteins for plasmid replication is currently unknown since the proteins have not been studied so far. An analysis of the amino acid sequence of these putative replication proteins, which are between 300 and 1100 amino acids long, does not reveal sequence similarity to known and characterized proteins, making an *in silico* prediction of the protein function impossible. However, the 808 amino acid *repH* homolog is the only large protein encoded by the 6.4 kb plasmid pHV2 from *H. volcanii*, suggesting that the RepH homologues are in fact involved in plasmid replication. Furthermore, the RepH protein of the megaplasmid pNRC100 from *Halobacterium* sp. NRC-1 is part of the 3.9 kb minimal replication origin (Ng & DasSarma, 1993). RepH homologs are also found in several megaplasmid/minichromosomes from *H. marismortui*, in the viruses φCh1 (*Natrilba magadii*), φH (*H. salinarium*) and in the plasmid pφHL (Gropp et al., 1992; Klein et al., 2002). Clearly, further studies are needed to characterize the *repH* replication proteins, which appear to be quite abundant within the halophilic Archaea.

Several *repH* replicons, e.g. the related plasmid pHH1 and pNRC100 from two *Halobacterium* strains, contain the gas vesicle genes, which are required for gas vesicle synthesis and enable the cells to float. Mutants unable to synthesize gas vesicles arise with extremely high frequency (up to 10^{-2}) (Pfeifer et al., 1981). Insertion, deletions, and rearrangements of the plasmids are responsible and are mostly caused by highly active insertion elements (DasSarma et al., 1983; Pfeifer & Blaseio, 1989). The plasmid pNRC100 contains 27 insertion elements (Ng et al., 1998). Moreover, this plasmid contains a large inverted repeat of about 35 kb. Both inversion isomers are observed in cultures and the plasmid appears to be inverted rapidly *in vivo* (Ng et al., 1991).

The megaplasmids from *Halobacterium* sp. have a different GC content than the main chromosome and are more prone to mutagenesis by insertion element. It is possible that the "plasmid" fractions of the genomes of *Halobacterium* species are evolving. The plasmids might disappear, fuse with each other or with the main chromosome, or develop into new chromosomes (Ng et al., 1998).

In Table 9.6 further archaeal plasmids, which are only poorly studied, are listed. For most of them the prediction of a replication protein is impossible due to the lack of sequence similarity to characterized proteins.

The plasmid pC2A from *Methanosarcina acetivorans* and the plasmid pURB500 from *Methanococcus maripaludis* encode a protein with the conserved domain of XerD/C integrases. The role of the respective proteins for plasmid replication has not, however, been further investigated. Both plasmids have served as backbones for the construction of shuttle vectors (Metcalf et al., 1997; Tumbula et al., 1997a).

The closely related plasmids pFV1 and pFZ1 from *Methanothermobacter thermautotrophicus* encode for a Cdc6-like protein. Cdc6 proteins are AAA$^+$-ATPases that participate in the replication initiation in Archaea and eukaryotes. It could be demonstrated that the Cdc6 proteins from *Pyrococcus abyssi* and *S. solfataricus* bind specifically to the chromosomal

Table 9.5 Plasmids with a *repH* replication protein.

Plasmid	Size	Original host	Remarks	Reference
pHV2	6354 bp	*Haloferax volcanii*	Shuttle	Charlebois et al., 1987
pHH1	143,000 bp	*Halobacterium salinarium*	Shuttle, only partially sequenced, gas vesicle genes	Pfeifer & Ghahraman, 1993
pNRC100	191,346 bp	*Halobacterium* sp. NRC-1	Shuttle, gas vesicle genes	Ng et al., 2000
pNRC200	365,425 bp	*Halobacterium* sp. NRC-1	Gas vesicle genes	Ng et al., 2000
pNG200	33,452 bp	*Haloarcula marismortui*		Baliga et al., 2004
pNG300	39,521 bp	*Haloarcula marismortui*		Baliga et al., 2004
pNG400	50,060 bp	*Haloarcula marismortui*		Baliga et al., 2004

Table 9.6 Plasmid with uncharacterized replication proteins.

Plasmid	Size	Original host	Remarks	Reference
pC2A	5467 bp	*Methanosarcina acetivorans*	Shuttle, XerDC integrase	Metcalf et al., 1997
pURB500	8285 bp	*Methanococcus maripaludis*	Shuttle, XerDC integrase	Tumbula et al., 1997
pFV1	13,514 bp	*Methanothermobacter thermautotrophicus*	Cdc6, DNA glycosylase, restriction endonuclease	Nolling et al., 1992
pFZ1	11,014 bp	*Methanothermobacter thermautotrophicus*	Cdc6, DNA glycosylase	Nolling et al., 1992
pME2200	6205 bp	*Methanothermobacter thermautotrophicus*		Luo et al., 2001
pME2001	4439 bp	*Methanothermobacter thermautotrophicus*		Luo et al., 2001
pZMX101	3918 bp	*Halorubrum saccharovorum*		
pSCM201	3463 bp	*Haloarcula* sp. AS7094		
large ECE*	58,407 bp	*Methanocaldococcus jannaschii*	Restriction/modification type I, MCM, ParA	Bult et al., 1996
small ECE	16,550 bp	*Methanocaldococcus jannaschii*	Restriction/modification type II	Bult et al., 1996
pHH205	16,341 bp	*Halobacterium salinarum*		Ye et al., 2003

*ECE: extrachromosomal element.

replication origins and could play a crucial role in origin firing (Matsunaga et al., 2001; Robinson et al., 2004). It is therefore possible that the plasmidal Cdc6 homologs initiate the replication of both plasmids.

Concluding remarks

The genome sequences of an increasing number of archaeal plasmids have been determined in the past few years. However, our knowledge of the biology, biochemistry, and molecular biology is lagging behind. Most of the archaeal plasmids appear to be cryptic. For a majority of the plasmids their replication and its regulation have not been addressed yet. We can expect, though, that the study of archaeal plasmids will deepen our knowledge of DNA replication and of archaeal conjugation and will allow us to assess the impact of plasmids on genome plasticity.

10

Integration mechanisms: possible role in genome evolution

Qunxin She, Haojun Zhu and Xiaoyu Xiang

Introduction

Enzymes that facilitate site-specific DNA recombination are widely present in Archaea and Bacteria. These enzymes play important roles in several molecular processes: some are involved in generating genetic variation or in regulating gene expression; some control the copy number of plasmids or resolve chromosome dimers; others facilitate integration of genetic elements into host chromosomes. This last group generates integrated plasmids or proviruses; therefore, the encoded enzymes are also called integrases (Ints). The integration process consists of multiple steps that are catalyzed by the integrase alone. No other proteins are absolutely required for the biochemical reaction but some facilitate the process (Craig et al., 2002). Characteristically an intermediate of a covalently linked integrase–DNA complex (Int-DNA) is formed. Some integrases form Int-DNA via a serine residue, whereas others use a tyrosine. This difference provides the basis for the classification of two major groups: serine and tyrosine recombinases. All the enzymes belonging to the latter category are conserved in the catalytic domain and generate a superfamily of tyrosine DNA recombinases (Esposito et al., 1997; Nunes-Duby et al., 1998).

There are two types of Int-encoding genetic elements. The first group consists of episomal integrative plasmids and viruses. The bacteriophage lambda is a well characterized representative of this group because it exists either in an episomal form or as an integrated prophage in its *Escherichia coli* host. The second category includes the Int-encoding genetic elements that are present in many archaeal and bacterial chromosomes. They are often referred to as integrated elements (IE) in Archaea (She et al., 2002) and as cryptic prophages, genomic islands, symbiotic islands, or pathogenicity islands in bacteria (Dobrindt et al., 2004). Both types of genetic elements have been implicated strongly in mediating horizontal gene transfer (HGT) during microbial speciation and genome evolution. There is now a general agreement that HGT constitutes an important evolutionary event in microbes and it has been suggested that 5–30% genes in microbial genomes have been transferred horizontally (Koonin et al., 2001). In this chapter we summarize the current knowledge relating to archaeal integrases and integration mechanisms as well as archaeal IEs and their roles in HGT during genome evolution.

Archaeal integrases belong to a superfamily of tyrosine recombinases

An archaeal tyrosine integrase was first identified in the genome of *Sulfolobus shibatae* virus SSV1 more than a decade ago (Palm et al., 1991). This virus can form a provirus in host chromosome via site-specific integration within the downstream half of a tRNA^{Arg} gene (Reiter et al., 1989; Schleper et al., 1992). Moreover, SSV1-encoded protein, SSV1-Int, can effect site-specific integration *in vitro* (Muskhelisvili et al., 1993). Taken together, these studies have demonstrated clearly that SSV1 carries an integrase gene (*int*). In analogy to the terminology of bacterial integration, the attachment site of SSV1 and that of the archaeal host are termed *attP* and *attA*, respectively, whereas the left and right hybrid attachment sites of the provirus are referred to as *attL* and *attR*. Figure 10.1 illustrates integration and excision of SSV1 and the sequence motifs involved in the recombination.

SSV1 integration has an exceptional feature: the *int* gene is partitioned upon integration because the *attP* site is located within the first part of the gene, corresponding to the N-terminal region of the integrase protein. This contrasts with the integration by all other known integrases. The resulting *int* gene fragments encode a smaller N-terminal fragment *intN* (70 amino acids) and a larger C-terminal fragment *intC* (270 amino acids) that border the provirus (Fig. 10.1a). At present it is not known whether SSV1 *intN* and *intC* can generate an active integrase, although *intC* is unlikely to be expressed in the first instance since it lacks a promoter for transcription. Thus, excision of the SSV1 provirus relies on the whole-length *int* gene present on the circular SSV1 viral genome. Interestingly, both episomal and integrated forms of SSV1 are always present in the host cell simultaneously (see Chapter 6 in this volume). A plausible explanation of this phenomenon is that the circular SSV1 genome provides the intact *int* gene required for dynamic integration/excision. This

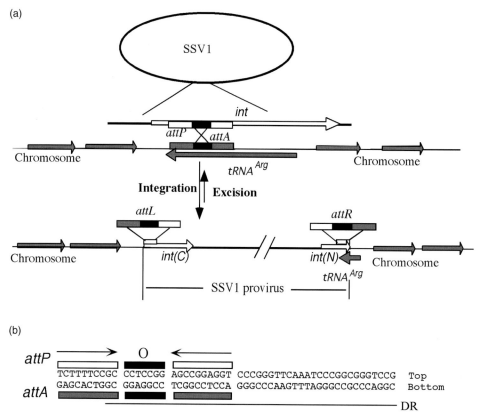

Fig. 10.1 Integration and excision of *Sulfolobus shibatae* virus SSV1. (a) The SSV1 integration scheme. (b) Sequences of the SSV1 *attP* and *attA* elements defined by Serre et al. (2002). Arrows illustrate genes and relevant genes are labeled. All four attachment sites are composed of an imperfect inverted repeat flanking the overlap sequence. O denotes the overlapping sequence within the attachment sites and it is illustrated with dark-filled boxes; unfilled boxes represent the imperfect inverted repeat in *attP*, whereas gray-filled bars indicate the imperfect repeat in *attA*. The direct repeat (DR) is underlined (this was previously called the attachment site by Muskhelishvili et al., 1993). *attA* and *attP* denote the attachment sites of SSV1 and its archaeal host, whereas *attL* and *attR* stand for the hybrid attachment sites. *int*, integrase gene; *intN* and *intC* indicate the partitioned SSV1 *int* gene fragments. Top, top strand of *attP*; bottom, bottom strand of *attA*. DR, direct repeat.

integrase represents the SSV1-type, since SSV1 is the first integrative genetic element discovered for Archaea (Schleper et al., 1992).

Recently, archaeal integrases that retain an intact *int* gene after integration have been found in the genomes of *Sulfolobus* species and this integrase type is also encoded by *Sulfolobus* conjugative plasmids (Garrett et al., 2004). Furthermore, it has been demonstrated that the pNOB8 plasmid can integrate site-specifically into host chromosome (She et al., 2004). Thus, this integrase is referred to as the pNOB8-type, after the first archaeal conjugative plasmid that was characterized (Schleper et al., 1995; She et al., 1998).

Thus far, a total of 43 whole-length *int* and 21 gene fragments thereof have been identified in 17 archaeal genomes (Table 10.1). This indicates that tyrosine integrases are widely present in the archaeal domain. Moreover, many archaeal genomes encode multiple integrases and at least one Xer protein (Table 10.1). As for bacterial recombinases (Nunes-Duby et al., 1998), the archaeal enzymes are very diverse in sequence and show only limited sequence similarity within the catalytic domains carrying 120–160 amino acids. This similarity enables us to classify archaeal integrases into subfamilies by multiple sequence alignment (She et al., 2002). Whereas archaeal XerD and Xer-like integrases form one major clade together with their bacterial counterparts, all other archaeal integrases form another clade that is distinctive from that of bacterial integrases (Fig. 10.2). Apparently archaeal tyrosine integrases constitute a major group by themselves. Moreover, the two different types of archaeal integrase, SSV1-type and pNOB8-type, also form distinctive groups (Fig. 10.2), reinforcing the mechanistic differences of their integration.

Crystal structures of six different tyrosine recombinases have revealed the presence of six highly conserved catalytic residues, which characterize the active site of tyrosine integrase (van Duyne, 2001). *Sulfolobus* integrases of both SSV1 and pNOB8 types are aligned with some well studied integrases to investigate their sequence conservation in Archaea. As shown in Fig. 10.3, the motifs in Box I, Kβ, and Box II, as well as all known catalytic residues thereof,

Table 10.1 Integrase genes and remnants in archaeal genomes.

Organism	pNOB8-type			SSV1-type	
	Full length *int*	Truncated *int*	*xerD* and *xer*-like *int*	*intN*	*intN+intC*
Aeropyrum pernix	1	0	1	n.d.	1
Archaeoglobus fulgidus	1	0	1	n.d.	0
Haloarcula marismortui	12	3	2	n.d.	0
Halobacterium salinarium	6	0	4	n.d.	0
Natronomonas pharaonis	4	2	1	n.d.	0
Methanococcus jannaschii	0	1	1	n.d.	0
Methanococcus maripaludis	1	1	1	n.d.	0
Methanosarcina acetivorans	7	3	2	n.d.	0
Methanosarcina mazei	1	0	1	n.d.	0
Picrophilus torridus	1	0	1	n.d.	0
Pyrococcus horikoshii	1	0	1	n.d.	2
Sulfolobus acidocaldarius	1	4	1	8	3
Sulfolobus solfataricus	6	2		6	2
Sulfolobus tokodaii	3	4	1	7	1
Thermococcus kodakaraensis	1	1	1	n.d.	4
Thermoplasma acidophilum	1	0	1	n.d.	0
Thermoplasma volcanium	1	4	1	n.d.	0

n.d., not determined; *int*, integrase gene, *intN* and *intC*, the partitioned N-terminal and C-terminal gene fragments of the SSV-type integrase gene.

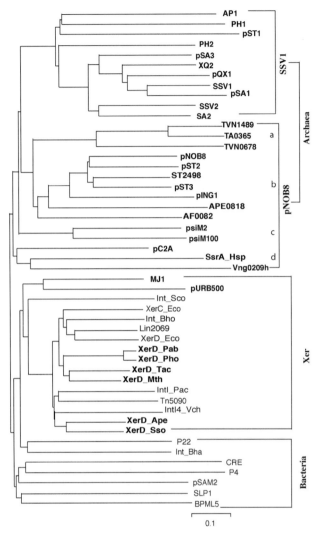

Fig. 10.2 Phylogenetic tree of archaeal tyrosine DNA recombinases. Sequences of the entire catalytic regions were used to generate the tree. Archaeal and bacterial integrases are shown in bold and normal type, respectively. Archaeal integrases are presented with the names of the genetic elements that carry them, including plasmids, integrated elements, and viruses. For those of an unknown origin, the annotations of the open reading frames (ORFs) are used, e.g. TA0365. Abbreviations: AP, *A. pernix*; AF, *A. fulgidus*; MJ, *M. jannaschii*; MTH, *M. thermautotrophicus*; PAB, *P. abyssi*; PH, *P. horikoshii*; Sso, *S. solfataricus*; ST, *S. tokodaii*; TA, *T. acidophilum*; TVN, *T. volcanium*; VNG, *Halobacterium* sp. NRC-1; IntI, bacterial integron integrases; SsrA, site-specific recombinase A. The sources of genetic elements are indicated in Table 10.2. SSV1- and pNOB8-type as well as Xer-like integrases are indicated.

are conserved within these archaeal members. However, important differences are found between the motifs of archaeal and bacterial enzymes. Whereas the bacterial consensus is R...K... H/KxxR...H/W...Y, the SSV1-type integrase has R...R...KxxR...R...Y and the pNOB8-type enzymes possess R...K...YxxR...R...Y. Thus, two differences have occurred for each group and these are indicated above the alignment of *Sulfolobus* integrases in Fig. 10.3.

SSV1-Int is the only archaeal integrase that has been characterized in detail. Using a recombinant SSV1-Int enzyme, Serre et al. (2002) showed that SSV1-Int forms a 3'-phosphotyrosine intermediate with *attA* or *attP* via the well conserved tyrosine "Y314" (Fig. 10.3), just as for bacterial tyrosine integrases. This work demonstrates that the general mechanism for integration by tyrosine enzymes is conserved in Archaea and Bacteria. Moreover, they have established that the minimal sequence required for SSV1 integration (*attP* and *attA*) is only 27 bp long and it overlaps the 44 bp direct repeat that was thought to be the SSV1 core attachment site (Muskhelisvili et al., 1993) (Fig 10.1b). Although this result remains to be tested *in vivo*, the identified core sequence for recombination is much shorter than that of bacterial *attP*s that normally span a few hundred base pairs (Craig et al., 2002). Further insight into archaeal integration was gained by mutagenesis of the putative catalytic residues of SSV1-Int and it has been shown that any substitution at these positions abolishes the catalytic activity (Letzelter et al., 2004). This is also true for the substitutions that have introduced the bacterial consensus sequence into the archaeal enzyme. This reinforces the idea that archaeal integrases constitute a distinctive group.

Two types of integrative genetic elements

During comparative analyses of the chromosome of *Sulfolobus solfataricus* and pHEN7 plasmid, an integrated plasmid pXQ1 was identified in the former (Peng et al., 2000). This plasmid is bordered by ORFs very similar to the IntN and IntC that are encoded by the SSV1 provirus. In the meantime, two putative proviruses were identified in the genome of

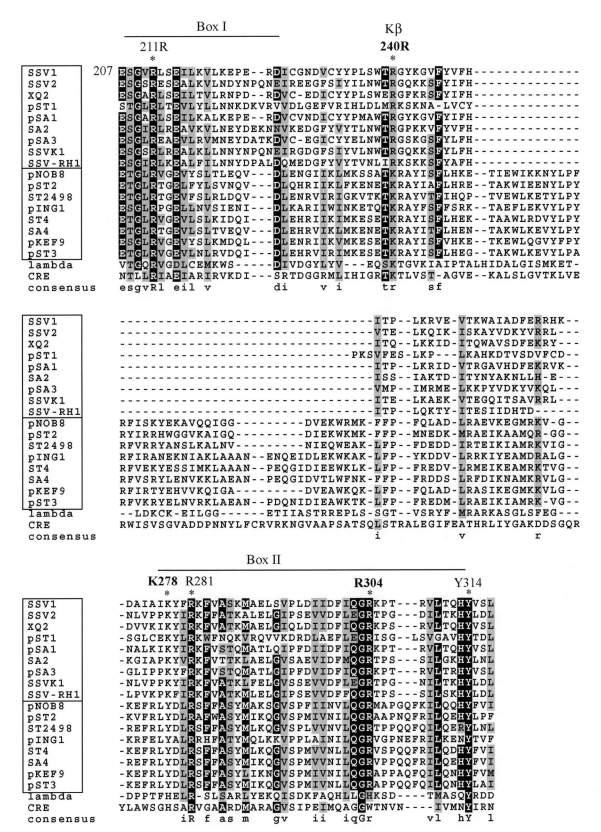

Fig. 10.3 A multiple sequence alignment of *Sulfolobus* integrases. The alignment was generated by using the Clustal X program and illustrated by BoxShader. Box I, Kβ, Box II, and Y are assigned according to Nunes-Duby et al. (1998), the consensus of the catalytic residues for SSV-type integrases is indicated above the alignment and the numbers indicate the position of amino acid in the SSV1 sequence. Catalytic residues that are shown in black type indicate the catalytic residues that are different between archaeal and bacterial consensuses. Residues with more than 80% identity and similarity are shaded black and gray, respectively.

Pyrococcus sp. OT3, again embedded by partitioned SSV1-like *intN*s and *intC*s (Makino et al., 1999). Subsequently, many *intN*s and *intC*s are identified in other archaeal genomes, including *Aeropyrum pernix*, *Pyrococcus horikoshii*, *Sulfolobus acidocaldarius*, and *Sulfolobus tokodaii* (She et al., 2001), and *Thermococcus kodakaraensis* (Fukui et al., 2005). For every *intC*, a corresponding *intN* is also present because each pair of *intC* and *intN* shares the same direct repeat. Thus, they border IEs. These IEs are referred to as the SSV1-type since they carry the partitioned *intN*s and *intC*s as for the SSV1 provirus. So far, 13 such elements have been identified in archaeal genomes (Table 10.2).

The IE encoding the pNOB8-type integrase pST2 was first found in the *S. tokodaii* genome (Kawarabayasi et al., 2001) on the basis of the following: (i) pST2 contains a pNOB8-type *int* gene that has remained intact after integration; (ii) it is bordered by a direct repeat with one end overlapping a tRNA^{Met} gene; and (iii) it encodes many ORFs highly similar to those of the conjugative plasmid pNOB8. This type of IE is referred to as pNOB8-type as for the SSV1-type. Only a subset of integrase-encoding archaeal genomes have been analyzed for integrated elements and eight pNOB8-type IEs have been found (Table 10.2).

The broad appearance of the *int* genes in most archaeal genomes suggests that archaeal integrative genetic elements occur widely in nature. Indeed, genetic elements encoding a pNOB8-type *int* have been isolated from diverse archaeal species. In addition to the *Sulfolobus* conjugative plasmids, many other genetic elements carry such an *int* gene. For Euryarchaea, three archaeal viruses encode an integrase, i.e. *Methanobacterium thermoautotrophicum* phage psiM2 (Pfister et al., 1998) and the haloviruses HF1 and HF2 (Tang et al., 2004). Two small cryptic plasmids carrying a putative *int* gene were isolated from methanogens. They are plasmid pURB500 from *Methanococcus maripaludis* (Tumbula et al., 1997a) and plasmid pC2A from *Methanosarcina acetivorans* (Metcalf et al., 1997). Furthermore, multiple plasmids covering a wide size range exist as separate replicons in haloarchaeal genomes, and many of them carry at least one gene encoding DNA recombinase, including plasmids pNRC100, pNRC200 of *Halobacterium* sp. (Ng et al., 1998), five large or megaplasmids of *Haloarcula marismortui* (Baliga et al., 2004b), and PL23 of *Natronomonas pharaonis* (Falb et al., 2005). For SSV1-type integrative elements, only *Sulfolobus* spindle viruses contain such an *int* gene (see Chapter 6 in this volume), and all other known genetic elements are devoid of such a gene. Thus, SSV1-type *int* genes are present in only six archaeal genomes, three of which are *Sulfolobus* species. Among the archaeal genomes that encode the SSV1-type integrase, only *Sulfolobus* species have systematically been screened for genetic elements (Zillig et al., 1994, 1998). Thus, more genetic elements carrying an SSV1-*int* gene remain to be isolated from other Archaea.

Recently, many genome sequences of archaeal hosts including genetic elements have become available and many are being sequenced. This has changed the perspective of the study of archaeal integration. Comparative sequence analysis and polymerase chain reaction are the preferred techniques for such investigations in order to exploit the genome sequence data. A first step is to identify the integration mechanism of the system under study. Four different mechanisms of integration have been found in Archaea thus far. The first is represented by SSV1, and all known SSV-type IEs belong to this type except TKV4. The difference between TKV4 and the other SSV-type IEs lies in the relative orientation of the *intN* and the tRNA target. Generally *intN* and tRNA are reverse oriented but in TKV4 they have the same orientation because integration has occurred upstream of the tRNA gene (Fig. 10.4).

Two integration mechanisms have been found for pNOB8-type IEs but integration occurs only in the downstream halves of the target tRNA genes. In one scheme, the target tRNA and *int* genes appear in a tandem arrangement at the *attL* end (Fig. 10.4c) and all known pNOB8-type IEs integrate in this fashion. However, for the PL23 plasmids integrated into the chromosome of *Natronomonas pharaonis* (Falb et al., 2005), the *int* gene is present at the *attR* end while the target tRNA site retains at the *attL* end (Fig. 10.4d). This reflects the fact that the PL23 *attP* site 3'-flanks the *int* gene (Falb et al., 2005).

tRNA genes are common targets for integration for both Archaea and Bacteria. Hundreds of tyrosine integrases are known to integrate within the downstream halves of tRNA genes and they very rarely integrate in the upstream halves (Williams, 2002). So far, only two IEs, TKV4 (Fig. 10.4b) and Mol38S from *Mesorhizobium loti* (Zhao & Williams, 2002), have been found to use this integration scheme. This raises a question as to why integration favors the

Table 10.2 Integrated elements characterized in Archaea.

Organism	Name	Size (kb)	Direct repeat (bp)	Overlapping tRNA gene	Gene content	References
SSV1-type						
(a) Euryarchaeota						
Pyrococcus sp. OT3	PT1	21.7	48	tRNA^{Ala}<	putative provirus	Makino et al., 1999
Pyrococcus sp. OT3	PT2	4.1	46	tRNA^{Val}<	putative provirus	Makino et al., 1999
Thermococcus kodakaraensis	TKV1	23.6	48	tRNA^{Val}<	MCM, transcriptional regulators	Fukui et al., 2005
Thermococcus kodakaraensis	TKV2	27.2	48	tRNA^{Glu}<	transcriptional regulators, ATPase	Fukui et al., 2005
Thermococcus kodakaraensis	TKV3	27.9	48	tRNA^{Arg}<	PCNA, enzymes	Fukui et al., 2005
Thermococcus kodakaraensis	TKV4	18.8	42	tRNA^{Leu}>	MCM	Fukui et al., 2005; this work
(b) Crenarchaeota						
Aeropyrum pernix	AP1	17.8	65	tRNA^{Leu}<	unknown	She et al., 2001
Sulfolobus acidocaldarius	pSA1	8.7	44	tRNA^{His}<	plasmid	She et al., 2004
Sulfolobus acidocaldarius	SA2	5.7	44	tRNA^{Arg}<	unknown	She et al., 2004
Sulfolobus acidocaldarius	pSA3	32.5	50	tRNA^{Glu}<	degenerated CP, 2 enzymes	Chen et al., 2005
Sulfolobus solfataricus	pXQ1	7.3	45	tRNA^{Val}<	pRN-like plasmid	She et al., 2001
Sulfolobus solfataricus	XQ2	67.7	45	tRNA^{Arg}<	11 Enzymes	She et al., 2001
Sulfolobus tokodaii	pST1	6.7	48	tRNA^{Ala}<	pRN-like plasmid	She et al., 2002
pNOB8-Type						
(a) Euryarchaeota						
Natronomonas pharaonis	PL23	36.9	25	tRNA^{Leu}>	three copies of degenerated PL23	Falb et al., 2005
Methanococcus jannaschii	MJ1	30.5	69	tRNA^{Ser}>	MCM	Makoni et al., 1999
(b) Crenarchaeota						
Sulfolobus tokodaii	pST2	44.8	39	tRNA^{Met}>	pNOB8-like	She et al., 2002
Sulfolobus tokodaii	pST3	6.9	42	tRNA^{Arg}>	pRN-like plasmid	She et al., 2002
Sulfolobus tokodaii	ST4	66.1	29	tRNA^{His}>	25 Enzymes	She et al., 2002
Sulfolobus tokodaii	ST5	nd	nd	tRNA^{Gly}>	highly degenerated	She et al., 2004
Sulfolobus acidocaldarius	SA4	7.5	43	tRNA^{Val}>	plasmid-related	Chen et al., 2005

>, < indicate the orientation of target tRNA genes relative to the integrase gene. GI, genomic island; CP, conjugative plasmid; IE: integrated element; MCM, minichromosome maintenance protein; nd: not determined; PCNA, proliferating cell nuclear antigen. ST5 is a degenerated integrated element for which only one integration border was identified.

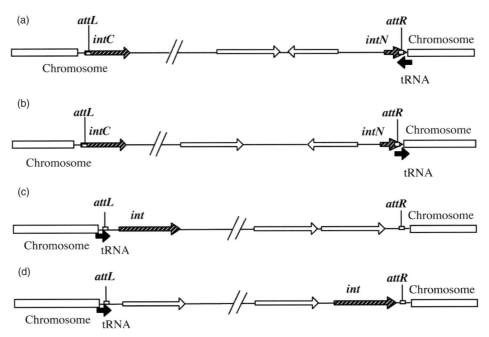

Fig. 10.4 Schematic illustration of the integration patterns of archaeal integrated elements. (a) The downstream of a tRNA gene as the integration target by an SSV-type integrase, resulting in reversely oriented *intN* and tRNA genes. (b) The upstream of a tRNA gene as the integration target by an SSV-type integrase, resulting in the same orientation of the target tRNA gene and *intN*. (c) Integration of a pNOB8-type integrase upstream of the *int* gene, resulting in the tandem arrangement of the tRNA and *int* genes at the *attL* end. (d) Integration of a pNOB8-type integrase downstream from the *int* gene, resulting in the tRNA target gene retained at the *attL* end, whereas the *int* gene is located at the *attR* end. *Int* represents the *int* gene of pNOB8-type; *intN* and *intC* denote the N-terminal and C-terminal parts of an intact SSV integrase gene; *attL* and *attR* denote the integration conjunctions; tRNA genes are illustrated as filled arrows.

downstream halves of tRNA genes. It is likely that a major disadvantage for integration at the upstream is that the inserted IEs will divide the target gene from its promoter. Although the target gene is fully restored after integration, the subsequent expression depends on whether an appropriate promoter occurs within the IE. In contrast, integration in the downstream region mainly results in the removal of the original termination signal. This can be replaced by a similar sequence motif in IEs because the termination sig-nals for transcription are much less conserved in sequence and position than the promoters.

Do archaeal IEs confer fitness on their hosts?

The archaeal IEs that have been identified so far are very different in size and gene content. For the relatively small IEs, some encode many ORFs carrying a viral or plasmid function. Others contain mainly unknown ORFs, presumably because the free integrative genetic elements have never been characterized. Remaining IEs are relatively large in size, often encoding enzymes, replication proteins, or transcriptional regulators in addition to the ORFs of plasmid or viral origin. The common appearance of IEs in archaeal genomes raises several questions such as what they are, where they are from, and what role they may play in Archaea.

The IEs on the three *Sulfolobus* chromosomes must have entered the present hosts via horizontal gene transfer. This is evident from the properties of these IEs as summarized in Table 10.3, including: (i) they exhibit an anomalous nucleotide composition; (ii) they contain plasmid-related ORFs; and (iii) they encode ORFs exhibiting exceptional codon usage. Three elements, pXQ1, pST1, and pST3, are apparently derived from pRN-like plasmids (see Chapter 9 in this volume), because they all encode at least two putative replication proteins characteristic of the pRN

Table 10.3 Properties and gene contents of *Sulfolobus* integrated elements.

Name	Size (kb)	G+C%*	ORF content						
			Total	HGT	Plasmid	Enzymes	Fragmented	IS	Other
pXQ1	7.3	38.2 (+2.4)	10	0	8	0	4	1	1
XQ2	67.7	36.8 (+1.0)	64	13	1	11	8	10	43
pST1	6.7	41.1 (+9.0)	9	9	7	0	2	0	2
pSA1	8.7	39.2 (+2.5)	7	n.d.	5	0	0	0	2
SA2	5.7	41.6 (+4.9)	7	n.d.	1	0	0	0	6
pSA3	32.5	39.0 (+2.3)	35	n.d.	11	1	7	3	16
pST2	44.8	36.9 (+4.8)	59	8	40	0	0	0	19
pST3	6.9	36.0 (+3.9)	6	>1	4	0	0	0	2
ST4	66.1	32.1 (−0.7)	86	10	14	25	5	3	50
SA4	7.5	41.0 (+4.3)	8	n.d.	3	0	0	0	5

*Numbers in parentheses indicate the difference between the integrated elements and the genomes carrying them; (+/−), higher/lower than host.

n.d., not determined; HGT, horizontal gene transfer; IS, insertion sequence. Predicted HGT genes were obtained from the HGT database www.fut.es/~debb/HGT/).

plasmid family. Moreover, pST2 and pSA3 must have been derived from *Sulfolobus* conjugative plasmids. Although pSA3 contains fewer conserved conjugative plasmid ORFs than pST2, ORFs important for conjugation are present (Chen et al., 2005b). Thus, pSA3 may have derived from a conjugative plasmid that is only distantly related to known ones. Although it may have been degenerated to some extent, pSA3 has been implicated in the process of chromosomal conjugation in *Sulfolobus acidocaldarius* (Chen et al., 2005b). However, it remains to be determined whether pST2 and pSA3 facilitate such a process.

Three *Sulfolobus* IEs have brought new genetic information and properties to the current hosts. Whereas ST4 encodes 25 enzymes, pSA3 adds two new genes encoding enzymes with closest orthologues from *M. jannaschii* to its host (Table 10.3). XQ2, on the other hand, encodes 11 enzymes, 12 IS elements and remnants thereof, and 13 hypothetical ORFs as well as 29 ORFs of unknown function. Among the 11 encoded enzymes, eight are apparently redundant for *S. solfataricus*, but present only in one copy in *S. tokodaii*. The remainder are unique, and their closest homologs are from Euryarchaea, showing 34–44% identity (Table 10.4). However, it remains to be established whether these enzymes are an advantage for the host.

Several other IEs present in euryarchaeal genomes encode metabolic enzymes, transcriptional regulators, and replication factors of the minichromosome maintenance protein (MCM) and proliferating cell nuclear antigen (PCNA) (Table 10.2). It was previously suggested that the archaeal and eukaryal replication apparatus originated from plasmids or viruses (Forterre, 1999). These IEs that carry MCM and PCNA genes, as well as cell cycle division protein 6- (Cdc6) and MCM-encoding plasmids (see Chapter 9 in this volume), provide plausible evidence for the capturing of genes encoding replication proteins in Archaea, as suggested by Forterre (1999).

Stability and evolution of integrated elements

A critical step for a successful HGT is the stable maintenance of foreign DNA in the host genome. For SSV1-type integration, stably integrated SSV-type IEs are already produced upon integration, since SSV-*int* genes are partitioned into *intN*s and *intC*s, providing that an episomal form of the genetic elements is absent and the *intN*s and *intC*s do not produce a functional enzyme. This is discussed in detail for SSV1 integration in a previous section (see also Fig. 10.1). The stability of the integrated pXQ1 in *S. solfataricus* has been studied by a polymerase chain reaction

Table 10.4 Metabolic enzymes in the integrated element XQ2 of *Sulfolobus solfartaricus*.

ORF ID	Enzyme	Copies S. solf./S. tok.	Identity/similarity (%)
SSO786	amino acid specific permease	2/1	77/90
SSO790	acetylornithine deacetylase	2/1	57/74
SSO791	glycosyltransferase	3/1	61/76
SSO809	UDP-glucose 4-epimerase (galE-2)	1/0	44/64
SSO810	UDP-glucose 6-dehydrogenase (ugd)	1/0	42/60
SSO813	UTP-glucose-1-phosphate uridylyltransferase	1/0	34/48
SSO830	dTDP-glucose 4,6-dehydratase	3/1	79/93
SSO831	glucose-1-phosphate thymidylyltransferase	2/1	86/93
SSO832	dTDP-4-dehydrorhamnose reductase	2/1	80/93
SSO833	dTDP-4-dehydrorhamnose 3,5-epimerase	2/1	78/91
SSO837	polysaccharide biosynthesis related protein	3/1	76/88

S. solf., Sulfolobus solfataricus; S. tok., Sulfolobus tokodaii.

(PCR) approach to amplify attachment sites *attA*, *attP*, *attL*, and *attR*. Only the integrated pXQ1 is present and it appears in all host cells, since PCR amplification of an *attA* does not yield a product (She et al., 2001). This indicates that the plasmid is stably maintained.

The mechanism for capturing a pNOB8-type IE was studied recently with the conjugative plasmid pNOB8. *Sulfolobus* sp. NOB8H2 mutants carrying only the integrated pNOB8 plasmid have been isolated, and characterization of the mutants revealed that they exhibit deletion mutations in the *int* gene region, which must have occurred after integration (Bao Shen & Q. She unpublished). Thus, it appeared that stabilization of pNOB8-type IEs relies on mutating at least one of the key elements that are required for excision.

It is widely accepted that IEs are subject to decay by deletion and degeneration as a result of persistent evolutionary pressure, yielding remnants of the genetic elements. Only genes that confer a selective advantage to the host will be retained and evolved. Thus, the extent of degeneration of IEs reflects the sequential HGT events that contributed to their being encaptured in the chromosome. In the *S. tokodaii* genome, pST3 is a selfish integrated pRN-like plasmid and it still has an intact *int* and a perfect direct repeat of 44 bp. This indicates that this plasmid was captured relatively recently. For pST2, the direct repeat is 41 bp with one mismatch in the middle, and its *int* gene remains intact, suggesting that pST2 represents an "older" IE than pST3. Furthermore, ST4 is an IE that resulted from an HGT event earlier than those of pST2 and pST3, since there are extensive changes in both its repeat sequence and *int* gene. The direct repeat is only 29 bp long and the putative overlap sequence, represented by "O" within the SSV1 attachment sites in Fig. 10.1b, no longer exists. Moreover, at least three mutations had occurred in the *int* gene, as was revealed by analysis of the *int* remnants of STO255 and STO256 (She et al., 2004). There is another *int* remnant in the *S. tokodaii* genome, ST5, which is indicative of a more ancient HGT via site-specific integration, since one border cannot be defined due to the lack of a direct repeat. Its *int* gene is so degenerated that only a small ORF of 74 amino acids (STS021) is recognizable. These demonstrate that degeneration of *int* genes and attachment sites can be used as a molecular evolutionary marker for IEs.

In summary, three molecular events occurred in the evolution of the IEs present in the *S. tokodaii* genome, indicating that: (i) multiple HGT events occurred via site-specific integration; (ii) IEs were subject to deletion and/or degeneration; and (iii) only acquired fitness genes will be retained and ameliorated. These findings provide a basis for a further study of mechanisms of HGT events conferred by IEs.

Concluding remarks

Two types of integrases (SSV1 and pNOB8-type) are present in Archaea in addition to the universal Xer homologues. Whereas the pNOB8-type genetic elements integrate into their host genomes by a similar mechanism to bacterial systems, integration by those of SSV1-type produces partitioned *intN*s and *intC*s. Most archaeal integrases form one group, which, together with sister groups of bacterial integrases and Xer enzymes, forms the superfamily of tyrosine recombinases (Fig. 10.2). Biochemical characterization of SSV1-Int revealed that this archaeal enzyme forms a covalent bond to the DNA substrate, reinforcing the hypothesis that all tyrosine recombinases share the same general mechanism of integration. Furthermore, six catalytic residues revealed by crystal structural analyses are well conserved in bacterial integrases. These motifs are also highly conserved within these archaeal members, but three substitutions occurred relative to the bacterial consensus. These changes in sequence motifs may reflect special features of archaeal integrases, since mutation of any archaeal consensus inactivates the SSV1-Int protein.

Acquiring genetic information via site-specific integration represents an important means of HGT in Archaea. Many genes encoding known enzymes or proteins, as well as a lot of unknown ORFs, are transferred via site-specific integration. Furthermore, the evolution of the acquired DNA can be estimated by examining the extent of degeneration of the *int* genes and the attachment sites of IEs. However, the evolution of the acquired genes cannot be assessed directly at present. A much larger dataset, including many genomes of closely related archaeal species, is required for such an analysis. In Bacteria, genes encoding physiological traits can be transferred from one species to another and the DNA enters a new host via site-specific integration. Whether Archaea can do the same is not yet clear and suitable archaeal physiological traits should be identified and then used for this study. Thus, outstanding questions such as the origin, function, and evolution of IEs remain to be answered. This will have to be done through concerted efforts in genomics, genetic, and biochemical studies of Archaea.

11
Genetics

Moshe Mevarech and Thorsten Allers

Introduction

In an Indian parable, six blind men are asked to identify an elephant. The first, touching only its side, proclaims that it is a wall. The second man feels the tusk and believes it to be a spear, while the third mistakes the elephant's trunk for a snake. In this fashion, its knee is mistaken for a tree, its ear for a fan, and its tail for a rope. The moral of the story is that by examining the parts of the elephant in isolation, each man "was partly in the right, and yet all were in the wrong" (from "The Blind Men and the Elephant" by John Godfrey Saxe, 1816–87).

We face a similar problem in our research on Archaea. These organisms lend themselves to biochemistry and structural biology, as their proteins are often very stable under normal laboratory conditions. Significant advances in our understanding of Archaea have therefore been made on account of elegant biochemical experiments. However, without genetics we risk being like the six blind men – by examining the individual parts of Archaea in exquisite detail, we will neglect to look at the whole.

A cell is more than the sum of its parts, for many processes have overlapping functions that can only be teased apart by studying them in the living organism. Historically, the most successful approach has combined genetics and biochemistry. While this partnership has recently been joined by various systems-level techniques (identifiable by the suffix "omics"), genetics remains the most effective technique for examining the role of a protein *in vivo*. Specifically, the ability to examine the phenotype of the cell in the absence of a protein, through the generation of mutants, has proven most fruitful in bacterial and eukaryotic model organisms.

There are several prerequisites for genetics, not least of which is the ability to grow the organism in isolation, under conditions that can easily be replicated in the laboratory. It is at this first hurdle that most potential archaeal genetic systems fall. Since these organisms are renowned as extremophiles, or are extremely intolerant of oxygen, it can prove difficult to cultivate them. Archaea that grow in moderate-temperature aerobic environments have recently risen to prominence, but thus far, attempts to grow them in isolation have not been successful.

Despite difficulties, a number of genetic systems have been developed for methanogens (*Methanococcus maripaludis*, *Methanococcus voltae*, and *Methanosarcina* spp.), halophiles (*Halobacterium* spp. and *Haloferax volcanii*), and thermophiles (*Sulfolobus* spp., *Pyrococcus* spp., and *Thermococcus kodakaraensis*). All of these species can be grown on solid media, can be transformed with foreign DNA, and have selectable markers for genetic manipulation. In the following sections, we describe the tools and techniques available for archaeal genetics in these species.

Random mutagenesis

Similarities between archaeal, bacterial, and eukaryotic genes can be used to predict the function of only half of the genes in archaeal genomes. Mutagenesis provides the means to uncover the function of the other half, and thereby characterize genes and reaction pathways that are unique to Archaea. Later in this review, we describe reverse genetics methodologies for directed gene-knockouts, which have been developed for a number of Archaea. In this section, we describe random mutagenesis strategies.

Chemical mutagenesis, using ethylmethanesulfonate (EMS), has been used to mutagenize *Hfx. volcanii* (Mevarech & Werczberger, 1985), *M. maripaludis* (Ladapo & Whitman, 1990), and *P. abyssi* (Watrin & Prieur, 1996). UV radiation mutagenesis has been used to mutagenize *M. voltae* (Bertani & Baresi, 1987), *S. acidocaldarius* (Woods et al., 1997), and *P. abyssi* (Watrin & Prieur, 1996). Although mutations induced by chemicals or UV radiation are usually point mutations and therefore are difficult to map, they have the advantage that one can randomly mutate essential genes to produce temperature-sensitive conditional mutations. One strategy to isolate such mutations, once a phenotype has been identified, is by complementation with genomic libraries.

Enrichment of mutants

Point mutants for "loss of function" often do not have phenotypes that are easy to select, and therefore are difficult to isolate. Ampicillin enrichment, which makes use of the fact that only growing cells are sensitive to ampicillin, is common for Bacteria. Since Archaea do not have peptidoglycan cell walls, this method cannot be used. Nevertheless, several enrichment protocols for loss of function mutants have been developed for Archaea. The 5-bromo-2′-deoxyuridine (BrdU) enrichment protocol for halophilic Archaea (Soppa & Oesterhelt, 1989) makes use of the fact that growing cells can incorporate BrdU instead of thymidine into the chromosome, making these cells extremely sensitive to UV irradiation. Similarly, 6-azauracil (zUra) and 8-azahypoxanthine (zHyp) base analogs were used to enrich for auxotrophs of *Methanococcus maripaludis* (Ladapo & Whitman, 1990). The bacitracin enrichment protocol is based on the fact that bacitracin inhibits the synthesis of the glycoprotein S-layer of Archaea (Mescher et al., 1976). This protocol was used for enrichment for formate auxotrophic mutants of *Methanobacterium thermoautotrophicum* (Tanner et al., 1989).

Other methods of random mutagenesis

Random insertional mutagenesis is possible as long as the species is proficient for homologous recombination. In this approach, a genomic library of small fragments is created in a "suicide vector" carrying a selectable genetic marker (see section on vectors). The library is transformed into the archaeal cell and transformants that carry the genetic marker are selected. Since the vector cannot replicate in the cell, only cells in which the cloned fragment have been recombined into the chromosome will survive. If the size of the cloned fragment is smaller than a gene, there is a good chance that it will disrupt the gene into which it has been recombined. The disrupted gene can be easily isolated by cutting and self-ligating genomic DNA fragments from the mutant, followed by introduction into *E. coli*. Such random insertional mutagenesis has been used to isolate acetate auxotrophs of *M. maripaludis* (Kim & Whitman, 1999).

Transposon mutagenesis has been attempted, but its usefulness has been limited. Heterologous transposon mutagenesis was used to study the *nifH* gene of *M. maripaludis* (Blank et al., 1995), using the Mudpur transposon. This is a miniMu transposon containing the chloramphenicol acetyltransferase and puromycin transacetylase genes between the left and right ends of Mu, thus providing selectable markers for *E. coli* and *Methanococcus* species. The *M. maripaludis nifH* gene was cloned into bacteriophage chromosome, and transposition of the Mudpur transposon was performed in *E. coli* cells. Phage DNA carrying the *nifH* gene, into which the transposon had been inserted, was isolated and transformed into the *M. maripaludis* cells. The transposon containing the *nifH* gene was integrated into the *M. maripaludis* genome by homologous recombination and mutant cells were selected on puromycin-containing plates. Although the transposition was random, it was restricted to a defined region of the chromosome rather the whole genome.

An elegant *in vivo* transposition system for *Methanosarcina acetivorans* was developed in the laboratory of Bill Metcalf (Zhang et al., 2000), using a modified version of the insect *mariner*-family transposon *Himar1*. This transposon carries a puromycin-resistance marker as well as features that allow easy cloning of transposon insertions. Unfortunately, this system is restricted to methanogens, as eukaryotic or bacterial transposons cannot function in the hypersaline interior of halophiles, or at the high temperatures that are required by hyperthermophiles.

Genetic exchange

The role of lateral gene transfer (LGT) as a driving force in archaeal evolution has only recently been appreciated, and is illustrated by the observation that the fraction of "foreign" genes in Archaea has been estimated to be as high as 20–30% (Koonin et al., 2001). The best examples for possible large-scale LGT in Archaea are the inflated genome of *Methanosarcina mazei*, comprising 4.1 Mb of which 30% is bacterial in origin, and the halophilic archaeon *Haloarcula marismortui*. The latter has three complete rRNA operons, containing extensive sequence differences, which are located on two separate replicons (Baliga et al., 2004). Whereas evidence is accumulating for LGT, the mechanisms that underlie this transfer are largely unexplored. Below, we describe novel genetic exchange systems discovered in the halophilic archaeon *Haloferax volcanii* and the hyperthermophilic archaeon *Sulfolobus acidocaldarius*.

Hfx. volcanii is a prototrophic halophilic archaeon that grows in a simple defined medium. Auxotrophic mutants of this archaeon were isolated after chemical mutagenesis (Mevarech & Werczberger, 1985). To test whether genetic exchange occurs, auxotrophic mutants were grown separately, and paired combinations of the mutants were mixed and grown together on the solid surface of rich medium. The mixed cultures were then resuspended and plated on minimal plates. Prototrophic colonies were obtained at a frequency of about 10^{-6}, which is about 1000 times higher than the frequency of spontaneous reverse mutation. Since recombinants were obtained in each cross of two auxotrophic mutants, it was not possible to distinguish between donor strains and recipient strains. The efficiency of transfer of genetic information was dependent on the establishment of physical contact between the mutant cells and the viability of each mutant. No transfer was observed when the two mutants were grown together in shaking culture. The transfer was not sensitive to DNase, and no transfer was observed when the filtrate of conditioned medium of one mutant was supplied to the other. Kinetic studies have shown that even a brief contact between *Hfx. volcanii* cells is sufficient for genetic exchange to occur, but the efficiency of this process increases with time of incubation (Rosenshine & Mevarech, 1991).

One possible explanation for these findings is that the genetic exchange system of *Hfx. volcanii* is a novel form of bidirectional conjugation. Alternatively, the lack of classical donor and recipient strains can be explained by cellular fusion of neighboring cells, reminiscent of eukaryotic mating. Some genetic evidence suggests that the latter mechanism is correct:

1 In genetic analyses of crosses of mutant strains, each carrying several auxotrophic mutations, stable diploids that carry the entire genomes of the two parental strains were obtained. These diploids segregated into the two parental strains upon removal of the selection pressure (Rosenshine & Mevarech, 1991).

2 When an auxotrophic mutant carrying a plasmid that confers resistance to one antibiotic is crossed with another auxotrophic mutant that carries a plasmid that confers resistance to another antibiotic, it is possible to demonstrate near-100% linkage between the transfer of the auxotrophic markers and the transfer of the plasmids that encode the antibiotic resistance markers (Ortenberg et al., 1998).

By attempting to transfer selectable plasmids between different halophilic Archaea, it was found that the genetic exchange system is genus-specific. *Hfx. volcanii* can transfer plasmids to *Hfx. mediterranei* but not to *Haloarcula marismortui* or *Halobacterium salinarum* (Tchelet & Mevarech, 1994). It seems, therefore, that a specific adherence factor is required for a successful genetic exchange process. Unlike Bacteria that are encapsulated by rigid cell walls, most Archaea are surrounded by flexible surface S-layers made of glycoproteins. It is possible that some Archaea have specific surface protein receptors that recognize the S-layer proteins of similar Archaea, leading to adherence and fusion of the two cells.

Electron micrographs of *Hfx. volcanii* have shown that cells in colonies are connected by cytoplasmic bridges, and this cytoplasmic continuity is established *de novo* between cells grown together on solid surfaces (Rosenshine et al., 1989). Based on these and other findings, summarized in Ortenberg et al. (1998), a four-step mechanism was proposed for the genetic exchange system of *Hfx. volcanii*: (i) genus-specific cell–cell interaction that occurs when two cells establish physical contact; (ii) creation of narrow cytoplasmic bridges; (iii) expansion of the bridges, thus fusing the neighboring cells and creating diploids (occurs in about 1% of the cells that are connected by cytoplasmic bridges); (iv) recombination and segregation.

The genetic exchange system of *S. acidocaldarius* was discovered by Dennis Grogan. Details of this system are described in Chapter 12 in this volume, but some comparative aspects of this system can be mentioned here. As in the case of *Hfx. volcanii*, cell–cell contact seems to be required for the exchange process, donor–recipient relationships cannot be inferred, and all strains seem generally fertile with each other. However, unlike *Hfx. volcanii*, *S. acidocaldarius* is able to exchange genetic material in liquid culture. It was hypothesized, therefore, that *S. acidocaldarius* cultures contain cells that are able to form cell–cell associations in dilute liquid suspensions (Ghane & Grogan, 1998). The difference between the mating systems of *Hfx. volcanii* and *S. acidocaldarius* might be only quantitative. Namely, it is possible that the surface interactions in the latter are much stronger and require only a brief encounter of the cells. In *Hfx. volcanii*, on the other hand, the surface layer interactions might be much weaker and require long exposure of the cells to each other. If this model is true and the existence of specific surface receptors can lead to cell fusion, this phenomenon might be widespread among Archaea and can have an important evolutionary role, similar to mating in eukaryotes.

Transformation

The development of selectable markers and transformation protocols are intimately linked. Without a selectable phenotype, it is impossible to quantify transformation efficiency. On the other hand, the development of selectable genetic markers and their validation requires efficient transformation protocols. To circumvent this problem, Cline and Doolittle (1987) developed a protocol for introduction of foreign DNA into the halophilic archaeon *Hbt. halobium*, by transfection with the naked halophage ΦH DNA and scoring for the appearance of plaques on *Hbt. halobium* lawn. Similarly, Schleper et al. (1992) have demonstrated the efficient uptake of the SSV1 viral DNA by transfecting *Sulfolobus solfataricus* cells using electroporation. A different approach was employed by Bertani and Baresi (1987) for *Methanococcus voltae*. Auxotrophic mutants of this archaeon were isolated after irradiation with UV light or gamma rays. The transformation protocols were then developed by transforming the cells with purified chromosomal DNA of the wild-type strain, in order to complement the chromosomal mutations. It was shown that *M. voltae* is naturally competent for transformation.

In evaluating transformation protocols, we consider the following variables: (i) the organism; (ii) transformation of spheroplasts versus whole cells; (iii) transformation of linear DNA versus circular DNA; (iv) the use of auxiliary materials or physical treatments.

Halophilic Archaea

The first transformation protocol for an archaeon was established in halophilic Archaea. This protocol was based on the observation that the cell surface of these organisms is made of a paracrystalline glycoprotein layer, termed the surface (S) layer, which depends on Mg^{++} ions for stability. Spheroplasts can be readily generated by removal of Mg^{++} ions from the medium, and these spheroplasts will regenerate to form normal cells upon incubation in rich media containing Mg^{++} ions. It was found (Cline & Doolittle, 1987) that the spheroplasts of *Hbt. halobium* could be efficiently transfected by the phage DNA when low molecular weight polyethylene glycol (PEG 600) was added. Subsequently, this protocol was adapted to the transformation of plasmids into other halophilic Archaea, such as *Hfx. volcanii*, *Har. vallismortis*, and *Har. hispanica* (Cline & Doolittle, 1992). PEG-mediated spheroplast transformation of halophilic Archaea is a very efficient method enabling the introduction of plasmid DNA or linear DNA, yielding around 10^6 transformants/μg DNA.

In considering transformation experiments in halophilic Archaea, one should be aware of the existence of restriction/modification systems in some members of this group. DNA that is isolated from *Hfx. volcanii* is resistant to cleavage by restriction enzymes at 5'-CTAG-3' sites, indicating that this site is modified (Charlebois et al., 1987). *Hfx. volcanii* also has a restriction system that recognizes adenine methylated 5'-GATC-3' sites (which occur frequently in DNA that is isolated from *E. coli*), resulting in a reduction of transformation efficiency of methylated DNA. This can be circumvented by passaging the DNA through an *E. coli dam⁻* strain that is deficient in GATC methylation (Holmes et al., 1991).

Methanogenic Archaea

The first demonstration of transformation in methanogens was made by Bertani and Baresi (1987), who found that *Methanococcus voltae* is naturally competent to incorporate naked chromosomal DNA. Subsequently, it was shown (Gernhardt et al., 1990) that circular plasmid could also be incorporated by this natural transformation system. The main disadvantage with natural transformation is its extremely low efficiency of transfer (~8 transformants/µg DNA). A substantial increase in transformation efficiency of plasmid DNA (~700 transformants/µg DNA) was obtained by natural transformation of protoplasts rather than whole cells (Patel et al., 1994). Electroporation of protoplasts with plasmid DNA yielded fewer recombinants than natural transformation (~170 transformants/µg DNA), but was much higher when linear DNA was electroporated (~7400/µg DNA).

Much higher transformation yields were obtained by PEG-mediated transformation of *Methanococcus maripaludis* (Tumbula et al., 1994). This method is reminiscent of the PEG-mediated spheroplast transformation of halophilic Archaea, and yields ~10^5 transformants/µg DNA. In both methods, the addition of PEG is important (though in the *M. maripaludis* case PEG 8000 is used), high Mg^{++} concentration inhibits the transformation, and spheroplasts are produced prior to the regeneration of native cells.

An extremely efficient liposome-mediated transformation protocol was developed for *Methanosarcina* (Metcalf et al., 1997). As in the PEG-mediated transformation mentioned above, best results are obtained after stripping off the glycoprotein S-layer, by suspending the cells in isoosmotic, Mg^{++}-free medium. The DNA:liposome complexes are formed by mixing the DNA with DOTAP (Roche), and the complexes are added to the cells. This method was successfully applied to different *Methanosarcina* species, with the best results obtained for *M. acetivorans* (>10^7 transformants/µg DNA). For some reason, no transformants have been obtained with *M. mazei*.

Thermophilic Archaea

Heat-shock dependent DNA transformation has been demonstrated in *Thermococcus kodakaraensis* (Sato et al., 2003), *Sulfolobus acidocaldarius*, and *Pyrococcus furiosus* (Aagaard et al., 1996). Initially, 80 mM $CaCl_2$ was included in the transformation of the three species, but it was shown that *T. kodakaraensis* is naturally competent for DNA uptake and $CaCl_2$ is not necessary. The selectable marker used in the *T. kodakaraensis* transformation was the *pyrF* gene and selection was initially made in liquid, because the Gelrite used as a solidifier appears to contain traces of pyrimidine. A selectable marker that can be used for *P. furiosus* is the alcohol dehydrogenase gene from *S. solfataricus*, which confers resistance to butanol and benzyl alcohol, and selection was made in liquid culture (Aravalli & Garrett, 1997).

PEG-mediated spheroplast transformation has been demonstrated for *Pyrococcus abyssi* (Lucas et al., 2002). As in the halophilic archaeal transformation protocol, spheroplasts were made by adding EDTA to cell suspension, and the uptake of the DNA by spheroplasts was dependent on the addition of PEG 600. The *S. acidocaldarius pyrE* gene was used for selection by complementation of a *pyrE* mutation in the *P. abyssi* cells. Transformants were selected on Phytagel (Gellan gum) plates, and the transformation efficiency was 10^2–10^3/µg of plasmid.

Electroporation was applied successfully for transfection of *S. solfataricus* with the bacteriophage SSV1 DNA (Schleper et al., 1992). Transfectants were obtained at yields of up to 10^6 transfectants/µg DNA. *Methanococcus voltae* is another species that has also been transformed using electroporation (Patel et al., 1994), but in general this protocol has not been used widely in Archaea. An obvious reason is that it is not appropriate for Haloarchaea, which do not survive in salt concentrations below ~1 M NaCl.

Selectable markers

Cell growth inhibitors

Cell growth inhibitors, and their corresponding resistance genes, are widely used for genetic selection in Bacteria and eukaryotic cells. The successful application of these inhibitors depends on their ability to inhibit processes that are essential for cells. Most bacterial antibiotics inhibit one of the processes of the genetic information flow – DNA replication, transcription, and translation – or cell wall synthesis. Since Archaea have cell walls that are distinct from those found in Bacteria, and their components of genetic information processing are more similar to eukaryotic ones, most bacterial cell growth inhibitors are harmless to Archaea. Another problem in the application of bacterial cell growth inhibitors for archaeal genetics is the scarcity of bacterial resistance genes, and their inefficiency when expressed in archaeal cells.

Resistance to cell growth inhibitors can be obtained in one of the following ways: (i) cleavage or modification of the inhibitor; (ii) modification of the target molecule to reduce binding of the inhibitor; (iii) overproduction of the target molecule; (iv) an alternative pathway that is insensitive to the inhibitors; (v) active efflux pumping of the inhibitor. Resistance genes that code for enzymes that cleave or modify cell growth inhibitors are widely used for selection against antibiotics. These genes are usually isolated from naturally occurring bacterial plasmids or from antibiotic-producing bacterial strains. Therefore, they can be used for construction of selectable makers for most methanogens. However, many Archaea are either hyperthermophiles that thrive at temperatures well above those that are suitable for growth of most Bacteria, or extreme halophiles that have a hypersaline cytoplasm. For these extremophilic Archaea, the enzymes encoded by antibiotic resistance genes are inactive under the physiological conditions existing inside these archaeal cells. Nevertheless, several cell growth inhibitors have been discovered for Archaea, and their use in the development for selection of recombinant DNA will be described.

Selectable genes that code for proteins that render the inhibitor inactive The most widely used inhibitor in this group is puromycin. Puromycin has been shown to inhibit growth in *M. voltae* (Possot et al., 1988) and other methanogens. The resistance marker (puromycin transacetylase) is encoded by a bacterial gene from *Streptomyces alboniger*. Due to differences between bacterial and archaeal gene regulation, it is transcribed using an *M. voltae* promoter (Gernhardt et al., 1990). Similarly, the bleomycin-resistance *ShBle* gene from *Streptoalloteichus hindustanus* was cloned downstream of haloarchaeal promoter sequences and used to select for bleomycin-resistant *Hfx. volcanii* cells (Nuttall et al., 2000). In order to be able to use hygromycin B to select recombinants of the thermophilic archaeon *S. solfataricus*, the *E. coli hph* gene encoding for hygromycin B phosphotransferase gene was randomly mutated and a thermostable variant was isolated (Cannio et al., 1998). Similarly, the *S. solfataricus adh* gene that encodes for alcohol dehydrogenase was used as a selectable marker, which enables growth of *Pyrococcus furiosus* and *S. acidocaldarius* in media containing butanol or benzyl alcohol (Aravalli & Garrett, 1997).

Selectable genes that code for modified targets The most useful genetic marker in this group is the mutated haloarchaeal *gyrB* gene that confers resistance of Haloarchaea towards novobiocin, a potent inhibitor of DNA gyrase. The original novobiocin-resistant *gyrB* mutant was isolated from *Haloferax* strain Aa2.2 and conferred resistance of *Hfx. volcanii* to novobiocin (Holmes & Dyall-Smith, 1991). Since the novobiocin-resistant *gyrB* gene is very similar to the chromosomal *Hfx. volcanii gyrB* gene, this selectable marker can integrate into the chromosomal allele by homologous recombination or replace it by gene conversion. This problem does not occur when the novobiocin-resistant *gyrB* is used to select for transformants of *Halobacterium* species. Pseudomonic acid is a cell growth inhibitor that inhibits the enzyme isoleucyl tRNA synthase (IleS). A mutated *Methanosarcina barkeri ileS* gene was selected and cloned into a shuttle vector, conferring resistance of various *Methanosarcina* spp. to pseudomonic acid (Boccazzi et al., 2000). However, the fact that pseudomonic acid is not commercially available limits its use. Mutations in the 23S rRNA of *Halobacterium* spp. confer resistance of this haloarchaeon to anisomycin and thiostrepton. However, these resistance markers were not further developed as selectable markers.

Resistance to cell growth inhibitors by overexpression of the target gene The best example of this system is the resistance of *Haloferax volcanii* to mevinolin. Mevinolin and other compounds of the statin family inhibit the enzyme 3-hydroxy-3-methylglutaryl coenzyme A (HMG-CoA) reductase, which is involved in mevalonic acid biosynthesis (an essential precursor for lipid biosynthesis in Archaea). A spontaneous *hmgA* mutation was isolated in *Hfx. volcanii* that conferred resistance to this inhibitor (Lam & Doolittle, 1989). It was found that this mutation is in the promoter of the *hmgA* gene, and causes the overexpression of the gene (Lam & Doolittle, 1992). As in the case of the novobiocin-resistance gene mentioned above, the applicability of this mutated gene for *Hfx. volcanii* is limited due to frequent homologous recombination with the chromosomal gene. This drawback can be alleviated by using markers from distantly related species, such as the mevinolin-resistant *hmgA* mutant allele from *Haloarcula hispanica* (Wendoloski et al., 2001).

Resistance due to an alternative pathway An interesting example of this mechanism is the trimethoprim resistance marker of *Hfx. volcanii*. Trimethoprim is an efficient inhibitor of bacterial dihydrofolate reductase (DHFR), an enzyme that catalyzes the last step of the biosynthesis of the important C_1 carrier tetrahydrofolate. Among all Archaea so far examined, *Hfx. volcanii* is unique in being sensitive to trimethoprim. This archaeon has two different bacterial type DHFRs, one that is highly expressed and is very sensitive to trimethoprim, and another that is not expressed under physiological conditions and is resistant to this drug (Ortenberg et al., 2000). Other Archaea either have an alternative pathway for tetrahydrofolate synthesis (Levin et al., 2004) or use other C_1 carriers instead of tetrahydrofolate. It was possible to use the trimethoprim-resistant *Hfx. volcanii hdrA* gene as a selectable marker for *Hfx. volcanii* (Ortenberg et al., 2000).

Auxotrophic selectable markers

Since most bacterial antibiotics are ineffective in eukaryotic cells, yeast geneticists have adopted the "auxotrophic approach," where genes that are involved in amino acid biosynthesis and other metabolic pathways are used to complement easily obtainable deletion mutations of these genes. This powerful methodology is becoming increasingly popular in archaeal genetics. The most straightforward approach for the development of metabolic selectable markers is to select first for uracil auxotrophs by growing the cells in the presence of the uracil analogue 5-fluoroorotic acid (5-FOA). This compound inhibits the growth of cells that can produce uracil. On the other hand, cells that are mutated in either the gene encoding for orotate phosphoribosyl transferase (*pyrE*) or orotidin-5'-phosphate decarboxylase (*pyrF*) are auxotrophic for uracil, and therefore resistant to 5-FOA. These auxotrophic mutations can be complemented by cloned wild-type *pyrE* or *pyrF* genes from a heterologous organism to avoid homologous recombination, or from the same organism after deleting the corresponding genes by a method that is described in detail below. This approach was applied successfully in the construction of selectable makers for *Halobacterium* spp. (Peck et al., 2000), *Hfx. volcanii* (Bitan-Banin et al., 2003), *S. solfataricus* (Jonuscheit et al., 2003), *Thermococcus kodakaraensis* (Sato et al., 2003), and *P. abyssi* (Watrin et al., 1999). Thymidine, tryptophan, and leucine auxotrophs were created in *Hfx. volcanii* and could be complemented by the *hdrB*, *trpA*, and *leuB* respectively (Allers et al., 2004). Proline auxotrophs were created in *M. acetivorans* and could be complemented by the *E. coli proC* gene (Pritchett et al., 2004), and *trpE* and *hisD* mutants were created in *T. kodakaraensis* (Sato et al., 2005). Of the above markers, thymidine, tryptophan, and uracil auxotrophs are particularly useful, as thymidine is absent from some commercial yeast extracts, and tryptophan and uracil are absent from commercial casamino acids. The principle drawback with auxotrophic markers is that they cannot be developed easily in obligatory autotrophs, which include most methanogens.

Vectors

Following the description of transformation protocols and selectable markers, it remains to describe the vectors that are available for archaeal molecular genetics. As a complete list of plasmids is available in the Supplementary Online Table of Allers and Mevarech (2005), we describe here only a few key examples. All of these vectors are based on *E. coli* plasmids, in order that all the cloning can be carried out in this bacterium, before the construct is introduced into the corresponding archaeon. Two strategies have been used for the construction of archaeal vectors: (i) construction of shuttle vectors, by cloning indigenous archaeal replicons into *E. coli* plasmids and introducing selectable genetic markers; (ii) construction of integrative plasmids, consisting of *E. coli* plasmids with a selectable genetic marker only, which in the absence of an archaeal replicon must integrate into the host chromosome by homologous or site-specific recombination.

Shuttle vectors

Most, but not all, shuttle vectors use origins of replication taken from indigenous archaeal plasmids. For example, a series of shuttle vectors were constructed for the halophilic archaeon *Hfx. volcanii*, using the replication origin of pHV2, a naturally occurring 6.4 kb *Hfx. volcanii* plasmid. In order to avoid incompatibility problems, *Hfx. volcanii* was cured of pHV2, either by incubating cells with ethidium bromide to generate the widely used WFD11 strain (Charlebois et al., 1987) or by chasing out pHV2 with a selectable plasmid using the same origin of replication, to generate the DS70 strain (Wendoloski et al., 2001). The first shuttle vector of this series is pWL102, which confers resistance to mevinolin (Lam & Doolittle, 1989). However, this plasmid-borne mevinolin resistance marker is unstable due to frequent homologous recombination with the chromosomal *hmg* gene, as the two sequences are virtually identical. A more stable mevinolin-resistance marker has therefore been isolated from *Haloarcula hispanica* (Wendoloski et al., 2001). More recently, a series of vectors were constructed that contain the metabolic selectable markers *pyrE*, *leuB*, *trpA*, or *hdrB*; these plasmids can be selected by complementing the corresponding auxotrophic deletion mutants, which cannot grow in the absence of uracil, leucine, tryptophan, or thymidine, respectively (Allers et al., 2004). Another *Haloferax* replicon is derived from the plasmid pHK2 (Holmes & Dyall-Smith, 1990), and has been used to construct a series of shuttle vectors

carrying the novobiocin-resistance marker (Holmes et al., 1991, 1994). Interestingly, plasmids that are based on pHV2 fail to replicate in recombination-deficient *radA* mutants of *Hfx. volcanii*, but pHK2-based replicons do not have this problem (Woods et al., 1997; Woods & Dyall-Smith, 1997). pHV2-based plasmids can also replicate in *Halobacterium* spp., but they are not very stable in this halophilic archaeon. Therefore, alternative shuttle vectors have been derived from the *Halobacterium* plasmid pHH1 (Blaseio & Pfeifer, 1990).

Shuttle vectors for methanogens are less common. A series of shuttle vectors were constructed for *Methanosarcina* spp., using the replicon of the naturally occurring plasmid pC2A of *Methanosarcina acetivorans* and the *E. coli* plasmid R6K; the latter replicon permits manipulation of the copy number by choice of an appropriate *E. coli* host strain. These shuttle vectors carry the puromycin-resistance gene *pac* and provide some useful features, such as blue-white screening for recombinant clones and an assortment of polylinkers (Metcalf et al., 1997). Shuttle vectors for *Methanococcus maripaludis* have been derived from the cryptic plasmid pURB500 of *M. maripaludis* C5 (Tumbula et al., 1997). For example, plasmid pDLT44 is composed of the pURB500 *Methanococcus* replicon and the *E. coli* pUC replicon, and contains the ampicillin resistance gene *bla* for selection in *E. coli* and the puromycin resistance gene *pac* for selection in *Methanosarcina*. Other vectors featuring a gene expression cassette for *M. maripaludis* and a *lacZ* gene for blue-white screening in *E. coli* of recombinant clones have since been constructed (Gardner & Whitman, 1999).

Shuttle vectors for hyperthermophilic Archaea have been constructed using the replicon of the cryptic pGT5 plasmid from *Pyrococcus abyssi*. For instance, the pAG21 shuttle vector is based on the *E. coli* pBR322 plasmid, and contains the *Pyrococcus* pGT5 replicon and the *S. solfataricus adh* gene. The latter encodes for alcohol dehydrogenase, and has been used to select for transformants in liquid media containing butanol or benzyl alcohol (Aravalli & Garrett, 1997). Since alcohol-resistant recombinants cannot be selected on plates, the pYS2 vector was constructed (Lucas et al., 2002), which uses the pGT5 replicon and the *S. acidocaldarius pyrE* gene as a selectable genetic marker. A mutant strain of *P. abyssi* GE9 (which is devoid of pGT5) was isolated that requires uracil for growth, thereby allowing pYS2 transformants to be selected on minimal plates without uracil.

Unlike the vectors described above, which rely on replication origins derived from indigenous plasmids, several shuttle vectors have been constructed using the autonomously replicating sequence of the *S. solfataricus* virus particle SSV1. The plasmid pEXSs was constructed by cloning the 1700 bp viral region containing the putative origin sequence for viral replication into the *E. coli* pGEM5Zf(-) plasmid, together with a temperature-resistant mutant *E. coli hph* gene that confers resistance to hygromycin (Cannio et al., 1998). It was found that this plasmid could be stably maintained in the cell without integrating into the chromosome. Exposing the cells to UV light or to mitomycin C has been reported to increase the copy number of the plasmid.

Integrating plasmids

Integrating plasmids circumvent the need for an autonomous origin of replication in order to be maintained in archaeal cells, and instead rely on their ability to integrate into the chromosome by homologous recombination. These plasmids are very useful for Archaea that lack indigenous plasmids, or where indigenous plasmids cannot be modified to accommodate the bacterial replicon and the selection marker. Integrating plasmids are often used in gene knockout protocols (see the section on reverse genetics), but were popular at the beginning of the archaeal molecular genetic era as a means to clone and maintain foreign genes in Archaea. One of the first integrating plasmids was the pMip1 plasmid for *M. voltae* (Gernhardt et al., 1990), and was based on the *E. coli* plasmid pUC18, into which the puromycin resistance gene *pac* and the *M. voltae hisA* genes were cloned. The purpose of the *hisA* gene was to enable integration of the plasmid on the *M. voltae* chromosome by means of homologous recombination.

Integrating plasmids are useful only in strains where homologous recombination is efficient. An elegant solution has been found for *S. solfataricus* P1 and P2 isolates that appear to lack homologous recombination. In these strains, vectors based on the SSV1 virus have been used to integrate into the chromosome by site-specific recombination. One such plasmid is pSSV64, which is made of the entire SSV1 DNA cloned into pUC18 for propagation in *E. coli*, and the *pyrEF* genes as selectable genetic markers that complement pyrimidine auxotrophic mutants (Jonuscheit et al., 2003). An intriguing feature of the SSV1-derived vectors is that they spread efficiently in cultures without lysis of the cells, thus eliminating the need for efficient transformation.

Reverse genetics

As more and more archaeal genome sequences become available, site-specific chromosomal mutagenesis, or reverse genetics, becomes increasingly popular. The various strategies for directed chromosomal mutagenesis are presented in Fig. 11.1. The simplest strategy is a direct replacement of a gene with a selectable marker, by recombination between a linear DNA fragment, which comprises the selectable marker flanked by DNA fragments from the upstream and downstream regions of the gene, and the chromosomal target (Fig. 11.1a). This strategy is widely used in yeast genetics, where linear DNA fragments introduced by transformation are relatively stable. Owing to the lack of knowledge about the stability of linear DNA fragments in various Archaea, this method has been used only rarely (Sato et al., 2005). Since the number of useful selectable markers for Archaea is very limited, this method has the disadvantage that the selectable marker cannot be reused. Furthermore, direct replacement of genes with a selectable marker is not possible where the gene is essential.

In order to overcome these disadvantages, it is possible to use the "pop-in/pop-out" strategy (Fig. 11.1b). In this strategy, the upstream and downstream flanking regions of the gene to be deleted are PCR amplified and cloned together into a "suicide plasmid," which carries a selectable genetic marker but cannot be autonomously replicated in the target organism. The plasmid is introduced into the cells and transformants, in which the plasmid has been integrated into the chromosome by homologous recombination, are selected. Upon relieving the selection pressure, spontaneous intrachromosomal homologous recombination will lead to excision of the integrated plasmid; this leaves behind either the wild-type allele or the deletion allele. The "pop-in/pop-out" method was used successfully in *Hfx. volcanii* to delete the two dihydrofolate genes

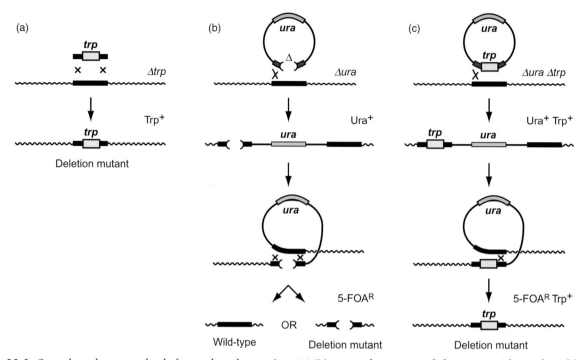

Fig. 11.1 Gene knockout methods for archaeal genetics. (a) Direct replacement of the gene with a selectable marker (gene for tryptophan biosynthesis shown here). Linear DNA, where the marker is flanked by upstream and downstream regions of the gene, is used to recombine with the chromosome. (b) Pop-in/pop-out method, which uses circular DNA to recombine with the chromosome. A plasmid carrying a gene deletion construct (comprising the upstream and downstream regions of the gene), and a selectable marker for uracil biosynthesis, is allowed to integrate at the wild-type locus by recombination. The plasmid is then allowed to excise by intrachromosomal recombination, resulting in either a wild-type cell or a deletion mutant. These can be selected on 5-flourootic acid (5-FOA) media, which is toxic to cells containing the uracil biosynthesis marker. (c) Variant of the pop-in/pop-out method, where the gene is replaced with a selectable marker (here tryptophan, as in a). This allows the direct selection of deletion mutants (as opposed to wild-type cells) in the final step, as only the former will be 5-FOA-resistant and able to grow without added tryptophan.

and the thymidylate synthase gene (Ortenberg et al., 2000).

In order to facilitate the excision of the integrated plasmid, it is possible to use counter-selectable genetic markers. The most popular counter-selectable markers are the *pyrE* or *pyrF* genes involved in uracil biosynthesis. Uracil auxotrophs are resistant to 5-fluoroorotic acid (5-FOA), owing to their inability to convert 5-FOA to the toxic analogue 5-fluoro-UMP, whereas uracil prototrophs are usually sensitive to this compound. The "pop-in/pop-out" strategy (Fig. 11.1b) is carried out in a ura⁻ strain, using a "suicide plasmid" that carries the complementing *pyrE* or *pyrF* gene and the flanking sequences of the gene to be deleted. Transformants in which the plasmid has been integrated into the chromosome are selected on plates that are deficient in uracil (such as casamino acid plates). Intramolecular recombinants that have lost the plasmid are then counter-selected by growing the cells on plates containing 5-FOA (and a small amount of uracil). The counter-selectable "pop-in/pop-out" method has enabled the establishment of gene-knockout systems in *Halobacterium* spp. (Peck et al., 2000; Wang et al., 2004), *Hfx volcanii* (Bitan-Banin et al., 2003), and *T. kodakaraensis* (Sato et al., 2003, 2005). A slightly modified version of this method, in which the gene to be knocked out is replaced by a selectable marker, enables direct selection of the deletion allele (Allers et al., 2004) (Fig. 11.1c).

An alternative counter-selectable gene that works for some Archaea, such as *M. maripaludis* (Moore &

Leigh, 2005), is the *hpt* gene that encodes for hypoxanthine phosphoribosyl transferase. The wild-type allele of this gene confers sensitivity to the base analogue 8-azahypoxanthine. This gene was used to knock out the genes encoding for alanine dehydrogenase, alanine racemase, and alanine permease in this archaeon. However, unlike the "pop-in/pop-out" methods using the *pyrE* or *pyrF* genes, gene knockout with the *hpt* gene requires an additional marker for positive selection of plasmid integration.

Additional recombinant DNA tools for archaeal genetics

To complete the arsenal of molecular genetic tools available for Archaea, we should consider two useful tools: (i) reporter genes, which are used to study the structure of regulatory elements that control gene expression; (ii) high-level expression systems that enable the production of recombinant proteins in the different archaeal cells, and controllable promoters that enable the expression or repression of genes at will.

Reporter genes

Reporter genes encode products that can be easily detected and quantitatively assayed. These genes are used to measure the strength of promoters and to analyze the different elements that control gene expression. The most familiar reporter gene is the *lacZ* gene of *E. coli*, which encodes the enzyme β-galactosidase. The activity of this enzyme can be easily assayed *in vitro*, and its expression *in vivo* can be detected by including in the medium the chromogenic substrate 5-bromo-4-chloro-3-indoyl-β-D-galactopyranoside (X-gal). In the presence of β-galactosidase, X-gal is converted into an insoluble blue dye that colors the colonies of the *lacZ*$^+$ cells. The *E. coli lacZ* gene can be used directly in mesophilic methanogens, as was demonstrated for *M. maripaludis* (Cohen-Kupiec et al., 1997) and *M. acetivorans* (Apolinario et al., 2005). Similarly, the *E. coli uidA* gene encoding β-glucuronidase has been used in *M. voltae* (Beneke et al., 1995) and *M. acetivorans* (Pritchett et al., 2004), and the *Bacillus subtilis treA* gene encoding for trehalase was used to monitor gene expression in *M. voltae* (Sniezko et al., 1998). Since *E. coli* enzymes are not active in the high salt concentrations existing in halophilic Archaea, and at the high temperatures necessary for growth of hyperthermophilic Archaea, analogous genes have to be isolated that can function under these extreme conditions.

The *bgaH* gene, which encodes for a halophilic β-galactosidase, was isolated from *Haloferax alicantei* (Holmes & Dyall-Smith, 2000) and was used to monitor gene expression in the halophilic Archaea *Halobacterium salinarum* (Patenge et al., 2000) and *Haloferax volcanii* (Gregor & Pfeifer, 2001). Similarly, the *S. solfataricus lacS* was used to monitor β-galactosidase activity in a mutant strain of *S. solfataricus* (Jonuscheit et al., 2003), where the chromosomal *lacS* is inactivated due to an insertion element (Schleper et al., 1994). Recently, a modified derivative of GFP (Green Fluorescent Protein) was developed that is soluble and active in the hypersaline cytosol of the halophilic archaeon *Hfx. volcanii* (Reuter & Maupin-Furlow, 2004).

High expression system and controllable promoters

Studies of protein structure and function require the production of substantial amounts of native proteins. The expression of archaeal proteins in heterologous bacterial systems is not always trivial. First, many Archaea are extremophiles that grow at temperatures, salt concentrations, and O_2 partial pressures that are very different from those that exist in bacteria. The intracellular environment in bacteria may affect the proper folding of the proteins and their susceptibility to oxygen. Second, when expressed in the native organism, proteins can be isolated that are complexed with cofactors or other proteins not found in Bacteria. It is therefore advantageous to express archaeal proteins in their native hosts, if possible, or in closely related hosts when no expression system is available for the native host. In a high expression vector, the gene to be expressed is put under the control of a strong (preferably controllable) promoter.

An expression vector for *M. maripaludis* was constructed by combining the *M. voltae* histone promoter with multiple cloning sites, in either an integrative vector or a shuttle vector (Gardner & Whitman, 1999). Similarly, the *Hbt. salinarum* P2 promoter of the rRNA operon enabled the construction of haloarchaeal expression vectors (Jolley et al., 1996; Kaczowka et al., 2005). Recently, high-level expression vectors were constructed for the hyperthermophilic archaeon *S. solfataricus* (Albers et al., 2006). The expression of genes cloned in these vectors is controlled either by the heat-inducible promoter of the major chaperonin TF55 or by the arabinose-inducible promoter of the arabinose-binding protein AraS of *S. solfataricus*. These vectors have enabled the high-level production of homologous and heterologous proteins, and of soluble as well as membrane-bound proteins. This arabinose-induced promoter is, so far, the only archaeal controllable promoter whose inducer has been identified.

Summary

As we have shown above, archaeal genetics is no longer in its infancy. By now, many of the techniques are in place for relatively sophisticated experiments. However, compared to the tools available for *E. coli* or yeast, archaeal genetics is still a blunt instrument. More work needs to be done, and efforts should be focused where they are most likely to benefit the greatest number of researchers.

The most significant limitation of reverse genetics, where a known gene is deleted and the resultant phenotype is analyzed, is in the study of essential genes. With traditional methods of chemical mutagenesis, it is possible to generate temperature-sensitive or hypomorphic point mutations (where the gene product is not completely inactivated). However, isolating such mutants requires laborious screening, and it seems that today's scientists have lost the appetite for this style of research. We should therefore follow the lead of eukaryotic geneticists, who have exploited RNA interference (RNAi) to the full. While RNAi is unlikely to work in Archaea (the requisite RNase III-like enzyme Dicer is not present), the potential exists to use the related technique of antisense RNA, where translation is blocked by formation of an RNA duplex. Evidence is accumulating from studies of non-coding RNAs, which suggests that this method of gene regulation is common in Archaea (see Chapter 20 in this volume). By harnessing antisense sequences to inducible/repressible promoters, it should be possible to create a transient inactivation of even the most essential gene.

Finally, it would be prudent to consider the early days of bacterial genetics, in particular the contributions made by studies on bacteriophages. The list of genetic tools that emerged directly from work carried out by phage geneticists is impressive – transduction, site-specific recombination, and strong promoters (such as those found in common gene expression vectors) are just some of techniques still in use today. The study of viruses that can infect (and better still lysogenize) archaeal hosts such as those discussed here will lead to the development of better genetics tools, and to a deeper understanding of the third domain of life.

12
Genetic properties of *Sulfolobus acidocaldarius* and related Archaea

Dennis W. Grogan

Microbiology of the genus *Sulfolobus*

In 1972, Thomas Brock and co-workers described a novel prokaryote cultured from acidic hot springs in Yellowstone National Park (USA) and similar geothermal sites around the world. Their various isolates shared an array of unusual properties, including a "lobate" cell morphology, optimal growth at extremely high temperature and low pH, and a proteinaceous cell wall (Brock et al., 1972). One of these isolates, "98-3," was deposited in culture collections as the type strain of the new species *Sulfolobus acidocaldarius*. This species was later identified as one of the Archaea, and was joined in later years by other thermoacidophiles (see Chapter 4 in this volume), including *S. solfataricus*, described by Zillig and co-workers (Zillig et al., 1980), *S. tokodaii*, described by Oshima and co-workers (Suzuki et al., 2002), and others documented by the groups of K. O. Stetter and others. The practical advantages of their heterotrophic, aerobic growth made *Sulfolobus* species popular experimental subjects for molecular, biochemical, evolutionary, and genetic studies on Archaea.

As is true for any microorganism, the genetics of *Sulfolobus* species relate closely to their cellular and metabolic properties. All *Sulfolobus* spp. are obligate aerobes and utilize a number of carbon sources for heterotrophic growth; many can grow on synthetic media with single carbon and nitrogen sources, thus enabling the identification and analysis of auxotrophic mutants. The cells also withstand considerable osmotic stress at room temperature, which enables diverse transformation methods and other experimental treatments to be applied. *Sulfolobus* cells are completely covered by a highly ordered, stable layer of glycoprotein subunits held out from the cytoplasmic membrane by thin "spacer elements" (Baumeister et al., 1989). This "subunit cell wall," or "S-layer," appears to mediate occasional cell–cell attachments seen in electron micrographs of *S. acidocaldarius* (Fig. 12.1). Polar lipids of the *S. acidocaldarius* cytoplasmic membrane have two C_{40} isoprenoid chains linked to glycerol at one end and a nine-carbon backbone at the other end (Sugai et al., 1995). The membrane contains additional glycoproteins and a sophisticated energy-conserving system, documented in considerable detail by Günter Schäfer and co-workers.

In *S. acidocaldarius*, cell division is followed almost immediately by chromosome replication, so that most cells in a population contain two complete copies of the chromosome (Bernander & Poplawski, 1997). A variety of transient stresses cause *S. acidocaldarius* cells to arrest in this "G2" phase, and release from this arrest can be used to synchronize cultures (Hjort & Bernander, 2001). Measurements of gene abundance in growing cultures have revealed that the *S. acidocaldarius* and *S. solfataricus* chromosomes replicate by simultaneous initiation at three widely separated origins of replication (Lundgren et al., 2004). The *S. acidocaldarius* cytoplasm contains several small, basic, and abundant DNA-binding proteins, which stabilize the duplex DNA against strand separation at the extremely high growth temperature. These proteins include Sac7d, which binds dsDNA and induces sharp bending (Robinson et al., 1998), and Sac10b, whose *S. solfataricus* homolog Sso10b ("alba") is reversibly acetylated *in vivo*, which modulates its DNA-binding affinity (Bell et al., 2002).

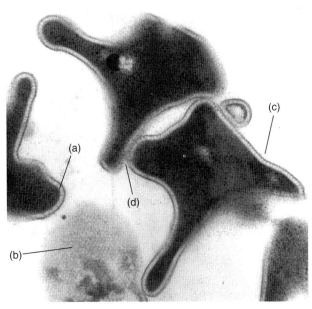

Fig. 12.1 Structural features of *S. acidocaldarius* cells. Cells were fixed, embedded, stained with uranyl acetate, and sectioned; area shown is about 1.5× 1.5 μm. Features indicated: (a) cytoplasmic membrane; (b) region of glycoprotein cell wall (S-layer) sectioned in-plane, showing the hexagonal symmetry of its periodic structure; (c) lateral section of cell envelope, showing the thin, regular spacer elements and the constant distance between glycoprotein layer and cell membrane; (d) area in which two cells have become associated.

Native genetic properties of *S. acidocaldarius*

The restriction-modification (R-M) system of *S. acidocaldarius*, *Sua*I, was reported in 1985 by Prangishvili et al., and remains one of the few R-M systems of Archaea to be analyzed in functional terms. The *Sua*I endonuclease cleaves the tetranucleotide GGCC, and the only effective protection against cleavage so far identified is N^4-methylation of the first C (Grogan, 2003). Based on its chemical properties, this modification should be less mutagenic than C5 methylation, and the threat of mutation is further decreased by underrepresentation of the *Sua*I recognition sequence in the *S. acidocaldarius* genome (Chen et al., 2005).

Another distinctive genetic property of *S. acidocaldarius* is the ability to transfer chromosomal DNA by a process of conjugation. The first evidence of this phenomenon was reported by Aagaard et al. (1995), who electroporated a self-splicing intron from the 23S tRNA of another hyperthermophilic archaeon, *Desulfurococcus mobilis*, into various *Sulfolobus* spp. In the case of *S. acidocaldarius*, the intron spread rapidly throughout the culture, and was found to have transferred from the original recipient into a different, co-cultured strain (Aagaard et al., 1995). At about the same time, attempts to differentiate between *S. acidocaldarius* auxotrophs by cross-streaking detected the spontaneous formation of prototrophic colonies from mixtures of stable auxotrophs. This was confirmed to reflect a mode of genetic exchange and recombination that required cell–cell contact and was resistant to DNase (Grogan, 1996; Ghane & Grogan, 1998).

This mode of genetic exchange provides a potentially interesting comparison to two other examples of conjugation in Archaea: chromosomal exchange in the halophile *Haloferax volcanii*, and transfer of certain plasmids among other *Sulfolobus* species. Conjugation in *H. volcanii* (described in detail in Chapter 11 in this volume) involves the transfer of chromosomal markers between two strains, but the two strains appear to contribute equally to the exchange, in the sense that neither is specialized to serve as the sole donor or recipient. In contrast, the conjugal plasmids that have so far been found in isolates of *S. solfataricus* and "*S. islandicus*" transmit themselves from their initial host to a plasmid-free recipient species. This transfer is documented experimentally by liquid co-culture, followed by extensive growth of the recipient and recovery of the plasmid in association with its genomic DNA (Schleper et al., 1995). This phenomenon thus appears to be analogous to bacterial conjugation, in the sense that it is initiated between a plasmid-containing donor and a plasmid-free recipient. Comparative sequence analysis suggests, however, that the *Sulfolobus* conjugative plasmids have a simpler transmission mechanism than seen in most conjugative plasmids of bacteria (Greve et al., 2004). Whereas interspecific conjugation promoted by pNOB8 is accompanied by formation of mixed cell clusters (Schleper et al., 1995), recombinant frequencies from *S. acidocaldarius* matings are too low (typically less than 10^{-4}) to make microscopic measurements meaningful. However, the genetic selections available in *S. acidocaldarius* enable various manipulations to be evaluated for their effects on conjugation. Data from this approach indicate that conjugation can initiate rapidly upon cell mixing and can occur efficiently in liquid medium, i.e. without stabilization of cell–cell contact by adsorption to a surface (Ghane & Grogan, 1998). More rigorous investigation of the transfer mechanism awaits efficient, nonlethal methods of interrupting the transfer process and blocking reinitiation.

Another question that warrants further study is the relationship between the system that transfers chromosomal DNA between S. acidocaldarius cells and the systems that transfer conjugal plasmids in other Sulfolobus species. The generation of prototrophic recombinants from S. acidocaldarius pyr mutants has not been observed with pyr mutants of S. solfataricus and "S. islandicus," reinforcing the idea that the S. acidocaldarius mode of transfer is not common among Sulfolobus species. However, homologs of genes on conjugal Sulfolobus plasmids have been documented in the S. acidocaldarius genome, raising the possibility that these genes mobilize the chromosome (Chen et al., 2005), in analogy to Hfr conjugation. It may be possible to test this explanation by similar evaluation of S. tokodaii, since at least one pNOB8-related plasmid has been detected in this species's genome (Kawarabayasi et al., 2001), making it analogous to the situation in S. acidocaldarius. Also, if appropriate mutagenesis techniques can be applied to S. acidocaldarius, it may be possible to disrupt the putative transfer genes found in its genome and determine whether this yields a "sterile" phenotype.

In the meantime, genetic properties suggest that much of the S. acidocaldarius chromosome can be transmitted between strains, and that much of the donor DNA is ultimately incorporated into the recipient as small fragments. This hypothesis is prompted by several experimental observations: (i) phenotypically diverse markers, which could represent up to ten different regions of the chromosome, were transmitted at reasonable frequencies by conjugation, and no marker was found that did not transfer (Grogan, 1996); (ii) unselected markers exhibited negligible genetic linkage to selected markers, even when separated by only 500–600 bp (Hansen et al., 2005); and (iii) when one parental strain was forced to serve as the recipient by irradiating the other parental strain with lethal doses of gamma radiation, the yield of recombinants was very low if the recipient's marker was a large deletion mutation rather than a point mutation. Since this last effect was not evident in standard (symmetrical) matings and could not be attributed to fragmentation of the donor chromosome, it argued that a large proportion of the DNA segments yielding recombinants under these conditions were less than about 400 bp long (Hansen et al., 2005).

Whereas S. acidocaldarius has restriction and DNA transfer systems not evident in other Sulfolobus species, it also lacks some capabilities that other species exhibit. For example, S. acidocaldarius apparently fails to propagate the diverse viruses and plasmids that have been recovered from other Sulfolobus spp. (Schleper et al., 1995; Zillig et al., 1996). In the case of the small, multicopy plasmids pRN1 and pRN2, this cannot be attributed to restriction by SuaI, because these plasmids lack the GGCC recognition site. Perhaps more striking is S. acidocaldarius's apparent lack of functional insertion sequences, which are otherwise abundant in Bacteria and Archaea. Although S. solfataricus and "S. islandicus" have been shown by various criteria to contain a diversity of transpositionally active ISs, the few intact ISs in the S. acidocaldarius genome detected by sequence analysis are not thought to be active (Chen et al., 2005), and none have been observed to cause mutation in a sensitive assay (described below).

Molecular-genetic properties

Culture media containing uracil plus 5-fluoro-orotic acid (FOA) select Sulfolobus mutants lacking either of the last two enzymatic steps of de novo pyrimidine biosynthesis. In the absence of reliable antibiotic selections for Sulfolobus genetics, the FOA selection provides two important capabilities. First, because resistance results from loss of enzyme function, the selection recovers the widest range of mutations possible. Second, the auxotrophic mutants selected by FOA provide a reverse selection for genetic events that restore growth in the absence of uracil. This reversibility is not shared by most resistance phenotypes, and allows the corresponding Sulfolobus genes (designated pyrE and pyrF) to be restored for subsequent rounds of selection.

Combined with appropriate controls, the FOA selection has shown that the S. acidocaldarius pyrE and pyrF genes are replicated with a maximum error rate of about 7.8×10^{-10} bp during exponential growth

(Grogan et al., 2001). The corresponding genomic error rate, about 0.0018 mutational events per genome replication, falls slightly below the remarkably tightly clustered values from diverse microbial DNA genomes, which span a 10^4-fold range of size and error rate per bp (Table 12.1). This result is significant in at least two respects: (i) it corroborates that a basic evolutionary force selects this approximate level of accuracy as optimal for all microbial genomes (Drake et al., 1998), regardless of environmental growth conditions; and (ii) it implies that *S. acidocaldarius* successfully compensates for the increased rates of depurination, deamination, and other spontaneous decomposition reactions expected at its intracellular temperatures and pH values. In fact, the *S. acidocaldarius* genomic mutation rate, which was calculated so as to avoid underestimation, is the lowest that has been measured for any microbial genome, despite the fact that this archaeon's optimal growth conditions rapidly kill the other microorganisms used for comparison. There is no firm evidence that other *Sulfolobus* species do not have similar accuracies of DNA replication, although this remains difficult to show experimentally, because frequent IS transposition dominates the mutational spectra of some other species and elevates the forward mutation rate above that seen in *S. acidocaldarius* (Martusewitsch et al., 2000; Blount & Grogan, 2005).

The ability to achieve such accurate replication under harsh conditions gives *S. acidocaldarius* and other *Sulfolobus* species special significance for understanding molecular mechanisms of mutation avoidance. In this context, the observation that the genomes of these and other hyperthermophilic Archaea lack important DNA repair genes remains especially perplexing (see Chapter 15 in this volume). Although hyperthermophilic Archaea may protect their DNA passively (through association with small, basic, proteins, for example) the potential for this strategy to replace active DNA repair seems limited. For example, (i) several other prokaryotic lineages encode such DNA-binding proteins but have not abandoned conventional MMR and NER pathways (White, 2003; Grogan, 2004); (ii) some of the most effective protection proteins, which occur in bacterial spores, decrease the rate of spontaneous decomposition only about 20-fold, compared to the approximately 1000-fold increase predicted in *Sulfolobus* cells; and (iii) the spore proteins impede vegetative growth and are normally destroyed early in the process of germination.

While quantitative fluctuation tests measure the accuracy of genome replication, sequencing of randomly chosen mutants documents the "mutational spectrum," which reveals the kinds of errors that escape the cell's accuracy-enforcing mechanisms. The available data (Grogan et al., 2001) allow *S. acidocaldarius* to be compared to model bacterial and eukaryotic systems in molecular terms. As shown in Table 12.2, the mutational spectrum of *S. acidocaldarius* resembles that of *E. coli* in being dominated by frameshift mutations. However, the *E. coli* frameshift affects a single block of four base pairs whereas the *S. acidocaldarius* frameshifts are primarily −1 events in mononucleotide runs throughout the target genes. The pattern in *S. acidocaldarius* is reminiscent of bacterial and yeast mutants deficient in DNA mismatch repair (MMR), thereby raising questions about the MMR capabilities of *S. acidocaldarius* and other hyperthermophilic archaea (see Chapter 15 in this volume). In contrast, the extremely low frequency of larger deletions within the *pyrE* and *pyrF* genes, and the absence of short direct repeats and inverted repeats at the endpoints of the few deletions that have been recovered, suggests that the dominant mechanisms that promote deletion in other microorganisms do not operate effectively in *S. acidocaldarius* (Grogan & Hansen, 2003). Finally, as mentioned above, the absence of any IS-induced mutation corroborates the sequence evidence that IS elements are inactive in *S. acidocaldarius* (Chen et al., 2005).

Table 12.1 Spontaneous mutation rates in microbial genomes.

Genome*	Rate per base pair†	Rate per genome†
phage M13	7.2×10^{-7}	0.0046
phage λ	7.7×10^{-8}	0.0038
phage T_2, T_4	2.4×10^{-8}	0.0040
Sulfolobus acidocaldarius	7.8×10^{-10}	0.0018
Escherichia coli	5.3×10^{-10}	0.0025
Saccharomyces cerevisiae	2.2×10^{-10}	0.0027
Neurospora crassa	7.2×10^{-11}	0.0030

*See Grogan et al. (2001) and other references cited therein; genomes are listed in order of increasing size.

†Rates are expressed as mutational events per generation.

Table 12.2 Comparison of spontaneous mutations in different microorganisms.

	Sulfolobus acidocaldarius pyrE	Escherichia coli lacI	Saccharomyces cerevisiae URA3
Frameshift			
+1	17.8	1.1	<2.9
−1	61.4	4.0	5.7
Block*	3.0	72.0	<2.9
Total frameshifts	82.2	77.1	5.7
Substitution			
G:C to A:T	7.9	6.2	14.3
A:T to G:C	1.0	0.7	<2.9
Transversions	3.0	4.1	74.3
Total substitutions	11.9	11.0	88.6
Duplications	5.0	1.0	2.9
Deletions	1.0	9.9	2.9
TE insertions	<1	1.1	<2.9

Data are percentages of independent, spontaneous mutants, representing 101 *pyrE* mutants (Grogan et al., 2001), 729 *lacI* mutants (Halliday & Glickman, 1991), and 35 *URA3* mutants (Lee et al., 1988).

*Includes all mutations in which a short (3–5 bp) repeating unit is added to, or subtracted from, a naturally occurring repeated sequence; in the case of trinucleotide repeats, these do not change the reading frame.

Thus, the experimental data predict that the *S. acidocaldarius* genome should be among the most stable with respect to the preservation of gene function and order, not only of hyperthermophilic Archaea, but of all microorganisms characterized so far. The different rates for individual events further imply a characteristic pattern for natural genetic variation in this species. Specifically, the most frequent mutations (frameshifts, base-pair substitutions, and tandem duplications) tend to inactivate genes but can be reversed by subsequent mutation. This reversibility, combined with the low rate of mutational events overall, implies a mode of genome evolution in *S. acidocaldarius* that could be described as "cautious" or "tentative"; in this scheme, genes must escape selection for many generations before they become permanently inactivated (Grogan & Hansen, 2003).

This situation contrasts sharply with other *Sulfolobus* spp., which exhibit active IS transposition (Martusewitsch et al., 2000). For example, the *S. solfataricus* P2 genome has the highest content of insertion sequences of any archaeal or bacterial genome yet reported. About 10% of this 2.99 Mb genome consists of small transposable elements, representing at least ten families present in an average of 25 copies each. The majority of the IS copies appear to be intact (Brugger et al., 2002), and a number transpose actively, as indicated by the frequency of spontaneous rearrangements detected in the course of genome sequencing (She et al., 2001). The situation also appears to apply to *S. solfataricus* strain P1 (Martusewitsch et al., 2000) and a related species not yet formally described, "*Sulfolobus islandicus*." In particular, experimental data from "*S. islandicus*" indicate that insertional inactivation of host genes by most *Sulfolobus* ISs is permanent, i.e. not reversed by precise excision events even when selection for restoration of gene function is applied (Blount & Grogan, 2005).

Homologous recombination

The conjugational mechanism of *S. acidocaldarius* so far provides the only natural genetic system that quantifies homologous recombination between chromosomal DNA segments of hyperthermophilic Archaea. It has been used to demonstrate that the DNA transfer or recombination is enhanced by prior UV-irradiation, and that this enhancement is countered by photo-reactivation or prolonged incubation of cells in the dark at physiological temperatures (Wood et al., 1997; Schmidt et al., 1999). These observations suggest that UV photoproducts are converted first into recombinogenic lesions that are repaired by enzymatic processes in the dark (Schmidt et al., 1999); this is consistent with the dark repair of UV photoproducts detected biochemically in *S. solfataricus* (Napoli et al., 2004).

Measurements of recombination frequencies following conjugation have also been used to investigate the nature of the recombination process in *S. acidocaldarius*. The classical (reciprocal) mode of homologous recombination in Bacteria and eukaryotes follows a characteristic pattern of frequency versus distance between markers, resulting in four regions that could be termed "threshold," "proportional," "transition," and "plateau" (Fig. 12.2a). The threshold region (I) reflects the mechanistic inefficiency of generating a crossover between two markers (i.e. mutations) that are extremely close together. It can be defined in practice by extrapolating recombination frequencies back to zero; the resulting x-intercept approximates a "minimal efficient processing segment," or MEPS (Shen & Huang, 1986), below which recombination is extremely inefficient, and above which it increases in proportion to the additional increment of separation. At larger separations, defining the "proportional" region (II), recombination is still relatively rare, so that the frequency of recombinants increases in proportion to the effective cross-section of the detector, i.e. the interval between the two markers. At greater separations, however, second crossovers between markers become increasingly frequent, resulting in the region of transition (III). Eventually, recombination between the two markers becomes so frequent that the probabilities of even and odd numbers of crossovers are about equal, and further increases in separation have no significant effect (IV). Reciprocal recombination in diverse systems fits the pattern depicted in Fig. 12.2a, provided that allowances are made for different scaling in different systems. For example, MEPSs observed in bacteriophage, bacteria, and eukaryotic cells are about 40, 70, and 250 bp, respectively, whereas the midpoint of Region II corresponds roughly to 2, 20, and 30 kbp, respectively.

Using a number of *pyr* mutations that have been mapped precisely by DNA sequencing, it has been possible to analyze homologous recombination in the *S. acidocaldarius* chromosome similarly as a function of distance (Hansen et al., 2005). No statistical support for an MEPS (Region I) was seen in the *S. acidocaldarius* recombination data, and the region of proportionality (Region II) extended to only about 50 base pairs, beyond which recombination remained fairly constant as a function of marker separation (Fig. 12.2b). It should be emphasized that this analysis focused on intragenic homologous recombination events triggered by conjugation. However, the fact that this unusual frequency-versus-distance behavior was observed with over 50 combinations of mutations, and was not altered by eightfold stimulation of recombination by prior UV irradiation (Hansen et al., 2005), argues that it is a robust property of homologous recombination in *S. acidocaldarius* under these conditions.

Absence of an MEPS, high frequencies of recombination between closely spaced markers, and a relative ineffectiveness of increasing separation beyond 50 base pairs are not characteristic of crossing over by homologous recombination, as illustrated in Fig. 12.2a. These properties seem easier to reconcile with non-reciprocal recombination mechanisms, in which base-pairing allows relatively short patches of single-stranded donor DNA simply to replace corresponding ssDNA segments of the recipient molecule (Fig. 12.2c). Non-reciprocal recombination is more difficult to analyze in haploid microorganisms than classical crossing-over, but it underlies a variety of genetic phenomena, including fine-structure map expansion (also called localized negative interference) in bacteriophages, homogenotization in bacteria, and gene conversion in fungi. These recombination mechanisms vary in their molecular features, but they generally involve DNA repair reactions. For example, fine-structure map expansion begins with removal of G:T mispairs in a cognate sequence by a specialized endonuclease, whereas gene conversion involves repair of heteroduplex DNA by MMR. The possible contribution of DNA repair proteins provides an additional incentive to clarify the molecular nature of the intragenic recombination observed in *S. acidocaldarius* (Hansen et al., 2005). It must also be emphasized, however, that the non-reciprocal modes of homologous recombination hypothesized for *S. acidocaldarius* exclude neither a role for the RadA protein nor the operation of a conventional, reciprocal mode of recombination in addition. For example, reciprocal recombination remains the best explanation for the successful transfer of marked genes to the chromosome of an *S. solfataricus*-like isolate and *Thermococcus kodakarensis* using non-replicating circular DNA constructs (Sato et al., 2003;

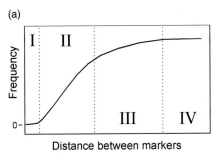

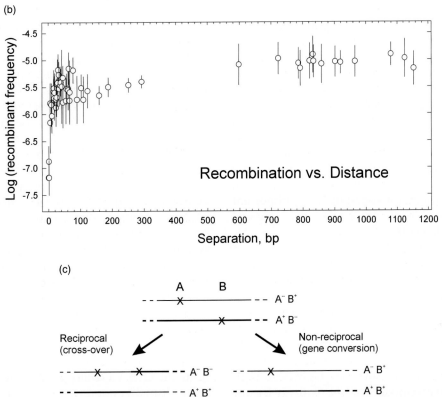

Fig. 12.2 Reciprocal and non-reciprocal modes of recombination. (a) Idealized frequency versus distance curve for reciprocal recombination when the length of DNA available for recombination is not limiting. The widths of regions I–IV generally increase as shown, but have not been drawn to scale, and differ among different organisms (see text for description). (b) Experimental data for *S. acidocaldarius* conjugation. Pairs of various *pyrE* and *pyrF* mutants were mated; Pyr$^+$ recombinants were enumerated by plating, and the resulting frequencies plotted as a function of the distance separating the mutations in the two parental strains. Data are means (open circles) and standard deviations (vertical bars) (Hansen et al., 2005). (c) Simplified depiction of the differences in products of reciprocal (left) versus non-reciprocal (right) recombination events. Since many prokaryotic systems use selections that detect only one recombination product (e.g. A$^+$ B$^+$), the mode of the recombination event is often not established.

Worthington et al., 2003). It should also be noted that another example of homologous recombination in *S. acidocaldarius* has been observed in which linear DNAs electroporated into *S. acidocaldarius* cells replace mutations in the host chromosome, thereby creating a selectable phenotype. This recombination seems reasonably efficient, and occurs even when the sequence homology flanking the selected marker is short (Kurosawa & Grogan, 2005).

Areas for future study

Recent analysis and publication of the complete genome sequence of *S. acidocaldarius* (Chen et al., 2005) provides a new perspective on the basic biology of this and other *Sulfolobus* spp. It has already identified a number of sequence features that seem relevant for the function and evolution of *Sulfolobus* genomes (Chen et al., 2005), and will greatly facilitate the functional study of individual proteins and genes. Methods to engineer specific changes in *Sulfolobus* genomes promise to open up many important questions of *Sulfolobus* biology to experimental analysis for the first time.

Genome engineering by recombination

Over the past decade, a number of experienced research groups have invested considerable effort into constructing selectable plasmid vectors for propagating cloned DNA extrachromosomally in *Sulfolobus* species. Various successes have been published but not reproduced, and the wider scientific community remains unaware of the pitfalls of plasmid cloning, which has worked so readily for many Bacteria, in *Sulfolobus* species. Tools for genetic analysis documented more recently involve the integration of circular, nonreplicating DNAs into the host chromosome. For example, a selectable vector derived from the virus SSV1 allows genes to be cloned *in vitro* and integrated by site-specific recombination at the viral attachment site (a tRNA gene) in the chromosome of *S. solfataricus* strain P1 (Jonuscheit et al., 2003). This construct can be selected via complementation of chromosomal *pyr* mutations, and it can be scored via the β-glycosidase encoded by the *lacS* gene included on the construct. In another system, loci cloned in *E. coli* and disrupted by the corresponding *lacS* gene of a related *Sulfolobus* strain have been transferred to the chromosome by homologous recombination (Worthington et al., 2003). The host strain bears a spontaneous deletion of *lacS*, allowing the cloned gene to provide a selection (growth on lactose), or visual scoring (colony staining with X-gal). In this system, the gene of interest is cloned into a bacterial plasmid and interrupted by *lacS*; transformation of the Δ*lacS* host followed by growth on lactose yields the desired recombinants (Worthington et al., 2003).

In addition to providing a way to determine the biological function of *Sulfolobus* genes, these techniques should focus renewed attention on the mechanisms of recombination that occur in hyperthermophilic Archaea. For example, evidence is emerging that *Sulfolobus* strains may differ with respect to their level of homologous recombination. Whereas recombination between an SSV1 construct and the *S. solfataricus* chromosome via *pyrE* and *pyrF* sequences seems to be extremely rare (Jonuscheit et al., 2003), events of this type appear to be relatively common in the strain used for in the *lacS* targeted mutation method (Worthington et al., 2003). A third situation may be represented by *S. acidocaldarius*, in which homologous recombination seems frequent and requires only short regions of sequence homology (Hansen et al., 2005; Kurosawa & Grogan, 2005).

Mutational analysis of complex processes

Analysis *in vitro* of individual proteins of hyperthermophilic archaea continues at a steady pace, thanks to published genome sequences, PCR, and efficient expression of foreign genes in *E. coli*. Putting the resulting biochemical information into the framework of archaeal cell function, however, will eventually require genetic dissection of complex cellular processes *in vivo*, such as chromosome replication and partitioning. In addition, the enzymatic activities of pure proteins identified *in vitro* may miss, or even obscure, unanticipated biological functions. It must also be stressed that isolating the appropriate mutants represents only the first part of a functional analysis; determining the nature of the cellular defect is equally important. The relative ease of cultivation under a variety of experimental conditions, combined with the various cellular and genetic processes that can be assayed quantitatively in *S. acidocaldarius* and *S. solfataricus*, provides a strategic advantage for these studies.

Certain classical genetic methods can be applied to *Sulfolobus* species and have strategic value for investigating complex cellular processes in archaea. For example, chemical mutagenesis of *S. acidocaldarius* followed by replica-plating has generated a number of thermosensitive (Ts) mutants, which, to the author's knowledge, represent the only conditional-lethal mutants of any Archaea that have been isolated. Fluorescence microscopy and flow cytometry of cultures shifted to the restrictive temperature identified several mutants in which the cell-division cycle stopped, or normal cellular structure degenerated, after the temperature up-shift. One mutant, for example, continued to grow but failed to divide, such that large cells formed which contained multiple, well-separated nucleoids. In other mutants, nucleoid structure was lost, such that the DNA appeared to expand to fill most of the cell interior (Bernander et al., 2000). If methods can be developed to identify the genes affected, mutants like these may provide strategic insight into the proteins that define cell structure in hyperthermophilic Archaea, and how the strategy of cell division compares to the bacterial and eukaryotic strategies.

The question of DNA repair

At the time of writing, the experimental evidence for and against the presence of DNA mismatch repair (MMR) and nucleotide excision repair (NER) in *Sulfolobus* spp. remains rather limited and remarkably balanced (see also Chapter 15 in this volume). The question of MMR and NER in hyperthermophilic Archaea presents particularly interesting opportunities for *Sulfolobus* genetics, for at least two reasons. First, the intrigue surrounding these two DNA repair strategies in hyperthermophilic Archaea stems from the fact that they may employ proteins that are fundamentally different from those known to operate in Bacteria or eukaryotes. As a result, any such putative DNA repair protein will require functional confirmation *in vivo*, which will, in turn, require isolation of the corresponding mutant and analysis of its phenotype. Second, DNA repair has genetic consequences. Its elucidation and manipulation in *Sulfolobus* accordingly promises to enhance our understanding and control of genetic processes in these species, with corresponding benefits for studying the molecular biology of hyperthermophilic Archaea.

Acknowledgments

I am indebted to the late Wolfram Zillig and members of his laboratory for helping me into the world of hyperthermophilic Archaea. I thank K. O. Stetter for personal interest and encouragement, and R. Garrett for discussions regarding genes of conjugative plasmids. Research in my laboratory was supported by grants from the US Office of Naval Research and the National Science Foundation.

13
Chromatin and regulation

John N. Reeve and Kathleen Sandman

Introduction

The information in DNA is linearly encoded, and genomic DNAs are therefore very long molecules that must be compacted to fit within the confines of the prokaryotic nucleoid or eukaryotic nucleus. Many different chromatin proteins (or nucleoid proteins) have evolved that now facilitate this process in Bacteria, Archaea, and eukaryotes that must also prevent the DNA from collapsing into a gel *in vivo*, as would occur *in vitro* with purified DNA in solution at the concentrations present *in vivo*. These proteins have unrelated structures but their convergent evolution for a common function has apparently resulted in common features, often generically designated as "histone-like," namely small size, positive charge, abundant and sequence-independent DNA binding.

In Bacteria, the HU-family of chromatin proteins has received most attention but typically the HU homolog(s) present share the DNA compaction role with several additional chromatin proteins; for example, with HNS, FIS, and IHF in *Escherichia coli* (Dame, 2005). These proteins vary in abundance under different growth conditions. Individually, they are not essential for viability but instead seem to have overlapping and compensating properties in terms of genome compaction, although some do play essential roles in regulating the expression of specific genes. In addition, after bacterial differentiation into metabolically inactive survival-stage cells, such as spores and elementary bodies, the chromatin proteins present are different from those present in vegetative cells. In stark contrast to the diversity of chromatin proteins in Bacteria, all eukaryotes (except dinoflagellates) employ the same four histones to compact and package their nuclear DNA into nucleosomes that further polymerize to form chromatin (Luger et al., 1997; Schlach et al., 2005). These histones, designated H2A, H2B, H3, and H4, form (H2A+H2B) and (H3+H4) heterodimers that assemble into a histone octamer [(H2A+H2B)+(H3+H4)$_2$+(H2A+H2B)] core around which ~165 bp of DNA are wrapped in ~1.7 circles to form a nucleosome. All four core histones have a histone fold (HF), a structural motif formed by three α-helices separated by two β-strand loop regions (α1-βL1-α2-βL2-α3) that directs and requires dimer formation for stability. The DNA between nucleosomes is bound by proteins, apparently misnamed histones H1 and H5, as they do not have HFs but seem to share a common ancestry with bacterial HU proteins, as do the HCC chromatin proteins that replace histones in the dinoflagellate *Crypthecodinium cohnii* (Wong et al., 2003). The HFs of H2A, H2B, H3, and H4 assemble to form the globular core of the nucleosome, and these histones also have N-terminal sequences (known as tails) that extend from the HFs outside of the nucleosome. The tails facilitate higher order nucleosome polymerization into chromatin and also carry the targets for the post-translation histone acetylation, methylation, phosphorylation, and ubiquitinylation events that regulate eukaryotic chromatin structure and gene expression (Briggs et al., 2002; Mellor, 2005). Only one HF-containing protein had been identified in Bacteria (Qiu et al., 2005), but many Archaea have DNA binding proteins, designated archaeal histones (see www.biosci.ohiostate.edu/~microbio/Archaealhistones; Reeve, 2003), with HFs structurally very similar to the HFs in the eukaryotic histones. However, unlike the almost universal use of histones to package nuclear DNA in eukaryotes, not all Archaea have histones, and the data available from biochemical studies and archaeal genome sequencing argue that most Archaea, like Bacteria, probably employ different amounts of different chromatin proteins to compact their genomic DNA depending on growth and gene expression needs. The predominant families of archaeal chromatin proteins are described individually below, although it should be remembered that they may have overlapping functions *in vivo*. An apparent example of the exchangeability of prokaryotic chromatin proteins is provided by the presence of HTa, a divergent member of the HU-family of "bacterial" chromatin proteins, in *Thermoplasma* and its close euryarchaeal relatives (Reeve, 2003). Presumably, this was acquired by lateral gene transfer and subsequently replaced the archaeal histone(s) that should be present in these Archaea based on their 16S rRNA phylogenetic positions.

The following sections describe the structures, DNA binding properties, and any evidence for the archaeal chromatin protein participating in gene expression, but it is important to emphasize that many of these proteins have also been studied in detail as model systems to identify and quantify molecular interactions that contribute to protein structure and stability. For such investigations, these proteins have the very attractive features of small size, established high-resolution structures, reversible unfolding–refolding and the availability of natural variants from Archaea that grow from ice-water to boiling-water temperatures. The results obtained with natural and laboratory-generated variants with different inherent structural stabilities are impressive and such structure–function–stability studies currently constitute much of the ongoing research focused on archaeal chromatin proteins. This work is beyond the scope of this review but this literature can be readily accessed through citations in recent reports (e.g. Clark et al., 2004; Topping & Gloss, 2004; Bedell et al., 2005).

Archaeal histones

Nomenclature

Archaeal histones were first discovered in *Methanothermus fervidus* and, as the histone preparations from *M. fervidus* contained two very similar polypeptides, they were designated HMf-1 and HMf-2, as Histones 1 and 2 from *M. fervidus*. Later, to conform to standard bacterial gene-protein nomenclature, the encoding genes were designated *hmfA* and *hmfB* and the proteins HMfA and HMfB. When archaeal histones were then identified and characterized in other species, they were named HXxA or HXxB with Xx being the first letters of the genus and species, and A and B indicating a closer sequence relationship to HMfA or HMfB. This became cumbersome with the discovery of multiple archaeal histones in some species – for example, there are three histones in *Methanobacterium thermoautotrophicum* (now *Methanothermobacter thermautotrophicus*), two of which are HMfA-like, resulting in HMtA1, HMtA2, and HMtB – and this nomenclature became essentially impractical when sequencing revealed that the *Methanococcus jannaschii* (now *Methanocaldococcus jannaschii*) genome encodes six different archaeal histones. These histones have always been identified as MJXXXX, where XXXX is the numerical designation of the encoding gene in the *M. jannaschii* genome, and with the rapid accumulation of genome sequences, from cultured Archaea and environmental DNA preparations, this now seems to be the only practical nomenclature. In this review, if both a genome sequence and a historical name exist, both are provided, e.g. HMtA1 (MTH0821), HMtA2 (MTH1696), and HMtB (MTH0254).

Sequences and structures

HMfB is the most thoroughly studied and the archetype of the most frequently reported family of archaeal histones, proteins with 65–69 residues that form just one HF. Regardless of their origin, all members of the HMfB-family have similar sequences (Plate 13.1) and can homodimerize or form heterodimers with different partners. High-resolution structures have been established from crystals of homodimers (Decanniere et al., 2000; Li et al., 2003) that confirm that these archaeal proteins have HFs that are almost identical to the eukaryotic HFs that form the core of the eukaryotic nucleosome. All HFs are stabilized by an intramolecular salt bridge formed between an arginine in L2 (R52 in HMfB) and an aspartate in α3 (D59 in HMfB), and by intermolecular bonds between hydrophobic residues located along the buried anti-parallel aligned faces of the two α2s in a HF dimer (Plate 13.1). Almost all members of the HMfB-family of archaeal histones have two proline residues (P4 and P7 in HMfB) conserved near the N-terminus of α1 that form a proline tetrad in a dimer. This motif positions surface-located residues appropriately for DNA binding but, as the archaeal histones recently identified in marine Crenarchaea (see HCs in Plate 13.1) do not have prolines at these locations (Čuboňová et al., 2005), this structure is not essential for archaeal histone-DNA binding.

In addition to the HMfB-family, three archaeal histones have been identified in *Methanococci* that have about 25 amino acid C-terminal extensions from their HFs that are predicted to fold to form an additional alpha-helix (see MJ1647 in Plate 13.1). The sequences of these archaeal histone tails are ~50% identical but they have no detectable homology to the sequences of the eukaryotic histone tails. Removal of the tail from recombinant MJ1647 resulted in decreased thermostability but a protein that formed more stable complexes with DNA (Li et al., 2000). Apparently, the presence of multiple C-terminal extensions impedes the higher-order assembly of $(MJ1647)_2$ homodimers needed for DNA binding and argues that MJ1647 most likely exists *in vivo* in HF heterodimers, partnered with one or more of the five HMfB-family members also present in *M. jannaschii*. The eukaryotic histones form only (H2A+H2B) and (H3+H4) heterodimers and the asymmetry of these heterodimers seems necessary to position DNA binding residues appropriately, and also to limit dimer oligomerization and so define the structure of the nucleosome core (Luger et al., 1997). In contrast, archaeal histone homodimers are symmetric, and it remains to be determined how such homodimers position the structurally homologous archaeal HF residues for DNA binding and if continued homodimer oligomerization is structurally limited. In this regard, intriguingly, one of the two archaeal histones in *Nanoarchaeum equitans* (see NEQ0288 in Plate 13.1) has four more residues (KKVG) in L1 than are present in L1 of the HMfB-family. Five additional residues (QDFKT) are similarly present in L1 in the eukaryotic histone H3 when compared with L1 of histone H4, and to accommodate this length difference (H3+H4) heterodimers have an asymmetric curved structure (Luger et al., 1997). If NEQ0288 forms heterodimers with NEQ0348, the second archaeal histone in *N. equitans* that belongs to the HMfB-family, they should also be curved and asymmetric. One of the additional residues in L1 of H3 is K69, the only residue within a eukaryotic HF, as opposed to a histone tail, that is known to be subject to regulatory post-translation modification. The extent of methylation of K69 in H3 has been correlated with regulation of eukaryotic gene expression and gene silencing (Briggs et al., 2002). The discovery of NEQ0288 with a lysine-containing peptide insert in L1 provides the first hint of a possible origin for this L1 insert and this regulation.

Two archaeal histones, HMk (MK1677) in *Methanopyrus kandleri* and VNG0134G in *Halobacterium* NRC1, contain two HFs covalently linked in tandem by short peptides. The crystal structure of HMk confirms that the two HFs interact, as predicted, and that HMk is therefore an archaeal histone heterodimer but with both HFs present in the same polypeptide chain (Fahrner et al., 2001). This linkage presumably guarantees that only HF "heterodimers" form and structural asymmetry that limits higher order oligomerization and the architecture of the core of a HMk–DNA complex (Pavlov et al., 2002a). HMk molecules exist in solution as $(HMk)_2$ dimers, and therefore as HF tetramers, a property that they share with eukaryotic histone $(H3+H4)_2$ tetramers (Alilat et al., 1999). By using the HMk structure as a model, HMfB–HMfB, HMfA–HMfA, HMfA–HMfB, and HMfB–HMfA fusions were generated *in vitro* with linker peptides based on the peptide that links the two HFs in VNG0134G. The genetically engineered archaeal HF homodimer-fusions had essentially the same DNA-binding properties *in vitro* as their natural unlinked homodimer counterparts (Marc et al., 2002). The fusions also allowed the properties of HMfA–HMfB heterodimers to be determined, and facilitated investigations of the roles of HF dimer–dimer interactions in determining DNA affinity, DNA sequence preferences and the differences in DNA compaction observed for $(HMfA)_2$ versus $(HMfB)_2$ homodimers (Marc et al., 2002).

DNA binding and complex formation

Archaeal histones bind DNA *in vivo* and *in vitro* to form structures that, when visualized by electron microscopy, resemble the classical "beads-on-a-string" structure described for nucleosomes in eukaryotic chromatin. The histone to DNA stoichiometry (Bailey et al., 1999), AFM measurements (Tomschik et al., 2001), and the length of the DNA protected from micrococcal nuclease digestion by archaeal histone binding (Pereira et al., 1997; Xie & Reeve, 2004), however, all argue that the structure formed at physiological archaeal histone to DNA ratios, designated an archaeal nucleosome (Pereira et al., 1997), has a histone tetramer core. The four HF present make direct contacts with DNA over ~60 bp and are fully circumscribed by ~90 bp of DNA (Plate 13.1). This results in a structure very similar to the tetrasome (Alilat et al., 1999) formed at the center of a eukaryotic nucleosome, where a similar length of DNA is circularized around the (H3+H4)$_2$ tetramer (Luger et al., 1997). The C-terminal regions of the two α2s and two α3s from two H3 monomers form a four α-helix bundle (4HB) that constitutes and stabilizes the interface between the two (H3+H4) dimers in a (H3+H4)$_2$ tetramer (Luger et al., 1997). Isolated tetrasomes, in common with archaeal nucleosomes, have the ability to wrap DNA in either a negative or positive supercoil, and *in vitro* can switch between these alternative configurations (Alilat et al., 1999; Musgrave et al., 2000). For archaeal histones, this property is dependent on the salt concentration and histone to DNA ratio *in vitro*, and on HF residues located near the C-terminus of α2 and within α3 (Musgrave et al., 2000; Bailey et al., 2002; Marc et al., 2002). These residues are the structural homologues of the H3 residues that stabilize the 4HB formed at the center of a (H3+H4)$_2$ tetramer (Malik & Henikoff, 2003). Some HMfB variants with amino acid substitutions at these locations can only wrap DNA in one direction, and so resemble HMk and EAG3, two natural archaeal histones that constrain DNA *in vitro* only in a negative (Musgrave et al., 2000) or positive (Čuboňová et al., 2005) supercoil, respectively.

Based on HF homology and the crystal structure of the eukaryotic nucleosome (Luger et al., 1997), the residues in an archaeal HF responsible for dimer and tetramer assembly, and for DNA binding, can be predicted. Many archaeal histone variants have been generated and assayed to test these predictions (Higashibata et al., 2003; Reeve, 2003) and the results obtained are fully consistent with the conclusion that an archaeal nucleosome has a structure very similar to that of the (H3+H4)$_2$ tetrasome. In addition, all four eukaryotic histones have a lysine in L2 that directly contacts the DNA in the nucleosome (Luger et al., 1997), and a lysine is also present at this location in the *M. jannaschii* archaeal histones, but a glycine residue fills this position in almost all other HMfB-family members. Replacing this glycine with a lysine resulted in an HMfB variant (G51K) that had increased affinity for DNA but formed less flexible complexes (Soares et al., 2003), arguing for positive selective pressure to retain archaeal histones that can form archaeal nucleosomes with topological flexibility.

Participation in regulating gene expression

Histones bind DNA primarily through interactions with the sugar-phosphate backbone, and virtually all DNA sequences can be incorporated into nucleosomes. But nucleosome assembly does occur at specific locations *in vivo*, and such positioned nucleosomes do participate directly in regulating gene expression (Mellor, 2005). Nucleosome positioning is facilitated by DNA sequences that more readily accommodate the structural distortions inherent in DNA wrapping around a histone core than does the adjacent DNA, and sequence "rules" for positioning motifs have been established. These rules apply to both eukaryotic and archaeal histone–DNA assembly (Bailey et al., 2000) and the genome sequences of histone-containing Archaea are enriched for sequences that should facilitate their compaction by HF-based DNA wrapping (Schieg & Herzel, 2004). The abundances and relative amounts of the different archaeal histones present in cells change with growth conditions, arguing for different but probably overlapping functions *in vivo* (Reeve, 2003). The *M. fervidus* genome has been shown to contain sequences that do localize HMfA and HMfB assembly *in vitro* (Pereira & Reeve, 1999) and most, but significantly not all, *M. thermautotrophicus* genomic DNA sequences were cross-linked to histones *in vivo* by formaldehyde treatment (Pereira et al., 1997).

A SELEX procedure was used to select DNA molecules to which HMfB binds with above average affinity (Bailey et al., 2000). All of these molecules have sequences that conform to the nucleosome-positioning rules, but every molecule has a different sequence and affinity for HMfB. Some of these molecules do not have a similarly high affinity for HMfA and this was shown to result from a difference in the structure of the 4HB formed at the center of 2(HMfA)$_2$ versus 2(HMfB)$_2$ homotetramers (Bailey et al., 2002; Marc et al., 2002). This is consistent with the results of investigations of eukaryotic nucleosomes assembled with the histone H3.3 variant. Nucleosomes containing H3.3 assemble specifically on centromeric DNA, and the residues that differ in H3.3 versus H3, presumably the residues responsible for this localized assembly (Malik & Henikoff, 2003), are the HF homologues of those that differ in HMfA versus HMfB that direct the assembly of the 4HB at the center of 2(HMfA)$_2$ versus 2(HMfB)$_2$ tetramers.

Most Archaea have more than one histone and therefore have the potential to assemble archaeal nucleosomes with different combinations of homodimers and heterodimers. If these have different sequence-dependent affinities they could be positioned to regulate gene expression, and archaeal histone binding to the promoter region of a template DNA does inhibit transcription *in vitro* by archaeal RNA polymerases (Soares et al., 1998; Xie & Reeve, 2004). In eukaryotes, genes assembled into chromatin are generally not expressed and, when necessary, this inhibition is relieved by the activities of chromatin remodeling and histone modification complexes (Mellor, 2005). Archaeal genomes do not, however, encode homologs of the eukaryotic complexes, and archaeal histones are not post-translationally modified *in vivo* (Forbes et al., 2004). Transcript elongation by eukaryotic RNA polymerase II (Pol II) through nucleosomal DNA similarly requires the aid elongation factors that also have no homologs in Archaea. In contrast, eukaryotic Pol III does have an inherent ability to transcribe through nucleosomes and *M. thermautotrophicus* archaeal RNA polymerase (RNAP) was shown to transcribe through an archaeal nucleosome *in vitro* without the aid of elongation factors (Xie & Reeve, 2004). The presence of the archaeal nucleosome decreased the rate but did not block transcription and, as illustrated in Fig. 13.1, this is also the case for *M. thermautotrophicus* RNAP transcription *in vitro* through tandemly assembled archaeal nucleosomes (Xie, 2005).

DBNP-B/Sso10b/Alba

Nomenclature

Sulfolobus species contain several families of DNA-binding proteins that have been designated helix-stabilizing nucleoid proteins (HSNP-A, -C, and -C′), DNA binding nucleoid protein (DBNP-B), and Sac/Sso/Ssh 7a–e, 8a,b, and 10a,b to identify the origin as *S. acidocaldarius* (Sac), *S. solfataricus* (Sso), or *S. shibatae* (Ssh) and the monomer size as ~7, 8, or 10 kDa. Based on amino acid compositions, HSPN-C′ and DBNP-B are apparently Sac7d and Sac10b, respectively, and Sso7d/Sac7d related proteins have also been isolated and characterized as RNases designated as p2, p3, and SaRD (Fusi et al., 1995; Kulms et al., 1995). p3 has the p2/Sso7d sequence plus a ten residue C-terminal extension with a sequence related to sequences present in some bacterial RNases. As described in later sections, considerable research has since been focused on the Sac10b/Sso10b and Sac7d/Sso7d families and structures have also been established for Sac10a/Sso10a proteins, but further studies of the Sac8/Sso8 proteins have not been reported.

In 2002, it was proposed that the Sac10b/Sso10b proteins should be renamed Alba, for acetylation lowers DNA binding affinity, based on native Sso10b, which is acetylated *in vivo* at the N-terminus and at lysine-16 (K16), having an ~30-fold lower affinity for DNA than non-acetylated recombinant Sso10b, or deacetylated native Sso10b (Bell et al., 2002). Unfortunately for this name, it has since been established that acetylation of Sso10b has a much less dramatic effect on DNA affinity, reducing its affinity for DNA only about two- or threefold (Jelinska et al., 2005; Marsh et al., 2005), and although Sso10b/Alba proteins are present in many archaeal lineages, acetylation has to date only been documented for Alba homologues purified from *Sulfolobales*. Some Sso10b/Alba family members do not have a lysine at the structural position homologous to K16 in Sso10b (Plate 13.2) and a second Alba paralog present in *S. solfataricus*, Sso10b2, is acetylated at only the N-terminus (Jelinska et al., 2005). Currently, these proteins are alternatively designated in the literature as Sso10b, Ssh10b, Mja10b, Afu10b, etc., or SsoAlba, SshAlba, MjaAlba, AfuAlba, etc., which is further complicated by the existence of paralogs. In

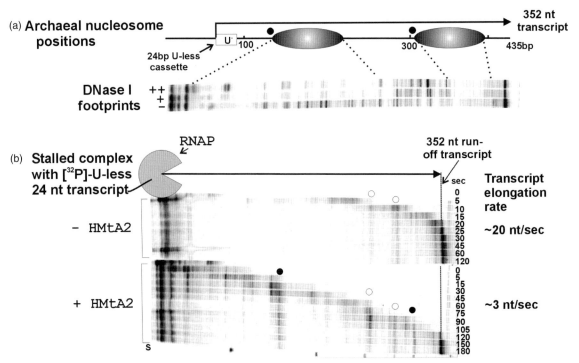

Fig. 13.1 Archaeal nucleosome positioning and transcription through archaeal histone-bound DNA. (a) The diagram shows the positions at which two archaeal nucleosomes (shaded ovals) assembled on the transcription template, based on the DNase I footprints shown below. The template DNA was ^{32}[P]-end-labeled, and incubated with HMtA2 at archaeal histone to DNA molar ratios of 1:1 (+) and 4:1 (++), or without the histone (−) and then exposed to DNase I. The DNA fragments generated were separated by denaturing gel electrophoresis. (b) Transcription by RNA polymerase (RNAP) purified from *M. thermautotrophicus* was initiated in a reaction mixture that contained only ATP, GTP and [^{32}P]-CTP on the template shown in (a). The first 24 bp (U-less cassette) were transcribed in the absence of UTP and stalled elongation complexes accumulated that contained the template DNA, RNAP and a 24-nucleotide ^{32}P-labeled transcript. These were washed and an aliquot was incubated with HMtA2 at a histone to DNA molar ratio of 1. All four unlabeled NTPs were then added and the transcripts synthesized by elongation of the [^{32}P]-labeled nascent transcript at the times indicated (seconds) were separated by electrophoresis and visualized by phosphorimaging. As noted, the rate of transcript elongation was reduced by the presence of the archaeal nucleosomes and the RNAP paused at multiple locations, predominantly at the 5′-boundaries (●) of these nucleosomes. However, full length run-off transcripts (352 nt) were generated. The RNAP also paused at two common sites (○) in both the presence and absence of archaeal nucleosome assembly. This figure is based on the results of research undertaken by Yunwei Xie (2005).

most cases, these have very similar sequences and appear to be products of recent gene duplication events, but the two paralogs in *Sulfolobus* species have sequences sufficiently different to warrant being designated Sso10b1 and Sso10b2, or Alba1 and Alba2 (see Plate 13.2). Here we use Alba1 and Alba2 together with a numerical genome designation when available, e.g. Alba1 (SSO0962), to identify the archaeal origin.

Sequences and structure

Crystal structures have been established for homodimers of Alba1 (SSO0962), Alba1 (MJ0212), Alba1 (AF1956), and Alba2 (SSO6877), and for [Alba1(SSO0962)+Alba2 (SSO6877)] heterodimers (Wardlesworth et al., 2002; Chou et al., 2003; Wang et al., 2003; Zhao et al., 2003; Jelinska et al., 2005). They are almost identical, consistent with the ~50%

sequence identity of all Alba1 proteins, and their ~40% identity with Alba2 proteins. The monomer has β-strand1–α-helix1–β-strand2–α-helix2–β-strand3–β-strand4 regions separated by loops (Plate 13.2) and dimer formation is stabilized primarily through interactions between the anti-parallel aligned α2-β3 regions. Dimer:dimer polymerization has also been identified in the crystal structures directed by α1:α1 interactions (Zhao et al., 2003) that is consistent with (Alba)$_2$ dimers assembling end-to-end to form a linear polymer when bound along the long axis of a DNA molecule (Zhao et al., 2003; Jelinska et al., 2005). Alba1 (SSO0962) and Alba2 (SSO6877) differ most noticeably in the loop region between β3 and β4, one of three regions that directly interact with DNA (Cui et al., 2003). In Alba2, the loop is shorter and contains an arginine-rich sequence, RDRRR, as is commonly found in nucleic acid binding proteins and particularly in RNA-binding proteins. The βαβαββ fold of Alba1 and Alba2 is present in many contemporary RNA-binding proteins, and a bioinformatics analysis concluded that archaeal Alba proteins share a common ancestry with these RNA-binding proteins (Aravind et al., 2003). Furthermore, it was argued that the ability to bind DNA is likely a recent acquisition and may not be a property of all archaeal Alba proteins. The genes encoding Alba1 and Alba2 in *S. solfataricus* and *S. shibatae* are located on the opposite DNA strand but directly adjacent to and partially overlapping, respectively, the gene that encodes reverse gyrase. Reverse gyrase apparently evolved with hyperthermophily, and perhaps having these three genes that encode DNA-structure modulating proteins organized as a functional unit was then advantageous.

DNA/RNA binding and complex formation

Purified Alba1 (SSO0962) binds *in vitro* to double-stranded DNA, single-stranded DNA, and RNA with similar affinity, and acetylation of recombinant Alba1 reduced the affinity about two- to threefold for both DNA and RNA (Jelinska et al., 2005; Marsh et al., 2005). Surprisingly, however, whether the primary target for Alba binding *in vivo* is DNA and/or RNA remains controversial. Cells of *S. solfataricus* have been exposed to UV-irradiation (Marsh et al., 2005), and cells of *S. shibatae* to formaldehyde (Guo et al., 2003), to covalently attach Alba *in situ* to the nucleic acid(s) to which it was bound *in vivo*. The results obtained have been interpreted as demonstrating that Alba is bound *in vivo* to both DNA and RNA (Marsh et al., 2005) or only to large RNA molecules and ribosomes (Guo et al., 2003). Given that Alba1 is a major component of the cell, initially reported as constituting ~4% (Xue et al., 2000) and then more precisely 1.7% of the total soluble protein in *S. shibatae* (Guo et al., 2003), it may be very difficult to distinguish between meaningful and spurious cross-linking of such an abundant protein to nucleic acids *in vivo* by these techniques. Consistent with Alba binding to ribosomes, electron microscopy of *S. solfataricus* cells that were immunogold stained using anti-DBNP-B (Sso10b) antibodies located this protein throughout the cytoplasm and not enriched in the nucleoid (Bohrmann et al., 1994).

Regardless of the *in vivo* target, considerable effort has been invested in characterizing and modeling the complexes formed by Alba1 and Alba2 *in vitro* with DNA. Residues in three regions of Alba1 directly contact the DNA based on changes in their NMR signals when measured in solution in the presence versus absence of DNA (Cui et al., 2003). These are located at the center and at each end of an Alba dimer, which, given the dimer dimensions, should facilitate DNA binding at three separate sites that extend over one complete helical turn (Plate 13.2). Consistent with this, at low Alba to DNA ratios, one Alba dimer binds per 10–12 bp and in one model (Wardlesworth et al., 2002; Wang et al., 2003), but not the only model for Alba binding (Zhao et al., 2003), the central DNA binding site makes contact with the major groove of the DNA and the two end-located sites extend to bind within the flanking minor grooves (Plate 13.2). At higher Alba to DNA ratios, about two Alba dimers bind per helical turn, leading to the proposal that Alba dimers are then assembled end-to-end (Zhao et al., 2003) as parallel linear polymers that bind along the long axis and essentially coat the DNA (Jelinska et al., 2005), but do not compact the DNA (Xue et al., 2000). In an early electron microscopy study, complexes formed at sub-saturating Alba concentrations appeared to link two DNA molecules but, with additional Alba binding, the complexes observed contained only one DNA molecule apparently coated by Alba (Lurz et al., 1986). In a more recent study, the complexes formed by (Alba1)$_2$ and (Alba2)$_2$ homodimers and by (Alba1+Alba2) heterodimers were compared (Jelinska et al., 2005). Recombinant (non-acetylated) (Alba1)$_2$ bound DNA with an affinity ~20-fold higher than recombinant (non-acetylated) (Alba2)$_2$, but only about threefold higher than recombinant (Alba1+Alba2) heterodimers or native (acetylated) (Alba1)$_2$. When visualized by EM, the complexes formed by (Alba1)$_2$ at different protein to DNA ratios were consistent with the earlier report that used native protein (Lurz et al., 1986) but, at a high protein to DNA ratio, (Alba1+Alba2) heterodimers did not simply coat one DNA molecule but instead formed more compact complexes that contained two, apparently branched, DNA molecules. As suggested for archaeal histone heterodimer formation, forming (Alba1+Alba2) heterodimers may limit dimer oligomerization and prevent the monotonous polymerization of (Alba1)$_2$ homodimers *in vivo* that apparently results in DNA coating *in vitro*.

Participation in the regulation of gene expression

Consistent with Alba participating in regulating gene expression, the abundance of Alba1 in *Thermococcus zilligii* (Dinger et al., 2000) and of transcripts encoding Alba1/Sso10b1 in *S. solfataricus* (see Jelinska et al., 2005) cells decreases substantially as cultures enter the stationary phase, although this was not observed for Alba1/Ssh10b1 in *S. shibatae* cells (Guo et al., 2003). Binding deacetylated Alba1 to the template DNA inhibited transcription *in vitro*, although the step in transcription inhibited was not determined (Bell et al., 2002). The presence of enzymes in *S. solfataricus* that specifically and reversibly acetylate (Pat; Marsh et al., 2005) and deacetylate (Sir2; Bell et al., 2002) the K16 residue in Alba1 (SSO0962), and the physical association of Alba1 with the deacetylase *in vivo*, argue convincingly that acetylation–deacetylation of this Alba1 must be biologically meaningful. When the acetylation of Alba1 was first found to reduce its affinity for DNA substantially, it was suggested that this could be the basis of an archaeal analog of the well established eukaryotic mechanism of regulating gene expression by histone-tail acetylation (Bell et al., 2002). This could still be the case, but the regulation of gene expression by Alba acetylation may be more indirect than originally imagined, with acetylation regulating (Alba1)$_2$ dimer:dimer polymerization, and the extent of (Alba)$_2$ polymerization modulating DNA binding (Zhao et al., 2003). The K16 residue acetylated in Alba1/Sso10b is located in a DNA binding region (Cui et al., 2003), within the loop that separates β1 and α1 (Plate 13.2), but this is also adjacent to the sites of α1:α1 interactions that are proposed to stabilize (Alba)$_2$:(Alba)$_2$ oligomerization (Zhao et al., 2003). Consistent with both K16 and the adjacent residue K17 playing direct roles in DNA binding by Alba1/Sso10b, K16A, K16E, K17A, and K17E, variants of this protein were reported to have ~7-, ~40-, ~14-, and ~42-fold reduced affinities for DNA *in vitro*, respectively (Bell et al., 2002), although these values must now be considered within the context of acetylation of Alba1/Sso10b resulting in only a two- to threefold, rather than 30-fold, reduction in DNA affinity (Jelinska et al., 2005; Marsh et al. 2005). The structural homolog of K16 in Alba1/Afu10b1 is not a lysine but N10, and an N10A variant of this protein has only about a twofold lower affinity for DNA than the wild-type protein (Zhao et al., 2003). The homolog of K17 in Alba1/Sso10b1 is a lysine (K11) in Alba1/Afu10b1 and K11R, K11Q, and K11M variants of Alba1/Afu10b1 do have about sevenfold reduced affinities for DNA, but this is also the case for the L18R and F54R variants of this protein, which were designed specifically to be defective in dimer:dimer interactions (Zhao et al., 2003). The enzyme that deacetylates K16 in Alba1/Sso10b1 is a member of the Sir2 family, and this residue follows the sequence P(X)$_6$G that is reminiscent of the P(X)$_4$G sequence that precedes a lysine residue that is acetylated by a bacterial Sir2 family member in acetyl-CoA synthetase. If this is a conserved recognition motif for Pat/Sir2 acetylation/deacetylation partnerships (Marsh et al., 2005), its absence from some Alba family members (see Plate 13.2) presumably argues against their acetylation/deacetylation *in vivo* by such a partnership.

Sso10a/Sac10a

Structures have been established for Sso10a (Kahsai et al., 2005) and Sac10a (Edmondson et al., 2004), two members of the second family of abundant ~10 kDa DNA binding proteins present in *Sulfolobus* species that, like the Sso10b1/Alba family, are also widely distributed in Archaea (COG3432). However, otherwise these protein families have only their ~10 kDa monomer size in common. The Sso10a/Sac10a proteins have about 95 amino acid sequences that form an α-helix1-β-strand1-α-helix2-α-helix3-β-strand1-β-strand3-α-helix4 fold. They are dimers in solution in which the two long C-terminal α4s are intertwined and form a two-stranded, anti-parallel, coiled-coil rod that separates the two N-terminal regions that fold into winged-helix DNA-binding domains (Chen et al., 2004). In the one biochemical investigation reported (Edmondson et al., 2004), Sac10a was found to bind poly(dAdT) in vitro with high affinity, at a stoichiometry of 1 dimer/17 bp, and compacted this DNA consistent with EM observations of the complexes formed by Sac10a (Lurz et al., 1986). But, under the same solution conditions, Sac10a did not bind poly(dA)-poly(dT), poly(dGdC), or poly(dC)-poly(dG) with such high affinity, leading to the proposal that Sac10a/Sso10a proteins may exhibit sequence-specific DNA binding and function as transcription regulators rather than chromatin proteins. When the sequences of many Sso10a/Sac10a family members are aligned, the winged-helix DNA-binding domains are conserved but some have long and unrelated N- and C-terminal sequences that extend from the COG3432 fold, consistent with different regulator-sensing domains fused to the same DNA-binding motif (Edmondson et al., 2004).

Sso7d/Sac7d/Ssh7d/HSPN-C′

Nomenclature and sequences

Sso7/Sac7/Ssh7/HSPN-C′ are small (~7 kDa), very abundant DNA binding proteins that, unlike the Sso10a and Sso10b/Alba families, are only present in *Sulfolobus* species so that Sul7 has been adopted as the family name. These proteins are not related to 7kMk, a DNA binding protein with a calculated mass of 8699 Da, but an electrophoretic mobility indicative of ~7 kDa, that has only been characterized from *Methanopyrus kandleri* (Pavlov et al., 2002b). Sac7 preparations, when first isolated from *S. acidocaldarius*, were shown to contain five closely related polypeptides that were designated Sac7a through 7e based on increasing positive charge. The sequences of Sac7a, b, d, and e (about 65 amino acids) are now known and are aligned in Plate 13.3, together with the sequence of Sso7d, the most studied member of this family from *S. solfataricus*.

Structure and DNA binding

High resolution structures have been established for Sac7d, Sso7d, Sac7d-DNA, and Sso7d-DNA complexes (Su et al., 2000; Ko et al., 2004; Bedell et al., 2005; Chen et al., 2005). All Sul7d proteins almost certainly have the same fold that resembles the Src homology 3 (SH3)/chromodomain fold with five β-sheet regions preceding one α-helix. DNA binding is primarily through contacts made by residues Y8, W24, V26, M29, and R42 within the minor groove over just a ~4 bp region. This results in a sharp ~60° DNA bend, helix unwinding, and an increase in the melting temperature of the DNA. The roles of W24, V26, and M29 in DNA binding and bending have been probed by mutagenesis (Bedell et al., 2005; Peters et al., 2005), and crystal structures have been established for the complexes formed with DNA by Sac7d variants with both single and combined V26A, M29A, and M29F substitutions (Chen et al., 2005). Five of the 14 lysine residues (K5, K7, K60, K62, and K63) in Sso7d are ε-amino-monomethylated *in vivo* to different extents and the extent of methylation increases *in vivo* with increasing growth temperature (Baumann et al., 1994), but this methylation does not affect DNA binding. These N- and C-terminally located lysine residues are remote from the protein:DNA interface, and methylation does not reduce the affinity of Sul7d for DNA (Baumann et al., 1994; Su et al., 2000). As confirmation that the C-terminal region is not required for DNA binding, a variant (L54Δ) with the C-terminal α1-region deleted has been generated and although this variant is structurally less stable, it still binds DNA (Shehi et al., 2003). To further probe the requirements for DNA binding, a Sac7d dimer has been constructed that retains but has modified DNA binding properties (Yang & Wang, 2004), and Sso7d-Taq and Sso7d-Pfu fusions have been shown to retain both high catalytic activity and structural stability and to have increased processivity (Wang et al., 2004). A DNA polymerase based on the Sso7d-Pfu construct has been commercialized as Phusion™.

RNase and chaperone activities

In addition to DNA binding, Sso7d has been shown to facilitate the re-annealing of complementary single strands of DNA at high temperatures, to have RNase activity, and to function as an ATP-dependent protein chaperone that disassembles and reactivates insolublized β-galactosidase (Guagliardi et al., 2002, 2004). The regions of Sso7d that bind protein aggregates and DNA apparently overlap, as these activities are mutually exclusive, but the residues responsible for ATP hydrolysis and RNase activity are located in regions not involved in DNA binding (Guagliardi et al., 2002). Two residues, K12 and E35, have been identified that are required for RNase activity and these, as illustrated in Plate 13.3, are located on the protein surface distant from the DNA binding face. Sso7d variants with K12L and E35L substitutions retain the native fold but they no longer have RNase activity (Shehi et al., 2001).

Participation in gene expression

Given their intracellular abundance, the Sul7d proteins must contribute to the overall architecture of the genome *in vivo*, and Sso7d has been shown to compact relaxed and positively supercoiled DNAs *in vitro* (Napoli et al., 2002). Not surprisingly, there are hints that these proteins also participate in DNA replication, repair, and/or expression. Addition of Sso7d repressed uracil DNA glycosylase activity (Dionne & Bell, 2005) and prevented the auto-inhibition of a Holliday junction-resolving enzyme (Kvaratskhelia et al., 2002) but did not inhibit reverse gyrase or transcription *in vitro*. Sso7d (and Sso10b/Alba) binding sites have been identified in the regulatory region upstream of a gene encoding alcohol dehydrogenase in *S. solfataricus* (Fiorentino et al., 2003).

MC1

By archaeal standards, *Methanosarcina* species have very large genomes (>5 Mbp) that encode only one archaeal histone and no Sso10b/Alba homologs. Histone synthesis was not detected in *Methanosarcina* CHTI55 (De Vuyst et al., 2005) and the genomic DNA of this methanogen, and related Methanosarcinae, is bound and compacted predominantly by members of the MC1 chromatin protein family. Based on genome sequences, *Methanosaeta*, *Halococcus*, *Halobacterium*, and *Haloferax* species have the capacity to synthesize two or three MC1 paralogs but members of this protein family are not otherwise widely distributed. A solution structure has been established for the MC1 from *Methanosarcina thermophila* (Plate 13.4). This 93 residue protein folds into β-β-α-β-β-loop-β regions that constitute a novel fold, although with features reminiscent of the Sul7d structure (Paquet et al., 2004), and MC1 apparently does also bind to the minor groove. It was calculated that one monomer interacts with ~11 bp of DNA (Paradinas et al., 1998) but the results of a SELEX selection argue for interactions over a longer 15 bp region. The DNA molecules selected, to which MC1 binds with about 50-fold higher affinity than to random DNA, have a consensus sequence of $A_5CACACAorCC_4$ (De Vuyst et al., 2005). MC1 binds non-cooperatively, causing a sharp kink, and binds preferentially to negatively supercoiled DNA and to four-way Holliday structures but not to bulged DNA (Paradinas et al., 1998). Based on photochemical crosslinking of MC1 to 5-bromouracil-substituted DNA, the DNA binding region is located near the C-terminus, between residues K62 and K87, and the surface locations of W74, K81, K85, and K86 are consistent with their predicted roles in contacting DNA (Paquet et al., 2004). MC1 binding results in negative supercoiling, and protects the DNA against radiolysis and heat denaturation. Addition of MC1 stimulated *E. coli* RNAP activity *in vitro* at low protein to DNA ratios, but inhibited transcription at higher ratios (Chartier et al., 1989). There are no reports of the effects of MC1 on transcription by archaeal RNAPs but, consistent with the idea that DNA binding and therefore regulation by archaeal chromatin proteins might be modulated by post-translational modifications, a lysine methyltransferase has been characterized from *Methanosarcina mazei* Göl that specifically methylates K37 of MC1 *in vitro*. This protein has homology to proteins that methylate lysines in eukaryotic histones but did not methylate the archaeal histone from *M. mazei* (Manzur & Zhou, 2005). The effects of MC1 methylation on DNA binding were not reported.

Summary

The almost universal use of histones and the regular organization of eukaryotic nuclear DNA into nucleosomes and higher order chromatin are attractive and seductive as models, but are clearly not applicable to prokaryotes. As described above, to date all Archaea investigated have the capacity to synthesize one or more members of more than one family of chromatin protein. These different families of proteins have apparently evolved independently, but with selection to accomplish essentially the same functions, namely to constrain and compact genomic DNA while not impeding and probably contributing to DNA replication, repair, and expression. The same selective pressures have apparently also been imposed in Bacteria, so that many different proteins now exist in both prokaryotic domains that qualify generically as "histone-like" proteins. By using sophisticated bacterial genetics, it has been possible to identify different and often overlapping roles for different chromatin proteins in Bacteria and this is likely also to be the case in Archaea. But, to date, there has been only one genetic investigation of chromatin proteins in Archaea, which, not surprisingly, revealed that neither the histones nor the Sso10b/Alba homolog present in *Methanococcus voltae* are essential for viability (Heinicke et al., 2004). A major difference between prokaryotes and most eukaryotes, in terms of genome organization, is that prokaryotes need to be able to access and transcribe essentially any gene at any time, whereas most nuclear DNA in a differentiated eukaryotic cell is not expressed and so can be archived in chromatin. This raises the question of whether genome compaction, as understood in terms of eukaryotic chromatin, is really a meaningful concept for prokaryotes. To date, there is no definitive demonstration that the genomic DNA in any growing prokaryotic cell is organized by chromatin protein(s) into a regular chromatin structure. Prokaryotic genomes must be compacted to meet size constraints, but their compaction has to be flexible and dynamic, and they may never be constrained into one definitive chromatin structure.

Acknowledgments

To meet space constraints many primary publications could not be cited directly. We apologize to the authors for their omission, and have cited Reeve (2003) to direct readers to many of these pre-2000 reports. Our research on archaeal chromatin proteins and gene expression at the Ohio State University is supported by the US National Institutes of Health and Department of Energy.

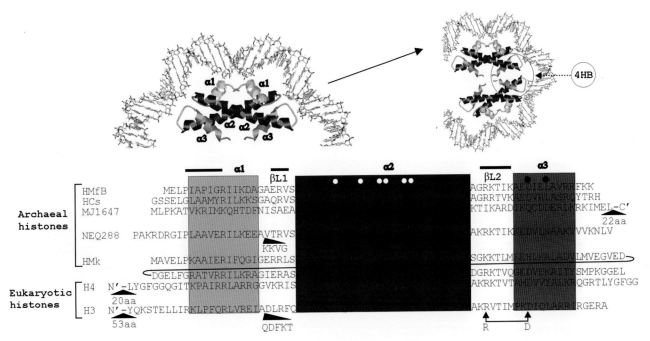

Plate 13.1 Histone fold sequences and structure. The complete sequences of representative archaeal histones are shown aligned with the sequences of the histone fold regions of eukaryotic (*Xenopus*) histones H3 and H4. As illustrated, the HMk sequence forms two histone folds. The α-helical regions of the histone folds are colored correspondingly in the alignment and structures. Black bars overline the regions that directly contact the DNA. Residues that stabilize histone dimer (○) and tetramer (●) formation are identified. As indicated, the histone fold monomer structure is universally stabilized by an arginine (R)–aspartate (D) interaction, and histone fold tetramers by the assembly of a four α-helix bundle (4HB).

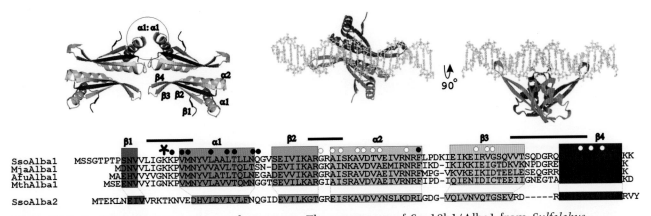

Plate 13.2 Sso10b/Alba sequences and structure. The sequences of Sso10b1/Alba1 from *Sulfolobus solfataricus* (SsoAlba1), *Methanococcus jannaschii* (MjaAlba1), *Archaeoglobus fulgidus* (AfuAlba1), and *Methanothermobacter thermautotrophicus* (MthAlba1) are shown aligned with the sequence of Sso10b2/Alba2 from *S. solfataricus* (SsoAlba2). The β-strand and α-helical regions are colored correspondingly in the sequences and in the dimer:dimer assembly modeled above to the left. The black bars identify regions that interact directly with DNA, as modeled and illustrated above to the right. Residues that stabilize dimer (○) and tetramer (●) formation, and the lysine residue (★) in Sso10b1/Alba1 (K16) that is acetylated *in vivo* in *S. solfataricus* are identified.

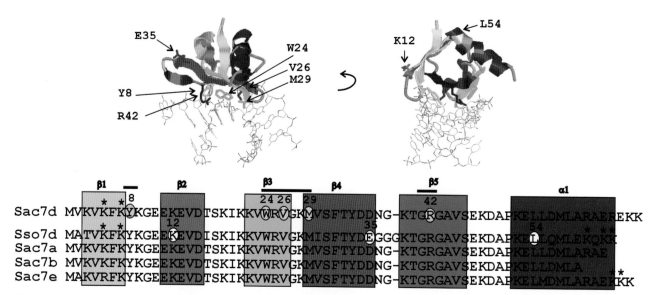

Plate 13.3 Sul7d sequences and structure. The sequences of the Sul7d proteins from *S. acidocaldarius* (Sac) and *S. solfataricus* (Sso) are aligned with the β-strand and α-helical regions colored correspondingly in the sequences and in the Sac7d–DNA complexes shown above. The black bars identify regions that interact directly with DNA, and residues that specifically bind DNA (Y8, W24, V26, M29, and R42) are circled in the Sac7d sequence. Residues required for the RNase activity (K12 and E45) of Sso7d, and the site of α1 truncation in an Sso7d deletion mutant (L54Δ) that retains DNA-binding activity are circled in the Sso7d sequence. Lysine residues that are ε-amino-monomethylated *in vivo* are identified (★).

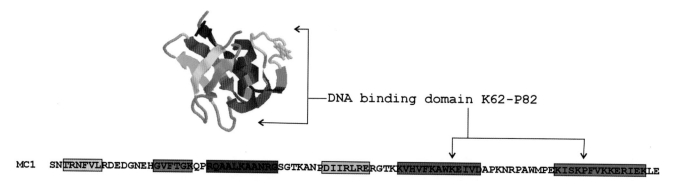

Plate 13.4 MC1 sequence and structure. The sequence of the MC1 protein from *Methanosarcina thermophila* is shown below the solution structure determined for this protein (Paquet et al., 2004). The α-helix and β-strand regions are colored identically in the sequence and structure, with the location of the DNA binding domain, identified by cross-linking studies, indicated.

14
DNA replication and the cell cycle

Victoria L. Marsh and Stephen D. Bell

Introduction

In the past few years there has been a huge increase in interest in, and knowledge of, the archaeal DNA replication system. With the availability of complete archaeal genomes in the late 1990s, it became apparent that the archaeal DNA replication apparatus was, in essence, a stripped down or ancestral version of that found in present day eukarya (Table 14.1). Thus, archaeal DNA replication is a particularly exciting field, as it addresses the inherently fascinating archaeal DNA replication machinery and manner in which Archaea divide their cells and also serves as a valuable model system for the fundamentally homologous, yet vastly more complicated, eukaryal machinery.

As we attempt to summarize in this chapter, there is now a large body of data available on the form, function, and identity of key players in the archaeal DNA replication machinery. However, far less is known about the temporal and physical context in which these proteins function during the archaeal cell cycle.

Cell cycle in Archaea

The process of cell division is a highly complicated and orchestrated process. At its very simplest, in order for a cell to pass on its full genetic complement to its progeny, chromosome segregation requires duplication of the chromosome. Eukaryotic cells and many bacteria have clearly defined periods of DNA synthesis (S-phase in the eukaryotic nomenclature) and of cell division (M-phase in eukaryotes). These phases are separated by two gap phases, G_1 between M and S phases and G_2 between S and M. Thus, the cell cycle proceeds G_1, S, G_2, M. The relative length of time organisms spend in the G_1 and G_2 phases vary, for example, G_1 dominates in budding yeast, whereas the majority of cells in asynchronous cultures of fission yeast are in G_2. In some Bacteria, for example rapidly growing *E. coli*, the situation becomes more complicated with multiple, overlapping rounds of DNA replication occurring within a single cell cycle. Thus, rapidly growing *E. coli* can actually segregate sister chromosomes that have already started their next round of DNA replication.

There is a fundamental difference in the manner in which bacterial cells and eukaryotic cells segregate sister chromosomes. In Bacteria, this event occurs concomitant with DNA replication; thus during the G_2 period of the bacterial cell cycle, newly replicated nucleoids are segregated and appropriately positioned so that septation can occur between them to generate daughter cells (Sherratt, 2003). In contrast, in the eukaryotic mitotic cell cycle, sister chromatids are paired post-replicatively in the G_2 phase of the cell cycle and this cohesion is only lost in the M phase.

The situation in Archaea remains very poorly understood. Indeed, the cell cycle parameters have only been established for a limited number of species, largely through the pioneering work of the Bernander laboratory. The best characterized species to date are the hyperthermophilic acidophiles of the *Sulfolobus* genus. It has been shown that *Sulfolobus* cells

Table 14.1 Summary of the replication proteins required across the three domains of life.

	Bacteria (*E. coli*)	Archaea (*S. solfataricus*)	Eukarya
Origin recognition protein	Dna A	Orc1/Cdc6 (3 homologs)	ORC
Helicase loading	Dna C	Orc1/Cdc6? (3 homologs)	Ccd6 and Cdt1
Replicative helicase (melts and unwinds the DNA)	Dna B	MCM homohexamer	MCM heterohexamer
Single stranded binding protein	SSB	RPA like SSB	RPA
Primase (makes RNA primers for replication)	Dna G	2 subunits, PriL and PriS	Primase complex, PriL, PriS, B subunit, DNA polymerase α
Polymerase	DNA pol III family A polymerase	DNA Pol B family polymerases	DNA Pol α, δ, ε family B polymerases
Sliding clamp (confers processivity to polymerase)	β clamp dimer	PCNA heterotrimer	PCNA homotrimer
Clamp loader (opens the ring of the sliding clamp)	γ complex, 5 different subunits	RFC, 1 large subunit and 4 small identical subunits	RFC, 1 large subunit and 4 small subunits
Ligase	DNA ligase NAD dependent	DNA ligase ATP dependent	DNA ligase ATP dependent
Primer removal	DNA Pol I	FEN1/ Rad2	FEN1/Rad 2

have a cell cycle with a short G_1 period, containing a single copy of the chromosome, and that the dominant period of the cell cycle is G_2, where cells have two copies of the chromosome (Bernander & Poplawski, 1997; Hjort & Bernander, 1999). Interestingly, microscopic studies have suggested that nucleoids are not clearly segregated in G_2 *Sulfolobus* cells. In contrast, studies of the euryarchaeon *Methanothermobacter thermautotrophicus* (Majernik et al., 2005) have shown that this organism has a more variable number of chromosomes (either two, four, or eight copies). Significantly, the majority of cells show a number of nucleoids that correlates with the number of genome copies, suggesting that genome segregation occurred either concomitant with, or immediately following, DNA replication. Studies of another euryarchaeon, *Halobacterium*, suggested that this organism also demonstrated a tight temporal linkage between DNA replication and chromosome segregation (Herrmann & Soppa, 2002).

While it is clear that significantly more studies are required before general principles can be derived, it is nevertheless tantalizing that the two Euryarchaea described above have an apparently bacterial-like mode of coupled replication and segregation. In contrast, the single crenarchaeote studied thus far, *Sulfolobus*, may be hinting at a distinct (and conceivably more eukaryotic) mode in which visible segregation of nucleoids occurs after a significant post-replicative period. It will be of considerable interest to determine whether sister chromatids in *Sulfolobus* are paired at the molecular level. Indeed, two-dimensional agarose gel analysis of replicating *Sulfolobus* chromosomes (see below) has revealed the presence of species corresponding to joint molecules between sister chromatids in this species (Robinson et al., 2004). Such species are also detected in eukaryotic chromosomes, where they have been proposed to play a role in establishing sister chromatid cohesion in S phase (Liberi & Foiani, 2004). Clearly, important goals of ongoing research will be to establish the molecular nature of these joint molecules; to determine how these structures are generated; to elucidate the longevity of the structures (are they maintained in G_2 cells?); and to perform high resolution studies of the location of genetic loci on sister chromatids in G_2 phase *Sulfolobus* cells.

The nuts and bolts of archaeal DNA replication

While there is much to be learnt about the mechanisms and control of the cell cycle in archaeal species, considerable detail is now known about the nature and function of the key players in the archaeal DNA replication apparatus. Sites of initiation of replication and the replication origins have been identified in several species. Some molecular detail is known about the manner by which these origins are defined. There have been some particularly exciting advances in the determination of the structures of a number of the replication-associated proteins and considerable insight has been gained into the architecture of the replication fork. In this section, we describe what is known about the various stages in the DNA replication process, starting with the definition of replication origins and leading to the establishment of replication forks.

Origins of replication in the Archaea

Until recently, it was generally perceived that there was a fundamental divide in the organization of DNA replication between Prokarya and Eukarya. Studies in Bacteria had revealed that they replicate their genomes from single origins of replication. In contrast, Eukarya use multiple origins per chromosome and these origins are typically spaced between 30 and 100 kb apart (for a recent review see Robinson & Bell, 2005). The pioneering work of Forterre and colleagues revealed that *Pyrococcus* species conformed to the prokaryotic paradigm, having a single origin of replication per circular chromosome (Myllykallio et al., 2000). However, bioinformatics analyses using the Z-curve methodology of Zhang and Zhang (http://tubic.tju.edu.cn/zcurve/) began to supply hints that some archaeal genomes may have a more complex replicon architecture. More specifically, Zhang and Zhang (2003) predicted that the main chromosome of *Halobacterium* had two origins of replication. However, a targeted genetic screen for autonomously replicating sequences by Berquist and DasSarma (2003) found evidence for only one of these origins of replication. Nevertheless, due to the relatively low resolution of the Z-curve methodology and the nature of the genetic screen, it remains possible that a second replication origin is present in the *Halobacterium* main chromosome.

Zhang and Zhang (2003) also revealed that the genome sequence of *Sulfolobus solfataricus* had a complex Z-curve plot that was interpreted as being indicative of three origins of replication. Direct experimental proof of the location and identity of two of these origins was obtained by 2D agarose gel electrophoresis analysis of DNA replication intermediates in *S. solfataricus* cells (Robinson et al., 2004). These origins, *oriC1* and *oriC2*, were mapped at high resolution and found to be upstream of the *S. solfataricus cdc6-1* and *cdc6-3* genes. The *cdc6-1*, *cdc6-2*, and *cdc6-3* genes encode homologs of eukaryal Orc1 and Cdc6 (see below). No origin activity was detected within 10 kb either side of the *cdc6-2* gene. A subsequent marker frequency analysis revealed that there was indeed a third origin of replication, *oriC3*, approximately 80 kb removed from the *cdc6-2* gene (Lundgren et al., 2004). Thus, it appears that *Sulfolobus* does not adhere to the prokaryotic paradigm of one replication origin per chromosome, but instead uses three origins. Replication initiation point mapping indicated that *oriC1* and *oriC2* were used in the majority of cells (Robinson et al., 2004). Additionally, the profile of replication derived from the marker frequency analysis fit with a model in which all three origins were being used in all cells (Lundgren et al., 2004). However, both these studies examined origin usage in cells growing at a high rate. It will be of considerable interest to determine whether there is any degree of differential origin usage either within the cell population or as a function of growth rate, at the level of either doubling time or position in early, middle, or late exponential growth.

In principle, the presence of multiple origins of replication could provide the cell with a mechanism for controlling the time spent to replicate the genome. Assuming no modulation of the rate of replication fork movement, in principle a cell with a single origin of replication functioning would take considerably longer to replicate its DNA than a cell with all three origins firing. In the simplest case, if the three origins were equidistant, this could result in a threefold modulation of the length of S-phase. In fact, the origins are somewhat asymmetrically distributed around the chromosome and so would allow a variability of about twofold.

If there is indeed differential origin usage in *Sulfolobus*, this may have some interesting implications for mechanisms of termination of replication. Bacterial cells have defined termination sites in the chromosome, at approximately half way round from the origin (Fig. 14.1a). These sites act as polar barriers to replication fork movement, trapping forks coming from one side, but permitting the passage of forks coming from the other. Termination sites in bacteria are often located near *dif* sites for the XerC/D-based site-specific recombination, a procedure that acts to convert dimeric chromosomes generated by homologous recombination to monomers, thereby facilitating segregation (Sherratt, 2003). In Eukarya, it is currently unclear whether there are specific mechanisms to ensure termination or if converging replication forks simply collide stochastically and terminate. In an archaeon like *Pyrococcus* with a circular chromosome and a single origin, it is possible that a bacterial-like termination system will exist (Fig. 14.1a). In this light it is important to note that many archaeal genomes encode clear homologs of the bacterial XerC/D recombinases. Presumably in *Pyrococcus*, as in bacteria, the *dif* site will be located across the chromosome from *oriC*. However, in an organism, like *Sulfolobus*, with multiple origins, it is unclear whether there will be active termination sites. Indeed, if the origins are used differentially, as suggested above, then the presence of active termination sites between all three origins could present a problem in ensuring complete replication of the chromosome (Fig. 14.1b, c). An alternative might be to have a single active termination site between two of the origins and rely on stochastic fork-collision mechanisms to ensure termination in the other two inter-origin regions (Fig. 14.1d). Depending on the positioning of this putative "master terminator," this could greatly influence the time taken to replicate the genome (Fig. 14.1e).

Origin definition

The replicon hypothesis posits the existence of a *cis*-acting sequence – the origin – that is acted on by *trans*-acting initiator factors (Jacob et al., 1963). The likeliest candidates for archaeal initiators are the homologs of eukaryotic Orc1 and Cdc6 encoded by almost all archaeal genomes. Eukaryal origins of replication are defined by their interaction with the six-member origin recognition complex (ORC; for review see Bell, 2002). One component of ORC, Orc1, is related to a second member of the eukaryal pre-replicative complex, Cdc6, and presumably derived from a common ancestor (Bell & Dutta, 2002). Present day Archaea encode what appears

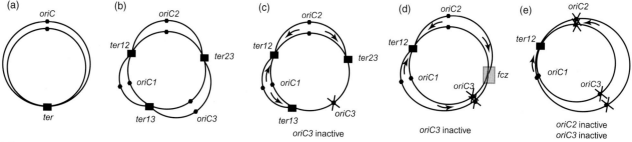

Fig. 14.1 Cartoons of replicon architecture and consequences for termination mechanism. (a) A simple replicon with a single origin (*oriC*) and a terminus (*ter*) positioned 180° around the circular chromosome. (b) A more complex situation with three origins and three active termination sites placed midway between the *ori*s. Forks moving away from the origins encounter the *ter* sites and arrest. However, (c) illustrates a situation in which *oriC3* does not fire. This would result in *ter13* and *ter23* blocking forks coming from *oriC1* and *oriC2* and thereby preventing replication of the *oriC3*-containing portion of the genome. (d) A situation where *oriC3* does not fire and the forks coming out simply collide at *fcz* (fork collision zone) and are resolved. (e) A situation where only *oriC1* is active and there is a single, asymmetrically positioned, active termination site (*ter12*). As can be seen from this diagram, the fork coming clockwise from the *oriC1* will rapidly reach *ter12* and terminate; in contrast the anticlockwise fork has to travel around over 80% of the genome before terminating. Clearly, the relative position of a *ter12* will influence the time taken to replicate the genome.

to be the descendant of that progenitor molecule. Intriguingly, the number of Orc1/Cdc6 homologs varies between species. *Pyrococcus* has a single Orc1/Cdc6, *Sulfolobus* has three homologs and halophilic archaeal species have at least ten. The origins of replication identified on *Pyrococcus* and *Halobacterium*, and two of the three *Sulfolobus* origins, are located beside the genes for Orc1/Cdc6 homologs. This is reminiscent of the situation in many Bacteria where there is tight linkage between the origin of replication and the gene for the initiator protein DnaA (Messer, 2002). The first evidence for binding of the origin of replication by archaeal Orc1/Cdc6s came from studies using chromatin immunoprecipitation (ChIP) in *Pyrococcus*. These revealed that the Orc1/Cdc6 in that species bound in the vicinity of the origin *in vivo* (Matsunaga et al., 2001). DNase footprinting studies performed *in vitro* with purified *Sulfolobus* Orc1/Cdc6 homologs (termed Cdc6-1, Cdc6-2, and Cdc6-3) revealed that these proteins bound to specific sites in the replication origins (Robinson et al., 2004). More specifically, Cdc6-1 and Cdc6-2 bound to all three origins and Cdc6-3 bound to *oriC2* and *oriC3*. These *in vitro* studies were supported by ChIP analyses *in vivo*. The binding sites for the proteins displayed complex cooperative or competitive interactions. For example, Cdc6-2 binding sites overlapped with, and competed for access to, binding sites for Cdc6-1 and/or Cdc6-3. This initially puzzling and complex situation was partly explained by the observation that there appears to be temporally regulated expression of the Orc1/Cdc6 homologs during the cell cycle. Cdc6-1 and Cdc6-3 levels are highest in G_1 and S-phase cells; in contrast, Cdc6-2 peaks in G_2. A simple prediction arising from these data is that Cdc6-1 and Cdc6-3 act to promote replication and Cdc6-2 may act as a negative regulator. Furthermore, Cdc6-1 and Cdc6-3 are encoded by genes located adjacent to *oriC1* and *oriC2*. It is possible, therefore, that the differential occupancy of the origins by Orc1/Cdc6 homologs during the cell cycle may modulate the transcription of the neighboring gene. The availability of purified Cdc6-1, 2, and 3 and the existence of a defined *in vitro Sulfolobus* transcription system (Qureshi et al., 1997) should permit testing of this hypothesis in a reconstituted system. Additionally, with the availability of genetic tools for manipulation of *Sulfolobus* it should be possible to test the effects of deleting or over-expressing Cdc6-2. One clear prediction is that deletion of Cdc6-2, if it does indeed serve as a negative regulator, would result in over-replication of the genome. Conversely, over-expression of Cdc6-2 would be predicted to result in a post-replicative arrest of cells in G_2.

While it was not possible to discern a clear consensus sequence for origin recognition by Cdc6-2 and Cdc6-3, it was apparent that conserved elements were bound by *Sulfolobus* Cdc6-1 at *oriC1* and *oriC2* (Robinson et al., 2004). The elements at *oriC1*, termed origin recognition boxes (ORBs), contained an inverted repeat motif flanked on one side by a G-rich sequence. Interestingly, clear ORB elements were also identified at the mapped origins in *Halobacterium* and *Pyrococcus* and at predicted origins in other archaeal species. The *Sulfolobus oriC2* did not possess full-length ORB elements, but instead had sequences, bound by Cdc6-1 that corresponded to the core ORB dyad symmetric element; these smaller (and lower binding affinity) motifs were termed mini-ORBs or mORBs. mORBs, but not full-length ORBs, have been also detected in the putative origin of *M. thermautotrophicum* (Capaldi & Berger, 2004).

Thus, it is clear that archaeal Orc1/Cdc6 homologs bind to, and thereby define, origins of replication. However, the mechanisms by which they lead to the recruitment of the replicate helicase (widely believed to be MCM, see below) remains unclear. Likewise the role of ATP binding and hydrolysis by the archaeal Orc1/Cdc6s is poorly understood. It is clear from sequence and structural analyses that the Orc1/Cdc6s have an N-terminal AAA$^+$ ATPase domain and a C-terminal winged helix (WH) DNA binding domain (Liu et al., 2000). Despite the presence of an ATP binding site, there is no evidence that nucleotide binding is required for the DNA binding activity of the Orc1/Cdc6s (Robinson et al., 2004). However, it has been observed that *M. thermautotrophicus* Orc1/Cdc6s inhibit the helicase activity of MCM and do so in a nucleotide-dependent manner (Shin et al., 2003). It is widely believed that the Orc1/Cdc6s will facilitate recruitment of MCM helicases to the origin(s) of replication. However, to date this reaction has not been recapitulated *in vitro*. Whether this is due to trivial reasons regarding experimental approach and conditions or indicative of the requirement for additional factors and/or modifications to the proteins remains unknown.

The replicative helicase

The best candidate for the eukaryal replicative helicase is the heterohexameric MCM complex composed of six related subunits, MCM2-7 (Labib & Diffley, 2001). The sequence similarity between these six subunits suggests that they evolved from a common, presumably homomultimeric, ancestor and such a molecule is found in present day Archaea. Archaeal MCM is a multimer of an approximately 70 kDa monomer. The precise stoichiometry varies with species, with hexamers present in *Archaeoglobus fulgidus* and *S. solfataricus* (Grainge et al., 2003; McGeoch et al., 2005), and double hexamers appearing to be the dominant form in *M. thermautotrophicus* (Mt) (Chong et al., 2000). Archaeal MCMs have been shown to be (d)ATP-dependent DNA helicases *in vitro* and are capable of melting at least 500 base pairs of duplex DNA. The crystal structure of the N-terminal 30 kDa of MtMCM has been solved and reveals a double hexameric organization with individual hexamers organized in a head-to-head orientation (Fletcher et al., 2003). These double hexamers of the MtMCM N-terminal region are extremely stable, being resistant to even 1 M urea, leading to the suggestion that this region of the MCM may play a role in facilitating multimerization of the protein. In this light, parallels have been drawn with the pseudo-hexameric sliding clamp, PCNA (proliferating cell nuclear antigen), a processivity factor for DNA polymerases (see below). An extension of this analogy is that the N-terminal region of MCM may act to tether MCM to DNA and thereby facilitate processive helicase activity. The circular nature of MCM presents a topological problem: how is this molecule loaded onto DNA? Again conceptual parallels might be drawn with PCNA. As discussed below, PCNA is loaded onto DNA by the clamp loader, RFC, in an ATP-dependent manner. It is apparent that the ATPase domain of the Orc1/Cdc6 homologs is closely related to the analogous domains in RFC, leading to the possibility that a complex of Orc1/Cdc6 molecules may act as ring-opening loaders for MCM.

Recent work has begun to shed some light on the mechanisms by which MCM moves along DNA and separates strands. It is proposed that two sets of β-hairpin motifs in the N terminal and AAA⁺ domains of MCM act to facilitate DNA binding in the central cavity of the ring-shaped MCM hexamer. Interestingly, mutational analyses have revealed that while mutation of a conserved lysine in the C-terminal hairpin to alanine only modestly affects DNA binding by MCM, it nevertheless abrogates the helicase activity of the protein. This suggests that this hairpin may effect the power stroke of the enzyme, coupling ATP hydrolysis to motion of the protein complex along DNA (McGeoch et al., 2005; Fig. 14.2).

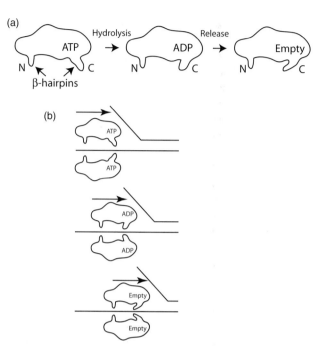

Fig. 14.2 (a) Cartoon of a monomer of MCM with the N-terminal and C-terminal putative β-hairpins indicated. The positioning of the C-terminal hairpin is shown to vary during the ATP binding, hydrolysis and release cycle. (b) A cutaway of an MCM hexamer bound to a flapped DNA substrate is shown, with two MCM monomers depicted. Motion of the C-terminal β-hairpins is depicted as driving the MCM complex along DNA, unwinding duplex ahead of it.

Primase

The first step in the synthesis of a new strand of DNA is the production of an oligoribonucleotide primer. Primers are synthesized from a DNA template by the activity of the primase enzyme. The bacterial primase (DnaG) is a single subunit protein, in contrast, eukaryotic core primase is a heterodimeric protein found in complex with DNA Pol α and polymerase B subunit (Frick & Richardson, 2001). The heterodimeric core primase has a large subunit of 58 kDa and a smaller subunit of 48 kDa. The smaller subunit contains the catalytic RNA primer synthesis ability, while the large subunit confers regulatory and stabilizing properties to the enzyme.

Initially archaeal primase was thought to be a monomeric enzyme, as only homologs of the small eukaryotic subunit were identified, via sequence similarity, in *Methanococcus jannaschii* (Desogus et al., 1999). Biochemical analysis of this subunit identified a requirement for divalent cations, in particular Mn^{2+}, for primase activity akin to the eukaryotic catalytic subunit (Desogus et al., 1999). Subsequent genome sequence searches of the euryarchaeal species *M. jannaschii*, *M. thermautotrophicus*, *A. fulgidus*, *Halobacterium* sp. and *Pyrococcus* sp. identified open reading frames homologous to both the large and small subunits of eukaryotic primase (Bocquier et al., 2001; Matsui et al., 2003). Biochemical analysis of the *P. horikoshii* primase established distinct functions for each of these subunits, with the small subunit being the catalytic subunit for primer synthesis, the large subunit C-terminal half being responsible for the DNA binding affinity of the enzyme, and the large subunit N-terminal half for the correct tethering of the catalytic subunit to the DNA binding domain (Matsui et al., 2003). Surprisingly, these biochemical investigations also highlighted the unique ability of archaeal primase to utilize dNTPs as well as NTPs in primer synthesis; this has not been observed in eukaryotic primases (Matsui et al., 2003). The crystal structure of *P. horikoshii* primase established that the active site pocket of the primase was of sufficient size to accommodate two NTPs for dinucleotide formation and primer initiation (Ito et al., 2003), as had previously been speculated for eukaryotic primases (Frick & Richardson, 2001), and that this NTP binding was DNA template-independent (Ito et al., 2003). It was noted that the positioning of aspartate residues in the active site closely resembles that of family X DNA polymerases, and that these are essential for the nucleotidyl transfer reaction (Ito et al., 2003).

The first crenarchaeal primase was identified in *S. solfataricus*, and was also identified as a heterodimeric protein (Lao-Sirieix & Bell, 2004). The biochemical analysis indicated that the capability of this protein to incorporate both dNTPs and NTPs for dinucleotide formation was most likely a characteristic of all archaeal primases and not just a feature of those found in the euryarchaeal sub-domain. However, it is most likely that RNA primers will be synthesized *in vivo* due to the higher affinity for NTPs than dNTPs, which was observed, and the presumed greater abundance of NTPs in the cell (Lao-Sirieix & Bell, 2004). Uniquely, terminal transferase activity was also observed for this primase. Recently, the crystal structure of the heterodimeric *Sulfolobus* enzyme has been solved, revealing that the large subunit is physically distant from the active site of the enzyme and suggesting a model whereby this subunit may act to count the length of the primer synthesized (Lao-Sirieix et al., 2005a).

It was postulated that, as there is an absence of DNA Pol α in Archaea, the primase may be required for both primer formation and elongation (Lao-Sirieix & Bell, 2004). Family X DNA polymerases are known to possess terminal transferase activity and the aspartate conformation in primases resembles that of DNA Pol X (Ito et al., 2003; Lao-Sirieix et al., 2005a). As there are no known homologs of family X DNA polymerases in this species, these data taken together lead to the further speculation that primase might be of importance in DNA repair pathways (Lao-Sirieix & Bell, 2004; Lao-Sirieix et al., 2005b).

Short Okazaki fragments, akin to those found in eukaryotes, and much shorter than those found in Bacteria, have been identified in Archaea (Matsunaga et al., 2003). These fragments are generated on the lagging strand of replication, and are initiated by the primase enzyme that primes DNA sites at regular intervals. Representatives of the euryarchaeal sub-domain *P. abyssi* and the crenarchaeal sub-domain *S. acidocaldarius* were utilized in the identification of these fragments. In both cases the fragment was seen to be approximately 100 bases in length and, in *P. abyssi*, to occur at a frequency of 40–60 bases between fragments (Matsunaga et al., 2003). The structure of the RNA/DNA fragment for *P. abyssi* was further studied and revealed a striking similarity to those previously observed in *Drosophila* cells and *Saccharomyces cerevisiae* (Matsunaga et al., 2003). Nucleosomal structure and chromosomal organization vary significantly between these two sub-domains yet Okazaki fragment length and the number of fragments initiated per second are very similar indicating that nucleosomal structure does not affect this processing (Matsunaga et al., 2003).

DNA polymerase

Primers, generated by primase complexes, are extended by DNA polymerases. DNA polymerases, based on phylogenetic relationships, are divided into six families: A, B, C, D, X, and Y (Savino et al., 2004). Eukaryotic DNA synthesis requires three B family polymerases: Pol α, Pol δ and Pol ε (Hubscher et al., 2002). Pol δ and Pol ε are the major replicative polymerases, with Pol δ acting on the lagging strand, and both acting on the leading strand of DNA replication (Fukui et al., 2004), while Pol α is involved in lagging strand Okazaki fragment synthesis, with switching of Pol α and Pol δ taking place (Maga et al., 2001).

It was initially thought that the archaeal replicative polymerases were family B enzymes. It was later established that, whilst multiple B family polymerases are a feature of Crenarchaea, such as *Aeropyrum pernix*, which has had two homologs identified from cell extracts (Cann et al., 1999a), euryarchaeal DNA replication is thought to require both a B family polymerase and another, unique, heterodimeric polymerase, Pol D (Cann et al., 1998). Pol D comprises two subunits, DP1 and DP2, DP2 being the polymerase domain and DP1 the 3′–5′ exonuclease domain.

Recent work has demonstrated a novel function associated with the archaeal B family DNA polymerases. They were found to possess a "read ahead" recognition pocket for uracil bases produced as a result of cytosine deamination. This recognition pocket is located in the N-terminal region of the polymerase and, upon binding the uracil base, halts DNA replication four bases from the primer/template junction, and prevents the mutation of C–G base pairing to T–A pairing (Fogg et al., 2002). This feature would be particularly important for those Archaea that inhabit high temperature environments, as deamination is more prevalent under these conditions, and genetic integrity would be rapidly lost by these organisms (Fogg et al., 2002).

The functional interplay between Pol B and Pol D in the Euryarchaea has been the subject of a recent study. Both are candidate replicative polymerases, but they possess different biochemical properties and may perform specific roles at the replication fork (Henneke et al., 2005). Pol D can perform primer extension of both RNA and DNA primers and addition of PCNA (see below) stimulates this extension (Henneke et al., 2005). In contrast, Pol B only extends DNA primers, and these can be extended to the full length of the template in the absence of PCNA. The inability of Pol B to extend RNA primers cannot be overridden by the addition of PCNA (Henneke et al., 2005). Other differences between the polymerases lie in their abilities to perform strand displacement synthesis; Pol D readily achieves strand displacement of DNA primers, but RNA primers can only be displaced by Pol D in the presence of PCNA. Conversely, Pol B is incapable of achieving strand displacement unless PCNA is present, in which case only DNA primers can be displaced (Henneke et al., 2005). These data led the authors to predict a model in which Pol B functions as the leading strand polymerase, extending initial products generated by Pol D, and Pol D is also the lagging strand polymerase.

As the Crenarchaea contain more than one B family homolog it is tempting to speculate that, as is seen in the eukaryotic domain, the euryarchaeal subdomain, and some members of the bacterial domain, these homologs may also have specific roles at the replication fork.

PCNA

Eukaryotic PCNA is a DNA polymerase processivity factor for the two major replicative DNA polymerases, Pol δ and Pol ε. This protein has been described as the "sliding clamp" and serves to tether DNA polymerase to the replication fork and stimulate DNA polymerase activity, as well as to act as a docking platform for several other accessory proteins involved in DNA replication such as FEN 1 (flap endonuclease) and DNA ligase (Warbrick, 2000). PCNA also has a role outside of DNA replication: it is required for DNA repair pathways and acts as a docking platform for accessory proteins in much the same manner.

In Bacteria, the analogous sliding clamp is the β subunit of the Pol III holoenzyme. The β clamp is a homodimeric protein complex, while eukaryotic PCNA is homotrimeric. These complexes both share a striking superficial structural resemblance, however, and the crystal structures can, in fact, be superimposed (Matsumiya et al., 2001). These doughnut shaped complexes have a positively charged central channel large enough to accommodate a double helix of DNA (Fig. 14.3).

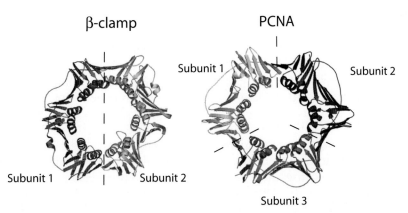

Fig. 14.3 Crystal structures of bacterial and archaeal/eukaryal sliding clamps. Images were generated using Pymol (www.pymol.org) and PDB coordinates 1UNN and 1VYM. Dashed lines indicate the approximate positions of inter-subunit interfaces.

The archaeal "sliding clamp" is a homolog of eukaryotic PCNA in all sequenced archaeal genomes (Cann & Ishino, 1999). Biochemical analysis of the *P. furiosus* homolog revealed that this protein stimulates primer extension *in vitro* by the two replicative DNA polymerases in this species, Pol B and Pol D (Cann et al., 1999b). Interactions were observed between PCNA and Pol B, and also between PCNA and the DP2 catalytic subunit of Pol D. An interaction between PCNA and the clamp loader RFC (see below) was also detected. Further analysis via dimerization suggested that the active form of this protein is a trimer, homologous to eukaryotic PCNA (Cann et al., 1999b). The sequence and structural similarities between the eukaryotic and archaeal PCNA homologs are so well conserved that *P. furiosus* PCNA is able to stimulate the primer extension activity of eukaryotic Pol δ from calf thymus using linear DNA as a template (Ishino et al., 2001). Furthermore, on the addition of human RFC, circular DNA can be used as a template for primer extension by Pol δ and enhances the *pfu*PCNA-dependent DNA synthesis activity of Pol δ (Bocquier et al., 2001). This indicates that human RFC works as the clamp loader for archaeal PCNA. The primer extended products are shorter than would be produced with human PCNA, but this is probably due to weaker interactions between the proteins, and a higher dissociation rate as a result (Ishino et al., 2001).

Phylogenetic analysis of archaeal PCNA homologs revealed that Euryarchaea contain a single homolog of PCNA, while the Crenarchaea possess more than one homolog; this duplication is hypothesized to have occurred by gene duplication after the Euryarchaea and Crenarchaea lineages split. In fact, three homologs of PCNA are encoded by some members of the crenarchaeal lineage (Daimon et al., 2002; Dionne et al., 2003). Interactions between the three PCNA homologs encoded by *A. pernix* and the two replicative DNA B family polymerases Pol I and Pol II were investigated (Daimon et al., 2002). PCNAs 1 and 2 coprecipitate, but no interaction was seen between PCNA3 and either of the other homologs. Individually all three PCNA homologs were able to stimulate polymerase activity in this species; notably PCNA3 stimulated primer extension approximately 200-fold. There is no additional improvement on the stimulation levels previously observed when both PCNA1 and 2 are assayed simultaneously (Daimon et al., 2002). From these data the authors hypothesized that PCNA3 is the replicative PCNA and that the PCNA1 and PCNA2 complex might be required in other cellular processes such as the stimulation of repair polymerase activity. Remarkably, the three PCNA homologs encoded by *S. solfataricus* are active solely in a heterotrimeric complex (Dionne et al., 2003). It was determined that the formation of this complex required PCNA1 to bind to PCNA2 initially and only then was PCNA3 able to bind (Dionne et al., 2003). Specific interactions were also detected between PCNA1 and 3 and other known replicative accessory factors, FEN1 and ligase, respectively. Sequence analysis of both Pol B1 and FEN1 protein residues detected the presence of

PIP (PCNA interacting protein) motifs, which are highly conserved compared to those found in analogous eukaryotic accessory proteins (Dionne et al., 2003). Furthermore, PCNA3 was seen to interact specifically with the large subunit of RFC, RFC_L, and the PCNA1/2 complex with the RFC heterotetrameric small subunits, RFC_S. Thus a model was derived whereby RFC_L binds PCNA3, RFC_S binds PCNA1/2, and ATP binding to RFC leads to a conformational change within the RFC, which in turn opens the PCNA ring; the PCNA ring then closes around the DNA and hydrolysis of the ATP allows recycling of the RFC (Dionne et al., 2003). The interactions between PCNA and RFC are discussed in greater detail below.

The archaeal PCNA toroidal structure is maintained via intermolecular hydrogen bonds, and eukaryotic toroidal PCNA structure is maintained by the use of hydrophobic contacts in addition to the hydrogen bonds. The *P. furiosus* PCNA crystal structure revealed conformational flexibility in each domain, and this is presumably linked to the loading and unloading of PCNA onto DNA (Matsumiya et al., 2001).

RFC

RFC, replication factor C (γ complex in Bacteria), is the clamp loader that loads PCNA (β clamp) onto a primer-template junction of DNA. In all three domains of life this is a five-subunit complex. This complex possesses ATPase activity, which is stimulated when bound to both the clamp and DNA; as such, clamp loaders are members of the AAA^+ superfamily of proteins.

Bacterial γ complex comprises three different subunits, γ, δ, and δ' at a stochiometry of $\gamma_3{:}\delta{:}\delta'$ (Jeruzalmi et al., 2001). Eukaryotic RFC comprises one large subunit, RFC1, and four smaller subunits, RFC2–5 (O'Donnell et al., 2001).

Archaeal RFC also comprises one large and four small subunits, but the four subunits are identical (Seybert et al., 2002). There is a high sequence similarity between the five eukaryotic RFC subunits and archaeal RFC subunits, and the most highly conserved regions, denoted RFC boxes I–VIII in eukaryotic RFC subunits, are also very highly conserved in Archaea (Cann & Ishino, 1999).

Clamp loader structural and mechanistic information was initially derived from the bacterial homologs. The δ subunit of the γ complex is the only subunit that contacts the β clamp. It was hypothesized that, by an induced fit mechanism, the δ subunit binds between domains II and III of the β clamp, leading to a conformational change in the β clamp resulting in the opening of the clamp ring (Jeruzalmi et al., 2001). ATP hydrolysis by the other subunits was suggested to render the δ subunit inaccessible to the β clamp, resulting in the release of the clamp onto the DNA. The δ subunit was therefore proposed to be the "wrench" in the bacterial homolog and RFC1 was hypothesized to perform the same function in eukaryotes (Jeruzalmi et al., 2001). Subsequent crystal structure analysis of the γ complex revealed that the five subunits bind, forming a closed ring structure at their C-terminals, while the ring remains open at the N-terminus, with the δ subunit being displaced the most from this structure (Jeruzalmi et al., 2001). It was surmised that sequential binding of ATP molecules to the γ subunits would lead to conformational changes in the γ complex, causing the δ subunit to swing out to allow binding to the clamp. Subsequent hydrolysis, triggered by DNA, would reset the closed structure of the γ complex and release the β clamp onto the DNA (Jeruzalmi et al., 2001). The structural information from the γ complex was used to predict the architecture of the eukaryotic RFC complex and highlighted particular structural similarities between the δ subunit and RFC1 (the PCNA binding subunit), the $\gamma1$–$\gamma3$ subunits and RFC2–4 (the ATP binding domains), and the δ' stator subunit and RFC5 (O'Donnell et al., 2001). Molecular modeling supports this hypothesis and further infers that the PCNA is contacted by three of the RFC subunits, RFC1, RFC3, and RFC5, with decreasing affinity on ATP binding.

The structure of the archaeal RFC small subunit can be superimposed on that of the δ subunit from the γ complex. Seven of the eight RFC boxes, common to clamp loaders, are highly conserved and are distributed around the nucleotide binding site (Oyama et al., 2001). Mutational analysis determined that archaeal RFC_S oligomerize via the C terminal regions akin to the crystal structure of bacterial γ complex (Jeruzalmi et al., 2001). Biochemical characterization of *Archaeoglobis fulgidus* RFC determined that ATP binding stimulated the interaction between

RFC and PCNA, and that subsequent ATPase activity of this protein complex is increased by PCNA, which, in turn, strongly depends on the presence of DNA, but not RNA. Thus, at a primer template junction it is likely that the RFC contacts the DNA moiety of the RNA/DNA primer (Seybert et al., 2002). Electron microscopy was employed for the visualization of the RFC–PCNA–DNA structure. Overall, the archaeal homologs strongly resemble the yeast clamp/clamp loader structure. However, contrary to the *E. coli* γ structure and the eukaryotic RFC structure, it is not a closed ring that is formed by the C-terminal regions of the RFC complex but a horseshoe shape (Miyata et al., 2004). This structure allows for the release of DNA from the RFC after successful PCNA loading. This structure was determined in the presence of non-hydrolyzable ATPγS yet a closed PCNA ring is visualized, and therefore ATP hydrolysis is not required for PCNA ring closure (Miyata et al., 2004). Distinct roles for ATP binding and hydrolysis for archaeal RFC were later attributed to specific subunits (Seybert & Wigley, 2004). In the presence of DNA and PCNA the *Af*RFC binds four ATP molecules: one binds to the large subunit and one ATP molecule is bound to each of three of the small subunits. ATP hydrolysis of the ATP molecules bound to the RFC_S subunits is required for PCNA release from RFC, while hydrolysis of the ATP molecule bound to RFC_L is required for the recycling of the RFC complex (Seybert & Wigley, 2004).

Concluding remarks

We now have a wealth of mechanistic knowledge about the function and form of individual archaeal DNA replication proteins. It is clear that it will be important to begin to integrate these various proteins into higher order systems to begin to attempt the *in vitro* reconstitution of defined replication systems. This will be a considerable technical challenge further complicated by the fact that it is currently unclear whether we even know the identities of all the players involved. Beyond the technical complexities of the biochemical analyses of DNA replication, it will be of great importance to begin to unravel the regulatory networks that govern and drive progression of the archaeal cell cycle.

Acknowledgments

Work in SDB's laboratory is funded by the Medical Research Council. We would like to apologize to our colleagues whose work we have been unable to cite because of space constraints.

15
DNA repair

Malcolm F. White

Introduction

The chapters of this book that deal with DNA replication and transcription are a clear testament to the fundamental conservation of the machinery of information processing in the Archaea and Eukarya. Indeed, this conservation is so strong at the level of protein structure that archaeal DNA replication and transcription proteins have been prime targets for structural biology, and have yielded many insights relevant to the equivalent eukaryal proteins. Studies of DNA repair have tended to lag behind those of replication and transcription. Historically, the issue of DNA damage was not appreciated, even though the structure of the DNA duplex suggested how an undamaged strand could act as a template for repair, as well as replication. The DNA repair field is now a fiercely competitive and highly productive one, initially focusing on *E. coli*, followed by yeast and latterly higher eukaryotes. There are important goals in the treatment and prevention of cancer, as well as more basic research prerogatives. Understandably, archaeal DNA repair has played Cinderella to her two sister domains. For one thing, an appreciation of the Archaea as a distinct domain of life only impinged on the consciousness of the general scientific community with the advent of whole genome sequencing in the past ten or so years. Second, a lack of defined genetic systems in the Archaea has retarded studies of DNA repair – a particular problem, as genetic studies in Bacteria and yeast have been so enlightening in this area. As we shall see, much of the recent progress has highlighted both the similarities of Archaea and Eukarya, and the unique nature of Archaea, and there are many unanswered questions that promise interesting and surprising answers.

Although the DNA duplex is rather chemically inert, it is subject to many types of modification and decay that arise from interactions with its chemical environment, in addition to external factors such as UV and ionizing radiation. It is now well understood that every cell in the human body suffers many thousands of damage events every day, and the same is true for microorganisms (Lindahl, 1993). Clearly this background rate of largely unavoidable DNA damage must be resisted, so all organisms devote a considerable portion of their coding potential to proteins that repair damaged DNA. These pathways can be classified in a variety of ways. Commonly they are classified as excision pathways, comprising base excision repair, mismatch repair, and nucleotide excision repair; direct reversal and error bypass; double strand break repair, and pathways for the rescue of stalled or collapsed replication forks. Overlying these pathways are the control mechanisms that coordinate the response to DNA damage at a transcriptional or post-translational level (Fig. 15.1). These are dealt with in turn.

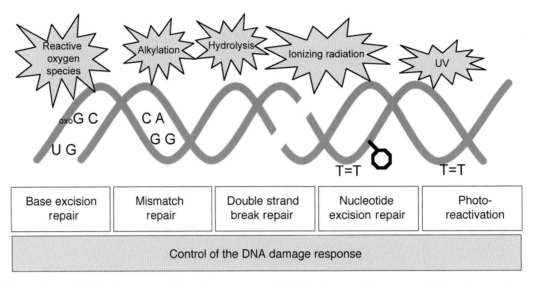

Fig. 15.1 Overview of DNA damage and repair processes. DNA is damaged in many different ways by environmental and chemical factors. DNA lesions and modifications are recognized and repaired by a variety of repair pathways with overlapping specificities. In most organisms the response to DNA damage is coordinated at the transcriptional and post-translational level.

Base excision repair

The base excision repair (BER) pathway repairs specific DNA bases modified or damaged as a result of simple chemical processes such as oxidation, hydrolysis, and methylation (Fig. 15.2). These forms of damage are frequent and unavoidable in all organisms from humans to microbes, and are likely to occur at increased frequency in hyperthermophiles as a consequence of elevated temperatures. Archaea in general possess a spectrum of specific BER glycosylases for modified bases such as uracil and 8-oxoguanine (Table 15.1). These enzymes recognize and remove modified DNA bases to generate an abasic site. Sometimes they also cleave the DNA backbone next to the damage site. These repair glycosylases are often coupled with an AP (apurinic or apyrimidinic) endonuclease that processes the resultant abasic sites and modified 3′ termini in DNA. This generates DNA duplexes with a small gap in one strand that are closed by DNA synthesis by DNA polymerase and finally nick ligation by a DNA ligase. In the crenarchaeote *Pyrobaculum aerophilum*, this pathway has been reconstituted *in vitro*. Repair of a G–U mismatch in DNA to a G–C base pair was observed upon incubation of the mispaired DNA with purified recombinant uracil DNA glycosylase,

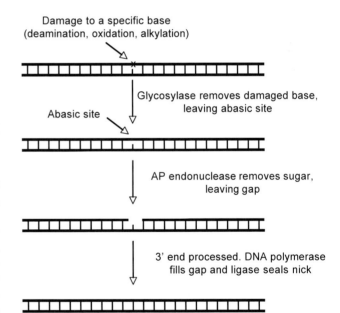

Fig. 15.2 Schematic representation of base excision repair. Damage to specific bases is recognized by specific DNA glycosylases, which remove damaged bases, often creating abasic sites and/or modified 3′ DNA ends that are in turn processed to produce a substrate for DNA polymerase to read the undamaged template strand and repair the duplex DNA structure.

Table 15.1 Distribution of base excision repair proteins in sequenced archaeal genomes.

Species	Endo IV	Endo V	OGT	UDG IV	HhH glycos	Temp (°C)
A. pernix	1	1	1	1	3	95
P. aerophilum	1	1	1	2	4	95
S. solfataricus	1	1	1	2	3	80
S. tokodaii	1	1	1	2	4	80
S. acidocaldarius	1	1	1	2	2	75
M. kandleri		1	1		2	100
P. furiosus	1	1	1	1	2	100
P. horikoshii	1	1	1	1	2	100
P. abyssi	1	1	1	1	2	100
T. kodakaraensis	1	1	1	1	2	85
N. equitans	1	1	1	1	1	85
M. jannaschii	1		1		3	85
A. fulgidus		1	1	1	3	85
M. thermoauto	1		1		4	70
P. torridus	1		1	3	3	60
T. acidophilum	2	1	1	1	4	60
T. volcanium	2	1	1	1	4	60
M. acetivorans	1		1	1	5	<40
M. mazei	1		1	1	5	<40
H. marismortui		1		4	6	<40
Halobacterium		1		3	5	<40

Crenarchaea are at the top, followed by euryarchaeal species, ranked by growth temperature. The number of each repair gene in each genome is indicated. Enzymes shown are: endonuclease IV (AP lyase and oxidative damage); endonuclease V (deoxyinosine); OGT (O6-methylguanidine cysteine methyltransferase); UDG IV (archaeal specific uracil DNA glycosylase) and HhH superfamily members (wide specificity, includes exonuclease III, UDG, 8-oxoguanine glycosylases).

AP endonuclease IV, DNA polymerase, and DNA ligase (Sartori & Jiricny, 2003). This suggests that BER in Archaea conforms to the paradigm observed in Bacteria and Eukarya, and that BER is essentially a universally conserved repair pathway across the three domains. The observation that hyperthermophiles appear to get along with a fairly standard set of BER enzymes probably reflects the very high efficiency with which these enzymes recognize and remove damaged bases.

Recent studies, however, have uncovered some quirks in archaeal BER where novel enzymes or enzyme combinations have been identified. One example is a protein from the euryarchaeote *Ferroplasma acidarmanus* that couples an O^6-alkylguanine-DNA alkyltransferase (AGT) domain with a domain homologous to *E. coli* endonuclease V (*nfi* gene product). The resultant protein (AGTendoV) can remove both O^6-alkylguanine lesions and a variety of deaminated base products such as uracil and hypoxanthine from DNA *in vitro* (Kanugula et al., 2005). This unusual combination of activities has been detected in the genomes of *Picrophilus torridus* and the *Thermoplasma*s and may thus reflect an adaptation to aerobic growth at extremely low pH. A second example is the discovery of an archaeal-specific 8-oxoguanine DNA glycosylase (AGOG) (Sartori et al., 2004). 8-Oxoguanine is an important premutagenic lesion formed by the oxidation of guanine. Many archaeal genomes, including the Sulfolobales and *Thermoplasma*s, encode an ortholog of the well characterized eukaryal OGG1 glycosylase, which removes 8-oxoguanine mispaired with cytosine. However, other Archaea, including *Pyrobaculum aerophilum* and the *Pyrococcus* species, lack this gene. Using a biochemical approach, Jiricny and coworkers identified a novel 8-oxoguanine glycosylase (AGOG) as the primary glycosylase for removal of 8-oxoguanine in

Pyrobaculum aerophilum. AGOG is present in many of the sequenced archaeal genomes that lack OGG1 (Sartori et al., 2004).

The structure of AGOG has revealed a new class of the helix–hairpin–helix superfamily of DNA repair glycosylases (Lingaraju et al., 2005). Many of the glycosylases involved in BER are members of this superfamily, whose members have diverse specificity for oxidized, deaminated, alkylated, and mismatched bases, as well as AP lyase activity (Lingaraju et al., 2005). A glance at Table 15.1 shows that there is no obvious correlation between growth temperature and the number of these genes in archaeal genomes. For example, *Methanopyrus kandlerii* has only two representatives of the glycosylase superfamily, while the mesophilic *Methosarcina*s and *Haloarcula* have five and six, respectively. While there is some correlation with genome size, it is possible that species living in surface environments are exposed to a more diverse array of base modifications, and therefore require a larger complement of BER glycosylases.

Uracil scanning polymerases

Archaea do have a very clever trick up their (figurative) sleeves when it comes to replication of DNA containing uracil. Uracil arises in DNA due to deamination of cytosine residues, which creates a promutagenic U–G base pair. As uracil closely resembles the structure of thymine, bacterial polymerases replicate happily past uracil by incorporating an adenine in the new strand, thus fixing a CG to TA mutation. This is also one of the most common types of mutation found in eukaryal organisms. However, the archaeal replicative B family DNA polymerases stubbornly refuse to pass a uracil in the template DNA and instead replication is stalled. An explanation for this phenomenon has now arisen from structural and biochemical studies. It turns out that archaeal polymerases have a pocket specific for uracil in the N-terminal domain. This functions as a "read ahead" domain, and when uracil is present in the template it binds in the pocket, stalling the polymerase. Mutagenesis of the pocket abolishes stalling (Fogg et al., 2002). It must be assumed that the stalled polymerase participates in some form of "hand-over" of the uracil lesion to a uracil glycosylase for repair, in a similar way to the hand-overs seen in other members of the BER pathway. Although the molecular details of this pathway are not yet defined, the observation that uracil glycosylase UDG1 from *Sulfolobus solfataricus* interacts specifically with the sliding clamp PCNA, which also interacts with the DNA polymerase, suggests a mechanism for the hand-over (Dionne & Bell, 2005).

Conclusions and future perspectives

BER is probably the most ubiquitous and best characterized repair pathway. While there are some peculiarities in Archaea such as fused genes, novel subfamilies of the HhH glycosylase superfamily and unusual substrate specificities, the Archaea seem to conform in general to the paradigm established from studies of BER in Eukarya and Bacteria. One exception is the emerging evidence for a link between DNA replication and the removal of uracil from DNA, where the Archaea appear to have evolved a unique proofreading mechanism involving a uracil-detecting read-ahead function in the replicative polymerase. The links between this novel uracil-detection mechanism and the downstream processing of lesions to allow replication to resume are likely to uncover interesting new examples of interactions between replicative and repair proteins.

Mismatch repair: desperately seeking proteins

The mismatch repair (MMR) pathway corrects misincorporated bases introduced primarily during DNA replication, and involves steps of mismatch detection, nicking of the newly synthesized strand, excision of a stretch of DNA including the mismatch, and gap filling by replication and ligation (reviewed in Kunkel & Erie, 2004). In *E. coli* the pathway is well defined, but the details are considerably less clear in Eukarya and even in other Bacteria. Common to all, though, are homologs of the mismatch detection proteins MutS and MutL. These proteins are absent from most archaeal genomes (Table 15.2). Their presence is restricted to low temperature Archaea, and like the UvrABC proteins with which they tend to coincide, they are likely to have reached the Archaea

Table 15.2 Distribution of putative archaeal nucleotide excision repair proteins (XPF, XPB, XPD & Fen1), error bypass polymerase Dpo4, photolyase, and homologs of bacterial UvrABC and MutSL (indicated "B").

Species	XPF	XPB	XPD	Fen1	Dpo4	Photolyase	Uvr ABC	MutS MutL
A. pernix	1	1	1	1				
P. aerophilum	1	2	1	1				
S. solfataricus	1	2	1	1	1	1		
S. tokodaii	1	2	2	1	1	1		
S. acidocaldarius	1	2	1	1	1	1		
M. kandleri	1			1				
P. furiosus	1	2	1	1				
P. horikoshii	1	2	1	1				
P. abyssi	1	2	1	1				
T. kodakaraensis	1	1	1	1				
N. equitans	1	1		1				
M. jannaschii	1		1	1				
A. fulgidus	2	1		1				
M. thermoauto	1			1		1	B	
P. torridus		1	1	1	1	1		
T. acidophilum		1	1	1				
T. volcanium		1	1	1				
M. acetivorans	1	1	1	1	1	1	B	B
M. mazei	1	1	1	1	1	1	B	B
H. marismortui	1	2		1	1	3	B	B
Halobacterium	1	2	1	1	1	2	B	B

Details are as for Table 15.1.

by lateral gene transfer from Bacteria. This has left the identity and even existence of an archaeal MMR pathway open to question. It has been suggested, for example, that the archaeon *Pyrobaculum aerophilum* lacks a functional MMR pathway (Fitz-Gibbon et al., 2002), while on the other hand studies of mutation rates in *Sulfolobus acidocaldarius* suggest a very low error frequency, consistent with functional MMR (Grogan et al., 2001; Chapter 12 in this volume). Recent data indicate that the *Sulfolobus* single-stranded DNA binding (SSB) protein can detect and melt mismatches *in vitro*, and a role for the protein has been proposed in the detection of a variety of different types of DNA damage (Cubbedu & White, 2005). However, there are currently no functional studies to underpin this observation. The controversy may not be simply resolved by genetic studies, as the genetically tractable organisms have bacterial MMR proteins. In this case a biochemical approach may be required to identify the archaeal MMR pathway, if indeed it exists.

Damage reversal and bypass

UV radiation is a potent DNA damaging agent for all organisms exposed to the sun (not least in humans; sunburn is by far the most common type of radiation damage we suffer). For the Archaea, there is a fundamental split between the deep sea species, which are never exposed to UV, and the surface organisms, which must cope with UV damage to a greater or lesser extent, depending on their lifestyle. UV radiation causes a wide spectrum of DNA damage including the two major lesions, cyclopyrimidine dimers (CPDs) and 6,4-photoproducts, which cross-link adjacent pyrimidines, resulting in a distortion of the DNA duplex and acting as a block to DNA replication and transcription. The most efficient way to remove photoproducts is to directly reverse the changes in bonding that create them, and this is achieved by the enzyme photolyase, which uses the energy from visible light to excite an electron that is used to reorganize bonding and reverse the damage caused by

UV radiation (Weber, 2005). This is sometimes called the "light repair pathway," and is notable as one of the few repair pathways that do not require DNA synthesis. Halobacterium NRC1 is much more resistant to UV radiation when allowed to recover in the light compared with the dark, suggesting that the light repair pathway predominates for UV damage repair in the organisms encoding photolyases (Baliga et al., 2004).

Unrepaired photoproducts prevent DNA replication by stalling replicative DNA polymerases, but a specialized family of error bypass polymerases can replicate past these lesions. The best characterized is Dpo4 (or DinB) from *Sulfolobus solfataricus*, a member of the Y family of error prone polymerases. Dpo4 can bypass UV photoproducts such as CPDs and abasic sites (Ling et al., 2003, 2004). This allows DNA damage to be bypassed and replication to proceed. However, translesion synthesis does come at a cost. Although Dpo4 is remarkably adept at incorporation of adenine residues opposite TT photoproducts in a template strand, misincorporation can occur opposite the 3' thymine of a CPD, resulting in mutation (Kokoska et al., 2002). As one might expect, photolyases are only found in surface dwelling organisms (Table 15.2) – after all, deep sea microbes will not experience photoproduct lesions nor have the means to catalyze light repair. There is also a very strong correlation in the presence of photolyase and a Dpo4-type polymerase, suggesting that the primary function of these specialized enzymes is in bypass of UV photoproducts rather than the other types of lesion that they can bypass *in vitro*.

Nucleotide excision repair

Photoproducts and other helix-distorting DNA lesions in Bacteria and Eukarya are removed by the nucleotide excision repair (NER) pathway, which excises a patch of DNA containing the damage, creating a gap that is filled by DNA polymerase using the undamaged strand as a template (Fig. 15.3). This is sometimes called the "dark repair pathway" to differentiate it from direct repair by photolyase. Bacterial and eukaryal NER proteins are unrelated: Bacteria utilize the UvrABC proteins, which remove a patch of about 12 nt (van Houten et al., 2005), whereas Eukarya have a much more complex machinery that creates a patch of about 28 nt. Some

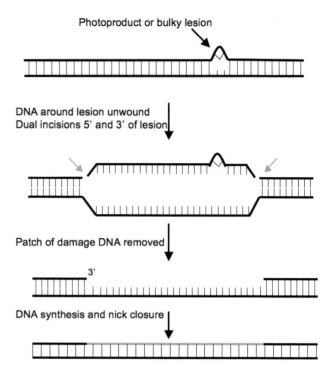

Fig. 15.3 Nucleotide excision repair. Helix-distorting DNA lesions such as photoproducts are removed by unwinding the DNA around the lesion, followed by cleavage on either side to release a patch containing the damage, allowing repair synthesis to proceed. In Bacteria, the UvrABC proteins accomplish this, with a patch size of around 10 nt, whereas in Eukarya many more proteins are required, and a patch size of around 30 nt is generated. The archaeal NER pathway is not yet known.

Archaea have clear orthologs of the UvrABC proteins, but these are confined to low temperature species, and are clear examples of lateral gene transfer (LGT) from Bacteria (Grogan, 2000) (Table 15.2). Regardless of the presence of UvrABC, most Archaea have a number of homologs of the eukaryal NER proteins, including the helicases XPB and XPD, and the nucleases XPF and Fen1/XPG (Table 15.2). This includes the case of *Nanoarchaeum equitans*, which has a highly streamlined genome and thus would be expected to encode a minimal set of repair proteins (Waters et al., 2003). This has led to the suggestion that the archaeal NER pathway may be a simpler version of the eukaryal one (Grogan, 2000; White, 2003). While this is an attractive hypothesis, it remains unproven at present, and there are significant caveats. First, there is no published experimental evidence reporting NER-like patch repair in Archaea lacking UvrABC. Second, there are no obvious archaeal homologs of the DNA damage-sensing proteins XPA and XPC. This may not be too great a hurdle, however, as these proteins appear to be absent from the genomes of higher plants, and may therefore be a more recent addition to the core NER pathway. Third, most of the proteins implicated in NER also perform other roles in the cell. For example, Fen1 is required for processing of Okazaki fragments during replication, and eukaryal XPB has dual roles in transcription and repair (Winkler et al., 2000). Further, not all Archaea encode all the potential NER proteins (Table 15.2). For example, the Thermoplasmatales lack an XPF homolog, and several species lack XPB, XPD, or both helicases.

The XPF endonuclease

One potential archaeal NER protein that has been studied extensively is the nuclease XPF. In Eukarya, XPF cuts splayed duplex and bubble substrates on the 5' side of DNA lesions. The eukaryal enzyme is a heterodimer of the nuclease XPF and the ERCC1 protein (Fig. 15.4). The XPF subunit consists of a degenerate helicase domain at the N-terminus, a central nuclease domain, and a C-terminal helix–hairpin–helix (HhH$_2$) DNA-binding domain. The ERCC1 subunit is shorter, with a degenerated nuclease domain for protein:protein interactions, coupled with a C-terminal HhH$_2$ domain. Structural and biochemical studies of the archaeal homologs have been very valuable in understanding the eukaryal enzyme. In Euryarchaea, the XPF homolog (also known as "Hef") is a homodimer with remarkable similarity to the eukaryal XPF subunit organization (Nishino et al., 2003), with the exception that the helicase domain is a functional helicase (Komori et al., 2004). In Crenarchaea, the helicase domain is absent, and the homodimeric enzyme consists of only the nuclease and HhH$_2$ domains (Roberts et al., 2003; Newman et al., 2005) (Fig. 15.4). This immediately suggests an evolution of the XPF enzyme from a homodimeric nuclease related to the Holliday junction resolving enzyme Hjc, with acquisition of an N-terminal helicase domain in Euryarchaea. Subsequent gene duplication and functional diversification of the second subunit, which is not required to be an active nuclease as the enzyme cleaves only a single

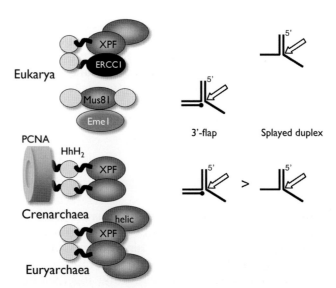

Fig. 15.4 XPF nuclease family domain organization and substrate preference. Eukarya have two related enzymes: XPF-ERCC1, a heterodimer that cleaves splayed duplexes and bubbles during NER; and Mus81-Eme1, a heterodimer that is specialized for 3'-flap substrates created during early recombination events. In Euryarchaea, XPF is a homodimer with a subunit organization very like eukaryal XPF, whereas the Crenarchaea lack an N-terminal helicase domain and show a requirement for the sliding clamp PCNA for activity. The archaeal enzymes cut both Mus81 and XPF-type substrates, with a preference for the former.

DNA strand, may have resulted in the heterodimeric eukaryal enzyme that makes interactions with other repair proteins. One further unexpected property of the crenarchaeal XPF enzyme has been observed: an almost absolute requirement for the sliding clamp PCNA for nuclease activity (Roberts et al., 2003). PCNA is regarded as a processivity factor for enzymes such as polymerases and the 5' flap endonuclease Fen1 (reviewed in Warbrick, 2000). In the case of *Sulfolobus solfataricus* XPF, the enzyme has no or only very limited activity in the absence of PCNA, which stimulates the enzyme by about four orders of magnitude (Roberts & White, unpublished observation).

Therefore in the case of XPF at least, there seems a prima facie case for a role in a prototypical NER machinery in Archaea. However, characterization of the substrate specificities of the crenarchaeal and euryarchaeal XPF enzymes have highlighted a preference for 3' flap over splayed duplex substrates

(Roberts & White, 2004) (Fig. 15.4). This suggests a closer affinity to the eukaryal Mus81 enzyme – a homolog of XPF that is specific for the types of substrates that arise during the rescue of stalled and collapsed replication forks (Haber & Heyer, 2001). Thus the actual function of archaeal XPF may be in a Mus81 type replication restart pathway, rather than an NER patch repair pathway.

Conclusions and future perspectives

Until more biochemical or genetic data are gathered, the existence and mechanism of archaeal nucleotide excision repair will remain highly uncertain. It is possible that the archaeal homologs of the eukaryal XP helicases and nucleases function together in a simplified version of the pathway. It is equally possible that a subset of the enzymes – for example, XPF and XPD – functions with as yet unknown archaeal-specific proteins to catalyze NER. However, we cannot rule out the possibility that Archaea accomplish NER by an alternative excision pathway. For example, a process involving damage recognition, nicking of the damaged strand, followed by an exonuclease activity to remove the DNA containing the damage could fit the bill. Such mechanisms are of course familiar for mismatch repair, and it is not outside the realms of possibility that mismatch repair and nucleotide excision repair are catalyzed by related machineries in Archaea. Recently we have demonstrated that the *Sulfolobus* XPF protein can function like an exonuclease, digesting DNA in a 3′ to 5′ direction from a nick, and that this activity can remove a variety of DNA lesions *in vitro* (Roberts & White, unpublished). What is certain is that biochemical and structural studies of the archaeal XP proteins over the next few years will provide valuable information on the equivalent proteins in Eukarya, which are so important for DNA repair and the avoidance of cancer.

Double strand break repair

Double strand breaks are one of the most serious types of DNA lesion experienced by a cell. The toxicity arises from the fact that both strands of the duplex are damaged, with the result that there is no undamaged strand to act as a template for repair, and the cut ends have no physical connection to keep them together. Despite the danger of a double strand break, they are often deliberately induced in order to initiate the process of homologous recombination (HR), which shuffles DNA to increase genetic diversity. The double strand break repair (DSBR) and HR pathways are equivalent in many situations, and can both proceed via the formation, branch migration, and finally resolution of a Holliday junction (four-way DNA junction) that provides both a physical linkage between two homologous DNA duplexes and a mechanism for exchange of information between them (reviewed in Liu & West, 2004). A further role for DSBR/HR in the rescue of stalled or collapsed replication forks has assumed much greater significance in recent years (reviewed in McGlynn, 2004). In fact, the most important role (and evolutionary origin) for the DSBR machinery may be in the resetting of replication forks that get into difficulty, and in this context DSBR could be considered a vital part of DNA replication. The pathway of DSBR is well understood in *E. coli* (Fig. 15.5), but surprisingly there are still very many uncertainties in the identities and roles of eukaryal recombination proteins. For example, it is not yet clear how double strand breaks are resectioned to produce 3′ overhangs suitable for strand invasion, and the identities of the Holliday junction branch migration and resolution enzymes are still unknown.

RadA, Rad50, and Mre11

As can be seen in Table 15.3, all sequenced archaeal genomes contain clear homologs of eukaryal DSBR proteins RadA/Rad51, Mre11, and Rad50, as well as the Holliday junction-resolving enzyme Hjc. The high degree of conservation of these proteins, and their clear similarity to the eukaryal equivalents, may reflect the linkage to DNA replication, which is strikingly conserved between these two domains. The archaeal RadA, Mre11, and Rad50 proteins are structurally very similar to their eukaryal counterparts, and have proven very useful for crystallographic and other biophysical studies (reviewed in Shin et al., 2004). The role of RadA/Rad51 (and the bacterial equivalent RecA) in catalyzing strand exchange is well defined and has been reviewed extensively elsewhere. Notably, it is the only repair protein found in every sequenced genome from all

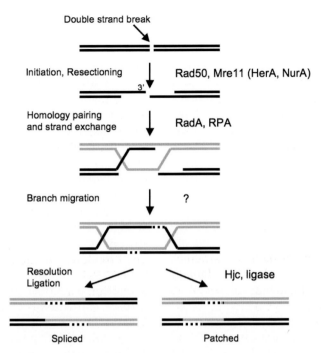

Fig. 15.5 Double strand break repair. An overview of the DSBR/homologous recombination pathway. Double strand breaks caused by environmental factors such as ionizing radiation or by cellular nucleases are processed to generate 3' ssDNA ends suitable for strand invasion of a homologous duplex. Strand exchange can lead to the formation of a mobile Holliday junction linking the two duplexes, which is ultimately resolved by a structure specific nuclease, regenerating two duplex DNA products. Other pathways for DSBR, including synthesis-dependent strand annealing, can function alongside the classical pathway shown here, and may predominate in certain organisms under certain conditions.

Table 15.3 Distribution of putative double strand break repair proteins in Archaea (proteins are described in the text).

Species	Hjc	RadA RadB	Rad50	Mre11	HerA	NurA
A. pernix	1	1	1	1	1	1
P. aerophilum	1	1	1	1	1	2
S. solfataricus	2	2	2	1	2	2
S. tokodaii	1	2	2	1	2	2
S. acidocaldarius	2	1	1	1	1	1
M. kandleri	1	2	1	1	1	1
P. furiosus	1	2	1	1	1	1
P. horikoshii	1	2	1	1	1	1
P. abyssi	1	2	1	1	1	1
T. kodakaraensis	1	2	1	1	1	1
N. equitans	1	1	1	1	1	1
M. jannaschii	1	2	1	1	2	2
A. fulgidus	1	2	1	1	1	1
M. thermoauto	1	2	1	1	1	2
P. torridus	1	2	1	1	1	1
T. acidophilum	1	2	1	1	1	1
T. volcanium	1	2	1	2	1	1
M. acetivorans	1	2	1	2		
M. mazei	1	2	1	1		
H. marismortui	1	2	1	1		
Halobacterium	1	2	1	1		

Details as for Table 15.1.

three domains of life (Aravind et al., 1999). There are still many unanswered questions about Mre11 and Rad50, which form a heterotetrameric complex (Hopfner et al., 2001). Rad50 is a member of the SMC family of proteins with globular ATPase domains at the N- and C-termini linked by a long coiled-coil domain (Hopfner et al., 2001). Mre11 is a nuclease with a 3'-5' polarity, which effectively rules out a direct role in resectioning DNA breaks to produce 3' overhangs for recombination. In Eukarya, Mre11–Rad50, together with Nbs1 and Xrs2, form a structure resembling "molecular callipers" and are implicated in many processes including DSBR, nonhomologous end-joining (NHEJ), DNA damage detection, and checkpoint signaling (reviewed in D'Amours and Jackson, 2002).

There are no clear homologs of Nbs1 or Xrs2 in Archaea, but Mre11 and Rad50 are commonly found in an operon in archaeal genomes that frequently includes the HerA and NurA genes. HerA (also known as MlaA) is a DNA-dependent ATPase (Manzan et al., 2004) and weak helicase *in vitro* (Constantinesco et al., 2004). HerA is a member of the hexameric AAA family ATPases. In *Methanothermobacter thermautotrophicus*, the HerA and Mre11 genes are fused, suggesting a close functional relationship (Constantinesco et al., 2004). These observations suggest a role for HerA in processing and manipulating DNA at double strand breaks. NurA is a 5'-3' nuclease and thus has the opposite polarity to Mre11 (Constantinesco et al., 2002). NurA may thus act together with Mre11 to process double strand breaks, and may be responsible for generation of 3' ssDNA tails (Constantinesco et al., 2002). An alternative suggestion is that HerA performs an analogous function to bacterial FtsK, to which it is distantly related, in archaeal cell division (Iyer et al., 2004). However, the lack of HerA in halophiles and mesophilic methanogens must then be addressed.

Holliday junction-resolving enzymes

The importance of the Holliday junction in DSBR and HR in Archaea is emphasized by the observation that the Holliday junction-resolving enzyme Hjc is absolutely conserved in Archaea, making it one of the few ubiquitous nucleases. Hjc was first identified in *Pyrococcus furiosus* (Komori et al., 1999), and two orthologs, Hjc and Hje, were subsequently identified in *Sulfolobus solfataricus* (Kvaratskhelia & White, 2000). Hjc is a small, dimeric nuclease with a very high degree of specificity for four-way DNA (Holliday) junctions over all other types of branched DNA structure. The structure of the enzyme revealed a nuclease superfamily fold that is present in a wide variety of nucleases including restriction enzymes (Bond et al., 2001; Nishino et al., 2001), and the same fold was subsequently observed in the XPF/Mus81 nuclease family (Nishino et al., 2003). Hjc interacts with the sliding clamp PCNA (Komori et al., 2000b), though the functional significance of this interaction is not yet clear. The Hje ortholog, which is restricted to *Sulfolobus solfataricus* and *Sulfolobus acidocaldarius*, is apparently of viral origin, with closest homologs present in the genome of archaeal SIRV viruses (Birkenbihl et al., 2001; Middleton et al., 2004). Hje is highly structure specific but sequence independent, with a turnover number that compares favorably with commonly used type II restriction enzymes such as *Eco*RI. These properties make Hje a useful enzyme for applications that require specific detection of Holliday junctions in DNA samples (Parker & White, 2005).

Conclusions and future perspectives

The HR/DSBR pathway is still not well understood in Eukarya, despite extensive investigation and the availability of highly developed genetic model systems. It is not clear why there has been such patchy progress in this area, but it may reflect the fact that the pathway has a variety of complex roles in eukaryotes, and the existence of overlapping specificities and redundancy for some of the proteins involved. The very strong relationship of archaeal and eukaryal DSBR is evident from a comparison of protein sequences and structures. In turn this may reflect the importance of DSBR for the rescue of DNA replication, which is itself very highly conserved between the two domains. These factors suggest that studies of archaeal DSBR/HR have the potential to shed light on the eukaryal pathway. The usual arguments about simplicity and robust, tractable proteins all apply. An interesting example is the archaeal helicase mth810, which is homologous to a human helicase of unknown function, hel308, and has been shown to target stalled replication forks *in vitro* (Guy & Bolt, 2005). This is an area in which archaeal genetic studies can provide a valuable role, as the core DSBR genes are present in the halophiles.

Although not yet identified, the eukaryal Holliday junction-resolving enzyme may yet turn out to be related to the archaeal Hjc nuclease.

Proteins seeking pathways: ORPs and RAMPs

While there are pathways such as MMR where we still have little idea of the proteins involved, Makarova and Koonin have suggested that a large group of archaeal proteins of unknown function play a role in DNA repair (Makarova et al., 2002). These proteins are found in clusters in the genomes of Archaea growing at elevated temperatures, leading to the suggestion that they may be an adaptation to hyperthermophilic conditions. There is reasonably convincing bioinformatic data suggesting that some of these proteins are nucleases, helicases, and possibly also a DNA polymerase. For the purposes of this discussion we designate these as orphan repair proteins (ORPs). However, many of these proteins have no obvious homologs of known function, and are therefore difficult to assign. This includes a family of repair associated mysterious proteins (RAMPs) that comprise COGs 1336, 1367, 1604, 1337, and 1332 (Makarova et al., 2002), which show detectable sequence similarity with one another. *Sulfolobus solfataricus* has the largest number of RAMPs and ORPs at around 90, while *Nanoarcheum equitans* has only two. To date, there is little or no published biochemical or genetic analysis of any RAMP or ORP in any organism, and a role in repair is still an untested prediction. Makarova et al. have suggested a role in translesion repair, but equally they could function in mismatch repair, recombination-dependent replication or non-homologous end joining. Makarova et al. recently withdrew their prediction of a role in DNA repair for these proteins, and suggested instead a role in a putative RNAi system (Makarova et al., *Biol Direct.*, 2006, 1–7).

Controlling DNA repair: the DNA damage response

In one of the first studies of DNA damage response in Archaea, *Pyrococcus furiosus* cells were exposed to 2500 Grays of ionizing radiation, resulting in fragmentation of the genomic DNA due to high numbers of double strand breaks (DiRuggiero et al., 1997). Following this level of radiation, 75% of the cells are estimated to survive, making *P. furiosus* approximately ten times more radioresistant than *E. coli*, but significantly less resistant than *Deinococcus radiodurans*. Chromosome fragmentation was reversed during incubation at 95°C, presumably by a DSBR process involving HR, resulting in reconstitution of intact genomes. However, in contrast to the situation in Eukarya and Bacteria, RadA was expressed constitutively in *Pyrococcus*, with little induction in response to DNA damage (Komori et al., 2000a; Jolivet et al., 2003). The transcriptional response to UV radiation in *Sulfolobus solfataricus* has also been studied for a subset of DNA binding and repair proteins (Salerno et al., 2003). Modest changes in mRNA levels were observed for most of the proteins studied, but a more dramatic effect was seen for the *Sulfolobus* XPB1 gene. XPB1 is a homolog of the eukaryal helicase XPB, which is a component of the transcription factor TFIIH and plays a dual role in DNA melting in both transcription initiation by RNA polymerase II and the eukaryal NER pathway (see section on NER). Following UV radiation, *Sulfolobus* XPB1 transcription increased dramatically, and initiated from a new start site about 130 nt upstream from the start codon (Salerno et al., 2003). These data suggest that Archaea do have a transcriptional response to UV-induced DNA damage, and therefore have damage-specific transcriptional control mechanisms.

The advent of whole genome microarrays has made large-scale analyses of transcriptional responses to external stimuli reasonably straightforward for microbial organisms. Microarray analysis of the UV damage response of *Halobacterium* NRC1 suggests there is no coordinated control of DNA repair proteins, analogous to the bacterial SOS response, following DNA damage (Baliga et al., 2004). Interestingly, a significant upregulation of the RNA polymerase subunit RpoM was observed following UV damage. RpoM is homologous to the TFIIS protein, which allows reinitiation of transcription by stalled RNA polymerase, suggesting a similar function for the protein in *Halobacterium*. Another study of the UV damage reponse of *Halobacterium* has identified the early induction of the single stranded DNA binding protein (McCready et al., 2005), and this has also been seen after UV irradiation of *Sulfolobus solfataricus* (Götz, Bernander, and White, unpublished). These observations may be consistent with a role for SSB in DNA damage detection (Cubbedu & White, 2005).

Conclusions and future perspectives

There is at present only limited information available on the response of Archaea to a variety of types of DNA damage. Partly this is due to the historical lack of robust genetic models. What is already clear is that the Archaea do not appear to have a coordinated SOS-type response to DNA damage. Instead, repair proteins in general seem to be expressed constitutively. This might be expected in hyperthermophiles, which experience relatively elevated levels of DNA damage continuously as a consequence of their growth temperature. However, it also appears to be the case for mesophilic Archaea such as the halophiles. There are no published studies yet on more eukaryal-type responses to DNA damage, which involve cascades of protein kinases and the use of protein phosphorylation as a signal, but there are no clear archaeal homologs of the ATM/ATR type kinases present in Eukarya. Nevertheless there are some clues that Archaea do respond to DNA damage at a transcriptional level: notably the induction of *Sulfolobus* XPB1 transcription in response to UV damage, and the elevated transcription of an ORP operon in response to ionizing radiation in *Pyrococcus*. These observations suggest a damage recognition signal links to the differential transcriptional control of specific genes or sets of genes. Possibilities for the damage signal or signals include binding of the archaeal single stranded binding protein SSB to damage sites, as observed for the eukaryal single stranded DNA binding protein RPA, or the presence of stalled RNA or DNA polymerases at DNA lesions. Microarray studies have shown induction of various hypothetical proteins and putative transcription factors in response to DNA damage. These leads must now be followed up using biochemical and genetic means to test their role in the archaeal damage response.

Concluding remarks

This review represents a snapshot of a field where the pace of research is accelerating rapidly, with new research groups and new experimental techniques being brought to bear. We can expect to see rapid progress in our understanding of DNA repair in the Archaea (and equally rapidly, this review will require updating!). We are assured of good progress at the biochemical and structural levels, where studies of archaeal repair proteins have played a very significant part in our understanding of the equivalent proteins in Eukarya. Equally, the microarray studies published and soon to be published will generate a host of new hypotheses that require testing. These are likely to address questions such as mechanisms of DNA damage recognition in the Archaea, and the links with transcriptional control of gene expression. Progress will depend on further developments of genetic systems for model archaeal species. At present, the halophiles are a very valuable model system with advanced genetics, but clearly these organisms are not always going to be representative of the Archaea as a whole. There are significant differences between repair protein content in the Crenarchaea and Euryarchaea, as well as between deep sea and surface dwelling organisms. In this context, recent progress in gene knockout and protein expression systems in *Sulfolobus solfataricus* are very encouraging.

Research effort in the latter part of the decade is likely to be focused on the following questions:
- Do Archaea have a eukaryal-type NER pathway, or a novel mechanism for NER?
- Is there a functional mismatch repair pathway in Archaea lacking MutS/MutL?
- Can we define the mechanism and protein components of the archaeal DSBR pathway?
- Do Archaea have a mechanism for transcription coupled repair?
- How do Archaea deal with RNA polymerase stalled at DNA lesions?
- Which proteins detect DNA damage (e.g. photoproducts) in Archaea?
- What is the basis for the transcriptional response to DNA damage in Archaea?
- What are the links between DNA replication and repair?
- If the RAMP and ORP families have a role in repair, what is it?

Answers to even some of these questions will expand our understanding of archaeal DNA repair greatly. We can expect to see more links with eukaryal DNA repair uncovered along the way, but also unique archaeal solutions to the problem faced by all cellular organisms in maintaining the genetic coding potential of DNA in response to frequent and varied DNA damage events.

Acknowledgments

Thanks are due to past and present members of the White lab for valuable discussions, and to colleagues who communicated information prior to publication. Thanks are due to the Biotechnology and Biological Sciences Research Council, Wellcome Trust, Association for International Canner Research, and Royal Society for funding.

16
Transcriptional mechanisms

Michael Thomm and Winfried Hausner

Introduction

The increasing amount of sequenced archaeal genomes revealed that Archaea contain a mosaic of bacterial and eukaryotic features. Genes involved in information processing are shared with Eukarya, whereas metabolic and structural proteins show similarities with bacterial ones. The striking similarities in information-processing systems between Archaea and Eukarya were first discovered by immunological analysis of purified archaeal and eukaryotic RNA polymerases (Huet et al., 1983). The development of *in vitro* transcription systems in *Methanococcus vannielii* and *Sulfolobus* led to the discovery of transcription factors and enabled the identification of archaeal promoter signals (Thomm, 1996; Soppa, 1999). In summary, accumulated data suggested that the archaeal transcriptional machinery represents a simple version of the eukaryotic one. These unexpected results encouraged other laboratories during the past few years to study archaeal transcription. Therefore, the transcription process is the best understood part of information-processing systems in Archaea. In the following review we summarize the latest research results in archaeal transcription. One main focus is the basal transcription apparatus itself and a comparison of it with the eukaryotic and the bacterial one. In the first part we describe the components involved in archaeal transcription. The second part illustrates the transcription cycle and in the third part we discuss some unique features of regulation of archaeal transcription.

Promoter structure

Archaeal promoter elements consist of three parts. A detailed mutational analysis using the tRNA^{Val} promoter of *Methanococcus vannielii* and the rRNA promoter of *Sulfolobus shibatae* in combination with *in vitro* transcription has defined an AT-rich sequence as major promoter signal, formerly designated as box A, which is located about 25 bp upstream of the transcription start site (Thomm, 1996). Similar experiments using an *in vivo* approach with a tRNA gene of *Haloferax volcannii* confirmed the essential role of this AT-rich element (Palmer & Daniels, 1995). The conserved element was renamed as the TATA box because it resembles with regard to position as well as sequence the eukaryotic TATA box, a core element of many RNA polymerase II promoters. The second part of the archaeal promoter element is the BRE (transcription factor B recognition element). It is located immediately upstream of the TATA box and is important for promoter strength as well as the orientation of the transcription initiation complex (Qureshi et al., 1997; Bell et al., 1999). Analysis of nucleotide frequencies in natural archaeal promoters confirmed a conserved element of 2–3 A residues at −35 to −33 (Soppa, 1999). This is in agreement with experimental data indicating an important role of an A residue at position −33 (Qureshi & Jackson, 1998). Substitution of the A residue at position −33 to a G residue of the tRNA^{Val} promoter of *Methanococcus* decreases transcriptional activity by a factor of two (Thomm, 1996). In the region around the transcription start site the initiator element (INR) motif exists as the third promoter element. The minimal requirement for an archaeal initiation site is a pyrimidine–purine dinucleotide in a proper distance to the TATA box. Mutational analysis revealed a strong preference for purines as starting nucleotides (Thomm, 1996; Soppa, 1999). Recently, an additional conserved A/T motif at −10 was suggested to play a role in archaeal transcription. The motif was derived from sequence analysis of upstream sequences of leaderless transcripts, but so far there are no confirming biochemical data available (Torarinsson et al., 2005).

The eukaryotic TATA box was the first identified core promoter motif of RNA polymerase II transcription and it was believed that this promoter element plays the central role in initiation of transcription. Meanwhile, at least four different eukaryotic core promoter motifs are known: BRE, TATA box, INR, and DPE (downstream promoter element). Furthermore, only a subset of these sequence elements is found in core promoters and a particular core promoter may contain some, all, or none of these elements (reviewed in Smale & Kadonaga, 2003). For instance, it was estimated that approximately 43% of 205 core promoters in *Drosophila* contain a TATA box (Kutach & Kadonaga, 2000). In humans, it was found that about 32% of 1031 potential promoter regions contain a putative TATA box motif (Suzuki et al., 2001), whereas in *Saccharomyces* genomes even 80% of the promoters are TATA-less promoters (Basehoar et al., 2004). These data indicate that different possibilities exist to initiate basal transcription in eukaryotes. In promoters without a TATA box, for instance, the INR together with the DPE can mediate initiation of transcription. There is also an apparent lack of TATA box elements in deep-branching eukaryotic lineages, such as the most ancient parasitic protists (Liston et al., 1999). The TATA box representing the archaeal type (with regard to sequence and location) is found in metazoans. It is typically located about 25–30 nucleotides upstream of the transcription start site, but variations are also known. In the yeast *Saccharomyces cerevisiae*, the position of the TATA box ranges from 40 to 100 nucleotides upstream of the start site, whereas in *Schizosaccharomyces pombe* the normal distance at −25 is found. Swapping experiments indicate that *S. cerevisiae* RNA polymerase and TFIIB dictate this unusual distance (Li et al., 1994).

TATA-box binding protein (TBP)

TBP is a general transcription factor that binds specifically to the TATA box. Consistent with the central role of the TBP–TATA box interaction, it is highly conserved in eukaryotes as well as in Archaea. Both archaeal and eukaryotic TBP molecules consist of two repeats of about 90 amino acids and adopt a symmetrical saddle-shaped form (Nikolov et al., 1992; Dedecker et al., 1996). The similarity of the first to the second repeat is much higher in archaeal TBPs (36–53% of identical amino acids) than in eukaryotic TBPs (22–26%; Soppa, 1999). Archaeal TBPs are acidic proteins with isoelectric points (IPs) within the range of 3.9–6.1, while the eukaryotic counterparts are basic, with IP ranging between 9.8 and 10.7 (Hickey et al., 2002). The structural similarity and the charge differences of a eukaryotic and an archaeal TBP are shown in a comparison of both molecules (Plate 16.1). The group of Ladbury demonstrated by mutational analysis that acidic residues of archaeal TBP are involved in site-specific cation binding and necessary to enable DNA-binding of *Pyrococcus* TBP at high salt concentrations (Bergqvist et al., 2001).

In eukaryotes, the conserved domain is located in the C-terminal region, whereas the N-terminal extension contains a highly diverged species-specific sequence (Hernandez, 1993). The function of this region is still not fully understood. Bondareva and Schmidt (2003) hypothesized that the N-terminus functions as a "signaling port" by which the basal transcription machinery receives specific regulatory signals, and that variation within this region represents commitment to gene-regulation processes that are unique to each phylum. So far no N-terminal extensions in archaeal TBPs are known. Furthermore, there is also no indication for TBP-associated factors (TAFs) in Archaea. By contrast, eukaryotic TBPs associate with several different TAFs. This results in the formation of at least three distinct complexes, SL1, TFIID, and TFIIIB, which dictate whether TBP functions in RNA polymerase I, II, or III transcription, respectively (Hernandez, 1993).

Beside the TAFs other proteins are known that could interact with TBPs. They are called TBP-interaction proteins (TIPs) and could be also found in some Archaea. In *Pyrococcus kodakaraensis* KOD1, TIP26 was described and is able to interact with a TBP/TFB/DNA complex (Matsuda et al., 2001). The cellular concentration of TIP26 is ten times higher than that of TBP. The protein seems to be present only in a confined subgroup of Archaea and the biological function of the protein is not known. A further archaeal TIP with unknown function is a homolog of the eukaryotic TIP49 gene family and is present in various Archaea, such as *Pyrococcus abyssi, Archaeoglobus fulgidus, Methanopyrus kandleri, Pyrobaculum aerophilum,* and *Sulfolobus solfataricus* (Kurokawa et al., 1999). Sequence similarities with bacterial RuvB indicate a potential ATP-dependent DNA helicase acitivity (Kanemaki et al., 1997). The eukaryotic homologs are possibly involved in transcription as bridging factors between the basic machinery and sequence-specific activators (Bellosta et al., 2005). Furthermore, the eukaryotic TIP gene family is also a component of the Ino80 chromatin remodelling complex in yeast (Shen et al., 2000).

TFB

The identification of a partial open reading frame of 152 amino acids with homology to the eukaryotic TFIIB in *Pyrococcus woesei* was the first indication that archaeal transcription factors are of the eukaryotic type (Ouzounis & Sander, 1992). Sequence analysis of the complete gene showed about 30% identity to eukaryotic *tfIIb* genes (Creti et al., 1993). In addition, this sequence exhibits distinct structural motifs characteristic of eukaryotic TFIIB such as an imperfect amino acid repeat or a zinc ribbon at the N-terminus. Experimental evidence that this putative TFIIB homolog in Archaea is indeed a transcription factor was provided by *in vitro* transcription experiments. In *Sulfolobus* as well as in *Methanococcus* it could be demonstrated that the archaeal TFIIB homolog (now called TFB) is essential to direct initiation of archaeal transcription (Hausner et al., 1996; Qureshi et al., 1997).

The structure of the N-terminal zinc ribbon of *Pyrococcus woesei* has been solved and it turned out that the zinc ribbon is a common motif in archaeal and eukaryotic transcription (Zhu et al., 1996; Chen et al., 2000). The N-terminal region is required for RNA polymerase recruitment in Archaea as well as in eukaryotes (Buratowski & Zhou, 1993; Ha et al., 1993; Bell & Jackson, 2000a). Recent data suggest that the archaeal recruitment function is redundant as deletions of the N-terminal zinc ribbon of TFB can be complemented by TFE (Werner & Weinzierl, 2005). The zinc ribbon is followed by the B-finger domain, which contains a short, highly conserved region ("conserved sequence block," CSB; Bushnell et al., 2004). A substitution of a conserved residue within this block increased the NTP dependence and suggested a role of the archaeal TFB in the late stages of transcription initiation, but the effect is influenced by the sequence context surrounding the transcription start site (Bell & Jackson, 2000a). Deletion of the B-finger domain in *Methanocaldococcus* also revealed a requirement of higher NTP concentrations (Werner & Weinzierl, 2005). Furthermore, the data also demonstrated that this domain can stimulate abortive and productive transcription in a recruitment-independent function. Mutational analysis of the corresponding eukaryotic TFIIB regions indicated functions in initiation efficiency and start site selection (Pinto et al., 1992; Zhang et al., 2002).

Multiple copies of TBP and TFB

With the increasing amount of sequenced archaeal genomes it becomes evident that most of the Archaea have only one copy of TBP and TFB. This indicates that in general binding of TBP to the promoter is the key step in initiation of archaeal transcription. But in halophiles multiple copies of transcription factors seem to be quite normal. The genome of *Halobacterium* NRC-1 has six *tbp* and seven *tfb* genes, and *Halobacterium marismortui* encodes a single TBP and eight different TFB molecules (Baliga et al., 2000). In *Haloferax volcanii* differential regulation of the TFB genes upon heat shock has been postulated for the regulation of genes involved in the appropriate responses (Thompson & Daniels, 1998). The multiplicity of these factors is hypothesized to play a role in transcription regulation through an assembly of pair-wise TBP–TFB combinations (Baliga et al., 2000).

Three *tbp* genes have also been published for *Methanosarcina mazeii* and *Methanosarcina acetivorans*. In *Pyrococcus horikoshii* and *Pyrococcus furiosus* two copies of TFB were found. By contrast, *Pyrococcus abyssi* contains a single copy of the *tfb* gene. In *Pyrococcus furiosus*, a Northern analysis revealed that the second copy of TFB seems to be also overexpressed upon heat shock (Shockley et al., 2003). But a specific interaction of this second TFB copy with heat shock promoters has so far not been demonstrated *in vitro* (M. Bartlett & M. Thomm, unpublished data). Crenarchaeal genome sequencing projects also indicate the presence of two copies of *tfb* genes. It is likely that the multiple copies of basal transcription factors are only involved in the regulation of a small subset of genes.

TFE

Archaeal genome sequencing projects have revealed that most likely all Archaea have a transcription factor that shows homology to the N-terminal region of the alpha subunit of eukaryotic transcription factor TFIIE. The eukaryotic protein is a heterotetramer composed of two 57 kDa and two 34 kDa subunits (Ohkuma et al., 1991; Ohkuma, 1997). Purified TFIIE has been found to possess no enzymatic activities, it stabilizes the preinitiation complex by binding to the complex as well as to the DNA, and it is involved in the transition from initiation to elongation (Watanabe et al., 2003; Forget et al., 2004). Mutational analysis revealed that the N-terminal half of hTFIIEα is sufficient for both basal and activated transcription (Ohkuma et al., 1995). Similar results were found in yeast using *in vivo* genetic experiments (Kuldell & Buratowski, 1997). Interestingly, this part of the alpha subunit is still conserved in Archaea. A recently determined crystal structure showed that the TFE domain from *Sulfolobus solfataricus* adopts an extended winged fold with unusual features that are consistent with a role of this domain as an adapter between RNA polymerase and general transcription factors (Meinhart et al., 2003). *In vitro* transcription experiments indicate that TFE is not absolutely required for transcription in a reconstituted archaeal *in vitro* system; it nonetheless plays a stimulatory role on some promoters and under certain conditions (Bell et al., 2001). Almost inactive promoters or reduction of TBP concentration in transcription reactions lead to sensitivity of a promoter to TFE addition (Bell et al., 2001; Hanzelka et al., 2001). Conversely, saturating reactions with TBP desensitize promoters to TFE. Since it is known that the alpha subunit of TFIIE can interact with TBP (Maxon et al., 1994) and can enhance binding of TBP to the promoter in the absence of other basal factors (Yokomori et al., 1998), it seems likely that this type of interaction was developed very early during evolution and may also play a role in archaeal transcription. The interaction of TFE with TBP does not interfere with the binding of Ptr2, a DNA-binding protein that activates transcription by direct recruitment of TBP, as recently demonstrated in a fully recombinant *in vitro* transcription system (Ouhammouch et al., 2004).

Archaeal RNA polymerase

The structural similarity of archaeal and eukaryotic RNA polymerases was the first evidence that the archaeal transcriptional machinery is more closely related to Eukarya than to Bacteria (Huet et al., 1983). Archaea share with Bacteria the presence of only one type of RNA polymerase, but by comparison of the subunit pattern it became immediately apparent that archaeal RNA polymerases contain about 10–14 subunits and therefore represent the eukaryotic type (reviewed by Langer et al., 1995). Most of these subunits have been shown to be homologs of subunits of eukaryotic RNA polymerases. The archaeal homolog of the largest subunit of eukaryotic RNA polymerases is split into two smaller subunits, A' and A", and the archaeal homolog of the second largest subunit exists as a single subunit B, or as split version B' and B", depending on the archaeal species (Langer et al., 1995). In contrast, bacterial RNA polymerases have a three-component core enzyme with the composition $\alpha_2\beta\beta'$. Nevertheless, the two largest eukaryotic subunits, Rbp1 and Rbp2 are also homologous to the bacterial β' and β subunits, indicating that all multisubunit RNA polymerases have a common universal precursor. Purified bacterial RNA polymerases are associated with specificity factors like σ^{70} and are able to direct *in vitro* transcription, whereas archaeal RNA polymerases – similar to the eukaryotic RNA polymerases – were unable to efficiently recognize promoter sequences on their own.

An active RNA polymerase of *Methanocaldococcus jannaschii* has been assembled *in vitro* from purified recombinant subunits (Werner & Weinzierl, 2002) and recently the *Pyrococcus furiosus* RNA polymerase was also reconstituted from single recombinant subunits (S. Naji, B. Goede, O. von Kampen & M. Thomm, manuscript in preparation). Although methods for reconstitution of bacterial RNA polymerases from recombinant subunits are well established, the attempts to reconstitute eukaryotic RNA polymerases have been unsuccessful (Zalenskaya et al., 1990; Kimura & Ishihama, 2000). Therefore, the reconstituted archaeal RNA polymerase may be helpful to address questions important for the mechanism of eukaryotic RNA polymerases. Furthermore, the availability of reconstituted RNA polymerases offers the possibility of analyzing the structure–function relationship in archaeal RNA polymerases in more detail on the molecular level. The first results of such experiments indicate that a minimal complex of RNA polymerase consisting of subunits A'-A"-B'-B"-D-L-N-P has a low, but distinct and reproducible, degree of catalytic activity and the subunits K, H, and the F/E complex are not absolutely essential for basal promoter-dependent transcription (Werner & Weinzierl, 2002; Ouhammouch et al., 2004). The K subunit has structural and functional homologs in both eukaryotic (RPB6) and bacterial (ω) RNA polymerases (Minakhin et al., 2001). Interestingly, experiments using two-hybrid and biochemical analysis revealed an interaction between subunit K and the N-terminal domain of TFB (Magill et al., 2001). In addition, biochemical data with RNA polymerase II suggest an interaction of TFIIB with the "dock" domain, which is located near the path of RNA exit and therefore also implies an interaction with RPB1 and RPB2, respectively (Chen & Hahn, 2003; Bushnell et al., 2004; Chen & Hampsey, 2004). The corresponding subunit ω in Bacteria seems to be involved in the assembly of the RNA polymerase (Ghosh et al., 2001b). It could be also demonstrated that ω is necessary to restore denatured RNA polymerase *in vitro*.

Subunit H is homologous to the eukaryotic RPB5, a subunit that is shared by all three eukaryotic RNA polymerases. The structure of RNA polymerase II shows that subunit RPB5 is located close to DNA downstream of the initiation site and possibly plays a role in activation of transcription (Miyao & Woychik, 1998; Cramer et al., 2000). Experiments using the reconstituted enzyme with and without subunit H revealed a tenfold increased activity in the presence of subunit H (Werner & Weinzierl, 2002). The function of the archaeal F/E complex is still unclear. The presence or absence of this complex in combination with the recombinant enzyme has no detectable effects (Werner & Weinzierl, 2002). The purified RNA polymerase from *Methanothermobacter thermoautotrophicus* seems to contain the F/E complex at sub-stoichiometric levels (Darcy et al., 1999). This is in agreement with the corresponding eukaryotic RPB4/RPB7 complex (Choder, 2004). Homologs of these subunits do not exist in Bacteria.

Formation of an initiation complex

Gel shift assays, template commitment experiments, and DNase I footprinting analyses have demonstrated that Archaea initiate transcription by the binding of TBP to the TATA box (Rowlands et al., 1994; Gohl et al., 1995; Hausner et al., 1996). Structural analysis of TBP–TATA box co-crystals revealed the mechanism of binding in more detail (Kosa et al., 1997; Littlefield et al., 1999). Like its eukaryotic counterparts, *Pyrococcus* TBP binds to the minor groove of the DNA and imposes a similar severe distortion on the DNA (Kosa et al., 1997). Several studies with eukaryotic TBPs suggest that TBP can bind in both orientations with only a minimal preference toward the correct orientation (Patikoglou et al., 1999; Liu & Schepartz, 2001). Due to the greater symmetry of archaeal TBPs it is most likely that archaeal TBPs cannot select the right orientation of binding to the TATA box. The polarity of the initiation complex is fixed in the next step by the binding of TFB (Bell et al., 1999b). It recognizes the distorted DNA–TBP complex and interacts with BRE. Bound TFB extends the DNase I footprint upstream as well as downstream of the TATA box (Hausner et al., 1996; Bell et al., 1999b). The geometry of bound archaeal TBP and TFB was also analyzed by photochemical cross-linking experiments (Bartlett et al., 2000, 2004; Renfrow et al., 2004). The data are summarized in Plate 16.2. These experiments confirmed the interaction of TFB with DNA upstream and downstream of the TATA box. Bound TFB enables the recruitment of the RNA polymerase and the formation of a pre-initiation complex. The RNA polymerase is positioned around the transcription start site and therefore extends DNase I footprints in the downstream direction (Hausner et al., 1996; Bell & Jackson, 1999). A more detailed picture resulted from the photochemical cross-linking experiments (Plate 16.2). Subunit B interacts with DNA around the transcription bubble on both DNA strands, whereas subunits A' and A" contact mainly downstream DNA. The downstream edge of interactions seems to be fixed by subunit H and is located at position +20. In comparison with eukaryotic RNA polymerase II and bacterial RNA polymerases it turned out that the geometry of interaction with transcribed DNA seems to be extremely conserved in all three domains (Naryshkin et al., 2000; Bartlett et al., 2004).

The next step in initiation of transcription is the transition of a closed complex to an open complex. RNA polymerase II needs for this step the DNA helicase activity of TFIIH. This activity hydrolyzes the β–γ phosphoanhydride bond of ATP and thereby triggers open complex formation. *In vitro* transcription experiments demonstrate that the TFIIH requirement could be bypassed by using a supercoiled template (Holstege et al., 1997; Yan & Gralla 1997). In contrast, archaeal RNA polymerases do not need ATP hydrolysis for open complex formation (Bell et al., 1998; Hausner & Thomm, 2001). This is independent of the topology of the template, but negative supercoiling of templates facilitates promoter melting at lower temperatures. These findings are also in agreement with the data of genome sequencing projects, as there is no indication for a TFIIH-like factor in Archaea. In Bacteria, ATP is also not required for open complex formation of σ^{70}-dependent promoters, but genes involved in nitrogen metabolism transcribed by a complex of σ^{54} with RNA polymerase require ATP for promoter melting.

Promoter escape

Recently, archaeal transcription complexes, stalled at different registers, were analyzed in more detail by exonuclease III and potassium permanganate footprinting techniques. It turned out that one major transition from initiation to elongation occurs between positions +6 and +7 (Spitalny & Thomm, 2003). At position +7 the upstream edge of the RNA polymerase resulted in a distinct exonuclease III footprint at position −7. The translocation of the downstream edge was found at register +10/+11, where too reclosure of the initially open complex occurred. Such a bubble collapse was very recently suggested to define the RNA polymerase II promoter clearance transition (Pal et al., 2005). The transition from initiation to elongation is most likely also combined with dissociation of TFB. Xie and Reeve (2004b) demonstrated that in *Methanothermobacter thermoautotrophicus* TFB is released during extension of the transcript from 4 to 24 nucleotides, whereas TBP remains bound to the template DNA. This is in agreement with results obtained with the eukaryotic RNA polymerase II transcription machinery. Recent data indicate that the TFIIB zinc ribbon domain overlaps with the RNA channel and exiting RNA beyond the DNA–RNA hybrid could facilitate the release of TFIIB (Chen & Hahn, 2003; Bushnell et al., 2004). Therefore it is tempting to speculate that archaeal TFB is released in a similar manner.

Elongation complex

After promoter clearance the RNA polymerase goes into the phase of formation of an early elongation complex. Figure 16.1 shows a comparison of elongation complexes from different domains on the basis of published footprinting data. The archaeal ternary complex covers a DNA segment of about 26–29 bp depending on the stalled register (Spitalny & Thomm, 2003). In complexes stalled at position +20, the enzyme protects the DNA region from position +4 to +32 from exonuclease III digestion (Fig. 16.1). Thus, the binding site of the enzyme extends over 29 bp of DNA. A similar size of the RNAP binding site was found in complexes of the *E. coli* RNAP (Metzger et al., 1989) stalled at position +20 (28 bp) and of RNA PolI (Kahl et al., 2000) stalled at position +12 (32 bp), indicating that the general extension of elongating RNAP-molecules is similar in the three domains of life. In the case of the eukaryotic RNA polymerase II complexes the size of the RNAP binding site determined by DNase I and exoIII footprintings varies between 26 and 55 bp (Linn & Luse, 1991; Samkurashivili & Luse, 1998; Kireeva et al., 2000; Fiedler & Timmers, 2001). However, most complexes analyzed by exoIII-footprinting also show a size of around 32 bp (Samkurashvili & Luse, 1998), indicating that polII complexes usually have a similar shape to other RNAPs. The larger complexes reported may be due to the presence of nonhomogeneous preparations of elongation complexes, the use of a different nuclease for footprinting (DNase I), and the existence of unusual conformations of elongating polII molecules, which are poorly understood. The front edge of the archaeal RNAP is located 12 bp downstream from the last incorporated nucleotide. This distance from the front edge of the footprint to the catalytic center is, at 8 bp, shorter in the bacterial and, at

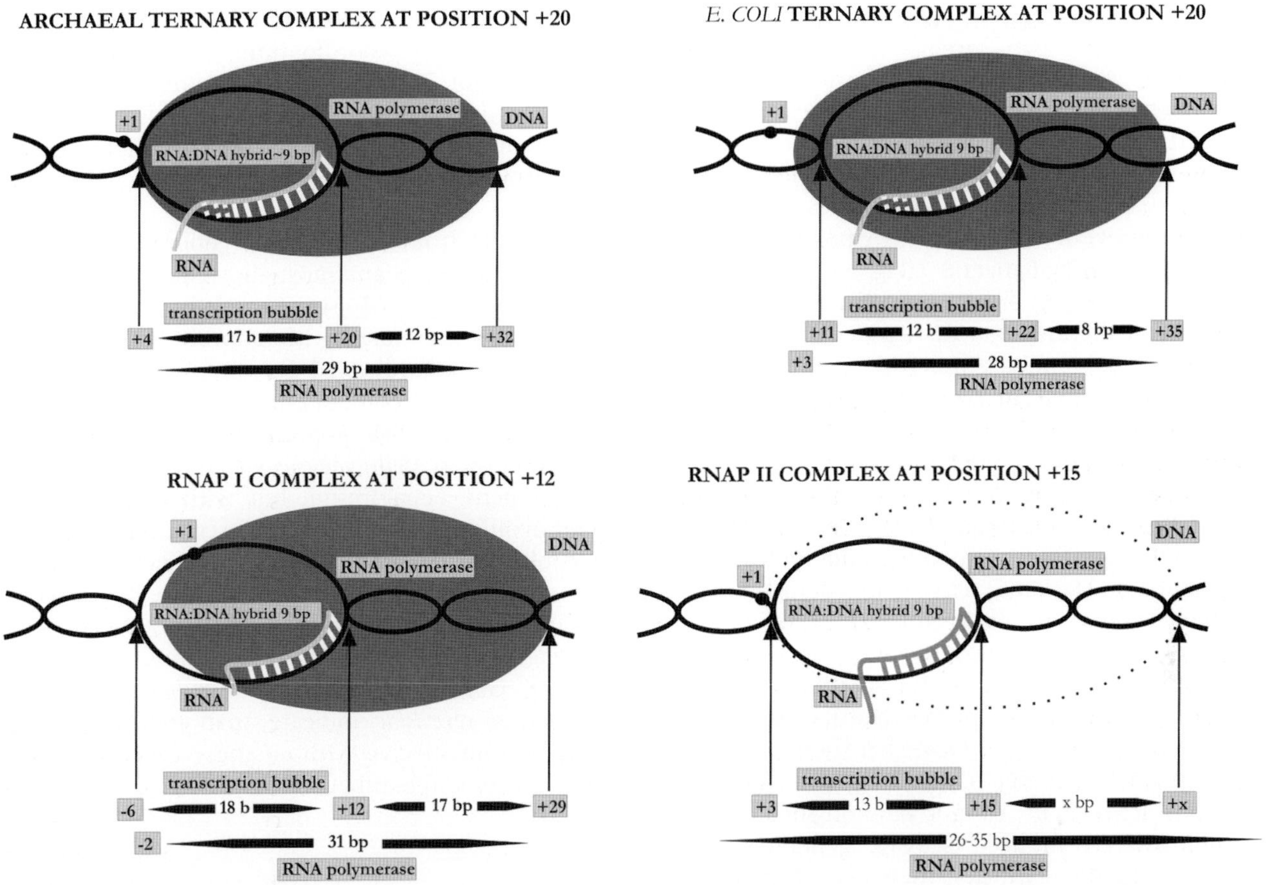

Fig. 16.1 Comparison of ternary elongation complexes. Footprinting data of ternary elongation complexes of all domains are shown. The data from the archaeal complex are from Spitalny and Thomm (2003) and the *Escherichia coli* complex is shown based on data published by Zaychikov et al. (1995). The eukaryotic RNA polymerase I complex was published by Kahl et al. (2000) and for the RNA polymerase II complex data are summarized from Linn and Luse (1991), Samkurashvili and Luse (1998), Kireeva et al. (2000), and Fiedler and Timmers (2001).

17 bp, longer in polI complexes (Fig. 16.1). The archaeal transcription bubble comprises 15–17 nucleotides, which is similar to the bubble formed by the polI enzyme (18 bp); the bacterial bubble (12 bp) and the polII bubble (13 bp) are shorter (Fig. 16.1). The DNA–RNA hybrid in stalled archaeal elongation complexes is 9 bp (Fig. 16.1). DNA–RNA hybrids of similar size were also found in bacterial and eukaryotic ternary complexes (Fig. 16.1; Zaychikov et al., 1995; Samkurashvili & Luse, 1998; Fiedler & Timmers, 2001). Analysis of the crystal structure of elongation polII complexes confirmed the existence of a 9 bp DNA–RNA hybrid (Fig. 16.1; Gnatt et al., 2001; Westover et al., 2004). Recent analyses of RNA polymerase II complexes indicate the formation of a stable elongation complex after synthesis of a RNA molecule of 23 nucleotides (Pal & Luse, 2003). Normally, such stable complexes are able to elongate transcription in a productive manner, but sometimes this process is made more difficult by misincorporation of nucleotides or certain DNA sequences, which promote pausing or backtracking of elongation complexes. Furthermore, elongating RNA polymerases have also to deal with DNA-binding proteins that could be involved in either gene regulation or wrapping the DNA. To overcome such difficulties during elongation of transcription additional factors are involved, such as TFIIS or GreA/B.

Cleavage induction factor TFS

Archaeal genomes encode several transcription factors that seem to be involved in elongation of transcription. One of these transcription factors, TFS, shows sequence similarity to the C-terminal domain of eukaryotic transcription factor TFIIS as well as to small subunits of eukaryotic RNA polymerases I, II, and III (Fig. 16.2a; Langer & Zillig, 1993; Kaine et al., 1994; Hausner et al., 2000). The corresponding subunits of the eukaryal RNA polymerases and the archaeal TFS homologs are, with about 120 amino acids, very similar in size and contain two zinc-binding domains. Western blot analysis demonstrated that archaeal TFS is a protein that does not copurify with the RNA polymerase and is therefore not a subunit of the archaeal RNA polymerase (Hausner et al., 2000). In addition, the protein is able to induce a cleavage activity in the RNAP in a manner similar to eukaryotic TFIIS or bacterial GreA/B (Fig. 16.2b). Backtracked or arrested elongation complexes are signals for TFS-induced hydrolysis. In backtracked complexes the catalytic center of the enzyme is no longer located at the 3'-end of the nascent RNA (Fig. 16.2b, second panel) and therefore the RNA chain cannot be elongated in these complexes. TFS-induced cleavage generates a new 3'-end of the RNA, which is then located again at the catalytic site of the enzyme (Fig. 16.2b, third panel). The catalytic activity of arrested elongation complexes is restored after cleaveage. In most cases, hydrolysis is combined with the release of dinucleotides (Fig. 16.2b, fourth panel; Hausner et al., 2000). Furthermore, TFS and functionally related proteins like GreA and GreB are able to improve the fidelity of transcription (Erie et al., 1993; Jeon & Agarwal, 1996; Thomas et al., 1998; Lange & Hausner, 2004). The molecular mechanism of GreA/B and TFIIS action and the structural models of GreA/B and TFIIS in combination with their cognate RNA polymerases were published by different groups (Kettenberger et al., 2003, 2004; Opalka et al., 2003; Sosunova et al., 2003). Although the corresponding proteins in Bacteria and Eukarya are completely different in primary sequence as well as three-dimensional structure, both use a very similar strategy to convert the active sites of their RNA polymerases into ribonucleases. Two highly conserved acidic residues, Asp and Glu, of TFS or GreA/B were placed into the active site of the RNA polymerase through the secondary channel and were involved in the coordination of Mg^{2+}, which is necessary for RNA hydrolysis (Kettenberger et al., 2003). Since these two residues are also conserved in archaeal TFS molecules, it is most likely that a similar mechanism for RNA cleavage is used in Archaea.

The data from the archaeal genome sequencing projects indicate that almost all Archaea contain one copy of the *tfs* gene. In some cases the corresponding genes were annotated with subunit M of the RNA polymerase according to the previous assumption that this gene is a component of the RNA polymerase (Kaine et al., 1994; Soppa, 1999; Hausner et al., 2000). Only in *Methanopyrus kandleri* has no copy of the *tfs* gene been found, but with respect to the high evolutionary rate of its transcription machinery it could also be possible that TFS function may be replaced by a non-homologous protein yet to be discovered (Brochier et al., 2004). It is also possible that this organism can survive without a TFS-like activity. Genetic analyses in *Escherichia coli* and *Saccharomyces cerevisiae* indicate that *greA/B* and *tfIIs* mutants can survive without these cleavage induction factors (Archambault et al., 1992; Orlova et al., 1995).

Termination

At present, knowledge of the process of termination of transcription in Archaea is rather poor. A comparison of 3'-ends of stable RNAs revealed the consensus sequence 5'-TTTTAATTTT-3' as a termination signal (Thomm, 1996). A mutational analysis of this sequence followed by *in vitro* transcription experiments confirmed the role of this sequence in termination of transcription (Thomm et al., 1994). Deletion of four residues from the 3'-end of the decameric sequence completely abolished termination. An internal deletion of the preceding stem-loop structure corresponding to the TΨC loop of a tRNA also significantly reduced termination efficiency. Furthermore, a Rho-independent terminator of *Escherichia coli* can perfectly replace the archaeal oligo-dT terminator (Thomm et al., 1994). Taken together, these findings indicate that archaeal transcription complexes can terminate by a mechanism similar to the intrinsic transcription terminators in Bacteria.

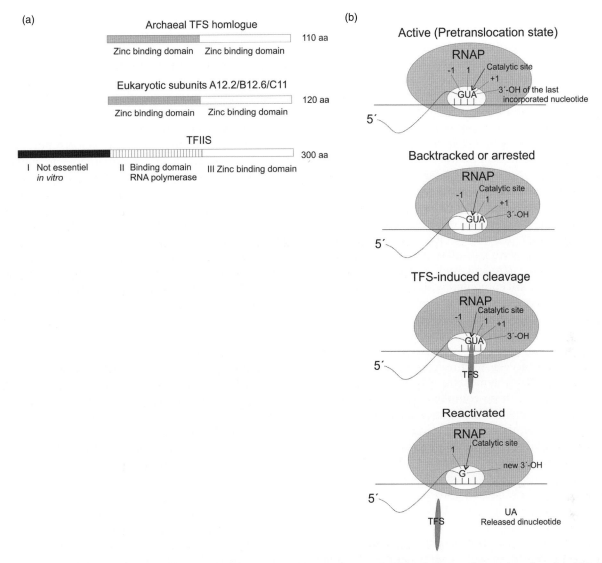

Fig. 16.2 Archaeal TFS. (a) Schematic drawing of domains present in archaeal TFS, eukaryotic subunits A12.2, B12.6, and C11, and eukaryotic TFIIS. The corresponding subunits of eukaryotic RNA polymerases and archaeal TFS are very similar in size and each consists of two zinc binding domains, whereas eukaryotic TFIIS contains only one zinc binding domain (III) and two other domains (I and II). Western blot analysis and *in vitro* transcription experiments demonstrate that TFS is no subunit of the archaeal RNA polymerase and catalyzes similar functions to the eukaryotic TFIIS (Hausner et al., 2000). (b) TFS is able to induce a cleavage activity in the RNA polymerase. Paused or arrested archaeal elongation complexes tend to backtrack (at least one nucleotide from the pretranslocation state). In the next step TFS interacts with the catalytic center of the backtracked RNA polymerase and induces the release of a dinucleotide from the 3′-end of the RNA (Kettenberger et al., 2003; Lange & Hausner, 2004). After cleavage the RNA polymerase could continue elongation.

At the moment it is not known whether there are further mechanisms to release RNA from elongation complexes. Genome data indicate that there is no Rho-like protein in Archaea. An Mfd homolog could not be found either. This protein is able to reactivate or recycle stalled or arrested elongation complexes in Bacteria and could therefore mediate a third mechanism of bacterial termination (Roberts & Park, 2004). It is structurally related to superfamiliy II helicases, but is unable to unwind DNA or RNA (Mahdi et al., 2003). Another helicase-like activity that is involved in elongation of transcription is the bacterial RapA (HepA) protein, which belongs to the SWI/SNF superfamily of helicase-like proteins (Sukhodolets et al., 2001). RapA has an ATPase activity and is able to stimulate *in vitro* transcription on supercoiled DNA and high salt concentrations, conditions that are combined with highly compacted DNA. Moreover, RapA can also promote RNA polymerase recycling (Sukhodolets et al., 2001). The corresponding eukaryotic SWI/SNF complex is an ATP-dependent chromatin-remodeling machinery that provides access to the compacted DNA for transcribing RNA polymerases (Martens & Winston, 2003). Although these complexes are very large in size and can be divided into different classes, it could be demonstrated that individual subunits containing the ATPase activity of such complexes can mediate chromatin-remodeling (Phelan et al., 1999; Saha et al., 2002). Interestingly, in most of the sequenced archaeal genomes, homologous proteins seem to be present. Therefore, it is possible that some Archaea can use a similar mechanism to remodel chromatin structure or to bypass DNA-bound protein.

In Archaea, there are several groups of small DNA-binding proteins that compact the genome; for instance, Sul7d, Alba, HTa, MC1, and histones (reviewed by Reeve, 2003). Archaeal histones are closely related to the eukaryotic histones H3 and H4. In contrast to eukaryotic octamers, in Archaea tetramers are responsible for wrapping and compacting DNA into nucleosomes (Bailey et al., 1999). Archaeal histones seemed to be a special invention of the Euryarchaeota, but the recent finding of histones in mesophilic Crenarchaeota supports the theory that histones evolved before the divergence of Archaea and Eukarya and were lost in hyperthermophilic crenarchaeotes (Cubonova et al., 2005). As histones or other DNA-binding proteins are involved in compacting of the DNA, the transcriptional machinery has to deal with the restricted access to the DNA. First *in vitro* experiments indicate that the RNA polymerase from *Methanothermobacter thermoautotrophicus* is able to traverse archaeal nucleosomes (Xie & Reeve, 2004a). The transcription rate was slowed down, but transcription was not blocked completely. This is in agreement with experiments using eukaryotic RNA polymerase III (Studitsky et al., 1997). In contrast, eukaryotic RNA polymerase II and the bacterial one were unable to extend a promoter-initiated transcript through a eukaryotic nucleosome *in vitro* under physiological salt conditions (Walter & Studitsky, 2001; Walter et al., 2003). In the case of the eukaryotic RNA polymerase II chromatin-remodeling factors, like FACT or SWI/SNF, are necessary to allow the RNA polymerase to pass the nucleosomes (reviewed by Sims et al., 2004). A recent publication indicates that the nucleosomes induce an arrest of backtracked RNA polymerase II molecules and TFIIS can reactivate the backtracked complexes and promote transcription through the nucleosomes (Kireeva et al., 2005). As the archaeal TFS can also reactivate arrested complexes (Hausner et al., 2000), it is most likely that this protein is also involved in transcription through archaeal nucleosomes. Furthermore, a genetic approach in *Saccharomyces cerevisiae* revealed an interaction between TFIIS and the SWI/SNF complex (Davie & Kane, 2000). Therefore, it is tempting to speculate that Archaea can apply in principle a similar mechanism to remodel the chromatin structure and to bypass DNA-bound protein in elongation complexes using TFS and an archaeal homolog of the SWI/SNF component.

Transcriptional regulation

Archaeal genome sequencing projects suggested that the basic eukaryotic-like transcription machinery is controlled to a great extent by bacterial-like transcriptional regulators. A detailed analysis of putative DNA-binding domains in archaeal genomes revealed that all Archaea encode a large number of proteins containing a helix–turn–helix motif, which is more similar to bacterial helix–turn–helix domains than to eukaryotic ones (Aravind & Koonin, 1999). Meanwhile, some negative transcriptional regulators of different Archaea were analyzed. On first sight many of them seemed to function in a similar way to bacterial repressors. But a closer look at the molecular detail revealed that unique mechanisms of gene regulation exist in Archaea, which we would like to emphasize in this review.

The basic concept of negative transcriptional regulation in Archaea uses a DNA-binding protein, whose binding site overlaps with the promoter or with the binding site of the RNA polymerase, and the DNA-bound protein represses transcription by preventing binding of the transcription factors TBP and TFB or the RNA polymerase. The various strategies of this concept used in Archaea have been reviewed by a number of authors (Reeve, 2003; Ouhammouch, 2004; Bell, 2005; Geiduschek & Ouhammouch, 2005). Therefore, here we summarize only some basic features. Most of the available data are derived from the Lrp family. Archaeal paralogs of the bacterial Lrps are present in almost all Archaea. LrpA from *Pyrococcus* is able to inhibit transcription of its own gene by binding to a DNA site that overlaps with binding of the RNA polymerase (Dahlke & Thomm, 2002). DNA-bound *Sulfolobus* Lrs14 inhibits binding of the archaeal transcription factors TBP and TFB (Napoli et al., 1999; Bell & Jackson, 2000b). Further examples of negative regulators are, for instance, NrpR or Mdr1. NrpR is an euryarchaeal nitrogen repressor that inhibits the expression of the *nif* operon in the presence of ammonia or alanine (Lie et al., 2005). Mdr1 is a metal-dependent repressor that is involved in transcriptional regulation of genes encoding metal-importing ABC transporters. Metal ion-dependent binding of Mdr1 to operator sequences that overlap the start site of transcription blocks RNA polymerase recruitment (Bell et al., 1999a).

Beside these strategies for negative regulation, which indeed represent bacterial repressor systems, there are some examples of archaeal-specific peculiarities. For instance, *Methanocaldococcus* Ptr2, a protein also belonging to the Lrp family, is able to recruit TBP to the promoter. Therefore, Ptr2 acts as a positive regulator and this kind of interaction between a bacterial-like regulator and an eukaryotic-like transcriptional machinery seems to be a novel invention of Archaea. A detailed biochemical analysis revealed that Ptr2 binds to specific DNA sequences, located in proper distances upstream of the promoter, and facilitates binding of TBP to the promoter, most probably by protein–protein interactions (Ouhammouch et al., 2003, 2005). A further transcriptional activator with archaeal-specific features, GvpE, was characterized in more detail by *in vivo* analysis and is the subject of Chapter 17 in this volume (see also Pfeifer et al., 2002).

In *Pyrococcus*, the regulation of two distinct ABC transporter systems for the uptake of maltose/trehalose and maltodextrins shows an archaeal speciality. The expression of the ABC transporters is controlled by the regulator TrmB. In the absence of maltose or trehalose TrmB binds to the promoter of the *mal* genes and inhibits transcription by preventing binding of the transcription factors. The repression can be counteracted by the presence of maltose or trehalose (Lee et al., 2003). The system described so far represents a bacterial repressor system, but TrmB can also bind to a completely different binding site overlapping the transcription start site of the maltodextrin promoter, and in this case bound TrmB can be only released by maltodextrin and not by maltose or trehalose (Lee et al., 2005). Therefore, the behavior of TrmB in responding in a different way to the presence of maltose in dependence on the bound target sequence seems to be a novel invention of Archaea, without any comparable parallel in Bacteria or Eukarya.

Phr, a putative regulator of the heat shock response, is present in most Euryarchaeota and was analyzed in more detail in *Pyrococcus furiosus*. The protein is able to repress the transcription of genes involved in the heat shock response by binding to a palindromic sequence overlapping the transcription start site (Vierke et al., 2003). Analysis of the crystal structure of Phr showed that it consists of a winged helix–turn–helix–protein containing four helices in the N-terminal domain. A mutational analysis demonstrated that amino acids in three helices and the wing-region are required for DNA binding (W. Liu, G. Vierke, A.-K. Wenke, M. Thomm & Rudolf Ladenstein, in preparation). Therefore, this kind of binding seems to be a novel mode of DNA–protein interactions because well described winged helix–turn–helix proteins contact the DNA only with one helix or with the wing-region (Gajiwala & Burley, 2000).

Analysis of the heat shock system in *Pyrococcus* revealed that heat shock regulation also differs from those in Bacteria and Eukarya. The basic principles of heat shock regulation are shown in Fig. 16.3. The main goal of the heat shock response is to switch on the expression of a special group of genes, the heat shock genes, which contain different classes of chaperones, like Hsp70/DnaK and

(a) Possible model of archaeal heat shock regulation

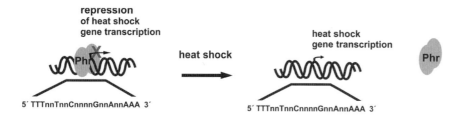

(b) Model of eukaryotic heat shock regulation

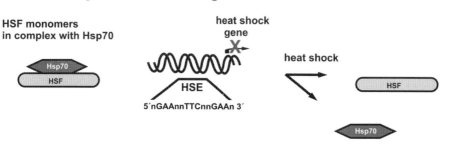

(c) Model of bacterial heat shock regulation

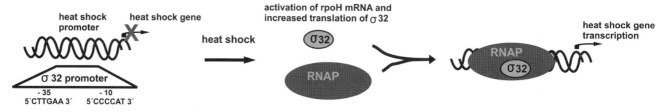

Fig. 16.3 Comparison of heat shock response. Schematic representation of archaeal, eukaryotic, and bacterial heat shock responses. The archaeal model is based on data of Vierke et al. (2003), the eukaryotic one on data of Morimoto (2002), and the bacterial one on data of Yura and Nakahigashi (1999).

Hsp60/GroEL. These proteins are necessary to refold or to degrade denatured proteins. The heat shock response seems to be extremely conserved in nature, but the data so far available in *Pyrococcus* indicate some differences. First of all, most Archaea and all hyperthermophiles lack the Hsp70/DnaK system (Hickey et al., 2002).

The major archaeal heat shock genes are a small heat shock protein Hsp20, an aaa$^+$ ATPase, and the Hsp60/GroEL homolog, the thermosome (Laksanalamai et al., 2001). Furthermore, *Pyrococcus* heat shock genes have an archaeal consensus promoter sequence and transcription of these genes is switched off *in vitro* due to the action of the repressor Phr (Vierke et al., 2003) binding to a conserved sequence typical for *Pyrococcus* promoter sequences located at the transcription start site (the consensus sequence is indicated in Fig. 16.3a). After heat shock, the repressor is released and transcription of these genes is possible, most likely by the normal transcriptional machinery (Fig. 16.3a; G. Vierke & M. Thomm, unpublished data).

Eukarya need an activator, HSF, for the expression of the Hsp70/DnaK system, and Bacteria need a special σ-factor. In Eukarya, the heat shock transcription factor HSF forms under physiological growth conditions a complex with Hsp70 and is released after heat shock, as Hsp70 has chaperone activity and interacts with heat-denatured proteins (Fig. 16.3b). After trimerization and phosphorylation of released HSF, the modified factor is able to recognize a specific DNA-binding site at heat shock promoters, the heat shock element (HSE) (the sequence of the HSE is shown in Fig. 16.3b). Binding of the heat shock factor to the HSE sequence element leads to activation of transcription of heat shock genes (Fig. 16.3b, to the right; reviewed by Morimoto, 2002).

In Bacteria, there are at least two main groups of heat shock systems. One is also negatively controlled with a repressor and is mainly found in *Firmicutes*. The other common system in Bacteria uses a special heat shock promoter sequence for the coordinate expression of these genes, and transcription of the corresponding genes requires σ^{32} instead of σ^{70} (Fig. 16.3c). The promoters recognized by the complex of the core enzyme of RNAP with σ^{32} have a consensus promoter differing from the bacterial consensus promoter sequence recognized by σ^{70} (the σ^{32} consensus sequence is shown in Fig. 16.3c, to the left). By contrast, in Archaea the heat shock promoters contain canonical TATA-box and BRE-consensus sequences but are characterized by a conserved inverted repeat sequence at the transcription start site recognized by the archaeal heat shock regulator Phr (Vierke et al., 2003). The concentration of σ^{32} increases after heat shock as a result of multiple modes of regulation at the transcriptional, translational, and post-translational levels. The complex of σ^{32} with RNAP recognizes the heat shock promoters, and as a consequence, transcription of the heat shock genes is dramatically intensified (reviewed by Yura & Nakahigashi, 1999).

Acknowledgments

The authors appreciate the preparation of figures by Christoph Reich, Patrizia Spitalny, and Gudrun Vierke.

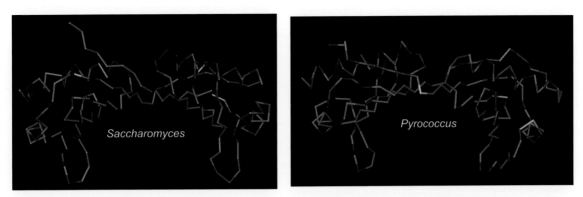

Plate 16.1 Structural comparison of archaeal and eukaryotic TBP. Tube models of TBP monomers from *Saccharomyces cerevisiae* (PDB: 1TBP) and *Pyrococcus woesei* (PDB: 1PCZ) are shown. Acidic residues are indicated in red and basic amino acids in blue. Both molecules show a symmetrical saddle-shaped form. Archaeal TBPs are more acidic, whereas eukaryotic TBPs contain more basic residues. The eukaryotic TBP was crystallized without the N-terminal extension and the archaeal TBP crystal does not contain the C-terminal acidic tail. The acidic residue in the *Pyrococcus* TBP (not present in *Saccharomyces*) that is involved in site-specific cation binding is shown in yellow (Bergqvist et al., 2001).

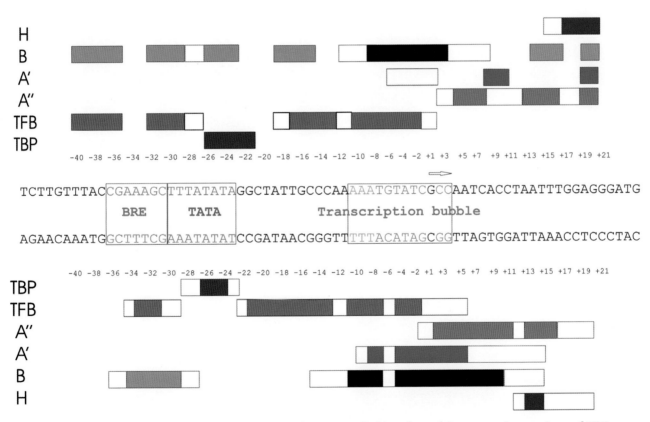

Plate 16.2 Summary of site-specific protein–DNA photo cross-linking data of *Pyrococcus*. Interactions of TBP, TFB, and RNA polymerase with the promoter region of the *gdh* promoter are shown in different colors. The colored regions indicate strong interactions and the boxed regions indicate weak interactions. Moderate DNA interactions with subunit B are shadowed in gray. The plate summarizes data shown in Bartlett et al. (2000, 2004) and Renfrow et al. (2004). The transcription start site is labeled by an arrow; the TATA box, BRE region, and the transcription bubble are indicated.

17
Transcriptional regulation in Haloarchaea

Felicitas Pfeifer, Torsten Hechler, Sandra Scheuch and Simone Sartorius-Neef

Introduction

Halophilic Archaea thrive in hypersaline environments such as natural salt lakes or salterns at salt concentrations up to the saturation of sodium chloride. The family Halobacteriaceae includes neutrophilic halophiles (such as *Halobacterium*, *Haloferax*, *Haloarcula*, and *Haloquadratum*), and haloalkaliphiles such as *Halorubrum* and *Natronobacterium*. The genome sequences of *Halobacterium salinarum* NRC-1 (Ng et al., 2000) and *Haloarcula marismortui* (Baliga et al., 2004) have been determined, and a comparison of the 2.01 Mbp genome of *Hbt. salinarum* and the 4.27 Mbp genome of *Hac. marismortui* suggests that *Hbt. salinarum* has a streamlined smaller genome possibly arising from a common ancestor to *Hac. marismortui*. In general, Haloarchaea contain circular chromosomes of relatively high G+C content (62–68% G+C), and several smaller replicons of 100–600 kbp with significantly lower G+C contents of 54–60% (Pfeifer, 1986; Ng et al., 2000; Baliga et al., 2004).

Two characteristic features of *Hbt. salinarum* are the purple membrane, synthesized under conditions of high light intensities and low oxygen tensions with bacteriorhodopsin as major constituent, and gas vesicles that enable the strain to move toward the surface of the brine. The *Hbt. salinarum* group includes the former species *Hbt. halobium*, *Hbt. cutirubrum*, and *Hbt. salinarium*, but also isolates derived from salterns in the San Francisco Bay (SB3), the Mediterranean coast in France (GRB), and the Baja California (GN101). All these strains are related, but the latter three isolates harbor different plasmid populations and possess a significantly smaller number of insertion elements (ISH elements) compared to the *Hbt. salinarum* strains PHH1 and NRC-1 that are commonly used in the lab and deposited in culture collections (Pfeifer, 1986; Ng et al., 2000). PHH1 and NRC-1 originate from the same *Hbt. halobium* isolate, but genomic rearrangements due to the action of a large number of insertion (ISH) elements occurred, especially in AT-rich islands in the chromosome and in the plasmid population (Pfeifer & Betlach, 1985; Pfeifer & Ghahraman, 1993; Ng et al., 2000).

The genes involved in purple membrane (mutation rate 10^{-4}), bacterioruberin (10^{-4}) or gas vesicle synthesis (10^{-2}) are affected by ISH elements at high frequencies (Pfeifer et al., 1981). The gene clusters involved in gas vesicle (*gvp*) or purple membrane (*bop*) formation have been identified in both strains (Betlach et al., 1983; DasSarma et al., 1983, 1988; Horne et al., 1991). Both clusters are regulated by a transcriptional activator protein, and environmental factors such as oxygen, light, and salt concentration appear to influence their expression.

The initiation of basal transcription in Archaea involves a multicomponent RNA polymerase (RNAP), a TATA-box binding protein (TBP), and the transcription factor TFB that is homologous to the eukaryal TFIIB (Bell & Jackson, 2001). Both gene clusters were investigated *in vivo*, since efficient genetic transformation systems are available for *Halobacterium* and *Haloferax* species. Vector plasmids have been developed that contain either the mevinolin or the novobiocin resistance gene as a selective marker. Together with the expression vector pJAS35 containing the strong ferredoxin gene promoter P_{fdx} (Pfeifer et al., 1994), and the reporter gene *bgaH* encoding a halophilic enzyme with β-galactosidase activity (Holmes & Dyall-Smith, 2000), these tools are suitable for *in vivo* studies. *Hbt. salinarum* or the moderately halophilic *Haloferax volcanii* is used to transform. A close relative of *Hfx. volcanii* is *Hfx. mediterranei*, a gas vesicle producer isolated from the salterns near Alicante, Spain.

This chapter summarizes our current knowledge regarding different *gvp* gene clusters of Haloarchaea, and of the *bop* gene cluster involved in purple membrane synthesis of *Hbt. salinarum*.

Regulation of gas vesicle formation in *Hbt. salinarum* and *Hfx. mediterranei*

Gas vesicles and vac regions in halophilic Archaea

Gas vesicles are gas-filled proteinaceous particles of spindle or cylindrical shape that are produced constitutively in *Hbt. salinarum* PHH1 and NRC-1, as well as in *Haloquadratum walsbyi* (Fig. 17.1). In contrast, *Hbt. salinarum* PHH4 (a pHH1-deletion strain) and the *Halobacterium* isolates SB3, GN101, and GRB, as well as *Hfx. mediterranei* and *Halorubrum vacuolatum*, produce gas vesicles in the stationary growth phase only. Gas vesicle formation in *Hfx. mediterranei* and the haloalkaliphilic *Hrr. vacuolatum* also depends on the salt concentration in the medium, i.e. cells grown in media containing 15–17% NaCl are gas vesicle free (Rodriguez-Valera et al., 1983; Englert et al., 1990; Mayr & Pfeifer, 1997). The possession of gas vesicles enables the cells to move to the surface of the brine where oxygen and higher light intensity are available. This is especially important at high salt concentrations, where the oxygen solubility is low. In cultures standing on the lab bench, *Hbt. salinarum* wild type cells form a pellicle at the surface, whereas all other Haloarchaea use gas vesicles to control their position in the water column rather than floating at the surface.

Gas vesicles are composed of at least two structural proteins. The major 8 kDa structural protein GvpA forms a rigid, ribbed envelope that allows the diffusion of dissolved gases, but prevents the passage of liquid water due to the hydrophobic inner surface formed by GvpA. The second structural protein, GvpC (31–42 kDa), is located on the outside of the gas vesicle and stabilizes the ribbed structure formed by GvpA, but also has a function in determining the constant diameter of the cylindrical portion. The genes encoding GvpA in *Hbt. salinarum* PHH1 and NRC-1 have been identified using a cyanobacterial *gvpA* gene probe, and all additional *gvp* genes involved in gas vesicle formation were detected through gas vesicle negative mutants that are usually caused by the integration of an ISH element in or near the *gvpA* gene in plasmid pHH1 (or the related pNRC100 of NRC-1) (Pfeifer et al., 1981; DasSarma et al., 1988; Horne et al., 1991). Transformation of the gas vesicle negative *Hfx. volcanii* with fragments containing the *gvpA* gene and surrounding sequences finally determined the boundaries of the *gvp* gene clusters and led to the identification of the genes that are required for their formation (Englert et al., 1992b; Offner et al., 2000). In *Hbt. salinarum* PHH1 and *Hfx. mediterranei*,

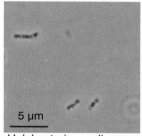
Halobacterium salinarum

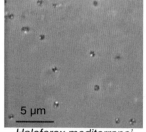
Haloferax mediterranei

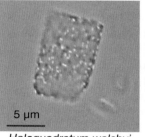
Haloquadratum walsbyi

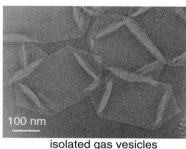
isolated gas vesicles

Fig. 17.1 Gas vesicle containing cells viewed by phase contrast microscopy, and isolated gas vesicles viewed by electron microscopy.

the gas vesicle gene clusters (= vac region) are composed of 14 *gvp* genes, arranged as two oppositely oriented clusters, *gvpACNO* and *gvpDEFGHIJKLM* (Englert et al., 1992a). Two distinct vac regions (p-vac located on plasmid pHH1, and c-vac on a larger mini-chromosome) are present in *Hbt. salinarum* PHH1 that are related but not completely identical to the *gvp1* (p-vac) and *gvp2* (c-vac) gene clusters of *Hbt. salinarum* NRC-1 (Ng et al., 2000). A single vac region (mc-vac) is found in the chromosome of *Hfx. mediterranei*. Eight of these genes are the minimal requirement for gas vesicle formation as determined for the p-vac region (Fig. 17.2). The p-vac region (and also the *gvp1* gene cluster in NRC-1) is frequently affected by ISH elements (Fig. 17.2) (Horne et al., 1991), whereas in the case of the slightly smaller c-vac region leading to gas vesicles in the p-vac deletion strain *Hbt. salinarum* PHH4 only, an integrated insertion element has never been observed in gas vesicle mutants. The related *gvp2* cluster of NRC-1 lacks the essential *gvpM* gene found in c-vac and is thus not functional in gas vesicle formation. The haloalkaliphilic *Hrr. vacuolatum* contains only 12 of the *gvp* genes arranged as the *gvpACNOFGHIJKLM* gene cluster (= nv-vac region) that is transcribed from a single promoter, P_{nvA} (Fig. 17.2) (Mayr & Pfeifer, 1997; Pfeifer, 2004). Genes related to *gvpD* and *gvpE* have not yet been identified.

Regulation of *gvp* promoters and promoter structures

The P_A promoter located in front of *gvpACNO* drives the synthesis of the major (GvpA) and minor (GvpC) gas vesicle structural proteins in all four vac regions. The 340-nt *gvpA* mRNA contains a 20–21 nt leader region and is produced as dominant *gvp* transcript in large amounts, with a relatively long half-life of 80 minutes determined for *Hfx. mediterranei* (Jäger et al., 2002). All other transcripts (including the longer *gvpACNO* mRNA) are only present in minor amounts (Offner & Pfeifer, 1995; Krüger & Pfeifer, 1996; Röder & Pfeifer, 1996). The relatively low stability determined for the latter *gvp* transcripts is not affected by the salt concentration in the medium, but the mc-*gvpA* mRNA appears to decay about twice as fast in cultures grown in 18% salt compared to cultures grown at 25% salt, contributing at least in part to the salt-dependent gas vesicle formation in *Hfx. mediterranei* (Jäger et al., 2002). The P_D promoter separated by 100 bp (p-vac and mc-vac) or 122 bp (c-vac) from P_A is responsible for the formation of the two endogenous regulatory proteins GvpD and GvpE, and additional proteins involved in the formation of the gas vesicle structure (Shukla & DasSarma, 2004). GvpE is a transcriptional activator, whereas GvpD is involved in the repression of gas vesicle formation.

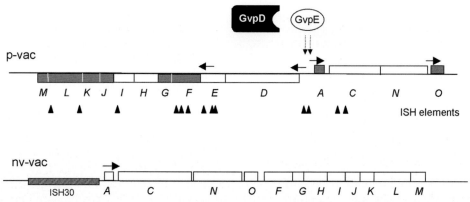

Fig. 17.2 Genetic map of the gas vesicle genes of *Hbt. salinarum* PHH1 (p-vac) and *Hrr. vacuolatum* (nv-vac). The *gvp* genes are depicted as boxes labeled A and C through O. The eight *gvp* genes constituting the minimal p-vac region are shaded in gray. Arrows above the map indicate start sites and the direction of transcription. The interaction of GvpD and GvpE, and of the transcriptional activator GvpE at P_{pA} and P_{pD}, are indicated above. Triangles below the map depict insertion sites of ISH elements in gas vesicle negative mutants. The 1443 bp ISH30 element is inserted immediately upstream of the TATA box in the P_{nvA} promoter in nv-vac.

The p-vac region expressed throughout growth is transcribed from four promoters (P_{pA}, P_{pD}, P_{pF}, and P_{pO}), two of which (P_{pA} and P_{pD}) are activated by GvpE (Fig. 17.2) (Hofacker et al., 2004). P_{pA} is the strongest promoter, followed by P_{pF}, P_{pD}, and P_{pO} (relative BgaH activities in mU/mg: 104 > 5 > 1.3 > 1 for $P_{pA} > P_{pF} > P_{pD} > P_{pO}$; Hofacker et al., 2004; Sartorius-Neef & Pfeifer, unpublished). The two promoters P_{pF} and P_{pO} are not described for the homologous $gvp1$ gene cluster of Hbt. salinarum NRC-1. A scanning mutagenesis encompassing 50 bp upstream of the transcription start site in P_{pA} demonstrates that the promoter strength is affected by the TATA box and surrounding sequences, since mutations in this region lead to a lack of basal transcription (Hofacker et al., 2004). The sequence AACCA located 8 bp upstream of the P_{pA}-TATA box appears to be crucial for the GvpE-mediated activation of the P_{pA} promoter (Fig. 17.3a). This sequence is part of the GvpE-responsive element that has been determined in P_{mcA} (see below). Only two promoters (P_A and P_D) drive the expression of c-vac and mc-vac, and all except P_{cD} are activated by GvpE (Gregor & Pfeifer, 2001; Zimmermann & Pfeifer, 2003). The reason for the lack of P_{cD} activaton is the 22 bp insertion upstream of the P_{cD}-TATA box that separates the putative GvpE-responsive element from this c-vac promoter.

A strong gvp promoter activity is only achieved in the presence of GvpE, and the largest promoter activity is found with P_{mcA} of mc-vac induced by cGvpE, that reaches a similarly high activity as found for the P_{fdx} promoter of the halobacterial ferredoxin gene (Gregor & Pfeifer, 2005). The lack of GvpE in ΔE transformants (that carry a deletion of the $gvpE$ gene in p-vac) results in gas vesicle negative mutants, indicating that the activation of P_{pA} and P_{pD} by GvpE is required for the formation of gas vesicles (Offner et al., 2000). The weakest of all gvp promoters is P_{cA} of the c-vac region that lacks a basal promoter activity and is only active in the presence of the homologous cGvpE (Gregor & Pfeifer, 2001).

A 4 bp scanning mutagenesis done throughout the 50 bp region separating the respective TATA-box of P_{mcD} and P_{mcA} in mc-vac determined the sequence TGAAACGG-n4-TGAACCAA as being important for the GvpE-mediated activation of P_{mcA} (Fig. 17.3b) (Gregor & Pfeifer, 2005). This sequence is located 40 nt upstream of the transcription start and 8 nt upstream of the P_{mcA}-TATA box. A similar sequence element is found in all promoters affected by GvpE,

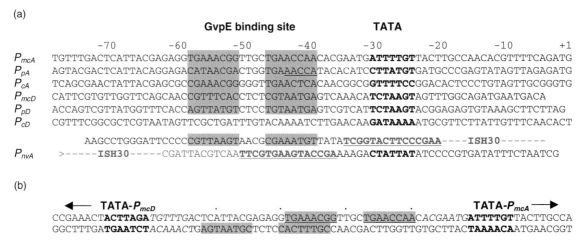

Fig. 17.3 Alignment of the promoter regions of various $gvpA$ and $gvpD$ genes (a) and intergenic region between P_{mcD} and P_{mcA} in mc-vac (b). The transcription start site in (a) is labeled +1, the TATA box are indicated in bold, and the GvpE-responsive elements located 40 bp upstream of the transcription start site are shaded in gray. The AACCA sequence in P_{pA}, determined as important for GvpE-mediated activation, is underlined. The partial sequence of the ISH30 element inserted in P_{nvA} given here is in gray, and the inverted repeats at the termini are underlined. (b) The intergenic region between P_{mcD} and P_{mcA} is given as double-stranded DNA to highlight the overlapping UAS elements shaded in gray in one strand only. Arrows mark the direction of the two TATA boxes. Dots above the sequence indicate distances of 10 nt.

whereas P_{cD}, P_{pF}, and P_{pO} (i.e. promoters not affected by GvpE) are different. Most likely this sequence constitutes the GvpE-responsive element. Altering the space between the TATA box and this sequence element leads to the loss of GvpE-induced gvpA transcription, suggesting that it is an upstream activator sequence (UAS). The distance between the respective TATA box of P_{mcD} and P_{mcA} is 50 bp and thus very short, and the putative divergent UAS element of P_{mcD} overlaps the UAS of P_{mcA} in the center (Fig. 17.3b). A search in the genome sequence of Hbt. salinarum NRC-1 revealed that the GvpE-responsive element is unique to the P_A and P_D promoters of the gvp gene clusters. The P_{nvA} promoter region of Hrr. vacuolatum is distinct in that a 1443 bp ISH-element is inserted adjacent to the TATA box (Fig. 17.3a) (Mayr & Pfeifer, 1997; and unpublished results). However, the promoter region upstream of the ISH30 shows a similar sequence element, implying that the GvpE would be able to activate the original P_{nvA} promoter.

The regulatory proteins GvpD and GvpE

The transcriptional activator GvpE is a 21 kDa protein that dimerizes in solution and resembles a basic leucine zipper protein as suggested by molecular modeling of the C-terminal AH6 helix and mutation analyses (Krüger et al., 1998; Plößer & Pfeifer, 2002). The GvpE protein appears in late exponential growth in Hbt. salinarum PHH4 and Hfx. mediterranei, prior to the gas vesicle formation (Krüger & Pfeifer, 1996; Zimmermann & Pfeifer, 2003). Mutants of cGvpE of the c-vac region with alterations of the leucine zipper helix AH6, or of the two putative DNA binding sites DNAB and AH4 (see Fig. 17.4), are unable to stimulate the expression of the reporter construct P_{cA}-bgaH in Hfx. volcanii transformants. Additional motifs that resemble functional domains of known bacterial regulator proteins have not been detected in GvpE; the protein contains neither a PAS/PAC nor a GAF domain as found in the Bat activator protein (see below).

The larger GvpD protein (54–60 kDa) is present during the very early growth phase of Hbt. salinarum PHH4 (Krüger & Pfeifer, 1996), whereas in Hfx. mediterranei mcGvpD appears shortly after mcGvpE in late exponential growth (Zimmermann & Pfeifer, 2003). GvpD has a function in the repression of gas vesicle formation: transformants containing an mc-vac region with an in-frame deletion in mc-gvpD (= ΔD) are gas vesicle overproducers, and in ΔD/D transformants the overproduction is complemented by the wild type gvpD gene (Englert et al., 1992b; Pfeifer et al., 1994). A p-loop motif found near the N-terminus of GvpD is essential for this repressing function (Fig. 17.4) (Pfeifer et al., 2001).

GvpE and GvpD are able to interact, since GvpE can be purified from the lysate of Hfx. mediterranei using a Ni-NTA matrix tagged with mcGvpD$_{his}$ (Zimmermann & Pfeifer, 2003). This interaction could be part of the negative P_A promoter control. Interestingly, GvpE is absent in DE^{ex} transformants (which contain mc-gvpDE expressed under P_{fdx} control in DE^{ex} pJAS35), whereas D^{ex}, E^{ex}, or ΔDE^{ex} transformants (carrying the in-frame deletion in mc-gvpD) contain large amounts of these proteins (Zimmermann & Pfeifer, 2003). This apparent

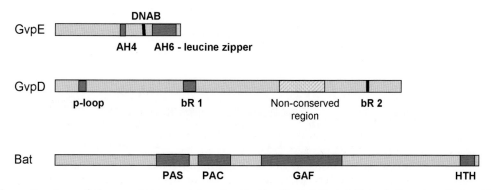

Fig. 17.4 Schematic view of the regulator proteins GvpD, GvpE, and Bat. Abbreviations used are: AH, amphiphilic helix; DNAB, DNA binding site; bR, basic region; HTH, helix–turn–helix motif. Further explanations are given in the text.

instability of GvpE in the presence of GvpD could be the mechanism underlying the GvpD-mediated reduction of the gas vesicle formation.

To date, the following model is proposed for the regulation of the *gvp* gene cluster: GvpE activates the P_A promoter by binding to the GvpE-responsive element (UAS sequence) and contacts proteins of the basal transcription apparatus (RNAP, TBP, or TFB). It is also possible that the binding of GvpE enables an improved binding of transcription factors as demonstrated for the activator protein Ptr2 of *Methanococcus jannaschii*, which contacts and enhances the binding of TBP (Ouhammouch et al., 2003). The presence of GvpD results in a reduced GvpE-mediated activation, most likely induced by the GvpD–GvpE interaction. Since a functional p-loop is required for the repression, this interaction could lead to a modified GvpE protein that has lost its activator function (due to a loss of the DNA binding or of another function), or results in an enhanced proteolytic degradation of GvpE. On the basis of the uniqueness of the UAS, GvpE appears to exclusively regulate the *gvp* gene clusters.

Regulation of the purple membrane synthesis in *Hbt. salinarum*

Structure of the *bop* gene cluster

The *bop* gene cluster located in the chromosome of *Hbt. salinarum* encodes the structural and regulatory proteins required for purple membrane synthesis. This specialized membrane contains the retinal protein bacteriorhodopsin (BR) as sole protein constituent. BR acts as a light-driven proton pump, and growth conditions of high light intensity and low oxygen tension strongly induce BR synthesis. The *bop* gene encodes the apoprotein bacterioopsin, and was one of the first archaeal genes to be cloned. The *bop* mRNA starts only two nucleotides upstream of the ATG start codon. Purple membrane mutants of *Hbt. salinarum* NRC-1 and *Hbt. salinarum* PHH1 occur with frequencies of 10^{-4} and usually contain an ISH element inserted either in the *bop* gene itself or further upstream (DasSarma et al., 1983; Pfeifer et al., 1984). The latter mutants led to the identification of additional genes involved in the purple membrane synthesis (Fig. 17.5). The three genes located here affect *bop* gene expression: *brp* (bacterioopsin-related protein gene), *bat* (bacterioopsin activator gene), and *blp* (bacterioopsin-linked protein gene) (Betlach et al., 1984; Leong et al., 1988b; Gropp et al., 1994). The *bop* and *brp* genes are oppositely oriented and separated by 526 nucleotides, and the *brp* transcript starts without an mRNA leader at the AUG start codon. A fourth gene (*crtB1*) located further upstream of *blp* has been identified in the genome of *Hbt. salinarum* NRC-1 on the basis of a common regulatory site in the promoter (Baliga et al., 2001). This gene seems to encode a phytoene synthase homolog required for the first step of the carotene biosynthesis.

Regulation of the *bop* gene cluster and the proteins encoded

The transcript level of *bop* remains low under conditions of normal aeration, but increases under conditions of low oxygen tension and high light intensities

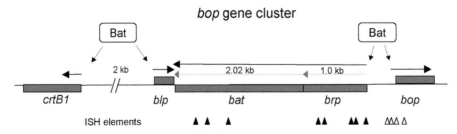

Fig. 17.5 Genetic map of the *bop* gene cluster of *Hbt. salinarum*. The genes are indicated by boxes, and arrows depict the length and direction of transcripts. Arrows in gray above *brp* and *bat* show the transcripts detectable by Northern analysis. Triangles below the map indicate the integration sites of various ISH elements in purple membrane mutants. Open triangles indicate mutants with orange phenotypes, whereas black triangles denote mutants with white or pale yellow phenotype. The Bat protein activates all four promoters as indicated.

(Shand & Betlach, 1991). The increase in transcripts is due to the product of the *bat* gene. This gene appears to be co-transcribed with *brp*, although single transcripts can be detected for *brp* and *bat* (Fig. 17.5) (Betlach et al., 1984; Leong et al., 1988a). In *Hbt. salinarum* wild type, the amount of the 2.0 kb *bat* mRNA increases 40-fold under low oxygen tension, whereas the 1.1 kb *brp* transcript remains unaltered despite the proposed co-transcription of both genes (Shand & Betlach, 1991). This difference could be due to different mRNA stabilities, with an enhanced degradation of the *brp* part of the *brp-bat* co-transcript.

The product of *brp* is a membrane-bound protein originally thought to be required for the activation of *bop* under high light intensities (Shand & Betlach, 1991). This suggestion was derived from insertion mutants in *brp* that are devoid of bacterioopsin and form white (when possessing gas vesicles) or pale yellow (when lacking gas vesicles) colonies. However, the lack of *bop* gene expression in these *brp* insertion mutants is most likely due to a polar effect of the ISH element in *brp* on *bat* transcription, resulting in a lack of the Bat activator. An additional analysis of a mutant containing an in-frame deletion in *brp* implies that Brp is involved in the formation of BR rather than in the production of the apoprotein Bop (Peck et al., 2001). The *bop* gene expression is not affected in the mutant, but this mutant contains an increased β-carotene content and forms orange red colonies. From these results Peck et al., conclude that Brp might be involved in the retinal synthesis and regulates or catalyzes the final conversion of β-carotene to retinal (Peck et al., 2001). However, this has never been demonstrated on a biochemical basis.

The 74 kDa Bat activator protein is most likely an oxygen and light sensor as suggested by the redox-responsive PAS/PAC domains and a putative photo-sensitive cGMP-binding domain, GAF, found in the protein sequence (Fig. 17.4) (Gropp & Betlach, 1994; Baliga et al., 2001). A helix–turn–helix motif near the C-terminus could be involved in DNA binding, although an interaction with the promoter region of *bop* has not yet been shown. Also, the functions of the postulated PAS/PAC and the GAF domains have not yet been confirmed *in vivo*. However, a four amino acid alteration in the PAC domain of Bat is thought to be the reason for the purple membrane overproducer phenotype of S9 (Fig. 17.4).

The fourth gene of the *bop* gene cluster, *blp* (bacterioopsin-linked protein gene), is co-regulated with *bop* under low oxygen tension, but the function of this protein is still not known (Gropp et al., 1994).

Regulatory sites in the *bop* gene promoter

The putative Bat binding site in P_{bop} has been determined by deletion and saturation mutageneses. The initial deletion analysis performed in Mary Betlach's lab confined the sequences responsible for oxygen- and light-mediated regulation to a minimal 54 bp region upstream of the transcriptional start site of the *bop* mRNA, and a 16 nt conserved sequence element located here has been postulated as Bat binding site (Gropp et al., 1995). This sequence is conserved between all promoters of the *bop* gene cluster (P_{bop}, P_{brp}, P_{blp}) and is also located upstream of *crtB1* encoding a putative phytoene synthase homolog (Fig. 17.6). Saturation mutagenesis confirmed the importance of this sequence element and defined the nucleotides ACC located near the 5′ end and TTnG at the 3′ end as most important (Baliga & DasSarma, 1999). Altering the spacing of this element to the TATA box of P_{bop} leads to the loss of *bop* transcription (Baliga & DasSarma, 2000).

Global transcriptome and proteome analyses underline the coordinated coregulation of the *bop* cluster genes: *bop*, *brp*, and *blp* are suppressed in a *bat*::ISH mutant, whereas they are overexpressed in the purple membrane overproducer strain S9 (Baliga et al., 2002). These global analyses also demonstrate a coordinate regulation of these genes under low oxygen tension and light. Interestingly, the *arcABC* operon encoding the enzymes required for arginine fermentation is induced in the *bat*::ISH mutant, whereas these genes are downregulated in the purple membrane overproducer S9 (Baliga et al., 2001). Thus, these two alternative energy producing pathways, purple membrane and arginine fermentation, might be inversely regulated by Bat.

Two additional sequence characteristics are found with the P_{bop} promoter. Saturation mutagenesis indicates that the 7 bp region between the TATA box and UAS in P_{bop} is not important for *bop* gene transcription, suggesting that P_{bop} lacks the TFB recognition element BRE that binds TFB (Baliga & DasSarma,

```
                                        UAS                    TATA   RY-box
                                        ACCcnactagTTnGG        tyTT/ATa
bop     ATCGGTTCTAAATTCCGTCACGAGCGTACCATACTGATTGGGTCGTAGAGTTACACACATATCCTCGTTAGGTACTGTTG
brp     GTTCCTTAAATATTCTCCAGTTCTGATACCCCACTAGCTTGGGTCTTTTTTTGATGCTCGGTAGTGACGTGTGTATTCATA
blp     GAGCGTGCCGCGCATGTGGGAGTTTCTACCAAATTGATTGGGGGCTGTAGTTACCCACCCACGCCACCATCATCCACACG
crtB1   TACCACGAGTACTAGCGCACCTAAAGCACCCAAAGTGATGGGGTGTCGGTTTGAGGCGCGCGAAACGAGTGTAGGTCCACA
             -70       -60       -50       -40       -30       -20       -10       +1
```

Fig. 17.6 Promoter regions of the four genes constituting the purple membrane cluster. The Bat-responsive UAS is shaded in gray, the TATA box is given in bold, and the RY box consisting of alternating purine/pyrimidines in *bop* and *brp* is underlined. The transcription start site is at position +1. The consensus sequences of UAS and TATA given on top have been determined by saturation mutagenesis.

2000). An additional regulatory element in P_{bop} appears to be an 11 bp alternating purine–pyrimidine sequence (RY box) that overlaps the TATA box by 4 bp and is able to adopt a non-B-DNA structure under conditions of high DNA supercoiling (Fig. 17.6). A saturation mutagenesis study demonstrates that the RY box in P_{bop} indeed modulates *bop* gene transcription (Baliga & DasSarma, 2000). Because lowered oxygen availability results in increased DNA supercoiling, the *bop* gene could be regulated in response to changes in dissolved oxygen concentration. An RY box is also found in the P_{brp} promoter with a similar effect on the transcription, whereas the activity of P_{blp} is not affected by DNA supercoiling.

Conclusions

Regulation of the *gvp* and *bop* gene clusters occurs at the transcriptional level and both depend on a particular activator protein that binds at the respective UAS element located 40 bp upstream of the transcription start site. In the case of the gas vesicle gene cluster, the 21 kDa GvpE activates the two oppositely oriented P_A and P_D promoters, and the activation is "repressed" by the anti-activator GvpD. The distance between these two *gvp* promoters is 50 bp (from TATA box to TATA box) and thus very short, leading to an overlap of the oppositely oriented UAS sequences in the center. Apart from a leucine zipper at the C-terminus and two DNA binding sites, GvpE does not indicate other domains that could enable a response to oxygen or light. In contrast, the two oppositely oriented P_{bop} and P_{brp} promoters are 470 bp apart, and the 74 kDa Bat activator appears to act as light and oxygen sensor, due to PAS/PAC and GAF domains found in the protein sequence. Bat seems to regulate a larger regulon, including the activation of all genes involved in the purple membrane and carotene biosynthesis, but also affects negatively the gene cluster required for arginine fermentation.

Acknowledgments

This work presented was supported by the Deutsche Forschungsgemeinschaft (DFG; Pf 165/1 through 165/9). Wolfram Zillig hosted the group of Felicitas Pfeifer at the MPI für Biochemie in Martinsried from 1985 to 1995, and F. Pfeifer expresses her special thanks for his generosity and continuous support during this time.

18
Aminoacyl-tRNAs: deciphering and defining the genetic message

Alexandre Ambrogelly, Juan Carlos Salazar, Kelly Sheppard, Carla Polycarpo, Hiroyuki Oshikane, Yuko Nakamura, Shuya Fukai, Osamu Nureki and Dieter Söll

Introduction

Aminoacyl-tRNAs are substrates for translation and are crucial in defining how the genetic code is interpreted as amino acids. The ancient process of aminoacyl-tRNA synthesis (Ibba & Söll, 2000; Woese et al., 2000) precisely matches an amino acid with its cognate tRNA species, which contains the corresponding anticodon. This is primarily achieved by the direct attachment of an amino acid to the corresponding tRNA by an aminoacyl-tRNA synthetase (see the end of this chapter for abbreviations and other notes). The presence of a set of 20 of these enzymes (one for each canonical amino acid), verified in *Escherichia coli* or in the eukaryotic cytoplasm, was taken as confirmation of the prediction of this fact by Crick's Adaptor Hypothesis (Crick, 1958). This led to the view that aa-tRNA synthesis was the almost exclusive domain of the aaRSs. The inadequacies of this view became obvious with the publication of the genome sequence of *Methanocaldococcus jannaschii*, which contained identifiable orthologs for only 16 of the 20 known aaRSs (Ibba & Söll, 2001). We now know that this apparent shortfall can be attributed to the existence of previously uncharacterized aaRSs and of novel pathways for aa-tRNA synthesis. Moreover, studies showed that some of these unexpected routes of aa-tRNA synthesis are not confined to Archaea, but are instead found in a wide variety of organisms (Ibba & Söll, 2001). Thus, investigations with Archaea had a major role in uncovering the unexpected diversity found in aa-tRNA synthesis in nature.

Aminoacyl-tRNA synthesis proceeds by two pathways

Two routes to aa-tRNA exist: direct esterification of the amino acid onto the 3'-terminus of the tRNA, or a tRNA-dependent amino acid conversion of a non-cognate amino acid charged to tRNA (Fig. 18.1). The latter pathway is mainly used for the synthesis of the amide aa-tRNA species Gln-tRNA and Asn-tRNA. Present in most Bacteria, the Gln-tRNA biosynthesis route takes advantage of a non-discriminating GluRS to synthesize Glu-tRNA^{Gln} and a tRNA-dependent amidotransferase (Glu-AdT) that converts the misacylated aa-tRNA by amidation to the cognate Gln-tRNA^{Gln} (Curnow et al., 1997). A similar pathway exists to form Asn-tRNA^{Asn} in many Archaea and Bacteria using a non-discriminating AspRS and a tRNA-dependent amidotransferase (Asp-AdT) (Curnow et al., 1996; Min et al., 2002). A synopsis of aa-tRNA formation can be found in more detailed reviews (Ibba & Söll, 2000). Below are two examples of discoveries in Archaea that were very important for the elucidation of processes occurring in Archaea and in other domains.

Existence of two lysyl-tRNA synthetases in nature

One of the most surprising outcomes of the sequencing of the *M. jannaschii* genome was the apparent lack of the expected class II LysRS (Bult et al., 1996).

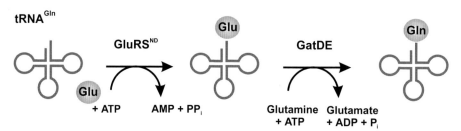

Fig. 18.1 Gln-tRNA^{Gln} synthesis in Archaea. The two required steps for synthesis of Gln-tRNA^{Gln} are catalyzed by a non-discriminating GluRS (GluRSND) and GatDE.

Purification to homogeneity of the protein responsible for Lys-tRNA formation in a *Methanococcus maripaludis* cell-free extract revealed a LysRS with all the features of a class I synthetase (Ibba et al., 1997a). LysRS offers the first and still unique example of functional convergence between the two evolutionary and structurally unrelated classes of aaRSs (Terada et al., 2002). While it was initially thought that this novel class I LysRS might be restricted to a small subset of archaeal organisms, the ever increasing number of sequenced genomes revealed the presence of this second LysRS in the majority of Archaea as well as in a significant number of Bacteria (Ibba et al., 1997b). Notably, both class I and class II LysRSs were found to coexist rarely; they have been found so far only in the Methanosarcinaceae and in *Bacillus cereus*. In these organisms the concomitant action of the class I and class II LysRSs is required for the charging of interesting new tRNA species involved in either suppression (Polycarpo et al., 2003; see below) or growth-phase-specific protein synthesis events (Ataide et al., 2005). The ability of the class I and class II LysRS proteins to form a ternary complex with a single tRNA molecule may be a remnant of the evolutionary process that generated the division of the aaRSs into two classes of approximately ten members each (Ribas de Pouplana & Schimmel, 2001).

tRNA-dependent formation of amide aminoacyl-tRNAs

Gln-tRNA and Asn-tRNA synthesis in most Bacteria, Archaea, and chloroplasts proceeds by the tRNA-dependent amidation pathway (reviewed in Ibba & Söll, 2004). Two tRNA-dependent amidotransferases have been discovered (Tumbula et al., 2000). The first is a heterotrimeric enzyme (GatCAB) that is capable of converting Glu-tRNA^{Gln} into Gln-tRNA^{Gln}, as well as converting Asp-tRNA^{Asn} into Asn-tRNA^{Asn}. This enzyme is found in Bacteria, chloroplasts, and Archaea. The second amidotransferase is a strictly archaeal heterodimeric enzyme (GatDE) responsible for Gln-tRNA^{Gln} formation only. Both enzymes contain subunits with activities not normally associated with aa-tRNA synthesis. GatA is an amidase, while GatD resembles a type I asparaginase. While their activities make sense in relation to their transamidation mechanisms (see below), it was unexpected to find that nature co-opted amino acid metabolizing enzymes to the process of protein synthesis. A further example of the unforeseen interrelationship of the processes of protein synthesis and amino acid formation is the finding that a large fraction of Bacteria as well as the archaeon *Sulfolobus* form asparagine solely by the tRNA-dependent conversion of aspartate to asparagine by GatCAB (Min et al., 2002).

Since the biochemical and structural analysis of these original enzymes is much more advanced for the Archaea-specific GatDE enzyme, we discuss in more detail below our first understanding of the *Methanothermobacter thermautotrophicus* Glu-tRNA^{Gln} amidotransferase GatDE.

The mechanism of Gln-tRNA formation

The strictly archaeal heterodimeric Glu-AdT GatDE enzyme is responsible for Gln-tRNA synthesis (Tumbula et al., 2000). Like the other heterotrimeric GatCAB amidotransferase, the GatDE enzyme catalyzes three separate reactions: (i) the enzyme hydrolyzes ATP to form and activate the γ-carboxyl group of the tRNA-bound glutamate (Wilcox, 1969); (ii) the enzyme hydrolyzes glutamine or asparagine to generate ammonia; (iii) the enzyme uses the liberated ammonia to amidate the activated Glu-tRNA^{Gln} to form Gln-tRNA^{Gln}. As expected from the similarity to L-asparaginase, the GatD subunit was expected to catalyze glutamine hydrolysis; this was recently verified (Feng et al., 2005). GatE (together with GatB) belongs to an isolated protein family (Tumbula et al., 2000); this subunit has kinase activity and activates Glu-tRNA^{Gln} by phosphorylation, resulting in ATP cleavage and generation of γ-phosphoryl-Glu-tRNA^{Gln} (P-Glu-tRNA^{Gln}) (Feng et al., 2005). GatE possesses an insertion domain (lacking in GatB) that structurally resembles an insertion domain found in bacterial AspRS proteins, and this may be the reason why GatDE functions only as a Glu-AdT (Schmitt et al., 2005). The recently solved structure of the *M. thermoautotrophicus* GatDE complexed with tRNA^{Gln} (Oshikane et al., in press) (Plate 18.1) shows that GatE possesses a cradle domain in which tRNA binds. The amino acids that make up the cradle's base are highly conserved between GatB and GatE, suggesting their importance in the kinase and transamidase activities of the AdTs (Schmitt et al., 2005). Biochemical studies implicated the *M. thermoautotrophicus* GatE as the subunit involved in tRNA binding and formation of the activated intermediate (Feng et al., 2005). With the use of pure enzymes the P-Glu-tRNA^{Gln} intermediate could not be trapped; however, this was achieved earlier using a crude *Bacillus subtilis* extract (Wilcox, 1969).

Similarity of sequence and structure, as well as biochemical evidence, point to GatD being related to L-asparaginases (Tumbula et al., 2000; Feng et al., 2005; Schmitt et al., 2005). L-Asparaginases catalyze the conversion of Asn or Gln to Asp or Glu while liberating ammonia. Like other L-asparaginases, GatD dimerizes (Schmitt et al., 2005). The *Pyrococcus abyssi* GatD homodimer and the homologous type I L-asparaginase are remarkably superimposable (Schmitt et al., 2005). The four catalytically important residues in L-asparaginases are conserved in GatD (T101, T177, D178, and K254 in *M. thermoautotrophicus* GatD) (Tumbula et al., 2000; Feng et al., 2005). Mutational studies in the *M. thermoautotrophicus* GatD have confirmed the importance of these residues in the glutaminase activity of GatD (Feng et al., 2005).

As the two subunits carry out different functions, it is clear that the ammonia liberated by GatD (asparaginase subunit) needs to be transported to the active site of the GatE (amidotransferase subunit) (Schmitt et al., 2005). In the structure (Plate 18.1) a molecular tunnel connecting the two subunits is discernable. This structure would also explain why exogenous ammonia is such a poor substrate for transamidation (Feng et al., 2005). All the observations made with the GatDE amidotransferase should be easily extended to unraveling the mechanism of the heterotrimeric GatCAB enzyme.

Why GatDE evolved in Archaea remains a mystery. It is known is that in archaeal proteins, Asn and especially Gln are underrepresented relative to their levels in proteins from mesophilic eubacteria and eukaryotes (Michelitsch & Weissman, 2000). The underusage of Gln in archaeal proteins may provide a partial explanation as to why GlnRS is not encoded in any archaeal genome. Differences between identity elements in tRNA^{Gln} from archaeal and eukaryotic or bacterial sources may also explain why *glnS* did not transfer to Archaea (Tumbula et al., 2000); for instance, *E. coli* and yeast GlnRS enzymes were unable to glutaminylate *M. thermautotrophicus* tRNA^{Gln}. Yet when the *E. coli* identity elements were transplanted into *M. thermautrophicus* tRNA^{Gln} the resulting tRNA could be charged by *E. coli* GlnRS (Tumbula et al., 2000). Differences in amino acid metabolism might be another reason why GlnRS is not found in Archaea. Much work, especially genetic replacement experiments, remains to be done to gain a complete understanding of the role of the unique archaeal GatDE amidotransferase in cell metabolism and protein synthesis.

Biosynthesis and incorporation of cysteine into proteins

Cysteine is one of the 20 canonical amino acids found in proteins. CysRS is a class I aaRS responsible for the formation of Cys-tRNA. While cysteine is present in archaeal proteins to a similar extent as found in other organisms, the *M. jannaschii*, *M. thermautotrophicus*, and *Methanopyrus kandleri* genomes do not encode a canonical CysRS (Bult et al., 1996; Smith et al., 1997; Slesarev et al., 1998). This left open the question as to how Cys-tRNA^{Cys} might be synthesized in these organisms. The solution to this puzzle remained elusive for more than a decade (Ambrogelly et al., 2004; Ruan & Söll, 2006). A combination of anaerobic biochemical purification and proteomic analysis of the chromatographic fractions finally led to success (Sauerwald et al., 2005). Starting from a cell-free *M. jannaschii* extract and employing a rigorous identification of aa-tRNA by acid gel electrophoresis and Northern blot, two proteins and two low-molecular weight factors were discovered to be essential for Cys-tRNA^{Cys} formation (Fig. 18.2a).

The first protein, an unusual aaRS homologous to the PheRS α-subunit, selectively acylates tRNA^{Cys} with phosphoserine (Sep). The second enzyme subsequently converts Sep-tRNA^{Cys} to Cys-tRNA^{Cys} in the presence of PLP and a still unidentified sulfur donor. The two novel enzymes were named phosphoseryl-tRNA synthetase (SepRS, encoded by *sepS*) and Sep-tRNA:Cys-tRNA synthase (SepCysS), respectively. Analysis of the phylogenetic repartition of these enzymes revealed their presence not only in *M. jannaschii*, *M. kandleri*, and *M. thermautotrophicus*

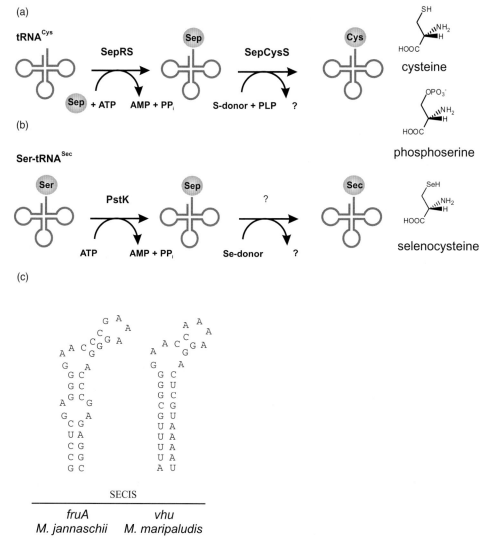

Fig. 18.2 Parallel between the pathways for Cys-tRNA^{Cys} synthesis (a) and the hypothetical route for Sec-tRNA^{Sec} formation (b) in *M. jannaschii*, *M. kandleri*, and *M. thermautotrophicus*. Structures of the SECIS element present in the *fruA* and *vhu* mRNAs of *M. jannaschii* and *M. maripaludis*, respectively, for recoding of the in-frame UGA stop codon as selenocysteine (c).

but also in genomes of other Archaea such as *M. maripaludis*, *Methanococcoides burtoni*, *Archaeoglobus fulgidus*, and the *Methanosarcinaceae*, organisms which already possess a canonical class I CysRS (Sauerwald et al., 2005). The redundancy of the Cys-tRNA^{Cys} biosynthetic route in *M. maripaludis* was foreseen, since the inactivation of the *cysS* gene encoding the canonical CysRS had no effect on cell growth in different media (Stathopoulos et al., 2001). The dispensability of the canonical CysRS strongly suggests that the SepRS/SepCysS pathway is fully capable of synthesizing Cys-tRNA^{Cys} *in vivo*. In contrast to the *cysS* deletion stain, a *M. maripaludis* strain with a *sepS* deletion displayed cysteine auxotrophy (Sauerwald et al., 2005). This conditional lethality establishes that in the presence of exogenous cysteine, CysRS and the SepRS/SepCysS pathway are operative and functionally equivalent. The auxotrophic nature of the *M. maripaludis* SepS knockout strain also demonstrates that the indirect route to Cys-tRNA^{Cys} is the sole source of free cellular cysteine in this organism (Sauerwald et al., 2005). Reliance on a tRNA-dependent route for cysteine biosynthesis is reminiscent of what has been shown for asparagine in many Bacteria (Min et al., 2002). Amino acid release from Cys-tRNA^{Cys} may require the presence of an enzyme selectively hydrolyzing this aa-tRNA. This is plausible as aa-tRNA hydrolases with different substrate specificities are known to exist throughout the living world (reviewed in Ibba & Söll, 2004). Furthermore, protein turnover in the cell would be expected to contribute significantly more to the cellular cysteine pool in an organism lacking CysRS, as this enzyme would be required to activate free cysteine for further protein synthesis.

Unlike for *M. maripaludis*, the presence of free cysteine in *M. jannaschii* is not established and a number of facts seem to indicate that there is none. First, paralogs of genes involved in bacterial or eukaryal cysteine biosynthesis that are observed in the genomes of other Archaea are absent in *M. jannaschii*, *M. thermautotrophicus*, and *M. kandleri* (Ambrogelly et al., 2004). Second, is cysteine required for other cellular functions in methanogens? In bacteria, cysteine is a general sulfur source for a number of metabolic processes. Formation of 2-thiouridine and 4-thiouridine present in tRNAs relies on activation of the sulfur group of cysteine by a cysteine desulfurase enzyme (IscS) and transfer of the sulfur atom to an iron cluster protein before its use by downstream enzymes (Lauhon, 2002). While *E. coli* contains three cysteine desulfurase activities (belonging to the NIF, ISC, and SUF pathways) to meet the different cellular metabolic needs, no similar proteins can be identified in the genomes of *M. jannaschii*, *M. thermautotrophicus*, and *M. kandleri* (Bult et al., 1996; Smith et al., 1997; Slesarev et al., 1998). The presence of thionucleotides in *M. jannaschii* tRNA (McCloskey et al., 2001) and of many iron cluster proteins (White, 2003b) may suggest that while sulfur gets activated and transferred to metabolites, the main sulfur donor might not be cysteine but an inorganic source instead. Fixation of sulfur from inorganic sulfide or sulfite was also shown to allow synthesis of other metabolites in *M. jannaschii*, such as homocysteine and phosphosulfolactate, key intermediates in methionine and coenzyme M biosynthesis, respectively (Graham et al., 2002a; White, 2003b). Cysteine and cysteine derivatives (e.g. glutathione) act as a powerful redox buffer. In contrast, methanogens might not need free cellular cysteine or glutathione to maintain a proper redox environment. Coenzyme M, the terminal methyl carrier in methanogenesis, was shown to be present in high levels in the cytoplasm and could potentially carry out this function in *M. jannaschii* (Balch & Wolfe, 1976; Sauerwald et al., 2005). Third, free cellular cysteine may even be toxic for the *M. jannaschii* cells. *M. jannaschii* ProRS was shown *in vitro* to mischarge very efficiently cysteine onto tRNA^{Pro}, hence potentially compromising translational fidelity and subsequently cell viability (Ambrogelly et al., 2004). While organisms from the bacterial and eukaryotic domains have dealt with the danger of a promiscuous ProRS by acquiring an editing mechanism able to clear mischarged Cys-tRNA^{Pro} (An & Musier-Forsyth, 2004; Ruan & Söll, 2006), no such mechanism can be identified in Archaea and in *M. jannaschii* in particular. The absence of selective pressure to acquire a Cys-tRNA^{Pro} editing mechanism could be explained by the absence of formation of the mischarged tRNA species due to a low cysteine:proline concentration ratio in the cell. Maintaining a low cellular cysteine concentration is then crucial. A cysteine desulfidase encoded by the MJ1025 open reading frame was recently shown to degrade cysteine into pyruvate, ammonia, and sulfide (Tchong et al., 2005). This protein would then serve the dual purpose of recycling the free cysteine released in the cell due to natural deacylation of Cys-tRNA^{Cys} and/or protein proteolysis, and maintain translation accuracy. Taken together these facts suggest that in *M. jannaschii*, *M. thermautotrophicus*, and *M. kandleri*, the SepRS/SepCysS pathway is only present for Cys-tRNA^{Cys} formation, as free cysteine may not be required.

All the other Archaea possess a presumably laterally transferred canonical CysRS and therefore need a means to produce free cellular cysteine. Most of these organisms have homologs to proteins known to be part of the cysteine biosynthesis metabolism in Bacteria and Eukarya (Ambrogelly et al., 2004). However, the actual involvement of these proteins in cysteine biosynthesis in Archaea has not yet been experimentally established. The *Methanosarcinaceae* are unique, as they were shown to have acquired from bacteria *cysE* and *cysK*, the two genes required for cysteine biosynthesis (Ambrogelly et al., 2004). Since these organisms possess a SepRS/SepCysS pathway, a CysRS, and a tRNA independent cysteine biosynthetic route, both tRNA cysteinylation and cysteine production are duplicated. Methanosarcinaceae genetics might in the future help us to understand the physiological role of each of these components.

The discovery of the SepRS/SepCysS pathway gives possible keys to a better understanding of contemporary phylogenetic repartition of tRNA cysteinylation enzymes and possibly how it came to be throughout evolution. CysRS is ubiquitous in Bacteria and Eukarya and may have been transferred to the Archaea in several instances from bacterial progenitors (Woese et al., 2000; Ambrogelly et al., 2004). Phylogenetic repartition of bacterial CysRS is consistent with accepted bacterial taxonomy deduced from 16S RNA sequence. It is therefore reasonable to think that CysRS is a bacterial invention that was already present in the common bacterial ancestor. The SepRS/SepCysS pathway may be the ancient route to either or both Cys-tRNA^{Cys} and cysteine that is still present in a few methanogens. Replacement of the SepRS/SepCysS pathway by a CysRS in most Archaea was presumably correlated with the establishment or acquisition of tRNA-independent cysteine biosynthesis. This process may have been driven by the higher efficiency of direct aminoacylation and by better regulation of gene expression offered by separate Cys-tRNA and cysteine biosynthetic routes. The coexistence of independent routes for cysteine biosynthesis and Cys-tRNA formation in some of the Archaea might correspond to a transient evolutionary stage before the final displacement of the SepRS/SepCysS pathway, possibly due to a later CysRS transfer.

Selenocysteine in Archaea

Selenocysteine is the twenty-first amino acid co-translationally inserted into proteins. Its biosynthesis and specific incorporation in response to an in-frame UGA codon has been worked out genetically and biochemically over the past 15 years in Bacteria (reviewed in Böck et al., 2004). However, much less is known about this process in eukaryotes and Archaea. Selenoproteins have been identified in organisms from all three domains of life (Hatfield & Gladyshev, 2002; Böck et al., 2004). While they are ubiquitous in Bacteria and present in most Eukarya (except yeast), their known occurrence to date in Archaea is restricted to only three genomes (*M. jannaschii*, *M. maripaludis*, and *M. kandleri*; Bult et al., 1996; Hendrickson et al., 2004; Slesarev et al., 1998). Sec insertion into proteins is a multistep process as demonstrated in *E. coli* (Böck et al., 2004). A specific Sec suppressor tRNA (tRNA^{Sec}) is first acylated by SerRS with serine to form Ser-tRNA^{Sec}. This misacylated tRNA species will then be converted to Sec-tRNA^{Sec} by the PLP-dependent enzyme selenocysteine synthase (SelA), using as selenium donor selephosphate, generated by selenophosphate synthetase (SelD) (reviewed in Böck et al., 2004). The incorporation of Sec-tRNA^{Sec} during translation requires a recoding of the UGA codon from chain termination to Sec. This is accomplished in *E. coli* by the new elongation factor SelB specific solely for the delivery of Sec-tRNA^{Sec} to the ribosome at the suppression site. Finally, the particular UGA codon in the mRNA is indicated to the translation machinery by a stem/loop structure named selenocysteine insertion sequence (SECIS).

Most steps in this process are still not understood in Archaea and in eukaryotes. As in Bacteria, formation of Ser-tRNA^{Sec} by SerRS occurs in Archaea (Fig. 18.2b) (Bilokapic et al., 2004; Kaiser et al., 2005). However, the conversion of Ser-tRNA^{Sec} to Sec-tRNA^{Sec} in Archaea might be different, as the *M. jannaschii* protein MJ0158, a putative SelA ortholog, is unable to produce Sec-tRNA^{Sec} *in vitro* (Kaiser et al., 2005). Conversion of Ser-tRNA^{Sec} into Sec-tRNA^{Sec} might need an additional step. Enzymatic conversion of Ser-tRNA to Sep-tRNA has been documented in eukaryotic cell extract since the 1970s (Maenpaa & Bernfield, 1970) and later the tRNA^{Ser} species was characterized as tRNA^{Sec} (reviewed in Böck et al., 2004). However, the enzyme that actively catalyzes the conversion, phosphoseryl-tRNA$^{Ser(Sec)}$ kinase (PSTK) has been identified in mouse (Carlson et al., 2004). Based on this result the *M. jannaschii* ortholog (MJ1538) was shown to carry out the same reaction (Kaiser et al., 2005) (Fig. 18.2b). The presence of PSTK in Archaea and eukaryotes suggests that Sep-tRNA^{Sec} may be an intermediate in Sec synthesis (Carlson et al., 2004). However, the enzyme that catalyzes the conversion of Sep-tRNA^{Sec} into Sec-tRNA^{Sec} remains to be identified. The demonstration (described above) that cysteine biosynthesis in *M. jannaschii* proceeds via Sep-tRNA^{Cys} and a subsequent Sep → Cys conversion by a PLP-dependent Sep-tRNA:Cys-tRNA synthase suggests that the same (or a similar) enzyme may carry out the Sep → Sec transformation, the missing step in Sec formation (Sauerwald et al., 2005). The striking similarity between the chemistry underlying the conversion of Sep to either Cys or Sec leaves open the possibility that SepCysS might be able to catalyze both reactions in the presence of the appropriate sulfur or selenium donors. Cooperative binding of Sep-tRNA^{Cys} and the sulfur source on one hand and Sep-tRNA^{Sec} and the selenium donor on the other hand would ensure selectivity of the reaction.

Our understanding of the UGA recoding process in Archaea is still incomplete. In Archaea and in Eukarya an mRNA structure similar to bacterial SECIS is present (Fig. 18.2c), but this element is located in the 3' noncoding region of the mRNA (Böck et al., 2004). The annotated SelB ortholog in *M. maripaludis* is essential for selenoprotein formation as shown by the growth properties of a strain in which SelB was inactivated (Rother et al., 2003). However, the archaeal SelB protein does not have all the functions of the bacterial one. Bacterial SelB has a 272 amino acid carboxy-terminal domain that binds the SECIS element (Böck et al., 2004). The archaeal and eukaryotic SelB orthologs lack this domain and are unable to bind their homologous SECIS elements (Böck et al., 2004). Thus, other components essential for the archaeal Sec recoding event are still unidentified. Further research in Archaea may be guided by the knowledge that the SECIS Binding Protein 2 (reviewed in Böck et al., 2004) and the ribosomal protein L30 (Chavatte et al., 2005) were shown to be involved in eukaryotic selenoprotein formation.

Pyrrolysine: an unexpected amino acid discovered in *Methanosarcina*

In the methanogenic Archaea *Methanosarcina barkeri*, *M. acetivorans*, *M. mazei*, and *M. thermophila*, as well as in the psychrophilic organism *Methanococcoides burtonii*, an in-frame UAG codon is located in the genes encoding mono-, di-, and tri-methylamine transferases (Mtmb, MtbB, and MttB), enzymes that are part of a methylamine-dependent methanogenic pathway in these organisms (Srinivasan et al., 2002; Goodchild et al., 2004). This UAG codon is read as a sense codon in the translation of presumably all these genes. Purification of the native MtmB enzyme from *M. barkeri* and subsequent determination of the enzyme's crystal structure led to the exciting finding that the UAG codon is translated with a new amino acid, the lysine derivative pyrrolysine (Hao et al., 2002) (Fig. 18.3a). The *M. barkeri* genome encodes an unusual suppressor tRNA (pylT) and a class II aminoacyl-tRNA synthetase like enzyme, PylS (Srinivasan et al., 2002), which presumably are involved in the UAG suppression mechanism.

In contrast to Sec-tRNA^{Sec} synthesis, where Sec is made via tRNA-dependent chemical modification of serine, Pyl-tRNA^{Pyl} is generated by direct charging of Pyl onto tRNA^{Pyl} in the presence of pyrrolysyl-tRNA synthetase (Fig. 18.3a), the product of the *pylS* gene (Blight et al., 2004; Polycarpo et al., 2004). PylRS, which is homologous to PheRS and LysRS, is a proposed class II enzyme with its classical three degenerate motifs. While the carboxy-terminus of the enzyme displays high sequence similarity to the aaRS class II catalytic core, the amino terminal part of the enzyme (anticipated to be an anticodon binding domain) is not homologous to known proteins and does not contain an RNA binding motif. All archaeal PylRS proteins contain this amino-terminal domain; yet the only bacterial PylRS homolog, found in *Desulfitobacterium hafniense*, is devoid of it (Srinivasan et al., 2002). *D. hafniense* also contains a pylT tRNA (Srinivasan et al., 2002). Cloning and production of the recombinant *M. barkeri*, *M. thermophila*, *M. acetivorans*, and *D. hafniense* PylRS enzymes shows

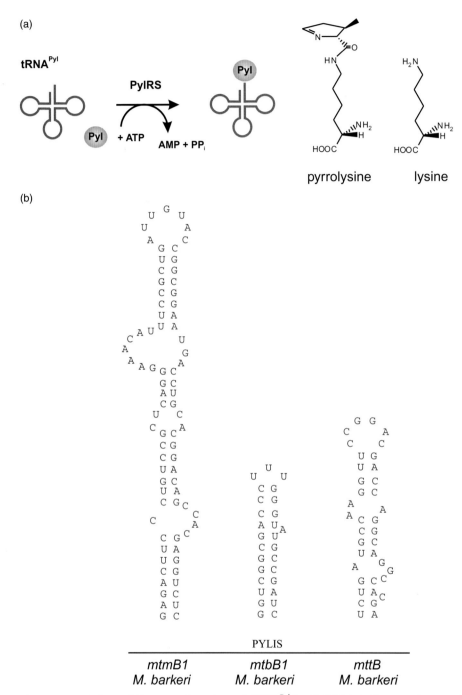

Fig. 18.3 Schematic representation of the pyrrolyzation of tRNA^{Pyl} by PylRS (a). Proposed secondary structure of a PYLIS element in *M. barkeri* mRNA (b).

that all these enzymes are able to activate and transfer selectively pyrrolysine onto their cognate tRNA^{Pyl} transcripts (C. Polycarpo & A. Ambrogelly, unpublished observations). Current research on Pyl-tRNA^{Pyl} formation is limited by the scarcity of Pyl, whose structure (Fig. 18.3a) was elucidated by a low-yield chemical synthesis (Blight et al., 2004; Polycarpo et al., 2004).

tRNA^{Pyl} has an unusual secondary structure reminiscent of the one seen for the bovine mitochondrial tRNA^{Ser} (Hayashi et al., 1998); it folds into a clover leaf structure but its anticodon stem is longer by one base pair, the variable loop with only three nucleotides is shorter, and only one nucleotide is present in the hinge region between the acceptor stem and D-stem. Only two modified nucleosides have been detected in native tRNA^{Pyl} (Polycarpo et al., 2004). The unusual secondary and likely tertiary structure of tRNA^{Pyl} together with its low level of modifications should ensure the flexibility needed for optimal Pyl charging and decoding of the UAG codon.

Research on the mechanism that leads to Pyl incorporation at a UAG codon in Mtmb, MtbB, and MttB is still in its infancy. In the absence of data showing that suppression occurs randomly at UAG codons, we assume that the Pyl incorporation in the three methylamine methyltransferase enzymes requires an active recoding process. It remains to be determined whether Pyl will be inserted via PYLIS stem/loop structures (Fig. 18.3b), found in the corresponding mRNAs analogous to the ones seen for Sec incorporation (Ibba & Söll, 2004; Namy et al., 2004; Theobald-Dietrich et al., 2005), or by other possible mechanisms (Zhang et al., 2005).

Outlook

Studies on aaRSs in Archaea have led to an exciting expansion of our knowledge of how the genetic code is translated *in vivo*. They also further shed light on the process of translation in Bacteria and Eukarya. Since some of the new aaRSs may have preceded pathways currently in use in eukaryotic cells (e.g. cysteine and Cys-tRNA^{Cys} formation), further studies with Archaea may uncover more surprises. The exciting discovery of novel aminoacyl-tRNA synthetases specific for noncanonical amino acids draws additional attention to the very inspired studies on the expansion of the genetic code. The use of designed orthogonal aminoacyl-tRNA synthetases/tRNA pairs (Furter, 1998) has led to exciting achievements in the incorporation of unusual amino acids into proteins (summarized in Wang & Schultz, 2004). While such studies have involved the use of *in vitro* modified archaeal LysRS (Anderson et al., 2004) and TyrRS (Zhang et al., 2005) as orthogonal enzymes in bacteria and mammalian systems, it is likely that the "naturally evolved" PylRS:tRNA^{Pyl} or SepRS:tRNA^{Cys} pairs may prove far more efficient. Additional studies on PylRS or SepRS mediated recoding in Archaea may be beneficial for devising more efficient systems for incorporation of non-natural amino acids in other systems (Köhrer et al., 2004).

Acknowledgments

Work in the authors' laboratories was supported by National Institute of General Medical Sciences (to DS), the Department of Energy (to DS), and the Ministry of Education, Culture, Sports, Science and Technology (to ON).

Notes

Abbreviations: aaRS, aminoacyl-tRNA synthetase; aa-tRNA, aminoacyl-tRNA; Pyl, pyrrolysine; Sep, *O*-phosphoserine; Sec, selenocysteine. Specific aaRSs are denoted by their three-letter amino acid designation, e.g. GlnRS for glutaminyl-tRNA synthetase. Glutamine tRNA or tRNA^{Gln} denote uncharged tRNA specific for glutamine; glutaminyl-tRNA or Gln-tRNA denote tRNA acylated with glutamine. In a misacylated tRNA the amino acid specificity does not match the tRNA anticodon, e.g. Glu-tRNA^{Gln}.

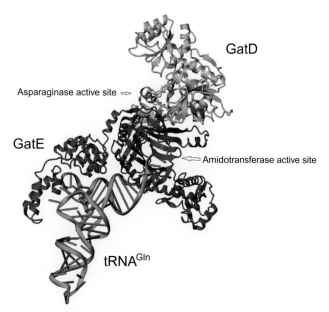

Plate 18.1 Crystal structure of the *M. thermautotrophicus* GatDE complexed with tRNA^{Gln}. The dimer of the heterodimeric GatDE (thus forming a heterotetramer) binds two tRNA molecules. The asparaginase active site of GatD and the kinase/amidotransferase active site of GatE are distantly separated, and are connected with a "molecular tunnel."

19
Translational mechanisms and protein synthesis

Paola Londei

Introduction

Translation is the last step of the gene expression pathway, whereby the genetic message carried over from the DNA in the form of a messenger RNA (mRNA) is finally converted into the final product, a polypeptide chain. Given its fundamental importance for all cells, translation is well conserved in its basic aspects across the three domains of life. Nevertheless, it has been known for a long time that the protein synthetic machineries of the Bacteria and the Eukarya present clear-cut differences in composition and complexity. Until recently, conventional wisdom had it that there were two different versions of the translational apparatus: the simple, streamlined one possessed by the Bacteria and the more complex one found in eukaryotic cells. This dichotomy seemed to be the logical result of the different organization and lifestyles of eukaryotic and prokaryotic cells. The "simpler" Bacteria, whose basic evolutionary strategy consisted of maximizing the velocity of growth and multiplication, had gene expression machineries made of fewer and, often, smaller components. The "complex" eukaryotic cells, many of which live in a highly integrated environment, had to add much sophistication to the basic design of their gene expression apparatus.

The discovery of the third domain of life, the Archaea, has challenged in many ways the classical textbook dichotomy between prokaryotes and eukaryotes. As far as cellular organization is concerned, all known Archaea are unicellular prokaryotes. However, a host of phylogenetic, molecular, and genomic studies have now shown that the Archaea are clearly separated from the other prokaryotic kingdom, the Bacteria. Moreover, the rooting of the universal tree of life has revealed that the Archaea have shared a tract of evolutionary path with the eukaryotes before branching out as a separate domain (Iwabe et al., 1989). This is reflected in the presence, at the molecular level, of many intriguing similarities between the Eukarya and the Archaea, regarding especially the structure and composition of the cellular machineries for gene expression. In recent years, for instance, an elegant series of experimental studies have shown that the archaeal transcriptional apparatus can be regarded as a "basic" or simplified version of the eukaryal one. The structure of the archaeal translational apparatus also exhibits an unexpected complexity, including several components, especially translational factors, that are found in Eukarya but have no counterparts in Bacteria (Bell & Jackson, 1998). Therefore, the study of the translational machinery in Archaea is not only interesting in its own right, but promises to yield new and exciting insights into the evolutionary history of the protein-synthetic mechanism. This chapter presents a survey of the most recent advances in the understanding of archaeal translation, focusing especially on the data that highlight the novel and Eukarya-like features of the archaeal translational apparatus.

Archaeal ribosomes

It has been known for a long time that archaeal ribosomes are composed of subunits that sediment at 30S and 50S and contain 16S, 23S, and 5S rRNAs, thus resembling their bacterial counterparts. However, the existence of certain similarities between archaeal and eukaryal ribosomes was suggested over two decades ago. Compositional analyses of the ribosomes of various archaeal species revealed that several of them were protein-richer than the bacterial particles, especially as far as the small subunit was concerned. Moreover, the ribosomes of certain sulfur-dependent thermophiles were shown to have morphological characteristics reminiscent of the eukaryal particle (Amils et al., 1993, and references therein).

Nowadays, the composition of archaeal ribosomes can be analyzed extensively taking advantage of the over 20 complete genomic sequences representative of archaeal diversity. Genomic studies have essentially confirmed the early surmise that archaeal ribosomes, although being closer in size to bacterial ones, have specific affinities with their eukaryal counterparts. A survey of the composition of archaeal ribosomes is presented in Table 19.1.

Archaeal ribosomal RNAs generally resemble bacterial ones in both size and structure. Most small subunit (SSU) rRNAs are 1400–1500 nucleotides in size; the smallest one, with 1344 nucleotides, is found in the parasitic species *Nanoarchaeum equitans* (Table 19.1). The large subunit (LSU) rRNAs have sizes comprising between 2850 and 3100 nucleotides, while 5S rRNAs have 119–132 nucleotides. Interestingly, the genome of the crenarcheon *Aeropyrum pernix* is unique in containing a 167 nucleotide long homolog of 5.8S rRNA, an rRNA specific to eukaryal LSU. However, there are no experimental observations available to confirm the presence of a separate 5.8S rRNA in *A. pernix* large ribosomal subunits. Both LSU and SSU archaeal rRNAs may contain introns (Lykke-Andersen et al., 1997).

Unlike the majority of Bacteria, most Archaea are endowed with single copies of the rRNA-coding genes. Exceptions are found among the methanogens and halophiles, which may have up to three copies of 16S and 23S RNA genes and up to four copies of 5S RNA genes (Table 19.1). The 16S and 23S rRNA genes are adjacent and co-transcribed in most Archaea; the exceptions are the Thermoplasmales and *N. equitans*, where these genes are located far apart in the genome and are transcribed independently. Also, the Archaea are unlike the Bacteria (and like the Eukarya) in having 5S RNA genes normally unlinked from the genes encoding the 16S and 23S rRNAs. Only the halophiles and some methanogens have Bacteria-like rRNA operons, including the 5S RNA genes and often also tRNA genes inserted as spacers between the rRNA genes. However, extra copies of the 5S RNA genes may be found unlinked from the rRNA operons in several species (e.g. *Methanococcus maripaludis*).

Where the similarity of archaeal ribosomes to the eukaryal ones becomes most apparent is in their complement of proteins. Archaeal genomes include a total of 68 r-protein families, 28 belonging to the SSU and 40 to the LSU (Lecompte et al., 2002). Thirty-four of these (15 in the SSU and 19 in the LSU) are universal proteins, having identifiable homologs in the ribosomes of all three domains of life. Another 33 (13 SSU and 20 LSU) are specifically shared by archaeal and eukaryal ribosomes and are not found in Bacteria, while no r-proteins are shared by the Bacteria and the Archaea to the exclusion of Eukarya. Only one r-protein (LXa in the LSU) is unique to the Archaea (Lecompte et al., 2002). These data show that, as far as their protein composition is concerned, archaeal ribosomes can be envisaged as a somewhat smaller version of the eukaryotic particles, which contain 78 r-protein families, including all of the 68 found in Archaea plus another ten Eukarya-specific ones. By contrast, bacterial ribosomes have altogether 57 r-protein families, 23 of which are exclusive of the Bacteria (Lecompte et al., 2002).

Interestingly, the protein composition of archaeal ribosomes is not constant over the domain, but presents a certain heterogeneity that correlates with the branching order of the various species (Lecompte et al., 2002). Thus, the deep-branching Crenarcheota tend to have protein-richer ribosomes than the Euryarchaeota (Table 19.1). The protein-richest ribosome is that of the crenarcheon *A. pernix*, which is endowed with the full complement of 68 r-proteins. On the other hand, the ribosomes of the late-branching halobacteria and Thermoplasmales have only 58 proteins, thus coming closer to the bacterial size. This evolutionary trend toward a "lighter" ribosome is not observed in the Bacteria or the Eukarya, where the protein composition of ribosomes remains essentially

Table 19.1 Ribosome composition in Archaea.

	23S	16S	5S	SSU prot.	LSU prot.
Crenarcheota					
*A. pernix**	4413†	1423	19/132	28	39
P. aerophylum	3024	2210‡	130	28	38
S. solfataricus	3049	1496	121	28	37
S. tokodaii	3012	1445	125	28	37
Euryarchaeota					
Thermococcales					
P. furiosus	3048	1446	121/125	25	37
P. horikoshii	3857	1494	121 (2)	25	37
P. abyssi	3017	1502	121 (2)	25	37
T. kodakaraensis	3028	1497	125 (2)	25	37
Methanopyrales					
M. kandleri	3097	1511	132	25	37
Methanobacteriales					
M. thermoautotrophicus	3028/3034	1478 (2)	126/128	25	36
Methanococcales					
M. jannaschii	2889/2948	1474/1477	119 (2)	25	36
M. maripaludis	2956 (3)	1391 (3)	114 (4)	25	36
Methanosarcinales					
M. mazei	2892 (3)	1473 (3)	134 (2)	25	34
			132 (1)		
M. acetivorans	2831	1429 (3)	134 (2)	25	34
	2848		132 (1)		
	2948				
Archeoglobales					
A. fulgidus	2931	1491	123	25	34
Halobacteriales					
H. marismortui	2923 (3)	1471 (3)	121 (3)	25	32
Halobacterium sp. NRC-1	2905	1472	122	25	32
Thermoplasmata					
T. acidophilum	3044	1470	122	25	32
T. volcanium	2906	1469	122	25	32
Nanoarcheota					
N. equitans	2861	1344	122		

The figures indicate the number of nucleotides in each rRNA gene. The number of genes, if more than one, is indicated in parentheses, except when the genes are of different length, in which case their size is shown in full.

*Has a 5.8S of 167 nt; †gene containing two introns; ‡gene containing one intron.

constant, excepting only some parasitic species where a few proteins may be missing (Lecompte et al., 2002). The functional significance of ribosomal protein loss in late-branching Archaea is unclear at present. However, it should be observed that none of the dispensable proteins belongs to the set of universally conserved ones, which are probably essential for an efficient ribosomal function in all cells.

The organization of ribosomal protein (r-protein) genes in archaeal genomes presents very interesting aspects. In Bacteria, about half of the r-protein genes are clustered in the two large spectinomycin (spc) and S10 operons, whose structure is largely conserved in most species. Likewise, in the Archaea over one-third of the r-protein genes are included in a few large clusters closely resembling the bacterial spc and

S10 operons in the type and order of genes. Most of the proteins belonging in these clusters are universal ones, suggesting that this genetic organization was already present in the last universal common ancestor (LUCA) of extant cells and predated the radiation of the three primary domains. The alternative hypothesis is that similar gene clustering is due to convergent evolution, namely positive selection because of some functional advantage independently operating in both prokaryotic domains. However, any such advantage is not immediately apparent, at least in present-day organisms, since the clusters can be broken, and frequently are, in both Bacteria and Archaea. Moreover, no information is available about the transcription patterns of the archaeal spc- and S10-like clusters, making it difficult to tell to what extent they are organized into functional operons that may resemble the bacterial ones.

However, other clusters of ribosomal protein genes are known to be organized and also regulated in a way similar to that observed in Bacteria. A well studied case is the *Methanococcus vannielii* L1 ribosomal protein operon, encoding the r-proteins L1, L10, and L12, which is transcribed as a single polycistronic mRNA (Kraft et al., 1999). This operon is subjected to autogenous translational regulation, namely its translation can be repressed by the protein encoded by the first cistron (L1). The regulatory protein L1 is a 23S rRNA-binding protein that under normal conditions interacts preferentially with its binding site on the ribosomal RNA. However, when in excess because of blocked or reduced ribosome synthesis, L1 also binds to a specific regulatory target site of its mRNA, thereby inhibiting translation of all three cistrons in the operon. The regulatory mRNA site, a structural mimic of the rRNA binding site for L1, is located within the L1 gene about 30 nucleotides downstream of the ATG initiation codon (Kraft et al., 1999). A similar regulatory mechanism also exists in *Methanococcus thermolitotrophicus* and *Methanocaldococcus jannaschii*; however, its presence in other Archaea is more doubtful.

The three-dimensional architecture of archaeal ribosomes has also been explored in some detail. Early electron microscopy observations showed that the ribosomes of sulfur-dependent thermophiles displayed morphological characteristics similar to those of the eukaryotic particles (Lake, 1985). This was especially true for the small ribosomal subunits, which possessed a "bill" on the head and "lobes" on the body similar to those observed in their eukaryal, but not bacterial, counterparts. These features were, however, absent in the ribosomes of halophiles and some methanogens, a fact that correlates well with the larger protein content of the ribosomes of sulfur-dependent themophiles (Table 19.1).

More recently, the three-dimensional structure of the large ribosomal subunit of the halophilic archaeon *Haloarcula marismortui* has been solved at high resolution by crystallographic analysis (Ban et al., 2000). Due to the lack of comparative data, however, it is difficult to discriminate between features that may be specific to archaeal (or halophilic) ribosomes and features that are common to all large ribosomal subunits. As an example, the exceptionally compact and "monolithic" quaternary packing observed in *H. marismortui* 50S particles (Ban et al., 2000) might be due, at least in part, to adaptation to a hypersaline cellular environment. Moreover, *H. marismortui* ribosomes have an RNA and protein content comparable to that of the bacterial ribosomes, making it difficult to detect archaeal-specific features as the position of the extra proteins in the three-dimensional structure. For instance, the one archaeal-specific protein (LXa) is missing from the genome of *H. marismortui* and other halophiles. Nevertheless, some specific observations can be made, such as that concerning the protein L7ae, shared by the Eukarya and the Archaea but not present in Bacteria. L7ae was initially identified as a ribosomal protein; however, it was later found to behave also as a component of the machinery for rRNA post-transcriptional modification. In fact, L7ae has homology with the eukaryal protein snu13p, which is the RNA-binding element of the snoRNPs involved in post-transcriptional modification of the rRNA transcripts (Kuhn et al., 2002). L7ae is clearly seen in the three-dimensional structure of *H. marismortui* 50S subunit, showing that it is a bona fide ribosomal protein; however, in agreement with its multifunctional character, it is located at the periphery of the subunit and is one of the few r-proteins that make contact with only one rRNA domain. Its function in the ribosome is uncertain (Ban et al., 2000). It is to be expected that the unraveling of more structures of whole ribosomal subunits will allow us in the future to learn more about any architectural features that may be unique to archaeal ribosomes.

Archaeal mRNAs

The structure and organization of mRNAs is another aspect of translation that presents clear-cut differences in Bacteria and Eukarya. Bacterial mRNAs are mostly polycistronic, always uncapped and devoid of long poly(A) tails. In addition, bacterial mRNAs are endowed with *cis*-acting sequences, the Shine–Dalgarno (SD) motifs that enhance the efficiency of ribosome binding to the translation initiation regions of the various cistrons in a polycistronic message. In contrast, eukaryal mRNAs are monocistronic, have 5′ cap structures and long poly(A) tails, and lack SD-motifs for mRNA/ribosome recognition. As the Archaea are prokaryotes with compact genomes, they were expected to have mRNAs similar in organization to the bacterial ones. In fact, the first studies of archaeal transcripts had uncovered the presence of polycistronic mRNAs endowed with SD sequences and coordinately translated into several polypeptides (Shimmin et al., 1989). However, recent *in silico* genome-wide studies analyzing the position of transcription start signals, initiation codons, and potential ribosome-binding motifs such as SD sequences have revealed interesting unique aspects of mRNA structure in Archaea.

Several years ago, genomic analyses of the thermophilic Archaea *Sulfolobus solfataricus* and *Pyrobaculum aerophilum* revealed that in these species a large proportion of mRNAs were predicted to be leaderless, i.e. to lack entirely, or almost so, a 5′ untranslated region ahead of the translation start codons (Sensen et al., 1996; Slupska et al., 2001). This constituted an unexpected unique feature of archaeal mRNAs, as leaderless messengers are rarely encountered in both the Bacteria and the Eukarya.

More recent surveys including a larger number of species have extended and refined those initial observations. It has been found that the archaeal genomes so far sequenced form two distinct groups as far as the structure of transcripts is concerned (Torarinsson et al., 2005). Group A genomes, including several (but not all) Crenarcheota, Euryarchaeota such as the Thermoplasmales, halobacteria and *N. equitans*, putatively produce a high proportion (about 50% on average) of leaderless transcripts. In some of these genomes the genes located internally in (presumably) polycistronic transcripts are preceded by clearly identifiable SD motifs. However, in other group A genomes, such as those of *N. equitans* and *P. aerophilum*, the internal cistrons do not appear to possess evident SD-like sequences.

Group B genomes, on the other hand, produce few leaderless transcripts and, accordingly, usually possess SD motifs ahead of the initiation codons of both the first and the internal genes in operons (or of genes in monocistronic transcripts). Group B genomes include a diverse array of species, mostly methanogens but also Crenarcheota as *A. pernix* and *Hyperthermus butylicus* and the Pyrococcales.

The presence of different types of mRNA organization in Archaea – "leaderless" messages as well as messages of more conventional bacterial type – has prompted several speculations about their respective evolutionary status. Some investigators have proposed that leaderless mRNAs are an evolutionary relic, i.e. represent the ancestral kind of mRNA, possibly the one prevalent at the LUCA stage. The most compelling evidence in favor of this hypothesis is that leaderless mRNAs are universally translatable (at least *in vitro*) by archaeal, bacterial, and eukaryotic ribosomes (Grill et al., 2000). Since "normal" eukaryotic mRNAs are poorly, if at all, translated in bacterial systems (and vice versa) this is a remarkable fact that argues for a common conserved mechanism for leaderless translation, predating the branching of the primary domains. Another recent observation supporting the ancestral nature of leaderless mRNAs is that in the protozoan *Giardia lamblia* most mRNAs are leaderless (Li & Wang, 2004). However, the "primitive" status of *G. lamblia* is uncertain: the species occupies a deep branch of the eukaryal evolutionary tree, but this may be an artefact of evolutionary analysis, due to an abnormally fast mutation rate, a frequent occurrence in parasitic organisms like *G. lamblia*.

Further evidence in favor of the ancestral nature of leaderless mRNAs is that their translation, at least in Bacteria, seems to have no stringent requirement for initiation factors, especially if carried out by preformed 70S ribosomes (Udagawa et al., 2004). Since only a very restricted set of IFs is common to all three primary domains (see below), translational initiation in the absence or semi-absence of accessory factors can be envisaged as a primitive condition.

However, there are also data arguing against the "primitivity" of leaderless mRNAs. First, there is the study mentioned above showing that most leaderless-mRNA-rich group A genomes are found in late-branching archaeal species, while early-branching species tend to have a prevalence of leadered mRNAs with SD motifs (Torarinsson et al., 2005). The latter would therefore represent the likely "ancestral" mRNA structure. If so, the prevalence of leaderless mRNAs in later-evolved, and especially in extremely thermophilic, archaeal species may have a physiological reason that currently escapes our understanding.

Second, there is reason to think that the polycistronic arrangement of genes and the SD motifs predate the branching of the primary domains from the common root of the tree of life (Londei, 2005). In fact, as observed previously for the case of certain ribosomal protein genes, groups of genes clustered (and sometimes transcribed) in the same or a similar order are frequently observed in Bacteria and Archaea. It is very unlikely, albeit not impossible, that this situation is the result of convergent evolution. Interestingly, short-branch Archaea in which most cistrons are endowed with SD motifs, such as *A. pernix* and *H. butylicus*, use AUG, GUG, and UUG as initiation codons in roughly the same proportion, while in most other species AUG is by far the prevalent initiation signal (Torarinsson et al., 2005). This suggests that a "primitive" function of the SD motifs may have been that of ensuring a correct ribosome positioning on the translation initiation region independently of the presence of an optimal initiation codon.

A better understanding of the evolutionary status of leaderless mRNAs may be reached when the mechanism for their translation will be unraveled. This mechanism is likely to be quite distinct from that operating on the leadered mRNAs (Tolstrup et al., 2000). As illustrated in the next section, the notion that the Archaea normally employ two different mechanisms for translating leaderless and leadered mRNAs has recently received some experimental confirmation. These studies are uncovering a very interesting aspect of archaeal translation that may have profound implications for the understanding of the evolution of the mechanism for mRNA/ribosome recognition.

Translational mechanism in Archaea

Initiation

mRNA–ribosome interaction Initiation is the step of translation that has incurred the greatest evolutionary divergence. Eukaryotic ribosomes normally identify the translational start site by a "scanning" mechanism, whereby the 40S subunits, aided by many protein factors and carrying pre-bound met-tRNAi, slide along the message until the initiation codon is found and codon–anticodon interaction takes place. In contrast, bacterial 30S ribosomes can bind directly to the translation initiation regions of individual cistrons by the pairing between the SD sequence on the mRNA and the anti-SD sequence on the SSU rRNA. Only three initiation factors (compared to over a dozen in eukaryotes) participate in bacterial initiation.

The Archaea, as prokaryotes endowed with polycistronic mRNAs, were expected to have a translation initiation mechanism of bacterial type. However, as explained above, the discovery of the abundance of leaderless mRNAs in the third domain of life has suggested the existence of two different mechanisms for archaeal translational initiation.

The first experimental data supporting the notion of two distinct modes of mRNA/ribosome recognition in Archaea were obtained from *in vitro* studies of translation in the crenarcheon *Sulfolobus solfataricus*. The functional relevance of the SD motifs for the decoding of *S. solfataricus* leadered mRNAs was demonstrated by the fact that the disruption of such motifs by site-directed mutagenesis completely inhibited the translation of the following cistrons (Condo et al., 1999). Moreover, it was shown that in a polycistronic mRNA each ORF can be translated independently of the others if preceded at the correct distance by an SD motif.

However, the most interesting fact revealed by the work in *S. solfataricus* was that the *in vitro* translation of the mRNAs whose SD motifs had been disrupted could be rescued by deleting entirely the 5′ untranslated region, thereby rendering the mRNA leaderless (Condo et al., 1999). These results strongly support the notion that a second specific mechanism exists for initiating translation on leaderless messages, operating independently of the SD-motif-based one.

The mechanistic details of ribosome interaction with leadered and leaderless mRNAs are as yet poorly understood. Recent *in vitro* studies carried out with purified translational components of *S. solfataricus* (Benelli et al., 2003) revealed that the 30S ribosomal subunits can interact directly and strongly with leadered mRNAs possessing SD motifs even in the absence of any other translational component, including initiator tRNA. The leaderless mRNAs, by contrast, were unable to form binary complexes with 30S subunits unless met-tRNAi was added to the samples (Benelli et al., 2003). These results suggest that ribosomal recognition of an initiation codon at the 5′ end of a leaderless mRNA requires codon–anticodon pairing, as previously observed for leaderless mRNA translation in *E. coli* (Grill et al., 2000). It should be pointed out that eukaryotic 40S ribosomes also need to carry met-tRNAi in order to recognize the initiation codon at the end of the scanning process.

The reason why the Archaea should keep two distinct mechanisms for mRNA/ribosome recognition is unclear at present, also because there are not enough data on the molecular details of either of them. An especially important task for future research will be to determine the function of the protein factors involved in archaeal translational initiation.

Translation initiation factors Translation initiation factors (IFs) are very interesting from an evolutionary point of view, since they differ to a large extent in the Bacteria and the Eukarya. Only three IFs exist in Bacteria. The principal one, called IF2, is an RNA-binding G-protein of about 90 kDa that performs the essential task of promoting the correct binding of the initiator tRNA (f-met-tRNA) to the ribosomal P site. The other two factors, IF1 and IF3, assist initiation by hindering premature subunit association (both) and by discouraging recognition of non-optimal initiation codons (IF3) (Gualerzi & Pon, 1990).

The Eukarya, in contrast, have an elaborate set of IFs. The cap-binding factor (termed eIF4F), absent in Bacteria, recognizes the 5′ cap structure and unwinds secondary structures in the mRNA, thus allowing ribosome binding. The preinitiation complex "scanning" the mRNA in quest of the initiator AUG codon is composed of the 40S subunits, met-tRNAi, and the proteins eIF1, eIF1A, and eIF3 (Pestova & Kolupaeva, 2002). eIF1 and eIF1A are both required for the correct identification of the start codon, while eIF3, among other things, connects the ribosome with the cap-binding factor eIF4F. Met-tRNAi binding to the 40S subunits is promoted by the G-protein eIF2, a hetero-trimeric complex not homologous to the bacterial factor IF2 (Kyrpides & Woese, 1998a). Adaptation of met-tRNAi in the P site is accompanied by the hydrolysis of the eIF2-bound GTP, whereupon the factor dissociates from the ribosome. However, eIF2 has no spontaneous GTPase activity and needs a GTPase activator factor, called eIF5, to trigger GTP hydrolysis. Moreover, the reactivation of eIF2-GDP obligatorily requires a GTP/GDP exchange factor, the pentameric protein eIF5B (Kimball, 1999). After the establishment of the codon–anticodon interaction, the factor eIF5B, also a G-protein and a homolog of bacterial IF2, stimulates subunit joining and thereby the formation of the monomeric ribosome 80S (Pestova et al., 2000).

The elaborate mechanism for translation initiation in eukaryotes is usually explained by invoking the greater complexity of eukaryotic over prokaryotic cells. Therefore, it was very surprising to discover that archaeal genomes contain genes encoding homologs of most eukaryotic factors, only excepting those involved in cap recognition (Bell & Jackson, 1998).

A summary of the putative translation IFs identified in Archaea on the basis of sequence homologies with known proteins in the other primary domains is shown in Table 19.2. Of the six proteins listed in the table, four are universal, i.e. have homologs in all domains of life, while another two are shared by the Archaea and the Eukarya to the exclusion of the Bacteria. Remarkably, no factor is shared by the Archaea and the Bacteria to the exclusion of Eukarya, once more stressing the evolutionary closeness of the gene expression machineries in the Archaea and the Eukarya. No archaeal-specific IF has been found so far, but it is possible that some will be identified following a more accurate biochemical and genetic analyses of archaeal initiation.

At present, very little is known about the function of the putative initiation factors in Archaea. The one archaeal factor to which a definite function can be assigned on the score of experimental data is the trimeric protein homologous to the eukaryal factor eIF2, here termed a/eIF2.

eIF2 has a central importance in eukaryal translation, as it interacts specifically with the initiator tRNA (met-tRNAi) and carries it to the 40S ribosomal subunits (Kimball, 1999). In Bacteria, the same essential function is carried out by the monomeric protein also called IF2, which, however, has no homology with any of the eIF2 subunits (Kyrpides & Woese, 1998a). The Eukarya do have a homolog of bacterial IF2 (termed eIF5B), which, however, does not interact with met-tRNAi but promotes the joining of the large ribosomal subunit to the

Table 19.2 Translation initiation factors in Archaea.

Factor name	E homolog	B homolog	Structure	Function in A	Function in other domains
aIF1A	eIF1A	IF1	Li & Hoffman, 2001	not determined	B: stimulates IF2 E: assists scanning
aIF2	eIF5B	IF2	Roll-Mecak et al., 2000	not determined	B: binds fmet-tRNA E: subunit joining
aSUI1	eIF1/SUI1	YCiH (some phyla)	Cort et al., 1999	not determined	B: unknown E: fidelity factor
a/eIF2 (αβγ)	eIF2 (αβγ)		γ, Roll-Mecak et al., 2004 β, Cho & Hoffman, 2002 α, Yatime et al., 2005	binds met-tRNAi	binds met-tRNAi
aIF6	eIF6		Groft et al., 2000	not determined	inhibits subunit association
aIF5A	eIF5A	EFP	Kim et al., 1998	not determined	B: formation 1st peptide bond E: undetermined

A, Archaea; B, Bacteria; E, Eukarya.

preinitiation complex (Pestova et al., 2000). The divergence of the tRNAi binding factors in Bacteria and Eukarya has been customarily attributed to the greater sophistication of translational regulation in Eukarya. In fact, eIF2 is central player in the regulation of eukaryal protein synthesis; the phosphorylation of its α-subunit, triggered by various stress signals, inhibits GTP/GDP exchange, thereby blocking the recycling of the factor and shutting off translation (Colthurst et al., 1987). However, the fact that the Archaea resemble the eukaryotes in having both eIF2-like and IF2-like factors shows that cellular complexity probably has nothing to do with the usage of these translation initiation factors.

Like eIF2, a/eIF2 is composed of three subunits that associate to form a hetero-trimeric complex (Yatime et al., 2004, 2005; Pedulla et al., 2005). The γ-subunit (about 45 kDa) and the α-subunit (about 30 kDa) have sizes comparable to those of their eukaryal homologs, while the archaeal β polypeptide (about 15 kDa) is much smaller than the eukaryal one, which is about 50 kDa in size and often the largest component of the trimeric complex. In fact, archaeal IF2-β is reduced to a conserved domain containing a zinc-finger motif while lacking the eukaryal-specific domains responsible for the interaction with the guanine nucleotide exchange factor eIF2B and with the GTPase activator eIF5. This agrees with the observation that all Archaea lack a homolog of eIF5 as well as a complete eIF2B. Archaeal genomes do include homologs of the α, β, and δ subunits of eIF2B, but lack counterparts of the γ and ε subunits that catalyze guanine nucleotide exchange on eIF2. Therefore, it is probable that the archaeal homologs of the eIF2B α, β, and δ proteins have a function unrelated to guanine nucleotide exchange (Kyrpides & Woese, 1998a).

Crystal or NMR structures are available for all three separate subunits of a/eIF2. As shown in Plate 19.1, the γ-subunit has a striking resemblance to the elongation factor 1A (formerly EF-Tu in Bacteria) (Schmitt et al., 2002; Roll-Mecak et al., 2004), in agreement with the fact that it contains the guanine-nucleotide binding domain and is principally involved in the interaction with met-tRNAi. a/eIF2-γ also contains a zinc-finger motif of uncertain function (Plate 19.1). The structures of the archaeal and eukaryal α-subunits are compared in Plate 19.2. Both proteins have a similar folding including three domains. The N-terminal domain (domain 1) has a β-barrel structure frequently observed in many RNA-binding proteins. The C-terminal domain (domain 3) contains an αββαβ module that is found in a large number of proteins and has been proposed to be an ancestral RNA binding motif (Yatime et al., 2005). An interesting feature of a/eIF2-α is the exposed loop in domain 1 (Plate 19.2), which is conserved in structure and contains the serine residue phosphorylated in the eukaryal factor. The archaeal α-polypeptides also contain a Ser residue in the same loop, although it occupies a slightly different position with respect to its eukaryal counterpart.

The function of a/eIF2 from *Pyrococcus abyssi* (Yatime et al., 2004) and *S. solfataricus* (Pedulla et al., 2005) has been explored by *in vitro* biochemical assays using the factor reconstituted from the cloned recombinant subunits. These studies have revealed that a/eIF2, like its eukaryal counterpart, binds specifically met-tRNAi and carries it to the ribosomes. However, a number of features differentiate functionally the archaeal and the eukaryal proteins. One regards the nature of the tRNA binding site: an αγ dimer of a/eIF2 is necessary and sufficient to achieve a stable interaction with met-tRNAi, while in the case of eIF2 met-tRNAi binding seems to involve principally the γ and β subunits (Das et al., 1982). The α-polypeptide of the eukaryal factor appears to have mainly a regulatory function.

Another very relevant difference is that a/eIF2 has a similar affinity for GDP and GTP and therefore does not require a guanine nucleotide exchange factor to be reactivated (Pedulla et al., 2005). This finding is consistent with the lack of a complete homolog of eIF2B in archaeal genomes (Kyrpides & Woese, 1998a). According to this observation, a/eIF2 should not be subjected to a eukaryal-type functional regulation based on the inhibition of guanine nucleotide exchange upon phosphorylation of the α-subunit. However, it has been reported recently that the α-subunit of *Pyrococcus horikoshii* a/eIF2 is phosphorylated by a specific protein kinase (Tahara et al., 2004). The function of this modification is unknown, but it cannot be related to the regulation of GTP/GDP exchange. Perhaps it controls the function of the factor by regulating the formation of the trimer or its interaction with the ribosome. The solution of the problem may help us to understand why the archaeal/eukaryal branch, unlike the bacterial one, originally evolved a trimeric tRNAi binding factor.

Finally, unlike its eukaryal counterpart, a/eIF2 probably does not require a companion GTPase activator factor. GTP hydrolysis on eukaryal eIF2 is triggered by the helper factor eIF5, which has no recognizable homolog in Archaea. It is therefore likely that a/eIF2 has an intrinsic, ribosome-triggered GTPase activity, although this has not yet been demonstrated experimentally. Alternatively, a/eIF2 may be helped by a new and still unidentified GTPase activator.

The function of all of the other putative archaeal IFs remains undetermined, although crystal or NMR structures are available for most of them. A particularly interesting protein is aIF2, homologous to bacterial IF2 and eukaryal eIF5B and therefore one the few universally conserved IFs (Kyrpides & Woese, 1998b). Despite its conservation in all primary domains, this factor seems to have diverged in function, since in Bacteria it binds f-met-tRNAi and carries it to the ribosome, while in Eukarya it appears to promote the joining of the ribosomal subunits in a late stage of initiation (Pestova et al., 2000).

To date, there are few published experimental data about the function of the archaeal IF2-like factor. The only study performed *in vivo* has shown that *M. jannaschii* aIF2 can partially rescue yeast mutants lacking eIF5B (Lee et al., 1999), thus demonstrating that aIF2 is to some extent functionally homologous to eIF5B. On the other hand, preliminary data have been obtained *in vitro* suggesting that *S. solfataricus* aIF2 promotes the binding of met-tRNAi to the ribosome (Londei, 2005). Thus, aIF2 would seem to have properties somewhat intermediate between those of the bacterial and the eukaryal protein, but more data are needed to understand its function.

Structurally, archaeal IF2 proteins are smaller than their eukaryal and bacterial homologs, since they lack the long and poorly conserved N-terminal tracts of uncertain function present in both IF2 and eIF5B. Crystallographic studies on the *Methanothermobacter thermoautotrophicus* aIF2 (Roll-Mecak et al., 2000) have revealed that it is characteristically shaped as a chalice (Plate 19.3). The globular "cup" of the chalice (N-terminal region) includes the guanine-nucleotide-binding domain and a β-barrel domain probably involved in the interaction with the ribosomes. The "stem" of the chalice is a long α-helix, while the globular "base" (domain IV) corresponds to the C-terminal domain known to bind f-met-tRNA in bacterial IF2 (Guenneugues et al., 2000). The aIF2 (and eIF5B) domain IV has, however, lost the capacity for tRNA binding because of some critical amino acid substitutions in the relevant region.

Despite the evident divergence in their tRNA-binding capacity, the universal conservation of the

IF2-like proteins suggests that they still have some common function in all cells. This function may consist of promoting the interaction of the ribosomal subunits, but this has yet to be proven for the archaeal protein. An interesting common feature of all IF2-like factors is their ability to interact with another universal initiation factor, the protein termed eIF1A/aIF1A in Eukarya and Archaea and IF1 in Bacteria. Experimental evidence for a direct interaction of the eukaryal proteins eIF5B and eIF1A has been obtained (Marintchev et al., 2003). In contrast, bacterial IF2 and IF1 do not form a complex in solution, but may interact on the surface of the ribosome, as suggested by earlier cross-linking data (Boileau et al., 1983) and by a more recent cryo-electron microscopy study (Allen et al., 2005). Experimental data on archaeal IF1A indicate that it resembles its eukaryal counterpart in being able to interact stably with aIF2 in solution (Londei, 2005). The complex between the universal factors IF1/IF1A and IF2/IF5B is likely to be an ancestral feature of translation initiation, whose significance will be fully understood once more data are available on the archaeal proteins.

The small protein termed aSUI (or aIF1) has homologous counterparts in all Eukarya (where it is known as SUI1 in yeast and as eIF1 in vertebrates) and in a limited number of bacterial species, including *E. coli*, where it is called YciH (Cort et al., 1999). A phylogenetic analysis has shown that SUI1 is very likely an ancestral factor that has been lost secondarily by most bacteria, possibly because its function has been replaced by another protein (Londei, 2005). In Archaea, aSUI1 interacts with the 30S subunits but its precise function in translation initiation has yet to be determined. In Eukarya, SUI1/eIF1 is an essential protein that controls the fidelity of initiation codon recognition and probably also of elongation (Cui et al., 1998).

A very interesting factor shared specifically by the Archaea and the Eukarya is the 25 kDa protein called aIF6 (eIF6). The function of this factor in the Eukarya has been studied in some detail, but remains somewhat enigmatic. In yeast, eIF6 is an essential protein that is found both in the nucleolus and in the cytoplasm, where it associates with the 60S ribosomal subunits (Basu et al., 2003). The main phenotype observed in conditional mutants lacking the factor is a defect in the synthesis of 60S ribosomes, specifically a block in the processing of the rRNA 26S precursor (Basu et al., 2001). However, the cytoplasmic, 60S-bound eIF6 behaves as a ribosome anti-associating factor, preventing the formation of 80S particles and thereby inhibiting protein synthesis. According to a recent report, the dissociation of eIF6 from mammalian ribosomes requires the phosphorylation of the factor, which takes place when certain environmental cues activate a specific kinase (Ceci et al., 2003). Thus, eIF6 would resemble eIF2 in being a general regulator of protein synthesis. However, it remains unclear whether the two functions described for eIF6 indeed coexist, and which is the relationship between them, if any. Clearly, the functional study of the archaeal factor will be of great help in advancing our understanding of the cellular role of this interesting protein.

A last protein generally included in the translation initiation factors is the universally conserved polypeptide known as aIF5A/eIF5A in Archaea and Eukarya and as EFP in Bacteria. As the bacterial name implies, this protein can be regarded as a specialized elongation factor since it appears to catalyze the formation of the first peptide bond at the end of the initiation process in Bacteria (Glick et al., 1979). It seems probable that the function of this factor is conserved in all cells; however, the structure of archaeal IF5A differs to some extent from that of bacterial EFP. The latter is composed of three β-barrel domains and has an L-shaped structure reminiscent of a tRNA; it seems to bind both ribosomal subunits and to stimulate the peptidyl transferase center on the 50S particle (Hanawa-Suetsugu et al., 2004). The archaeal factor (structures are available for *M. jannaschii*, *P. aerophylum*, and *Pyrococcus horikoshii* aIF5A) is somewhat shorter than its bacterial homolog. It includes only two β-barrel domains and has a rod-like shape rather than a L-shape, and therefore it may interact preferentially with the 50S subunit (Kim et al., 1998). On the basis of structural comparisons, it has been proposed that archaeal/eukaryal IF5A evolved from an EFP-like common ancestor following the deletion of one of its three domains, and that perhaps another still unidentified protein has replaced functionally the missing domain in Archaea and Eukarya (Hanawa-Suetsugu et al., 2004). A remarkable feature of aIF5A, shared with its eukaryal homolog, is the presence of a uniquely modified lysine known as hypusine (N-ε-(4-aminobutyl-2-hydroxy) lysine), whose functional role is poorly understood.

Elongation and termination

The process of elongation is the basic biochemical core of protein synthesis and as such is extremely well conserved in evolution (see Table 19.3). All cells make use of two elongation factors, EF1 and EF2 (also known as EF-Tu and EFG in Bacteria). Like the majority of the components of the translational apparatus, the archaeal elongation factors have the closest homology with the eukaryal ones. Indeed, elongation factors-based evolutionary trees have first allowed us to place the root of the universal tree between the Archaea and the Bacteria, identifying the Archaea and the Eukarya as sister domains (Iwabe et al., 1989).

Termination, like initiation, has incurred a certain divergence in the primary domains of cell descent (see Table 19.3). In Bacteria and Eukarya, the specific task of recognizing the stop codons is performed by the class-1 termination factors, which release the completed polypeptide by promoting the hydrolysis of the ester bond anchoring it to the tRNA in the P-site. Bacteria possess two class-1 termination factors: RF1, recognizing UAA and UAG, and RF2, recognizing UAA and UGA. By contrast, the Eukarya appear to employ a single factor (eRF1) to recognize all three stop codons (Kisselev & Buckingham, 2000). All archaeal genomes include genes encoding a polypeptide homologous to eRF1 (termed aRF1), while no counterparts of bacterial RF2 have been detected. Therefore, the Archaea appear to resemble the Eukarya in using a single factor for stop codon recognition. That this is in fact the case has been demonstrated by the observation that *M. jannaschii* aRF1 can promote termination on eukaryotic ribosomes (Dontsova et al., 2000). Despite exhaustive *in silico* analyses, no meaningful similarity has ever been detected between the bacterial and the archaeal/eukaryal class-1 RF, which therefore seem to belong to two distinct protein families (Kisselev & Buckingham, 2000). Given the functional similarity between aRF1 and eRF1, the archaeal proteins were expected to have a structure comparable to that of their eukaryal counterpart. However, aRF1 appears to lack entirely a C-terminal domain present in both bacterial and eukaryal class-1 RF (Kisselev & Buckingham, 2000). Very likely, this reflects the fact that the archaeal genomes do not include any homolog of the class-2 RF, present in both the Bacteria and Eukarya, where they are termed respectively RF3 and eRF3. Class-2 RF are G proteins that do not participate in the peptide release reaction itself. The function of bacterial RF3 has been analyzed in some detail: briefly, its main task seems to be to accelerate the recycling of class-1 RFs after translational termination (Zavialov et al., 2002). Class-1 RFs interact with class-2 RFs by means of the C-terminal domains that are lacking in archaeal class-1 RFs. The Archaea also lack any apparent homolog of a bacterial-specific termination factor called RRF.

Thus, the data so far available suggest that the Archaea are endowed with a simplified version of the eukaryal translation termination mechanism, based on a single class-1 RF and dispensing with both the RF3 and the RRF proteins. Obviously, detailed experimental studies are needed to tell whether the Archaea possess unique termination factors that may take up the role played by the RF3s and/or RRF in the other two domains.

Conclusion and prospects

The study of the translational apparatus and of the protein synthesis mechanism in Archaea is still in its

Table 19.3 Translation elongation and termination factors in the primary domains.

	Archaea	Eukarya	Bacteria	Function
Elongation factors	aEF1A	eEF1A	EFT	Adapts aa-tRNA in ribosomal A site
	aEF2	eEF2	EFG	Promotes translocation
Termination factors	aRF1	eRF1	RF1/RF2	Stop codon recognition
		eRF3	RF3	Recycling of RF1
			RRF	Ribosome recycling

infancy, but the relatively few data available are revealing an interesting scenario of "hybrid" features whose detailed understanding will give new and exciting insights to the evolutionary history of the protein synthetic machinery. Foremost questions to be addressed in the near future regard the understanding of the mechanism for translation of leaderless mRNAs and the unraveling of the function of the putative translation initiation factors. A particularly interesting task will be to determine whether the Archaea make use of translational regulation mechanisms based on the phosphorylation of the translational factors that the Archaea share with the Eukarya, i.e. a/eIF2 and aIF6. The unraveling of the functional and regulatory role of these proteins, besides being interesting in itself, will be important to clarify some still obscure aspects of their function in eukaryotes.

A further subject almost completely unexplored is the mechanism of archaeal translational termination, which seems to be based on a single, eukaryal-like termination factor. Once more, the study of archaeal termination is likely to shed light on the unclear aspects of the corresponding eukaryal process, especially on the function of eRF3, a factor that in Archaea is apparently missing. Thus, translational studies in Archaea will not only improve our knowledge of the basic workings of the gene expression machinery, but also be precious to orient future research in the eukaryotic field.

Acknowledgments

Thanks are due to the people in the author's laboratory who worked on archaeal translation in the past few years: Dario Benelli, Enzo Maone, Maria Grazia Pantano, Nadia Pedullà, Rocco Palermo, Laura Peri, and Alessandra Barbazza. Thanks are also due to Udo Blaesi and David Hasenhörl of Vienna University Biozentrum for fruitful collaboration and helpful discussions. The work in the author's laboratory was supported by MURST projects PRIN 2000 "Translational initiation of leadered and leaderless mRNAs in Bacteria and Archaea: comparative analysis of the functions of IF2, an essential and universally conserved translation initiation factor" and PRIN 2002 "Evolution of the gene expression machinery: characterization of a set of universally conserved factors modulating translational initiation."

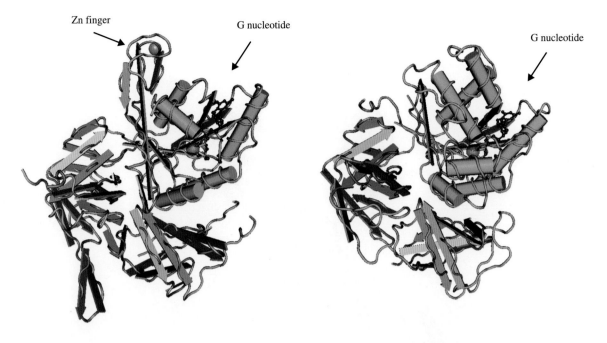

Plate 19.1 a/eIF2-γ is a structural homolog of elongation factor 1. The three-dimensional structures of archaeal IF2-γ subunit (left) and of bacterial elongation factor Tu (right) with a bound guanine nucleotide are compared. α-Helix elements are shown in green, while β sheets are shown in yellow. The G-nucleotide binding site in both proteins, and the a/eIF2-γ specific zinc-finger, are indicated by arrows. (Taken from the NCBI structure databank.)

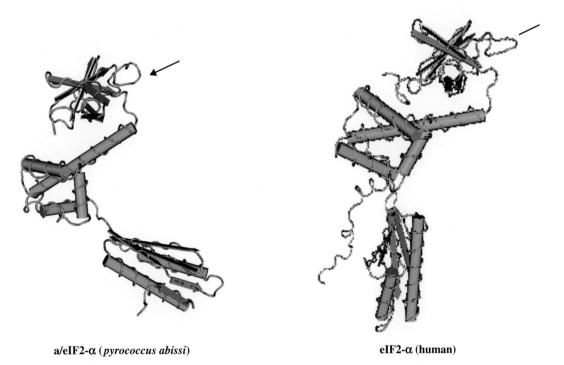

Plate 19.2 The IF2 α-subunit in Archaea and Eukarya. The similarity in the three-dimensional folding between the eIF2 and a/eIF2 α-subunits is shown. Colouring is as in Plate 19.1. The arrows indicate the exposed loop containing the serine residue, which is phosphorylated in the eukaryal protein. (Taken from the NCBI structure databank.)

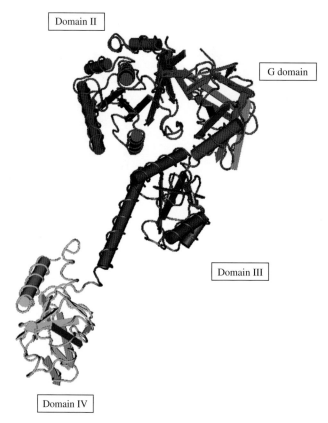

Plate 19.3 The "chalice" structure of aIF2. Three-dimensional folding of *Methanococcus thermoautotrophicus* aIF2. The four domains that compose the proteins are shown in different colors: domain I (G domain), magenta; domain II, blue; domain III, brown; domain IV (the f-met-tRNA binding domain in the bacterial protein), green. (Taken from the NCBI structure databank.)

20
Expanding world of small noncoding RNAs in Archaea

Arina Omer, Maria Zago and Patrick P. Dennis

Introduction

A Herculean effort by a large cadre of investigators during the 1950s and 1960s defined the ribosome (and its rRNA and protein components), tRNA, and mRNA as the core machinery of the information transfer system – the system that translates nucleotide sequence information stored in DNA into the amino acid sequence of proteins. In the decades that followed few biologists recognized or appreciated the possibility that the role of RNA in biology could extend much beyond the translation process. This prevailing attitude began to change in the 1980s, both with the discovery of introns (review in Witkowski, 1988) and with the discovery that some RNAs possess catalytic activity (Cech et al., 1981; Guerrier-Takada & Altman, 1984). These observations coincided with the discovery and characterization of a number of different cellular RNAs that exist, like the ribosome, as ribonucleoprotein complexes. Examples of such complexes included RNaseP (processing of tRNAs), the signal recognition particle (translocation of protein), the splicosome (removal of introns from eukaryotic mRNAs), telomerase (restoration of telomeric sequences at the ends of eukaryotic chromosomes), and small nucleolar (sno)RNPs (processing and modification of pre rRNA). In the past few years, whole new families of small regulatory RNAs or RNA elements that are believed to modulate gene expression have been discovered and described in both prokaryotic and eukaryotic organisms (Huttenhoffer et al., 2001; Wassarman et al., 2001; Eddy, 2002; Gottesman, 2002; Tang et al., 2002b; Nelson et al., 2003; Mattick, 2004; Storz et al., 2004; Zago et al., 2005).

The advent of the genomics era over the past decade has provided complete sequences for hundreds of microbial genomes. Although protein coding genes (ORFs) in these genomes are generally easy to identify using various informatics tools, the regions encoding small RNAs (other than tRNAs) are generally much more difficult to identify, particularly where there are no conserved box sequence elements or RNA structural motifs and where the function of the RNA is poorly understood. This chapter reviews the recent progress that has been made in identifying and characterizing the structure and function of the plethora of small noncoding RNAs in Archaea.

The eukaryotic nucleolous

In eukaryotic cells the nucleolus functions as a factory for the production and assembly of ribosomes. The factory contains scores of distinct RNP machines, each working along the ribosome assembly line and participating in the processing, modification, and folding of the nascent rRNA. Based on the sequence and structural features of their RNA components, the machines divide into two major classes – the C/D box snoRNPs that guide 2′O-ribose methyl modifications and the H/ACA box snoRNPs that guide pseudouridine modifications in the nascent rRNA (Plate 20.1) (Bachellerie et al., 1995, 2000; Maxwell & Fournier, 1995; Balakin et al., 1996; Ganot et al., 1997; Tollervey & Kiss, 1997).

Both the C/D box and the H/ACA box snoRNPs use guide sequences that are complementary to sequences in the rRNA to direct the modifications to specific locations (Kiss-Lazslo et al., 1996; Tollervey, 1996; Kiss, 2001). Biochemical and genetic studies with various eukaryotic organisms have led to the identification of four protein components of C/D box RNPs (Fib (the methyl transferase), two paralogous proteins Nop56 and Nop58, and the 15.5 kDa protein) and four protein components of H/ACA box RNPs (Cbf5 (the pseudouridine synthetase), Gar1, Nop10, and Nhp2 (a paralog of the 15.5 kDa protein)). In addition to guide-directed modification, both types of complexes appear to participate in the endonucleolytic processing and mediate the folding of the nascent rRNA at critical points along the assembly pathway (Steitz & Tycowski, 1995; Ofengand & Fournier, 1998).

Biochemical and informatics identification of C/D box RNAs in Archaea

Seminal work over the past few years has revealed that the ribosomes of *Sulfolobus acidocaldarius* (and presumably other archaeal species) contain a large number of sites of 2'O-ribose methyl modification and that these modifications are mediated by RNP complexes that are homologous to eukaryotic C/D box snoRNPs (Noon et al., 1998; Omer et al., 2000). These discoveries imply that nucleolar function originated prior to the divergence of Archaea and eukaryotes. The initial biochemical discovery of archaeal C/D box RNAs was made by sequencing entries in a *S. acidocaldarius* cDNA library that was prepared from small RNAs that were co-immunoprecipitated using antibodies against the archaeal Fib or Nop5 (also referred to as Nop56/58) protein (Omer et al., 2000). Based on the presence of the four conserved box (C, D', C', and D) sequences and the spacing between the boxes in the cloned RNAs, search programs were designed to help to identify the genes encoding these sRNAs in sequenced archaeal genomes. The results of these searches revealed that genes encoding C/D box sRNAs are abundant in archaeal genomes, particularly the genomes of organisms that grow at high temperatures (Gaspin et al., 2000; Omer et al., 2000). This correlation with growth temperature may mean: (i) that the base pairing occurring between the guide regions of the C/D box RNAs and the nascent rRNA assists in the proper folding of the rRNA during assembly; or (ii) that the deposition of methyl groups along the rRNA backbone contributes to the stabilization of higher order structure of the RNA within the ribosome. In a number of instances, the presence of a methyl modification predicted by the complementarity between the sRNA guide and rRNA target sequences has been confirmed (reviewed by Dennis et al., 2001; Omer et al., 2000).

Archaeal C/D box RNP complexes

The archaeal methylation guide C/D box RNPs consist of a single small RNA (sRNA) about 50–60 nucleotides in length and two copies of each of three proteins: L7Ae (a homolog of the eukaryotic 15.5 kDa protein), Fib, and Nop5 (Omer et al., 2002; Rashid et al., 2003). The RNAs are characterized by moderately conserved C (UGAUGA) and D (CUGA) box sequence motifs located near the 5' and 3' ends of the molecules and reiterated D' and C' boxes located near the center of the molecule (Plate 20.1). This imparts a bipartite structure to the RNA. The C/D, and D'/C' boxes form two separate K-turn motifs in the bipartite RNA that are each stabilized by association with the L7Ae protein (Kuhn et al., 2002). The K-turn motif is found in many RNAs and is characterized by an asymmetric internal (or terminal) loop in a helical region of the RNA where the loop is partially closed by two adjacent sheared AG base pairs. The nucleotide adjacent to the second AG base pair (usually U) is protruded from the loop by the sharp 45–63° bend in the RNA backbone (Klein et al., 2001). The molecular details of this interaction are provided by a number of X-ray crystal structures that include the *Haloarcula marismortui* 50S subunit (Ban et al., 2000), the L30e (Mao et al., 1999), 15.5 kDa (Vidovic et al., 2000), or L7Ae (Moore et al., 2004; Suryadi et al., 2005) proteins in complex with various K-turn containing RNA fragment. All of the structures are highly similar and generally equivalent. In the L7Ae–C/D box RNA co-crystal, the protein makes contact with six RNA nucleotides: U5, G6, and G19 at the top of stem II and C16, G17, and U18 in the loop region (Plate 20.1) (Moore et al., 2004; reviewed by Dennis & Omer, 2005). At the present time there is no crystal structure of the fully assembled methylation guide RNP complex.

The Fib protein component of C/D box RNPs has an S-adenosyl methionine binding domain and has been implicated in the methyl transferase function of the complex, whereas the Nop5 protein is believed to serve as a bridge between the sRNA and the catalytic Fib subunit. The Fib, Nop5, and Fib–Nop5 heterodimer do not show appreciable affinity for either the naked C/D box sRNA or for the L7Ae protein (Omer et al., 2002). Nevertheless, two copies of the Fib–Nop5 heterodimer rapidly assemble onto the L7Ae–sRNA complex in vitro to produce a larger complex that is active in guide-directed methylation. The explanation of this anomaly seems to be that the binding of the L7Ae protein stabilizes the structure of the K-turn within the RNA, revealing features that can then be recognized by the Fib–Nop5 complex (Omer et al., 2002).

Mutational analyses along with a crystal structure of the Fib–Nop5 heterodimer have provided additional clues relating to the overall structure of the fully assembled RNP complex. The positively charged surface on the C-terminal domain of Nop5 has been implicated as the region that interacts with the accessible features of the tightly kinked K-turn RNA that are revealed by binding of the stabilizing L7Ae protein. These features are believed to include the sharply bent phosphate backbone of the box C, including the major groove edge of the protruding U base and stem II. The N-terminal domain of Nop5 forms a complementary surface to Fib that interfaces the two subunits around a stabilized binding pocket for the S-adenosyl methonine cofactor. The coiled coil domain in the center of the Nop5 protein mediates self-dimerization and optimally positions the two Fib–Nop5 heterodimers into the bipartite, fully assembled, and active RNP complex (Plate 20.1; Aittaleb et al., 2003).

Probing the structure and function of archaeal C/D box RNPs

The C/D box sR1 sRNA from *S. acidocaldarius* uses the D box guide to direct methylation to position U52 in 16S rRNA (Fig. 20.1a). A major step toward understanding of the methylation function of archaeal C/D box RNPs came with the reconstitution of active complexes using *in vitro* transcribed sR1 RNA and purified recombinant proteins from *S. acidocaldarius* (or similar systems from other Archaea) (Omer et al., 2002). The initial study demonstrated that the L7Ae protein binds directly to the C/D box RNA and nucleates the sequential addition of the Nop5 and Fib to the complex. When the complex was supplied with a fragment of rRNA that was complementary to the guide region of the sRNA along with S-adenosyl methionine as a methyl donor, methylation was directed to the predicted N plus five position in the target RNA that base pairs with the guide region in the sRNA. The methylation activity of the complex was dependent on the presence of active fibrillarin protein and on a Watson–Crick base pair between the guide and target at the site of methylation (Omer et al., 2002). This *in vitro* assembly and methylation system has been used extensively to examine aspects of the structure and function of C/D box RNPs and the nature of the interaction of the complex with the target RNA.

Structure–function constraints in C/D box sRNAs

A number of analyses that use mutant sRNAs to establish structure–function constraints have been carried out. In circularly permuted RNAs the wild type 5′ and 3′ termini are connected and new extremities are created at alternative sites. Circular permutations have been used as a tool to probe the folding pathway and tertiary structure of various RNAs. We constructed circular permutation mutants of sR1 and examined the ability of these altered sRNAs to assemble into RNP complexes and to catalyze methylation of a complementary fragment of rRNA containing position U52 (Omer et al., in the press). When the 5′ and 3′ ends were relocated to the region connecting the D′ and C′ boxes (to give an sRNA where the order of the boxes was 5′–C′–D–C–D′–3′), the sRNA was fully active in both assembly and methylation, whereas when the 5′ and 3′ ends were relocated to the connector region between the C and D′ box (to give sRNA where the order of the boxes was 5′–D′–C′–D–C–3′), the sRNA was able to assemble into a RNP complex but the complex was inactive in methylation (Fig. 20.1b, c). In a separate study, a systematic analysis of the spacing between the C and D′

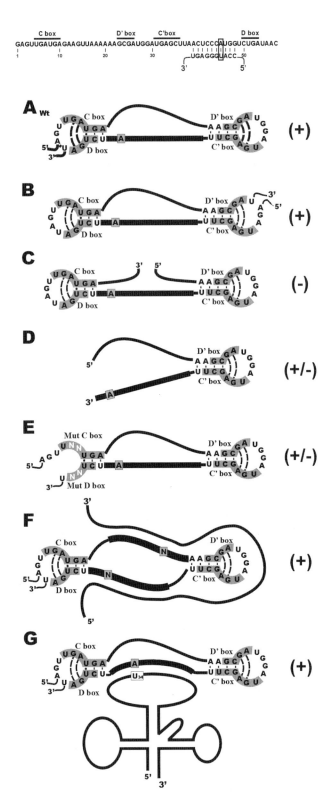

and the C′ and D in archaeal sRNAs indicated that the length of the spacer regions is generally narrowly constrained to between 11 and 13 nucleotides (most often 12 nucleotides) (Tran et al., 2005). To understand the importance of this spacing, deletion and insertion mutants that alter the length of both spacers by increments of two nucleotides without altering the guide–target interaction were constructed. Although the spacer mutant sRNAs were efficiently assembled into sRNPs, their *in vitro* methylation activity was drastically impaired (Tran et al., 2005). Together these observations indicate that although individual C/D and the C′/D′ K-turns are flexible and dynamic structures in the context of symmetrically assembled RNPs, the two juxtaposed K-turn motifs need to be spatially constrained with respect to each other in order for the methylation reaction to occur efficiently.

Fig. 20.1 Structure of mutationally altered C/D box sRNAs. The structure–function relationships of archaeal C/D box sRNAs have been probed using mutated RNAs in *in vitro* assembly and activity assays. The sequence of the sR1 sRNA from *S. acidocaldarius* is illustrated at the top of the figure along with the partial sequence of the fragment of rRNA that is recognized as target. The C, D′, C′, and D boxes are overlined and the base pair at the site of methylation (U52) is boxed. The structures illustrated are: (a) the structure of the typical wildtype sR1 sRNA from *S. acidocaldarius*; (b) circular permutation of sR1 where the 5′ and 3′ ends have been relocated to the connector loop between the D′ and C′ boxes; (c) circular permutation where the 5′ and 3′ ends have been relocated to the connector regions between the C and D′ boxes; (d) deletion mutant where the C and D boxes have been removed to give a half sRNA; (e) missense mutant where the critical A and G residues in the C and D boxes required for the formation of the K-turn have been replaced by Cs; (f) structure of *P. horikoshii* sR24 sRNA containing D and D′ box antisense elements that guide methylation respectively to positions C1221 and C1243 in 23S rRNA; (g) structure sR14 sRNA from *S. acidocaldarius* that guides methylation to position U34 in tRNA^{Gln} (UUG). The activity of the complexes on directing methylation *in vitro* is indicated on the right: (+), normal or near normal activity; (+/−), low residual activity; (−), inactive.

A number of studies have used targeted mutagenesis to determine the function of the two separate K-turns formed by the C/D and C'/D' sequences (Tran et al., 2003; Omer et al., in press). Deletions that remove either the C/D or the D'/C' K-turns and generate half sRNAs retain the ability to bind L7Ae, Nop5, and Fib (Fig. 20.1d) (Tran et al., 2003). Similarly, RNAs that contain one mutant (misfolded) K-turn are not able to bind L7Ae efficiently at the aberrant site but maintain the capacity to assemble L7Ae with at least one Nop5–Fib heterodimer at the normal K-turn site to form asymmetric particles (Fig. 20.1e) (Rashid et al., 2004). Both the half sRNP complexes and the asymmetric sRNP complexes possess low residual methylation activity. Although the results are not always equivocal, it is generally the case that disruption of the K-turn associated with the guide function (box C/D K-turn and D guide function or box C'/D' K-turn and D' guide function) is more detrimental to the activity of the complex than is disruption of the nonguiding K-turn motif (Tran et al., 2003; Omer, in preparation). Merged together these observations provide strong support for a symmetrical sRNP structure that contains two copies of each of the three proteins organized around the C/D and D'/C' K-turns as depicted in Plate 20.1. The integrity of both K-turns, particularly the one associated with guide function, is required for full activity of the complex.

Double guide RNPs

The fact that archaeal C/D box sRNA can use either the D- or the D'-associated guide sequence to direct methylation to a suitable target RNA has been well established. The symmetrical nature and the presence of predicted D- and D'-associated guides in many C/D box sRNAs suggests that at least some of these may have "double guide" enzymatic activity and be capable of directing methylation to two separate locations within the 16S or 23S rRNA target. The predicted sites of methylation for many of these double guide sRNAs are located in close proximity, suggesting further that the two guide–target interactions may occur simultaneously and perhaps influence the localized folding of the rRNA and its assembly into ribosomal subunits (Dennis et al., 2001; Zieche et al., 2004). The importance of this double interaction and its impact on the efficiency of the methylation reaction has been examined using the *Pyrococcus horikoshii* sR24 RNA (Fig. 20.1f). The D and D' box anti-sense elements of this sRNA are predicted to guide methylation respectively to positions C1221 and C1243 in 23S rRNA (Gaspin et al., 2000; Omer et al., 2000). When assembled into a RNP complex with L7Ae, Nop5, and Fib, the complex efficiently catalyzed methylation to both of the predicted sites in a longer fragment of 23S rRNA-containing regions of complementarity to both the D and D' guide sequences. In contrast, the catalytic activity of the complex was significantly attenuated when the D and D' regions of complementarity were presented on two separate shorter rRNA fragments (Zieche et al., 2004). This suggests that in at least some double guide sRNAs, the two guide regions work in concert to select the complementary target and to catalyze efficient methylation on the target RNA.

Methylation of tRNA

The D- and D'-associated guide regions of most archaeal sRNAs exhibit a 9 to 12 nucleotide long region complementarity to either 16S or 23S rRNA. However, in many instances, the guide regions of some sRNAs that lack sequence complementarity to rRNA instead exhibit complementarity to tRNA sequences. The predicted sites of methylation are confined exclusively to the known positions of 2'-O-ribose methyl modification in at least some tRNAs (Dennis et al., 2001). These methyl modifications function to enhance the accuracy of the codon–anticodon interaction or to stabilize higher order structure in the tRNA. In eukaryotes and bacteria, methyl transferase enzymes that recognize sequence and structure within the tRNA mediate all of the known ribose methylations in tRNA and none of these modifications is known to use a RNA guide.

The ability of archaeal C/D box sRNAs to carry out the modification of full-length tRNA sequences has been examined *in vitro* (Zieche et al., 2004). The D box guide in *S. acidocaldarius* sR14 was predicted to methylate the wobble base U34 in tRNA^{Gln} (UUG), but not the corresponding position C34 in the isoacceptor tRNA^{Gln} (CUG) (Fig. 20.1g). The ability of sR14 to guide tRNA modification at position U34 has been tested by assembling the guide RNA into an RNP complex and determining the degree of methylation of either tRNA^{Gln} (UUG) or tRNA^{Gln} (CUG) substrates. The complex was enzymatically active only when the base at position 34 of the tRNA^{Gln} was U. The specificity of the complex was changed by introducing an A to G mutation into the D box guide sequence of sR14 opposite the site of methylation; the altered sRNA as predicted was now active in methylation of tRNA^{Gln} (CUG) and inactive on tRNA^{Gln} (UUG) (Zieche et al., 2004).

In a second example, the D' box guide of sR11 from *S. solfataricus* is predicted to target methylation to position G18 in tRNA^{Gln}. The residues G18 and G19 are highly conserved in all tRNAs (Giege et al., 1993; Dirheimer et al., 1995), and are responsible for the formation of the L-shaped three-dimensional structure by D-loop/T-loop interaction through the tertiary base pairs G18-Ψ55 and G19-C56 (Kim et al., 1974; Robertus et al., 1974). In *E. coli* and yeast, the G18 methyltransferase enzymes are believed to be SpoU and Trmp3 respectively (Persson et al., 1997; Cavaille et al., 1999). When assembled into complexes with L7Ae, Nop5, and Fib, the *S. solfataricus* sR11 was able to direct methylation to the predicted location in the RNA. Mutational alteration of the sRNA guide or the tRNA target sequence at the site of modification abolished the activity of the complex.

These studies demonstrate the unique function of archaeal C/D box sRNAs in the methyl modification of tRNAs (Zieche et al., 2004). The importance of this reaction may be related to the fact that many Archaea grow at high temperatures. It has been suggested that under these conditions the tRNA primary sequences may sometimes lack sufficient structure to be recognized by conventional methyl transferase enzymes. The introduction of the methyl group by the sRNP guide complex, which presumably does not require tRNA structure, likely contributes to stabilization and folding of tRNA into the conventional tertiary structure (Dennis et al., 2001; Renalier et al., 2005).

Introns in archaeal tRNA genes

The genomes of many Archaea contain one or a small number of tRNA genes that are disrupted by short introns, usually within the anticodon loop of the tRNA (Kaine et al., 1983; Daniels et al., 1985). The most interesting example of this is the tRNA^{Trp} gene found in most species of Euryarchaea. In *Haloferax volcanii*, the gene contains a 142 nucleotide long intron located between positions 37 and 38 within the anticodon loop of the mature tRNA. The 5'exon–intron and intron–3'exon boundaries are partially complementary and are capable of forming a short helical region containing a bulge–helix–bulge (BHB) structural motif that is characterized by two three-base bulges, positioned on opposite strands and separated by four base pairs. This motif is the substrate for the intron excision endonuclease that cleaves within the bulges to excise all archaeal introns having the motif at the intron–exon boundaries (Thompson et al., 1988; Thompson & Daniels, 1990). In addition, the BHB motif is generally present in the processing stems of the rRNA operon transcript (see below) and the endonuclease activity is used to excise pre16S and pre23S rRNAs from these precursor transcripts (Dennis et al., 1998) and possibly unstable pre-mRNAs (Lykke-Andersen et al., 1997).

The tRNA^{Trp} intron is unusual in that it contains a sequence with all of the hallmark features of a C/D box sRNA and the two guide regions are predicted to guide methylation to positions C34 and U39 in mature tRNA^{Trp}. It has been suggested that this arrangement might be used to facilitate *cis*-directed methylation and to link methyl modification of positions C34 and U39 to intron excision and tRNA maturation (Thompson et al., 1988; Thompson & Daniels, 1990; Clouet d'Oval et al., 2001). However, recent experiments suggest that guide-directed methylation occurs most efficiently in *trans* between two intron-containing (unprocessed) transcripts or between an excised intron and a unprocessed transcript (Singh et al., 2004). Methylation at position C34 precedes methylation at U39. To date there has been no identification of the RNA ligase enzyme and some investigators have speculated that ligation may be catalyzed by the RNA bulge–helix–bulge motif.

Circular C/D box sRNAs

The complexity of C/D box sRNA biogenesis and function has been highlighted by the recent discovery that these RNAs can exist as circular molecules (Starostina et al., 2004). It was noticed that in a cDNA library of small RNAs from *Pyrococcus furiosus* there were a number of unusual C/D box sRNA clones that appeared to be circular permutations of the corresponding gene sequence. To explain the existence of these C/D box variants, a model involving the circularization of C/D box RNAs that are randomly cleaved during the RNA purification and cloning procedure was proposed. Using Northern blotting and RT-PCR analysis, it was demonstrated that in *Pyrococcus* linear C/D box RNAs coexist with single circles or multiple circularized forms that resulted from concatemerization and ligation of the normal 5' to 3' ends of the sRNA. The RNA end ligation activity in this case is clearly distinct from the ligase activity that is involved in the tRNA intron excision/ligation pathway and in the rRNA processing pathway (see below). Interestingly, both linear and circular C/D box RNAs were able to associate with all three core C/D box proteins, as indicated by their presence in anti-L7Ae and anti-Nop5 immunoprecipitates. At this point it is unclear what role these circular sRNAs might play in sRNA biogenesis and function, or if the RNP complexes assembled with circular sRNAs are active in methylation.

Structure and function of archaeal H/ACA box RNP complexes

In contrast to eukaryotic organisms where both C/D and H/ACA box RNPs are abundant, archaeal organisms appear to contain only a few H/ACA RNPs and pseudouridine modifications in rRNA are infrequent. Identification of H/ACA RNAs in Archaea was accomplished by sequencing of entries in cDNA libraries prepared from small RNAs from several species of Archaea (Tang et al., 2002b; Zago et al., 2005). The low abundance of these H/ACA RNAs and their poorly conserved features have precluded the development of an effective H/ACA gene finding program that can be widely applied to other archaeal genomes. Again, the presence of pseudouridine modifications in rRNA at sites predicted from guide sequences of the cloned sRNAs have been confirmed by biochemical analysis. In eukaryotic organisms four proteins are associated with H/ACA snoRNAs and these have homologs that are encoded in archaeal genomes: Cfb5, the putative pseudouridine synthetase, Gar1, Nop10, and Nhp2. Interestingly, in a subset of Archaea, the gene encoding the Cfb5 enzyme contains introns. In addition, the exon–intron junctions of this gene are predicted to fold into a bulge–helix–bulge motif, suggesting that splicing of the Cfb5 pre-mRNA relies on the same endonuclease activity involved in the processing of tRNA introns and rRNA precursors (Watanabe et al., 2002). The Nhp2 protein is a member of the 15.5 kDa family of proteins and in archaeal H/ACA sRNPs this protein is replaced by the homologous L7Ae protein that is a component of the large ribosome subunit and of C/D box RNPs (see above). Analysis of archaeal H/ACA sRNAs reveals the presence of an RNA K-turn structural motif that serves as the binding site for the L7Ae protein (Plate 20.1; Rozhdestvensky et al., 2003).

The majority of information relating to the structural organization of archaeal H/ACA box RNAs comes from the reconstitution of an active complex using protein components and Pf9 sRNA from *P. furiosus* and Pab91 sRNA from *P. abyssi* (Baker et al., 2005; Charpentier et al., 2005). The functional hairpin unit of the H/ACA pseudouridine guide RNA is about 60–75 nucleotides in length. The Pf9 sRNA consists of a single hairpin structure containing: (i) a bipartite guide sequence located within a central loop; (ii) a K-turn near the top of the stem; (iii) a conserved GAG sequence that is part of the terminal loop; and (iv) a conserved box ACA at the 3' base of the stem. The Pab91 sRNA differs somewhat in that the K-turn is located in the loop at the top of the hairpin and the GAG sequence is not present in the terminal loop (Charpentier et al., 2005). In the target RNA the two components that base pair with the bipartite guide sequence are usually separated by two nucleotides, one of which is the uridine to be modified. The guide target interaction positions the uridine in the pseudouridylation pocket (Plate 20.1). The distance between the pseudouridine pocket that interrupts the bipartite guide and the ACA box sequence at the base of the helix is generally about 14 nucleotides. The H/ACA sRNAs can contain between one and three repeats of this basic unit structure. Each unit appears to contain one copy of each of the four proteins (Tang et al., 2002b).

Gel shift assays using wild type or mutant Pf9 guide RNAs indicate that the Cbf5 and the L7Ae bind independently to the sRNA, whereas neither Gar1 nor Nop10 binds directly to the RNA (Baker et al., 2005). The Cfb5 determinants on the RNA are complex and include the box ACA sequence, the pseudouridylation pocket and the conserved GAG trinucleotide in the terminal loop. Protein interactions and gel shift experiments indicate that Gar1 and Nop10 are both able to interact with Cbf5 but not with each other or with the L7Ae protein. This has led to the model in which the Cbf5 protein binds to the RNA through at least three widely spaced contact points and independently recruits Gar1 and Nop10 to the complex. The complex remains inactive in pseudouridylation until L7Ae binds to the K-turn and introduces a sharp bend into the backbone of the RNA hairpin that is presumably required for activation of the archaeal H/ACA complex. Inclusion of a fragment of rRNA complementary to the bipartite guide with an appropriately positioned uridine results in target uridine isomerization. It has been suggested that Gar1 and Nop10 may be involved in interaction with or release of the rRNA substrate (Baker et al., 2005).

The second study with Pab91 has provided a slightly modified view of the *in vitro* structure, assembly, and activity of archaeal box H/ACA sRNAs (Charpentier et al., 2005). The Pab91 sRNA differs from Pf9 in that it is somewhat shorter and the K-turn is located in the terminal loop of the hairpin; there is no conserved GAG sequence in this terminal loop. In contrast to the earlier work, the results of Charpentier et al. (2005) suggest that: (i) the Cbf5 protein shows rather low affinity for the sRNA; (ii) only the Cfb5 and Nop10 proteins were required for activity; and (iii) the box ACA was not required for activity (or for Cbf5 binding). These minor discrepancies may be the result of differences in the *in vitro* structure, assembly, and activity of different archaeal box H/ACA sRNA or in the sensitivities of the assays used in the two studies. Both studies indicate that the maximal pseudouridylation activity occurred when all four proteins and box ACA were present, and the efficient binding of the target RNA to the RNP complex required a uridine at the site of pseudouridylation. At this point most of the evidence suggests that the interactions enumerated in the archaeal system are likely to be conserved in eukaryotic systems (Baker et al., 2005; Charpentier et al., 2005).

L7Ae binding and the RNA K-turn motif are common features of archaeal small RNAs

An informatic analysis of RNA K-turn motifs based on a consensus sequence suggested that K-turns might be common to many types of RNAs. In Archaea the single multifunctional L7Ae protein, which binds to the K-turn motif, is a component of the ribosome (Ban et al., 2001), as well as C/D box (Omer et al., 2002) and H/ACA box RNP complexes (Rozhdesventsky et al., 2003). To survey the frequency of the motif more carefully and to begin to characterize the diversity of RNAs containing the motif, we constructed a cDNA library from the small RNAs that were immunoprecipitated from *S. solfataricus* cell extracts with antibodies against the multifunctional L7Ae protein (Zago et al., 2005). The sequencing of 128 insert-containing clones revealed a remarkable diversity of RNAs that extended far beyond the canonical C/D box and H/ACA box sRNAs. Single or multiple clones representing overlapping regions of 45 different core sequences were recovered. Most of these RNAs represented by these clones had recognizable K-turn motifs and were able to form a stable association with the L7Ae protein. The remaining RNAs that did not bind L7Ae directly were presumed to be present in the library because they were in larger complexes that contained somewhere a K-turn and were able to bind the L7Ae protein. A second more general library, constructed from a pool of unfractionated small RNAs, resulted in the identification of 57 novel sequences (Tang et al., 2005). The clones recovered using the two different library construction strategies exhibit numerous common RNA entries or RNA entries with similar features. Based on characteristic features, the RNAs can be grouped into six categories (Plate 20.2): (i) RNAs that exhibited some or all of the features of canonical C/D box or H/ACA box sRNAs; (ii) sense strand RNAs that were encoded within or overlapping annotated ORFs; (iii) RNAs that were derived from intergenic regions; (iv) antisense RNAs that were partially or completely complementary to the mRNAs of protein encoding ORFs or to C/D box sRNAs; (v) RNAs that corresponded to fragments of 7S RNA; and (vi) stable tRNAs and fragments from the 5′ and 3′ ends of 16S and 23S rRNA. It is notable that no internal fragments of 16S or 23S rRNA were recovered (Zago et al., 2005).

Pre-rRNA processing complex

Of particular interest were the four rRNA fragments obtained from the L7Ae library representing the 5′ and 3′ ends of 16S and 23S rRNA. None of the four fragments exhibits binding affinity to the L7Ae protein and only the fragment from the 5' end of 16S rRNA contains a known or predicted site of 2'O-ribose methylation (position U52). Why were these specific RNA fragments recovered in the L7Ae library, whereas no other internal rRNA fragments were recovered? The answer to this question is likely to be related to the machinery that cleaves the precursor rRNA transcript to release the 16S and 23S rRNA sequences. In *Sulfolobus*, as in most other Archaea, the pre-rRNA contains long bulge–helix–bulge (BHB) processing stems that surround the respective 16S and the 23S rRNA sequences (Dennis et al., 1998). The BHB RNA motif is found not only in the processing stem of archaeal pre-rRNA but also at the intron–exon junction of archaeal intron-containing RNA transcripts. This motif is the substrate for an intron excision endonuclease that is homologous to the eukaryotic tRNA intron excision endonuclease (Thompson & Daniels, 1988). Another interesting group of RNAs were recovered from both the small RNA and the L7Ae libraries; these RNAs contained 5'ETS, ITS, and 3'ETS sequences from the pre-rRNA that had been ligated to each other at the site of endonuclease cleavage in the BHB motif (Tang et al., 2002a; Zago et al., 2005). Interestingly, the ITS spacer sequence contains a well defined K-turn motif and exhibits high affinity for L7Ae protein binding. Together these results suggest that a large processing complex forms on the pre-rRNA and contains the L7Ae protein, the intron excision endonuclease, the yet to be identified exon ligase, and probably other components. Antibodies against the L7Ae protein apparently precipitated the complex and fragments of the RNA near the center of the complex (i.e. the spacer sequences and the 5' and 3' ends of the mature rRNAs) were recovered in the L7Ae library. It should also be noted that in eukaryotes a large complex containing a number of different C/D box and H/ACA box snoRNPs is assembled and mediates essential endonucleolytic cleavage events within pre-rRNA (Granneman & Baserga, 2005).

The signal recognition particle

Both small RNA libraries also contained fragments of 7S RNA, the RNA component of the signal recognition particle. In eukaryotes, the complex contains, in addition to the 7S RNA, six different proteins (Keenan et al., 2001). Homologs to only two of these proteins (SRP19 and the GTPase SRP54) have been identified in Archaea (Eichler & Moll, 2001). Biochemical and mutational analysis identified a K-turn motif within the large asymmetric loop in the middle of helix 5 of the *Sulfolobus* 7S RNA (Zago et al., 2005). In the eukaryotic SRP, this region is the binding site for SRP68/72; these proteins are responsible for bending the RNA so that it can interact simultaneously with the A site and the exit tunnel on the surface of the large ribosome subunit (Halic et al., 2004; Iakhiaeva et al., 2005). The binding of the L7Ae protein to the K-turn motif in the *Sulfolobus* 7S RNA would be expected to stabilize the tightly kinked structure in the RNA backbone and might therefore eliminate the need for the SRP68/72 proteins to provide this function. It is also interesting to note that a full-length antisense 7S transcript has been detected in *Sulfolobus*; the role of this transcript in SRP function remains unclear (Zago et al., 2005).

Sense and antisense RNAs

Interest in small RNAs has burgeoned in recent years with the discovery and characterization of: (i) small noncoding RNAs in bacteria; and (ii) small inhibitory and micro RNAs in eukaryotes (Wassarman et al., 2001; Nelson et al., 2003; Bartel, 2004; Stortz et al., 2004). Many of these RNAs are antisense, exhibit full or partial complementarity to mRNAs, and have been implicated in the processes that regulate mRNA translation and stability. Others are sense strand elements such as riboswitches or other *cis* acting elements that regulate translation, attenuation, processing, or degradation of the mRNA.

A number of fragments of mRNA that contain stable K-turns were recovered from the small RNA and the L7Ae libraries (Tang et al., 2005; Zago et al., 2005). The K-turn motifs in these sense strand fragments are most often derived from regions of complex secondary structure that overlap either the translation initiation or translation termination sites and suggest an important role for the L7Ae in the post-transcriptional regulation of gene expression (Plate 20.3a). One example of an mRNA fragment clone (sR110) corresponds to the 5' UTR and translation initiation region of the ORF that encodes subunit four of formate dehydrogenlyase. The 5' UTR forms a long hairpin structure with a K-turn at the base; the Shine–Dalgarno and AUG translation initiation codon are sequestered in the descending side of the hairpin above the K-turn. Another example of an mRNA fragment clone (sR114) corresponds to the translation termination and 3′ UTR region of a hypothetical transposase-related protein (Plate 20.3b). The region exhibits a hairpin structure that contains the translation termination codon near the base of the descending strand, immediately under a K-turn motif. The genome of *S. solfataricus* contains two additional highly similar copies of this element and it is conserved in the genome of several related archaeal species (Zago et al., 2005). This suggests that the sequence may represent a module or controlling element that can be recruited to function at the 3′ end of different ORFs.

Both *Sulfolobus* small RNA libraries contain small fragments of antisense RNA that are complementary to 5′, 3′, or internal regions of mRNAs that most often (but not exclusively) encode transposon related proteins (Tang et al., 2005; Zago et al., 2005). The synthesis of these proteins is expected to be tightly regulated because of the genomic instability that results from uncontrolled transposition. The detected antisense RNAs are predicted to be important components in this regulation. Many of these antisense RNAs contain K-turn motifs and exhibit high affinity binding to the L7Ae protein. One example of an antisense RNA fragment (sR129) is complementary to the last 15 nucleotides of the transposase-related gene and the first 49 nucleotides of the transposase 3′ UTR (Plate 20.3c). The fragment contains a K turn motif in the region complementary to the 3′ UTR. This 64 nucleotide fragment may have been derived from a longer RNA since we also recovered a second antisense RNA fragment (sR126) that is complementary to the 5′ end of the same transposase mRNA but does not contain a K-turn motif (Zago et al., 2005). Indeed, a number of the antisense clones do not contain a recognizable K-turn or bind the L7Ae protein. Some or all of these may be fragments of larger antisense transcripts that do contain the motif and do bind the protein. These results suggest that K-turn motifs and their interaction with the L7Ae protein may in at least some cases be involved in the antisense regulation of gene expression in *S. solfataricus*. How these antisense RNAs might function is currently unknown.

Conclusion

It is no longer feasible to think of RNA within the narrow context of translation. The discoveries over the past 25 years demonstrate that RNAs are ubiquitous and pervasive, and capable of participating in a large number of important biological processes. In almost all cases the RNAs are associated with proteins to form dynamic ribonucleoprotein complexes. The structure and function of these dynamic complexes depend on the ability of RNAs to form unique inter- and intramolecular secondary (complementary base pairing) and higher order structures that in terms of dynamic flexibility are generated generally far beyond the range that can be achieved by structurally more rigid protein complexes.

The characterization of the entries in small RNA libraries from *Sulfolobus* has revealed an unexpected number and diversity of small cellular RNAs. Many of these RNAs exhibit some of the features of the small interfering or micro RNAs that are found in eukaryotes. Equally surprising was the prevalence of the K-turn structural motif in these RNAs and their apparent propensity to assemble with L7Ae and probably other proteins into RNP complexes. Virtually every clone points toward novel and interesting translational regulation, biochemistry, or physiology. The challenge for the future will be to elucidate the structures, compositions, and functions of these ncRNA-containing RNP complexes and to identify their roles in information transfer and control of gene expression. Every indication suggests that our current understanding of the ncRNA world in prokaryotic systems is only a small part of a much larger picture that is likely to penetrate into virtually every aspect of microbial physiology.

Acknowledgments

This work has been supported by the National Science Foundation (PPD) and by the Natural Sciences and Engineering Research Council of Canada (ADO). Any opinions, findings, and conclusions expressed in this chapter are those of the authors and do not necessarily reflect the views of the National Science Foundation or of the Natural Sciences and Engineering Research Council of Canada.

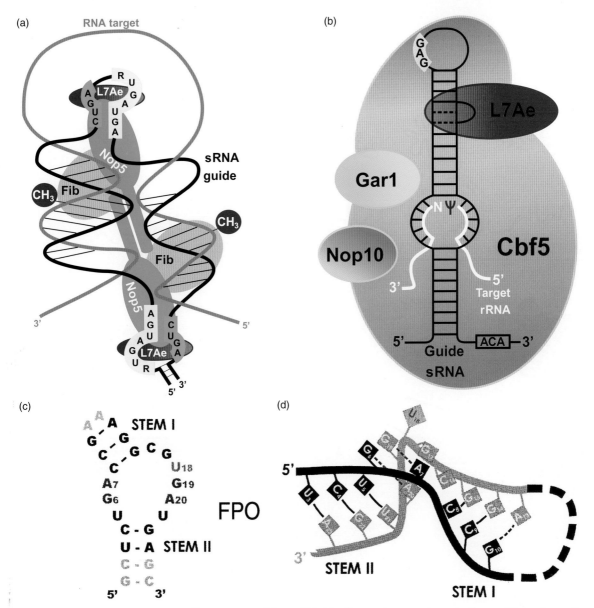

Plate 20.1 Structural organization of box C/D and box H/ACA ribonucleoprotein complexes. The schematic structure of archaeal C/D box and H/ACA box RNPs are depicted in (a) and (b) respectively and the structure of the K-turn motif is represented in (c). (a) The archaeal C/D box sRNAs are bipartite molecules about 50–60 nucleotides in length that guide 2′-O-ribose methylation to target RNAs. They contain the conserved C (RUGAUGA) and D (CUGA) box sequences near their respective 5′ and 3′ ends and second copies, termed C′ and D′, located near the center of the molecule. The C and D and C′ and D′ sequences form K-turn structural motifs within the RNA. The active RNP complex contains three proteins, L7Ae, Nop5, and Fib (Fibrillarin), each present in two copies per complex. This contributes to the symmetrical structure of the complex. The guide sequences are located immediately preceding the D or D′ boxes; when base paired with a complementary target RNA, methylation is directed to the nucleoside in the target molecule that is base paired five nucleotides in front of the beginning of the D′ or D box sequence; this is the N plus five rule. Many archaeal C/D box sRNAs contain two guide sequences and direct methylation to closely linked target sites within the small or large subunit rRNAs. (b) The core RNA component of box H/ACA RNPs is a helical structure with an internal loop containing the bipartite guide sequence and the pseudouridine (ψ) pocket, a K-turn loop and terminal loop containing the conserved GAG sequence, and the conserved box ACA located at the 3′ base of the stem. In some cases the stem is truncated at the K-turn loop and the terminal loop containing the GAG sequence is missing. The Cbf5 and the L7Ae proteins bind separately to the RNA. Cfb5 recognizes determinants in the box ACA, the pseudouridine pocket, and GAG terminal loop (when present) for binding to the RNA, and through protein–protein interactions recruits Gar1 and Nop10 to the complex. The guide sequence is bipartite and located in the large internal loop containing the pseudouridine pocket. Binding of the target RNA positions the uridine to be isomerized into the pocket (adapted from Baker et al., 2005). The primary nucleotide sequence of a typical box C/D RNA motif used for co-crystallization with the L7Ae protein is illustrated in the open form (c) and the closed tightly kinked form (d) that occurs when the RNA is bound by the protein in the co-crystal (adapted from Moore et al., 2004). The K-turn motif exhibits a sharp bend of between 45 and 63° and is characterized by two sheared A:G base pairs that close the loop and cause the adjacent U at position 18 to protrude from the loop.

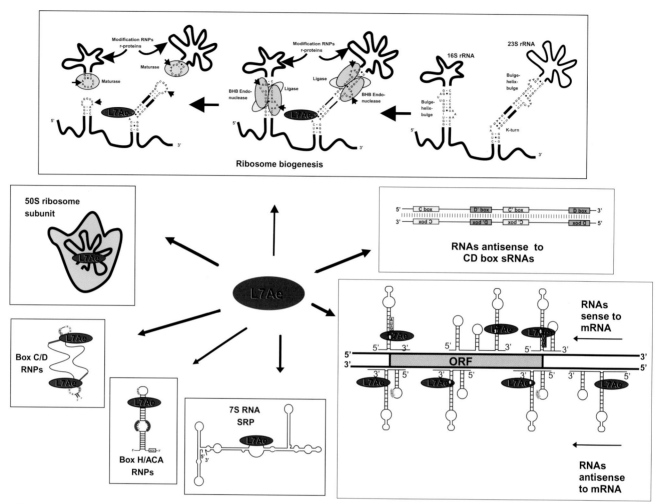

Plate 20.2 Ribonucleoprotein complexes that contain the L7Ae protein. The *S. solfataricus* RNA K-turn motif is common to many different RNA transcripts and is the binding site for the L7Ae protein. Examples of all of the RNAs have been isolated from the two small RNA libraries that have been characterized (see text). The different complexes include: (i) the 50S subunit of the ribosome; (ii) box C/D methylation guide sRNPs; (iii) box H/ACA pseudouridylation guide sRNPs; (iv) the RNP complex that mediates processing of the precursor rRNA transcript; (v) the 7S RNA containing signal recognition particle; and (vi) sense, antisense, and intergenic RNAs. In addition, *S. solfataricus* also contains antisense RNAs to at least some box C/D sRNAs and to 7S RNA. These RNAs do not bind L7Ae directly but are believed to be at some time in RNP complexes that contain the protein; the function of these antisense RNAs CD is not known. The scheme depicting the early stages of ribosome biogenesis is illustrated in three steps at the top of the plate. The first step (right) shows the rRNA primary transcript. The 16S and 23S rRNA sequences are depicted as loops extending from the top of the respective processing stems containing the bulge–helix–bulge motif that is the substrate for the tRNA intron excision endonuclease. The 23S rRNA processing stem also contains a K-turn motif. The 5′ETS, ITS and 3′ETS are depicted as solid lines. In the second step (center) the binding of the L7Ae to the K-turn and the intron excision endonuclease, and the putative ligase to the BHB motifs, is illustrated. In step three (left), the endonuclease cleaves in the loops on opposite sides of the processing stem and the ligase connects the 5′ and 3′ ends so that the pre 16S and pre 23S rRNAs are released from the primary transcript as circular intermediates. The ligase also joins the 5′ETS to the ITS and the ITS to the 3′ETS. A maturase activity is required to remove precursor sequences from the circular pre 16S and pre 23S RNAs and convert them to the linear mature forms. The ligase and maturase activities have not been identified or characterized.

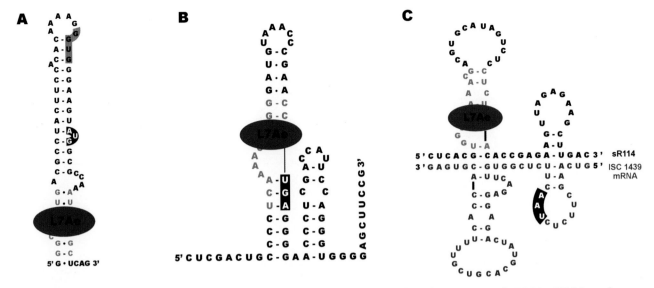

Plate 20.3 Sense and anti-sense RNAs associated with L7Ae. The predicted structures of sR110, sR114, and sR129 are illustrated. The initiation and termination codons are highlighted in black. The nucleotides forming the predicted K-turn motifs are in blue and the L7Ae protein bound to the RNA is depicted. (a) sR110 is overlapping the 5′UTR and the first 21 nucleotides of the protein coding region of the formate dehydorgenlyase subunit 4. The Shine–Dalgarno motif present in the 5′UTR of the mRNA transcript is highlighted in green. (b) sR114 overlaps the 3′end and 3′UTR of a hypothetical transposase-related protein and the predicted L7Ae binding site of this RNA is overlapping the termination codon. (c) sR129 RNA is complementary to the 3′ end and part of the 3′UTR of the mRNA coding for transposase 1439.

21
Transcriptomics, proteomics, and structural genomics of *Pyrococcus furiosus*

Michael W. W. Adams, Francis E. Jenney Jr,
Chung-Jung Chou, Scott Hamilton-Brehm,
Farris L. Poole II, Keith R. Shockley,
Sabrina Tachdjian and Robert M. Kelly

Introduction

Pyrococcus furiosus is a hyperthermophilic archaeon first isolated from shallow hydrothermal marine vents near Vulcano, Italy (Fiala & Stetter, 1986). It is an obligately anaerobic heterotroph that can grow on a variety of glucan- and peptide-based media, generating acetate as a primary metabolic product as well as either H_2 or H_2S depending on whether elemental sulfur is present. At the time of its isolation in 1986, *P. furiosus* was one of few hyperthermophilic Archaea that could be grown to significant cell densities (Verhagen et al., 2001), making it an important source of thermostable proteins and thereby emerging as a model microorganism within this group of extremophiles. Indeed, over 600 literature citations on PubMed exist for *P. furiosus* to date. The focus on *P. furiosus* as a model hyperthermophile was further reinforced by the availability of its genome sequence (Robb et al., 2001). In addition, *P. furiosus* has been the target of a recent structural genomics study (Adams et al., 2003) and was the first member of the Archaea to be studied by DNA microarrays (Schut et al., 2001). The aim of this chapter is to summarize these recent developments.

The *P. furiosus* genome and problems with genome annotation

P. furiosus has a small, well defined genome of 1.9 Mb comprising at least 2065 predicted genes (Robb et al., 2001). Genome sequences of two other members of the genus, *P. horikoshii* and *P. abyssi*, are available and these enable issues such as genome evolution and horizontal gene transfer to be explored (see Cohen et al., 2003). However, the extent to which definitive annotations are available for genes encoded in these genomes or any other is difficult to gauge. In the case of *P. furiosus*, the original genome annotation (2065 genes) was subsequently annotated in two public databases by the Institute for Genomic Research (TIGR) and the National Center for Biotechnology Information (NCBI). Remarkably, more than 500 of the originally annotated genes differ in size in the two databases, many very significantly (Poole et al., 2005). For example, more than 170 of the predicted proteins differ at their N-termini by more than 25 amino acids. Similar discrepancies were observed in the TIGR and NCBI databases with the other archaeal and bacterial genomes examined (Poole et al., 2005). In addition, the two databases contain 60 (NCBI) and 221 (TIGR) genes not present in the original annotation of *P. furiosus* (Poole et al., 2005). Experimental confirmation of the precise sizes of the genes in the *P. furiosus* genome is available for only a small fraction of them. However, use of newly available tools in transcriptomics, proteomics, and structural genomics is allowing a better definition of the genome and the genes that are transcribed. Examples of such efforts are discussed below.

Table 21.1 Transcriptional response experiments with *P. furiosus*.

Condition	Reference
Maltose ± sulfur	Schut et al., 2001
Maltose/peptides	Schut et al., 2003
Heat shock	Shockley et al., 2003
Glucan polysaccharide utilization	Lee et al., 2006
Cold shock	Weinberg et al., 2005
Growth rate	Chou & Kelly, unpublished
Growth phase	Chou & Kelly, unpublished

The *P. furiosus* transcriptome

The development of DNA microarrays to assess microbial transcriptomes represented a quantum leap in our ability to understand the genetic components of microbial biochemistry, physiology, and ecology. However, it is becoming clear that meaningful differential gene expression information requires effective experimental design, experimental technique, processing of cellular material to obtain RNA, microarray slide preparation and scanning, and statistical analysis of data. These steps become more complicated in the case of organisms such as *P. furiosus*, given the need to maintain anaerobic growth conditions at high temperatures and the relatively low biomass yields characteristic of hyperthermophilic Archaea (Kelly, 2004). Even with the full complement of global transcriptional response information, there is the additional challenge of sorting through such data to glean meaningful insights into physiology and metabolism. Nonetheless, transcriptomics have been utilized to investigate *P. furiosus* and these have led to new perspectives into microbial life at extremely high temperatures. Table 21.1 lists experimental conditions for which genome-wide transcriptional response information has been collected for this organism and the results are summarized here.

The number of genes in the *P. furiosus* genome

Transcriptome analyses using DNA microarrays containing representatives of all genes that are annotated in the genome provide a means of assessing what fraction is expressed under a given growth condition. Hence, in peptide-grown *P. furiosus* at 95°C, approximately 80% of the 2065 genes were expressed at a detectable level (Schut et al., 2001). In addition, DNA microarrays were used to assess the validity of previously unannotated genes (Poole et al., 2005). Of the 61 genes examined, 11 were expressed in *P. furiosus* when the organism was grown (on maltose or peptides) at either 95 or 72°C. In the same study (Poole et al., 2005) a structural genomics approach (see below) was used to assess the validity of 54 previously unannotated genes. It was reasoned that artifactual genes would yield unfolded proteins that would be degraded in *Escherichia coli*, and would also not be stable to a heat treatment step. Seven of the 54 genes yielded heat-stable recombinant proteins when they were expressed in *E. coli* (although only one of the seven genes was expressed in *P. furiosus*). Hence, it was concluded that, based on transcriptional and recombinant approaches, the *P. furiosus* genome contains at least 17 genes not previously recognized in the original annotation (Poole et al., 2005).

The response of *P. furiosus* to growth phase and growth rate

When examining the mechanisms underlying transcriptional responses, it is important to account for those that stem from growth phase and/or growth rate effects rather than the stress or environmental change that is administered. Growth-related effects can impact data interpretation, confounding interpretation of "cause-and-effect" experiments. To illustrate this, batch and continuous cultures of *P. furiosus* were grown at 90°C on media containing maltose and yeast extract as carbon and nitrogen sources, respectively. For a 1.5 L batch culture, samples were taken at 5, 8, and 16 hours, corresponding to mid-exponential, early stationary, and late stationary phase, respectively, of batch growth. Approximately 5% of the genes in the *P. furiosus* genome were differentially expressed as a function of growth phase (see Table 21.2). Over 100 genes were differentially expressed twofold or more during the transition from exponential to early stationary phase. The observed up-regulation of many genes involved in amino acid and thiamine biosynthesis likely reflects nutrient-limiting conditions that play a role in growth rate reduction. The gene encoding a universal stress protein (PF1557) was up-regulated 11-fold in stationary phase; the corresponding gene in *E. coli* has been observed to be an indicator of stationary phase onset (Kvint et al., 2003). Genes responsive to growth phase changes need to be considered when the influence of environmental factors on the transcriptome is examined.

In addition to the growth phase, growth rate can also affect transcription. Using the growth conditions that were used for the batch culture, *P. furiosus* was grown in a chemostat at three dilution rates, 0.15 (A), 0.25 (B), and 0.45/h (C), to examine the influence of growth rate on the transcriptome (Chou & Kelly, unpublished data). Perhaps not surprisingly, differential gene expression was growth rate-dependent. After each change of dilution rate (A to B, and B to C), more than 100 genes were differentially expressed twofold or more. These included some of the same amino acid and cofactor biosynthesis genes that had already been observed to be up-regulated in the transition from exponential to stationary phase in batch culture. Furthermore, genes related to cellular repair mechanisms, such as heat shock and DNA repair and recombination, were higher in A-type cells, while oxidative stress genes were highest in C-type cells. Many genes annotated as "hypothetical," but that encoded putative membrane proteins and had signal peptides, were also affected by changing dilution rates.

Table 21.2 illustrates the sensitivity of the *P. furiosus* transcriptome to growth phase and growth rate. Similarly, organisms can respond to what might appear to be minor variations in growth medium. For example, Plate 21.1 shows a volcano plot illustrating differential expression of genes in *P. furiosus* when yeast extract was added to batch cultures growing on tryptone. This plot reports statistical significance on the y-axis ($-\log_{10}$ p-value) and $\log_2$ fold change on the x-axis. A total of 165 genes (95 up-regulated, 70 down-regulated) were differentially expressed twofold or more (Shockley & Kelly, unpublished data). This level of differential expression was unanticipated and points to the sensitivity of *P. furiosus* to medium composition. For example, the results show that thiamine pyrophosphate biosynthesis is required in the absence of yeast extract, based on genes encoding a thiamine biosynthetic enzyme (PF1530), phosphomethylpyrimidine kinase

Table 21.2 Differential gene expression (more than twofold) during batch and continuous cultures of *P. furiosus* grown on maltose at 90°C.

Batch	Comparisons between different growth phases		
	8 v. 5 hour	16 v. 8 hour	16 v. 5 hour
Genes up-regulated	67	28	95
Genes down-regulated	37	0	39

Continuous	Comparisons between different growth rates		
	0.25 v. 0.45/h	0.15 v. 0.25/h	0.15 v. 0.45/h
Genes up-regulated	94	66	130
Genes down-regulated	51	62	78

(PF1333), hydroxyethylthiazole kinase (PF1335), thiamine phosphate phosphorylase (PF1334), and phosphoribosylaminoimidazole carboxylase (PF0426). Each of these enzymes has been implicated in thiamine pyrophosphate biosynthesis (Rodionov et al., 2002), and all were expressed at elevated levels in cells grown with tryptone only. Two genes (PF1337 and PF1338) were also expressed at significantly higher levels without yeast extract in the medium. These genes are predicted to be the transcriptional activator TenA but recent advances in understanding the crystal structure of PF1337 suggest that these TenA homologs may be biosynthetic enzymes (Benach et al., 2005). These examples serve to demonstrate that medium composition must be carefully controlled if meaningful information is to be gleaned from transcriptional responses.

The response of *P. furiosus* to the presence of elemental sulfur

The first investigation of the transcriptome of *P. furiosus* utilized DNA microarrays that contained a representative subset of 271 genes (Schut et al., 2001). These were used to investigate the effects of elemental sulfur (S°) in cells grown at 95 °C with maltose as the carbon source. Cells were grown in batch culture and were harvested at the same phase of growth (mid-exponential). The 271 genes on the array included those that are proposed to encode proteins mainly involved in the pathways of sugar and peptide catabolism, in the metabolism of metals, and in the biosynthesis of various cofactors, amino acids, and nucleotides. The expression of 21 of the genes decreased by more than fivefold when cells were grown with S° and, of these, 18 encoded subunits associated with the three different hydrogenase enzymes in this organism (Sapra et al., 2003). The remaining three genes encode homologs of ornithine carbamoyltransferase and HypF, both of which appear to be involved in hydrogenase biosynthesis, as well as a conserved hypothetical protein. On the other hand, two previously uncharacterized genes (PF2025, PF2026) were up-regulated by more than 25-fold in S°-grown cells and these were proposed to encode a novel S°-reducing, membrane-associated, iron-sulfur cluster-containing complex (SipAB). This complex is thought to involve a putative flavoprotein and a second FeS protein, as their genes were also up-regulated in S°-grown cells. Genes encoding homologs of proteins involved in amino acid metabolism were similarly up-regulated, a finding consistent with the fact that growth on peptides is a S°-dependent process. While these results represented the first derived from the application of DNA microarrays to either an archaeon or a hyperthermophile, the precise mechanism by which S° is reduced by *P. furiosus* and the role of the novel SipAB complex remain unknown.

The response of *P. furiosus* to changes in the primary carbon source

The first complete-genome DNA microarray to be constructed for a hyperthermophile or a nonhalophilic archaeon contained the 2065 genes that have been annotated in the *P. furiosus* genome. It was used to determine relative transcript levels in cells grown at 95 °C with either peptides or maltose as the primary carbon source (Schut et al., 2003). Of the 1667 genes that were expressed at a significant level, 125 of them (8%) differed in expression by more than fivefold between the two cultures, and 82 of the 125 (65%) appeared to be part of co-regulated operons, indicating extensive coordinate regulation. Surprisingly, five of the 27 operons encode (conserved) hypothetical proteins. Eighteen operons were up-regulated in maltose-grown cells, including those responsible for maltose transport and for the biosynthesis of 12 amino acids as well as citric acid cycle intermediates. Nine operons are up-regulated in peptide-grown cells, including those encoding enzymes involved in the production of acyl and aryl acids and 2-ketoacids, all of which are used for energy conservation. Consistent with this, spent medium from peptide-grown cells was found to contain branched-chain and aromatic acids. In addition, six nonlinked enzymes in the pathways of sugar metabolism were regulated more than fivefold. Three of them are unique to the unusual glycolytic pathway and were up-regulated in maltose-grown cells, while the other three are unique to gluconeogenesis and were up-regulated in peptide-grown cells. In addition, the catalytic activities of 16 metabolic enzymes, whose expression appeared to be highly regulated in the two cell types, were measured and the results correlated very well with the microarray data. This study demonstrated that *P. furiosus* readily adapts to changes in its primary carbon source with quite dramatic and highly coordinated changes in gene expression. While many of the responses were in accordance with known pathways, five of 27 highly regulated (greater than fivefold) operons encode primarily conserved hypothetical proteins and their functions remain unknown.

The response of *P. furiosus* to cold shock and cold adaptation

Whole genome transcriptional profiling was used to assess how gene expression differs in *P. furiosus* when it is grown at 95 °C, near its optimum growth temperature (doubling time ~1 hour), compared to 72 °C, the lower end of the temperature range for significant growth (doubling time ~5 hours) (Weinberg et al., 2005). In addition, in separate experiments, cultures were shocked by rapidly dropping the temperature from 95 to 72 °C, which resulted in a 5 hour lag phase, during which time little growth occurred. From the transcriptional data it was evident that cells undergo three very different responses to the "cold" stress (72 °C): there is an early shock response over the first 1–2 hours, a late shock response up to 5 hours, and an adapted response that occurs when cells are grown for many generations at 72 °C. Each of these responses involved up-regulation in the expression of more than 30 genes, and these were unique to that response. In general, these included genes encoding proteins involved in translation, solute transport, amino acid biosynthesis, and tungsten and intermediary carbon metabolism, as well as numerous conserved-hypothetical and/or membrane-associated proteins. The up-regulation of genes involved in amino acid biosynthesis by cold shock is particularly intriguing since this occurs when cells are either not growing (in the acclimation phase) or growing very slowly (at 72 °C rather than 95 °C). Consequently, this is clearly not an example of the general growth phase response that was discussed above, but indicates a high rate of "cold-specific" protein synthesis, independent of cell growth and culture condition. In addition, examination by conventional 1D-SDS gel electrophoresis of highly washed membranes of cold-adapted cells revealed two major membrane proteins that were not evident in cells grown at 95 °C. Unexpectedly, staining revealed that these were glycoproteins. Mass spectrometric analysis of the excised gel bands showed that the two proteins corresponded to PF0190 and PF1408, and they were termed CipA and CipB (for cold-induced proteins), respectively. Their cold-induced expression was confirmed by the DNA microarray data and by quantitative PCR. CipA and CipB both appear to be solute-binding proteins. While the Archaea do not contain members of the bacterial cold shock protein (Csp) family, they all contain homologs of CipA and CipB. These proteins are also related phylogenetically to some cold-responsive genes recently identified in certain Bacteria. It was speculated (Weinberg et al., 2005) that the Cip proteins may represent a general prokaryotic-type cold response mechanism, although their function has yet to be elucidated.

The response of *P. furiosus* to heat shock

The effect of supra-optimal temperatures on gene expression in *P. furiosus* was assessed using a targeted DNA microarray that contained a representative subset of 201 genes (Shockley et al., 2003). Their products were known or thought to be involved in proteolysis, stress response, proteolytic fermentation, and glycoside hydrolysis. When a culture growing with peptides as the carbon source was shifted from 90 to 105 °C for one hour, differential expression (fourfold or higher) of several stress genes was noted, including the thermosome (PF1974), small heat shock protein (PF1883), and two other putative molecular chaperones (PF0963 and PF1882, CDC-48 homologs, VAT-related). Of the 42 protease-related genes monitored, only two were differentially expressed more than fivefold pyrolysin (PF0287) was repressed, while a subtilisin-like protease (PF0688) was induced. The two ATP-dependent proteases in *P. furiosus* responded to heat shock in different ways. The Lon protease (PF0467) was down-regulated (threefold), and while both proteasome β-subunits (PF0159 and PF1404) were slightly stimulated (twofold), the α-subunit (PF1571) was repressed (fourfold), and the proteasome ATP-dependent regulatory subunit (PAN PF0115) was not affected by heat shock. This result raises questions concerning 20S proteasome assembly and function during thermal stress, and cellular abundance of this complex. Genes related to proteolytic fermentation were among the most highly expressed under normal growth conditions and were either not affected or down-regulated to varying extents upon heat shock.

Compatible solute formation under thermal stress was indicated by the induction of genes encoding a putative trehalose synthase (PF1742) and L-myo-inositol 1-phosphate synthase (PF1616), the latter result being obtained through Northern analysis. Mannosyl glycerate formation (e.g. PF0591) was not induced. Several glycosidase genes were significantly induced upon heat shock (e.g. PF0073, PF0076, PF0442), as were genes encoding maltose/trehalose-binding proteins (e.g. PF1739 and PF1938). Saccharide recruitment involving glycosidases during thermal stress may serve to meet the increased bioenergetic needs of the cell or be recruited for compatible solute synthesis. Subsequent transcriptome analyses of heat shock using a whole genome microarray have revealed that the response of *P. furiosus* to thermal stress is a transient process, including immediate and long-term effects. Hence, the initial response immediately upon reaching 105 °C is significantly different from that observed at

60 minutes (Shockley, 2004). Plate 21.2 shows the changes observed 5 minutes after switching a culture from 90 to 105 °C, and indicates that the expression of some heat shock proteins, such as the thermosome (PF1974) and small heat shock protein (PF1983), responds more dramatically in the early stages of thermal stress.

Use of a P. furiosus DNA microarray to assess species relationships

The DNA microarray representing the complete genome of P. furiosus was used to assess its evolutionary relationship to Pyrococcus woesei. Like P. furiosus, this organism grows optimally near 100 °C and, like P. furiosus, was isolated from a shallow marine volcanic vent system off the coast of Vulcano Island, Italy (Zillig et al., 1987). Hybridization of genomic DNA from P. woesei to the DNA microarray, in combination with PCR analysis, indicated that homologs of 105 genes present in P. furiosus were absent from the uncharacterized genome of P. woesei (Hamilton-Brehm et al., 2005). Pulsed field electrophoresis indicated that the two genomes are of comparable size and the results were consistent with P. woesei lacking the 105 genes found in P. furiosus. The missing genes included one gene cluster (Mal I, PF1737-PF1751) involved in maltose metabolism, and another (PF0691-PF0695) whose products are thought to remove toxic reactive nitrogen species. P. woesei, in contrast to P. furiosus, was subsequently shown to be unable to utilize maltose as a carbon source for growth, and its growth (on starch) was inhibited by the addition of a nitric oxide generator. In P. furiosus the ORF clusters not present in P. woesei are bracketed by or close to either insertion sequences or long chain terminal repeats. It is, therefore, possible that P. woesei acquired more than 100 genes by lateral gene transfer from other organisms living in the same thermophilic environment to produce the type strain P. furiosus (Hamilton-Brehm et al., 2005).

The P. furiosus proteome

P. furiosus was one of the first archaeal species whose proteomes were examined using mass spectrometry (MS) for protein identification, and more than 60 proteins were identified (Holden et al., 2001). Moreover, this proteomics effort was one of the first for any prokaryote that was directed toward membrane-bound proteins (Holden et al., 2001). Prior to the experimental efforts, the complete genome was analyzed for signal peptides using the prediction programs SignalP, TargetP, and SOSUI, and for transmembrane-spanning α-helices, using TSEG, SOSUI, and PRED-TMR2 programs. The former predicted that 270–465 (12–21%) of the genes have signal sequences, and that 452–563 (20–25%) have membrane-spanning domains. The consensus was that 533 (24%) of the genes in P. furiosus encode proteins located in the membrane.

To investigate the validity of these approaches, cell extracts of P. furiosus were separated into soluble cytoplasmic and insoluble membrane fractions and each was analyzed by 2D-gel electrophoresis (Holden et al., 2001). Major spots were subjected to trypsin digestion and the peptides were identified by what is now the standard MALDI-TOF approach, as well as a tandem MS approach coupled with microcapillary chromatography (μ-LC-ESI-MS/MS). A total of 32 membrane proteins and 34 cytoplasmic proteins were identified. From the bioinformatic analyses it was concluded that 23 of the 32 proteins (72%) from the membrane fraction should be membrane-associated, and that all of the proteins from the cytoplasmic fraction should be in the cytoplasm. Two membrane-associated proteins predicted to be cytoplasmic were also predicted to consist primarily of transmembrane-spanning β-sheets. On the other hand, a predicted cytoplasmic ATPase subunit homolog was found in the membrane fraction, and is most likely complexed with other membrane-bound ATPase subunits. The membrane-prediction programs used in this study therefore appear quite accurate when used in a consensus fashion. The P. furiosus proteome was also utilized to systematically compare the efficiency of the MALDI-TOF and the μ-LC-MS/MS approaches in identifying proteins after their separation by 2D-gel electrophoresis (Lim et al., 2003). A total of 62 cytoplasmic proteins were analyzed. The tandem MS approach gave better protein sequence coverage than MALDI-TOF over a large range of protein masses, although MALDI-TOF identified some peptides not identified by μ-LC-MS/MS. A combination of both techniques was recommended, and these yielded in excess of 50% sequence coverage for all proteins identified.

Structural genomics of *P. furiosus*

The concept of "structural genomics" (SG) emerged in the post-genomic era of the late 1990s. The objective was to carry out large-scale, rapid determination of the three-dimensional structures of proteins using nuclear magnetic resonance (NMR) and X-ray crystallography with the goal of filling structure space or, rather, determining the structure of a representative fold from all possible families of protein folds in the living world (see Protein Structure Initiative, 2005). It was estimated that approximately 5000–15,000 different structures may be necessary to cover the estimated 1000–3000 unique folds that may be represented in living organisms, though this estimate is often revised (Liu et al., 2004). It was recognized that such an undertaking required extensive development of novel high-throughput (HTP) technologies at the levels of bioinformatics, gene cloning, heterologous protein expression, crystallization, structure determination, data processing, and data normalization and distribution. In the United States, the National Institute of General Medical Sciences (NIGMS/NIH) created in 2000 a five-year Protein Structure Initiative (PSI; Protein Structure Initiative, 2005) devoted to the development of new SG tools required for HTP production of protein structures. Most of the resulting SG groups focused on groups of organisms, both prokaryotic and eukaryotic, as gene sources for protein production. At the Southeast Collaboratory for Structural Genomics (SECSG; Southeast Collaboratory for Structural Genomics, 2005), the three target organisms were human, the worm *Caenorhabditis elegans*, and *P. furiosus* (Adams et al., 2003).

The initial goal of the SG effort with *P. furiosus* was to heterologously express all of its 2065 genes and to obtain the recombinant proteins in a fully folded, functional form. However, there would obviously be difficulties in obtaining recombinant forms of membrane proteins, proteins containing both simple and complex cofactors (especially those not synthesized by *E. coli*), and proteins which may only be stable when co-expressed with their partners to form a multiprotein complex. We therefore estimated that only about 20% of genes in the complete genome would probably be expressed as a stable, properly folded protein, often referred to as "low-hanging fruit" (Adams et al., 2003). This prediction so far holds true based on the data available from multiple SG groups (Protein Data Bank, 2005). In fact, there is a remarkable attrition rate from gene target to structure, which is independent of research group, protocols used, and target list (e.g. see Acton et al., 2005). Consequently, some attempts were made with *P. furiosus* to ensure that proteins incorporate the correct cofactor, and/or are part of multiprotein complexes. For example, this might require growth of the recombinant host (*E. coli*) in the presence of excess metals for certain cofactors, or the co-expression of multiple genes to produce multiprotein complexes. However, no special attempt was made to produce membrane proteins.

For *P. furiosus*, 1909 genes have been cloned into an expression vector containing an amino-terminal affinity tag (MAHHHHHGS-) (Southeast Collaboratory for Structural Genomics, 2005). This allows purification by immobilized metal affinity chromatography, as well as detection using a commercial antibody against the affinity tag (Sugar et al., 2005). Since screening of thousands of clones at a preparative scale (1 liter) is prohibitive in terms of time and cost, the initial goal was to screen heterologous protein expression in a simple *E. coli* system, but in a number of different growth conditions known to affect protein expression, such as culture medium, temperature, and host strain. These were screened in a small scale (1 ml) expression system (SSE) using robotics to automate screening for protein expression via the His-tag in 96-well plates (Adams et al., 2003; Sugar et al., 2005). To date, 2373 1-liter cultures representing 1006 unique *P. furiosus* genes have been grown, 575 (57%) of which have successfully expressed some protein (as determined by SDS-polyacrylamide gel electrophoresis after one affinity column). A total of 385 unique proteins have been purified, 259 of which appear to be correct by mass spectrometry analysis (or at least not degraded, or subjected to some unknown modification in *E. coli*), and 240 have been submitted for X-ray crystallography screening, 137 for NMR screening. The results so far are indicated in Table 21.3 (Southeast Collaboratory for Structural Genomics, 2005). Current summary statistics for all SG groups can be found at the Protein Data Bank (Protein Data Bank, 2005), but at the time of this writing (August 2005), 89,721 targets have resulted in 54,530 clones, 11,632 purified proteins, and 1926 crystal and 782 NMR structures.

In the truly short time of five years, tremendous strides have been made in developing HTP protocols

Table 21.3 Statistics for *Pyrococcus furiosus* proteins from gene to structure in the structural genomics project.

Genes cloned	Express attempt	Express success	Pure	Crystal	Diffracts	X-ray structure	NMR folded	NMR structure
1909	1106	575	385	108/259	59	29	112/137	2

for every step in the structure determination pipeline, from bioinformatics, through cloning, expression, screening, and protein purification, to actual structure determination. The resulting dramatic reduction in time and cost for determination of high resolution structures (at least for the low-hanging fruit category of proteins) makes structure determination much more accessible. All the techniques currently in use, however, are still essentially empirical in nature and no complete predictive rules have yet emerged at any practical level in the production scheme, though some relationships between predictions and protein properties have emerged (e.g. Canaves et al., 2004; Acton et al., 2005). During the next few years the US Protein Structure Initiative will involve several large production centers where protein families are the targets, not individual organisms. Smaller research-based centers will tackle the "high-hanging fruit" represented by membrane and post-translationally modified proteins (Protein Structure Initiative, 2005). Consequently, the genome of *P. furiosus* is just one of almost 300 genome sequences currently available that can provide homologous members of a particular protein family. The next five years will doubtless provide even more tools, structures, and new insights into the all-important protein folding problem.

Conclusions

P. furiosus was isolated almost 20 years ago by Karl Stetter and coworkers and has become arguably one of the, if not the, best studied of the nonhalophilic Archaea. Tremendous progress has been made in understanding the physiology, metabolism, and biochemistry of this organism, and how it responds to a range of environmental stresses. Such studies have been facilitated and greatly invigorated by the advent of genome-based technologies to examine its transcriptome and proteome. At present the Achilles heel of the *P. furiosus* field is the absence of a genetics system. Fortunately, there have recently been some very encouraging developments along these lines with other members of the Thermococcales (see Sato et al., 2005). One can therefore confidently envision carrying out genetic manipulations with *P. furiosus* at some point in the near future to fully capitalize on the tremendous amount of progress that has already been made from molecular and biochemical studies of this remarkable organism.

Acknowledgments

Research on *Pyrococcus furiosus* is supported in the authors' laboratories by grants from the National Science Foundation, the Department of Energy, the National Institutes of Health, and the National Aeronautics and Space Administration.

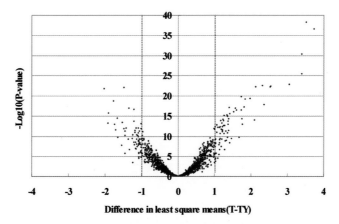

Plate 21.1 Volcano plot (statistical significance on *y*-axis, $\log_2$ fold change on *x*-axis) for *P. furiosus* grown in the presence of elemental sulfur on tryptone compared to tryptone/yeast extract-based media. Significant changes in the transcriptome (165 genes differentially expressed twofold or more) were observed upon addition of yeast extract.

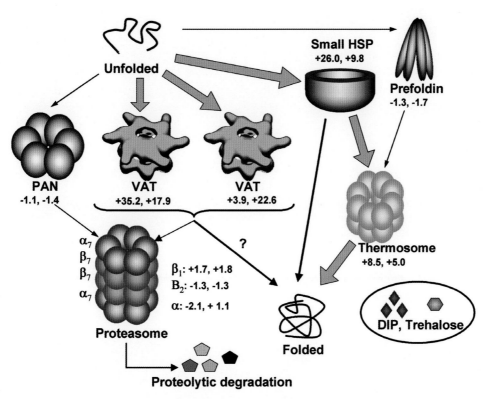

Plate 21.2 Differential expression of genes encoding protein folding and degradation complexes in *P. furiosus* upon heat shock. Immediate response after reaching 105 °C fold change values are indicated for the tryptone and maltose grown cells, respectively, following a shift from 90 to 105 °C. Genes encoding the thermosome, the small heat shock protein, and probably VAT (Rockel et al., 2002) were differentially expressed to a greater extent in the early going compared to the 60 minute case reported previously (Shockley et al., 2003).

22
The glycolytic pathways of Archaea: evolution by tinkering

John van der Oost and Bettina Siebers

Introduction

The evolution of life on earth has resulted in three distinct types of living systems: the Eukarya and two distinct types of Prokarya: the Archaea and the Bacteria (Woese et al., 1990). Like the metabolic diversity that has been found in Bacteria, the overall metabolic potential of Archaea is very versatile, ranging from photo- and chemo-autotrophy to heterotrophy, and from fermentation to anaerobic and aerobic respiration (Schönheit & Schäfer, 1995). Representatives of the archaeal classes that were discovered several decades ago (methanogens, halophiles, and thermophiles) have become model organisms for studying archaeal metabolism. In agreement with their monophyletic evolutionary status, Archaea harbor many unique properties compared to Bacteria as well as Eukarya. The aim of this chapter is to recapitulate some remarkable discoveries on archaeal metabolism in general, and to more extensively review the well characterized archaeal glycolytic pathways. Comparison of the archaeal glucose catabolism with bacterial and eukaryotic counterparts reveals several variant pathways, with unprecedented conversions, novel enzymes, and distinct regulatory mechanisms.

Evolution of metabolic pathways

Reconstruction of metabolic pathways is an exciting exercise that allows for speculation on the evolutionary history of the route in general and of its enzymes in particular. Initial comparative genome studies on archaeal genomes have revealed a "conserved archaeal core" of genes (Makarova et al., 1999). This core, which in later studies has been found to consist of approximately 300 clusters of orthologous groups of proteins (COGs), mainly consists of housekeeping genes involved in information processing, many of which do not resemble bacterial systems but share similarity with eukaryal counterparts (Makarova & Koonin, 2003a). In addition to this stable core, archaeal genomes were also found to contain a "flexible shell" that is sensitive to evolutionary events like lineage-specific gain and loss of genes; the flexible shell contains the majority of the metabolic genes, obviously reflecting the wide diversity of archaeal metabolism. Based on this observation, in both Archaea and Bacteria, the evolution of central metabolic pathways has been referred to as "the playground of non-orthologous gene displacement" (Koonin & Galperin, 2003).

Apart from relatively rare cases of real novel inventions that obviously played an important role in the "early evolution" of life, the "modern evolution" appears to proceed merely via the adaptations of available polypeptides, i.e. from one function to another. The latter process of genomic drift is expected to occur most frequently in the case of gene redundancy, as a result of either a vertical (homologous) or a horizontal (heterologous) recombination event. In the vertical process a gene duplication results in two paralogs, one of which is stable and maintains the original function encoded by the ancestral gene, whereas the other is relieved from selective pressure and can rapidly evolve and eventually reach another function. In the horizontal process a gene is introduced into a recipient genome (which already contained a gene encoding a protein with a similar function) via lateral transfer from a distinct donor genome; again due to the generated redundancy, there will be no selective pressure on one of the resulting analogous genes, and the gene is allowed either to be lost (when the original gene is lost this would be an example of non-orthologous gene replacement) or to evolve into some distinct function. Hence, the evolution of apparent novel functionalities is often the result of recombination from old materials, a process that has previously been compared to "tinkering" (Jacob, 1997), i.e. using available leftovers and spare parts to restore or adjust a certain function, or to create a novel one. This phenomenon can be recognized at different levels of complexity: (i) domains in proteins (e.g. domains involved in ligand binding and allosteric regulation of transcriptional regulators as well as enzymes (Aravind & Koonin, 1999; Ettema et al., 2002); (ii) subunits in protein complexes (e.g. subunits that contain crucial redox centers in complexes involved in anaerobic or aerobic respiration; van der Oost et al., 1994); and (iii) enzymes in pathways (e.g. mosaic-like composition of enzymes in glycolysis; Plate 22.1).

Apart from the development of a new function by the sequence drift of one member of a pair of functional analogs, the reverse may occur as well. In some instances it has become clear that a single polypeptide is responsible for more than one function ("moonlighting proteins"), replacing the simple idea of one gene/one protein/one function (Jeffery, 1999). The function of a moonlighting protein can vary as a consequence of changes in cellular localization, cell type, oligomeric state, or the cellular concentration of a ligand, substrate, cofactor, or product. Several dual- and multiple-function metabolic proteins have been described, including: (i) the maltose binding protein, which is a subunit of the high-affinity ABC-like transport system, and also functions as a sensor of a signal transduction pathway (chemotaxis) to control the mode of rotation of the flagellum (Ehrmann et al. 1998); (ii) the eukaryotic phosphoglucose isomerase, which not only catalyzes the second step in glycolysis, but also appears to function as extracellular cytokine (Jeffery, 1999); (iii) GAPDH, which is a key glycolytic enzyme, but in Eukarya also plays a role as a potential redox-sensing subunit of a transcription regulation complex that is involved in cell cycle control (Zheng et al., 2003); and (iv) enolase, which is a glycolytic enzyme, as well as a subunit of the bacterial degradasome complex responsible for RNA processing/hydrolysis (Carpousis, 2001). In analogy to GAPDH, the role of enolase in the latter complex may be that of a regulatory switch, sensing the metabolic state of the cell; glycolytic intermediates are known to regulate the activity of this enzyme (see also below). It is anticipated that many more examples of dual- or even multiple-function proteins will be discovered in the course of ongoing holistic, multidisciplinary research projects.

Archaeal metabolism

The most extensive recent overview of archaeal metabolism has been provided by Koonin and Galperin (2003), by means of a comparative genomics analysis using the COG classification, with reference to experimental verification whenever available. Extensive comparative genomics has revealed that non-orthologous gene displacement is a frequently observed phenomenon at the level of central metabolic pathways. Often, missing steps in metabolic pathways have been the goal of directed experimental studies. In several instances, non-orthologous enzymes have been identified that catalyze a similar metabolic conversion, most likely reflecting a convergent evolutionary event.

It appears to be a general trend that catabolic and anabolic pathways are well conserved in Bacteria and Eukarya, whereas Archaea possess a relatively high number of distinct features. Comparative genomics analyses of archaeal genomes have revealed many gaps in classical metabolic pathways. Several metabolic missing links have since been identified (Galperin & Koonin, 2000; Makarova & Koonin, 2003b). The most remarkable examples of unique archaeal enzymes have been reviewed by Ettema et al. (2005), including distantly related or non-homologous variants of PEP carboxylase (linking the tricarboxylic acid (TCA) cycle and the glycolysis), aconitase (TCA cycle), the α-aminoadipic acid (AAA) pathway (also present in yeast and some thermophilic Bacteria), a unique shikimate synthase (amino acid biosynthesis), and 2-methyl-4-amino-5-

hydroxymethylpyrimidine kinase (vitamin biosynthesis). Additional features of archaeal metabolism, such as sulfur metabolism and methanogenesis, are discussed in detail in Chapters 23 and 24 of this book.

Glycolytic pathways

To illustrate the variability within the central metabolic pathways of Archaea, the well characterized example of glucose degradation is reviewed. Because of some overview articles (Ronimus & Morgan, 2003; Verhees et al., 2003; Sakuraba et al., 2004; Siebers et al., 2004; Ahmed et al., 2005; Siebers & Schönheit, 2005; Snijders et al., 2006) in which the characteristics of the archaeal glycolytic enzymes have been discussed in great detail, we restrict ourselves here to some remarkable features of the archaeal glycolytic pathways, some recently gained insights, and expected future developments.

Although some unique archaeal conversions have been demonstrated to occur in archaeal glycolytic pathways (Ronimus & Morgan, 2003; Verhees et al., 2003; Siebers et al., 2004; Ahmed et al., 2005) (Fig. 22.1a), the overall series of chemical conversions from glucose to pyruvate is generally very well conserved in saccharolytic species of all three domains, due to chemical and thermodynamic constraints (Melendez-Hevia et al., 1997). However, the enzymes catalyzing identical steps in some instances differ dramatically; as described below, some spectacular examples of convergent evolution have been identified in the glucose-degrading pathways of Archaea. Biochemical and genomics-based analyses have revealed that Archaea possess variants of the Embden–Meyerhof–Parnas (EMP) and Entner–Doudoroff (ED) pathway that are constituted by different combinations of classical (bacterial/eukaryal) and novel (archaeal) enzymes (Fig. 22.1, Table 22.1). The recently gained insight in some relevant steps of the central pentose metabolism is described as well.

Embden–Meyerhof–Parnas pathway

The EMP pathway can be considered as a highway of the central metabolism in all three domains of life. In organisms with a versatile (saccharolytic) metabolism, the pathway is bidirectional, and the complete set of enzymes that catalyze the stepwise conversions via either glycolysis or gluconeogenesis is present. In autotrophs and non-saccharolytic heterotrophs, however, the pathway is only used to convert pyruvate (or phosphoenol pyruvate) to at least fructose-6-phosphate; subsequently, the latter intermediate may be further converted to ribose-5-phosphate, an essential building block for nucleotides and precursor of some amino acids (see below). Examples of Archaea that appear to have an incomplete glycolysis are the methanogens *Methanothermobacter thermautotrophicus* and *Methanopyrus kandleri* and the hyperthermophile *Archaeoglobus fulgidus* VC-16; all three lack obvious homologs of glucokinase, glucose-6-phosphate isomerase, phosphofructokinase, and pyruvate kinase (Verhees et al., 2003). On the other hand, several methanogens from the genera *Methanococcus*, *Methanocaldococcus*, and *Methanosarcina* have the capacity to store carbon as glycogen. For the degradation of this glucose polymer they need a glycolytic pathway that starts with the conversion of glucose or glucose-6-phosphate; indeed, genes encoding all required enzymes have been identified (Verhees et al., 2003). The few organisms that do not possess the complete set of glycolytic genes are the organisms that have minimal-size genomes, reflecting their parasitic or endosymbiotic lifestyle (e.g. the bacterium *Rickettsia*, and the archaeon *Nanoarchaeum*) (Siebers & Schönheit, 2005).

The glycolytic enzymes that are completely conserved in all three domains of life are triose-phosphate isomerase (TIM), enolase, and pyruvate kinase (PYK). The former two have a phylogenetic pattern that resembles that of the ribosomal sequences, with monophyletic distribution of the archaeal sequences (Ronimus & Morgan, 2003) (Fig. 22.1a, b; Table 22.1). Pyruvate kinase has been divided into two related subclasses, one that is allosterically regulated and the other (present in Archaea) that appears not to be controlled by allosteric effectors (Schramm et al., 2000); as stated above, PYK is not present in the non-glycogen forming methanogens, which is in line with the enzyme's role in glycolysis. Interestingly, however, analysis of modulated gene expression in *P. furiosus* reveals (differential display, Robinson & Schreier, 1994; DNA microarray, Schut et al., 2003) on the one hand a slight down-regulation of the gene encoding

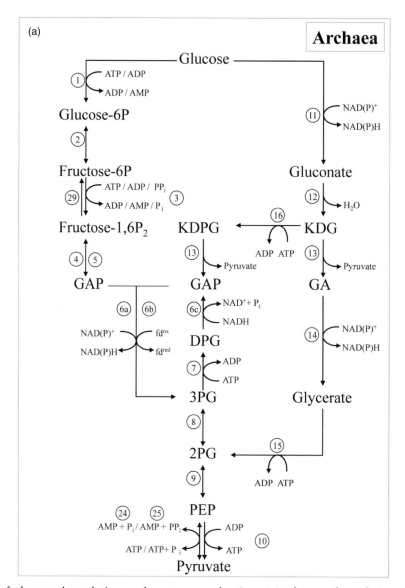

Fig. 22.1 Variation of glucose degradation and pentose production. (a) Glucose degradation in Archaea via variants of Embden–Meyerhof (EM) and Entner–Doudoroff (ED) pathways. (b) Bacterial/eukaryal routes of glucose catabolism: EM, ED, and oxidative pentose phosphate (PP) pathway. (c) Pentose-phosphate pathway/cycle in Bacteria and Eukarya. (d) Pathways that lead to the generation of ribulose-5P in Archaea. Abbreviated metabolites are: KDG, 2-keto-3-deoxygluconate; KDPG, 2-keto-3-deoxy-6-phosphogluconate; GAP, glyceraldehyde-3-phosphate; GA, glyceraldehyde; DPG, 2,3-diphospho-glycerate; 3PG, 3-phosphoglycerate; 2PG, 2-phosphoglycerate; PEP, phosphoenolpyruvate. Numbers refer to the following enzymes: (1) GLK, glucokinase (two types, ATP- or ADP-dependent); (2) PGI, phosphoglucose isomerase; (3) PFK, phosphofructokinase (three types, ATP-, PPi-, or ADP-dependent); (4) FBA, fructose-1,6-bisphosphate aldolase; (5), TIM, triose-phosphate isomerase; (6a) GAPN, non-phosphorylating GAP dehydrogenase; (6b) GAPOR, non-phosphorylating GAP oxidoreductases; (6c) GAPDH, classical GAP dehydrogenase; (7) PGK, phosphoglycerate kinase; (8) PGM, phosphoglycerate mutase; (9) ENO, enolase; (10) PYK, pyruvate kinase; (11) GlcDH/Glc-lac; glucose dehydrogenase/gluconolactonase; (12) G-hydr, gluconate dehydratase; (13) KD(P)G-ald, KD(P)G-aldolase; (14) Ald DH, glyceraldehyde dehydrogenase; (15) Gly-kin, glycerate kinase; (16) KDG-kin, KDG kinase; (17) KDPG-ald, KDPG aldolase; (18) G6PDH/6PG-lac, glucose-6-posphate dehydrogenase/6-phospho-gluconolactonase; (19) GlcA-kin, gluconate kinase; (20) PG-hydr, 6-phospho-gluconate kinase; (21) PG-DHDC, 6-phospho-gluconate dehydrogenase (decarboxylase); (22) RP-epi, ribulose-5-phosphate 3-epimerase; (23) Tr.ketolase, transketolase; (24) PEPS, PEP synthetase (see text); (25) PPDK, pyruvate phosphate dikinase (see text); (26) HPI, 3-hexulose-6-phosphate isomerase, and HPS, 3-hexulose-6-phosphate synthase (may occur as fusion, see text); (27) R5P iso, ribose-5-phosphate isomerase; (28) DERA, 2-deoxyribose-5-phosphate aldolase; (29) FBA, fructose bisphosphatase.

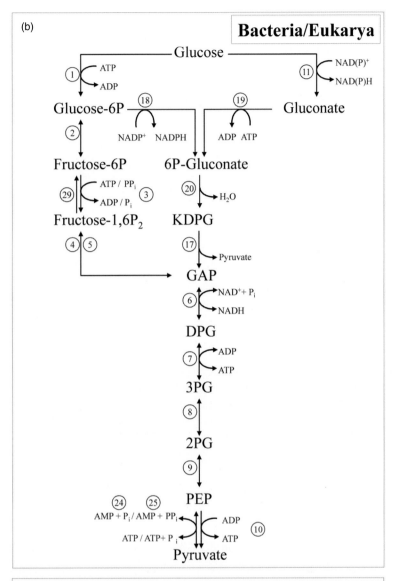
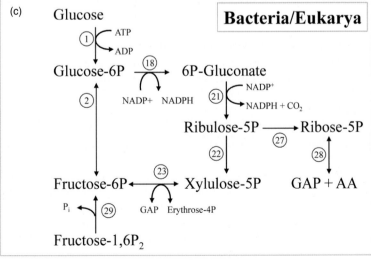

Fig. 22.1 *Continued*

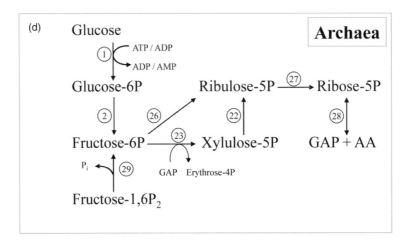

Fig. 22.1 Continued

PYK, and on the other hand an up-regulation of the gene encoding PEP synthetase. The latter would agree very well with the isolated AMP-dependent kinase from *Pyrococcus*, which was reported to catalyze the ATP-yielding conversion to pyruvate (Sakurabe et al., 1999, 2004), and would imply an unprecedented, very efficient site of ATP generation in pyrococcal glycolysis.

In *T. tenax* the interconversion of phosphoenolpyruvate and pyruvate is catalyzed by three different enzymes: pyruvate kinase (PYK), phosphoenolpyruvate synthetase (PEPS), and pyruvate phosphate dikinase (PPDK). Whereas PYK and PEPS in *T. tenax* catalyze the unidirectional glycolytic and gluconeogenic reaction, respectively, the PPDK catalyzes the reversible interconversion of phosphoenolpyruvate and pyruvate (ATP+P_i+pyruvate ↔ AMP+PP_i+PEP) (Tjaden et al., 2006). However, recent enzymatic characterization of the recombinant *T. tenax* PPDK suggests a glycolytic function (Tjaden et al., 2006) in addition to a pyruvate kinase without allosteric properties (Schramm et al., 2000).

Many examples of non-orthologous gene displacements have been established for the archaeal glycolytic enzymes: kinases (GLK and PFK), isomerase, aldolase, GAP oxidoreductase, and mutase (Plate 22.1, Table 22.1). For each case, relevant features are discussed below. It should be noted that none of these appear to be novel inventions, but instead result from evolutionary adjustments of substrate specificity of existing enzymes.

The classical glycolytic pathway is well conserved among Bacteria and Eukarya (Fig. 22.1b). A significant difference in the glycolytic pathways of some Euryarchaea concerns the involvement of ADP-dependent kinases (COG4809, "Archaeal ADP-dependent phosphofructokinase (ADP-PFK)/glucokinase (ADP-GLK)"; remotely similar to the ribokinase family, that also includes the "minor PFK" of *E. coli*) (Fig. 22.1a) (Verhees et al., 2003). Some interesting observations have been made that may reveal some details on the evolution of this class of enzymes. In *Methanocaldococcus* there is a single gene that encodes ADP-PFK. The latter enzyme has been reported to have a broad substrate specificity (Sakuraba et al., 2002); a physiological role of the proposed bifunctional ADP-GLK/PFK may be the conversion of glucose that is the product of glycogen hydrolysis; apart from a gene coding for a glycogen phosphorylase, two genes potentially encoding glucoamylase and α-amylase are linked to the cluster encoding the ADP-kinase and the phosphoglucose isomerase (Table 22.1). In *Methanosarcina* spp. two adjacent paralogous genes are present (Verhees et al., 2003), obviously as a result of a gene duplication; the corresponding proteins have not yet been characterized. In *Pyrococcus* spp. and *Thermococcus kodakaraensis* two specialized enzymes (ADP-GLK and ADP-PFK) are present at distinct genomic loci; most likely they are the result of a gene duplication, as has been observed in *Methanosarcina*. Single copies of the ADP-kinases are also encoded by the genomes of several

Table 22.1 Phylogenetic distribution of genes involved in glucose and pentose metabolism in Archaea.

						Euryarchaea						Crenarchaea			Bacteria
Enzyme	Remark	EC	COG	PF (PH/PAB)	MJ	MA (MM)	MT (MK)	AFU	VNG	TA (TV)	SSO (STO SACI)	APE	PAE	TTX	TM

Embden-Meyerhof-Parnas (EMP) pathway-specific enzymes

Enzyme	Remark	EC	COG	PF	MJ	MA	MT	AFU	VNG	TA	SSO	APE	PAE	TTX	TM
GLK	ADP/glucose	2.7.1.147	4809*	0312	1604ª‡										
	ATP/hexose	2.7.1.2	1940						2629	0825		2091	3437	AJ510140	1469
PGI		5.3.1.9	2140	0196		0821§		(1494)	1992	1419	(2281)	0768	1610	AJ621272	1385
		5.3.1.9	0166												
PFK	ADP	2.7.1.146	4809*	1784	1605ª‡	3562						0012ª	0835	Y14655ª	0209ᵇ
	ATP	2.7.1.11	0205		1604ª‡										0289
	PPi	2.7.1.90													
FBA		4.1.2.13	1830	1956	1585	0439	0579	0108	0683		3226	0011ª		AJ310483ª	0273
		4.1.2.13	0191												

EMP and Entner-Doudoroff (ED) general enzymes

Enzyme	Remark	EC	COG	PF	MJ	MA	MT	AFU	VNG	TA	SSO	APE	PAE	TTX	TM
TIM		5.3.1.1	0149	1920†	1528	4607	1041ª	1304	1027	0313	2592	1538	1501	AJ012066	0689ª#
GAPOR	Fd	1.2.7.6	2414	0464	1185								1029	AJ621330	
GAPN	NAD(P)	1.2.1.9	1012	(0755)	(1411)	(1355)	(0978)	1732	(2513)	(0809)	3194	(1786)		Y10625	
GAPDH	NADP	1.2.1.13	0057	1874†	1146	1018	1009		0095	1103	0528ª	0171ᵇ	1740*	Y10126ᵇ	0688ª#
	NAD	1.2.1.12													0689ª#
PGK	ADP	2.7.3.2	0126	1057†	0641	2669	1042ª	1146	1216	1075	0527ª	0173ᵇ	1742*	Y10126ᵇ	1374
PGM	2,3-PG	5.4.2.1	0406							1347	2236	1616	2326	AJ621333	1774
	independent	5.4.2.1	3635	1959	1612	0132	1591	1751	1887	0413	0417				
		5.4.2.1	0696												
ENO		4.2.1.11	0148	0215	0232	1672	0043	1132	1142	0882	0913	2458	0812	AJ621325	0877
PYK	ADP	2.7.1.40	0469	1188	0108	3890	1118		0324	0896ª	0981	0489	0819	AF065890	0208ᵇ
PEPS	AMP + Pi	2.7.9.2	0574	0043	0542	3408		0710	0330		0883	0650	2423	AJ515537	0272
						2667									
PPDK	AMP + PPi	2.7.9.1				0608				0886					

ED pathway-specific pathways

Enzyme	Remark	EC	COG	PF	MJ	MA	MT	AFU	VNG	TA	SSO	APE	PAE	TTX	TM
Glc DH	NAD(P)	1.1.1.47	1063						0446ª	0897ª	3003ᵇ		3383	AJ621346	0298
											3042ᵇ				
											3041ᵇ				
Glc-lac		3.1.1.17													
G-hydr		4.2.1.39	4948**						0442ª	0085	3198ᶜ			AJ621281ᶜ	(0006)
KD(P)G ald		4.1.2.-	0329***						0444ª	0619	3197ᶜ			AJ621282ᶜ	
Ald DH	NAD(P)	1.2.1.3	1012							(0809)	(1629)			ns††	
Gly kin	ATP	2.7.1.31	2379							0453	0666			AJ621354	
KDG kin		2.7.1.45	0524						0158		3195ᶜ			AJ621283ª	0067ᶜ
G6P DH	NAD(P)	1.1.1.49	0364												1155ᵈ
6PG-lac		3.1.1.31	0363												1154ᵈ
GlcA-kin	ATP	2.7.1.12													(0006)
PG-hydr		4.2.1.12	4948**												0066ᶜ
KDPG ald		4.1.2.14	0800												

(Continued)

Table 22.1 Phylogenetic distribution of genes involved in glucose and pentose metabolism in Archaea.

Enzyme	Remark	EC	COG	PF (PH/PAB)	MJ	MA (MM)	MT (MK)	AFU	VNG	TA (TV)	SSO (STO/SACI)	APE	PAE	TTX	TM
Pentose phosphate (PP) metabolism-specific enzymes															
Oxidative PP pathway															
G6P DH	NAD(P)	1.1.1.49	0364												1155[d]
6PG-lac		3.1.1.31	0363												1154[d]
PG DHDC		1.1.1.44	1023	(0716)					(2553)		(1560)		(1145)		0438
			0362												
Non-oxidative PP pathway															
R5P iso		5.3.1.6	0120	1258	1603	1683	0608	0943	2272	0878	0978	0665	1027	AJ621331	1080
		5.3.1.6	0698												
RP epi		5.1.3.1	0036		0680[b]										1718
Tr.ketolase	N-term SU	2.2.1.1	3959	1688[a]	0679[b]					0617[a]	0299[d]	0583[c]	1929[c]	AJ621132[d]	0953[e]
	C-term SU	2.2.1.1	3958	1689[a]	0681[b]					0618[b]	0297[d]	0586[c]	1927[b]	AJ621131[d]	0954[e]
Tr.aldolase		2.2.1.2	0176		0960					0616[b]					0295
Ribulose monophosphate pathway															
PHI		5.-.-.-	0794	0220#‡	1247	1384	1546 249	1796			0151	1940	1489	1049	
HPS		4.1.2.-	0269	0220#‡	1447§§	4608§§	1474§§	1305§§ 0861¶¶			0202	0952	1679	1521	

Numbering of the genes is according to www-archbac.u-psud.fr/projects/sulfolobus/. Seventeen Archaea have been clustered (based in similar genomic profiles) for the sake of clarity; the genes of the species between brackets have not been included. For *T. tenax* the central carbohydrate metabolism was reconstructed by genomic and biochemical data; the respective accession numbers are given (Siebers et al. 2004). PF, *Pyrococcus furiosus* (PH, *Pyrococcus horikoshii*; PAB, *Pyrococcus abyssi*); MA, *Methanosarcina acetivorans* (MM, *Methanosarcina mazei*); MJ, *Methanocaldococcus jannaschii*; MT, *Methanobacterium thermoautotrophicum* (MK, *Methanopyrus kandleri*); AFU, *Archaeoglobus fulgidus* VC-16; VNG, *Halobacterium* NRC-1; TA, *Thermoplasma acidophilum*; SSO, *Sulfolobus solfataricus* (STO, *Sulfolobus tokodaii*; SACI, *Sulfolobus acidocaldarius*); APE, *Aeropyrum pernix*; PAE, *Pyrobaculum aerophilum*; TTX, *Thermoproteus tenax*. The hyperthermophilic bacterium *Thermotoga maritima* (TM) is included to allow comparison of the archaeal genomes with a bacterial genome. Experimentally confirmed gene products are shown in bold. †Characterized from *Pyrococcus woesei*; ‡bifunctional ADP-glucokinase/phosphofructokinase; §characterized from *Methanosarcina mazei*; ¶no homolog was identified in SACI; ††ns, not specified, different gene homologs for aldehyde dehydrogenase and aldehyde oxidoreductase were identified in *T. tenax* (Siebers et al. 2004); ‡‡characterized from *P. horikoshii*; §§fused to formaldehyde-activating enzyme; ¶¶fused to demethylmenaquinone methyltransferase. It should be stressed that the proposed gene functionality requires experimental verification; brackets indicate uncertain substrate specificity. Enzyme abbreviations are as given in Fig. 22.1. EC, Enzyme Commission (http://www.genome.ad.jp/kegg); COG, Clusters of Orthologous Groups (http://www.ncbi.nlm.nih.gov/COG). Asterisks (*) in the COG-column indicate homology; in the species-columns gray shading and superscript letters ([a], [b], [c], [d], [e]) indicate clustering of respective genes, and (#) indicates a fusion of two enzyme-encoding genes.

higher Eukarya (from worm to man); the ortholog from mouse has been characterized as an ADP-dependent glucokinase, the role of which remains to be established (Ronimus & Morgan, 2004).

In some Crenarchaea, such as *Aeropyrum pernix* and *T. tenax*, the first phosphorylation step in their variant EMP pathway is catalyzed by a hexokinase-like ATP-dependent glucokinase (ATP-GLK) with broad substrate specificity. The enzyme is a member of the ROK (repressor protein, open reading frame, sugar kinase) family, which belongs to the actin-ATPase superfamily, and harbors different archaeal and bacterial sugar kinases, most of which have unknown substrate specificity (Siebers & Schönheit, 2005). The phosphorylation of fructose-6-phosphate to fructose-1,6-bisphosphate in *A. pernix* as well as *Desulfurococcus amylolyticus* is catalyzed by an ATP-dependent phosphofructokinase, which belongs to the PFK-B family of the ribokinase superfamily (Hansen & Schönheit, 2001). In *T. tenax* the reaction is catalyzed by a reversible PP_i-dependent phosphofructokinase (PP_i-PFK) (Fru $6-P+PP_i \leftrightarrow FBP+P_i$), which substitutes for the antagonistic enzyme couple ATP-PFK and fructosebisphosphatase (class V) (Siebers et al., 2004). Phylogenetic analyses revealed that the enzyme is, like bacterial and eukaryal PP_i-PFKs, distantly related to classical ATP-dependent phosphofructokinases, and belongs to the phosphofructokinase (PFK-A) family (Siebers et al., 1998).

The regulatory sites of the classical EMP pathway are the irreversible reactions of the route; the kinases are allosterically regulated by a range of metabolic effectors. Interestingly, all archaeal glucokinases, phosphofructokinases, and pyruvate kinases that have been characterized so far appear not to be regulated allosterically, meaning that the classical control points are absent in the archaeal pathways; the only exception is the *T. acidophilum* PYK, which is stimulated by AMP (Siebers & Schönheit, 2005) (see below, Regulation).

An archaeal-type phosphoglucose isomerase (PGI; COG2140, "Thermophilic glucose-6-phosphate isomerase and related metalloenzymes"; member of the cupin family that consists of a variety of ligand-binding domains, and some enzymes such as the bacterial phosphomannose isomerase) has been identified in the Thermococcales as well as in *Archaeoglobus* and *Methanosarcina* (Verhees et al., 2003; Hansen et al., 2005). This PGI-type has also been found in Bacteria (e.g. *Salmonella* spp.), probably introduced via horizontal transfer from Euryarchaea (Hansen et al., 2005). Another unusual bifunctional phosphoglucose/phosphomannose isomerase (PGI/PMI) has been characterized from *A. pernix*, *P. aerophilum*, *Thermoplasma acidophilum*, and *T. tenax* (Hansen et al., 2004; Siebers & Schönheit, 2005). Unlike all known PGIs, the enzyme shows similar catalytic efficiency on both substrates (glucose 6-phosphate, mannose 6-phosphate) and converts them to fructose 6-phosphate. The enzyme belongs to a new family within the PGI superfamily. In most saccharolytic organisms, including Bacteria, Eukarya, as well as some Archaea such as *Methanocaldococcus jannaschii*, *Haloarcula marismortui*, and *Halobacterium* NRC1, a "classical" type of PGI is present (COG0166) (Siebers & Schönheit, 2005). Recent crystallographic and mechanistic analyses reveal that the catalytic mechanisms of the two structurally unrelated PGIs, the PGI of the cupin family and the classical PGI, may be similar and based on an enediol intermediate (Berrisford et al., 2006) (Plate 22.2).

An archaeal-type aldolase (aFBPA; COG1830, DhnA-type fructose-1,6-bis-phosphate aldolase and related enzymes; distantly related to the *E. coli* DhnA type-I fructose bisphosphate aldolase) is present in most Archaea (Siebers et al., 2001). Although the enzyme uses the same reaction mechanism as classical Schiff base class I enzymes, the aFBPA shows no obvious sequence similarity to either classical class I or metal-dependent class II FBPA, which are predominantly found in higher Eukarya and Bacteria as well as a few unicellular Eukarya, respectively. An archaeal phophoglycerate mutase (COG3635) was originally predicted on the basis of distant relatedness to bacterial counterparts (Graham et al., 2002; van der Oost et al., 2002).

Archaea have alternative enzymes that catalyze the oxidation of glyceraldehyde-3-phosphate (GAP) to 3-phospho-glycerate (3PG). In the bacterial/eukaryal system two reversible enzymes catalyze the two-step conversion of GAP via 1,3-bisphospho-glycerate to 3PG: "classical" NAD^+-/P_i-dependent GAP dehydrogenase (GAPDH) and phosphoglycerate kinase (PGK, ADP-dependent, ATP generating) (Fig. 22.1b). Strikingly, although the GAPDH and PGK enzyme couple is present in all archaeal genomes sequenced so far (Ettema et al., unpublished), biochemical and transcript analyses in *P. furiosus* (Schäfer & Schönheit, 1993; van der Oost et al., 1998; Schut et al., 2003) and

T. tenax (Brunner et al., 1998, 2001) strongly suggest a gluconeogenetic function of GAPDH/PGK (Fig. 22.1a). In the archaeal glycolytic variant, a single enzyme catalyzes the (either ferredoxin- or NAD(P)$^+$-dependent, non-phosphorylating, irreversible) single-step conversion of GAP to 3PG: GAPOR (COG2414, "Aldehyde-ferredoxin oxidireductase") has first been characterized in *Pyrococcus furiosus*, and GAPN (COG1012, "NAD-dependent aldehyde dehydrogenase") in *Thermoproteus tenax* (Fig. 22.1a). Whereas GAPOR exhibited no allosteric properties, GAPN is regulated by the energy charge of the cell and intermediates of the central carbohydrate metabolism. Interestingly, both organisms possess the genes encoding both GAPOR and GAPN in addition to the GAPDH/PGK enzyme couple (Table 22.1); the gene disruption system recently developed for *Thermococcus* might reveal further insights into the possible role of these three GAP-converting enzymes, at least in the Themococcales (see also below, Regulation).

Phylogenetic analysis of enolase (see the dendrogram presented by Ronimus & Morgan, 2003) shows two paralogs of enolase in several Euryarchaea, including the Thermococcales. One of the paralogs clusters with the group that has been demonstrated to encode the glycolytic enzyme. Interestingly, as noted by Rogozin et al. (2002), the other paralog has a conserved context: it is part of a gene cluster that encodes proteins whose functions appear to be directly linked to translation (ribosome structure and biogenesis) and transcription (RNA polymerase subunits). It is tempting to speculate that the role of this paralog is analogous to the aforementioned examples: enolase in the bacterial degradosome, and GAPDH in the eukaryal histon-associated transcription regulatory system. The Archaeal enolase-like proteins may have lost their catalytic capacity, but most likely have retained their allosteric sensing mode (2-phosphoglycerate, Mg^{2+}). The latter might play a crucial role as a sensor of the metabolic state, as a means to adjust the rate of transcription or translation.

With respect to the glycolytic energetics, the net yield of the classical EMP pathway is 2 mole ATP per mole glucose. In the archaeal EMP variants, in which the enzyme couple glyceraldehyde-3-phosphate dehydrogenase and phosphoglycerate kinase is replaced by non-phosphorylating GAPN or GAPOR, the formal net ATP yield is zero. Only in the hyperthermophile *T. tenax* is the calculated energy gain 1 mole ATP per mole glucose, due to the recycling of anabolically formed PP$_i$ by the reversible PP$_i$-PFK. PP$_i$ is generally regarded as a waste product of the cell, which is formed during anabolic processes (e.g. DNA synthesis) and is directly cleaved by a cytoplasmatic pyrophosphatase in order to drive the anabolic reaction. In *P. furiosus* two alternative sites of ATP generation have been reported. The ATP formation by a reversible "glycolytic" phosphoenolpyruvate synthetase, which would imply that the yield (ATP/glucose) of the pyrococcal glycolysis (with substrate phosphorylation loss at the level of the (GAP → 3PG) conversion by glyceradehyde:ferredoxin oxidoreductase (GAPOR), and ATP gain from AMP at the (PEP → PYR) conversion (AMP + P$_i$ + PEP → ATP + PYR)) would be similar to that of the classical pathway; indeed, the latter conclusion has previously been reported on the basis of a yield analysis (Kengen & Stams, 1994). The longstanding debate on this remarkable feature of the archaeal glycolysis has recently been solved by the disruption of the *pps* gene from *Thermococcus kodakaraensis*, which confirms a crucial role of PEPS in glycolysis (H. Atomi & T. Imanaka, personal communication). In addition, in *P. furiosus* the formation of H$_2$ by a ferredoxin-dependent hydrogenase has been demonstrated to be coupled to proton translocation and thus to ATP synthesis (Sapra et al., 2003).

Entner–Doudoroff pathway

The ED pathway is an alternative route for glucose degradation. Whereas the classical phosphorylative ED pathway seems to be restricted to Bacteria (Fig. 22.1b), modifications of the pathway have been identified in all three domains of life (Fig. 22.1a, b). The classical ED pathway is initiated by the phosphorylation of glucose to glucose-6-phosphate, whereas in the modified pathways the phosphorylation step is shifted to the level of either 2-keto-3-deoxygluconate (KDG) or glycerate in the semi- and non-phosphorylative variant, respectively. Studies in halophilic Archaea revealed the presence of the semi-phosphorylative ED involving KDG phosphorylation by KDG kinase, 2-keto-3-deoxy-6-phosphogluconate (KDPG) cleavage by KDPG aldolase, and GAP oxidation via a "classical" GAP dehydrogenase (*Halobacterium saccharovorum*, *Halococcus saccharolyticus*, *Haloferax volcanii*). This pathway has previously been shown to operate in several species of *Clostridium*. In addition, the non-phosphorylative variant has been proposed for the hyperthermophiles *S. solfataricus* and *T. tenax* as well as the thermophile *T. acidophilum*, on the basis of metabolite analysis and enzymatic activities in crude extracts. This variant is characterized by direct cleavage of KDG to pyruvate and glyceraldehyde via KDG aldolase and phosphorylation of glycerate via glycerate kinase. The non-phophorylative pathway has also been reported for several species of the fungus *Aspergillus* (reviewed in Ahmed et al., 2005; Siebers & Schönheit, 2005).

A recent genomics-based approach suggested the presence of a "branched non-phosphorylative and semi-phosphorylative ED variant" in *S. solfataricus* as well as *T. tenax* (Ahmed et al., 2005). This branched ED pathway is characterized by KDG kinase, a bifunctional KD(P)G aldolase (converting both KDG and KDPG), and a non-phosphorylating glyceraldehyde-3-phosphate dehydrogenase (GAPN). *Sulfolobus solfataricus* is one of the Archaea that rely on its ED-like pathway for sugar degradation. Since the variant ED pathways converge with the EMP pathway at the level of either glyceraldehyde-3-phosphate (semi-phosphorylating) or 2-phosphoglycerate (non-phosphorylating), the lower/C3 part of the EMP glycolysis is present. The incomplete upper/C6 part of the EMP pathway in *S. solfataricus* and aerobic halophiles most likely plays a role in gluconeogenesis (conversion of pyruvate to glucose-6-phosphate; Verhees et al., 2003; Snijders et al., 2006) (Fig. 22.1a); in halophiles (and perhaps in Sulfolobales) some of the latter enzymes are responsible for fructose catabolism (reviewed in Siebers & Schönheit, 2005). Interestingly, the ED variant of *S. solfataricus* is promiscuous for glucose and galactose degradation (Lamble et al., 2003), whereas the pathway seems to be specific for glucose in *T. tenax* (Ahmed et al., 2005). It should be noted that *Thermoproteus tenax* is the only archaeon to date that has been demonstrated to operate its glycolytic variants of both the EMP and ED pathways in parallel (Siebers et al., 2004; Ahmed et al., 2005; Siebers & Schönheit, 2005).

The net yield of the classical ED pathway is 1 mole ATP per mole glucose. In halophilic Archaea, which use the semi-phosphorylative ED variant involving the glyceraldehyde-3-phosphate dehydrogenase, phosphoglycerate kinase enzyme couple, the calculated ATP gain is also 1 mole ATP per mole glucose. However, when the non-phosphorylative ED pathway (e.g. *T. acidophilum*), the semi-phosphorylative ED pathway (e.g. *S. solfataricus*, *T. tenax*), or a combination of the two is operational, the energy yield is zero (Ahmed et al., 2005) (Fig. 22.1a).

Pentose phosphate pathway

A classical oxidative pentose phosphate pathway (PPP) is generally present in Bacteria and Eukarya, where it is essential for the generation of reducing power (NADPH), as well as of precursors for nucleotide/histidine biosynthesis (ribose-5-phosphate) and for aromatic amino acids (erythrose-4-phosphate) (Fig. 22.1c). The PPP is divided into an oxidative phase (OPPP) during which two molecules of NADPH are generated, and a non-oxidative phase (NOPPP) that results in biosynthetic building blocks (for review see Soderberg, 2005) (Fig. 22.1c). Strikingly, no orthologs of the OPPP were identified in archaeal genomes, with the exception of an ortholog of 6-phosphogluconate dehydrogenase (COG 1023) in *Halobacterium* sp. NRC-1 (Soderberg, 2005) (Table 22.1).

Comparative genomics-based approaches and some biochemical studies indicate the presence of either the NOPPP or the reversed ribulose monophosphate pathway (RuMP) in Archaea (Verhees et al., 2003; Soderberg, 2005) (Table 22.1). Only in *M. jannaschii* do both pathways seem to be active in parallel; in *Halobacterium* sp. NRC-1 orthologs of the NOPPP as well as RuMP are absent and an alternative OPPP is proposed via a predicted unusual enzyme to generate gluconate 6-phosphate (Soderberg, 2005). Orthologs of the complete NOPPP were identified in *Thermoplasma acidophilum*, *T. volcanium*, and *M. jannaschii*, and are probably responsible for the generation of pentoses as well as erythrose 4-phosphate. In most Archaea the generation of pentoses seems to proceed via the reversed RuMP, which was originally described as a pathway for formaldehyde fixation in methylotrophic Bacteria. However, orthologs of the encoding key enzymes were identified in Bacteria and Archaea and in some cases the respective activities were confirmed. The key enzymes of the pathway are 3-hexulose-6-phosphate isomerase (PHI; catalyzing the isomerization of fructose 6-phosphate to 3-hexulose-6-phosphate) and 3-hexulose-6-phosphate synthase

(HPS; forming ribulose-5-phosphate and formaldehyde). In contrast to the NOPPP no erythrose-4-phosphate is generated by the RuMP. An interesting observation is the fusion (*hps-phi*) of the responsible genes in the Thermococcales, and in the bacterium *Methylococcus capsulatus*. The HPS-PHI fusion protein from *Pyrococcus horikoshii* has been characterized and proposed to be involved in the fixation of formaldehyde via the RuMP pathway, as in methylotrophic bacteria (Orita et al., 2005). In addition, fusions to the formaldehyde-activating enzyme (e.g. *M. jannaschii*, *A. fulgidus*) or fusions to demethylmenaquinone methyltransferase (e.g. *A. fulgidus*) are found (see Table 22.1). However, the reversed conversion of fructose-6-phosphate to ribulose-5-phosphate has also been demonstrated (Fig. 22.1d), albeit that under the conditions used this reaction appeared less efficient than the forward reaction; genetic analyses should be applied to definitely demonstrate the role of the reversed RuMP pathway in the synthesis of pentoses as the precursor of nucleotides and amino acids in Archaea.

In some Archaea (*P. abysii*, *A. pernix*, *P. aerophilum*, *S. solfataricus*) the precursor for aromatic amino acids, erythrose 4-phosphate, has been proposed to be formed from fructose 6-phosphate and glyceraldehyde 3-phosphate by transketolase (incomplete NOPPP) (Verhees et al., 2003). Interestingly, the presence of transketolase in Archaea correlates with the need for erythrose 4-phosphate as precursor in aromatic amino acid biosynthesis. In Archaea that utilize the classical erythrose 4-phosphate-dependent chorismate pathway, orthologs of transketolase (N- and C-terminal subunit; COG 3959/3958) were identified. On the other hand, some Archaea have recently been demonstrated to use a novel erythrose 4-phosphate-independent chorismate pathway, which differs in the initial steps of 3-dehydroquinate formation (Tumbula et al., 1997; White, 2004); indeed, the transketolase genes are missing in these organisms (*M. kandleri*, *M. thermoautotrophicum*, *A. fulgidus*, *M. acetivorans*) (Verhees et al., 2003; Soderberg, 2005) (Table 22.1).

Another metabolic link between the central carbohydrate metabolism and (deoxy)nucleosides biosynthesis/degradation pathways in Bacteria and Eukarya is established via phosphopentomutase (COG1109) and 2-deoxyribose-5-phosphate aldolase (DeoC/DERA; COG0274). Homologs were identified in some Archaea, and recently the two enzymes from *T. kodakaraensis* were characterized (Rashid et al., 2004). Nucleoside phosphorylases release the pentose moiety ((deoxy)ribose-1-P) from (deoxy)ribonucleosides, after which phosphopentomutase catalyzes the isomerization to (deoxy)ribose-5-phosphate, and subsequently DERA converts the latter to glyceraldehyde-3-phosphate and acetaldehyde. The pathway is reversible and a function in catabolism and biosynthesis has been proposed (Rashid et al., 2004).

Regulation of carbohydrate metabolism

Regulation of metabolism in Bacteria and Eukarya is executed at all levels: DNA (transcription initiation), RNA (transcription elongation, attenuation), and protein (post-translational modification, allosteric regulation). Regulation at protein/enzyme level is certainly expected to be commonly present in Archaea, based on protein domains that are involved in allosteric regulation (ACT, RAM) (Aravind & Koonin, 1999; Ettema et al., 2002). However, in one of the few well characterized archaeal metabolic pathways, the aforementioned EMP pathway, regulation by metabolic effectors is rare as compared to the glycolytic enzymes in Bacteria and Eukarya (Verhees et al., 2003; Siebers & Schönheit, 2005).

The classical regulation sites of the EMP pathway are the irreversible reactions catalyzed by the antagonistic enzyme couples: hexokinases/glucose-6-phosphatase (Eukarya), phosphofructokinase/fructose-1,6-bisphosphatase and pyruvate kinase/phosphoenolepyruvate synthetase. Archaeal kinases, such as ADP- and ATP-dependent glucokinases, ADP-, ATP-, and PP_i-dependent phosphofructokinases, and pyruvate kinase, with the only exception being the AMP-stimulated pyruvate kinase of *T. acidophilum*, exhibit no allosteric properties and thus the classical control sites seem to be absent in Archaea (Siebers & Schönheit, 2005). The only glycolytic enzyme identified so far that is under allosteric control is the non-phophorylating GAPN. GAPN has an important function in the regulation of the EMP variant of *T. tenax* (Brunner et al., 1998, 2001; Lorentzen et al., 2004), as well as the semi-phosphorylative ED pathway of *S. solfataricus* (Ahmed et al., 2005; Ettema et al., in preparation).

In addition, a second control point seems to emerge at the level of phosphoenolpyruvate/pyruvate interconversion by PEPS, PYK, and PPDK in *T. tenax* (Tjaden et al., 2006).

Regulation at translational level (attenuation, anti-termination) has not yet been demonstrated in Archaea. A relatively important site of regulation of archaeal metabolism appears to be at transcriptional level. After the finding that the archaeal transcription machinery is not like the bacterial system (four subunit RNA polymerase (RNAP) and a series of Sigma factors that interact with specific −10/−35 promoter elements), but instead resembles the core of the eukaryal PolII system (10–12 subunit RNAP, TATA-binding protein (TBP) and transcription factor B (TFB)) (Bell, 2005), different types of transcriptional regulators have been identified in Archaea:

1 Bacterial-like regulators that often block transcription initiation by competing with the transcription factors (TBP, TFB) or the RNAP for their respective binding site (e.g. Lrp-like regulators) (Brinkman et al., 2000, 2003); bacterial-like activators have also been described, that stimulate the formation of the TBP-TFB-RNAP-promoter complex by affecting the DNA structure, by a direct interaction, or via an additional bridging protein (see below).
2 Eukaryal-like regulators are not abundant, but interestingly one has been reported to be present in all archaeal genomes, the multi-protein bridging factor (MBF) (Makarova & Koonin, 2003a); whereas the function of eukaryal MBF is to connect transcriptional regulators (generally containing bZIP-type leucine zipper domains) with the pre-initiation complex (TBP-TFB-RNAP), the role of the archaeal MBF is currently under investigation.
3 Unique archaeal regulators, such as the transcriptional activator of gas vesicle formation in haloarchaea (GvpE), a proposed leucine zipper that has no obvious homolog in Bacteria and Eukarya (Kruger et al., 1998).

Recently a few archaeal functional genomics studies have been published (reviewed by van der Oost et al., 2006). In a few studies the control of glycolysis has been addressed. Microarray analysis in *Pyrococcus furiosus* reveals coexpression of genes encoding the glycolytic EMP enzymes, as well as key enzymes of amino acid biosynthesis during growth on maltose; growth on peptides induces the expression of genes that code for gluconeogenetic enzymes (Schut et al., 2003; reviewed by Verhees et al., 2003). Surprisingly, however, a combined microarray/proteomics study in *Sulfolobus solfataricus* indicated significantly less fluctuation (Snijders et al., 2006).

The DNA microarray analysis of *Pyrococcus* suggests coregulation of the expression of the EMP genes. A careful analysis of the promoter regions of *P. furiosus* indeed reveals a conserved palindromic site between the TATA box and the translation start site. This site is missing in the corresponding promoters in *Pyrococcus abyssi* and *P. horikoshii*, but was found to be present in the promoter regions of all glycolytic genes in *Thermococcus kodakaraensis* (van de Werken et al., 2006). Interestingly, in both *P. furiosus* and *T. kodakaraensis*, the domain is present downstream the TATA box of the *pps* gene encoding PEP synthetase, but not of the *pyk* gene; this again is in agreement with a role of PEPS in the glycolysis in the Thermococcales. The glycolytic regulatory protein still remains to be identified. Recently, Lee et al. (2005) reported on a sugar-sensing regulator (TrmB) that differentially controls the expression of the genes encoding two ABC-type sugar transporters in *P. furiosus* and *Thermococcus litoralis*. A link with the glycolysis, although tempting, has not yet been made.

Conclusions

We have described some examples of mosaicity and non-homologous gene replacement in the well characterized archaeal glycolytic pathways. The unraveling of the complete metabolic network is anticipated to occur much more efficiently in the near future, due to implementation of holistic, genomics-based technology. It will be exciting to see whether the pathway mosaicity described here is a general phenomenon, and to find that tinkering of metabolism is a not an exception but a rule.

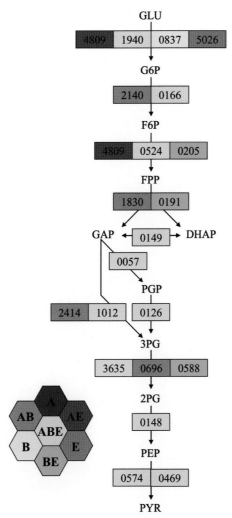

Plate 22.1 Non-orthologous gene displacement. COG analysis of the EMP pathway; COG numbers are given (see Table 22.1) and the distribution of orthologous genes in the three domains of life Archaea (A), Bacteria (B), and Eukarya (E) is given by the color code indicated. Although the COG classification is very useful for comparative studies, one has to keep in mind that a single COG may contain homologs with different physiological roles (substrate specificities); for example, the paralogous archaeal GLK and PFK kinases (COG4809).

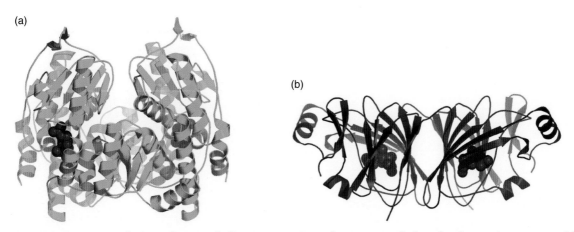

Plate 22.2 Convergent evolution of a metabolic enzyme. Crystal structure of phosphoglucose isomerase with bound 5-phospho-D-arabinonate from (a) *Pyrobaculum aerophilum* and (b) *Pyrococcus furiosus*.

23
Metabolism of inorganic sulfur compounds in Archaea

Arnulf Kletzin

Introduction

Oxidation and reduction of inorganic sulfur compounds (ISCs) was associated with the Archaea from the start: one of the three – physiologically defined – groups of prokaryotes in the newly created urkingdom was the so-called "thermophilic and sulfur-dependent Archaebacteria" (Woese et al., 1978; Huber et al., 2002). Sulfur metabolism became soon a hallmark of (hyper-) thermophiles and of the Archaea, from both the Crenarchaeota and Euryarchaeota kingdoms: the first hyperthermophilic isolate, *Sulfolobus acidocaldarius*, and also the first organism growing optimally above the boiling point of water, *Pyrodictium occultum*, were sulfur-dependent Crenarchaeota (Brock et al., 1972; Stetter, 1982). In particular the reduction of elemental sulfur (S°) is one of the most common physiological properties (Table 23.1; Amend & Shock, 2001), which is not surprising given the abundance of sulfur in volcanic environments. This chapter covers the current knowledge of the physiology and biochemistry of **dissimilatory** sulfur metabolism in Archaea, i.e. reactions used for energy conservation. I do not describe **assimilatory** reactions used for synthesis of cell proteins and cofactors (see, for example, Chapter 18 in this volume).

Sulfur (geo-) chemistry and physiology of sulfur-dependent Archaea

Geochemistry and physiology

Sulfur is the fourteenth most abundant element in the earth's crust. The bulk consists of sulfidic metal ores or sulfate sediments (Middelburg, 2000). S° and ISCs are also prevalent in volcanic exhalations and can amount to up to 10% of the dry volume. Their main constituents are water vapor (usually 60–90%; Plate 23.1) and CO_2 (90% of the dry volume). SO_2 is the most abundant sulfur species; however, the proportion of SO_2, SO_3, H_2S, and S° varies (Stoiber, 1995). SO_2 and SO_3 are characteristic for exhalations from water intrusions reacting with magma and overheated igneous rocks. H_2S dominates in evolved terrestrial hydrothermal fields and submarine vents. Elemental sulfur is deposited from H_2S following air oxidation or comproportionation with SO_2.

Sulfate is the predominant sulfur species and the second most abundant anion in seawater (Middelburg, 2000). Mixing of superheated, H_2S-reduced hydrothermal fluids with cold, oxidized seawater in deep-sea hydrothermal vents results in the formation of the spectacular "black" or "white smokers" (Hannington et al., 1995). The smoker chimneys consist of metal sulfides, sulfates, and silica. Their porous walls are loaded with hyperthermophilic methanogens, sulfate and S° reducers. The microorganisms are stratified into in several temperature zones within the chimney walls according to their optimal growth conditions (Reysenbach & Cady, 2001).

Anaerobic enrichment cultures with organic carbon sources and S° poised to a near-neutral pH (5–8) will result in the growth of *Pyrococcus* or

Table 23.1 Electron donors, acceptors, and growth products of sulfur-dependent archaeal genera.

Electron donor(s)	Electron acceptor	Product(s)	Genera
H_2	$S°$	H_2S	*Acidianus, Caldivirga, Ignicoccus, Metallosphaera, Pyrobaculum, Pyrodictium, Sulfurisphaera, Stygiolobus, Thermocladium, Thermoproteus*
H_2	$S_2O_3^{2-}$	H_2S	*Archaeoglobus, Ferroglobus, Pyrobaculum, Pyrodictium*
H_2	HSO_3^-	H_2S	*Pyrobaculum*
H_2	SO_4^{2-}	H_2O	*Archaeoglobus*
H_2	O_2	H_2O	*Acidianus, Metallosphaera, Pyrobaculum, Sulfolobus*
Organics	$S°$	H_2S	*Acidilobus, Caldisphaera, Caldivirga, Desulfurococcus, Hyperthermus, Palaeococcus, Pyrobaculum, Pyrococcus, Staphylothermus, Stetteria, Thermococcus, Thermophilum, Thermoplasma, Thermoproteus, Vulcanisaeta*
Organics	$S_2O_3^{2-}$	H_2S	*Archaeoglobus, Caldivirga, Pyrobaculum, Pyrodictium, Pyrolobus, Thermocladium, Vulcanisaeta*
Organics	HSO_3^-	H_2S	*Archaeoglobus, Pyrobaculum, Pyrodictium*
Organics	SO_4^{2-}	H_2S	*Archaeoglobus, Caldivirga*
Organics	H^+	H_2	*Pyrococcus*
$S°$	O_2	SO_4^{2-}	*Acidianus, Metallosphaera, Sulfolobus*
MeS, MeS_2	O_2	$Me^{n+}SO_4^{2-}$	*Acidianus, Metallosphaera, Sulfolobus*
$S_2O_3^{2-}$	O_2	$S_4O_6^{2-}$	*Natronorubrum*
$S_2O_3^{2-}$	O_2	SO_4^{2-}	*Aeropyrum*
$S_4O_6^{2-}$	O_2	SO_4^{2-}	*Sulfolobus*
H_2S	NO_3^-	$S°, NO_2^-, NO$	*Ferroglobus*

Compiled from Stetter (1999), Amend & Shock (2001), and the abstracts accessible via the taxonomy browser in Pubmed (www.ncbi.nih.gov).

Thermococcus-like organisms, which are easily isolated. Others like *Pyrodictium* and *Thermoproteus* are chemolithoautotrophs: redox reactions of inorganic substances are the sources of energy and reducing equivalents. The oxidation of H_2, which is a natural component of volcanic gasses, with sulfur as electron acceptor is probably the most common energy-yielding process (Stetter, 1999; Amend & Shock, 2001). The carbon source is CO_2. Most sulfur metabolizers are anaerobes; others use O_2 as an electron acceptor, sometimes in concentrations as low as 10 ppm (Stetter, 1999).

Solfataras are the typical habitats of sulfur-oxidizing hyperthermophiles. They are mostly small, steam-heated pools of boiling surface water or mud, named after the Solfatara caldera near Naples, Italy. The cell densities are sometimes in excess of 1×10^8/ml; the pH values are between 1 and 4 and will lower with increasing cell densities due to the microbial sulfuric acid production. Most of the cells in these habitats show a uniform morphology: lobed, irregular cocci of 1–2 μm in diameter, growing aerobically or anaerobically. These are Sulfolobales, six different genera of Crenarchaeota (Huber & Prangishvili, 2005). Species of the genus *Sulfolobus* are characterized by aerobic growth and a high metabolic versatility. *Stygiolobus* species are obligatory anaerobic chemolithoautotrophic sulfur reducers. *Acidianus* and *Metallosphaera* are facultative anaerobes and facultative heterotrophs, which also grow, like *Sulfolobus metallicus*, by oxidation of metal sulfides. They contribute to bioleaching of base and precious metal ores and to the formation of acidic drainage downstream of mines and slag heaps (Table 23.1; Huber & Prangishvili, 2005).

The first reports on non-thermophilic sulfur-dependent Archaea appeared only recently: a halophilic, heterotrophic, and sulfur-respiring strain, named *Haloferax sulfurifontis*, was isolated from a sulfur-rich spring in Oklahoma (USA; Elshahed et al., 2004). An extremely halophilic *Natronorubrum* strain (4 M NaCl) was isolated from neutral hypersaline lakes in Altai (Russia). It oxidized thiosulfate to tetrathionate, which was subsequently oxidized to sulfate by an extremely halophilic, chemolithoautotrophic bacterium isolated from the same source (Sorokin et al., 2005).

Inorganic chemistry

Sulfur is an element with a complex inorganic chemistry. S° is almost insoluble in water (5 µ/l at 25 °C, solubilities at higher temperatures are unknown; Boulegue, 1978). The oxidation state may adopt any value between −2 and +6; however, biologically relevant are H_2S, polysulfides, metal sulfides (MeS and MeS_2), S°, and the sulfur oxyanions sulfite, thiosulfate, polythionates, and sulfate (Table 23.2, Fig. 23.1; Roy & Trudinger, 1970). The high enthalpy of S–S bond formation as in S_8 (264 kJ/mol) has the consequence that numerous sulfur compounds form reactive homoatomic chains and rings. Thus, many reactions of ISCs will occur non-enzymatically (Steudel, 2000). Some examples:

- The nucleophile HS^- reacts with S° to soluble polysulfides at pH >5 (Table 23.2, eqn 23.1; Schauder & Kröger, 1993), S° and polysulfides react in a similar manner with thiols forming organic sulfane chains (eqn 23.2).
- S° disproportionates to thiosulfate and HS^- in neutral and alkaline aqueous solutions catalyzed by hydroxyl ions (eqn 23.3; Kletzin, 1989; Steudel, 2000).

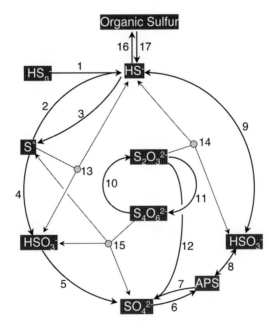

Fig. 23.1 The biological sulfur cycle shown by enzyme reactions involving inorganic sulfur compounds. Enzymes (cofactors in brackets): 1, polysulfide reductase (Mo/FeS/heme); 2, sulfur reductase (Mo/FeS); 3, sulfide:quinone oxidoreductase (flavin) or sulfide:cytochrome *c* oxidoreductase (flavin, heme); 4, sulfur oxygenase (GSH); 5, sulfite:acceptor oxidoreductase (Mo/FeS, acceptor is mostly cytochrome *c*, in some cases quinones) or sulfite oxidase (Mo, heme); 6, ATP sulfurylase; 7, ATP sulfurylase or adenylylsulfate:phosphate adenylyltransferase (APAT); 8, adenylylsulfate (APS) reductase (FAD/FeS); 9, sulfite reductase (siroheme, FeS); 10, tetrathionate reductase (Mo/FeS); 11, thiosulfate:acceptor oxidoreductase; 12, Sox complex (Mo, Mn, heme); 13, sulfur oxygenase reductase (Fe); 14, thiosulfate reductase (Mo/FeS); 15, tetrathionate hydrolase (there is also a trithionate hydrolase in addition); 16, O-acetylserin or O-phosphoserine sulfhydrolases (pyridoxal-5′-phosphate); 17, cysteine desulfurase (pyridoxal-5′-phosphate). Gray circles denote disproportionation reactions. Reactions of phosphoadenosinphosphosulfate formation and reduction and of cysteine breakdown are omitted. (Compiled from Takakuwa, 1992, and other references cited in the text.)

Table 23.2 Biologically relevant inorganic sulfur compounds and non-enzymic reactions of ISCs.

Sulfur species	Formula*	S oxidation state	Eqn no.	Non-enzymic reaction
(Metal) Sulfides	HS^-, $[MeS]$ $[MeS_2]$†	-2 -1	— —	
Polysulfides	$HS\text{-}S_n\text{-}S^-$‡	-1 and 0	23.1	Polysulfide formation§ $\quad S_n^\circ + HS^- \underset{pH<5}{\overset{pH>5}{\rightleftharpoons}} S_{n+1}^{2-} + H^+$
Organic thiols	R-SH	-1	23.2	Organic sulfane formation $\quad GSH + S_n^\circ \longrightarrow G_{n+1}^- + H^+$ $\quad GSH + S_n^{2-} \longrightarrow GS_{n+1}^- + HS^-$
Sulfur	S_8	0	23.3	S° disproportionation§ $\quad S_8 + 6H_2O \underset{pH\leq5}{\overset{pH\geq7}{\rightleftharpoons}} 2S_2O_3^{2-} + 4HS^- + 8H^+$
Sulfite	HSO_3^-	$+4$	23.4	Thiosulfate formation§ $\quad S_8 + 8HSO_3^- \underset{pH<4}{\overset{pH>5}{\rightleftharpoons}} S_2O_3^{2-} + H^+$
Polythionates	$O_3S\text{-}S_n\text{-}SO_3^{2-}$¶	$+5$ and 0	23.5	Disproportionation $\quad O_3S\text{-}S_n\text{-}SO_3^{2-} \overset{pH\geq4}{\longrightarrow} S^\circ + SO_4^{2-} + S_2O_3^{2-}$
Thiosulfate Sulfate	$S\text{-}SO_3^{2-}$ SO_4^{2-}	$+6$ and -2 $+6$		See eqns 23.3 and 23.4

*Predominant dissociation state at pH 7; †pure or mixed metal sulfides, e.g. FeS_2, $CuFeS_2$; ‡n = 2–3; §pH at $T \geq 75$ °C; ¶n ≤ 50.
Source: Roy and Trudinger (1970), Takakuwa (1992), Steudel (2000).

- Aqueous sulfite solutions incubated with S° at higher temperatures will form thiosulfate (eqn 23.4). In contrast, thiosulfate decomposes at pH <4 rapidly to sulfite and S° (Roy & Trudinger, 1970; Kletzin, 1989; Steudel, 2000).
- Polythionates can accumulate at acidic pH and to chain lengths up to 50 S. Reductants and/or a pH ≥6 will cause decomposition into thiosulfate, sulfite, and sulfur (eqn 23.5; Steudel, 2000).

For reasons of this complexity it is important to understand the behavior of ISCs under the conditions in which hyperthermophilic microorganisms will thrive. Many intermediates react rapidly to the thermodynamically most stable product, impairing the interpretation of enzyme assays and other biochemical measurements and cultivation in defined media. For example, reports claiming to grow thermoacidophiles on thiosulfate (e.g. Sun et al., 2003) should therefore be regarded with care, as the resulting substrate might have been S° (eqn 23.4).

Dissimilatory sulfate and sulfite reduction: *Archaeoglobus* and *Pyrobaculum*

Archaeal model organisms for dissimilatory sulfate reduction are *Archaeoglobus* spp. and *Pyrobaculum islandicum* (Dahl et al., 2001; Dahl & Trüper, 2001; Sperling et al., 2001). These organisms use organic substrates and/or H_2 as electron donors to reduce sulfate, thiosulfate, or sulfite to H_2S. The steps of dissimilatory and assimilatory sulfate reduction are the same, with the only difference being that the dissimilatory reactions are fully reversible and can be used for ISC oxidation (Fig. 23.1). Three enzymes are required (eqns 23.6–23.8), which have been purified from *Ag. fulgidus* and are localized in the cytoplasm. Therefore sulfate must be transported into the cell, a process that has not been studied in Archaea (Dahl & Trüper, 2001; Sperling et al., 2001).

ATP sulfurylase
$$ATP + SO_4^{2-} \leftrightarrow APS + PP_i$$
(eqn 23.6)

Adenylylsulfate (APS) reductase
$$APS + 2e^- + H^+ \leftrightarrow AMP + HSO_3^-$$
(eqn 23.7)

Sulfite reductase
$$HSO_3^- + 6e^- + 6H^+ \leftrightarrow HS^- + 3H_2O$$
(eqn 23.8)

ATP sulfurylases or, more correctly, **ATP:sulfate adenylyltransferases** (eqn 23.6) are key enzymes in sulfate reduction, which fall into two unrelated classes. Most are homooligomeric like the Archaeoglobus enzyme and have subunits of 41–69 kDa, while the heterodimeric enzymes consist of 27 and 62 kDa subunits and are exclusively assimilatory (e.g. from *E. coli*; Sperling et al., 2001). Crystal structures are available from three assimilatory and one dissimilatory ATP sulfurylase from Bacteria, all of the homooligomeric type. There is a difference between the enzymes from mesophiles and thermophiles and not between Archaea and Bacteria: the *Thermus* ATP sulfurylase contained a zinc site, which is supposed to be stabilizing against thermal denaturation (Taguchi et al., 2004). The coordinating residues are conserved in all homologs from hyperthermophiles supporting this notion.

Dissimilatory APS reductases (eqn 23.7) are bidirectional FAD and iron–sulfur (FeS) cluster-containing enzymes with a molecular mass of approximately 90,000 (Dahl & Trüper, 2001). The larger FAD-containing α-subunit is paralogous to the flavoprotein subunit of succinate dehydrogenases (SDH). The smaller FeS subunits are not homologous and contain two FeS clusters in the APS reductase and three in SDH. The *Archaeoglobus* enzyme became an important model for the homologous enzymes of SRB and sulfur-oxidizing bacteria. Their spectroscopic properties (EPR and UV/Vis) are virtually identical (Plate 23.2; Fritz et al., 2002a). Thus, the redox cofactors appear to be embedded into a conserved protein matrix.

The three-dimensional structure of the *Archaeoglobus* enzyme showed that the flavin is located at the bottom of a deep pocket of the large subunit, providing access for the substrates and products (Plate 23.2; Fritz et al., 2002b). The [4Fe–4S] cluster I located in the small subunit is deeply buried in the protein matrix in close proximity to the FAD, whereas [4Fe–4S] cluster II lies at the surface of the protein and may interact with the still unknown physiological electron donor/acceptor (Plate 23.2). Cluster I showed a reduction potential of −57 mV, cluster II of −520 mV, thus differing by 460 mV, whereas the FAD showed a potential of about

+50 mV (Fritz et al., 2002a). Upon incubation with AMP and sulfite a rather broad radical signal appeared, indicating the interaction of a flavosemiquinone with another paramagnet, most likely cluster I. The results allowed the unraveling of the reaction mechanism of the APS reductases, showing that an FAD N(5)-sulfite adduct is the intermediate during both the oxidative and the reductive reactions (Plate 23.2).

Dissimilatory, siroheme-containing sulfite reductases (DSRs, eqn 23.8) catalyze the six-electron reduction from sulfite to sulfide. They are present in all anaerobic sulfate or sulfite reducers and in some sulfide-oxidizing bacteria (Kappler & Dahl, 2001). Archaeal DSRs were purified from *Ag. fulgidus* and *Pb. islandicum* (Dahl & Trüper, 2001; Dahl et al., 2001). The core enzymes consist of two different subunits in an $\alpha_2\beta_2$ structure (Kappler & Dahl, 2001). The *dsrA* and *dsrB* genes have similar sizes, are adjacent to each other in genomes, and show distant similarity, suggesting that they probably arose from a gene duplication event (~30% amino acid identity between *Ag. fulgidus* DsrA and B). The enzymes are remarkably conserved between *Ag. fulgidus* and *Desulfovibrio vulgaris* (~60% identity in both subunits; Dahl & Trüper, 2001). Additional small subunits are encoded in archaeal *dsr* operons. The function is unknown and not even the resolution of the three-dimensional structures of two of these small subunits solved this problem (Mander et al., 2005).

The siroheme, which also occurs in the paralogous nitrite reductases (NirB), is a complex cofactor consisting of a heme molecule and an electronically coupled [4Fe–4S] cluster. The thiol sulfur of a bridging cysteine is the axial ligand of the heme iron and one of the ligands of the [4Fe–4S] cluster. The distal side of the heme iron is poised to bind the substrate and does not have a protein ligand. The available three-dimensional structures of assimilatory sulfite reductases revealed the siroheme geometry, but they belong to a family with a different domain structure compared to the *Archaeoglobus* enzyme (Schnell et al., 2005).

The *Archaeoglobus* DSR contained 2 mol/mol siroheme per heterotetramer as indicated by UV/Vis spectra. The quantitation of iron gave 22–24 mol/mol, of acid-labile sulfur 20 mol/mol heterotetramer, respectively. The dissimilatory DSRs thus have, compared to the *E. coli* assimilatory enzyme, numerous additional cysteine residues in their amino acid sequences (in both DsrA and B), which could coordinate a minimum of six FeS clusters in the $\alpha_2\beta_2$ heterotetramer in agreement with the iron quantitation. Two of them are part of the siroheme, leaving four clusters to create one or two intramolecular cluster chain(s) as found in the APS reductase (Plate 23.2; Dahl & Trüper, 2001). The *in vivo* electron donor and the mechanism of action are not fully resolved.

A novel DSR with different domain composition and cofactor specificity was recently isolated from *Methanocaldococcus jannaschii* (Johnson & Mukhopadhyay, 2005). *Mc. jannaschii* and *Methanothermobacter thermoautotrophicus* grew in the presence of sulfite as the sole sulfur source, although it is commonly toxic to methanogens. A 620 aa protein encoded in the genome had an N-terminal F_{420} dehydrogenase and a C-terminal DsrA domain. The purified protein was an oxygen-sensitive F_{420}-dependent sulfite reductase containing siroheme. Homologs were exclusively found in genomes of other strictly hydrogenotrophic and sulfite-resistant thermophilic methanogens. It is believed that the function of this novel enzyme is in sulfite detoxification and in assimilatory sulfite reduction.

Anaerobic sulfur reduction: *Pyrococcus*, *Pyrodictium*, and *Acidianus*

The reduction of S° coupled to the oxidation of organic compounds or molecular hydrogen is the most common pathway of energy metabolism in hyperthermophilic Archaea (Table 23.1; Stetter, 1999; Amend & Shock, 2001). Several metabolic types with S° as electron acceptor can be distinguished:

1 Many fermentatively growing heterotrophs require the addition of S° to the media. The main products are small organic compounds like acetate or alanine. H_2S is produced in the presence and H_2 in the absence of S° (Adams et al., 2001).

2 Some Archaea oxidize organic substrates completely to CO_2 and depend on membrane-bound electron transport chains with S° as the terminal electron acceptor (Stetter, 1999).

3 Numerous S° reducers gain energy from H_2 oxidation; ATP and reduction equivalents are used for CO_2 fixation (eqn 23.9).

$$H_2 + S° \rightarrow H_2S$$
(or HS^-, depending on the pH)

(eqn 23.9)

4 Some methanogens like *Methanopyrus* and *Methanobacterium* produce significant amounts of H$_2$S when grown in the presence of S° (Stetter & Gaag, 1983).

Anaerobic heterotrophic sulfur and proton reduction

Pyrococcus and *Thermococcus* species are easily enriched and grow fast with doubling times sometimes below 40 minutes. Many are sufficiently aerotolerant to be cultured without special precautions. *Pc. furiosus* has been studied as a model organism in great detail; however, several paradoxes remain. It ferments peptides as the optimal substrate, maltose, or both and requires the addition of S° to the medium when it grows on peptides alone, an observation that is also true for other *Pyrococcus* species (Adams et al., 2001). In contrast, the addition of S° to cells growing in maltose medium resulted in H$_2$S formation but had no beneficial effect on growth rates. Three different soluble proteins with S-reducing activity were purified from *Pc. furiosus*. All had other activities and none was the major S°-reducing enzyme in peptide-grown cells or was coupled to electron transport phosphorylation. Two of these enzymes were hydrogenases with additional S° or polysulfide reductase activity using H$_2$ or NADPH as electron donors (sulfhydrogenases; Ma & Adams, 2001). The third enzyme was a ferredoxin:NADP oxidoreductase with a broad substrate range, which oxidizes and/or reduces many different substrates, including FAD, polysulfides, NAD(P)(H), and O$_2$ (Ma & Adams, 1994). Paradoxically, their specific activities increased in response to the addition of maltose but not of S° to the medium (Adams et al., 2001).

A third hydrogenase from *Pc. furiosus* was membrane-bound and did not reduce S° or polysulfides (MBH; Sapra et al., 2003). MBH activities increased even more in the presence of maltose and the absence of S° (Adams et al., 2001). The MBH is part of a membrane-bound complex, which couples fermentation to proton translocation over the membrane driven by proton reduction with ferredoxin as electron donor. It constitutes a novel, simple, and so far unique type of anaerobic electron transport chain present in a single complex, resolving the question of energy conservation in *Pyrococcus* cells growing in the absence of S° (Sapra et al., 2003). It is still unknown whether S° reduction is coupled to an electron transport chain, as the known S°-reducing enzymes are apparently not involved and their sulfur reductase (SR) activity might be fortuitous (Adams et al., 2001; Sapra et al., 2003).

Inducible transcripts and proteins in sulfur- and peptide-grown cultures were identified in *Pc. furiosus* using microarray and two-dimensional gel techniques. The two genes with the highest increase in mRNA levels were *sipA* and *sipB*, which are transcribed from a common promoter region in opposite directions (Schut et al., 2001). The gene products were identified as membrane-associated proteins previously. Their function is not known and there is no biochemical evidence that they belong to a membrane-associated sulfur or polysulfide reductase complex (PSR; Adams et al., 2001; Sapra et al., 2003). The data suggested that *Pc. furiosus* should have some kind of SR or PSR. Unfortunately, the enzyme seems to be rather elusive.

Anaerobic sulfur metabolism in chemolithoautotrophic Archaea

H$_2$ oxidation with S° as electron acceptor is widespread in hyperthermophilic Archaea, including the *Pyrodictium* and *Acidianus* species (eqn 23.9, Table 23.1). The latter were the first to grow alternatively by anaerobic H$_2$ or aerobic S° oxidation without organic substrates (Segerer et al., 1985; Zillig et al., 1985). One of the major differences between these two genera is the pH of the medium: *Pyrodictium* spp. require pH 5–8 whereas *Acidianus* species grow at pH 1–4. However, their internal, cytoplasmic pH is supposed to be near neutrality. Polysulfides could theoretically form non-enzymically from S° powder added to the media during growth of *Pyrodictium* but not of *Acidianus* (Schauder & Kröger, 1993).

Pyrodictium sulfur reductase and hydrogenase

Sulfur-reducing chemolithoautotrophs require at least two membrane-bound enzymes for energy conservation: a hydrogenase and an SR or PSR coupled in a short electron transport chain. Both were characterized from *Pd. brockii* (Pihl & Maier, 1991; Pihl et al., 1992) and *Pd. abyssi* (Dirmeier et al., 1998). H$_2$-

dependent SR (H$_2$:S° oxidoreductase) activity of *Pd. brockii* depended on quinones (Pihl et al., 1992). The SR was not purified in contrast to the NiFe hydrogenase, which consisted of two subunits. More is known about the 520 kDa membrane-bound H$_2$:sulfur oxidoreductase multienzyme complex purified from the related hyperthermophile *Pd. abyssi*. It is composed of nine subunits and has hydrogenase and SR activities (Dirmeier et al., 1998). Fe, Ni, Cu, acid-labile sulfur, and hemes b and c were found, but no Mo or W. Quinones were not required for activity. The complex contains all the constituents necessary for the electron transport from H$_2$ to S°, thus being another example of an entire electron transport chain within a single complex. The N-termini of the 65 and 42 kDa subunits show similarity to the NiFe and the FeS subunits of the *Ac. ambivalens* but not to *Pyrococcus* hydrogenases (Dirmeier et al., 1998; Laska et al., 2003). The N-termini of the 24 and 85 kDa subunits showed similarity to the FeS and molybdopterin subunits of the *Ac. ambivalens* SR and the *Wolinella* PSR, respectively (Laska et al., 2003), suggesting phylogenetic relationship. The enzyme should therefore contain Mo or W, although none was found. It is not uncommon for these elements to be depleted beyond the detection limit upon protein purification.

It can be concluded that the *Pyrodictium* enzymes differ significantly from *Pyrococcus* enzymes. There are also differences to the *Wolinella* hydrogenase/PSR system (Fig. 23.2; Hedderich et al., 1999), especially due to the formation of the complex and the presence of c-type hemes in *Pyrodictium*. The two *Pyrodictium* systems have many properties in common, especially the heme b and c content and the sizes of the hydrogenase subunits. But there are also clear differences, most notably the quinone dependence of the *Pd. brockii* enzyme and the presence of a tightly bound complex in *Pd. abyssi*.

Acidianus ambivalens sulfur reductase and hydrogenase

An SR purified from membrane fractions of anaerobically grown *Ac. ambivalens* cells had a comparable H$_2$:S° oxidoreductase activity to the *Pyrodictium* sulfur reductase in the presence of a copurified hydrogenase and with either 2,3-dimethyl-1,4-naphthoquinone (DMN) or cytochrome *c* as electron carriers (Laska et al., 2003). The bidirectional and oxygen-stable hydrogenase purified separately did not have the sulfur-reducing side activity of the soluble *Pc. furiosus* hydrogenases. In contrast to the *Pyrodictium* SRs, horse heart cytochrome *c* was not reduced and, moreover, no c-type cytochromes have been found in any of the Sulfolobales or their genomes yet, excluding them as natural electron carriers. The activity of the basic cytochrome in the SR assay might originate from linking of the hydrogenase and SR non-specifically for direct electron transfer. The purified enzymes did not form a single complex as in *Pd. abyssi* (Laska et al., 2003).

The *A. ambivalens* hydrogenase encoded by a polycistronic cluster including genes for a NiFe and an FeS subunit was rather dissimilar to other hydrogenases (Laska et al., 2003). The most similar relatives were from a streptomycete and *Ralstonia* (~34% sequence identity). The Isp1 protein (Fig. 23.2) is probably the membrane anchor. The FeS subunit contained a leader peptide with a twin-arginine translocation motif (TAT) not present in the mature protein, suggesting a transport across the membrane (Fig. 23.2). Fe and Ni were present in membrane and in enriched hydrogenase fractions in accordance with the sequence analysis (Laska et al., 2003).

The *Ac. ambivalens* SR gene cluster consisted of five ORFs, *sre*ABCDE (Fig. 23.2). The deduced amino acid sequences of *sreA* encoding a 110 kDa subunit visible on SDS gels and *sreB* showed similarity to the molybdopterin and the FeS subunits of the DMSO/nitrate reductase family, respectively (Laska et al., 2003). SreA also contained a twin arginine motif, suggesting an export by the TAT pathway. The *sreC* gene encoded a protein with ten predicted transmembrane helices, *sreD* an unknown polyferredoxin with 26 cysteine residues, and *sre*E a small system-specific chaperone. Mo, but not W, was found in solubilized membrane fractions, suggesting that the SR is a molybdoprotein in accordance with the sequence analysis. The predicted orientation and the molecular composition of the *Ac. ambivalens* SR deduced from the biochemical results and the sequence analysis are similar to the *W. succinogenes* PSR (Fig. 23.2; Laska et al., 2003). Both enzymes consist of homologous catalytic and electron transfer subunits (SreA/PsrA and SreB/PsrB, respectively) and non-homologous membrane anchors (SreC/PsrC). The hydrogenases also have comparable quaternary

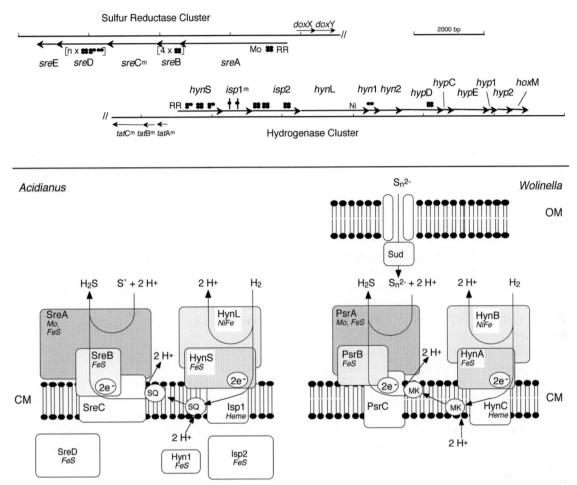

Fig. 23.2 Upper panel: gene clusters encoding the NiFe hydrogenase and the SR in *Ac. ambivalens*. See Laska et al. (2003) for the explanation of the gene names. Black dots, Cys-containing sequence motifs potentially coordinating FeS clusters; Ni, Mo, potential metal binding sites; m, predicted transmembrane protein; RR, twin arginine protein translocation motif; the *tat* cluster encodes proteins required for twin arginine translocation pathway; *doxX*, *doxY*, a thiosulfate:quinone oxidoreductase paralog (see below). Lower panel: schematic representation of the sulfur and polysulfide respiration in *Ac. ambivalens* and *Wolinella succinogenes*. The *Ac. ambivalens* model was developed in analogy to *Wo. succinogenes* from the results of the sequence comparison and the biochemical data (Laska et al., 2003). Homologous subunits are shown in identical shading. CM, cytoplasmic membrane; OM, outer membrane; MK, menaquinone; SQ; sulfolobus quinone.

structures; they are composed of at least three subunits each, the homologous Ni-containing catalytic (HynL/HynB; Fig. 23.2), the FeS-containing electron transfer subunits (HynS/HynA), and the non-homologous membrane anchors (Isp1/HynC).

The only other probably orthologous *sreABCDE* gene clusters were identified in the genome of *Su. solfataricus* (but not in *Su. acidocaldarius* and *Su. tokodaii*), suggesting that it should grow by heterotrophic anaerobic S° respiration; however, this has not been demonstrated yet (A. Kletzin, unpublished). Phylogenetic analysis showed that homologs of SreA are present in the genomes of other facultative sulfur reducers and that they form a branch in the DMSO reductase family tree together with other enzymes with unrelated substrates (Laska et al., 2003). It can be concluded that these organisms have a mutually similar set of membrane-bound SRs with related but diverse molybdopterin and electron transfer subunits, while additional subunits might vary. The membrane-bound hydrogenases diversify similarly, although the analysis is more complicated because NiFe hydrogenases fall into four subfamilies, which are all realized in different Archaea. It was interesting to see that the *Sulfolobus* species sequenced so far do not have hydrogenase genes, showing that they should not be able to grow lithotrophically with H_2.

Sulfur oxidation in *Acidianus*

Numerous (micro-) organisms oxidize ISCs for dissimilatory energy metabolism, including anaerobic phototrophic and aerobic chemolithotrophic, mesophilic, and (hyper-) thermophilic bacteria. Bacterial sulfur oxidizers are physiologically and phylogenetically diverse. In contrast, most of the known archaeal sulfur oxidizers belong to the acidophilic Sulfolobales (Huber & Prangishvili, 2005). The biochemistry was consequently studied most thoroughly in *Acidianus* species (Kletzin et al., 2004; Chen et al., 2005). The schemes of S° oxidation and electron transport developed for *Ac. ambivalens* (Fig. 23.3) differ significantly from the bacterial models.

The oxidation of S° to sulfuric acid proceeds in at least two steps, often more involving intermediates like sulfite, thiosulfate, tetrathionate, etc. (Fig. 23.1). Several pathways are distinguished depending on the organisms and growth conditions (reviewed, for example, in Takakuwa, 1992; Kelly et al., 1997; Friedrich et al., 2001, 2005).

1 Thermoacidophilic Archaea oxidize S° with a cytoplasmic sulfur oxygenase reductase (SOR), a sulfur-disproportionating enzyme. Thiosulfate and sulfite are oxidized by membrane-bound oxidoreductases (Fig. 23.3; Kletzin, 1989; Zimmermann et al., 1999; He et al., 2000; Müller et al., 2004).

2 Acidophilic Bacteria like *Acidithiobacillus* spp. oxidize S° to sulfite by a GSH-dependent soluble sulfur oxygenase. Sulfite is oxidized by a sulfite:cytochrome *c* oxidoreductase and thiosulfate by a tetrathionate-forming thiosulfate oxidoreductase. Tri- and tetrathionate hydrolases were found as well (Takakuwa, 1992; Visser et al., 1997; Rohwerder & Sand, 2003; Friedrich et al., 2005).

3 Neutrophilic bacteria oxidize S° and ISCs using the periplasmic Sox multienzyme complex. The complex binds S° and ISCs covalently to the thiol of a cysteine side chain and oxidizes the S atoms to sulfate without formation of free intermediates (Friedrich et al., 2001, 2005). The electrons are transferred via cytochrome *c* to the terminal oxidase. Sox genes were identified in genomes of numerous mesophilic and thermophilic Bacteria but not in Archaea or in *Acidithiobacillus* (Friedrich et al., 2005).

Acidianus ambivalens and *Ac. tengchongensis* sulfur oxygenase reductase

The initial enzyme in the archaeal S° oxidation pathway is unique in several aspects. The soluble protein was termed sulfur oxygenase reductase (SOR) because it is the only S° disproportionating enzyme. Sulfite, thiosulfate, and hydrogen sulfide are products. Oxygen is required for activity but no organic cofactors (Kletzin, 1989; He et al., 2000):

$$4S° + O_2 + 4H_2O \rightarrow 2HSO_3^- + 2H_2S + 2H^+$$
(eqn 23.10)

Thiosulfate is probably a non-enzymic product of sulfite condensation with excess S° (eqn 23.4). The glutathione-dependent periplasmic sulfur oxygenases of mesophilic *Acidithiobacillus* species do not have a reductase activity or bear any similarity to the SOR (sequences are not known; Rohwerder & Sand, 2003). Another protein with a SOR-like activity was described earlier from a phylogenetically uncharacterized isolate of "*Sulfolobus brierleyi*" as a sulfur oxygenase (Table 23.3). The enzyme largely resembles the *Acidianus* SORs (the isolate was most probably an *Acidianus* species; Kletzin et al., 2004).

Other *sor* genes were identified in the genomes of the crenarchaeote *Sulfolobus tokodaii*, of the Euryarchaeota *Ferroplasma acidarmanus* and *Picrophilus torridus*, and of the hyperthermophilic bacterium *Aquifex aeolicus*. The deduced amino acid sequences shared 35–88% identical residues and constitute a unique protein family without similarities to other enzymes (Urich et al., 2004). Interestingly, the *sor* gene was missing in *Su. solfataricus* and *Su. acidocaldarius*, which were described as facultative chemolithoautotrophic, sulfur-dependent aerobes (Brock et al., 1972; Zillig et al., 1980). They had been isolated by successive rounds of serial dilution and not by plating of single colonies. Other researchers were unable to grow both strains under the

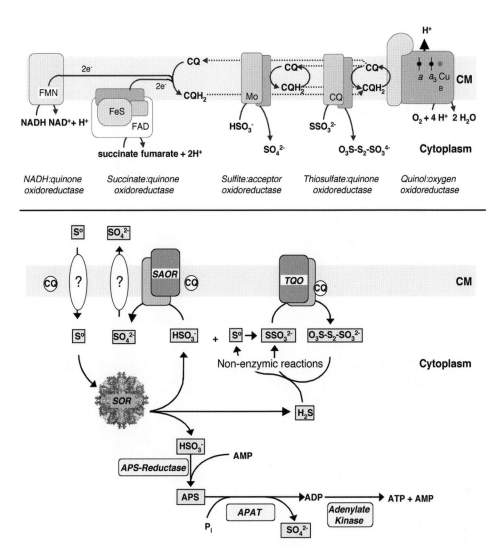

Fig. 23.3 Upper panel: scheme of *Ac. ambivalens* respiratory chain components and coupling points with sulfur metabolism. The activities of all enzymes depicted have at least been measured. a, heme a; CQ, caldariella quinone; CQH2, caldariella quinol; FeS, iron-sulfur centers (Kletzin et al., 2004; Pereira et al., 2004). Lower panel: hypothetical model of S° oxidation in *Ac. ambivalens* derived from known enzyme activities (in italics) and possible non-enzymic reactions: SAOR, sulfite:acceptor oxidoreductase; TQO, thiosulfate:quinone oxidoreductase; SOR, sulfur oxygenase reductase; APS, adenylylsulfate; APAT, adenylylsulfate:phosphate adenylyltransferase; CQ, caldariella quinone. See text for details (Kletzin et al., 2004).

chemolithoautotrophic conditions they were isolated upon, suggesting that they lost or never possessed the ability to oxidize S° (Norris & Johnson, 1998). The original cultures had probably not been pure, whereas those presently available were separated later with improved plating techniques.

The SOR activity was inhibited by thiol-binding reagents, pointing to the involvement of cysteines in catalysis (Kletzin, 1989; Urich et al., 2004; Chen et al., 2005). Three conserved cysteine residues are present in the SOR sequences. Site-directed mutagenesis showed that one of these (C_{31} in *Ac. ambivalens* numbering) was indispensable and could not be replaced by Ala or Ser, while mutagenesis of the other two resulted in markedly reduced activities (Chen et al., 2005; Urich et al., 2005b). EPR spectroscopy and redox titration showed that the SOR contains a mononuclear non-heme iron center in the high-spin Fe^{3+} state and with a low reduction potential ($E_o' = -268$ mV, protein as

Table 23.3 Properties of the SORs and sulfur oxygenase.

Source		Ac. ambivalens wild type SOR	Ac. ambivalens recombinant SOR (E. coli)*	"Su. brierley" S-oxygenase	Ac. tengchongensis recombinant SOR (E. coli)
Subunit mol. mass		35,187†	36,311†	35,000	35,172†
Holoenzyme mol. mass	Apparent	550,000‡	550,000‡/732,000§	550,000§	n.r.
	X-ray crystallography	844,488†	871,464†	n.r.	n.r.
pH range/pH$_{opt}$		4–8/7–7.4	n.r.	n.r./6.5–7.5	3.5–9/5
T$_{opt}$/T$_{max}$		85/108 °C	n.r.	65/>80 °C	70/>90 °C
Spec. oxygenase activity at optimal temperature ¶		10.6 U/mg	2.8 U/mg	0.9 U/mg††	753 U/mg§§ 29.7 U/mg‡‡,§§
Spec. reductase activity at optimal temperature**		2.6 U/mg	0.66 U/mg	n.r.	45.2 U/mg††,§§ 3.3 U/mg**,††
^{18}O-incorporation		n.r.	n.r.	+	n.r.
Diameter		~15.5 nm	~15 nm	n.r.	n.r.
Reference(s)		Kletzin, 1989, 1992; Urich et al., 2004	Urich et al., 2004, 2005a	Emmel et al., 1986	He et al., 2000; Sun et al., 2003; Chen et al., 2005

n.r., not reported; *including 10 aa C-terminal Streptag; †from sequence without N-terminal Met; ‡gel permeation chromatography; §non-denaturing PAGE; ¶one unit was defined as 1 μmol of sulfite plus thiosulfate formed per minute assuming that thiosulfate is formed non-enzymically; **one unit was defined as 1 μmol of H_2S per minute; ††wild type enzyme; ‡‡recombinant enzyme; §§values taken from He et al. (2000). One unit was defined as 1 mmol sulfite or H_2S formed per minute. The values reported by Chen et al. (2005) were: 4.85 U/mg for the heat-treated E. coli cell extract with recombinant SOR and 0.48 U/mg for the membrane fraction of Ac. tengchongensis cell extracts, respectively.

isolated). The signal disappeared upon reduction with dithionate or incubation of the SOR with S° at elevated temperature (Urich et al., 2004). It was intriguing to find that the reduction potential was more than 300 mV lower than usually found for this type of iron center and low enough to explain the S° reducing activity of the enzyme (E_o' [$H_2S/S°$] = −270 mV).

SOR structure and mechanism

Electron microscopic preparations of immunogold-labeled *Ac. tengchongensis* cells showed that the SOR is most likely associated with the cytoplasmic membrane (Chen et al., 2005). Hollow globular particles of 15.5 nm in diameter appeared in EM pictures of the purified *Ac. ambivalens* SOR (Kletzin, 1989; Urich et al., 2004).

X-ray crystallographic analysis to 1.7 Å resolution showed that the SOR is a spherical homoicosatetramer (i.e. 24 subunits) with a 432 point group symmetry and an external diameter of 150 Å. It surrounds an empty cavity with a diameter of 71–107 Å (Plate 23.3; Urich et al., 2005a, 2006). The resulting molecular mass of 844,000 for the native SOR (871,000 for the recombinant; Table 23.3) was higher than anticipated from biochemical analyses (550–730 kDa; Urich et al., 2004, 2005a). Each subunit consisted of a β-barrel core surrounded by α-helices. The iron center was coordinated by a bidentate glutamate and two histidine ligands completed by two water molecules in a structural motif known as "2-His 1-Carboxylate facial triad" (Plate 23.3; Urich et al., 2006; Costas et al., 2004). Mutation of any of the three iron ligands to alanine resulted in the loss of activity and iron-binding capabilities (Urich et al., 2005b). Residue C_{31} showed additional electron density, which proved to be a persulfide modification (Urich et al., 2006). The iron center and the three cysteines are buried in a pocket in the interior of each monomer, which is accessible only from the interior cavity. The low reduction potential of the iron site is probably the result of a surrounding hydrogen bonding network (Plate 23.3).

Some conclusions for the reaction mechanism could be derived from the structure. The core active site is composed of the iron site and the modified C_{31}. Substrate entry has to proceed through the hydrophobic channels along the fourfold axes of the sphere (Plate 23.3). The SOR thus provides an enclosed reaction compartment separated from the cytoplasm. The presence of a persulfide shows that S° is most probably bound covalently to C_{31} (eqn 23.2). The linear sulfur chain is aligned to the iron site and replaces the water ligands, poising the iron site for oxygen binding and activation. Further steps in the reaction sequence are currently not known, as no data of transition states are available; however, not even the non-enzymic sulfur disproportionation reaction is well understood (eqn 23.3; Steudel, 2000).

A question arises as to why a cytoplasmic and not a periplasmic or membrane-bound enzyme as in most Bacteria is used for the initial step of sulfur oxidation (Rohwerder & Sand, 2003; Friedrich et al., 2005). One probable answer is that the closed sphere allows use of the high reactivity of the S° disproportionation at elevated temperature and near-neutral pH, so that only little activation energy is required.

Electron transport chains coupled to the oxidation of H_2S, sulfite, and thiosulfate

The oxidation of sulfite, sulfide, and thiosulfate in *Ac. ambivalens* requires membrane-bound, proton-pumping oxidoreductases since the SOR does not couple S° oxidation to electron transport or substrate level phosphorylation. Bacterial H_2S oxidoreductases include the membrane-bound sulfide:quinone oxidoreductase (SQR), a flavocytochrome *c* sulfide dehydrogenase and oxidatively acting sulfite reductases (eqn 23.8; Theissen et al., 2003). SQR activity has been found in many Bacteria and even in higher Eukaryota but not yet in *Ac. ambivalens* or other Archaea, although homologs of the *sqr* gene are present in archaeal genomes (A. Kletzin, unpublished).

A tetrathionate-forming, membrane-bound **thiosulfate:quinone oxidoreductase** (TQO) was isolated from aerobically grown *Ac. ambivalens* cells. Ferricyanide, methylene blue and decyl ubiquinone were electron carriers of the reversible reaction. Caldariella quinone was bound to the protein. Optimal activity was observed at 85 °C and pH 5. The 102 kDa glycosylated holoenzyme consisted of 28 and 16 kDa subunits, suggesting an $\alpha_2\beta_2$ stoichiometry. Oxygen electrode measurements showed an electron transport from thiosulfate to molecular oxygen via the terminal heme copper quinol:oxygen oxidoreductase (Fig. 23.3). The TQO subunits were identical to DoxA and DoxD, previously described as parts of the *Ac. ambivalens* terminal oxidase (Müller et al., 2004). Interestingly, the recently isolated haloarchaeon *Natronorubrum* HG 1 (Sorokin et al., 2005) contained a membrane-associated tetrathionate synthase or thiosulfate:acceptor oxidoreductase, whose activity depended specifically on elevated concentrations of Cl^-. Its function is similar to the TQO described here but the protein has not been purified yet and the sequence is not known.

The fate of the tetrathionate formed by the *Ac. ambivalens* TQO has not been investigated yet. However, there is a possibility that a thiosulfate/tetrathionate cycle exists (Müller et al., 2004). Tetrathionate is unstable in the presence of strong reductants and is reduced to thiosulfate *in vitro* at high temperatures. Therefore, H_2S and sulfite might re-reduce tetrathionate formed by the TQO and thus feed electrons indirectly from the S° disproportionation reaction catalyzed by the SOR into the quinone pool (Fig. 23.3).

Sulfite:acceptor oxidoreductases or dehydrogenases (SAOR) and other sulfite-oxidizing enzymes are known from many organisms. Several pathways can be distinguished (Kappler & Dahl, 2001):

1 SAORs were described from several Bacteria; most were periplasmic enzymes, only one was membrane-bound (Kappler & Dahl, 2001). SAOR activity is also included in the SOX complex. The enzymes feed electrons typically via *c*-type cytochromes into the respiratory chain (Friedrich et al., 2001, 2005).

2 Two variants of a sulfite oxidation pathway coupled to substrate level phosphorylation were identified in Bacteria and Archaea involving the indirect sulfite oxidation via an APS reductase (eqn 23.7, Fig. 23.1) and either an ATP sulfurylase or an adenylylsulfate:phosphate adenylyltransferase (APAT, eqn 23.6; Brüser et al., 2000).

The activities of a SAOR (membrane fraction), APS reductase, APAT, and adenylate kinase (soluble fraction) were demonstrated in *Ac. ambivalens* (Fig. 23.3; Zimmermann et al., 1999), showing that both pathways are realized. The enzymes have not yet been purified. ATP sulfurylase activity was not found.

Conclusions

The sketches of the S° oxidation pathway and of electron transport in *Ac. ambivalens* presented in Fig. 23.3 were outlined from the presently available data; however, some gaps are still open. Sulfur is oxidized by the SOR; the products thiosulfate and sulfite are oxidized by membrane-bound oxidoreductases. There is also the APS oxidation pathway. Unsolved questions are how the sulfur gets into the cell, how it gets into the SOR, and how products get out. It is also unsolved whether a separate tetrathionate oxidation pathway and additional membrane-bound enzymes exist, catalyzing sulfur compound oxidation at the outer surface of the membrane or the S-layer. Another unsolved question is, how NAD^+ reduction is coupled to sulfur metabolism. It is to be hoped, however, that some of these gaps will be filled in the near future.

Acknowledgments

Special thanks are due to Wolfram Zillig, without whom the whole work would not have been possible. He was very special as a PhD adviser, because he gave me the chance and freedom of choice to do what I wanted with my project. I also wish to thank Felicitas Pfeifer for her support and Tim Urich for critically reading the manuscript. This work has been supported by grants from the Deutsche Forschungsgemeinschaft (Kl885/3-1, 3-2, and 3-3).

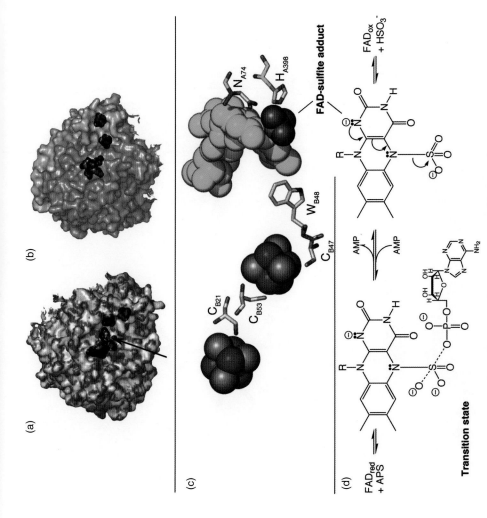

Plate 23.2 Structure and mechanism of *Archaeoglobus fulgidus* adenylylsulfate reductase. (a, b) Molecular surface representations of an αβ heterodimer in two different orientations with FAD and the two [4Fe–4S] clusters highlighted as spheres (Fritz et al., 2002b). The enzyme crystallized in an $\alpha_2\beta_2$ conformation in the asymmetric unit although it is assumed that its biologically active form is the heterodimer. (a) Active site pocket with access site to the FAD (arrow). Color codes: gray, carbon; red, oxygen; blue, nitrogen. (b) Back side view with the α subunit in blue and β in brown. (c) Enlargement of the sulfite-FAD adduct (green) and the FeS clusters (magenta and orange spheres) and amino acid residues (with chain and number), which provide putative electron transfer routes. Two polar amino acids positioning the sulfite are also highlighted. The figures were prepared with Pymol (DeLano, 2002) using the coordinates deposited at the protein database under the accession numbers 1JNR and 1JNZ (Fritz et al., 2002b). (d) Reaction intermediates of the APS reductase; note that all steps are reversible (modified after Fritz et al., 2002a).

Plate 23.1 Fumaroles in the crater of White Island off the cost of New Zealand, showing exhalation of ample amounts of water vapor and elemental sulfur. (Photo: A. Kletzin.)

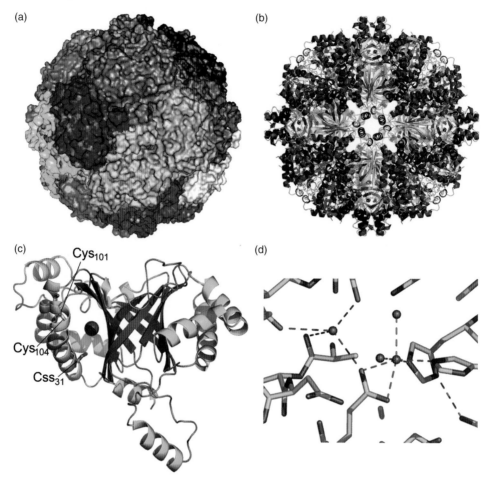

Plate 23.3 Structural model of the sulfur oxygenase reductase. (a) Molecular surface representation of the holoenzyme viewed from the crystallographic fourfold axis highlighting the subunits. (b) Cartoon representation shown in the same orientation. (c) Cartoon representation of a single subunit with cysteines, cysteine persulfide (Css) and iron (sphere). (d) Hydrogen-bonding network around the iron site (magenta), including the coordinating residues Glu_{114}, His_{86}, and His_{90}; ordered water molecules are given in red (Urich et al., 2006). The figures were prepared with Pymol (DeLano, 2002).

24

Methyl-coenzyme M reductase in methanogens and methanotrophs

Rudolf K. Thauer and Seigo Shima

Introduction

Methyl-coenzyme reductase (MCR) catalyzes the reduction of methyl-coenzyme M (CH_3-S-CoM) with coenzyme B (HS-CoB) to methane and CoM-S-S-CoB ($\Delta G^{o\prime}=-30\,kJ/mol$) (Fig. 24.1a). This is the methane forming reaction in all methanogenic Archaea. MCR contains nickel bound within cofactor F_{430} (Fig. 24.1b) (for review see Thauer, 1998). The nickel protein has been thought to be present only in methanogenic Archaea, which all contain this enzyme. Therefore, the *mcr* genes have been used as a taxonomic marker for methanogenic Archaea (Springer et al., 1995; Lueders et al., 2001; Luton et al., 2002; Chin et al., 2004; Banning et al., 2005). However, two years ago MCR, its nickel cofactor, and the *mcr* genes were also found in methanotrophic Archaea of the ANME1 and ANME2 cluster (Hallam et al., 2003, 2004; Krüger et al., 2003). These methanotrophic Archaea are phylogenetically most closely related to methanogenic Archaea of the order of Methanosarcinales and this phylogenetic relationship is reflected in the primary structures of the MCR from these organisms (Hallam et al., 2003; Krüger et al., 2003). Based on these and other findings it has been proposed that MCR is involved not only in methanogenesis but also in anaerobic oxidation of methane (AOM).

The most convincing evidence for an involvement of MCR in AOM comes from the biochemical analysis of microbial mats from the Black Sea, which catalyze AOM and are composed of more than 50% Archaea from the ANME 1 cluster (Pimenov et al., 1997; Thiel et al., 2001; Michaelis et al., 2002; Tourova et al., 2002; Pimenov & Ivanova, 2005). These mats were found to contain high concentrations (7% of the protein extracted from the mats) of an enzyme that, because of its primary structure, subunit composition, and nickel cofactor, is closely related to MCR from methanogenic Archaea (Krüger et al., 2003). These mats catalyzed sulfate-dependent AOM in the laboratory at a specific rate of 0.1 mU/mg protein (1 U = 1 μmol/min). The rate of methane formation from added methanogenic substrates (H_2/CO_2, formate, methanol, methylamines, and/or acetate) was less than 0.01 mU/mg protein, indicating that the mats are dedicated to AOM rather than to methanogenesis (Krüger et al., 2003).

Despite these findings there is a serious question. Methane formation from methyl-coenzyme M and coenzyme B is a strongly exergonic reaction ($\Delta G^{o\prime}=-30\,kJ/mol$). It is catalyzed by MCR from methanogenic Archaea with a maximal specific activity of 100 units (μmol/min) per mg protein (Thauer, 1998). Can this enzyme really catalyze the back reaction at sufficiently high rates?

In this chapter the properties of MCR from methanogenic Archaea are described and analyzed with respect to a possible involvement of this enzyme in AOM. Then the rates of AOM, the inhibition of AOM by bromoethane sulfonate, and the apparent restriction of AOM to sulfate as electron acceptor are discussed with respect to the properties of MCR. Finally, some thoughts are presented as to why AOM and the anaerobic oxidation of alkanes with more than one carbon appear to proceed via two completely different mechanisms.

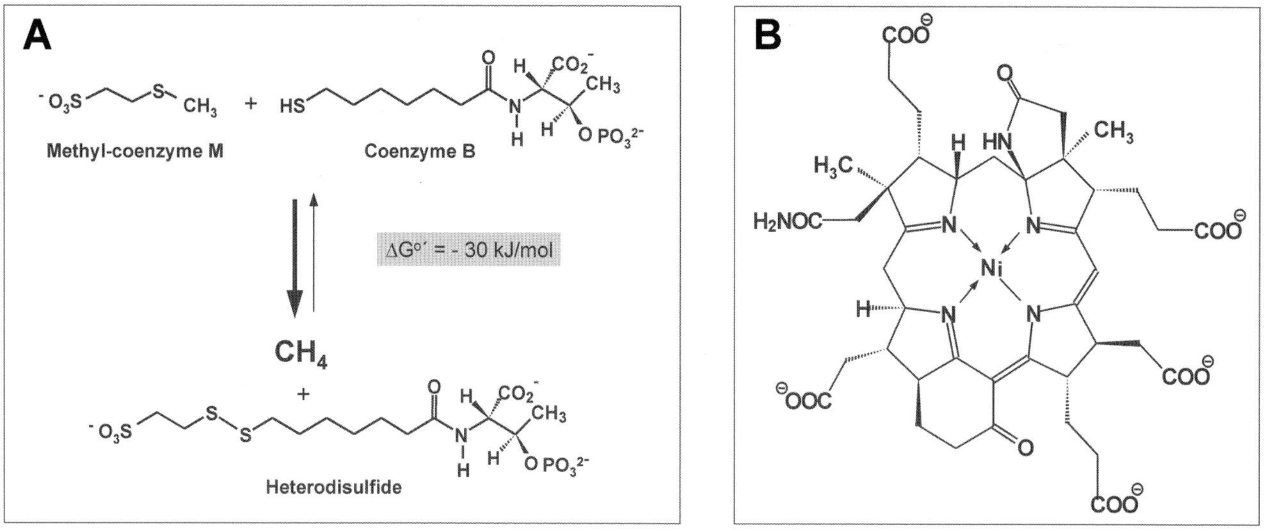

Fig. 24.1 Methyl-coenzyme M reductase from methanogenic archaea. (a) The reaction catalyzed by the enzyme; (b) the prosthetic group (coenzyme F_{430}) of the enzyme.

Reviews of different aspects of the subject can be found in Thauer (1998) (methyl-coenzyme M reductase), DeLong (2000), Valentine and Reeburgh (2000), Valentine (2002), Boetius and Suess (2004), Strous and Jetten (2004), Chistoserdova et al. (2005), Shima and Thauer (2005) (anaerobic oxidation of methane), Buffett and Archer (2004) (global inventory of methane clathrate), Spormann and Widdel (2000), and Widdel and Rabus (2001) (anaerobic oxidation of hydrocarbons). The reader is referred to these reviews for literature published before 2000.

Properties of methyl-coenzyme M reductase from methanogenic Archaea

Structure of MCR

MCR from methanogenic archaea is composed of three different subunits in an $\alpha_2\beta_2\gamma_2$ arrangement and contains 2 mol of a nickel porphinoid, designated coenzyme F_{430}, as prosthetic group, which has to be in the Ni(I) oxidation state for the enzyme to be active (for review see Thauer, 1998). The crystal structure of MCR with F_{430} in the Ni(II) oxidation state was resolved to 1.16 Å. It revealed that the subunits are intertwined such that they form two structurally interlinked active sites made up of the subunits α, $\alpha'\beta$, and γ, and α', α, β', and γ, respectively. Each active site harbors one F_{430} molecule buried deeply within the protein and accessible from the outside only via a 50 Å long channel made up of mainly hydrophobic amino acid residues. Near to the active site are five modified amino acids: a thioglycine, an N-methyl-histidine, an S-methyl cysteine, a 5-(S)-methyl arginine and a 2-(S)-methyl glutamine (Ermler et al., 1997; Grabarse et al., 2000, 2001a, b). Labeling studies have shown that the methyl groups are biosynthetically derived from the methyl group of methionine and not from the methyl group of methyl-coenzyme M (Selmer et al., 2000). The five modified amino acids are highly conserved. However, in MCR from *Methanosarcina* the respective glutamine is not methylated (Grabarse et al., 2000). Generally the genes coding for the three MCR subunits are coded in a transcription unit *mcrBDCGA*. The function of McrC and McrD, which do not copurify with MCR, is not known. In some methanogens the *mcrC* and/or *mcD* genes are not located together with the *mcrAGB* genes. *Methananothermobacter marburgensis*, and *Methanococcus jannaschii* contain two sets of *mcr* genes coding for two MCR isoenzymes, the synthesis of which is regulated (Thauer, 1998; Luo et al., 2002).

The redox state of F_{430} in the active enzyme

Within the past few years there was a major controversy with respect to the number of electrons required to reduce MCR from the inactive EPR silent Ni(II) state to the active Ni(I) state. Evidence was published that three electrons are required, indicating that besides the nickel the porphinoid ligand was also reduced (Tang et al., 2002). It was subsequently unambiguously shown that the stoichiometry is 1 (Piskorski & Jaun, 2003; Craft et al., 2004a). Likewise there was controversy over the oxidation state of nickel in the EPR active but enzymatically inactive MCRox forms (Mahlert et al., 2002a). Evidence was published that the EPR signal exhibited by MCRox could only be derived from Ni(I) (Telser et al., 2000, 2001; Singh et al., 2003). More detailed analyses revealed, however, that the nickel in the MCRox form can best be described as a Ni(III) thiolate in equilibrium with a Ni(II) thiyl radical complex (Duin et al., 2003, 2004; Craft et al., 2004b; Goenrich et al., 2004; Harmer et al., 2005).

The catalytic mechanism

The active site structure indicates that methyl-coenzyme M reduction to methane takes place in a hydrophobic pocket from which water is completely excluded. When entering the active site via the hydrophobic channel methyl-coenzyme M is stripped of water and after the reaction the heterodisulfide is expelled into the water phase. The reaction most probably starts with a conformational change within the active site that is induced upon binding of coenzyme B. This is indicated by the finding that upon addition of coenzyme B to active MCR in the presence of coenzyme M the enzyme is converted from the MCRred1c state exhibiting an axial EPR signal into the MCRred2 state exhibiting a highly rhombic EPR signal (Mahlert et al., 2002b; Finazzo et al., 2003a, b). In single turnover experiments methane formation from methyl-coenzyme M was found to be dependent on coenzyme B (Horng et al., 2001).

There is evidence that the two active sites are structurally and functionally interlinked. The two α subunits in the enzyme are intertwined such that a conformational change in the one active site (made up of the subunits α, α', β, and γ) can be transmitted to the other active site (made up of the subunits α', α, β', and γ') and vice versa. An indication for the coupling of the two active sites is the finding that at most 50% of the enzyme is converted from the MCRred1c state into the MCRred2 state upon addition of coenzyme B (Goenrich et al., 2005). MCR thus shows "half-of-the-sites" reactivity. Based on these findings it has been proposed that the enzyme operates according to a dual stroke engine mechanism. This would allow the coupling of endergonic steps of the catalytic cycle in one active site to the exergonic steps in the other site. The coupling is predicted to lower the activation energy for both the forward and the back reaction (Goenrich et al., 2005) (Fig. 24.2).

Two different mechanisms for methyl-coenzyme M reduction have been proposed. In mechanism 1 the methyl group of methyl-coenzyme M reacts with the Ni(I) in a nucleophilic substitution reaction, yielding methyl-Ni(III) and coenzyme M, which in turn react to methyl-Ni(II) and the thiyl radical of coenzyme M. Methyl-Ni(II) then reacts with a proton in an electrophilic substitution reaction to

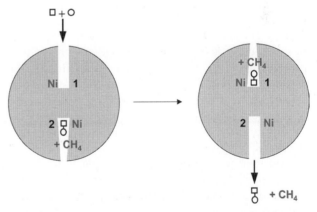

Fig. 24.2 Dual stroke engine mechanism proposed for methyl-coenzyme M reductase. The mechanism allows the coupling of endergonic steps of the catalytic cycle in the one active site to exergonic steps in the other active site. The coupling is predicted to lower the activation energy in both the forward and the back reaction.

methane and Ni(II) and the coenzyme M thiyl radical reacts with coenzyme B, yielding a disulfide anion radical, which is a strong reductant that reduces Ni(II) back to Ni(I), thus closing the catalytic cycle (Ermler et al., 1997; Grabarse et al., 2001a, b). This mechanism is mainly supported by the findings that MCR-catalyzed methyl-coenzyme M reduction proceeds with inversion of stereoconfiguration (Ahn et al., 1991), that enzyme-bound Ni(I)F_{430} reacts with 3-bromopropane sulfonate to alkyl Ni(III) (Goenrich et al., 2004), and that free Ni(I)F_{430} in aprotic solvents reacts with methylbromide to methyl-Ni(II)F_{430}, which subsequently can be protonolyzed to methane and Ni(II)F_{430} (Jaun, 1993). Also, the finding that MCR catalyzes the reduction of ethyl-coenzyme M with less than 1% of the catalytic efficiency of methyl-coenzyme M reduction is consistent with a nucleophilic substitution as the first step in the catalytic cycle (Goenrich et al., 2004).

However, density function calculations have revealed that mechanism 1 is energetically not favorable (Pelmenschikov et al., 2002; Pelmenschikov & Siegbahn, 2003), although the calculations have not taken into account that the two active sites could be energetically coupled. Based on their calculations Ghosh (Ghosh et al., 2001) and Siegbahn (Pelmenschikov et al., 2002; Pelmenschikov & Siegbahn, 2003) have proposed mechanism 2, in which as the first step in the catalytic cycle the methyl thioether bond in methyl-coenzyme M is reductively cleaved, yielding a Ni(II) thiolate and a methyl radical, which in turn reacts with HS-CoB, yielding methane and a CoBS thiyl radical. The latter reacts with coenzyme M thiolate to the disulfide anion radical, which, as in mechanism 1, is used to re-reduce the Ni(II)F_{430} in MCR to the Ni(I) oxidation state. Mechanism 2 is backed up by the experimental finding that in active MCR coenzyme M reversibly coordinates with its thiol sulfur to Ni(I) of F_{430} when coenzyme B is present (Finazzo et al., 2003a, b).

Considering an involvement of MCR in AOM, both mechanisms are not very likely. Assuming mechanism 1, methane oxidation would start by a nucleophilic attack of methane to Ni(II) of F_{430}. This can be excluded since Ni(II) of F_{430} is not electrophilic enough to be able to attack methane with a pKa of above 40. The low electrophilicity of F_{430}(Ni II) is reflected by the low redox potential $E^{o'} < -600$ mV of the Ni(II)F_{430}/Ni(I)F_{430} couple (Jaun, 1993; Piskorski & Jaun, 2003). Mechanism 2 is likewise problematic.

Methane oxidation would start by the reaction of methane with the CoB-$\dot{S}$ thiyl radical. The bond dissociation energy of the C–H bond in methane is 439 kJ/mol as compared to that of the S–H bond of only 365 kJ/mol, which makes a reaction of methane with a thiyl radical yielding a methyl radical and a thiol thermodynamically very unfavorable.

A third mechanism is therefore proposed, which is a modification of mechanism 1. Methyl-Ni(III)F_{430} generated from Ni(I)F_{430} and methyl-coenzyme M reacts with a proton, yielding methane and Ni(III)F_{430}. Free Ni(III)F_{430} is a very strong electrophile ($E^{o'} > +1$ V) (Jaun, 1993) and this is probably also true for enzyme-bound Ni(III)F_{430}, although the EPR spectrum and the UV/visible spectrum of free Ni(III)F_{430} and those of Ni(III)F_{430} in MCRox (see above) are very different, indicating major differences in the coordination sphere (Craft et al., 2004b; Duin et al., 2004). The back reaction, the oxidation of methane, would therefore start with the electrophilic metalation of methane by reaction of methane either end-on or side-on with the high-valent Ni(III) complex in MCR, as described for the activation of C–H bonds by other high-valent metal complexes (Shilov & Shul'pin, 1997).

The reversibility of the MCR catalyzed reaction

The MCR catalyzed reaction is exergonic (Fig. 24.1). Under standard conditions (non-gaseous substrates and products at 1 M concentration and CH_4 at 10^5 Pa pressure) the free energy change (ΔG^o) associated with the reaction is estimated to be -30 kJ/mol methane (see below). The free energy change under physiological conditions (ΔG) is obtained from $\Delta G = \Delta G^o + RT\ln$ [Products]/[Substrates] $= -30 + 5.7 \log$ [Products]/[Substrates]. The equation predicts that the back reaction becomes exergonic when the product to substrate concentration ratio is approximately 10^5. Such a ratio is physiologically not unrealistic. At 10^5 Pa methane the ratio is 10^5 when the intracellular concentration of CoM-S-SCoB is, for example, 1 mM (10^{-3} M) and that of CH_3-S-CoM and HS-CoB each 0.1 mM (10^{-4} M). Consistent with this, methanogenic Archaea have been shown to be capable of very slow methane oxidation (Shilov et al., 1999; Moran et al., 2004). In some of the earlier reports (Zehnder & Brock, 1979, 1980) it has to be considered, however, that the ^{14}C methane used to follow AOM was generated from ^{14}CO$_2$ by methanogens and was therefore most likely contaminated by ^{14}CO (Conrad & Thauer, 1983).

$\Delta G°$ for the MCR-catalyzed reaction is obtained from the difference in the free energy changes associated with several reactions (Keltjens & van der Drift, 1986; Thauer, 1998). One of these reactions is the reduction of CoM-S-S-CoB with H$_2$. $\Delta G°$ for this reaction was revised recently due to the finding that the redox potential of the CoM-SS-CoB/HS-CoB + HS-CoM couple is -143 ± 10 mV rather than -200 mV (Tietze et al., 2003). As a result, $\Delta G°$ for methyl-coenzyme M reduction with coenzyme B decreased from -45 to -33 kJ/mol. This value also has some uncertainty since it is in part based on $\Delta G° = -28$ kJ/mol associated with methyl-coenzyme M formation from methanol and coenzyme M, which was calculated from differences in bond energies and neglects differences in solvation energies (Keltjens & van der Drift, 1986). $\Delta G°$ for methyl-coenzyme M reduction is probably best given as being -30 ± 10 kJ/mol.

Specific activity of MCR

Methyl-coenzyme M reductase (McrI) from *Methanothermobacter marburgensis* catalyzes methane formation from methylcoenzyme M with a maximal specific activity of approximately 100 U/mg protein (Thauer, 1998). Exponentially growing *M. marburgensis* can produce methane at a specific rate of up to 5 U/mg protein. Consistently, such grown methanogenic Archaea contain MCR at concentrations of 5–10% of the soluble cell proteins. For experimental reasons (equilibrium already reached after a few turnovers) the rate of the back reaction catalyzed by the enzyme has not yet been determined experimentally. The specific rate can be estimated, however, using the Haldane equation, which correlates the equilibrium constant (K_{eq}) of a reaction with the catalytic efficiency (k_{cat}/K_M) of an enzyme to catalyze the forward and the back reaction: K_{eq} is catalytic efficiency (forward reaction) divided by catalytic efficiency (backwards reaction). K_{eq} for the MCR catalyzed reaction is calculated from $\Delta G° = -RT\ln K_{eq} = -30$ kJ/mol to be near 10^5. Assuming the K_M values of MCR for its substrates and products to be all e.g. 0.1 mM, the Haldane equation predicts that MCR with a V_{max} for methyl-coenzyme M reduction of 100 U/mg catalyzes methane oxidation at a maximal specific rate of 1 mU/mg ($10^{-5} \times 100$ U/mg). As indicated above $\Delta G° = -30$ kJ/mol of the MCR catalyzed reaction is only known with an uncertainty of ± 10 kJ/mol. Therefore, the maximal specific rate of methane oxidation could be as high as 10 mU/mg and as low as 0.01 mU/mg.

Anaerobic oxidation of methane and methyl-coenzyme M reductase

Anaerobic oxidation of methane by microorganisms has, until now, only been observed with sulfate as electron acceptor.

$$CH_4 + SO_4^{2-} + H^+ = CO_2 + HS^- + 2H_2O$$
$$\Delta G°' = -21 \text{ kJ/mol} \quad \text{(eqn 24.1)}$$

(The free energy change under standard conditions at pH 7.0 ($\Delta G°'$) is given for CH$_4$ and CO$_2$ in the gaseous state at 10^5 Pa pressure and SO$_4^{2-}$ and HS$^-$ at 1 M concentrations). The very slowly growing organisms involved have not yet been recovered in pure culture, which is why our knowledge of their physiology and biochemistry is still very limited. Most of what we know is based on indirect evidence obtained from the analysis of environmental samples. But there is hope, since there are first studies of growth and population dynamics of mixed cultures in continuous-flow reactors (Girguis et al., 2003, 2005).

Archaeal lipid biomarkers and isotopic evidence indicate that AOM is catalyzed by Archaea (Hinrichs et al., 1999; Bian et al., 2001; Pancost et al., 2001; Aloisi et al., 2002; Schouten et al., 2003; Zhang et al., 2003; Blumenberg et al., 2004; Wakeham et al., 2004). These are phylogenetically most closely related to methanogenic Archaea of the order Methanosarcinales as revealed by 16S RNA analyses (Hinrichs et al., 1999; Boetius et al., 2000; Orphan et al., 2001a, b). Three lineages have been identified that are referred to as ANME-1, ANME-2, and ANME-3 (Hinrichs et al., 1999; Boetius et al., 2000; Orphan et al., 2001a, 2002; Blumenberg et al., 2004; Knittel et al., 2005; Nauhaus et al., 2005; Stadnitskaia et al., 2005). Whether these Archaea also catalyze the reduction of sulfate is still a much-debated question. In their natural environment they are always associated with sulfate-reducing Bacteria belonging taxonomically to the delta group of Proteobacteria (Boetius et al., 2000; Orphan et al., 2001b; Elvert et al., 2003; Knittel et al., 2003). Whether the Archaea and Bacteria are syntrophically associated remains to be shown. H_2, formate, and acetate, which could be involved in interspecies electron transfer, have been shown not to inhibit AOM (Nauhaus et al., 2002, 2005), which for thermodynamic reasons they should if they were end products of AOM in the Archaea (Hoehler et al., 1994).

Syntrophic association of methanotrophic Archaea and sulfate-reducing Bacteria could also proceed by extracellular electron transfer involving a "nanowire" (Reguera et al., 2005) or a diffusible electron carrier. The former would require each archaeon to be in physical contact with a sulfate-reducing bacterium. In the microbial mats investigated so far this is not always the case. A present view, therefore, is that at least in some of the Archaea methane oxidation and sulfate reduction occur in the same cells, although this is not backed up by metagenomic data: it has not yet been possible to demonstrate the presence of gene homologs for the enzymes required for dissimilatory sulfate reduction in Archaea of the ANME-1 or -2 cluster.

In this respect it is of interest that methanogenic Archaea have been shown to require a free energy change under physiological conditions (ΔG) of at least $-10\,kJ/mol$ and sulfate-reducing Bacteria one of at least $-19\,kJ/mol$ to support their metabolism *in situ* (Hoehler et al., 2001). The limits of anaerobic metabolism have been discussed by Jackson and McInerney (2002).

Presence of MCR in methanotrophic Archaea

The microorganisms catalyzing AOM have been shown to contain gene homologs of *mcrBGA* and of most of the other genes involved in CO_2 reduction to methane in methanogenic Archaea. Only a gene homologue of *mer* coding for methylenetetrahydromethanopterin reductase was not found (Hallam et al., 2004). Methanotrophic Archaea present in the microbial mats catalyzing AOM in the Black Sea were found to contain at least two different MCRs, designated Ni-protein I and Ni-protein II, which could be separated by anion exchange chromatography (Krüger et al., 2003). Ni-protein II contained normal F_{430} with a molecular mass of 905 Da, whereas Ni-protein I contains a modified F_{430} with a molecular mass of 951 Da. Ni-protein I is present in a concentration of 7% of the extracted soluble proteins and Ni-protein II in a concentration of up to 3%. The N-terminal amino acid sequences of the three subunits of Ni-protein I were determined by Edman degradation and used to identify the encoding genes in a metagenome library of the mats. The codon usage and tetranucleotide signature of the three genes in the cluster *mcrBGA* revealed that the three genes are located on the genome of the dominant ANME-1 archaeon present in the mats. The deduced amino acid sequences show a high degree of sequence similarity to MCR from methanogenic Archaea but with some distinct differences: in the α-subunit the glutamine, which in MCR from methanogenic Archaea is post-translationally methylated at C2, is replaced by a valine (Plate 24.1). Two amino acids downstream of the valine there is a cysteine-rich sequence CCX_4CX_5C not present in MCR from methanogenic Archaea (Plate 24.1). This cysteine rich stretch in the α-subunit of MCR is also found in the DNA sequence of the gene homologs present in the metagenomic library of other microbial consortia catalyzing AOM (Hallam et al., 2003). Due to these differences and the presence of a modified F_{430} the catalytic properties of the enzyme from methanotrophic Archaea could differ significantly from those of MCR from methanogenic Archaea. Thus, the catalytic efficiency could be higher.

As indicated above, the most abundant MCR in microbial mats catalyzing AOM in the Black Sea contains a modified F_{430} with a molecular mass of 951 Da (Krüger et al., 2003). This 951 Da cofactor, which can easily be identified by its characteristic MALDI-TOF mass spectrum, was not found in any of the methanogenic Archaea analyzed in this respect or in microbial cells present in the anaerobic digesters of the waste treatment plant in Marburg. The modified cofactor was found, however, in all habitats with AOM. But besides the modified cofactor the normal F_{430} with a mass of 905 Da is always present. It has already been mentioned that Ni-protein II, isolated from the microbial mats catalyzing AOM in the Black Sea, contained only the 905 Da cofactor (Krüger et al., 2003), indicating that AOM is not restricted to MCR containing the 951 Da cofactor.

MCR and the rate of AOM

The rates reported for AOM are all very low. They are generally between 1 and 100 nmol/cm^3/day (Nauhaus et al., 2002, 2005; Treude et al., 2003, 2005; Wakeham et al., 2003; Joye et al., 2004; Kallmeyer & Boetius, 2004; Orcutt et al., 2004; Werne et al., 2004). Only a few direct estimates of the specific rates have been reported. Microbial mats from the Black Sea were found to catalyze AOM in the laboratory at a specific rate of 0.1 nmol/min and mg protein and (Krüger et al., 2003). The low specific rate of AOM is consistent with an involvement of MCR in this process. As outlined above, MCR from methanogenic Archaea has been estimated from the specific activity of methylcoenzyme M reduction (100 U/mg) and from the free energy change associated with the reaction ($\Delta G° = -30$ kJ/mol) to catalyze the oxidation of methane at a maximal specific activity of approximately 1 mU/mg protein. This is ten times the specific rate of AOM catalyzed by the microbial mats from the Black Sea, which were shown to contain MCR concentrations as high as 10% of the extracted proteins (Ni-protein I and II; see above) (Krüger et al., 2003). The observed specific rates of AOM and the estimated specific activity of MCR thus appear to match quite well. But there is a caveat. Neither the maximal specific rate of AOM nor the specific activity of MCR are at present known sufficiently well to be able to draw decisive conclusions from these results.

MCR and the inhibition of AOM by bromoethane sulfonate

MCR is known to be inhibited by 2-bromoethane sulfonate which is a substrate analog of methylcoenzyme M (Goenrich et al., 2004). Inhibition is competitive with respect to methylcoenzyme M. 2-Bromoethane sulfonate also inactivates MCR by quenching the Ni(I) oxidation state in a reaction yielding Ni(II)F$_{430}$, ethane and a sulfite radical (Goenrich et al., 2004). To inhibit and/or inactivate MCR *in vivo*, 2-bromoethane sulfonate has to be taken up by the cells, which is most probably achieved by a coenzyme M transporter present in many but not all methanogens (Santoro & Konisky, 1987). These findings explain why 2-bromoethane sulfonate is a specific inhibitor of methanogenesis. Inhibition of AOM by 2-bromoethane sulfonate, which has repeatedly been reported, is therefore an indication that MCR is involved in AOM (Nauhaus et al., 2005).

MCR and the restriction of AOM to sulfate as terminal electron acceptor

Until now AOM has only been observed with sulfate as electron acceptor despite the fact that, for example, methane oxidation with Fe(III) ($\Delta G°' = -782$ kJ/mol CH$_4$), nitrate ($\Delta G°' = -522$ kJ/mol CH$_4$), or nitrite ($\Delta G°' = -765$ kJ/mol CH$_4$) is thermodynamically much more favorable. This can possibly be explained on the basis of the properties of MCR.

As indicated above, MCR from methanogenic Archaea is only active when its prosthetic group F$_{430}$ is in the Ni(I) oxidation state. The redox potential (E$°'$) of the Ni(II)F$_{430}$/Ni(I)F$_{430}$ couple has been determined to be below -600 mV (Jaun, 1993; Piskorski & Jaun, 2003). Due to the negative redox potential, which is more than 200 mV below that of the hydrogen electrode at pH 7.0, and due to the fact that in the enzyme F$_{430}$ is not completely electrically insulated from electron acceptors present in the solvent, Ni(I)F$_{430}$ in methyl-coenzyme M reductase slowly oxidizes to Ni(II)F$_{430}$ even under strictly anoxic conditions. The rate of MCR inactivation increases with increasing redox potential in its environment. Considering that re-reduction of MCR in the cells requires ATP (Thauer, 1998), the negative redox potential of F$_{430}$ could preclude the operation of MCR in cells, in which the redox potential of the terminal electron acceptors is more positive than 0 Volt as in the case of the Fe(III)/Fe(II) couple (E$°'$ = +772 mV) or of the nitrate/nitrite couple (E$°'$ = +433 mV). The redox potentials of the adenylylsulfate (APS)/sulfite couple (E$°'$ = -60 mV) and of the sulfite/H$_2$S couple (E$°'$ = -116 mV) involved in dissimilatory sulfate reduction are probably sufficiently negative to allow the function of MCR in the presence of these electron acceptors. With respect to the redox potential, elemental sulfur with an E$°'$ of -270 mV would even be a more suitable electron acceptor, but anaerobic oxidation of methane with sulfur is an endergonic reaction ($\sim G^{o\prime}$ = +19 kJ/mol CH_4). However, the reaction becomes exergonic at the HS^- concentrations (<0.1 mM) prevailing in the natural habitats of the methanotrophic microorganisms ($E'(S°/HS^-)$ = −120 mV).

AOM and the inhibition of MCR by sulfite

All methanogenic Archaea known to date cannot use sulfate as soulfur source or as terminal electron acceptor and none of the known sulfate-reducing Archaea are capable of methanogenesis. Apparently, methanogenesis and dissimilatory sulfate reduction in one organism exclude each other. Is there an explanation for this?

It has been shown that MCR is inactivated by sulfite (Mahlert et al., 2002a). However, in the presence of methyl-coenzyme M, due to competitive binding, the enzyme is largely protected from inactivation. Thus some methanogens can grow with sulfite as a sulfur source. However, in microorganisms with dissimilatory sulfate reduction the intracellular sulfite concentration is expected to be higher than in organisms using sulfate or sulfite only as sulfur source. Therefore, the susceptibility of MCR to inactivation by sulfite could be an argument why methanogenesis and dissimilatory sulfate reduction do not occur in the same microorganism. On the same basis it could be explained why AOM appears to be preferably catalyzed by syntrophic associations of methanotrophic Archaea and sulfate-reducing microorganisms.

Anaerobic oxidation of alkanes with more than one carbon

In recent years sulfate-reducing Bacteria and denitrifying Bacteria have been characterized that can grow on alkanes with more than one carbon (Aeckersberg et al., 1991; Rueter et al., 1994). None of these organisms can metabolize methane. In denitrifying bacteria the first step in the anaerobic oxidation of alkanes has been shown to be the formation of 2-alkylsuccinate from the alkane and fumarate (Rabus et al., 2001; Wilkes et al., 2003). The functionalized alkane is then degraded via β-oxidation regenerating fumarate. Alkylsuccinate formation is predicted to be catalyzed by a glycyl radical enzyme (Spormann & Widdel, 2000; Widdel & Rabus, 2001), which is neither structurally nor mechanistically related to MCR. In the proposed catalytic mechanism the alkane is attacked by a thiyl radical, which is generated by reaction of a cysteine residue of the enzyme with the glycyl radical as deduced from other glycyl radical enzymes (Eklund & Fontecave, 1999; Frey, 2001; Knappe & Wagner, 2001; Krieger et al., 2001; Becker & Kabsch, 2002; Duboc-Toia et al., 2003; Logan et al., 2003; Andrei et al., 2004; Verfurth et al., 2004). The question is: could such a mechanism also work for AOM?

If methane is oxidized via the same mechanism as alkanes then as first intermediate 2-methylsuccinate would have to be formed. The formation of 2-methyl succinate from fumarate and methane is metabolically attractive, since fumarate can easily be regenerated from 2-methylsuccinate via mesaconate, citramalate, pyruvate plus acetate, oxaloacetate, and malate via known biochemical reactions. But such a metabolic pathway for AOM could be problematic for two independent reasons:

1 In the first step, the reaction of methane with fumarate to 2-methylsuccinate, a methyl radical would have to be formed at the expense of a glycyl radical. The dissociation energy of the C–H bond in methane (439 kJ/mol) is much higher than that of the C–H bond in glycine, which in glycyl radical enzymes is predicted to be even lower than the dissociation energy of the S–H bond (365 kJ/mol). The dissociation energy of the C–H bond in methane is also higher than that at C1 (409 kJ/mol) or at C2 (395 kJ/mol) of alkanes. The formation of a methyl radical at the expense of a glycyl radical is thus thermodynamically even more unfavorable than that of an alkyl radical.

2 From bond dissociation energy differences, the reaction of methane with fumarate to 2-methylsuccinate is predicted to proceed irreversibly and from the reaction mechanism not to be coupled with energy conservation. Therefore, most if not all of the free energy generated during methane oxidation with sulfate ($\Delta G^{o\prime}$ = −21 kJ/mol) would be dissipated as heat in the first step, leaving not enough energy to drive the phosphorylation of ADP. However, this argument holds true only for AOM with sulfate as electron acceptor. With Fe(III), nitrate, or nitrite, the free energy change associated with AOM would be more than sufficient (see above) to allow the formation of 2-methylsuccinate in the first step.

Recently evidence has been published that in some sulfate-reducing bacteria the anaerobic oxidation of alkanes could be initiated by the subterminal carboxylation of the alkane at the C3 position (So et al., 2003). Although this mechanism does not apply to AOM, it is a reminder that there could also be more than one way to anaerobically oxidize methane.

Conclusions

The presence of high concentrations of methyl-coenzyme M reductase (MCR) in methanotrophic Archaea strongly suggests that this enzyme catalyzes the first step in the anaerobic oxidation of methane (AOM). The thermodynamics of the MCR-catalyzed reaction and the catalytic properties of MCR from methanogenic Archaea appear to conform to this proposed function. But since the free energy change associated with the MCR-catalyzed reaction and the maximal specific rate of AOM are not yet known with certainty, it is too early for a decisive conclusion.

After completion of the script of this chapter a paper appeared describing the anaerobic oxidation of methane with nitrate or nitrite as terminal electron acceptor (Raghoebarsing et al., 2006). The article was highlighted by Thauer and Shima (2006).

Acknowledgments

This work was supported by the Max Planck Society and by a grant from the Fonds der Chemischen Industrie. We thank Martin Krüger and Fritz Widdel from the Max Planck Institute for Marine Microbiology in Bremen for the fruitful collaboration.

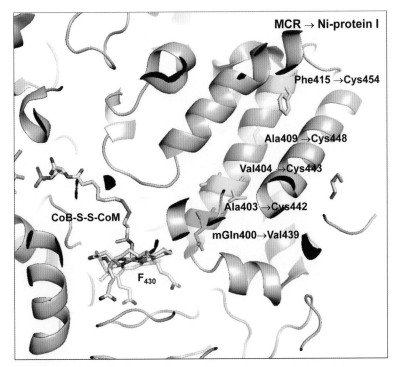

Plate 24.1 Active site structure of methyl-coenzyme M reductase (MCR) from *Methanothermobacter marburgensis*. The amino acids remarkably different in the enzyme (Ni-protein I) from methanotrophic Archaea are highlighted.

25

Methylation of metal(loid)s by Methanoarchaea: production of volatile derivatives with high ecotoxicological impact and health concern

Klaus Michalke, Jörg Meyer and Reinhard Hensel

Introduction

Enzymatically catalyzed methylation reactions in the living cell (so-called biomethylation), transferring methyl groups from a donor to an acceptor molecule, are common reactions featuring a high functional importance for the viability of the cell. They are integrated in various processes, such as signal cascades for controlling gene activities (e.g. by post-translational modifications of histones and/or by modifying bases of polynucleotides), and in the perception of gradients of chemical and physical agents (e.g. chemo-, photo-, and aero-taxis of bacteria), thus fulfilling regulation functions; they are involved in the central carbon and energy metabolism (e.g. reductive acetyl-CoA pathway, methanogenesis); and they are also essential for the synthesis of a number of essential cofactors and prosthetic groups of enzymes. The common methylated atoms of the acceptor molecules are H, C, N, S, and O, but they also include metal atoms such as Co and Ni.

More recently, interest in the methylation capacity of the cell was evoked by the observation that a plenitude of various biologically produced methylated compounds – mainly methyl derivatives of metals and metalloids – can be detected in the environment. This is of particular concern because, unlike with organometalloids of anthropogenic origin, whose entry into the environment can be controlled and regulated, the parameters affecting biomethylation are not as yet clearly understood.

Methylmetal(loid)s are generally volatile and, with only a few exceptions, they are more toxic than their inorganic educts owing to their greater hydrophobicity. This results in a higher solubility in, and permeability through, cell membranes, leading ultimately to a higher accumulation and greater potential for damage. One example is trimethylbismuth, for which considerable damaging effects on kidney and central nervous system have been described ($LD50_{Rabbit}$ 11 mg/kg (i.v.); Sollmann & Seifter, 1939). Others are monomethylmercury and dimethylmercury, which cause damage to the brain and induce chromosomal aberrations. Methylmercury is one of the strongest toxins known (methylmercury: $LD50_{Guinea\ Pig}$ 7 mg/kg (i.p.); http://physchem.ox.ac.uk/msds/).

Mercury and virtually all group IV, V, and VI elements, with the possible exception of lead, can be biomethylated (Gadd, 1993; Bentley & Chasteen, 2002; Thayer, 2002; Craig, 2003).

The first example of biomethylation of a metalloid was given by Gosio (1897), who described the production of a toxic, arsenical gas with a garlic-like odor from various "arsenic" fungi (*Scopulariopsis bervicaulis* (formerly *Penicillium brevicaule*) and various species of *Aspergillus, Cephalothecium, Fusarium, Mucor, Sterigmatocystis, Paecilomyces*) when these were grown on organic matter or soil samples containing arsenic compounds. This compound was later identified as trimethylarsine (Challenger et al., 1933).

It is generally accepted that microorganisms play a predominant role in the biotic synthesis of these compounds. Although biomethylation can also take place under aerobic conditions by the action of various fungi and Bacteria (*S. brevicaulis*, various species of *Aspergillus, Penicillium*; *Escherichia coli*, various species of *Bacillus*,

Flavobacterium, Aeromonas; Thayer, 2002), anaerobic habitats such as sediments, wetlands, waste deposits, and sludge treatment plants seem to be preferred locations for the biogenesis of the volatile metal(loid) derivatives (Feldmann et al., 1994; Feldmann & Hirner, 1995; Wickenheiser et al., 1998b; Michalke et al., 2000).

Several anaerobic Bacteria such as species of *Clostridium* and various sulfate reducers, as well as facultatively anaerobic bacteria such as species of *Pseudomonas* or *Alcaligenes* and Methanoarchaea, could be identified as versatile producers of methylated metal(loid) derivatives (Cheng & Focht, 1979; Michalke et al., 2000). First reports of methylation of metal(loid)s by Methanoarchaea go back some way: in 1968 Wood et al. showed that mercury was methylated by crude extracts of *Methanobacterium bryantii*, and a little later McBride and Wolfe (1971) reported an equivalent methylation reaction with arsenate, by the same organism, yielding dimethylarsine.

In the meantime, much progress in analytical techniques has been made, allowing the analysis of metal(loid) organic derivatives in environmental samples and laboratory assays to be performed more easily and with higher precision. Thus, especially in recent years, our knowledge about the formation and conversion of these compounds and about the identity of the producing organisms has increased significantly. Nevertheless, knowledge regarding the mechanism of organometalloid biosynthesis and the real hazards of these processes is still scarce.

In this chapter we focus on Methanoarchaea as one of the major groups in anaerobic microbial consortia, and we discuss their role in the biogenesis of volatile methylated metal(loid) derivatives under anaerobic conditions. We present data documenting the versatility of various species of Methanoarchaea with regard to transforming metal(loid)s into volatile derivatives and we give some details about the biomethylation of bismuth by these organisms, a metal with increasing importance in technological applications and consumer products. Finally, the potential of Methanoarchaea as producers of toxic organometal(loid) compounds in the human intestine – and thus their role as potential pathogens – is discussed.

Analytical techniques for identification and quantification of organometal(loid) compounds

Several analytical techniques have been developed to analyze and quantify metal(loid)-organic derivatives in environmental samples such as wastewater, soils, and sediments, as well as in laboratory assays. This allows one to study the formation, conversion, and degradation of volatile organometal(loid) compounds. Because of their excellent selectivity and sensitivity, combined ("hyphenated") techniques have been widely used to separate volatile metal(loid) species by coupling a gas chromatographic system to various detectors, such as atomic-absorption spectrometers (Dirkx et al., 1994), microwave-induced plasma atomic emission spectrometers (Pereiro et al., 1999), and mass spectrometers (Dunemann et al., 1999). These methods include inductively coupled plasma mass spectrometry (ICP-MS; Wickenheiser et al., 1998a). Non-volatile and ionic species of organometal(loid) compounds can be extracted from the sample matrix by solvent extraction or by derivatization by *in situ* hydride generation or alkylation before analysis (Bouyssiere et al., 2002).

We developed a purge-and-trap gas chromatographic system coupled to an inductively coupled plasma mass spectrometer (PT-GC/ICP-MS). With this method, the volatile metal(loid) derivatives in the headspace of the sample are purged with a helium flow and are trapped on a column (10% SP-2100, 80/100 mesh, Supelco) cooled to −196 °C with liquid nitrogen. To release matrix gases (such as CO_2 and CH_4), which would otherwise interfere with the subsequent analysis, the cryo-focusing trap is heated to −70 °C with ethanol that is cooled with liquid nitrogen. After refocusing of the analytes on the chromatographic column (10% SP-2100, 80/100 mesh, Supelco) cooled with liquid nitrogen to −196 °C, they are separated according to boiling point by heating the column to 250 °C and are then transferred via a heated transfer line to an ICP-MS (VG Elemental PQ2, Fisons) as an element-specific detector. Because of the complete destruction and ionization of the analytes in the argon plasma of the ICP-MS, they are detected as the corresponding metal(loid) and identified by boiling-point retention-time correlation. The identities of the species were confirmed by mass spectrometry of fragments, as described by Wickenheiser et al. (1998a).

Environmental samples – i.e. soil samples, sewage sludge samples, or human faecal samples – were incubated under strictly anaerobic conditions (N_2 or He atmosphere) at appropriate temperatures and by moderately shaking in the dark before ICP-MS analysis. Pure cultures of microorganisms were spiked with inorganic metal(loid) salts in the early exponential growth phase (1 μM to 1 mM of KH_2AsO_4, $SbCl_3$, $HgCl_2$, $Bi(NO_3)_3$, TeO_2, or SeO_2, respectively). The products were analyzed by ICP-MS from the headspace. As a control, samples of sterilized media were incubated under the same conditions.

Versatility of Methanoarchaea in methylating metal(loid)s

Up to now, the production of methylated metal(loid)s among the Archaea has only been reported for several species of Methanoarchaea. The general capacity of Methanoarchaea to methylate metals and metalloids was initially documented by Wood et al. (1968) and McBride and Wolfe (1971), who demonstrated the methylation of mercury and the transformation of arsenate to dimethylarsine, respectively, by cell extracts of the methanogen strain M.o.H. (now named *Methanobacterium bryantii*).

Screening experiments with growing cultures of various species of Methanoarchaea spiked with a mixture of various metal(loid) salts (KH_2AsO_4, $SbCl_3$, $HgCl_2$, $Bi(NO_3)_3$, TeO_2, SeO_2) suggested that antimony is generally susceptible to methylation by methanogens. As shown in Table 25.1, volatile species of antimony could be identified in all cultures of methanogens tested and confirm corresponding results already obtained for Methanoarchaea in earlier studies (Michalke et al., 2000, 2002).

Surprisingly, the methylation of mercury to volatile derivatives does not seem to be common in Methanoarchaea. None of the strains investigated by Michalke et al., nor the strains in the present study, produced any volatile methylated mercury derivatives. Thus, the *in vitro* production of methylmercury by cell extracts of *M. bryantii* (Wood et al., 1968) seems to be the only hint that Methanoarchaea may possibly be able to biomethylate mercury.

In conclusion, methanogens of various genera are very versatile in methylation – with the sole exception of the thermophilic *Methanothermobacter thermautotrophicus*, which is seemingly able to methylate only antimony under the conditions used (Table 25.1). However, we could not exclude the possibility that some volatile metal(loid) derivatives remained undetected because of their lability at the elevated growth temperature of this organism (65 °C). Since the production of certain derivatives may depend on the specific conditions of the assay and does not allow an absolute assessment of intrinsic biomethylation, Table 25.1 should be regarded as summarizing the minimum methylation capacities of the organisms tested.

An instructive example, which demonstrates the influence of extrinsic factors on biomethylation, is given by *Ms. barkeri*. As shown in Table 25.1, and explained in more detail below, the methylation of bismuth by this methanoarchaeon depends strongly on the presence of specific chemical compounds in its environment.

Present knowledge about the biochemistry of biomethylation in the three domains of life: Bacteria and Eukarya versus Methanoarchaea

Up to now, only a limited number of mechanistic investigations of the biomethylation of metal(loid)s have been reported in the literature, despite the importance of these reactions. Only for As, Se, Hg, Bi, and – with some restrictions – for Sb and Te are detailed data available. Whereas biochemical methylation mechanisms for As, (Sb), Se, (Te), and Hg were mainly investigated in Bacteria and Eukarya, the methylation of bismuth has only been analyzed in Methanoarchaea as yet. The most extensive dataset exists for arsenic. The arsenic methylation mechanism was deduced about 60 years ago from the investigation of *S. brevicaulis* by Challenger (1945). The so-called "Challenger mechanism" is still accepted today. It proposes a reaction sequence of alternating reduction and oxidative methylation steps. Starting from arsenate, four reduction and three methylation steps are necessary for the formation of trimethylarsine (Fig. 25.1). As reductants, thiols such as glutathione and lipoic acid are

Table 25.1 Versatility of Methanoarchaea in methylation of metals/metalloids: volatile metal/metalloid species produced by growing cultures of various Methanoarchaea.

	Metal/metalloid					Reference
	As	Bi	Se	Te	Sb	
Methanobacterium formicicum	AsH_3, CH_3AsH_2, $(CH_3)_2AsH$, $(CH_3)_3As$, X	BiH_3, CH_3BiH_2, $(CH_3)_2BiH$, $(CH_3)_3Bi$	$(CH_3)_2Se$, $(CH_3)_2Se_2$	$(CH_3)_2Te$	SbH_3, CH_3SbH_2, $(CH_3)_2SbH$, $(CH_3)_3Sb$	Michalke et al., 2000
Methanobrevibacter smithii	CH_3AsH_2, $(CH_3)_2AsH$, $(CH_3)_3As$	CH_3BiH_2, $(CH_3)_2BiH$, $(CH_3)_3Bi$	$(CH_3)_2Se$, $(CH_3)_2SeS$, $(CH_3)_2Se_2$, X	$(CH_3)_2Te$	$(CH_3)_3Sb$	This study
Methanococcus vanielli	CH_3AsH_2, $(CH_3)_2AsH$, $(CH_3)_3As$, X	CH_3BiH_2, $(CH_3)_3Bi$	$(CH_3)_2Se$, $(CH_3)_2SeS$	$(CH_3)_2Te$	$(CH_3)_3Sb$	This study
Methanolacinia paynteri	n.d.	$(CH_3)_2BiH$, $(CH_3)_3Bi$	$(CH_3)_2Se$, $(CH_3)_2SeS$, $(CH_3)_2Se_2$, X	$(CH_3)_2Te$	$(CH_3)_3Sb$	This study
Methanolobus tindarius	n.d.	$(CH_3)_3Bi$	$(CH_3)_2Se$, X	$(CH_3)_2Te$	CH_3SbH_2, $(CH_3)_3Sb$	This study
Methanoplanus limicola	$(CH_3)_3As$	$(CH_3)_3Bi$	$(CH_3)_2Se$, $(CH_3)_2SeS$, $(CH_3)_2Se_2$, X	$(CH_3)_2Te$, X	$(CH_3)_3Sb$	This study
Methanosarcina barkeri	AsH_3, X	$(CH_3)_3Bi^*$	$(CH_3)_2Se$, $(CH_3)_2Se_2$	n.d.	$(CH_3)_3Sb$	Michalke et al., 2000
Methanosarcina mazei	$(CH_3)_3As$	$(CH_3)_3Bi$	$(CH_3)_2Se$, $(CH_3)_2Se_2$	$(CH_3)_2Te$	$(CH_3)_3Sb$	This study
Methanosphaera stadtmanae	$(CH_3)_2AsH$, $(CH_3)_3As$	CH_3BiH_2, $(CH_3)_2BiH$, $(CH_3)_3Bi$	$(CH_3)_2Se$, $(CH_3)_2SeS$, $(CH_3)_2Se_2$, X	$(CH_3)_2Te$	$(CH_3)_3Sb$	This study
Methanothermobacter thermautotrophicus	AsH_3	n.d.	n.d.	n.d.	$(CH_3)_3Sb$	Michalke et al., 2000

X, unidentified volatile metal(loids); n.d., not detected; *mediated by addition of octamethylcyclotetrasiloxane and monensin (see below). Culture volume 50ml. Medium composition as proposed by DSMZ. Metal(loid) spiked at approx. 10^6 cells/ml. Head space of cultures was sampled after 24–72 hours and analyzed by PT-Gc/ICP-MS.

discussed. Labeling experiments using [$^{14}C_3$]methionine have demonstrated that SAM is the relevant methyl donor in *S. brevicaulis* and other fungi (Cullen & Reimer, 1989; Andrewes et al., 2000).

Most experimental evidence for the enzymatic catalysis of arsenic methylation comes from mammalian systems, where two different reaction pathways have been described for transforming arsenite to dimethylarsinic acid:

1 Methylation and reduction by two different enzymes identified in rabbit and human liver (Aposhian et al., 2004; Zakharyan et al., 2005) with a methyltransferase responsible for the methylation of arsenite to monomethylarsonate and monomethylarsonous acid to dimethylarsinic acid acting in cooperation with a reductase (glutathione S-transferase omega), which reduces not only monomethylarsonate to monomethylarsonous acid but also dimethylarsinic acid to dimethylarsinous acid, using reduced glutathione as substrate.

2 Methylation and reduction by a bifunctional enzyme, as found in mouse and rat but also in the human system (Thomas et al., 2004). This enzyme (an obvious Cyt19 homologue) not only converts arsenite to monomethylarsonate and monomethylarsonous acid to dimethylarsinic acid using SAM as methyl donor, but also catalyzes the reduction of monomethylarsonate to monomethylarsonous acid, probably using thioredoxin as reductant.

In contrast to eukaryotic arsenic methylation, our knowledge of the biochemistry of arsenic methylation by methanogens is only sparse. Despite this, the observation by McBride and Wolfe (1971) that methylcobalamin represents the most suitable methyl donor for the formation of dimethylarsine in *in vitro* assays with cell-free extracts of *M. bryantii* led to vigorous discussions as to whether Bacteria and Archaea in general use methylcobalamin instead of SAM for methylation (Bentley & Chasteen, 2002). Apparently, there is no evidence that methyl CoM could function as methyl donor in Methanoarchaea.

The mechanism of antimony biomethylation has been little studied up to now. Labeling experiments indicate that methylation in *S. brevicaulis* occurs via SAM, with a dimethylated antimony species as a possible intermediate product (Andrewes et al., 2000).

In the case of selenium and tellurium, methylation has been investigated only in Eukarya and Bacteria. There appear to be many similarities between transformations of these elements and sulfur in both domains, and the same enzymes seem to be involved in the reactions methylating sulfur and selenium/tellurium (Terry et al., 2000). The formation of both dimethylselenide and dimethyltellurium has been proposed to start from methylselenomethionine (methylmethionine), which is synthesized by an S-adenosylmethionine-dependent methionine methyltransferase. This yields dimethylselenide/dimethyltellurium (dimethylsulphide), either directly by the action of a methylmethionine hydrolase or indirectly by the combined action of a decarboxylase, transaminase, and aldehyde dehydrogenase through the formation of the intermediate dimethylselenonium propionate (dimethylsulfonium propionate). In mouse and rat tissues, a further methyl transfer by a S-adenosylmethionine:thioether S-methyltransferase has been reported to generate the trimethylselenonium ion, a

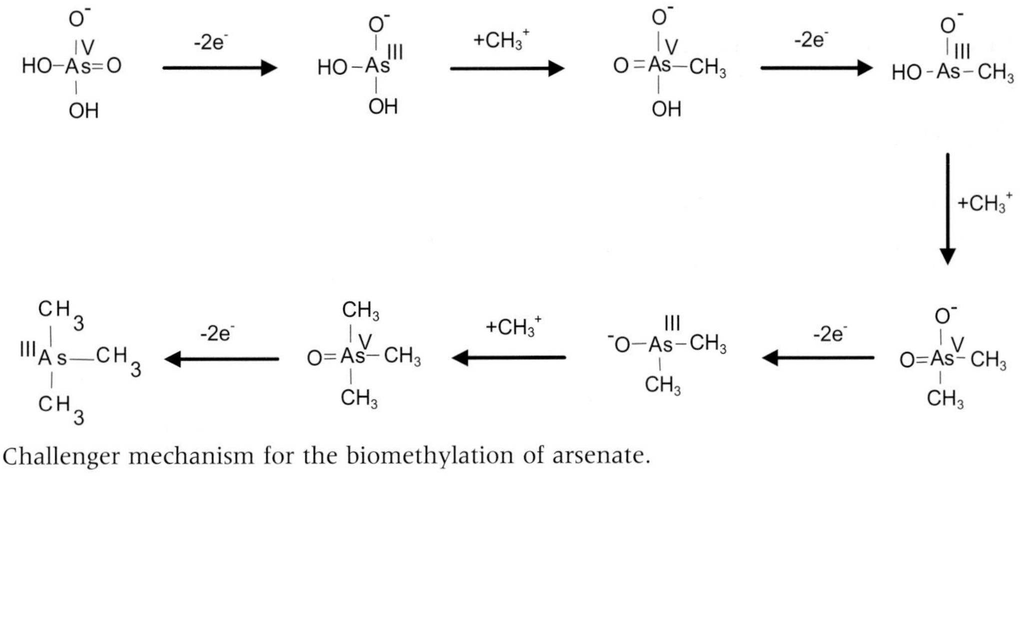

Fig. 25.1 Challenger mechanism for the biomethylation of arsenate.

transformation that seems to take place for excretion purposes (Ganther, 1966; Mozier et al., 1988).

Substantial effort has been invested in the study of mercury biomethylation, most probably motivated by the high toxicity of the methylated products, monomethylmercury and dimethylmercury, and by some fatal incidents with methylmercury, e.g. in Japan (Minamata bay; Westoo, 1966; Luke & Tedeschi, 1982), in Sweden (Borg et al., 1986), and in Iraq (Bakir et al., 1973).

Sulfate-reducing Bacteria have been the subject of particular study in this connection, since this physiological group of prokaryotes has been assumed to be mainly responsible for the formation of monomethyl- and dimethylmercury in the environment. Although previous studies suggest that the methylation of mercury is closely interlinked with the acetyl coenzyme A pathway, involving methyl transfer from methyltetrahydrofolate through methylcobalamin to Hg^{2+} (Choi et al., 1994), more recent investigations (Ekstrom et al., 2003) indicate that mercury methylation in sulfate-reducing Bacteria also takes place independently of the acetyl coenzyme A pathway and does not necessarily require methylcobalamin.

As mentioned above, the first report of the methylation of mercury by Methanoarchaea was made by Wood et al. (1968). These authors observed a methylated mercury derivative (presumably $(CH_3)_2Hg$) upon incubation of crude extracts of *M. bryantii* in the presence of methylcobalamin and concluded that the methyl transfer is mainly enzyme-catalyzed. However, no mechanistic details were given, and neither was the *in vivo* methyl donor definitively identified.

Mechanistic studies of the biomethylation of bismuth by Methanoarchaea

Interest in the biomethylation of bismuth has been evoked only recently, on account of its increasing importance in technological application as an additive in glass and ceramics, as well as in the cosmetic and pharmaceutical industries, resulting consequently in an increasing load of bismuth in waste water and in sewage sludge at concentrations of 1–5 mg/kg dry weight (Michalke et al., 2000).

As a further consequence of the frequent application of bismuth and its well known susceptibility to biomethylation, a methylation product, trimethylbismuth ($(CH_3)_3Bi$) occurs in rather high concentrations in sewage and landfill gases (up to $25 \mu g/m^3$; Feldmann et al., 1994). Of special ecotoxicological relevance is the high susceptibility of bismuth for methylation, as demonstrated by its conversion in sewage sludge. The turnover of bismuth in this rather complex habitat of sewage sludge has been reported to be some 4000 times faster than that of tin (Michalke et al., 2003). Since $(CH_3)_3Bi$ shows toxic effects in animal experiments (Sollmann & Seifter, 1939), the reaction of bismuth to its permethylated product is of high interest with regard to public health.

In vivo studies of the biochemistry of bismuth methylation have been performed using *M. formicicum* (Michalke et al., 2002), a known representative of the anaerobic sewage sludge microflora in municipal waste-water treatment plants. In *in vivo* experiments with batch cultures, the formation of the volatile and non-volatile reaction products was followed after addition of $0.1 \mu M$ $Bi(NO_3)_3$ in the early exponential growth phase, and the methane production was monitored as a measure of the physiological activity of the culture. As shown in Fig. 25.2, $(CH_3)_3Bi$ was identified as the dominant volatile compound – with a production of approximately 150 ng in a 50 ml culture over a 7-week incubation. The partially methylated volatile derivatives monomethyl- and dimethylbismuthine, together with bismuthine itself, evolved at significantly lower rates. Since these products occurred mainly in the stationary phase, when the $(CH_3)_3Bi$ production has already ceased, they probably do not represent intermediates of the reaction, but are instead the result of competing reactions occurring predominantly under the changed conditions of the stationary phase. As a hypothesis, we suppose that the formation of the partially methylated products, including bismuthine, may be favored by a higher supply of hydride ions during the stationary phase.

In addition to the volatile product $(CH_3)_3Bi$, the non-volatile ionic derivatives $[(CH_3)_2Bi^+]$ and $[CH_3Bi^{2+}]$ in addition to $[Bi^{3+}]$ were also observed (Fig. 25.3). We suppose that these ionic species represent intermediates that react further, either to bismuthine, to the partially methylated bismuthines, or to $(CH_3)_3Bi$, depending on the ratio between hydride ions and methyl groups available.

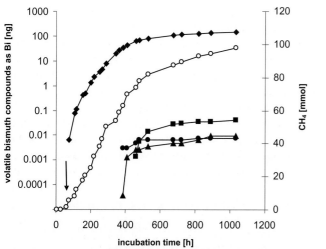

Fig. 25.2 *In vivo* assay for the biovolatilization of bismuth by *Methanobacterium formicicum*. The culture (50 ml culture medium, 70 ml head space, growth temperature 37 °C), was spiked with 0.1 μM Bi(NO₃)₃ at the beginning of the exponential growth phase (arrow), and the volatile bismuth derivatives (● BiH₃, ■ CH₃BiH₂, ▲ (CH₃)₂BiH, ◆ (CH₃)₃Bi) were analyzed with PT-GC/ICP-MS by purging the head space; methane production (○) was analyzed by GC-FID and used as a measure of the physiological activity of the culture.

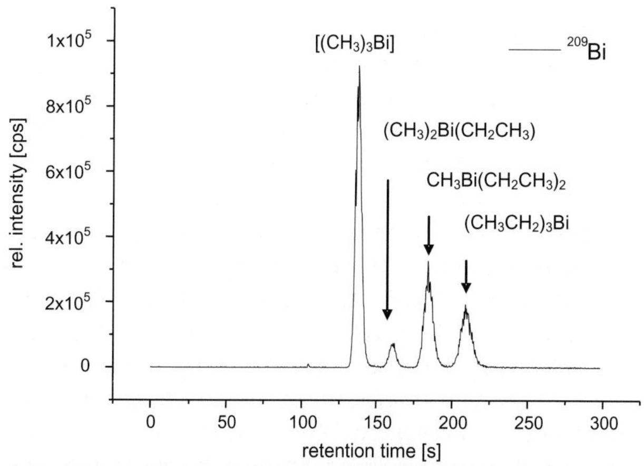

Fig. 25.3 Separation and identification of ionic bismuth species found in *M. formicicum* cultures spiked with Bi(NO₃)₃. The growth conditions of the cultures were the same as described in Fig. 25.2. Samples of the liquid phase (5 ml) were withdrawn at the end of the exponential growth phase and treated with 50 μl sodium tetraethylborate(III) (0.05% w/v in doubly distilled water) at pH 5. The ethylated derivatives were separated by gas chromatography and identified by fragment MS.

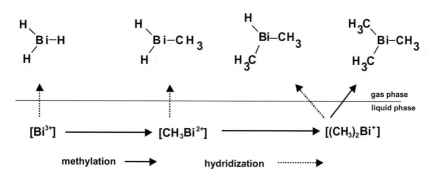

Fig. 25.4 Proposed mechanism of bismuth biovolatilization by *M. formicicum*.

Thus, the methylation pathway for bismuth can be described in the simplest case as a consecutive methylation of [Bi³⁺] up to volatile (CH₃)₃Bi or – including competing reactions with hydride ions as proposed – as a branched pathway resulting in the by-products bismuthine or mono- or dimethylbismuthine (Fig. 25.4). The proposed pathway would be an alternative to the Challenger mechanism characterized by alternating reduction and methylation steps (see above).

Influence of environmental conditions on biomethylation: the effect of cyclic polydimethylsiloxanes (cyclic PDMS) on the methylation of bismuth by *Ms. barkeri*

Motivated by the observation that the emission of $(CH_3)_3Bi$ in sludge stabilization plants is increased by the presence of certain chemicals such as cyclic polydimethylsiloxanes (cyclic PDMSs; data not shown), we investigated the influence of cyclic PDMSs on the biomethylation in more detail. These crown-ether-like molecules are high-volume chemicals and are heavily used in medicinal products, food, and personal care products. As a consequence of their frequent application, concentrations up to 500 mg/kg of these compounds can be found in sewage sludges (Fendinger et al., 1997).

As shown in pure-culture experiments, cyclic PDMSs – like octomethylcyclotetrasiloxane (OMCTS) – are indeed strong stimulators of biomethylation of bismuth. Especially striking effects could be found for *Ms. barkeri*, which was seemingly unable to produce $(CH_3)_3Bi$ without OMCTS (Table 25.1) but showed, in the presence of that compound, production rates similar to those of *M. formicicum*, which synthesizes $(CH_3)_3Bi$ without OMCTS. Thus, in a methylation assay with 5 µM $Bi(NO_3)_3$, a 50 ml culture of *Ms. barkeri* in the presence of 0.07 nM OMCTS produced 4.4 nmol $(CH_3)_3Bi$ in 35 days; this is comparable to the production of 3.6 nmol by *M. formicicum*.

We have shown previously that OMCTS does not serve as a methyl donor in bismuth biomethylation (Wickenheiser et al., 2000), indicating that the observed methylation stimulation by OMCTS is instead due to indirect effects.

As yet, the mechanism by which OMCTS stimulates bismuth methylation is not clear, but the structures of these cyclic compounds suggest that they can chelate metal ions, which could support the penetration of bismuth through the cell membrane. This assumption is supported by the observation that known ionophores, such as monensin or lasalocide, caused similar effects on *Ms. barkeri*. These ionophores increased the biomethylation to a comparable extent (Michalke et al., in preparation). Consequently, we assume that the methylation of bismuth is a direct response to the intracellular concentration of bismuth and thus represents some kind of self-defense by volatilizing and removing the toxic metal from the cytoplasm. Possibly, the differences between organisms methylating with or without adjuvants are caused by different membrane permeabilities for the respective metal(loid).

On the other hand, these observations also emphasize the general influence of environmental factors on biomethylation processes and the resultant ecotoxicological and medical hazards.

Health concerns regarding biomethylation: Methanoarchaea as pathogens?

As shown above, the transformation of metal(loid)s to volatile toxic methylated derivatives seems to be a common property of Methanoarchaea, and therefore this metabolic property should be considered in any evaluation of their impact on health and ecology. However, a reliable assessment of the hazards caused by these organisms remains difficult. Consequently, we here confine ourselves to some suggestions regarding the risks of the biomethylating methanogens in various scenarios. The main criteria for a risk assessment are: (i) the toxicity of the emitted volatile compounds themselves; (ii) the concentration of the compounds reached in the gas phase, which is dependent upon the production rate and affected by dilution (which occurs through convection) and the decay rate; (iii) accumulation of the compound in or by organisms; and (iv) biotic and abiotic conversion processes leading to a modification of their toxicity. We examine the special case of $(CH_3)_3Bi$, the dominant product of bismuth methylation by the majority of the Methanoarchaea investigated here. This compound is known to be toxic, causing encephalopathic symptoms in animal experiments. Fatal effects have been observed with rats and rabbits after exposure to a concentration of approximately 200–300 mg/m^3 of $(CH_3)_3Bi$ (Sollmann & Seifter, 1939). Considering, however, that (i) the concentration of $(CH_3)_3Bi$ found in gases evolved in sludge stabilization plants is rather low (around 25 µg/m^3; Feldmann et al., 1994), (ii) the compound, once formed, would immediately be diluted by convection flows, and (iii) the compound exhibits only moderate stability in air (half-life < 17 hours) (Michalke et al., 2002), one would expect that the biomethylation of bismuth under these conditions should present little hazard. Nevertheless, it should not be forgotten that we do not yet know whether $(CH_3)_3Bi$ can be bioaccumulated, or whether there are conversion processes that notably increase its toxicity.

Certainly, a more dangerous situation arises in closed reaction systems or in compartments or tissues of organisms, where significant accumulation of volatile compounds could take place and where absorption effects further increase their concentration. Such a promotion of biomethylation could take place in the (commonly) strictly anaerobic intestine of various animals, a well known habitat for Methanoarchaea in invertebrates and vertebrates, including humans (Wolin, 1981; Hackstein & Stumm, 1994). It has already been suggested that intestine microflora could be involved in the transformation reactions of bismuth to derivatives that undergo faster absorption by the human body; this suggestion was prompted by bismuth poisoning during medical therapy, causing mainly renal failures or mental disorders (Slikkerveer & Wolff, 1989) – symptoms similar to those observed following $(CH_3)_3Bi$ inhalation by cats and dogs (Sollmann & Seifter, 1939).

In *ex situ* studies of faecal samples of human volunteers we were able to show that bismuth from ingestion of the drug Telen® (containing bismuth subcitrate; bismuth content 215 mg/tablet) is in fact converted to $(CH_3)_3Bi$ under anaerobic conditions. Under the rather non-physiological conditions of these *ex situ* assays, a $(CH_3)_3Bi$ production of approximately 0.5 μg Bi per kg of faeces (wet weight) was observed; this corresponded to some 1–2% of the specific production in the pure-culture assays with *M. formicicum* (incubation in the presence of 0.1 μM $Bi(NO_3)_3$ over 35 days). Presumably, the biomethylation in the intestine, which is probably more extensive *in vivo*, is also the cause for the occurrence of $(CH_3)_3Bi$ in the exhaled breath of the volunteers. Although the observed concentration (up to 0.5 μg/m^3) seems fairly harmless, we cannot exclude the possibility that the measured content in breath signals an already critical accumulation in the body, which could result in more or less severe damage to health.

The organisms responsible for the $(CH_3)_3Bi$ production in the faeces samples have not been as yet identified. However, since *Mb. smithii* and *Msp. stadtmanae*, both typical inhabitants of the human gut (Miller & Wolin, 1982, 1985), are able to synthesize $(CH_3)_3Bi$ in pure-culture assays (Table 25.1), we suggest that Methanoarchaea are in general involved in the intestinal production of $(CH_3)_3Bi$ – and thus deserve the attribute "pathogenic."

Acknowledgments

The work was supported by grants from the Deutsche Forschungsgemeinschaft. Special thanks are due to Jan Koesters for analytical support.

26
Biotechnology

Ksenia Egorova and Garabed Antranikian

Introduction

The application of enzymes and microorganisms for the sustainable production of chemicals, biopolymers, and fuels from renewable resources, also defined as industrial (white) biotechnology, offers great opportunities for the chemical and pharmaceutical industries. White biotechnology aims at the reduction of waste, energy input, and raw materials to improve the environmental friendliness of processes. The majority of the industrial enzymes used to date have been derived from Bacteria and fungi. The global annual enzyme market has been estimated to be around 5 billion euros. In the case of Archaea, which represent the third domain of life, only a few enzymes have found their way to the market. Since many representatives of this group are able to grow optimally under extreme conditions, they are an interesting source of stable enzymes (extremozymes). Based on the unique stability of their enzymes at high temperature and pressure, at extremes of pH and salinity, and in the presence of organic solvents, detergents, and heavy metals, they are expected to be a powerful tool in industrial biotransformation processes that run in harsh conditions. Tailor-made enzymes are needed for various industrial applications, such as food, feed, textile, paper, pharmaceuticals, and fine chemistry. The recent development in the energy sector underlines the relevance of Archaea in industrial biotechnology, especially by using their enzymes for the utilization of renewable resources such as cellulose, hemicellulose, starch, and lignin. In order to meet future challenges, innovative technologies for the production of new generations of enzymes and bioprocesses are needed. In this chapter we focus on extremophilic Archaea and their relevance for industrial biotechnology.

Extremophilic Archaea as a source of enzymes and other compounds

Extreme environments, such as geothermal sites (80–121 °C), polar regions (−20 to +20 °C), acidic (pH <4) and alkaline (pH<8) springs, saline lakes (2–5 M NaCl), and the cold pressurized depths of the oceans (<5 °C), are promising sources of unique microorganisms and their enzymes. The majority of extremophiles, identified to date, belong to the archaeal domain (nearly 300 species), which consists of four kingdoms: Crenarchaeota, Euryarchaeota, Korarchaeota, and Nanoarchaeota. The recently discovered nanosized, parasitic, hyperthermophilic archaeon *Nanoarchaeum equitans* grows in co-culture with the crenarchaeon *Ignicoccus* sp. Another symbiotic psychrophilic crenarchaeon, *Cenarchaeum symbiosum*, inhabits a marine sponge. Thermophiles (growth 50–80 °C) and hyperthermophiles (80–113 °C) are widely distributed in the archaeal domain. Interestingly, methanogens are able to grow up to 110 °C (*Methanopyrus kandleri*) or below 0 °C (*Methanogenium frigidum*). Some members of Archaea can also survive under two extreme conditions, namely high temperature and low pH (50–90 °C, pH 0–4) (e.g. *Acidianus*, *Ferroplasma*, *Picrophilus*, *Sulfolobus*, and *Thermoplasma*). These thermoacidophiles provide interesting enzyme systems and cell components that are active at elevated temperatures and acidic conditions. Various archaeal strains can also tolerate other extreme conditions, such as high pressure (*Paleococcus ferrophilus*, *Thermococcus barophilus*), high levels of radiation or toxic compounds (*Pyrococcus furiosus*), or low water and nutrient supply (*Halobacterium* sp.).

Biocatalysts from Archaea

Starch processing enzymes: biochemistry at the boiling point of water

Many archaeal enzymes involved in carbohydrate metabolism, particularly those of the glucosyl hydrolase family, are of industrial interest. The starch-processing industry, which converts starch into more valuable products such as dextrins, glucose, fructose, maltose, and trehalose, can profit from thermostable enzymes. In all starch-converting processes, high temperatures are required to dissolve starch and make it accessible to enzymatic hydrolysis. The synergetic action of thermostable amylolytic enzymes results in an advantage to those processes, lowering the cost of sugar syrup production. Extremophilic Archaea have been shown to be a good source for the production of a number of starch-degrading hydrolases.

Amylases, glucoamylases, and α-glucosidases α-Amylase (α-1,4-glucan-4-glucanohydrolase; EC 3.2.1.1) hydrolyzes linkages in the interior of the starch polymer in a random fashion, which leads to the formation of linear and branched oligosaccharides. The sugar-reducing groups are liberated in the α-anomeric configuration. Most starch-hydrolyzing enzymes belong to the α-amylase family that contains a characteristic catalytic $(\beta/\alpha)_8$-barrel domain. Throughout the α-amylase family, only eight amino acid residues are invariant, seven at the active site and a glycine in a short turn. A variety of amylolytic enzymes has been detected in thermophilic and halophilic archaea (Table 26.1). Extremely thermostable α-amylases have been characterized from a number of hyperthermophilic Archaea belonging to the genera *Methanocaldococcus*, *Pyrococcus*, *Sulfolobus*, and *Thermococcus* (Antranikian et al., 2005). The thermostability of the enzymes is often enhanced in the presence of divalent metal ions. The optimal temperatures for the activity of these enzymes range between 80 and 100 °C. The high thermostability of the pyrococcal extracellular α-amylase (thermal activity even at 130 °C and after autoclaving for 4 hours at 120 °C) and α-amylase from *Methanocaldococcus jannashii* (temperature optimum 120 °C, half-life of 50 hours at 100 °C) makes these enzymes interesting candidates for industrial applications (Kim et al., 2001). α-Amylases with lower thermostability have been isolated from the Archaea *Thermococcus profundus*, *Thermococcus kodakaraensis*, and the halophilic Archaea of the genera *Haloarcula*, *Haloferax*, *Halothermothrix*, and *Natronococcus* sp. Ah-36. Amylases from halophiles could be used in processes running at high salt concentrations and hydrophobic organic solvents. The α-amylase from *Haloarcula* sp. S-1 exhibits its maximal activity at 4.3 M NaCl and is stable in benzene, toluene, and chloroform (Fukushima et al., 2005). The use of α-amylases in detergents for medium-temperature laundering demands enzymes with high stability and activity at alkaline conditions. The α-amylase from *Haloferax mediterranei* is highly stable, having a half-life of 240 days at pH 10 (Perez-Pomares et al., 2003). A maltotriose-forming amylase from *Natronococcus* sp. Ah-36 with maximal activity at pH 8.7, 55 °C, and 2.5 M NaCl was expressed in the halophilic archaeon *Haloferax volcanii* (Kobayashi et al., 1994).

Unlike α-amylase, the production of glucoamylase seems to be very rare in Archaea (Table 26.1). Glucoamylases (EC 3.2.1.3) hydrolyze terminal α-1,4-linked-D-glucose residues successively from non-reducing ends of the chains, releasing β-D-glucose. An ideal catalyst for starch liquification should be optimally active at 100 °C and pH 4.0–5.0 without requiring calcium for the stabilization of the enzyme. Therefore, intensive experiments have been carried out to identify glucoamylases in Archaea. Recently, it has been shown that the thermoacidophilic Archaea *Thermoplasma acidophilum*, *Picrophilus torridus*, and *Picrophilus oshimae* produce heat- and acid-stable extracellular glucoamylases. The purified archaeal glucoamylases are optimally active at pH 2 and 90 °C. Catalytic activity is still detectable at pH 0.5 and 100 °C. These enzymes are more thermostable than already described glucoamylases from Bacteria, yeasts, and fungi. This has been the first report on the production of

Table 26.1 Archaeal enzymes with potential applications in starch processing.

Enzymes	Strain	MW (kDa)	T_{opt} (°C)	pH_{opt}	Thermostability (half-life)	Possible applications	References
Starch-processing							
α-Amylase	*Desulfurococcus mucosus*		100	5.5		Bread and baking industry, starch liquefaction and saccharification, production of glucose, fructose for sweeteners, textile desizing, paper industry, synthesis of oligosaccharides, detergent application, gelling, thickening in food industry	Antranikian et al., 2005
	Haloarcula sp. S-1	70	50	7			Fukushima et al., 2005
	Halobacterium salinarum						Margesin & Schinner, 2001
	Haloferax mediterranei	58	50–60	7–8	10 h at 50 °C		Perez-Pomares et al., 2003
	Halothermothrix orenii	62	65	7.5			Mijts & Patel, 2002
	Methanocaldococcus jannaschii		120	5–8	50 h at 100 °C		Kim et al., 2001
	Natronococcus sp. Ah-36	74	55	8.7			Antranikian et al., 2005
	Pyrococcus furiosus (intracell.)	76	92	7			Antranikian et al., 2005
	Pyrococcus furiosus (extracell.)	100	100	5.5–6	13 h at 98 °C		Antranikian et al., 2005
	Pyrococcus woesei (struct.)	68	100	5.5			Antranikian et al., 2005
	Pyrodictium abyssi		100	5			Antranikian et al., 2005
	Staphylothermus marinus		100	5			Antranikian et al., 2005
	Sulfolobus solfataricus	240	95	6.5			Antranikian et al., 2005
	Thermococcus aggregans		90	5.5			Antranikian et al., 2005
	Thermococcus celer		75–85	5–5.5	4 h at 90 °C		Antranikian et al., 2005
	Thermococcus hydrothermalis	49	80	5–6	3 h at 80 °C		Antranikian et al., 2005
	Thermococcus profundus	43					Antranikian et al., 2005
	Thermococcus kodakaraensis (amyS)	49.5	90	6.5	24 h at 70 °C		Uotsu-Tomita et al., 2001
Glucoamylase	*Methanocaldococcus jannaschii*		80	6.5			Serour & Antranikian, 2002
	Picrophilus oshimae (extracell.)	140	90	2	20 h at 90 °C		Serour & Antranikian, 2002
	Picrophilus torridus (extracell.)	133	90	2	24 h at 90 °C		Antranikian, unpublished
	Picrophilus torridus (intracell.)	312	50	5	4 h at 55 °C		Serour & Antranikian, 2002
	Thermoplasma acidophilum (extracell.)	141	90	2	24 h at 90 °C		Antranikian, unpublished
	Thermoplasma acidophilum (intracell.)	140	75	5	40 h at 60 °C		Kim et al., 2004
	Sulfolobus solfataricus	250	90	5.5–6			

(Continued)

Table 26.1 Archaeal enzymes with potential applications in starch processing.—cont'd

Enzymes	Strain	MW (kDa)	T_{opt} (°C)	pH_{opt}	Thermostability (half-life)	Possible applications	References
α-Glucosidase	*Ferroplasma acidiphilum*	57	55–60	2.4–3.5			Ferrer et al., 2005
	Pyrococcus furiosus	125	115	5.5			Eichler, 2001
	Pyrococcus woesei	90	110	5–5.5			Antranikian et al., 2005
	Sulfolobus shibatae	313	85	5.5	6 h at 80 °C		Leveque et al., 2000
	Sulfolobus solfataricus	80	105	4.5	39 h at 85 °C		Rolfsmeier et al., 1998
	Thermococcus hydrothermalis	57	120	5.5			Antranikian et al., 2005
	Thermococcus sp. AN1	63	120	7			Eichler, 2001
	Thermococcus zilligii		75	7			Leveque et al., 2000
Pullulanase type II	*Desulfurococcus mucosus*	74	85	5	0.8 h at 85 °C		Antranikian et al., 2005
	Pyrococcus furiosus	90	105	6			Antranikian et al., 2005
	Pyrococcus woesei (struct.)	90	100	6			Antranikian et al., 2005
	Pyrodictium abyssi		100	9			Niehaus et al., 1999
	Thermococcus celer		90	5.5			Niehaus et al., 1999
	Thermococcus hydrothermalis	128	95	5.5			Antranikian et al., 2005
	Thermococcus litoralis	119	98	5.5			Antranikian et al., 2005
	Thermococcus sp. ST489		80–95				Leveque et al., 2000
Pullulan hydrolase	*Pyrococcus furiosus*	77.2	90	5	2 h at 95 °C		Yang et al., 2004
	Thermococcus aggregans	80	95	6.5	2.5 h at 100 °C		Niehaus et al., 2000
	Thermococcus profundus (amyL)	43	90	5.5	6.7 h at 90 °C		Antranikian et al., 2005
Cyclodextrin glucanotransferase	*Thermococcus kodakaraensis*	77	80	5.5–6	0.33 h at 100 °C		Rashid et al., 2002
	Thermococcus sp.	83	90–110	5–5.5	0.66 h at 110 °C		Tachibana et al., 1999
Amylomaltase	*Pyrobaculum aerophilum*		95	6.7			Kaper et al., 2005
	Thermococcus litoralis (struct.)	87	90	6			Imamura et al., 2001
Trehalosyl dextrin-forming enzyme	*Sulfolobus solfataricus*	87	75	5	2 h at 85 °C		Fang et al., 2004
Maltooligosyl trehalose synthase	*Sulfolobus acidocaldarius*	84	75	5	72 h at 80 °C		Gueguen et al., 2001

(struct.): the protein is crystallized and the three-dimensional structure is determined.

glucoamylase in Archaea. These enzymes are of interest for applications in the beverage industry (Serour & Antranikian, 2002). However, the lack of suitable genetic methods for thermoacidophiles has precluded structural studies aimed at discovering their adaptation to very low pH. Recently, the gene encoding a putative glucoamylase from *Sulfolobus solfataricus* was cloned and expressed in *E. coli*, and the properties of the recombinant protein were examined in relation to the glucose production process (Kim et al., 2004). This recombinant enzyme is extremely thermostable, with an optimal temperature of 90 °C; however, it is most active in a slightly acidic pH range from 5.5 to 6.0. The tetrameric enzyme liberates β-D-glucose from maltotriose, and the substrate preference for maltotriose distinguishes this enzyme from fungal glucoamylases. Genome analysis of other thermoacidophiles revealed further putative glucoamylases, which were cloned and expressed in *E. coli*. Thus, the recombinant intracellular glucoamylase from *Methanocaldococcus jannashii* is optimally active at pH 6.5 and 80 °C (Uotsu-Tomita et al., 2001). In our laboratory, the intracellular glucoamylases from the extreme thermoacidophiles *Picrophilus torridus* and *Thermoplasma acidophilum* have recently been cloned; the enzymes are optimally active at 50–75 °C and pH 5 (Antranikian, unpublished results).

α-Glucosidases (EC 3.2.1.20) attack the α-1,4 linkages of oligosaccharides that are produced by the action of other amylolytic enzymes. Unlike glucoamylase, α-glucosidase prefers smaller oligosaccharides (e.g. maltotriose) and liberates glucose with an α-anomeric configuration. Intracellular and extracellular α-glucosidases have been purified and characterized from Archaea belonging to the genera *Pyrococcus*, *Sulfolobus*, and *Thermococcus* (Table 26.1) (Antranikian et al., 2005). The enzymes exhibit optimal activity at pH 4.5–7.0 over a temperature range from 105 to 120 °C. The recombinant α-glucosidase from *Sulfolobus solfataricus* with a calculated size of 80.5 kDa hydrolyzes *p*-nitrophenyl-D-glucopyranoside (Rolfsmeier et al., 1998). At pH 4.5 it exhibits a pH optimum for maltose hydrolysis. Unlike maltose hydrolysis, glycogen was hydrolyzed efficiently at the intracellular pH of the organism (pH 5.5). The recombinant α-glucosidase exhibits greater thermostability than the native enzyme, with a half-life of 39 hours at 85 °C at a pH of 6.0. Recently, a novel α-glucosidase from the extreme acidophilic *Ferroplasma acidiphilum* was characterized (Ferrer et al., 2005). The highest α-glucosidase activity was detected at pH 2.4–3.5, with <74% activity remaining at pH 1.5. Iron was found to be essential for enzymatic activity and His30 was shown to be responsible for iron binding.

Thermoactive pullulanase, CGTase, and amylomaltase Enzymes capable of hydrolyzing α-1,6 glucosidic bonds in pullulan and branched oligosaccharides are defined as pullulanases. On the basis of substrate specificity and product formation, pullulanases have been classified into three groups: pullulanase type I, pullulanase type II, and pullulan hydrolases. Pullulanase type I (EC 3.2.1.41) specifically hydrolyzes the α-1,6-linkages in pullulan as well as in branched oligosaccharides (debranching enzyme); its degradation products are maltotriose and linear oligosaccharides, respectively. Pullulanase type I is unable to attack α-1,4 linkages in α-glucans. Pullulanase type II (amylopullulanase) attacks α-1,6-glycosidic linkages in pullulan and branched α-glucans. Unlike pullulanase type I, this enzyme also attacks α-1,4 linkages in branched and linear oligosaccharides. The enzyme has multiple specificity and is able to fully convert polysaccharides (e.g. amylopectin) to small sugars (e.g. glucose, maltose, maltotriose) in the absence of other enzymes, such as α-amylase or β-amylase (Antranikian et al., 2005).

Interestingly, pullulanase type I has not been identified in Archaea so far, whereas the enzyme has been characterized in several thermophilic Bacteria. Thermostable and thermoactive pullulanases type II from *Desulfurococcus mucosus*, *Pyrococcus furiosus*, *Pyrococcus woesei*, and *Thermococcus hydrothermalis* have been reported to have temperature optima between 85 and 105 °C (Table 26.1), as well as remarkable thermostability even in the absence of substrate and calcium ions (Antranikian et al., 2005). In the presence of calcium ions the pullulanase activity was also measured at 130–140 °C. On the basis of the amino acid sequences, the pullulanases from *Thermococcus hydrothermalis* and *Pyrococcus furiosus* belong to family 57 (GH-57) of the glycoside hydrolases. Five conserved regions were identified, which are postulated to be GH-57 consensus motifs by comparison to the 659 amino-acid-long 4-α-glucanotransferase from *Thermococcus litoralis*. These motifs correspond to 13_HQP (region I), 76_GQLEIV (region II), 120_WLTERV (region III), 212_HDDGEKFGVW (region IV), and 350_AQCNDAYWH (region V). The third and fourth conserved regions contain the hypothetical catalytic nucleophile E291 and the proton donor D394, respectively. To validate this prediction, the characterization of catalytic sites of certain members of GH-57 has been investigated. Site-directed mutagenesis performed on the pullulanase from *Thermococcus hydrothermalis* reveals that both residues are indeed critical for the pullulanolytic and amylolytic activities of the pullulanase (Zona et al., 2004). The crucial role of E291 as the catalytic nucleophile has been also confirmed for the pullulanase from *Pyrococcus furiosus* (Kang et al., 2004). The apparent catalytic efficiencies (k_{cat}/K_m) of mutants E291Q and D394N on pullulan were 123.0 and 24.4 times lower than that of the native enzyme. The activity of mutant E396Q on pullulan was too low to allow reliable determination of its catalytic efficiency. The apparent specific activities of these enzymes on starch also decreased by 91.0 times (E291Q), 11.7 times (D394N), and 37.2 times (E396Q). The hydrolytic patterns for pullulan and starch were the same, while the hydrolysis rates differed as reported. Therefore, these data strongly suggest that the bifunctionality of the pullulanase type II is determined by a single catalytic center. Due to the dual specificity of amylopullulanases to degrade both α-1,4- and α-1,6-glucosidic linkages they cannot be used as debranching enzymes in maltose and glucose syrup production. It is proposed that these archaeal enzymes are promising candidates to optimize starch liquefaction for producing fermentation syrups. Since the major products of the amylpullulanase action on starch are maltose, maltotriose, and maltotetraose, they can be applied as biocatalysts in a one-step liquification–saccharification process to obtain syrups of these molecules.

The third class of pullulan-hydrolyzing enzymes includes pullulan hydrolases type I, II, and III. Pullulan hydrolases I and II are active toward α-1,4 linkages of amylose, starch, aand pullulan, but are unable to hydrolyze α-1,6 linkages. An exception is pullulan hydrolase type III. This enzyme attacks α-1,4 as well as α-1,6 linkages of pullulan. The enzymes from *Pyrococcus furiosus* (AmyL), *Thermococcus aggregans*, and *Thermococcus profundus* (Table 26.1) exhibit maximal activity at 90 °C and pH 5.5–6.5 and are stable for several hours at 95–100 °C. In addition, the pullulan hydrolase from *Pyrococcus furiosus* degrades β-cyclodextrin.

Cyclodextrin glucosyltransferase (CGTase; EC 2.4.1.19) converts starch and oligodextrins into cyclodextrins, which are composed of 6 (α), 7 (β), or 8 (γ) α-1,4 linked glucose molecules. The internal cavities of cyclodextrins are hydrophobic and they can encapsulate hydrophobic molecules. Thermostable CGTases are generally found in Bacteria and were recently discovered in Archaea. The archaeal enzyme found in *Thermococcus* sp. is optimally active at 90–110 °C (Table 26.1). During an incubation for 24 hours at 96 °C and pH 4.5 with 30% corn-starch (wt/vol) the enzyme converts 34.4% of starch into cyclodextrins, and the production ratios are 69% α-cyclodextrin, 20% β-cyclodextrin, and 11% γ-cyclodextrin (Tachibana et al., 1999). On the other hand, the main cyclodextrin formed by the action of the CGTase from *Thermococcus kodakaraensis* is β-cyclodextrin (Rashid et al., 2002).

Amylomaltases (EC 2.4.1.25, 4-α-glucanotransferase) catalyze the transfer of a segment of an α-1,4-D-glucan to the reducing end of an acceptor, which may be glucose or another α-1,4-D-glucan. Acting upon starch, amylomaltases can produce products of commercial interest, such as cycloamylose and thermoreversible starch gel, which can be used as a substitute for gelatin. In combination with α-amylase, the amylomaltases produce syrups with reduced sweetness and low viscosity (Kaper et al., 2004). The only archaeal amylomaltases known to date are the enzymes from *Pyrobaculum aerophilum* and *Thermocacccus litoralis*. The thermostable amylomaltase from *Pyrobaculum aerophilum* produces a thermoreversible starch product with gelatin-like properties (Kaper et al., 2005). The enzyme from *Thermococccus litoralis* produces linear α-1,4-glucans and a cycloamylose (cyclic α-1,4-glucan) with a high degree (up to hundreds) of polymerization. The residue Glu-123 was identified as the catalytic nucleophile of the enzyme, suggesting that family 57 and 38 glycoside hydrolases may have a common ancestor (Imamura et al., 2001).

The finding of novel thermostable starch-modifying enzymes will be a valuable contribution to the starch-processing industry. By using robust enzymes from thermophiles, innovative and environmentally friendly processes can be developed, aiming at the formation of products of high added value for the food and pharmaceutical industries. At elevated temperatures starch is more soluble (30–35% wt/vol) and the risk of contamination is reduced. This is of advantage when starch is to be converted to high glucose and fructose syrups. Industrial production of fructose from starch consists of three steps: liquefication, saccharification, and isomerization. This multistage process (step 1, pH 6.5, 98 °C; step 2, pH 4.5, 60 °C; step 3, pH 8.0, 65 °C) leads to the conversion of starch to fructose, with concurrent formation of high concentrations of salts that have to be removed by ion exchangers. Furthermore, high energy is required for cooling from 100 to 60 °C in step 2. The application of thermostable enzymes such as amylases, glucoamylases, pullulanases, and glucose isomerases, which are active and stable above 100 °C and at acidic pH values, can simplify this complicated process. The use of the extremely thermostable amylolytic enzymes can lead to other valuable products, which include innovative starch-based materials with gelatin-like characteristics and defined linear dextrins that can be used as fat substitutes, texturizers, aroma stabilizers, and prebiotics. CGTases are used for the production of cyclodextrins that can be used as a gelling, thickening, or stabilizing agent in jelly desserts, dressings, confectionery, and dairy and meat products. Due to the ability of cyclodextrins to form inclusion complexes with a variety of organic molecules, cyclodextrins improve the solubility of hydrophobic compounds in aqueous solution. This is of interest for the pharmaceutical and cosmetic industries. Cyclodextrin production is a multistage process in which starch is first liquefied by a heat-stable amylase and in the second step a less thermostable CGTase from *Bacillus* sp. is used. The application of heat-stable CGTase from the *Thermococcus* species in jet cooking, where temperatures up to 105 °C could be achieved, will allow liquefication and cyclization to take place in one step (Biwer et al., 2002).

Another promising application of archaeal enzymes is the production of a disaccharide trehalose, a stabilizer of enzymes, antibodies, vaccines, and hormones. The use of thermoactive enzymes in the process would eliminate problems associated with viscosity and sterility. The process was developed to produce trehalose from dextrins using *Sulfolobus* enzymes at 75 °C in a continuous bioreactor, with a final conversion of 90% (Schiraldi et al., 2002). Recently, the trehalose biosynthetic pathway was identified in Sulfolobales and the responsible enzymes were cloned and expressed in *E. coli* (de Pascale et al., 2002; Schiraldi et al., 2002; Fang et al., 2004).

Cellulose-degrading enzymes

Cellulose is the most abundant organic biopolymer in nature, since it is the structural polysaccharide of the cell wall in the plant kingdom. It consists of glucose units linked by β-1,4-glycosidic bonds with a polymerization grade of up to 15,000 glucose units in a linear mode. The minimal molecular weight of cellulose from different sources has been estimated to vary from about 50,000 to 2,500,000 g/mol in different species, which is equivalent to 300–15,000 glucose residues. Although cellulose has a high affinity to water, it is completely insoluble in it. Natural cellulose compounds are structurally heterogeneous and have both amorphous and highly ordered crystalline regions. The degree of crystallinity depends on the source of the cellulose and the higher crystalline regions are more resistant to enzymatic hydrolysis. Cellulose can be hydrolyzed into glucose by the synergistic action of different enzymes: endoglucanase (cellulase), exoglucanase (cellobiohydrolase), and β-glucosidase (cellobiase). Endoglucanase (EC 3.2.1.4) hydrolyzes cellulose in a random manner as endohydrolase, producing various oligosaccharides, cellobiose, and glucose. Exoglucanases (EC 3.2.1.91) hydrolyze β-1,4-D-glycosidic linkages in cellulose and cellotetraose, releasing cellobiose from the non-reducing end of the chain. β-Glucosidases (EC 3.2.1.21) catalyze the hydrolysis of terminal, non-reducing β-D-glucose residues, releasing β-D-glucose.

Several cellulose-degrading enzymes have been identified in Archaea (Table 26.2). Thermostable endoglucanases, which degrade β-1,4 or β-1,3 linkages of β-glucans and cellulose, have been identified in *Pyrococcus furiosus*, *Pyrococcus horikoshii*, and *Sulfolobus solfataricus* (Antranikian et al., 2005). The purified recombinant endoglucanase from *P. furiosus* is active at 100 °C and hydrolyzes β-1,4 but not β-1,3 glycosidic linkages. It has the highest specific activity with cellopentaose and cellohexaose (Bauer et al., 1999). Another thermoactive glucanase (laminarinase) (T_{opt} 100 °C) from this strain catalyzes the hydrolysis of mixed-linked oligosaccharides with both β-1,4 and β-1,3 specificities (van Lieshout et al., 2004). The E170A mutant of the enzyme is

Table 26.2 Other archaeal polymer-degrading enzymes with industrial relevance.

Enzymes	Strain	MW (kDa)	T_{opt} (°C)	pH_{opt}	Thermostability (half-life)	Possible applications	References
Cellulose-degrading enzymes							
β-Glucosidase	*Pyrococcus furiosus*	232	102	5	13 h at 110 °C	Polymer degradation, color brightening, color extraction of juice, saccharification of agricultural and industrial wastes, animal feed, biopolishing of cotton products, synthesis of sugars, optically pure heterosaccharides	Lebbink et al., 2001
	Pyrococcus horikoshii		>100	6	15 h at 90 °C		Matsui et al., 2002
	Sulfolobus acidocaldarius	224		7–8			Antranikian et al., 2005
	Sulfolobus shibatae						Antranikian et al., 2005
	Sulfolobus solfataricus (struct.)		85				Antranikian et al., 2005
	Thermosphaera aggregans (struct.)						Chi et al., 1999
Endoglucanase	*Pyrococcus furiosus* (EglA)	36	100	6	40 h at 95 °C		Bauer et al., 1999
	Pyrococcus furiosus (LamA)	31	100	6–6.5	19 h at 100 °C		van Lieshout et al., 2004
	Pyrococcus horikoshii	43–52	97	5.6	>3 h at 97 °C		Kashima et al., 2005
	Sulfolobus solfataricus MT4 (struct.)	40	65	6			Limauro et al., 2001
	Sulfolobus solfataricus P2	37	80	1.8	8 h at 80 °C and pH 1.8		Huang et al., 2005
Chitin-degrading enzymes							
	Pyrococcus furiosus (ChiA)	40	6	90–95		Utilization of biomass of marine environment	Gao et al., 2003
	Pyrococcus furiosus (ChiB)	55	6	90–95			Gao et al., 2003
	Thermococcus chitonophagus (Chi70)	70	70	7	1 h at 120 °C		Andronopoulou & Vorgias, 2004b
	Thermococcus chitonophagus (Chi50)	50	80	6			
	Thermococcus chitonophagus (Chi90)	90					Andronopoulou & Vorgias, 2004b
	Thermococcus kodakaraensis (GlmA)	193	80	6			
	Thermococcus kodakaraensis (Tk-Dac)	160	75	8.5			Andronopoulou & Vorgias, 2004b
	Thermococcus kodakaraensis (ChiA)	134	85	5			Tanaka et al., 2003; Tanaka et al., 2004; Tanaka et al., 1999

Category	Organism	Value	pH	Stability	Application	Reference
Xylan-degrading enzymes	Halorhabdus utahensis	45	55–70	6.5		Paper bleaching	Waino & Ingvorsen, 2003
	Pyrodictium abyssi		110	5.5			Antranikian et al., 2005
	Sulfolobus solfataricus	57	100	7	0.8 h at 90 °C		Cannio et al., 2004
	Thermococcus zilligii AN1	95		6	4 h at 95 °C		Antranikian et al., 2005
Proteolytic enzymes							
Serine protease	Aeropyrum pernix	34	90	8–9	1 h at 100 °C	Detergents, baking, brewing, amino acids production	Catara et al., 2003
	Desulfurococcus mucosus	43–54	95	7.5	4.3 h at 95 °C		Ward et al., 2002
	Halobacterium halobium	66					Ryu et al., 1994
	Haloferax mediterranei		37				Antranikian et al., 2005
	Natrialba asiatica		37				Antranikian et al., 2005
	Natrialba magadii		37	8–10			Oren, 2002
	Natronococcus occultus		60				Horikoshi, 1999
	Natronomonas pharaonis		60	10			Stan-Lotter et al., 1999
	Pyrobaculum aerophilum						Ward et al., 2002
	Pyrococcus abyssi	60	95	9			Ward et al., 2002
	Pyrococcus furiosus	150	115	6–9	0.33 h at 105 °C		Antranikian et al., 2005
	Thermococcus aggregans		90	7			Eichler, 2001
	Thermococcus celer		95	7.5			Eichler, 2001
	Thermococcus kodakaraensis	44	80	9.5			Ward et al., 2002
	Thermococcus litoralis		95	9.5			Eichler, 2001
	Thermococcus stetteri	142	85	8.5–9	22 h at 90 °C		Ward et al., 2002
	Thermoplasma acidophilum (struct.)	120	65	8.5			Ward et al., 2002
	Staphylothermus marinus	150	90	9			Antranikian et al., 2005
	Sulfolobus solfataricus (struct.)	118	>90	6.5–8			Ward et al., 2002
Thiol protease	Thermococcus kodakaraensis	45	110	7	1 h at 100 °C		Antranikian et al., 2005
Acidic protease	Sulfolobus acidocaldarius	46–51	90	2			Ward et al., 2002
Metalloprotease	Aeropyrum pernix	52	100	6–8			Ward et al., 2002
	Pyrococcus furiosus	128	100	6.5			Ward et al., 2002
	Pyrococcus furiosus	79	75	7			Ward et al., 2002
	Pyrococcus horikoshii OT3	95	>95	7.5			Ishikawa et al., 2001
	Sulfolobus solfataricus	320	75	6.7			Ward et al., 2002
	Sulfolobus solfataricus	170	85	5.5–7			Ward et al., 2002

(struct.): the protein is crystallized and the three-dimensional structure is determined.

additionally active as a glycosynthase, catalyzing the condensation of α-laminaribiosyl fluoride to different acceptors at pH 6.5 and 50 °C (van Lieshout et al., 2004). Depending on the acceptor, the synthase generates either β-1,4 or β-1,3 linkage. Recently, a recombinant endoglucanase from *Pyrococcus horikoshii* was characterized; the enzyme is active even toward crystalline cellulose. Its activity was recently increased by protein engineering (Kashima et al., 2005). This enzyme is expected to be useful for industrial hydrolysis of cellulose, particularly in biopolishing of cotton products. A novel acid-stable endoglucanase from *Sulfolobus solfataricus* P2 was recently cloned and expressed in *E. coli* (Huang et al., 2005). The purified recombinant enzyme, with optimal activity at 80 °C and pH 1.8, hydrolyzes carboxymethylcellulose and cello-oligomers, with cellobiose and cellotriose as main products. The presence of a signal peptide indicates that this secreted protein enables *Sulfolobus solfataricus* to utilize cellulose as carbon source. This is the first description of the cellulolytic enzyme with a combination of high stability and activity at high temperatures and low pH. The enzyme could be applicable for the large-scale hydrolysis of cellulose under acidic conditions.

Unlike endoglucanases, several β-glucosidases have been detected in Archaea. These enzymes have been detected in strains of the genera *Sulfolobus*, *Pyrococcus*, and *Thermosphaera*. The β-glucosidase from *Pyrococcus furiosus* is very stable with optimal activity at 103 °C, and it also shows a β-mannosidase activity (Nagatomo et al., 2005). The β-glucosidase from *S. solfataricus* MT4 is very resistant to various denaturants, with activity up to 85 °C. The gene for this β-glucosidase has been cloned and expressed in *E. coli* and *Saccharomyces cerevisiae*. Using a mixture of both β-glucosidases from *P. furiosus* and *S. solfataricus*, an ultra high temperature process for the enzymatic production of novel oligosaccharides from lactose was developed (Schiraldi et al., 2002). The thermoactive β-glucosidase from *Pyrococcus horikoshii* is active in organic solvents and it synthesizes a heterosaccharide that has a high optical purity (Antranikian et al., 2005). For the production of glucose from cellobiose (dimer of β-1,4-linked glucose) a bioreactor system with the immobilized recombinant β-glucosidase from *S. solfataricus* was developed (Schiraldi et al., 2002). The system runs at a high flow rate and has a high degree of conversion, productivity, and operational stability.

There is a great demand for robust cellulolytic enzymes for various applications such as alcohol production, improvement of juice yield, and effective color extraction of juices. Other suitable applications of cellulases include the pre-treatment of cellulose biomass and forage crops to improve nutritional quality and digestibility. Furthemore, cellulases are useful tools for the saccharification of agricultural and industrial wastes and production of fine chemicals (Schulein, 2000).

Xylan-degrading enzymes

Xylan is a heterogeneous molecule that constitutes the main polymeric compound of hemicellulose, a fraction of the plant cell wall, which is a major reservoir of fixed carbon in nature. The main chain of the heteropolymer is composed of xylose residues linked by β-1,4-glycosidic bonds. Approximately half of the xylose residues have substitution at O-2 or O-3 positions with acetyl-, arabinosyl-, and glucuronosyl-groups. The complete degradation of xylan requires the combined action of endo-β-1,4-xylanase (EC 3.2.1.8), it hydrolyzes β-1,4-xylosydic linkages in xylans, and β-1,4-xylosidase (EC 3.2.1.37), hydrolyzing β-1,4-xylans and xylobiose by removing the successive xylose residues from the non-reducing termini.

To date only a few extreme thermophilic microorganisms have been able to grow on xylan and secrete thermoactive xylanolytic enzymes (Table 26.2). Among the thermophilic Archaea, a xylanase from *Pyrodictium abyssi* has been produced with an optimum temperature of 110 °C, one of the highest values reported for a xylanase (Antranikian et al., 2005). The crenarchaeon *Thermosphaera aggregans* was shown to grow on heat-treated, but not native, xylan (Eichler, 2001). The xylanase from *Thermococcus zilligii* AN1 is active up to 100 °C, can attack different xylans, and is not active toward cellulose. Recently, a thermoactive endoxylanase from xylan-degrading *Sulfolobus solfataricus* was purified and characterized (Cannio et al., 2004). Xylooligosaccharides are the products of the enzymatic hydrolysis of xylan, and the smallest degradation product was xylobiose. The production of an endo-β-1,4-xylanase and a β-xylosidase from the extremely halophilic archaeon *Halorhabdus utahensis* was also reported (Waino & Ingvorsen, 2003).

Xylanases from bacteria and fungi have a wide range of potential biotechnological applications. They are already produced on an industrial scale and are used as food additives for poultry, for increasing feed efficiency diets, and in wheat flour for improving dough handling and the quality of baked products. In the past decade, the major interest in thermostable xylanases was in enzyme-aided bleaching of paper. The chlorinated lignin derivatives generated by the bleaching process constitute a major environmental problem caused by the pulp and paper industry. Investigations have demonstrated the feasibility of enzymatic treatments as alternatives to chlorine bleaching for the removal of residual lignin from pulp. A treatment of craft pulp at elevated temperatures with cellulase-free thermostable xylanases facilitates xylan and lignin removal without undue loss of other pulp components and thus enhances the development of environmentally friendly processes.

Chitin-degrading enzymes

Chitin is a linear β-1,4 homopolymer of *N*-acetylglucosamine residues and it is one of the most abundant biopolymers on earth. It has been estimated that the annual worldwide formation rate and steady state amount of chitin is in the order of 10^{10} to 10^{11} tonnes per year. Particularly in the marine environment, chitin is produced in enormous amounts, and its turnover is due to the action of chitinolytic enzymes. Chitin is the major structural component of most fungi and some invertebrates, while for soil or marine bacteria chitin serves as a nutrient. Chitin degradation is known to proceed with the endo-acting chitin hydrolase (chitinase A; EC 3.2.1.14) and the chitin oligomer degrading exo-acting hydrolases (chitinase B), and *N*-acetyl-D-glycosaminidase (chitobiase; EC 3.2.1.52). Chitin and chitosan exhibit interesting properties that make them valuable raw materials for several applications.

Although a large number of bacterial chitin-hydrolyzing enzymes have been isolated and their corresponding genes have been cloned and characterized, only few thermostable chitin-hydrolyzing enzymes are known. So far, only three hyperthermophilic Archaea, *Thermococcus chitonophagus* (Andronopoulou & Vorgias, 2004a), *Thermococcus kodakaraensis* (Tanaka et al., 1999; Imanaka et al., 2001), and *Pyrococcus furiosus* (Gao et al., 2003) have been shown to grow on chitin and produce chitinolytic enzymes (Table 26.2). The extreme thermophilic anaerobic archaeon *Thermococcus chitonophagus* possesses a multicomponent enzymatic system, consisting of an extracellular exochitinase (Chi50), a periplasmic chitobiase (Chi90), and a cell-membrane-anchored endochitinase (Chi70) (Andronopoulou & Vorgias, 2004b). The chitinolytic system is strongly induced by chitin, although a low-level constitutive production of the enzymes in the absence of any chitinous substrates was detected. The archaeal chitinase (Chi70) was purified and characterized. It is a monomeric enzyme with an apparent molecular weight of 70 kDa and appears to be associated with the outer surface of the cell membrane. The enzyme is optimally active at 70 °C and pH 7.0 and is thermostable, maintaining 50% activity even after 1 hour at 120 °C. The enzyme was not inhibited by allosamidin and was also resistant to denaturation by urea and SDS. The chitinase has a broad substrate specificity for several chitinous substrates and derivatives and has been classified as an endochitinase due to its ability to release chitobiose from colloidal chitin (Andronopoulou & Vorgias, 2004a). The purified recombinant chitinase from the hyperthermophile *Thermococcus kodakaraensis* is optimally active at 85 °C and pH 5.0 and produces chitobiose as major end product (Tanaka et al., 1999). This unique multi-domain protein consists of two active sites with different cleavage specificities and three substrate-binding domains, which are related to two families of cellulose-binding domains (Tanaka et al., 2001). A chitin-degrading pathway involves the unique enzymes diacetylchitobiose deacetylase and exo-β-D-glucosaminidase. After the hydrolysis of chitin by chitinase, the product diacetylchitobiose was deacetylated and than successively hydrolyzed to glucosoamine (Tanaka et al., 2004). The thermostable chitinase from *T. kodakaraensis* is active in the presence of detergents and organic solvents and can be applied, for example, for the production of *N*-acetyl-chitooligosaccharides with biological activity (Imanaka et al., 2001). *Pyrococcus furiosus* was also found to grow on chitin, adding this polysaccharide to the inventory of carbohydrates utilized by this hyperthermophilic archaeon. Two open reading frames (ChiA and ChiB) were identified in the genome of *P. furiosus*, and encode chitinases with sequence similarity to proteins from the glycosyl hydrolase family 18 in less thermophilic organisms (Gao et al., 2003). The two chitinases share little sequence homology to each other, except in the catalytic region, where both have the catalytic glutamic acid residue that is conserved in all family 18 bacterial chitinases. The pH optimum of both recombinant chitinases is 6, with a temperature optimum between 90 and 95 °C. The chitinase A with a melting temperature of 101 °C exhibits no detectable activity toward chitooligomers smaller than chitotetraose, indicating that the enzyme is an endochitinase, whereas the chitinase B (melting temperature 114 °C) is a chitobiosidase, processively cleaving off chitobiose from the non-reducing end of chitin or other chitooligomers. The synergetic action of both thermoactive chitinases on colloidal chitin allows *P. furiosus* to grow on chitin as the sole carbon source (Gao et al., 2003).

Proteolytic enzymes

Proteases are involved in the conversion of proteins to amino acids and peptides. They have been classified according to the nature of their catalytic site in the following groups: serine, cysteine, aspartic, or metalloproteases. Proteases and proteasomes play a key role in the cellular metabolism of Archaea and a variety of heat-stable proteases have been identified in hyperthermophilic Archaea belonging to the genera *Aeropyrum, Desulfurococcus, Sulfolobus, Staphylothermus, Thermococcus, Pyrobaculum,* and *Pyrococcus* (Table 26.2; Antranikian et al., 2005). It has been found that most proteases from extremophilic Archaea belong to the serine type, and are stable at high temperatures even in the presence of high concentrations of detergents and denaturing agents (Ward et al., 2002). Those properties of extracellular serine proteases are reported in a number of *Thermococcus* species (Kannan et al., 2001; Morikawa & Imanaka, 2001) and are well illustrated by the extracellular enzyme from *T. stetteri*, which is highly stable (half-life of 2.5 hours at 100 °C) and resistant to chemical denaturation such as 1% SDS. A globular serine protease from *Staphylothermus marinus* was found to be extremely thermostable and is heat-resistant up to 125 °C in the stalk-bound form (Mayr et al., 1996). Proteases have also been characterized from the thermoacidophilic Archaea *Sulfolobus solfataricus* and *Sulfolobus acidocaldarius*. Serine proteases were also characterized from halophilic Archaea (Table 26.2). An extracellular serine protease from *Halobacterium halobium* is highly salt-dependent and active in dimethylformamide/water mixtures, and is expected to be an excellent tool for the synthesis of glycine-containing peptides (Ryu et al., 1994). A serine protease from *Natrialba magadii* has a broad pH profile (pH 6–12) with an optimum at pH 8–10, and a high dependence on salt, which is required for enzymatic activity and stability (Oren, 2002). The salt-dependent proteolytic activity has also been described in the halophile *Natronococcus occultus* (Horikoshi, 1999). A chymotrypsinogen B-like protease was isolated and characterized from another halophile, *Natronomonas pharaonis*, and acts optimally at 61 °C and pH 10. In contrast to other haloarchaeal enzymes, which lose their catalytic activity at low salt concentrations, this protease can function at salt concentrations lower than 3 mM. This property makes the enzyme a suitable detergent additive.

In addition to serine proteases other subclasses of proteases have been identified in Archaea: aminopeptidases (Dura et al., 2005), a thiol protease from *Thermococcus kodakaraensis* KOD1, an acidic protease from *Sulfolobus acidocaldarius*, and a propylpeptidase and a new type of protease from *Pyrococcus furiosus* (Antranikian et al., 2005). Indeed, *P. furiosus* contains at least 13 different proteins with proteolytic activity.

The amount of proteolytic enzymes produced worldwide on a commercial scale is the largest. Heat stable proteases are useful enzymes, especially for the detergent industry. Serine alkaline proteases from thermophilic Archaea could be used as additives for laundering, where they have to resist denaturation by detergents and alkaline conditions. Proteases are also applied for peptide synthesis using their reverse reaction, mainly because of their compatibility with organic solvents. A number of heat-stable proteases are now used in molecular biology and biochemistry procedures. The protease S from *Pyrococcus furiosus* is used to fragment proteins before peptide sequencing (TaKaRa Biomedicals). Carboxy- and aminopeptidases from *Pyrococcus furiosus* and *Sulfolobus solfataricus* are used for protein N- or C-terminal sequencing (Cheng et al., 1999; Vieille & Zeikus, 2001).

DNA-processing enzymes

DNA amplification DNA polymerases (EC 2.7.7.7) are the key enzymes in the replication of DNA and present in all life forms. They catalyze, in the presence of Mg^{2+}-ions, the addition of a deoxyribonucleoside 5'-triphosphate onto the growing 3'-OH end of a primer strand, forming complementary base pairs to a second strand. More than 100 DNA polymerase genes have been cloned and sequenced from various organisms, including thermophilic Bacteria and Archaea. Thermostable DNA polymerases play a major role in a variety of molecular biological applications, e.g. DNA amplification, sequencing, or labeling (Table 26.3). The DNA polymerase I from the bacterium *Thermus aquaticus*, called *Taq* polymerase, was the first thermostable DNA polymerase characterized and applied in PCR. Due to the absence of a 3'-5'-exonuclease activity, this enzyme is unable to excise mismatches and, as a result, the base insertion fidelity is low. The use of high fidelity DNA polymerases is essential for reducing amplification errors in PCR products. Several thermostable DNA polymerases with 3'-5'-exonuclease-dependent proofreading activity have been described and the error rates (number of misincorporated nucleotides per base synthesized) for these enzymes have been determined. Archaeal polymerases from *Pyrococcus* or *Thermococcus* species, which have stringent proofreading abilities, are of widespread use. Archaeal proofreading polymerases, such as Pwo pol from *Pyrococcus woesei*, Pfu pol from *Pyrococcus furiosus*, Deep Vent™ pol from *Pyrococcus* sp. and GB-D or Vent™ pol from *Thermococcus litoralis* have an error rate that is up to tenfold lower than that of *Taq* polymerase. The 9°N-7 DNA polymerase from *Thermococcus* sp. strain 9°N-7 has a 15-fold higher 3'-5'-exonuclease activity. However, *Taq* polymerase was not replaced by these DNA polymerases because of their low extension rates, among other factors. DNA polymerases with higher fidelity are not necessarily suitable for amplification of long DNA fragments because of their potentially strong exonuclease activity. The recombinant KOD1 DNA polymerase from *Thermococcus kodakaraensis* KOD1 has been reported to show low error rates, high processivity, and the highest known extension rate, resulting in an accurate amplification of target DNA sequences up to 6 kb (Hashimoto et al., 2001). Recently, the PCR technique has been improved to allow low error synthesis of long amplificates (20–40 kb) by the addition of small amounts of thermostable archaeal proofreading DNA polymerases, containing 3'-5'-exonuclease activity, to *Taq* or other non-proofreading DNA polymerases. The supplement of the PCR reaction mixtures with recombinant *P. woesei* dUTPase improves the efficiency of the reaction and allows amplification of longer targets (Dabrowski & Kiaer Ahring, 2003). Low fidelity mutants of *P. furiosus* polymerase were also created for performance in error-prone PCR (Biles & Connolly, 2004).

The ssDNA-binding proteins are known to be involved in eliminating DNA secondary structure, and are key components in DNA replication, recombination and repair. The archaeal ssDNA-binding proteins derived from *Methanocaldococcus jannashii*, *Methanothermobacter thermoautotrophicum*, and *Archaeoglobus fulgidus* are therefore useful reagents for genetic engineering and other procedures involving DNA recombination, such as PCR (Kowalczykowski et al., 2005).

DNA sequencing DNA sequencing by the Sanger method has undergone countless refinements in the past 20 years. A major step forward was the introduction of thermostable DNA polymerases, leading to the cycle sequencing procedure. This method uses repeated cycles of temperature denaturation, annealing, and extension with dideoxy-termination to increase the amount of sequencing product by recycling the template DNA. Due to this "PCR-like" amplification of the sequencing products several problems could have been overcome. Caused by the cycle denaturation, only fmoles of template DNA are required, no separate primer annealing step is needed, and unwanted secondary structures within the template are resolved at high temperature elongation. The first enzyme used for cycle sequencing was the thermostable DNA polymerase I from *Thermus thermophilus*. A combination of thermostable enzymes has been developed that produces higher quality cycle sequences. Thermo Sequenase DNA polymerase is a thermostable enzyme engineered to catalyze the incorporation of ddNTPs with an efficiency of several thousandfold better than other thermostable DNA polymerases. Since the enzyme also catalyzes pyrophosphorolysis at dideoxy termini, a thermostable inorganic pyrophosphatase is needed to remove the pyrophosphate produced during

Table 26.3 Application of archaeal DNA-modifying enzymes as molecular biology reagents.

Enzymes	Strain	MW (kDa)	T_{opt} (°C)	pH_{opt}	Thermostability (half-life)	References
DNA polymerase (family B)	*Aeropyrum pernix* (pol I)	108			0.5 h at 85 °C	Cann et al., 1999
	Aeropyrum pernix (pol II)	88			0.5 h at 100 °C	Cann et al., 1999
	Cenarchaeum symbiosum	96	38	7.5	0.17 h at 46 °C	Schleper et al., 1997
	Pyrobaculum islandicum	90	70–80		>5 h at 90 °C	Kahler & Antranikian, 2000
Pfu pol	*Pyrococcus abyssi*	90	70–80	7.3	5 h at 100 °C	Dietrich et al., 2002
Deep Vent pol	*Pyrococcus furiosus*	90	72–78	9.0	4 h at 95 °C	Antranikian et al., 2005
Pwo pol	*Pyrococcus* sp. GB-D	90.6	70–80	8–9	8 h at 100 °C	Antranikian et al., 2005
	Pyrococcus woesei	90				Antranikian et al., 2005
	Sulfolobus acidocaldarius	100	65–75		0.25 h at 87 °C	Antranikian et al., 2005
	Sulfolobus solfataricus	101	75	7.5	0.1 h at 90 °C	Nastopoulos et al., 1998
KOD1	*Thermococcus aggregans*	90	70–80	6.8	1.2 h at 90 °C	Bohlke et al., 2000
Vent pol	*Thermococcus kodakaraensis* (struct.)	90	75	7.5	12 h at 95 °C	Antranikian et al., 2005
9°N-7 pol	*Thermococcus litoralis*	98.9	70–80	6.5	2 h at 100 °C	Antranikian et al., 2005
	Thermococcus sp. 9°N-7	90	70–80	8.8	6.7 h at 95 °C	Antranikian et al., 2005
Phosphatase	*Haloarcula marismortui*	160	25	8.5		Marhuenda-Egea et al., 2002
	Halobacterium salinarum					Marhuenda-Egea et al., 2002
	Pyrococcus abyssi	108	70	11	18 h at 100 °C	Zappa et al., 2001
	Sulfolobus acidocaldarius	80	56	6.5		Wakagi et al., 1992
	Thermoplasma acidophilum (struct.)					van der Horn et al., 1997
Ligase	*Acidianus ambivalens*					Eichler, 2001
	Pyrococcus furiosus		45–80		>1 h at 95 °C	Stratagene
	Sulfolobus shibatae	62	50–70	6–7		Lai et al., 2002
	Thermococcus fumicolans		65	7		Rolland et al., 2004
	Thermococcus kodakaraensis	52	100	8	0.15 h at 90 °C	Nakatani et al., 2002
ssDNA-binding proteins	*Archaeoglobus fulgidus*					Kowalczykowski et al., 2005
	Methanocaldococcus jannashii					
	Methanothermobacter thermoautothrophicum					

sequencing reactions. *Thermoplasma acidophilum* inorganic pyrophosphatase (TAP) is thermostable and effective for converting pyrophosphate to orthophosphate. The combination of Thermo Sequenase polymerase and TAP for cycle sequencing yields sequence data with uniform band intensities and allows the determination of longer, more accurate sequence reads (Van der Horn et al., 1997). Highly thermostable alkaline phosphatases, which dephosphorylate linear DNA fragments, were also identified in Archaea (Table 26.2). The alkaline phosphatase from *Pyrococcus abyssi* dephosphorylates linear DNA fragments with efficiencies of 94 and 84% regarding cohesive and blunt ends, respectively (Zappa et al., 2001).

DNA Ligation A variety of analytical methods are based on the use of thermostable ligases. Of considerable potential is the construction of sequencing primers by high temperature ligation of hexameric primers, the detection of trinucleotide repeats through repeat expansion detection, or DNA detection by circularization of oligonucleotides (Landegren et al., 2004). Several archaeal DNA ligases, displaying nick joining and blunt-end ligation activities using either ATP or NAD^+ as a cofactor, have been identified and characterized in detail (Table 26.3). Unlike bacterial enzymes, the ligase from *Acidianus ambivalens* is NAD^+-independent but ATP-dependent, in a similar way to the enzymes from bacteriophages, eukaryotes, and viruses. The DNA ligase from a hyperthermophilic archaeon *Thermococcus kodakaraensis* is also ATP-dependent (Nakatani et al., 2002). Sequence comparison with previously reported DNA ligases and the presence of conserved motifs indicated that the ligase is closely related to the ATP-dependent DNA ligase from *Methanobacterium thermoautotrophicum* H, a moderate thermophilic archaeon, along with putative DNA ligases from Euryarchaeota and Crenarchaeota. The optimum pH of the recombinant monomeric enzyme is 8.0, the optimum concentration of Mg^{2+} is 14–18 mM, and of K^+ is 10–30 mM. The protein does not display single-stranded DNA ligase activity. At enzyme concentrations of 200 nM, a significant DNA ligase activity is observed even at 100 °C. Surprisingly, the protein also displays a DNA ligase activity when NAD^+ is added as the cofactor (Nakatani et al., 2002). The ability of DNA ligases to use either ATP or NAD^+ as a cofactor appears to be specific to DNA ligases from Thermococcales. A DNA ligase from *Thermococcus fumicolans* displays nick joining and blunt-end ligation activity using either ATP or NAD^+ as a cofactor (Rolland et al., 2004). The presence of $MgCl_2$ (optimally at 2 mM) is required for the enzymatic activity. In contrast to that, the recombinant ATP-dependent ligase from the thermoacidophilic crenarchaeon *Sulfolobus shibatae* is more active in the presence of Mn^{2+} ions than in the presence of other divalent cations such as Mg^{2+} or Ca^{2+} (Lai et al., 2002). A splicing ligase activity is characterized in cell extracts of the halophile *Haloferax volcanii* (Oren, 2002).

Pyrococcus furiosus is thought to be highly resistant to γ-radiation and therefore may have a unique method for removing damaged DNA. A thermostable flap endonuclease from *P. furiosus* is described, which cleaves the replication fork-like structure endo/exonucleolytically (Matsui et al., 1999). The O6-methylguanine-DNA methyltransferase is the most common form of cellular defense against the biological effects of O6-methylguanine in DNA. The thermostable recombinant O6-methylguanine-DNA methyltransferase from *Thermococcus kodakaraensis* is functional *in vivo* and complements the mutant phenotype, making the cells resistant to the cytotoxic properties of the alkylating agent *N*-methyl-*N'*-nitro-*N*-nitrosoguanidine (Leclere et al., 1998).

A thermostable type I group B DNA topoisomerase has been isolated and purified from the hyperthermophilic methanogen *Methanopyrus kandleri* (Slesarev et al., 2002). The enzyme is active over a wide range of temperatures and salt concentrations and does not require magnesium or ATP for its activity, which makes manipulations of DNA more convenient and more efficient. Exploitation of the common features and the differences of topoisomerases will be important for the modeling of novel drugs and for an understanding of the action of cancer chemotherapeutic agents.

Alcohol dehydrogenases

Dehydrogenases are enzymes belonging to the class of oxidoreductases. Within this class, alcohol dehydrogenases (EC 1.1.1.1, also named keto-reductases) represent a biotechnologically important group of biocatalysts due to their ability to stereospecifically reduce prochiral carbonyl compounds. Alcohol dehydrogenases (ADHs) can be used efficiently in the synthesis of optically active alcohols, which are key building blocks in the synthesis of chirally pure pharmaceutical agents. From a practical point of view, alcohol dehydrogenases that use NADH as cofactor are of particular importance, because they represent an established method to regenerate NADH efficiently. By contrast, for NADP-dependent enzymes the cofactor-recycling systems that are available are much less efficient (Radianingtyas & Wright, 2003). The secondary specific alcohol dehydrogenase, which catalyzes the oxidation of secondary alcohols and, less readily, the reverse reaction (the reduction of ketones), has a promising future in biotechnology. Although ADHs are widely distributed among microorganisms, only a few of them have been characterized from archaea (Table 26.4). The ADH from *Sulfolobus solfataricus* requires NAD as cofactor and contains Zn ions (Radianingtyas & Wright, 2003). In contrast, the enzyme from *Thermococcus litoralis* lacks metal ions and catalyzes preferentially the oxidation of primary alcohols, using NADP as cofactor. The enzyme is thermostable, having half-lives of 15 minutes at 98 °C and 2 hours at 85 °C (Ma & Adams, 2001). The pyrococcal ADH is the most thermostable short-chain ADH (half-life of 150 hours at 80 °C) known to date (van der Oost et al., 2001). The NADP-dependent ADH from *Thermococcus hydrothermalis* oxidizes a series of primary aliphatic and aromatic alcohols preferentially from C_2 to C_8 but is also active toward methanol and glycerol and is stereospecific for monoterpenes (Antoine et al., 1999). The enzyme structure is pH-dependent, being a tetramer (45 kDa per subunit) at pH 10.5 (pH optimum for alcohol oxidation), and a dimer at pH 7.5 (pH optimum for aldehyde reduction).

Esterases

In the field of industrial biotechnology, esterases are gaining increasing attention because of their application in organic biosynthesis. In aqueous solution, esterases catalyze the hydrolytic cleavage of esters to form the constituent acid and alcohol, whereas in organic solutions, transesterification reaction is promoted. Both the reactants and the products of transesterification are usually highly soluble in the organic phase and the reactants may even form the organic phase themselves. Several archaeal esterases were successfully cloned and expressed in mesophilic hosts (Table 26.4). Esterases from *Aeropyrum pernix*, *Pyrobaculum calidifontis*, and *Sulfolobus tokodaii* exhibit high thermoactivity and thermostability and are also active in a mixture of a buffer and water-miscible organic solvents, such as acetonitrile and dimethyl sulfoxide (Antranikian et al., 2005). The optimum activity for ester cleavage of the esterase from *Sulfolobus tokodaii* strain 7 is at 70 °C and pH 7.5–8.0. From the kinetic analysis, p-nitrophenyl butyrate is a better substrate than caproate and caprylate (Suzuki et al., 2004). The *Pyrococcus furiosus* esterase is the most thermostable (a half-life of 50 minutes at 126 °C) and thermoactive (temperature optimum 100 °C) esterase known to date (Ikeda & Clark, 1998). A carboxylesterase from *Pyrobaculum calidifontis*, stable against heating and organic solvents, is active toward tertiary alcohol esters, a very rare feature among previously reported lipolytic enzymes (Hotta et al., 2002). A novel thermostable esterase from *Aeropyrum pernix* K1 with an optimal temperature at 90 °C exhibits additionally a phospholipase activity (Wang et al., 2004). The esterases from *Archaeoglobus fulgidus* and *Sulfolobus shibatae* are of interest for the dairy industry (Schiraldi et al., 2002). In our laboratory two thermoactive esterases from the thermoacidophile *Picrophilus torridus* have been recently characterized after successful expression in *E. coli* (Antranikian, unpublished observations). Both esterases are active at 50–60 °C and neutral pH. A gene coding the esterase from *Archaeoglobus fulgidus* was subjected to error-prone PCR in an effort to increase the low enantioselectivity toward the racemic mixture of an intermediate in the synthesis of the optically pure herbicide mecoprop (Manco et al., 2002). The esterase from *Sulfolobus solfataricus* P1 has been studied in detail for the chiral resolution of 2-arylpropionic esters (Sehgal & Kelly, 2003). Thus, the application of the esterase toward *R,S*-naproxen methyl ester yields highly optically pure *S*-naproxen (ee(p) > 90%) (Sehgal & Kelly, 2002, 2003). The enzyme is activated by DMSO to various extents, due to small changes in the enzyme structure resulting in an increase in its conformational flexibility. Thus, added cosolvents, which are useful for solubilization of hydrophobic substrates in water, also serve as activators in applications involving thermostable biocatalysts at suboptimal temperatures (Sehgal et al., 2002). Interestingly, experimental data on kinetic resolution of α-arylpropionic acid revealed

Table 26.4 Enzymes from Archaea with industrial relevance for fine chemistry.

Enzymes	Strain	MW (kDa)	T_{opt} (°C)	pH_{opt}	Thermostability (half-life)	Possible applications	References
Alcohol dehydrogenase	Aeropyrum pernix (struct.)		90			Stereoselective transformation of ketones to alcohols	Guy et al., 2003
	Methanoculleus thermophilicus		70				Radianingtyas & Wright, 2003
	Pyrococcus furiosus	55	90	7.5	150 h at 80 °C		van der Oost et al., 2001
	Sulfolobus solfataricus (struct.)	71	95	7.5			Radianingtyas & Wright, 2003
	Thermococcus hydrothermalis	80.5	80	7.5	0.25 h at 80 °C		Antoine et al., 1999
	Thermococcus litoralis	192	80	8.8	2 h at 85 °C		Ma & Adams, 2001
	Thermococcus sp. AN1	200	85	7			Li & Stevenson, 2001
	Thermococcus sp. ES-1						Ma & Adams, 2001
	Thermococcus zilligii	184	85	7	0.25 h at 80 °C		Radianingtyas & Wright, 2003
Aldolase	Methanocaldococcus jannaschii	271	80	7–8.5	24 h at 80 °C	Synthesis of chiral carbohydrates	Soderberg & Alver, 2004
	Sulfolobus solfataricus (struct.)	133	100		2.5 h at 100 °C		Buchanan et al., 1999
Amidase	Sulfolobus solfataricus (struct.)	56	95	7.5	25 h at 80 °C	Synthesis of fine chemicals	Antranikian et al., 2005
Aminoacylase	Pyrococcus furiosus	170	100	6.5		Pharmaceutical industry (production of stereoisomers)	Story et al., 2001
	Pyrococcus horikoshii OT3	95	95	7.5	>48 h at 90 °C		Ishikawa et al., 2001
	Thermococcus litoralis	172	85	8	1.7 h at 85 °C		Taylor et al., 2004
Cysteine synthase	Aeropyrum pernix	65	>60	7.5–8	>6 h at 100 °C	Synthesis of sulfur-organic compounds	Ishikawa & Mino, 2004
β-Galactosidase	Haloferax alicantei	156	22	7.2		Synthesis of oligosaccharides	Holmes & Dyall-Smith, 2000
	Pyrococcus woesei	61	90	4	3.5 h at 100 °C		Dabrowski et al., 2000

(Continued)

Table 26.4 Enzymes from Archaea with industrial relevance for fine chemistry.—cont'd

Enzymes	Strain	MW (kDa)	T_{opt} (°C)	pH_{opt}	Thermostability (half-life)	Possible applications	References
Esterase	*Aeropyrum pernix* (struct.)	18	90		1 h at 100 °C	Biotransformation in organic solvents	Wang et al., 2004
	Archaeoglobus fulgidus (struct.)	35.5	70	7			Manco et al., 2000
	Methanocaldococcus jannaschii (struct)		70	9.5			Chen et al., 2004
	Picrophilus torridus	21	70–80	6.5			Antranikian, unpublished
	Pyrobaculum calidifontis	35	90	7	2 h at 100 °C		Hotta et al., 2002
	Pyrococcus furiosus		100		2 h at 120 °C		Sehgal & Kelly, 2003
	Sulfolobus acidocaldarius	128	90				Antranikian et al., 2005
	Sulfolobus shibatae		90	6	0.33 h at 120 °C		Antranikian et al., 2005
	Sulfolobus solfataricus P1	33	100	5–6			Sehgal & Kelly, 2003
	Sulfolobus solfataricus P2	34	80	7.4	0.66 h at 80 °C		Kim & Lee, 2004
	Sulfolobus tokadaii		70	7.5–8			Suzuki et al., 2004
Lycopene β-cyclase	*Sulfolobus solfataricus*					Accumulation of beta-carotene	Hemmi, 2003
N-Methyltransferase	*Pyrococcus horikoshii*	23	90–100	8.5	>2 h at 100 °C	Synthesis of phosphatidylcholine for medicine and food	Matsui et al., 2002
Nitrilase	*Pyrococcus abyssi*	60	60–90	6–8	6 h at 90 °C	Production of mononitriles	Müller et al., 2006

(struct.): the protein is crystallized and the three-dimensional structure is determined.

that a carboxylesterase from *Sulfolobus solfataricus* P2 hydrolyzes the *R*-ester of racemic ketoprofen methylester with enantiomeric excess of 80% (Kim & Lee, 2004).

C–C bond-forming enzymes

Synthetic building blocks bearing hydroxylated chiral centers are important targets for biocatalysis. C–C bond-forming enzymes, such as aldolases and transketolases, have been investigated for new applications, and various strategies for the synthesis of sugars and related oxygenated compounds have been developed (Fessner & Helaine, 2001). The use of aldolases in stereoselective C–C bond-forming reactions is applicable for asymmetric synthesis of carbohydrates, leading to the development of new therapeutics and diagnostics. However, many aldolases display a narrow specificity, and often prefer phosphorylated substrates, which can limit the product range of chiral aldols. In contrast, an extremely thermostable aldolase (half-life 2.5 hours at 100 °C) from *Sulfolobus solfataricus*, actively expressed in *E. coli*, possesses a broad specificity for non-phosphorylated substrates and has a great potential for use in asymmetric aldol reactions (Table 26.4) (Demirjian et al., 2001). This aldolase represents a rare example of an enzyme that exhibits no diastereocontrol for the aldol condensation of its natural substrates pyruvate and glyceraldehyde. Recently, it was demonstrated that the stereoselectivity of the enzyme has been induced by employing the substrate engineering procedure (Lamble et al., 2005). In another application thermostable pentose phosphate enzymes, e.g. the transaldolases from *Methanocaldococcus jannaschii*, could greatly increase the efficiency of an enzymatic hydrogen production that employs a novel archaeal hydrogenase (Soderberg & Alver, 2004).

Nitrile-degrading enzymes

Nitrile-degrading enzymes are of considerable importance in industrial biotransformations, and to date several processes have been developed for chemical and pharmaceutical industries for the production of optically pure compounds, drugs, acrylic, and hydroxamic acids (Antranikian et al., 2005). Nitrile-degrading enzymes also play a significant role in the protection of the environment due to their ability to eliminate highly toxic nitriles. Thermostable amidases and nitrilases are gaining more attention, especially in enzymatic processes in mixtures of organic solvent or in the formation of highly pure products with a concomitant reduction of wastes (Alcantara, 2000). A number of bacterial amidases and nitrilases have been purified and characterized. Very little, however, is known about the enzymes that are active at high temperatures. Amidases (EC 3.5.1.4) catalyze the conversion of amides to the corresponding carboxylic acids and ammonia. The only amidase derived from Archaea is that from the thermoacidophile *Sulfolobus solfataricus* (Table 26.4). This enzyme is *S*-stereoselective (*S*-enantioselectivity is common for amidases) with a broad substrate spectrum and is active at 95 °C (Antranikian et al., 2005). Nitrilases (EC 3.5.5.1) are thiol enzymes that convert nitriles directly to the corresponding carboxylic acids, with release of ammonia. The only thermoactive nitrilases described so far were isolated from the Bacteria *Acidovorax facilis* 72W and *Bacillus pallidus* Dac521. Recently, the first archaeal nitrilase from the hyperthermophile *Pyrococcus abyssi*, regiospecific toward aliphatic dinitriles, was cloned and characterized in our laboratory. The enzyme is highly thermostable, having a half-life at 90 °C of 6 hours (Müller et al., 2006).

Aminoacylases

Due to their chiral specificity in the synthesis of acylated amino acids, aminoacylases (EC 3.5.1.14) are attractive candidates for application in fine chemistry (Story et al., 2001). The L-aminoacylase from *Thermococcus litoralis* accepts a wide range of amino acid side chains and *N*-protecting groups and was recently commercialized (Toogood et al., 2002). The application of the thermostable enzyme reduces the process time, simplifies filtration procedure, improves substrate solubility, and increases the enantiomeric excess to 99%. In contrast to the chemical process, the reaction completes overnight at 70 °C, which avoids boiling in 20 equivalent volumes of 6 M HCl for 2 days (Taylor et al., 2004). Two thermostable zinc-containing aminoacylases were also characterized from *Pyrococcus* species (Ishikawa et al., 2001; Story et al., 2001; Table 26.4).

Other archaeal enzymes of practical use

Thermostable β-galactosidase is potentially useful for whey utilization and for the preparation of low-lactose milk and other dairy products, or it can be used as a catalyst in the synthesis of galactooligosaccharides, using lactose as substrate and a nucleophile. A β-galactosidase from the halophilic archaeaon *Haloferax lucentensis* was cloned and expressed in *Haloferax volcanii*, a widely used strain that lacks detectable β-galactosidase activity (Holmes & Dyall-Smith, 2000). The extremely halotolerant β-galactosidase is optimally active at 4 M NaCl, cleaves several different β-galactoside substrates, and does not exhibit β-glucosidase, β-arabinosidase, or β-xylosidase activities. A β-galactosidase from *Pyrococcus woesei* was also cloned and characterized (Dabrowski et al., 2000).

Carotenoids act as anti-oxidant agents, leading to their use as food additives and drugs. The majority of carotenoids are synthesized from lycopene. The β-carotene, the precursor of vitamin A, is biosynthesized directly from lycopene by β-cyclization at both termini, and the reaction is catalyzed by lycopene β-cyclase. Recently, lycopene β-cyclase was predicted in the carotenogenic gene cluster in the genome of the thermoacidophilic archaeon *Sulfolobus solfataricus* (Hemmi, 2003). The recombinant expression of the gene in *E. coli* resulted in the accumulation of lycopene β-carotene in the cells.

Sulfur-containing organic compounds have been synthesized mainly chemically. Due to the side reactions, the chemical synthesis of those molecules results in the unavoidable production of impurities in the product and environmental pollution by the formation of by-products such as sulfur oxides. In order to overcome these problems, a method for the synthesis of sulfur-containing organic compounds using O-acetylserine sulfhydrolase has been proposed (Ishikawa & Mino, 2004). A recombinant cysteine synthase from *Aeropyrum pernix* is highly stable within pH 6–10 and resistant to organic solvents. Due to its high heat resistance, the enzyme can act on highly concentrated substrate solutions compared to mesophilic thermolabile cysteine synthases.

Phosphatidylethanolamine N-methyltransferase plays a key role in the synthesis of phosphatidylcholine, a main component of liposomal membrane, which is present in various foods as digestible surfactant. It plays an important role in medicine as a component of microcapsules for drugs. A phosphatidylethanolamine N-methyltransferase from *Pyrococcus horikoshii* was cloned and expressed in *E. coli*. The enzyme is thermostable and is active in organic solvents. This opens the possibility of developing a new process for the synthesis of polar lipids with high optical purity (Matsui et al., 2002).

Whole-cell biocatalysis

Biomining

Industrial mineral processing has been developed in several countries, such as South Africa, Brazil, and Australia. Iron- and sulfur-oxidizing microorganisms are used to release occluded gold from mineral sulfides. Most industrial plants for biooxidation of gold-bearing concentrates have been operated at 40 °C with mixed cultures of mesophilic Bacteria of the genera *Thiobacillus* or *Leptospirillum*. In subsequent studies dissimilatory iron-reducing Archaea *Pyrobaculum islandicum* and *Pyrococcus furiosus* were shown to reduce gold chloride to insoluble gold (Lloyd, 2003). The potential of thermophilic sulfide-oxidizing Archaea in copper extraction has attracted interest due to the efficient extraction of metals from sulfide ores that are recalcitrant to dissolution (Schiraldi et al., 2002). The acidophilic Archaea *Sulfolobus metallicus* and *Metallosphaera sedula* tolerate up to 4% copper and have been exploited for mineral biomining (Norris et al., 2000). The efficiency of copper extraction from chalcopyrite by thermoacidophilic Archaea was influenced by the characteristics of mineral concentrates. Between 40 and 60% copper extraction was achieved in primary reactors and more than 90% extraction in secondary reactors with overall residence times of about 6 days (Norris et al., 2000).

The handling and recycling of spent tyres is a significant and worldwide problem. The reuse of rubber material is preferable from an economic and environmental point of view. The anaerobic sulfate-reducing thermophilic archaeon *Pyrococcus furiosus* was investigated for its capacity to desulfurize rubber. The tyre rubber treated with *P. furiosus* for 10 days was subsequently vulcanized with virgin rubber material (15% wt/wt). This results in the desulfurization of ground rubber and leads to a product with good mechanical properties (Bredberg et al., 2001). The thermoacidophilic archaeon *Sulfolobus acidocaldarius* has also been also tested for desulfurization of rubber material (Bredberg et al., 2001).

Decontamination and hydrogen production

Microorganisms, growing in the presence of toxic chemicals, heavy metals, halogenated solvents, and radionuclides, can be used to detoxify those compounds during treatment of waste. The archaeal strains *Thermococcus marinus* and *Thermococcus radiotolerans* are resistant to high levels of ionizing and ultraviolet radiation (Jolivet et al., 2004). The biological treatment of synthetic saline wastewater was investigated. High removal efficiencies were obtained at salt concentrations above 4% using immobilized cells of *Halobacterium halobium*. Extremely halophilic Archaea, such as *Haloferax mediterranei*, were found to utilize crude oil even at high salinities and therefore could be used in bioremediation of polluted sites. The strain *Halobacterium* sp. degrades *n*-alkanes with a C_{10}–C_{30} composition in the presence of 30% NaCl. Halophilic archaea belonging to genera *Haloarcula*, *Halobacterium*, and *Haloferax* degrade halogenated hydrocarbons, such as trichlorophenols or the insecticides lindane and DDT. It has been demonstrated that the hyperthermophilic anaerobic archaeon *Ferroglobus placidus* is capable of oxidizing aromatic compounds with the reduction of Fe^{3+}. Such microorganisms can be useful when heat treatment is employed to aid in the extraction of organic contaminants that are trapped in sediments. Textile wastewater is often of high salinity and therefore problematic for conventional biological treatment. The use of halophilic microorganisms for the biodegradation of the segregated dye bath has been reported (Oren, 2002).

There is an increasing interest in the utilization of renewable sources to satisfy exponentially growing energy needs. Research on biological hydrogen production became attractive due to the possible use of biohydrogen as a clean energy carrier and raw material. The production of hydrogen through photobiological or heterotrophic fermentation routes depends on a supply of organic substrates and could therefore be ideally suited for coupling energy production with treatment of organic wastes. A two-stage fermentation system was constructed for the production of biohydrogen from keratin-rich biowaste (Balint et al., 2005). First, the bacterial strain *Bacillus licheniformis* KK1 was employed to convert keratin-containing waste into a fermentation product that is rich in amino acids and peptides. In the next stage the thermophilic anaerobic archaeon *Thermococcus litoralis* was fermented on the hydrolysate and hydrogen was produced. Archaeal hydrogenases have also been the target of intensive research. A cytosolic NiFe-hydrogenase from the hyperthermophilic archaeon *Thermococcus kodakaraensis* is optimally active at 90 °C for hydrogen production, with methyl viologen as the electron carrier (Kanai et al., 2003).

Archaeal membrane lipids, proteins, and polymers

Lipids

Liposomes are artificial spherical closed vesicles consisting of one or more lipid bilayers. Liposomes made from ether phospholipids have been studied extensively over the past 30 years as artificial membrane models with remarkable thermostability and tightness against solute leakage. Considerable interest has been generated for applications of liposomes in medicine, including their use as diagnostic agents, as carrier vehicles in vaccine formulations, or as delivery systems for drugs, genes, or cancer imaging agents (Sprott et al., 2003). In general, archaeosomes (liposomes from Archaea) demonstrate higher stability to oxidative stress, high temperature, alkaline pH, attack by phospholipases, bile salts, and serum proteins. Some archaeosome formulations can be sterilized by autoclaving without problems of fusion or aggregation of the vesicles. The uptake of archaeosomes by phagocytic cells can be up to 50-fold higher than that of conventional liposomes (Patel & Sprott, 1999). Ether-linked lipids from halophilic Archaea have high chemical stability and resistance against esterases and thus a higher survival rate than liposomes based on fatty acid derivatives (Margesin & Schinner, 2001). Novel patented ether lipids, prepared by pressure extrusion from the halophile *Halobacterium cutirubrum*, were resistant to attack by phospholipases and could be stored for more than 60 days. Cyclic and acyclic dibiphytanylglycerol tetraether lipids were identified in non-thermophilic crenarchaeotes (DeLong et al., 1998). The immune stimulating activity of the lipid vesicles, prepared from the archaeon *Haloferax volcanii*, was investigated, and an increase in immune responses was observed. The unique ability of the archaeosomes, consisting of sulfoglycolipids, phosphoglycerols, and cardiolipins, to maintain antigen-specific T cell immunity may be attributed to a property of the archaeal 2,3-diphytanylglycerol lipid core (Sprott et al., 2003). Due to their bipolar tetraether structure, archaeal lipids have been also proposed as monomers for bioelectronics (De Rosa et al., 1994).

Crystalline cell surface layers (S-layers) that are composed of protein and glycoprotein subunits are one of the most commonly observed cell envelope structures of Bacteria and Archaea. S-layers could be produced in large amounts by continuous cultivation of S-layer-carrying microorganisms and used as isoporous ultrafiltration membranes or as matrices for immobilization of biologically active macromolecules such as enzymes, ligands, or mono- and polyclonal antibodies (Sleytr et al., 1997). S-layers have been shown to be excellent patterning structures in molecular nanotechnology due to their high molecular order, high binding capacity, and ability to recrystallize with perfect uniformity on solid surfaces, at the water/air interface, or on lipid films. The two-dimensionally organized S-layers of *Sulfolobus acidocaldarius* are suggested to be of practical use as biomimetic templates for material deposition and fabrication of advanced materials (Sleytr et al., 1997).

Proteins and peptides

Production of antibiotic peptides and proteins is a near-universal feature of living organisms regardless of phylogenetic classification. Antimicrobial agents from Bacteria and Eukarya have been studied for more than 50 years. However, archaeal strains are just at the beginning of investigation for the production of peptide antibiotics. A variety of halocins have been detected in halophilic Archaea. These antimicrobial agents are diverse in size, consisting of proteins as large as 35 kDa and peptide "microhalocins" as small as 3.6 kDa (O'Connor & Shand, 2002). Microhalocins with an unclear mechanism of action are hydrophobic and robust, withstanding heat, desalting, and exposure to neutral residues, and are not cationic. The microhalocins S8 and R1 lack the biochemical and structural properties demonstrated by other antibiotics, suggesting that their mechanisms of action should be novel. The halocin H7 has been suggested for reducing injury during organ transplantation. Archaeocins are also produced by a thermoacidophilic *Sulfolobus* strain. The 20 kDa protein antibiotics are not excreted and are associated with small particles apparently derived from the cell's S-layer (O'Connor & Shand, 2002).

Chaperones and chaperonins of extremophilic Archaea are useful agents in protein refolding, stabilization, and solubilization of recombinant proteins (Ideno et al., 2004; Maruyama et al., 2004). Peptidyl prolyl *cis-trans* isomerase (PPIase) is involved in many processes, such as the regeneration of denatured protein, stabilization of proteins, production of recombinant protein, and development of novel immunosuppressant and physiologically active substances. A novel cyclophilin type PPIase derived from *Halobacterium cutirubrum* was characterized (Margesin & Schinner, 2001).

Extremely halophilic Archaea often contain membrane-bound retinal pigments, bacteriorhodopsin and halorhodopsin, that enable them to use light energy directly for bioenergetic processes by the generation of proton and chloride gradients. Bacteriorhodopsin is a 26.5 kDa protein with seven helical protein segments in its transmembrane domain. The excellent thermodynamic and photochemical stability of bacteriorhodopsin has led to many technical applications based on its protonmotive, photoelectric, and photochemical properties. The applications comprise holography, spatial light modulators, artificial retina, neural network optical computing, and volumetric and associative optical memories. The optical properties of bacteriorhodopsin could be exploited to manufacture electronic ink for laptop displays, which will be an important contribution to the problem of battery lifetime in portable computing. Another application of bacteriorhodopsin is the renewal of biochemical energy by conversion of ADP to ATP. Such a solar-driven recycling system could be of interest for biotechnological processes that require large amounts of expensive ATP. A patented ATP-synthesizing device, useful for bioelements, has been obtained by using of bacteriorhodopsin and ATP synthase (Margesin & Schinner, 2001). Bacteriorhodopsin is commercially offered in the form of purple membrane patches, isolated from *Halobacterium salinarum* S9.

In recent years carotenoids have gained importance in the nutraceutical field. These pigments, including canthaxanthin, have been shown to possess physiological functions in the prevention of cancer and heart diseases, in enhancing *in vitro* antibody production, and as precursors for vitamins. Canthaxanthin was reported to have greater antioxidant activity than its non-oxygenated analog; it has been used as a food and feed additive, in cosmetics, and in pharmaceuticals. Recently, the production of this pigment by the archaeon *Haloferax alexandrinus* was investigated. The highest production of the pigment (2.19 µg/l) was obtained during batch fermentation of the strain on a medium with 25% salinity. Other carotenoids, such as 3-hydroxy echinenone and *trans*-astaxanthin, are also produced by Archaea (Asker, 2002).

Further archaeal properties could be used to improve fermentation processes. For the harvesting of cells after completion of fermentation, buoyancy of cells is desired. The genes coding for the synthesis of gas vesicles in *Halobacterium halobium* were cloned. The recombinant expression of these genes provided the cells with the property of floating and the biomass could be separated by skimming or decanting (Oren, 2002).

Biopolymers

Polyhydroxyalkanoates are microbial storage compounds with properties comparable to those of polyethylene and polypropylene. Such biodegradable plastics could replace oil-derived thermoplastics. The archaeon *Haloferax mediterranei* accumulates poly(β-hydroxybutyric acid) up to 60% of cell dry weight. Production can be enhanced to 6 g/l using phosphate limitation and starch as carbon source (Hezayen et al., 2000). The polymer can be easily recovered using cell lysis caused by an exposure of the cultures to low salt concentrations. The maximum production of poly(β-hydroxybutyric acid) (up to 53% of cell dry weight) by another extremohalophilic archaeon was reached after 11 days of fermentation on *n*-butyric acid and sodium acetate as carbon sources (Hezayen et al., 2000). In contrast to *Haloferax mediterranei*, phosphate does not affect the polyester production by the strain. The accumulated polyester was recovered at 87% using chloroform extraction. The exopolymer poly(γ-D-glutamic acid) can be used as a biodegradable thickener, humectant sustained-release material, or drug carrier in the food or pharmaceutical industries. The extreme halophilic archaeon *Natrialba aegyptiaca* starts to produce the polymer at 20% NaCl and the maximum production is reached at NaCl saturation. After 90 hours of growth in a corrosion-resistant bioreactor, 470 mg/l of the polymer was produced (Hezayen et al., 2000, 2001).

Microbial exopolysaccharides are high molecular mass polymers composed mainly of carbohydrates excreted by Bacteria and Archaea. Among the family Halobacteriaceae several species of the genus *Haloferax* have been described as producers of extracellular polysaccharides. The high viscosity at low concentrations, excellent rheological properties, and remarkable tolerance to high pH, temperature, and salinity make these polymers suitable as emulsifying agents and mobility controllers in microbial enhanced oil recovery. Furthermore, several biosurfactant producers such as *Methanothermobacter thermoautotrophicum* produce bioemulsifiers active over a wide range of pH (pH 5–10) and at high salinities (up to 20%) (Oren, 2002).

Compatible solutes

Accumulation of osmotically active substances, so-called compatible solutes, by uptake or *de novo* synthesis, enables microorganisms to reduce the difference between osmotic potentials of the cell cytoplasm and the extracellular environment. Those compounds are highly water-soluble sugars, alcohols, amino acids, or their derivatives. They gained increasing attention in biotechnology due to their action as stabilizers of enzymes, DNA, membranes, tissues, and stress-protecting agents (Borges et al., 2002). Additionally, compatible solutes support the high-yield periplasmic production of functional active recombinant proteins in different expression systems (Barth et al., 2000). Di-*myo*-inositol-1,1'-phosphate is the most widespread solute of hyperthermophilic Archaea and was never detected in a mesophile. This thermoprotective compound was found in a variety of archaeal strains (Table 26.5; Santos & da Costa, 2002). In most of these organisms, an increase of the solute concentrations is observed at growth temperatures above the optimum, reaching 20-fold in the case of *Pyrococcus furiosus* grown at 101 °C. In contrast, the concentration of

Table 26.5 Other applications of extremophilic Archaea.

Product	Strain	Application	Ref.
Bacteriorhodopsin	*Halobacterium salinarum*	Holography, color-sensors, neural network optical computing, spatial light modulators	Margesin & Schinner, 2001
Carotenoids (canthaxanthin)	*Haloferax alexandrinus*	Food and feed additives, cosmetics	Asker, 2002
Chaperones, chaperonins, maltodextrin-binding proteins, peptidyl-prolyl *cis-trans* isomerases	*Halobacterium cutirubrum* *Haloferax volcanii* *Methanothermococcus* sp. *Pyrococcus* spp. *Sulfolobus shibatae* *Thermococcus* spp.	Stabilization and solubilization of recombinant proteins	Iida et al., 2000; Fox et al., 2003; Laksanalamai et al., 2003; Lund et al., 2003; Suzuki et al., 2003; Ideno et al., 2004; Maruyama et al., 2004
Compatible solutes (mannosylglycerate, mannosylglyceramide, diglycerol phosphate, di-*myo*-inositol-phosphate, N-acetyl-β-lysine, trehalose, 2-sulfotrehalose, cyclic-2,3-bisphosphoglycerate)	*Aeropyrum pernix* *Archaeoglobus* spp. *Methanococcus igneus* *Methanopyrus kandleri* *Methanosarcina thermophila* *Methanothermus fervidus* *Natronobacterium* spp. *Natronococcus* sp. *Pyrobaculum aerophilum* *Pyrococcus* spp. *Pyrodictium occultum* *Pyrolobus fumarii* *Thermococcus* spp.	Cosmetics, biomolecules and tissue stabilizers, molecular biology	Desmarais et al., 1997; Barth et al., 2000; Borges et al., 2002; Santos & da Costa, 2002; Pfluger et al., 2003
Cytochrome P450	*Sulfolobus solfataricus*	Selective regio- and stereospecific hydroxylations in chemical synthesis	Vieille & Zeikus, 2001
DNA topoisomerase type I-group B	*Methanopyrus kandleri*	Modeling of novel drugs	Margesin & Schinner, 2001
Expression systems (vector/host)	*Haloarcula hispanica* *Halobacterium salinarum* *Haloferax volcanii* *Methanococcus* spp. *Sulfolobus solfataricus*	Production of recombinant proteins	Holmes et al., 1994; Aagaard et al., 1996; Cannio et al., 2001; Jonuscheit et al., 2003; Zhou et al., 2004; Kaczowka et al., 2005

	Organism	Application	References
Ni-Fe-hydrogenase	*Thermococcus kodakaraensis*	H_2 production	Kanai et al., 2003
Peptides, proteins Sulfolobicin	*Sulfolobus islandicus*		
Halocins	*Halobacterium salinarum*	Antibiotics	O'Connor & Shand, 2002
	Haloferax mediterranei		
	Haloferax gibbonsii		
	Strain S8a (Archaea)		
Polymers			
Exopolysaccharides	*Haloferax* spp., *Halobacterium* spp.	Emulsifiers	Margesin & Schinner, 2001
Poly(γ-D-glutamic acid)	*Natrialba aegyptiaca*	Food and pharmaceuticals	Hezayen et al., 2001
Poly(β-hydroxy butyric acid)	*Haloferax mediterranei*	Bioplastics	Hezayen et al., 2000
	Strain 56 (Archaea)		
S-layer proteins, lipids, liposomes	*Halobacterium* spp.	Vaccine development, diagnostics, biomimetics, drugs, nanotechnology	Sleytr et al., 1997; DeLong et al., 1998; Patel & Sprott, 1999; Eichler, 2003; Sprott et al., 2003; Upreti et al., 2003
	Haloferax volcanii		
	Haloarcula japonica		
	Methanobrevibacter smithii		
	Methanococcus spp.		
	Methanothermus spp.		
	Natronobacterium sp.		
	Staphylothermus marinus		
	Sulfolobus solfataricus		
Whole cell biocatalysis	*Palaeococcus ferrophilus*	Formation of gels and starch granules	Abe & Horikoshi, 2001; Gomes & Steiner, 2004
	Thermococcus barophilus		
	Methanogenic Archaea	Methane production, low temperature waste treatment	Scherer et al., 2000; Schiraldi et al., 2002
	Haloarcula spp.	Detoxification of halogenated organic compounds and toxic metals, nuclear waste treatment	Norris et al., 2000; Jolivet et al., 2003, 2004
	Halobacterium spp.		
	Haloferax spp.		
	Thermococcus gammatolerans		
	Thermococcus marinus		
	Thermococcus radiotolerans		
	Sulfolobus metallicus		
	Pyrococcus furiosus	Rubber recycling	Bredberg et al., 2001

mannosylglycerate, detected in the euryarchaeotes of the genera *Archaeoglobus*, *Pyrococcus*, *Thermococcus*, and *Methanothermus* and in the crenarchaeote *Aeropyrum pernix*, increases concomitantly with the salinity of the medium and serves therefore as a compatible solute under salt stress. Mannosylglycerate has also been observed to have a profound effect on thermoprotection and protection against desiccation on enzymes of mesophilic, thermophilic, and hyperthermophilic origin. The biosynthetic routes for the synthesis of mannosylglycerate in the archaeon *Pyrococcus horikoshii* and di-*myo*-inositol-1,1′-phosphate in *Pyrococcus woesei* and *Methanococcus igneus* have been investigated (Chen et al., 1998; Scholz et al., 1998; Empadinhas et al., 2001). The hyperthermophilic archaeon *Archaeoglobus fulgidus* accumulates a very rare compound diglycerol phosphate under salt and temperature stress. This solute demonstrated a considerable stabilizing effect against heat inactivation of various dehydrogenases and a strong protective effect on bacterial rubredoxins (with a fourfold increase in the half-lives) (Lamosa et al., 2000). A compatible solute, cyclic 2,3-bisphosphoglycerate, has been detected only in methanogenic Archaea such as *Methanopyrus kandleri*. The thermoprotective role of this solute was proven by *in vitro* studies showing that the solute protects selected enzymes from *M. kandleri* against thermal denaturation (Santos & da Costa, 2001). Another compatible solute, restricted to methanogenic Archaea, is a derivative of β-amino acid (N^ε-acetyl-β-lysine), which is synthesized in response to increasing osmotic stress in both marine and non-marine species of methanogenes. The first genes involved in the biosynthesis of the solute have been identified (Pfluger et al., 2003). A novel compound, 2-sulfotrehalose, was found to be the major organic solute accumulated by halophilic Archaea from the genera *Natronococcus* and *Natronobacterium*, which grow under high salt and high pH conditions (Desmarais et al., 1997).

Advantages and limitations of using extremophilic Archaea

Bioprocesses under extreme conditions

Running processes under extreme conditions such as elevated temperature has many advantages, including increased solubility of polymers, decreasing viscosity, increased bioavailability, and a decreased risk of contamination. Stable enzymes derived from extremophilic Archaea are active in organic solvents and detergents, and they are more resistant to proteolytic attack. The efficient separation of recombinant extremozymes from mesophilic production hosts can be achieved by simple treatment such as heat denaturation, extraction with organic solvents, or increased salinity. Consequently, the delivery of sufficient amounts of enzymes for industrial trials can be obtained. The availability of extremozymes capable of catalysis at high pressures will also offer a novel biotechnological alternative to currently running processes, especially in the food industry. During the processing and sterilization of food materials, high pressure can be used to induce the formation of gels and starch granules, the denaturation/coagulation of proteins, or the transition of lipid phases. The use of high pressure leads to better flavor and color preservation.

On the other hand, a number of limitations have so far prevented the broad application of extremophiles. These limitations include difficulties associated with large-scale cultivation of extremophiles, non-efficient systems for the overexpression of archaeal genes and unknown factors that confer enzyme stability under extremes of temperature, pH, and pressure.

Enzyme stabilization

Thermoactive enzymes have been investigated intensively and used as model systems to understand structure–function relationships. Three-dimensional structures of thermostable enzymes have been resolved and compared with their mesophilic counterparts. So far, no single stabilizing factor has been found that seems to be responsible for thermostability of enzymes from extremophiles. Several mechanisms are thought to confer the stability of proteins from extremophiles, such as an increased hydrophobicity and surface charge, an increased number of ion pairs and hydrogen bonds, decreased flexibility and decreased size of surface loops, reduced ratio of surface area to volume, less exposed thermolabile amino acids, and truncated amino and carboxyl termini. Other factors affecting the stability of extremozymes include the presence of salts, high protein concentrations, stabilizers such as sorbitol, thermamine, or cyclic polyphosphates, and ultrahigh pressure (up to 50 MPa) (Bruins et al., 2001). As described for other microorganisms, chaperones, which assist protein folding, are also found in thermophiles (Demirjian et al., 2001). However, stabilization mechanisms are still not clearly elucidated and further efforts are needed to understand structure–function relationships.

Heterologous gene expression

Molecular cloning of the archaeal genes and their expression in heterologous hosts circumvent the problem of insufficient expression in natural hosts. A variety of archaeal enzymes have been cloned and successfully expressed in mesophilic hosts such as *Escherichia coli*, *Bacillus subtilis*, and yeasts. In *E. coli* the overexpression of proteins was optimized using cotransformation of the host cells with a plasmid encoding tRNA synthases for low-frequency codons. Recombinant enzymes could be also obtained by cloning and expressing a synthetic gene with a codon usage optimized for the corresponding host. To overcome the limitations of different codon usage, the production of recombinant proteins can be increased using extremophilic microorganisms as hosts for autologous gene expression. Therefore, expression systems using extremophilic Archaea as production hosts have to be developed, including hyperthermophiles, thermoacidophiles, and halophiles. Expression systems in *Haloferax*, *Methanococcus*, and *Sulfolobus* species have been constructed (Gardner & Whitman, 1999).

Conclusions

Owing to their properties, such as activity over a wide temperature and pH range, substrate specificity, stability in organic solvents, diverse substrate range, and enantioselectivity, extremophiles and their enzymes will represent the choice for countless future applications in industry. The growing demand for more robust biocatalysts has shifted the trend toward improving the properties of existing proteins for established industrial processes and producing new enzymes tailor-made for entirely new areas of application. The new technologies, such as genomics, metanogenomics, gene shuffling, and mutagenesis, provide valuable tools for improving or adapting enzyme properties to the desired requirements. However, the success of these techniques demands the production of recombinant enzymes on a large scale, allowing experimental trials and application tests. Thus, modern methods of genetic engineering, combined with an increasing knowledge of the structure and function, and process engineering, will allow further adaptation to industrial needs, the exploration of novel applications, and protection of the environment.

Wolfram Zillig

Wolfram Zillig was a lively and enthusiastic researcher who vigorously promoted the concept of Archaea with his sharp intellect and creative experimentation. He greatly valued the interaction between molecular biology and the study of microbial evolution and approached important questions in these fields with an intellectual and physical energy that left a lasting impression on those of us who he trained and worked with.

Wolfram Zillig was a brilliant scientist who made a major impact on Archaea research for more than two decades with many outstanding contributions and he was highly active at the lab bench until the age of 78. The hyperthermophilic Archaea found in hot springs attracted him especially, as did their novel, and amazingly diverse, viruses. His lively, original, and entertaining discussions were a daily occurrence in the laboratory. This, combined with his charismatic and cheerful character, made him a wonderful mentor and friend. He will stay in the memory of his friends and co-workers for many years.

The picture was taken by F. Pfeifer in the lab at the Max-Planck-Institut für Biochemie. Wolfram Zillig is inspecting *Pelomyxa*, a methanogen-containing lower eukaryote that he collected from the Löschteich at the MPI in Martinsried and tried to grow in the lab.

References

Aagaard, C., Dalgaard, J. Z. & Garrett, R. A. (1995) Intercellular mobility and homing of an archaeal rDNA intron confers a selective advantage over intron-cells of *Sulfolobus acidocaldarius*. *Proc Natl Acad Sci USA* **92**, 12285–9.

Aagaard, C., Leviev, I., Aravalli, R. N., Forterre, P., Prieur, D. & Garrett, R. A. (1996) General vectors for archaeal hyperthermophiles: strategies based on a mobile intron and a plasmid. *FEMS Microbiol Rev* **18**, 93–104.

Abe, F. & Horikoshi, K. (2001) The biotechnological potential of piezophiles. *Trends Biotechnol* **19**, 102–8.

Achenbach-Richter, L., Gupta, R., Zillig, W. & Woese, C. R. (1988) Rooting the archaebacterial tree: the pivotal role of *Thermococcus celer* in archaebacterial evolution. *Syst Appl Microbiol* **10**, 231–40.

Acton, T. B., Gunsalus, K. C., Xiao, R. et al. (2005) Robotic cloning and protein production platform of the northeast structural genomics consortium. *Methods Enzymol* **394**, 210–43.

Adams, M. W., Dailey, H. A., DeLucas, L. J. et al. (2003) The Southeast Collaboratory for Structural Genomics: a high-throughput gene to structure factory. *Acc Chem Res* **36**, 191–8.

Adams, M. W., Holden, J. F., Menon, A. L. et al. (2001) Key role for sulfur in peptide metabolism and in regulation of three hydrogenases in the hyperthermophilic archaeon *Pyrococcus furiosus*. *J Bacteriol* **183**, 716–24.

Aeckersberg, F., Bak, F. & Widdel, F. (1991) Anaerobic oxidation of saturated hydrocarbons to CO_2 by a new type of sulfate-reducing bacterium. *Arch Microbiol* **156**, 5–14.

Ahmed, H., Ettema, T. J., Tjaden, B. et al. (2005) The semi-phosphorylative Entner–Doudoroff pathway in hyperthermophilic archaea: a re-evaluation. *Biochem J* **390**, 529–40.

Ahn, D.-G., Kim, S.-I., Rhee, J.-K., Kim, K. P. & Oh, J.-W. (2004) TTSV1, a novel globuloviridae family virus isolated from the hyperthermophilic crenarchaeote *Thermoproteus tenax*. GenBank AY722806.

Ahn, Y., Krzycki, J. A. & Floss, H. G. (1991) Steric course of the reduction of ethyl coenzyme M to ethane catalyzed by methyl coenzyme M reductase from *Methanosarcina barkeri*. *J Am Chem Soc* **113**, 4700–1.

Aittaleb, M., Rashid, R., Chen, Q., Palmer, J. R., Daniels, C. J. & Li, H. (2003) Structure and function of archaeal box C/D sRNP core proteins. *Nat Struct Biol* **10**, 256–63.

Akhmanova, A. S., Kagramanova, V. K. & Mankin, A. S. (1993) Heterogeneity of small plasmids from halophilic archaea. *J Bacteriol* **175**, 1081–6.

Albers, S.-V., Jonuscheit, M., Kletzin, A., Driessen, A. J. & Schleper, C. (2006) Production of recombinant and tagged proteins in the hyperthermophilic archaeon *Sulfolobus solfataricus*. *Appl Environ Microbiol* **72**, 102–11.

Alcantara, A.-R., Sanchez-Montero, J. M. & Sinisterra, J. V. (2000) Chemoenzymatic preparation of enantiomerically pure S(+)-2-arylpropionic acids with anti-inflammatory activity. *In:* R. Patel (ed.) *Stereoselective biocatalysis*. Marcel Dekker, New York and Basel, pp. 659–702.

Alilat, M., Sivolob, A., Révet, B. & Prunell, A. (1999) Nucleosome dynamics. Protein and DNA contributions in the chiral transition of the tetrasome, the histone (H3-H4)2 tetramer-DNA particle. *J Mol Biol* **291**, 815–41.

Allen, G. S., Zavialov, A., Gursky, R., Ehrenberg, M. & Frank, J. (2005) The cryo-EM structure of a translation initiation complex from *Escherichia coli*. *Cell* **121**, 703–12.

Allers, T. & Mevarech, M. (2005) Archaeal genetics – the third way. *Nat Rev Genet* **6**, 58–73.

Allers, T., Ngo, H., Mevarech, M. & Lloyd, R. G. (2004) Development of additional selectable markers for the halophilic archaeon *Haloferax volcanii* based on the *leuB* and *trpA* genes. *Appl Environ Microbiol* **70**, 943–53.

Allmang, C., Carbon, P. & Krol, A. (2002) The SBP2 and 15.5 kD/Snu13p proteins share the same RNA binding domain: identification of SBP2 amino acids important to SECIS RNA binding. *RNA* **8**, 1308–18.

Aloisi, G., Bouloubassi, I., Heijs, S. K. et al. (2002) CH$_4$-consuming microorganisms and the formation of carbonate crusts at cold seeps. *Earth Planet Sci Lett* **203**, 195–203.

Amann, R. I., Ludwig, W. & Schleifer, K. H. (1995) Phylogenetic identification and *in situ* detection of individual microbial cells without cultivation. *Microbiol Rev* **59**, 143–69.

Ambrogelly, A., Kamtekar, S., Sauerwald, A. et al. (2004) Cys-tRNA^{Cys} formation and cysteine biosynthesis in *Methanocaldococcus jannaschii*: two faces of the same problem? *Cell Mol Life Sci* **61**, 2437–45.

Amend, J. P. & Shock, E. L. (2001) Energetics of overall metabolic reactions of thermophilic and hyperthermophilic archaea and bacteria. *FEMS Microbiol Rev* **25**, 175–243.

Amils, R., Cammarano, P. & Londei, P. (1993) Translation in archaea. In: M. Kates, D. Kushner & A. Matheson (eds) *Biochemistry of Archaea*. New Comprehensive Biochemistry Series, Elsevier, Amsterdam, pp. 393–437.

An, S. & Musier-Forsyth, K. (2004) Trans-editing of Cys-tRNA^{Pro} by *Haemophilus influenzae* YbaK protein. *J Biol Chem* **279**, 42359–62.

Anderson, J. C., Wu, N., Santoro, S. W. et al. (2004) An expanded genetic code with a functional quadruplet codon. *Proc Natl Acad Sci USA* **101**, 7566–71.

Anderson, J. O., Sarchfield, S. W. & Roger, A. J. (2005) Gene transfer from *Nanoarchaeota* to an ancestor of Diplomonades and Parabasalids. *Mol Biol Evol* **22**, 85–90.

Andrei, P. I., Pierik, A. J., Zauner, S., Andrei-Selmer, L. C. & Selmer, T. (2004) Subunit composition of the glycyl radical enzyme p-hydroxyphenylacetate decarboxylase – a small subunit, HpdC, is essential for catalytic activity. *Eur J Biochem* **271**, 2225–30.

Andrewes, P., Cullen, W. R. & Polishchuk, E. (2000) Antimony biomethylation by *Scopulariopsis brevicaulis*: characterization of intermediates and the methyl donor. *Chemosphere* **41**, 1717–25.

Andronopoulou, E. & Vorgias, C. E. (2004a) Isolation, cloning, and overexpression of a chitinase gene fragment from the hyperthermophilic archaeon *Thermococcus chitonophagus*: semi-denaturing purification of the recombinant peptide and investigation of its relation with other chitinases. *Protein Expr Purif* **35**, 264–71.

Andronopoulou, E. & Vorgias, C. E. (2004b) Multiple components and induction mechanism of the chitinolytic system of the hyperthermophilic archaeon *Thermococcus chitonophagus*. *Appl Microbiol Biotechnol* **65**, 694–702.

Antoine, E., Rolland, J., Raffin, J. & Dietrich, J. (1999) Cloning and over-expression in *Escherichia coli* of the gene encoding NADPH group III alcohol dehydrogenase from *Thermococcus hydrothermalis*. *Eur J Biochem* **264**, 880–9.

Antranikian, G., Vorgias, C. & Bertoldo, C. (2005) Extreme environments as a resource for microorganisms and novel biocatalysts. *Adv Biochem Eng Biotechnol* **96**, 219–62.

Apolinario, E. E., Jackson, K. M. & Sowers, K. R. (2005) Development of a plasmid-mediated reporter system for in vivo monitoring of gene expression in the archaeon *Methanosarcina acetivorans*. *Appl Environ Microbiol* **71**, 4914–18.

Aposhian, H. V., Zakharyan, R. A., Avram, M. D. et al. (2004) A review of the enzymology of arsenic metabolism and a new potential role of hydrogen peroxide in the detoxication of the trivalent arsenic species. *Toxicol Appl Pharmacol* **198**, 327–35.

Aravalli, R. N. & Garrett, R. A. (1997) Shuttle vectors for hyperthermophilic archaea. *Extremophiles* **1**, 183–91.

Aravind, L., Anantharaman, V., Balaji, S., Babu, M. M. & Iyer, L. M. (2005) The many faces of the helix–turn–helix domain: transcription regulation and beyond. *FEMS Microbiol Rev* **29**, 231–62.

Aravind, L., Iyer, L. M. & Anantharaman, V. (2003) The two faces of Alba: the evolutionary connection between proteins participating in chromatin structure and RNA metabolism. *Genome Biol* **4**, R64.

Aravind, L. & Koonin, E. V. (1999) DNA-binding proteins and evolution of transcription regulation in the archaea. *Nucleic Acids Res* **27**, 4658–70.

Aravind, L., Walker, D. R. & Koonin, E. V. (1999) Conserved domains in DNA repair proteins and evolution of repair systems. *Nucleic Acids Res* **27**, 1223–42.

Archambault, J., Lacroute, F., Ruet, A. & Friesen, J. D. (1992) Genetic interaction between transcription elongation factor TFIIS and RNA polymerase II. *Mol Cell Biol* **12**, 4142–52.

Arnold, H. P., She, Q., Phan, H. et al. (1999) The genetic element pSSVx of the extremely thermophilic crenarchaeon *Sulfolobus* is a hybrid between a plasmid and a virus. *Mol Microbiol* **34**, 217–26.

Arnold, H. P., Ziese, U. & Zillig, W. (2000a) SNDV, a novel virus of the extremely thermophilic and acidophilic archaeon *Sulfolobus*. *Virology* **272**, 409–16.

Arnold, H. P., Zillig, W., Ziese, U. et al. (2000b) A novel lipothrixvirus, SIFV, of the extremely thermophilic crenarchaeon *Sulfolobus*. *Virology* **267**, 252–66.

Arp, D. J., Sayavedra-Soto, L. A. & Hommes, N. G. (2002) Molecular biology and biochemistry of ammonia oxidation by *Nitrosomonas europaea*. *Arch Microbiol* **178**, 250–5.

Asker, D. & Ohta, Y. (2002) Production of canthaxanthin by *Haloferax alexandrinus* under non-aseptic conditions and a simple, rapid method for its extraction. *Appl Microbiol Biotechnol* **58**, 743–50.

Ataíde, S. F., Jester, B. C., Devine, K. M. & Ibba, M. (2005) Stationary-phase expression and aminoacylation of a transfer-RNA-like small RNA. *EMBO Rep* **6**, 742–7.

Atomi, H., Matsumi, R. & Imanaka, T. (2004) Reverse gyrase is not a prerequisite for hyperthermophilic life. *J Bacteriol* **186**, 4829–33.

Bachellerie, J. P., Cavaille, J. & Qu, L. H. (2000) Nucleotide modifications in eukaryotic rRNAs: the world of small nucleolar RNA guides revisited. *In:* R. A. Garrett, R. S. Douthwaite, A. Liljas, A. T. Matheson, P. B. Moore & H. F. Noller (eds) *The Ribosome: Structure, Function, Antibiotics, and Cellular Interactions.* ASM Press, Washington, DC, pp. 191–203.

Bachellerie, J. P., Nicoloso, M., Qu, L. H. et al. (1995) Novel intron-encoded small nucleolar RNAs with long sequence complementarities to mature rRNAs involved in ribosome biogenesis. *Biochem Cell Biol* **73**, 835–43.

Bailey, K. A., Chow, C. S. & Reeve, J. N. (1999) Histone stoichiometry and DNA circularization in archaeal nucleosomes. *Nucleic Acids Res* **27**, 532–6.

Bailey, K. A., Marc, F., Sandman, H. & Reeve, J. N. (2002) Both DNA and histone fold sequences contribute to archaeal nucleosome stability. *J Biol Chem* **277**, 9293–301.

Bailey, K. A., Pereira, S. L., Widom, J. & Reeve, J. N. (2000) Archaeal histone selection of nucleome positioning sequences and the prokaryotic origin of histone-dependent genome evolution. *J Mol Biol* **303**, 25–34.

Baker, D. L., Youssef, O. A., Chastkofsky, M. I., Dy, D. A., Terns, R. M. & Terns, M. P. (2005) RNA-guided RNA modification: functional organization of the archaeal H/ACA RNP. *Genes Dev* **19**, 1238–48.

Bakir, F., Damluji, S. F., Amin-Zaki, L. et al. (1973) Methylmercury poisoning in Iraq. *Science* **181**, 230–41.

Balakin, A. G., Smith, L. & Fournier, M. J. (1996) The RNA world of the nucleolus: two major families of small RNAs defined by different box elements with related functions. *Cell* **86**, 823–34.

Balch, W. E. & Wolfe, R. S. (1976) New approach to the cultivation of methanogenic bacteria: 2-mercaptoethanesulfonic acid (HS-CoM)-dependent growth of *Methanobacterium ruminantium* in a pressurized atmosphere. *Appl Environ Microbiol* **32**, 781–91.

Balch, W. E. & Wolfe, R. S. (1979) Specificity and biological distribution of coenzyme M (2-mercaptoethanesulfonic acid). *J Bacteriol* **137**, 256–63.

Baliga, N. S., Bjork, S. J., Bonneau, R. et al. (2004a) Systems level insights into the stress response to UV radiation in the halophilic archaeon *Halobacterium* NRC-1. *Genome Res* **14**, 1025–35.

Baliga, N. S., Bonneau, R., Facciotti, M. T. et al. (2004b) Genome sequence of *Haloarcula marismortui*: a halophilic archaeon from the Dead Sea. *Genome Res* **14**, 2221–34.

Baliga, N. S. & DasSarma, S. (1999) Saturation mutagenesis of the haloarchaeal *bop* gene promoter: identification of DNA supercoiling sensitivity sites and absence of TFB recognition element and UAS enhancer activity. *Mol Microbiol* **36**, 1175–83.

Baliga, N. S., Goo, Y. A., Ng, W. V., Hood, L., Daniels, C. J. & DasSarma, S. (2000) Is gene expression in *Halobacterium* NRC-1 regulated by multiple TBP and TFB transcription factors? *Mol Microbiol* **36**, 1184–5.

Baliga, N. S., Kennedy, S., Ng, W. V., Hood, L. & DasSarma, S. (2001) Genomic and genetic dissection of an archaeal regulon. *Proc Natl Acad Sci USA* **98**, 2521–5.

Baliga, N. S., Pan, M., Goo, Y. et al. (2002) Coordinate regulation of energy transduction modules in *Halobacterium* sp. analyzed by a global systems approach. *Proc Natl Acad Sci USA* **99**, 14013–18.

Balint, B., Bagi, Z., Tóth, A., Rákhely, G., Perei, K. & Kovács, K. L. (2005) Utilization of keratin-containing biowaste to produce biohydrogen. *Appl Microbiol Biotechnol* **69**, 404–10.

Ban, N., Nissen, P., Hansen J., Moore, P. B & Steitz, T. A. (2000) The complete atomic structure of the large ribosomal subunit at 2.4 Å resolution. *Science* **289**, 905–20.

Banning, N., Brock, F., Fry, J. C., Parkes, R. J., Hornibrook, E. R. C. & Weightman, A. J. (2005) Investigation of the methanogen population structure and activity in a brackish lake sediment. *Environ Microbiol* **7**, 947–60.

Bapteste, E., Boucher, Y., Leigh, J. & Doolittle, W. F. (2004) Phylogenetic reconstruction and lateral gene transfer. *Trends Microbiol* **12**, 406–11.

Bapteste, E. & Brochier, C. (2004) On the conceptual difficulties in rooting the tree of life. *Trends Microbiol* **12**, 9–13.

Bapteste, E., Brochier, C. & Boucher, Y. (2005) Higher-level classification of the Archaea: evolution of methanogenesis and methanogens. *Archaea* **1**, 353–63.

Bapteste, E., Susko, E., Leigh, J., MacLeod, D., Charlebois, R. L. & Doolittle, W. F. (2005) Do orthologous gene phylogenies really support tree-thinking? *BMC Evol Biol* **5**, 33.

Barns, S. M., Delwiche, C. F., Jeffrey, J. D. & Pace, N. R. (1996) Perspectives on archaeal diversity, thermophily and monophyly from environmental rRNA sequences. *Proc Natl Acad Sci USA* **93**, 9188–93.

Bartel, D. P. (2004) MicroRNAs: genomics, biogenesis, mechanism, and function. *Cell* **116**, 281–97.

Barth, S., Huhn, M., Matthey, B., Klimka, A., Galinski, E. & Engert, A. (2000) Compatible-solute-supported periplasmic expression of functional recombinant proteins under stress conditions. *Appl Environ Microbiol* **66**, 1572–9.

Bartlett, M. S., Thomm, M. & Geiduschek, E. P. (2000) The orientation of DNA in an archaeal transcription initiation complex. *Nat Struct Biol* **7**, 782–5.

Bartlett, M. S., Thomm, M. & Geiduschek, E. P. (2004) Topography of the euryarchaeal transcription initiation complex. *J Biol Chem* **279**, 5894–903.

Basehoar, A. D., Zanton, S. J. & Pugh, B. F. (2004) Identification and distinct regulation of yeast TATA box-containing genes. *Cell* **116**, 699–709.

Basu, U., Si, K., Deng, H. & Maitra, U. (2003) Phosphorylation of mammalian eukaryotic translation initiation factor 6 and its *Saccharomyces cerevisiae* homologue Tif6p: evidence that phosphorylation of Tif6p regulates its nucleocytoplasmic distribution and is required for yeast cell growth. *Mol Cell Biol* **23**, 6187–99.

Basu, U., Si, K., Warner, J. R. & Maitra, U. (2001) The *Saccharomyces cerevisiae* TIF6 gene encoding translation initiation factor 6 is required for 60S ribosomal subunit biogenesis. *Mol Cell Biol* **21**, 1453–62.

Bauer, M. W., Driskill, L. E., Callen, W., Snead, M. A., Mathur, E. J. & Kelly, R. M. (1999) An endoglucanase, EglA, from the hyperthermophilic archaeon *Pyrococcus furiosus* hydrolyzes beta-1,4 bonds in mixed-linkage (1→3),(1→4)-beta-D-glucans and cellulose. *J Bacteriol* **181**, 284–90.

Baumann, H., Knapp, S., Lundback, T., Ladenstein, R. & Hard, T. (1994) Solution structure and DNA-binding properties of a thermostable protein from the archaeon *Sulfolobus solfataricus*. *Struct Biol* **1**, 808–19.

Baumeister, W. & Lembcke, G. (1992) Structural features of archaebacterial cell envelopes. *J Bioenerg Biomembr* **24**, 567–75.

Baumeister, W., Wildhaber, I. & Phipps, B. M. (1989) Principles of organization in eubacterial and archaebacterial surface proteins. *Can J Microbiol* **35**, 215–27.

Becker, A. & Kabsch, W. (2002) X-ray structure of pyruvate formate-lyase in complex with pyruvate and CoA – how the enzyme uses the Cys-418 thiyl radical for pyruvate cleavage. *J Biol Chem* **277**, 40036–42.

Bedell, J. L., Edmondson, S. P. & Shriver, J. W. (2005) Role of a surface tryptophan in defining the structure, stability, and DNA binding of the hyperthermophile protein Sac7d. *Biochemistry* **44**, 915–25.

Beiko, R. G., Harlow, T. J. & Ragan, M. A. (2005) Highways of gene sharing in prokaryotes. *Proc Natl Acad Sci USA* **102**, 14332–7.

Beja, O., Koonin, E. V., Aravind, L. et al. (2002) Comparative genomic analysis of archaeal genotypic variants in a single population and in two different oceanic provinces. *Appl Environ Microbiol* **68**, 335–45.

Beja, O., Suzuki, M. T., Koonin, E. V. et al. (2000) Construction and analysis of bacterial artificial chromosome libraries from a marine microbial assemblage. *Environ Microbiol* **2**, 516–19.

Belfort, M., Reaban, M. E., Coetzee, T. & Dalgaard, J. Z. (1995) Prokaryotic introns and inteins: a panoply of form and function. *J Bacteriol* **177**, 3897–903.

Bell, S. D. (2005) Archaeal transcriptional regulation – variation on a bacterial theme? *Trends Microbiol* **13**, 262–5.

Bell, S. D., Botting, C. H., Wardleworth, B. N., Jackson, S. P. & White, M. F. (2002) The interaction of Alba, a conserved archaeal chromatin protein, with Sir2 and its regulation by acetylation. *Science* **296**, 148–51.

Bell, S. D., Brinkman, A. B., van der Oost, J. & Jackson, S. P. (2001) The archaeal TFIIEalpha homologue facilitates transcription initiation by enhancing TATA-box recognition. *EMBO Rep* **2**, 133–8.

Bell, S. D., Cairns, S. S., Robson, R. L. & Jackson, S. P. (1999a) Transcriptional regulation of an archaeal operon *in vivo* and *in vitro*. *Mol Cell* **4**, 971–82.

Bell, S. D. & Jackson, S. P. (1998) Transcription and translation in Archaea: a mosaic of eukaryal and bacterial features. *Trends Microbiol* **6**, 222–8.

Bell, S. D. & Jackson, S. P. (2000a) The role of transcription factor B in transcription initiation and promoter clearance in the archaeon *Sulfolobus acidocaldarius*. *J Biol Chem* **275**, 12934–40.

Bell, S. D. & Jackson, S. P. (2000b) Mechanism of autoregulation by an archaeal transcriptional repressor. *J Biol Chem* **275**, 31624–9.

Bell, S. D. & Jackson, S. P. (2001) Mechanism and regulation of transcription in archaea. *Curr Opin Microbiol* **4**, 208–13.

Bell, S. D., Jaxel, C., Nadal, M., Kosa, P. F. & Jackson, S. P. (1998) Temperature, template topology, and factor requirements of archaeal transcription. *Proc Natl Acad Sci USA* **95**, 15218–22.

Bell, S. D., Kosa, P. L., Sigler, P. D. & Jackson, S. P. (1999b) Orientation of the transcription preinitiation complex in archaea. *Proc Natl Acad Sci USA* **96**, 13662–7.

Bell, S. P. (2002) The origin recognition complex: from simple origins to complex functions. *Genes Dev* **16**, 659–72.

Bell, S. P. & Dutta, A. (2002) DNA replication in eukaryotic cells. *Ann Rev Biochem* **71**, 333–74.

Bellosta, P., Hulf, T., Diop, S. D. et al. (2005) Myc interacts genetically with Tip48/Reptin and Tip49/Pontin to control growth and proliferation during *Drosophila* development. *Proc Natl Acad Sci USA* **102**, 11799–804.

Benach, J., Edstrom, W. C., Lee, I. et al. (2005) The 2.35 Å structure of the TenA homolog from *Pyrococcus furiosus* supports an enzymatic function in thiamine metabolism. *Acta Crystallogr D Biol Crystallogr* **61**, 589–98.

Beneke, S., Bestgen, H. & Klein, A. (1995) Use of the *Escherichia coli uidA* gene as a reporter in *Methanococcus voltae* for the analysis of the regulatory function of the intergenic region between the operons encoding selenium-free hydrogenases. *Mol Gen Genet* **248**, 225–8.

Benelli, D., Maone, E. & Londei, P. (2003) Two different mechanisms for ribosome mRNA interaction in archaeal translation initiation. *Mol Microbiol* **50**, 635–43.

Bentley, R. & Chasteen, T. G. (2002) Microbial methylation of metal(loid)s: arsenic, antimony, and bismuth. *Microbiol Mol Biol Rev* **66**, 250–71.

Bentley, S. D. & Parkhill, J. (2004) Comparative genomic structure of prokaryotes. *Annu Rev Genet* **38**, 771–91.

Bergqvist, S., O'Brien, R. & Ladbury, J. E. (2001) Site-specific cation binding mediates TATA binding protein-DNA interaction from a hyperthermophilic archaeon. *Biochemistry* **40**, 2419–25.

Bernander, R. & Poplawski, A. (1997) Cell cycle characteristics of thermophilic archaea. *J Bacteriol* **179**, 4963–9.

Bernander, R., Poplawski, A. & Grogan, D. W. (2000) Altered patterns of cellular growth, morphology, replication and division in conditional-lethal mutants of the thermophilic archaeon *Sulfolobus acidocaldarius*. *Microbiology* **146**, 749–57.

Berrisford, J. M., Akerboom, J., Brouns, S. et al. (2004) The structures of inhibitor complexes of *Pyrococcus furiosus* phosphoglucose isomerase provide insights into substrate binding and catalysis. *J Mol Biol* **343**, 649–57.

Berrisford, J. M., Hounslow, A. M., Akerboom, J. et al. (2006) Evidence supporting a *cis*-enediol-based mechanism for *Pyrococcus furiosus* phosphoglucose isomerase. *J Mol Biol* **358**, 1353–66.

Bertani, G. & Baresi, L. (1987) Genetic transformation in the methanogen *Methanococcus voltae* PS. *J Bacteriol* **169**, 2730–8.

Betlach, M., Friedman, J., Boyer, H. W. & Pfeifer, F. (1984) Characterization of a halobacterial gene affecting bacterio-opsin gene expression. *Nucleic Acids Res* **12**, 7949–59.

Betlach, M., Pfeifer, F., Friedman, J. & Boyer, H. W. (1983) Bacterio-opsin mutants of *Halobacterium halobium*. *Proc Natl Acad Sci USA* **80**, 1416–20.

Bettstetter, M., Peng, X., Garrett, R. A. & Prangishvili, D. (2003) AFV1, a novel virus infecting hyperthermophilic archaea of the genus *Acidianus*. *Virology* **315**, 68–79.

Bian, L. Q., Hinrichs, K. U., Xie, T. M. et al. (2001) Algal and archaeal polyisoprenoids in a recent marine sediment: Molecular isotopic evidence for anaerobic oxidation of methane. *Geochem Geophys Geosyst* **2**, doi:10.1029/2000GC000112.

Biles, B. D. & Connolly, B. A. (2004) Low-fidelity *Pyrococcus furiosus* DNA polymerase mutants useful in error-prone PCR. *Nucleic Acids Res* **32**, e176.

Bilokapic, S., Korencic, D., Söll, D. & Weygand-Durasevic, I. (2004) The unusual methanogenic seryl-tRNA synthetase recognizes tRNA^{Ser} species from all three kingdoms of life. *Eur J Biochem* **271**, 694–702.

Bintrim, S. B., Donohue, T. J., Handelsman, J., Roberts, G. P. & Goodman, R. M. (1997) Molecular phylogeny of Archaea from soil. *Proc Natl Acad Sci USA* **94**, 277–82.

Birkenbihl, R. P., Neef, K., Prangishvili, D. & Kemper, B. (2001) Holliday junction resolving enzymes of archaeal viruses SIRV1 and SIRV2. *J Mol Biol* **309**, 1067–76.

Bitan-Banin, G., Ortenberg, R. & Mevarech, M. (2003) Development of a gene knockout system for the halophilic archaeon *Haloferax volcanii* by use of the *pyrE* gene. *J Bacteriol* **185**, 772–8.

Biwer, A., Antranikian, G. & Heinzle, E. (2002) Enzymatic production of cyclodextrins. *Appl Microbiol Biotechnol* **59**, 609–17.

Blank, C. E., Kessler, P. S. & Leigh, J. A. (1995) Genetics in methanogens: transposon insertion mutagenesis of a *Methanococcus maripaludis nifH* gene. *J Bacteriol* **177**, 5773–7.

Blaseio, U. & Pfeifer, F. (1990) Transformation of *Halobacterium halobium*: development of vectors and investigation of gas vesicle synthesis. *Proc Natl Acad Sci USA* **87**, 6772–6.

Blight, S. K., Larue, R. C., Mahapatra, A. et al. (2004) Direct charging of tRNA^{CUA} with pyrrolysine *in vitro* and *in vivo*. *Nature* **431**, 333–5.

Blount, Z. D. & Grogan, D. W. (2005) New insertion sequences of *Sulfolobus*: functional properties and implications for genome evolution in hyperthermophilic archaea. *Mol Microbiol* **55**, 312–25.

Blum, H., Zillig, W., Mallok, S., Domdey, H. & Prangishvili, D. (2000) The genome of the archaeal virus SIRV1 has features in common with genomes of eukaryal viruses. *Virology* **281**, 6–9.

Blumenberg, M., Seifert, R., Reitner, J., Pape, T. & Michaelis, W. (2004) Membrane lipid patterns typify distinct anaerobic methanotrophic consortia. *Proc Natl Acad Sci USA* **101**, 11111–16.

Blumenthal, T. (2004) Operons in eukaryotes. *Brief Funct Genomic Proteomic* **3**, 199–211.

Boccazzi, P., Zhang, J. K. & Metcalf, W. W. (2000) Generation of dominant selectable markers for resistance to pseudomonic acid by cloning and mutagenesis of the *ileS* gene from the archaeon *Methanosarcina barkeri* fusaro. *J Bacteriol* **182**, 2611–18.

Böck, A., Thanbichler, M., Rother, M. & Resch, A. (2004) Selenocysteine, in Aminoacyl-tRNA Synthetases. In: M. Ibba, C. S. Francklyn & S. Cusack (eds) Landes Bioscience, pp. 320–7.

Bocquier, A. A., Liu, L. D., Cann, I. K., Komori, K., Kohda, D. & Ishino, Y. (2001) Archaeal primase: bridging the gap between RNA and DNA polymerases. *Curr Biol* **11**, 452–6.

Boetius, A., Ravenschlag, K., Schubert, C. J. et al. (2000) A marine microbial consortium apparently mediating anaerobic oxidation of methane. *Nature* **407**, 623–6.

Boetius, A. & Suess, E. (2004) Hydrate Ridge: a natural laboratory for the study of microbial life fueled by methane from near-surface gas hydrates. *Chem Geol* **205**, 291–310.

Bohlke, K., Pisani, F. M., Vorgias, C. E., Frey, B., Sobek, H., Rossi, M. & Antranikian, G. (2000) PCR performance of the B-type DNA polymerase from the thermophilic euryarchaeon *Thermococcus aggregans* improved by mutations in the Y-GG/A motif. *Nucleic Acids Res* **28**, 3910–17.

Bohrmann, B., Kellenberger, E., Arnold-Schulz-Gahmen, B. et al. (1994) Localization of histone-like proteins in thermophilic archaea by immunogold electron microscopy. *J Struct Biol* **112**, 70–8.

Boileau, G., Butler, P., Hershey, J. W. & Traut, R. R. (1983) Direct cross-links between initiation factors 1, 2, and 3 and ribosomal proteins promoted by 2-iminothiolane. *Biochemistry* **22**, 3162–70.

Bond, C. S., Kvaratskhelia, M., Richard, D., White, M. F. & Hunter, W. N. (2001) Structure of Hjc, a Holliday junction resolvase, from *Sulfolobus solfataricus*. *Proc Natl Acad Sci USA* **98**, 5509–14.

Bondareva, A. A. & Schmidt, E. E. (2003) Early vertebrate evolution of the TATA-binding protein, TBP. *Mol Biol Evol* **20**, 1932–9.

Borg, K., Wanntrop, H. & Erne, K. (1986) Alkyl mercury poisoning in terrestrial Swedish wildlife. *Viltrevy* **6**, 301–79.

Borges, N., Ramos, A., Raven, N. D., Sharp, R. J. & Santos, H. (2002) Comparative study of the thermostabilizing properties of mannosylglycerate and other compatible solutes on model enzymes. *Extremophiles* **6**, 209–16.

Boulegue, J. (1978) Solublity of elemental sulfur in water at 298 K. *Phosphorus Sulfur* **5**, 127–8.

Boult, Z. D. & Grogan, D. W. (2005) New insertion sequences of *Sulfolobus*: functional properties and implications for genome evolution in hyperthermophilic archaea. *Mol Microbiol* **55**, 312–25.

Bouyssiere, B., Szpunar, J. & Lobinski, R. (2002) Gas chromatography with inductively coupled plasma mass spectrometric detection in speciation analysis. *Spectrochim Acta B* **57**, 805–28.

Breathnach, R. & Chambon, P. (1981) Organization and expression of eukaryotic split genes coding for proteins. *Annu Rev Biochem* **50**, 349–83.

Bredberg, K., Persson, J., Christiansson, M., Stenberg, B. & Holst, O. (2001) Anaerobic desulfurization of ground rubber with the thermophilic archaeon *Pyrococcus furiosus* – a new method for rubber recycling. *Appl Microbiol Biotechnol* **55**, 43–8.

Breed, R. S., Murray, E. G. D., Smith, N. R. et al. (1957) *Bergey's Manual of Determinative Bacteriology*, 7th edn. Williams & Wilkins, Baltimore.

Breitbart, M., Wegley, L., Leeds, S., Schoenfeld, T. & Rower, F. (2004) Phage community dynamics in hot springs. *Appl Environ Microbiol*, **70**, 1633–40.

Briggs, S. D., Xiao, T., Sun, Z.-W. et al. (2002) Trans-histone regulatory pathway in chromatin. *Nature* **418**, 498.

Brinkman, A. B., Bell, S. D., Lebbink, R. J. et al. (2002) The *Sulfolobus solfataricus* Lrp-like protein LysM regulates lysine biosynthesis in response to lysine availability. *J Biol Chem* **277**, 29537–49.

Brinkman, A. B., Dahlke, I., Tuininga, J. E. et al. (2000) An Lrp-like transcriptional regulator from the archaeon *Pyrococcus furiosus* is negatively autoregulated. *J Biol Chem* **275**, 38160–9.

Brinkman, A. B., Ettema, T. J., De Vos, W. M. et al. (2003) The Lrp family of transcriptional regulators. *Mol Microbiol* **48**, 287–94.

Brinkmann, H. & Philippe, H. (1999) Archaea sister group of Bacteria? Indications from tree reconstruction artifacts in ancient phylogenies. *Mol Biol Evol* **16**, 817–25.

Brochier, C., Bapteste, E., Moreira, D. & Philippe, H. (2002) Eubacterial phylogeny based on translational apparatus proteins. *Trends Genet* **18**, 1–5.

Brochier, C., Forterre, P. & Gribaldo, S. (2004) Archaeal phylogeny based on proteins of the transcription and translation machineries: tackling the *Methanopyrus kandleri* paradox. *Genome Biol* **5**, R17.

Brochier, C., Forterre, P. & Gribaldo, S. (2005a) An emerging phylogenetic core of Archaea: phylogenies of transcription and translation machineries converge following addition of new genome sequences. *BMC Evol Biol* **5**, 36–42.

Brochier, C., Gribaldo, S., Zivanovic, Y., Confaloniere, F. & Forterre, P. (2005b) *Nanoarchaea*: representatives of a novel archaeal phylum or a fast-evolving euryarchaeal linage related to *Thermococcales*? *Genome Biol* **6**, R42.1–10.

Brock, T. D. (1978) *Thermophilic Microorganisms and Life at High Temperatures*. Springer-Verlag, Berlin.

Brock, T. D., Brock, K. M., Belly, R. T. & Weiss, R. L. (1972) *Sulfolobus*: a new genus of sulfur-oxidizing bacteria living at low pH and high temperature. *Arch Mikrobiol* **84**, 54–68.

Brügger, K., Redder, P., She, Q., Confalonieri, F., Zivanovic, Y. & Garrett, R. A. (2002) Mobile elements in archaeal genomes. *FEMS Microbiol Lett* **206**, 131–41.

Brügger, K., Torarinsson, E., Chen, L. & Garrett, R. A. (2004) Shuffling of *Sulfolobus* genomes by autonomous and non-autonomous mobile elements. *Biochem Soc Trans* **32**, 179–83.

Bruins, M., Janssen, A. & Boom, R. (2001) Thermozymes and their applications. *Appl Biochem Biotechnol* **90**, 155–86.

Brunner, N. A., Brinkmann, H., Siebers, B. et al. (1998) NAD^+-dependent glyceraldehyde-3-phosphate dehydrogenase from *Thermoproteus tenax*. The first identified archaeal member of the aldehyde dehydrogenase superfamily is a glycolytic enzyme with unusual regulatory properties. *J Biol Chem* **273**, 6149–56.

Brunner, N. A., Siebers, B. & Hensel, R. (2001) Role of two different glyceraldehyde-3-phosphate dehydrogenases in controlling the reversible Embden–Meyerhof–Parnas pathway in *Thermoproteus tenax*: regulation on protein and transcript level. *Extremophiles* **5**, 101–9.

Brüser, T., Selmer, T. & Dahl, C. (2000) "ADP sulfurylase" from *Thiobacillus denitrificans* is an adenylylsulfate: phosphate adenylyltransferase and belongs to a new family of nucleotidyltransferases. *J Biol Chem* **275**, 1691–8.

Buchanan, C. L., Connaris, H., Danson, M. J., Reeve, C. D. & Hough, D. W. (1999) An extremely thermostable aldolase from *Sulfolobus solfataricus* with specificity for non-phosphorylated substrates. *Biochem J* **343**, 563–70.

Buckley, D. H., Graber, J. R. & Schmidt, T. M. (1998) Phylogenetic analysis of nonthermophilic members of the kingdom crenarchaeota and their diversity and abundance in soils. *Appl Environ Microbiol* **64**, 4333–9.

Buffett, B. & Archer, D. (2004) Global inventory of methane clathrate: sensitivity to changes in the deep ocean. *Earth Planet Sci Lett* **227**, 185–99.

Bult, C. J., White, O., Olsen, G. J. et al. (1996) Complete genome sequence of the methanogenic archaeon, *Methanococcus jannaschii*. *Science* **273**, 1058–73.

Buratowski, S. & Zhou, H. (1993) Functional domains of transcription factor TFIIB. *Proc Natl Acad Sci USA* **90**, 5633–7.

Burggraf, S., Stetter, K. O., Rouviere, P. & Woese, C. R. (1991) *Methanopyrus kandleri*: an archaeal methanogen unrelated to all other known methanogens. *Syst Appl Microbiol* **14**, 346–51.

Burkhardt, R. W. Jr (1977) *The Spirit of System: Lamarck and Evolutionary Biology*. Harvard University Press, Cambridge, MA.

Bushnell, D. A., Westover, K. D., Davis, R. E. & Kornberg, R. D. (2004) Structural basis of transcription: an RNA polymerase II-TFIIB cocrystal at 4.5 Angstroms. *Science* **303**, 983–8.

Caetano-Anolles, G. & Caetano-Anolles, D. (2003) An evolutionarily structured universe of protein architecture. *Genome Res* **13**, 1563–71.

Caetano-Anolles, G. & Caetano-Anolles, D. (2005) Universal sharing patterns in proteomes and evolution of protein fold architecture and life. *J Mol Evol* **60**, 484–98.

Cambillau, C. & Claverie, J. M. (2000) Structural and genomic correlation of hyperthermostability. *J Biol Chem* **275**, 32383–6.

Canaves, J. M., Page, R., Wilson, I. A. & Stevens, R. C. (2004) Protein biophysical properties that correlate with crystallization success in *Thermotoga maritima*: maximum clustering strategy for structural genomics. *J Mol Biol* **344**, 977–91.

Cann, I. K. & Ishino, Y. (1999) Archaeal DNA replication: Identifying the pieces to solve a puzzle. *Genetics* **152**, 1249–67.

Cann, I. K., Ishino, S., Hayashi, I. et al. (1999a) Functional interactions of a homolog of proliferating cell nuclear antigen with DNA polymerases in Archaea. *J Bacteriol* **181**, 6591–9.

Cann, I. K., Ishino, S., Nomura, N., Sako, Y. & Ishino, Y. (1999b) Two family B DNA polymerases from *Aeropyrum pernix*, an aerobic hyperthermophilic crenarchaeote. *J Bacteriol* **181**, 5984–92.

Cann, I. K., Komori, K., Toh, H., Kanai, S. & Ishino, Y. (1998) A heterodimeric DNA polymerase: evidence that members of Euryarchaeota possess a distinct DNA polymerase. *Proc Natl Acad Sci USA* **95**, 14250–5.

Cannio, R., Contursi, P., Rossi, M. & Bartolucci, S. (1998) An autonomously replicating transforming vector for *Sulfolobus solfataricus*. *J Bacteriol* **180**, 3237–40.

Cannio, R., Contursi, P., Rossi, M. & Bartolucci, S. (2001) Thermoadaptation of a mesophilic hygromycin B phosphotransferase by directed evolution in hyperthermophilic Archaea: selection of a stable genetic marker for DNA transfer into *Sulfolobus solfataricus*. *Extremophiles* **5**, 153–9.

Cannio, R., Di Prizito, N., Rossi, M. & Morana, A. (2004) A xylan-degrading strain of *Sulfolobus solfataricus*: isolation and characterization of the xylanase activity. *Extremophiles* **8**, 117–24.

Capaldi, S. A. & Berger, J. M. (2004) Biochemical characterization of Cdc6/Orc1 binding to the replication origin of the euryarchaeon *Methanothermobacter thermoautotrophicus*. *Nucleic Acids Res* **32**, 4821–32.

Carlson, B. A., Xu, X. M., Kryukov, G. V. et al. (2004) Identification and characterization of phosphoseryl-tRNA[Ser]Sec kinase. *Proc Natl Acad Sci USA* **101**, 12848–53.

Carpousis, A. J., Leroy, A., Vanzo, N. et al. (2001) *Escherichia coli* RNA degradosome. *Methods Enzymol* **342**, 333–45.

Casjens, S. R. (2005) Comparative genomics and evolution of the tailed-bacteriophages. *Curr Opin Microbiol* **8**, 451–8.

Catara, G., Ruggiero, G., La Cara, F., Digilio, F. A., Capasso, A. & Rossi, M. (2003) A novel extracellular subtilisin-like protease from the hyperthermophile *Aeropyrum pernix* K1: biochemical properties, cloning, and expression. *Extremophiles* **7**, 391–9.

Cavaille, J., Chetouani, F. & Bachellerie, P. (1999) The yeast *Saccharomyces cerevisiae* YDL112w ORF encodes the putative 2'-O-ribose methyltransferase catalyzing the formation of Gm18 in tRNAs. *RNA* **5**, 66–81.

Cavalier-Smith, T. (2002) The neomuran origin of archaebacteria, the negibacterial root of the universal tree and bacterial megaclassification. *Int J Syst Evol Microbiol* **52**, 7–76.

Cech, T. R., Zaug, A. J. & Grabowski, P. J. (1981) In vitro splicing of the ribosomal RNA precursor of *Tetrahymena*: involvement of a guanosine nucleotide in the excision of the intervening sequence. *Cell* **27**, 487–96.

Ceci, M., Gaviraghi, C., Gorrini, C. et al. (2003) Release of eIF6 (p27BBP) from the 60S subunit allows 80S ribosome assembly. *Nature* **426**, 579–84.

Challenger, F. (1945) Biological methylation. *Chem Rev* **36**, 315–18.

Challenger, F., Higginbottom, C. & Ellis, L. (1933) The formation of organo-metalloid compounds by microorganisms. Part I. Trimethylarsine and dimethylethylarsine. *J Chem Soc* **5**, 95–101.

Charbonnier, F. & Forterre, P. (1994) Comparison of plasmid DNA topology among mesophilic and thermophilic eubacteria and archaebacteria. *J Bacteriol* **176**, 1251–9.

Charlebois, R. L. & Doolittle, W. F. (2004) Computing prokaryotic gene ubiquity: rescuing the core from extinction. *Genome Res* **14**, 2469–77.

Charlebois, R. L., Lam, W. L., Cline, S. W. & Doolittle, W. F. (1987) Characterization of pHV2 from *Halobacterium volcanii* and its use in demonstrating transformation of an archaebacterium. *Proc Natl Acad Sci USA* **84**, 8530–4.

Charpentier, B., Muller, S. & Branlant, C. (2005) Reconstitution of archaeal H/ACA small ribonucleoprotein complexes active in pseudouridylation. *Nucleic Acids Res* **33**, 3133–44.

Chartier, F., Laine, B., Belaiche, D., Touzel, J. P. & Sautiere, P. (1989) Primary structure of the chromosomal protein MC1 from the archaebacterium *Methanosarcina* sp. CHTI 55. *Biochim Biophys Acta* **1008**, 309–14.

Chatton, E. (1938) *Titres et travaux scientifiques (1906–1937)*. Edouard Chatton, Ste, France.

Chavatte, L., Brown, B. A. & Driscoll, D. M. (2005) Ribosomal protein L30 is a component of the UGA-selenocysteine recoding machinery in eukaryotes. *Nat Struct Mol Biol* **12**, 408–16.

Chen, B. S. & Hampsey, M. (2004) Functional interaction between TFIIB and the Rpb2 subunit of RNA polymerase II: implications for the mechanism of transcription initiation. *Mol Cell Biol* **24**, 3983–91.

Chen, C. Y., Ko, T. P., Lin, T. W., Chou, C. C., Chen, C. J. & Wang, A. H. (2005a) Probing the DNA kink structure induced by the hyperthermophilic chromosomal protein Sac7d. *Nucleic Acids Res* **33**, 430–8.

Chen, H. T. & Hahn, S. (2003) Binding of TFIIB to RNA polymerase II: mapping the binding site for the TFIIB zinc ribbon domain within the preinitiation complex. *Mol Cell* **12**, 437–47.

Chen, H. T., Legault, P., Glushka, J., Omichinski, J. G. & Scott, R. A. (2000) Structure of a (Cys3His) zinc ribbon, a ubiquitous motif in archaeal and eucaryal transcription. *Protein Sci* **9**, 1743–52.

Chen, L., Brügger, K., Skovgaard, M. et al. (2005b) The genome of *Sulfolobus acidocaldarius*, a model organism of the crenarchaeota. *J Bacteriol* **187**, 4992–9.

Chen, L., Chen, L.-R., Zhou, X. E. et al. (2004a) The hyperthermophile protein Sso10a is a dimer of winged helix DNA-binding domains linked by an antiparallel coiled coil rod. *J Mol Biol* **341**, 73–91.

Chen, L., Spiliotis, E. T. & Roberts, M. F. (1998) Biosynthesis of Di-myo-inositol-1,1′-phosphate, a novel osmolyte in hyperthermophilic archaea. *J Bacteriol* **180**, 3785–92.

Chen, S., Yakunin, A. F., Kuznetsova, E. et al. (2004b) Structural and functional characterization of a novel phosphodiesterase from *Methanococcus jannaschii*. *J Biol Chem* **279**, 31854–62.

Chen, Z. W., Jiang, C. Y., She, Q., Liu, S. J. & Zhou, P. J. (2005c) Key role of cysteine residues in catalysis and subcellular localization of sulfur oxygenase–reductase of *Acidianus tengchongensis*. *Appl Environ Microbiol* **71**, 621–8.

Cheng, C. N. & Focht, D. D. (1979) Production of arsine and methylarsines in soil and in culture. *Appl Environ Microbiol* **38**, 494–8.

Cheng, T. C., Ramakrishnan, V. & Chan, S. I. (1999) Purification and characterization of a cobalt-activated carboxypeptidase from the hyperthermophilic archaeon *Pyrococcus furiosus*. *Protein Sci* **8**, 2474–86.

Chi, Y. I., Martinez-Cruz, L. A., Jancarik, J., Swanson, R. V., Robertson, D. E. & Kim, S. H. (1999) Crystal structure of the beta-glycosidase from the hyperthermophile *Thermosphaera aggregans*: insights into its activity and thermostability. *FEBS Lett* **445**, 375–83.

Chin, K. J., Lueders, T., Friedrich, M. W., Klose, M. & Conrad, R. (2004) Archaeal community structure and pathway of methane formation on rice roots. *Microbial Ecol* **47**, 59–67.

Chistoserdova, L., Jenkins, C., Kalyuzhnaya, M. G. et al. (2004) The enigmatic planctomycetes may hold a key to the origins of methanogenesis and methylotrophy. *Mol Biol Evol* **21**, 1234–41.

Chistoserdova, L., Vorholt, J. A. & Lidstrom, M. E. (2005) A genomic view of methane oxidation by aerobic bacteria and anaerobic archaea. *Genome Biol* **6**, 208.

Cho, S. & Hoffman, D. W. (2002) Structure of the beta subunit of translation initiation factor 2 from the archaeon *Methanococcus jannaschii*: a representative of the eIF2beta/eIF5 family of proteins. *Biochemistry* **41**, 5730–42.

Choder, M. (2004) Rpb4 and Rpb7: subunits of RNA polymerase II and beyond. *Trends Biochem Sci* **29**, 674–81.

Choi, S. C., Chase, T. Jr & Bartha, R. (1994) Enzymatic catalysis of mercury methylation by *Desulfovibrio desulfuricans* LS. *Appl Environ Microbiol* **60**, 1342–6.

Chong, J. P., Hayashi, M. K., Simon, M. N., Xu, R. M. & Stillman, B. (2000) A double-hexamer archaeal minichromosome maintenance protein is an ATP-dependent DNA helicase. *Proc Natl Acad Sci USA* **97**, 1530–5.

Chou, C.-C., Lin, T.-W., Chen, C.-Y. & Wang, A. H. (2003) Crystal structure of the hyperthermophilic archaeal DNA-binding protein Sso10b2 at a resolution of 1.85 angstroms. *J Bacteriol* **185**, 4066–73.

Ciaramella, M., Napoli, A. & Rossi, M. (2005) Another extreme genome: how to live at pH 0. *Trends Microbiol* **13**, 49–51.

Clark, A. T., McCrary, B. S., Edmondson, S. P. & Shriver, J. W. (2004) Thermodynamics of core hydrophobicity and packing in the hyperthermophilic proteins Sac7d and Sso7d. *Biochemistry* **43**, 2840–53.

Clarke, G. D., Beiko, R. G., Ragan, M. A. & Charlebois, R. L. (2002) Inferring genome trees by using a filter to eliminate phylogenetically discordant sequences and a distance matrix based on mean normalized BLASTP scores. *J Bacteriol* **184**, 2072–80.

Cline, S. W. & Doolittle, W. F. (1987) Efficient transfection of the archaebacterium *Halobacterium halobium*. *J Bacteriol* **169**, 1341–4.

Cline, S. W. & Doolittle, W. F. (1992) Transformation of members of the genus *Haloarcula* with shuttle vectors based on *Halobacterium halobium* and *Haloferax volcanii* plasmid replicons. *J Bacteriol* **174**, 1076–80.

Clouet d'Orval, B., Bortolin, M. L., Gaspin, C. & Bachellerie, J. P. (2001) Box C/D RNA guides for the ribose methylation of archaeal tRNAs. The tRNATrp intron guides the formation of two ribose-methylated nucleosides in the mature tRNATrp. *Nucleic Acids Res* **29**, 4518–29.

Cobucci-Ponzano, B., Rossi, M. & Moracci, M. (2005) Recoding in Archaea. *Mol Microbiol* **55**, 339–48.

Coenye, T. & Vandamme, P. (2005) Organisation of the S10, spc and alpha ribosomal protein gene clusters in prokaryotic genomes. *FEMS Microbiol Lett* **242**, 117–26.

Cohen, G. N., Barbe, V., Flament, D. et al. (2003) An integrated analysis of the genome of the hyperthermophilic archaeon *Pyrococcus abyssi*. *Mol Microbiol* **47**, 1495–512.

Cohen-Kupiec, R., Blank, C. & Leigh, J. A. (1997) Transcriptional regulation in Archaea: in vivo demonstration of a repressor binding site in a methanogen. *Proc Natl Acad Sci USA* **94**, 1316–20.

Colthurst, D. R., Campbell, D. G. & Proud, C. G. (1987) Structure and regulation of eukaryotic initiation factor eIF-2. Sequence of the site in the alpha subunit phosphorylated by the haem-controlled repressor and by the double-stranded RNA-activated inhibitor. *Eur J Biochem* **166**, 357–63.

Condo, I., Ciammaruconi, A., Benelli, D., Ruggero, D. & Londei, P. (1999) Cis-acting signals controlling translational initiation in the thermophilic archaeon *Sulfolobus solfataricus*. *Mol Microbiol* **34**, 377–84.

Conrad, R., Klose, M. & Claus, P. (2002) Pathway of CH_4 formation in anoxic rice field soil and rice roots determined by 13C-stable isotope fractionation. *Chemosphere* **47**, 797–806.

Conrad, R. & Thauer, R. K. (1983) Carbon monoxide production by *Methanobacterium thermoautotrophicum*. *FEMS Microbiol Lett* **20**, 229–32.

Constantinesco, F., Forterre, P. & Elie, C. (2002) NurA, a novel 5'-3' nuclease gene linked to rad50 and mre11 homologs of thermophilic archaea. *EMBO Rep* **3**, 537–42.

Constantinesco, F., Forterre, P., Koonin, E. V., Aravind, L. & Elie, C. (2004) A bipolar DNA helicase gene, *herA*, clusters with *rad50*, *mre11* and *nurA* genes in thermophilic archaea. *Nucleic Acids Res* **32**, 1439–47.

Contursi, P., Cannio, R., Prato, S. et al. (2003) Development of a genetic system for hyperthermophilic archaea: expression of a moderate thermophilic bacterial alcohol dehydrogenase gene in *Sulfolobus solfataricus*. *FEMS Microbiol Lett* **218**, 115–20.

Cort, J. R., Koonin, E. V., Bash, P. A. & Kennedy, M. A. (1999) A phylogenetic approach to target selection for structural genomics: solution structure of YciH. *Nucleic Acids Res* **27**, 4018–27.

Costas, M., Mehn, M. P., Jensen, M. P. & Que, L. Jr (2004) Dioxygen activation at mononuclear nonheme iron active sites: enzymes, models, and intermediates. *Chem Rev* **104**, 939–86.

Craft, J. L., Horng, Y. C., Ragsdale, S. W. & Brunold, T. C. (2004a) Spectroscopic and computational characterization of the nickel-containing F_{430} cofactor of methyl-coenzyme M reductase. *J Biol Inorg Chem* **9**, 77–89.

Craft, J. L., Horng, Y. C., Ragsdale, S. W. & Brunold, T. C. (2004b) Nickel oxidation states of F_{430} cofactor in methyl-coenzyme M reductase. *J Am Chem Soc* **126**, 4068–9.

Craig, N. L., Craigie, R., Gellert, M. & Lambowitz, A. (2002) *Mobile DNA II*. ASM Press, Washington, DC.

Craig, P. (ed.) (2003) *Organometallic Compounds in the Environment*. John Wiley & Sons, Chichester.

Cramer, P., Bushnell, D. A., Fu, J. et al. (2000) Architecture of RNA polymerase II and implications for the transcription mechanism. *Science* **288**, 640–9.

Creevey, C. J., Fitzpatrick, D. A., Philip, G. K. et al. (2004) Does a tree-like phylogeny only exist at the tips in the prokaryotes? *Proc Biol Sci* **271**, 2551–8.

Creti, R., Londei, P. & Cammarano, P. (1993) Complete nucleotide sequence of an archaeal (*Pyrococcus woesei*) gene encoding a homolog of eukaryotic transcription factor IIB (TFIIB). *Nucleic Acids Res* **21**, 2942.

Crick, F. H. C. (1958) The biological replication of macromolecules. *Symp Soc Exp Biol* **12**, 138–63.

Cubbedu, L. & White, M. F. (2005) DNA damage detection by an archaeal single stranded DNA binding protein. *J Mol Biol* **353**, 507–16.

Čuboňová, L., Sandman, K., Hallam, S. J., Delong, E. F. & Reeve, J. N. (2005) Histones in crenarchaea. *J Bacteriol* **187**, 5482–5.

Cui, Q., Tong, Y., Xue, H., Huang, L., Feng, Y. & Wang, J. (2003) Two conformations of archaeal Ssh10b. *J Biol Chem* **278**, 51015–22.

Cui, Y., Dinman, J. D., Kinzy, T. G. & Peltz, S. W. (1998) The Mof2/Sui1 protein is a general monitor of translational accuracy. *Mol Cell Biol* **18**, 1506–16.

Cullen, W. R. & Reimer, K. J. (1989) Arsenic speciation in the environment. *Chem Rev* **89**, 713–64.

Curnow, A. W., Hong, K., Yuan, R. et al. (1997) Glu-tRNA^{Gln} amidotransferase: a novel heterotrimeric enzyme required for correct decoding of glutamine codons during translation. *Proc Natl Acad Sci USA* **94**, 11819–26.

Curnow, A. W., Ibba, M. & Söll, D. (1996) tRNA-dependent asparagine formation. *Nature* **382**, 589–90.

Dabrowski, S. & Kiaer Ahring, B. (2003) Cloning, expression, and purification of the His6-tagged hyperthermostable dUTPase from *Pyrococcus woesei* in *Escherichia coli*: application in PCR. *Protein Expr Purif* **31**, 72–8.

Dabrowski, S., Sobiewska, G., Maciunska, J., Synowiecki, J. & Kur, J. (2000) Cloning, expression, and purification of the His-tagged thermostable beta-Galactosidase from *Pyrococcus woesei* in *Escherichia coli* and some properties of the isolated enzyme. *Prot Exp Pur* **19**, 107–12.

Dahl, C., Molitor, M. & Trüper, H. G. (2001) Siroheme-sulfite reductase-type protein from *Pyrobaculum islandicum*. *Methods Enzymol* **331**, 410–19.

Dahl, C. & Trüper, H. G. (2001) Sulfite reductase and APS reductase from *Archaeoglobus fulgidus*. *Methods Enzymol* **331**, 427–41.

Dahlke, I. & Thomm, M. (2002) A *Pyrococcus* homologue of the Leucine-responsive regulatory protein, LrpA, inhibits transcription by abrogating RNA polymerase recruitment. *Nucleic Acids Res* **30**, 701–10.

Daimon, K., Kawarabayasi, Y., Kikuchi, H., Sako, Y. & Ishino, Y. (2002) Three proliferating cell nuclear antigen-like proteins found in the hyperthermophilic archaeon *Aeropyrum pernix*: Interactions with the two DNA polymerases. *J Bacteriol* **184**, 687–94.

Dalgaard, J. & Garrett, R. A. (1992) Protein-coding introns from the 23S rRNA-encoding gene form stable circles in the hyperthermophile archaeon *Pyrobaculum organotrophum*. *Gene* **121**, 103–10.

Dame, R. T. (2005) The role of nucleoid-associated proteins in the organization and compaction of bacterial chromatin. *Mol Microbiol* **56**, 858–70.

D'Amours, D. & Jackson, S. P. (2002) The Mre11 complex: at the crossroads of dna repair and checkpoint signalling. *Nat Rev Mol Cell Biol* **3**, 317–27.

Daniels, C. J., Gupta, R. & Doolittle, W. F. (1985) Transcription and excision of a large intron in the tRNATrp gene of an archaebacterium, *Halobacterium volcanii*. *J Biol Chem* **260**, 3132–4.

Darcy, T. J., Hausner, W., Awery, D. E., Edwards, A. M., Thomm, M. & Reeve, J. N. (1999) *Methanobacterium thermoautotrophicum* RNA polymerase and transcription *in vitro*. *J Bacteriol* **181**, 4424–9.

Darland, G., Brock, T. D., Samsonoff, W. & Conti, S. F. (1970) A thermophilic, acidophilic mycoplasma isolated from a coal refuse pile. *Science* **170**, 1416–18.

Darwin, C. (1859) *On the Origin of Species*. Cambridge University Press, Cambridge.

Das, A., Bagchi, M. K., Ghosh-Dastidar, P. & Gupta, N. K. (1982) Protein synthesis in rabbit reticulocytes. A study of peptide chain initiation using native and beta-subunit-depleted eukaryotic initiation factor 2. *J Biol Chem* **257**, 1282–8.

DasSarma, S., Halladay, J. T., Jones, J., Donovan, J., Giannasca, P. & Tandeau de Marsac, N. (1988) High frequency mutations in a plasmid-encoded gas vesicle gene in *Halobacterium halobium*. *Proc Natl Acad Sci USA* **85**, 6861–5.

DasSarma, S., RajBhandary, U. L. & Khorana, H. G. (1983) High-frequency spontaneous mutation in the bacterio-opsin gene in *Halobacterium halobium* is mediated by transposable elements. *Proc Natl Acad Sci USA* **80**, 2201–5.

Daubin, V., Gouy, M. & Perriere, G. (2002) A phylogenomic approach to bacterial phylogeny: evidence of a core of genes sharing a common history. *Genome Res* **12**, 1080–90.

Daubin, V., Moran, N. A. & Ochman, H. (2003) Phylogenetics and the cohesion of bacterial genomes. *Science* **301**, 829–32.

Davie, J. & Kane, C. M. (2000) Genetic interactions between TFIIS and the Swi-Snf chromatin-remodeling complex. *Mol Cell Biol* **20**, 5960–73.

Decanniere, K., Babu, A. M., Sandman, K., Reeve, J. N. & Heinemann, U. (2000) Crystal structures of recombinant histones HMfA and HMfB from the hyperthermophilic archaeon *Methanothermus fervidus*. *J Mol Biol* **303**, 35–47.

DeDecker, B., O'Brien, R., Fleming, P. J., Geiger, J. H., Jackson, S. & Sigler, P. B. (1996) The crystal structure of a hyperthermophilic archaeal TATA-box binding protein. *J Mol Biol* **264**, 1072–84.

DeLano, W. L. (2002) The PyMOL Molecular Graphics System, 0.97 edn (www.pymol.org). San Carlos, CA, DeLano Scientific.

DeLong, E. F. (1992) Archaea in coastal marine environments. *Proc Natl Acad Sci USA* **89**, 5685–9.

DeLong, E. F. (2000) Microbiology – resolving a methane mystery. *Nature* **407**, 577–9.

DeLong, E. F. (2005) Microbial community genomics in the ocean. *Nat Rev Microbiol* **3**, 459–69.

DeLong, E. F. & Karl, D. M. (2005) Genomic perspectives in microbial oceanography. *Nature* **437**, 336–42.

DeLong, E. F., King, L. L., Massana, R. et al. (1998) Dibiphytanyl ether lipids in nonthermophilic crenarchaeotes. *Appl Environ Microbiol* **64**, 1133–8.

DeLong, E. F., Taylor, L. T., Marsh, T. L. & Preston, C. M. (1999) Visualization and enumeration of marine planktonic archaea and bacteria by using polyribonucleotide probes and fluorescent in situ hybridization. *Appl Environ Microbiol* **65**, 5554–63.

Delsuc, F., Brinkmann, H. & Philippe, H. (2005) Phylogenomics and the reconstruction of the tree of life. *Nat Rev Genet* **6**, 361–75.

Demirjian, D. C., Moris-Varas, F. & Cassidy, C. S. (2001) Enzymes from extremophiles. *Curr Opin Chem Biol* **5**, 144–51.

Dennis, P. P. & Omer, A. (2005) Small non-coding RNAs in Archaea. *Curr Opin Microbiol* **8**, 1–10.

Dennis, P. P., Omer, A. & Lowe, T. (2001) A guided tour: small RNA function in Archaea. *Mol Microbiol* **40**, 509–19.

Dennis, P. P., Ziesche, S. & Mylvaganam, S. (1998) Transcription analysis of two disparate rRNA operons in the halophilic archaeon *Haloarcula marismortui*. *J Bacteriol* **180**, 4804–13.

Department of Chemistry, University of Oxford (n.d.) Chemical and other safety information (http://physchem.ox.ac.uk/msds/).

de Pascale, D., Di Lernia, I., Sasso, M. P., Furia, A., De Rosa, M. & Rossi, M. (2002) A novel thermophilic fusion enzyme for trehalose production. *Extremophiles* **6**, 463–8.

Deppenmeier, U., Johann, A., Hartsch, T. et al. (2002) The genome of *Methanosarcina mazei*: evidence for lateral gene transfer between bacteria and archaea. *J Mol Microbiol Biotechnol* **4**, 453–61.

De Rosa, M., Morana, A., Riccio, A., Gambacorta, A., Trincone, A. & Incani, O. (1994) Lipids of the archaea: a new tool for bioelectronics. *Biosens Bioelectron* **9**, 669–75.

Desmarais, D., Jablonski, P., Fedarko, N. & Roberts, M. (1997) 2-Sulfotrehalose, a novel osmolyte in haloalkaliphilic archaea. *J Bacteriol* **179**, 3146–53.

Desogus, G., Onesti, S., Brick, P., Rossi, M. & Pisani, F. M. (1999) Identification and characterization of a DNA primase from the hyperthermophilic archaeon *Methanococcus jannaschii*. *Nucleic Acids Res* **27**, 4444–50.

De Vuyst, G., Aci, S., Genest, D. & Culard, F. (2005) Atypical recognition of particular DNA sequences by the archaeal chromosomal MC1 protein. *Biochemistry* **44**, 10369–77.

Dietrich, J., Schmitt, P., Zieger, M. et al. (2002) PCR performance of the highly thermostable proof-reading B-type DNA polymerase from *Pyrococcus abyssi*. *FEMS Microbiol Lett* **217**, 89–94.

Dinger, M. E., Baillie, G. J. & Musgrave, D. R. (2000) Growth phase-dependent expression and degradation of histones in the thermophilic archaeon *Thermococcus zilligii*. *Mol Microbiol* **36**, 876–85.

Dionne, I. & Bell, S. D. (2005) Characterization of an archaeal family 4 uracil DNA glycosylase and its interaction with PCNA and chromatin proteins. *Biochem J* **387**, 859–63.

Dionne, I., Nookala, R. K., Jackson, S. P., Doherty, A. J. & Bell, S. D. (2003) A heterotrimeric PCNA in the hyperthermophilic archaeon *Sulfolobus solfataricus*. *Mol Cell* **11**, 275–82.

Dirheimer, G., Keith, G., Dumas, P. & Westhof, E. (1995) Primary, secondary, and tertiary structures of tRNAs. In: D. Söll & U. L. RajBhandary (eds) *TRNA: Structure, Biosynthesis and Function*. American Society of Microbiology, Washington, DC, pp. 93–126.

Dirkx, W. M. R., Lobinski, R. & Adams, F. C. (1994) Speciation analysis of organotin in water and sediments by gas chromatography with optical spectrometric detection after extraction separation. *Anal Chim Acta* **286**, 309–18.

Dirmeier, R., Keller, M., Frey, G., Huber, H. & Stetter, K. O. (1998) Purification and properties of an extremely thermostable membrane-bound sulfur-reducing complex from the hyperthermophilic *Pyrodictium abyssi*. *Eur J Biochem* **252**, 486–91.

DiRuggiero, J., Dunn, D., Maeder, D. L. et al. (2000) Evidence of recent lateral gene transfer among hyperthermophilic archaea. *Mol Microbiol* **38**, 684–93.

DiRuggiero, J., Santangelo, N., Nackerdien, Z., Ravel, J. & Robb, F. T. (1997) Repair of extensive ionizing-radiation DNA damage at 95 degrees C in the hyperthermophilic archaeon *Pyrococcus furiosus*. *J Bacteriol* **179**, 4643–5.

Dobrindt, U., Hochhut, B., Hentschel, U. & Hacker, J. (2004) Genomic islands in pathogenic and environmental microorganisms. *Nat Rev Microbiol* **2**, 414–24.

Dontsova, M., Frolova, L., Vassilieva, J., Piendl, W., Kisselev, L. & Garber, M. (2000) Translation termination factor aRF1 from the archaeon *Methanococcus jannaschii* is active with eukaryotic ribosomes. *FEBS Lett* **472**, 213–16.

Doolittle, W. F. (1999) Phylogenetic classification and the universal tree. *Science* **284**, 2124–9.

Doolittle, W. F. & Logston, J. M. Jr (1998) Archaeal genomics: do archaea have a mixed heritage? *Curr Biol* **8**, R209–11.

Drake, J. W., Charlesworth, B., Charlesworth, D. & Crow, J. F. (1998) Rates of spontaneous mutation. *Genetics* **148**, 1667–86.

Duboc-Toia, C., Hassan, A. K., Mulliez, E. et al. (2003) Very high-field EPR study of glycyl radical enzymes. *J Am Chem Soc* **125**, 38–9.

Duin, E. C., Cosper, N. J., Mahlert, F., Thauer, R. K. & Scott, R. A. (2003) Coordination and geometry of the nickel atom in active methyl-coenzyme M reductase from *Methanothermobacter marburgensis* as detected by X-ray absorption spectroscopy. *J Biol Inorg Chem* **8**, 141–8.

Duin, E. C., Signor, L., Piskorski, R. et al. (2004) Spectroscopic investigation of the nickel-containing porphinoid cofactor F_{430}. Comparison of the free cofactor in the +1, +2 and +3 oxidation states with the cofactor bound to methylcoenzyme M reductase in the silent, red and ox forms. *J Biol Inorg Chem* **9**, 563–74.

Dunemann, L., Hajimiragha, H. & Begerow, J. (1999) Simultaneous determination of Hg(II) and alkylated Hg, Pb, and Sn species in human body fluids using SPMEGC/MS-MS. *Fresen J Anal Chem* **363**, 466–8.

Dura, M. A., Receveur-Brechot, V., Andrieu, J. P., Ebel, C., Schoehn, G., Roussel, A. & Franzetti, B. (2005) Characterization of a TET-like aminopeptidase complex from the hyperthermophilic archaeon *Pyrococcus horikoshii*. *Biochemistry* **44**, 3477–86.

Dyall-Smith, M., Tang, S. L. & Bath, C. (2003) Haloarchaeal viruses: how diverse are they? *Res Microbiol* **154**, 309–13.

Ebert, K., Goebel, W. & Pfeifer, F. (1984) Homologies between heterogeneous extrachromosomal DNA populations of *Halobacterium halobium* and four new halobacterial isolates. *Mol Gen Genet* **194**, 91–7.

Eckburg, P. B., Bik, E. M., Bernstein, C. N. et al. (2005) Diversity of the human intestinal microbial flora. *Science* **308**, 1635–8.

Eckburg, P. B., Lepp, P. W. & Relman, D. A. (2003) Archaea and their potential role in human disease. *Infect Immun* **71**, 591–6.

Eddy, S. R. (2002) Computational genomics of noncoding RNA genes. *Cell* **109**, 137–40.

Edmondson, S. P., Kahsai, M. A., Gupta, R. & Shriver, J. W. (2004) Characterization of Sac10a, a hyperthermophile DNA-binding protein from *Sulfolobus acidocaldarius*. *Biochemistry* **43**, 13026–36.

Ehrmann, M., Ehrle, R., Hofmann, E. et al. (1998) The ABC maltose transporter. *Mol Microbiol* **29**, 685–94.

Eichler, J. (2001) Biotechnological uses of archaeal extremozymes. *Biotechnol Adv* **19**, 261–78.

Eichler, J. (2003) Facing extremes: archaeal surface-layer (glyco)proteins. *Microbiology* **149**, 3347–51.

Eichler, J. & Moll, R. (2001) The signal recognition particle of Archaea. *Trends Microbiol* **9**, 130–6.

Eklund, H. & Fontecave, M. (1999) Glycyl radical enzymes: a conservative structural basis for radicals. *Structure* **7**, R257–62.

Ekstrom, E. B., Morel, F. M. & Benoit, J. M. (2003) Mercury methylation independent of the acetyl-coenzyme A pathway in sulfate-reducing bacteria. *Appl Environ Microbiol* **69**, 5414–22.

Elshahed, M. S., Savage, K. N., Oren, A., Gutierrez, M. C., Ventosa, A. & Krumholz, L. R. (2004) *Haloferax sulfurifontis* sp. nov., a halophilic archaeon isolated from a sulfide- and sulfur-rich spring. *Int J Syst Evol Microbiol* **54**, 2275–9.

Elvert, M., Boetius, A., Knittel, K. & Jorgensen, B. B. (2003) Characterization of specific membrane fatty acids as chemotaxonomic markers for sulfate reducing bacteria involved in anaerobic oxidation of methane. *Geomicrobiol J* **20**, 403–19.

Embley, T. M. & Finlay, B. J. (1993–4) Systematic and morphological diversity of endosymbiotic methanogens in anaerobic ciliates. *Antonie Van Leeuwenhoek* **64**, 261–71.

Embley, T. M. & Hirt, R. P. (1998) Early branching eukaryotes? *Curr Opin Genet Dev* **8**, 624–9.

Emmel, T., Sand, W., König, W. A. & Bock, E. (1986) Evidence for the existence of a sulfur oxygenase in *Sulfolobus brierleyi*. *J Gen Microbiol* **132**, 3415–20.

Empadinhas, N., Marugg, J. D., Borges, N., Santos, H. & da Costa, M. S. (2001) Pathway for the synthesis of mannosylglycerate in the hyperthermophilic archaeon *Pyrococcus horikoshii*. Biochemical and genetic characterization of key enzymes. *J Biol Chem* **276**, 43580–8.

Engelhardt, H. & Peters, J. (1998) Structural research on surface layers: a focus on stability, surface layer homology domains and surface layer-cell wall interactions. *J Struct Biol* **124**, 276–302.

Englert, C., Horne, M. & Pfeifer, F. (1990) Expression of the major gas vesicle protein in the halophilic archaebacterium *Haloferax mediterranei* is modulated by salt. *Mol Gen Genet* **222**, 225–32.

Englert, C., Krüger, K., Offner, S. & Pfeifer, F. (1992a) Three different but related gene clusters encoding gas vesicles in halophilic archaea. *J Mol Biol* **227**, 586–92.

Englert, C., Wanner, G. & Pfeifer, F. (1992b) Functional analysis of the gas vesicle gene cluster of the halophilic archaeon *Haloferax mediterranei* defines the vac-region boundary and suggests a regulatory role for the *gvpD* gene or its product. *Mol Microbiol* **6**, 3543–50.

Enright, A. J., Iliopoulos, I., Kyrpides, N. C. et al. (1999) Protein interaction maps for complete genomes based on gene fusion events. *Nature* **402**, 86–90.

Erauso, G., Marsin, S., Benbouzid, R. N. et al. (1996) Sequence of plasmid pGT5 from the archaeon *Pyrococcus abyssi*: evidence for rolling-circle replication in a hyperthermophile. *J Bacteriol* **178**, 3232–7.

Erie, D. A., Hajiseyedjavadi, O., Young, M. C. & von Hippel, P. H. (1993) Multiple RNA polymerase conformations and GreA: control of the fidelity of transcription. *Science* **262**, 867–73.

Erkel, C., Kemnitz, D., Kube, D. et al. (2005) Retrieval of first genome data for rice cluster I methanogens by a combination of cultivation and molecular techniques. *FEMS Microbiol Ecol* **53**, 187–204.

Ermler, U., Grabarse, W., Shima, S., Goubeaud, M. & Thauer, R. K. (1997) Crystal structure of methyl-coenzyme M reductase: the key enzyme of biological methane formation. *Science* **278**, 1457–62.

Espinosa, M., del Solar, G. H., Rojo, F. & Alonso, J. C. (1995) Plasmid rolling circle replication and its control. *FEMS Microbiol Lett* **130**, 111–20.

Esposito, D. & Scocca, J. J. (1997) The integrase family of tyrosine recombinases: evolution of a conserved active site domain. *Nucleic Acids Res* **25**, 3605–14.

Ettema, T. J., Brinkman, A. B., Tani, T. H. et al. (2002) A novel ligand-binding domain involved in regulation of amino acid metabolism in prokaryotes. *J Biol Chem* **277**, 37464–8.

Ettema, T. J., De Vos, W. M. & van der Oost, J. (2005) Discovering novel biology by *in silico* archaeology. *Nat Rev Microbiol* **3**, 859–69.

Fahrner, R. L., Cascio, D., Lake, J. A. & Slesarev, A. (2001) An ancestral nuclear protein assembly: crystal structure of the *Methanopyrus kandleri* histone. *Protein Sci* **10**, 2002–7.

Falb, M., Pfeiffer, F., Palm, P. et al. (2005) Living with two extremes: conclusions from the genome sequence of *Natronomonas pharaonis*. *Genome Res* **15**, 1336–43.

Fang, T. Y., Hung, X. G., Shih, T. Y. & Tseng, W. C. (2004) Characterization of the trehalosyl dextrin-forming enzyme from the thermophilic archaeon *Sulfolobus solfataricus* ATCC 35092. *Extremophiles* **8**, 335–43.

Feldmann, J., Grümping, R. & Hirner, A. V. (1994) Determination of volatile metal and metalloid compounds in gases from domestic waste deposits with GC/ICP-MS. *J Anal Chem* **350**, 228–34.

Feldmann, J. & Hirner, A. V. (1995) Occurence of volatile metal and metalloid species in landfill and sewage gases. *Intl J Environ Anal Chem* **60**, 339–59.

Fendinger, N. J., Lehmann, R. G. & Mihaich, E. M. (1997) Polydimethylsiloxane. *In:* G. Chandra (ed.) *Organosilicon Materials*, Vol. 3. Springer-Verlag, Berlin, pp. 181–223.

Feng, L., Sheppard, K., Tumbula-Hansen, D. & Söll, D. (2005) Gln-tRNA^{Gln} formation from Glu-tRNA^{Gln} requires cooperation of an asparaginase and a Glu-tRNA^{Gln} kinase. *J Biol Chem* **280**, 8150–5.

Ferrer, M., Golyshina, O. V., Plou, F. J., Timmis, K. N. & Golyshin, P. N. (2005) A novel alpha-glucosidase from the acidophilic archaeon, *Ferroplasma acidiphilum* Y with high transglycosylation activity and an unusual catalytic nucleophile. *Biochem J* **391**, 269–76.

Fessner, W. D. & Helaine, V. (2001) Biocatalytic synthesis of hydroxylated natural products using aldolases and related enzymes. *Curr Opin Biotechnol* **12**, 574–86.

Fiala, G. & Stetter, K. O. (1986) *Pyrococcus furiosus* sp-nov represents a novel genus of marine heterotrophic archaebacteria growing optimally at 100-degrees C. *Arch Microbiol* **145**, 56–61.

Fiedler, U. & Timmers, H. T. (2001) Analysis of the open region of RNA polymerase II transcription complexes in the early phase of elongation. *Nucleic Acids Res* **29**, 2706–14.

Filee, J., Forterre, P., Sen-Lin, T. & Laurent, J. (2002) Evolution of DNA polymerase families: evidences for multiple gene exchange between cellular and viral proteins. *J Mol Evol* **54**, 763–73.

Finazzo, C., Harmer, J., Bauer, C. et al. (2003a) Coenzyme B induced coordination of coenzyme M via its thiol group to Ni(I) of F_{430} in active methyl-coenzyme M reductase. *J Am Chem Soc* **125**, 4988–9.

Finazzo, C., Harmer, J., Jaun, B. et al. (2003b) Characterization of the MCR$_{red2}$ form of methyl-coenzyme M reductase: a pulse EPR and ENDOR study. *J Biol Inorg Chem* **8**, 586–93.

Fiorentino, G., Cannio, R., Rossi, M. & Bartolucci, S. (2003) Transcriptional regulation of the gene encoding an alcohol dehydrogenase in the Archaeon *Sulfolobus solfataricus* involves multiple factors and control elements. *J Bacteriol* **185**, 3926–34.

Fitz-Gibbon, S. T., Ladner, H., Kim, U. J., Stetter, K. O., Simon, M. I. & Miller, J. H. (2002) Genome sequence of the hyperthermophilic crenarchaeon *Pyrobaculum aerophilum*. *Proc Natl Acad Sci USA* **99**, 984–9.

Fleischmann, R. D., Adams, M. D., White, O. et al. (1995) Whole-genome random sequencing and assembly of *Haemophilus influenzae* Rd. *Science* **269**, 496–512.

Fletcher, R. J., Bishop, B. E., Leon, R. P., Sclafani, R. A., Ogata, C. M. & Chen, X. S. (2003) The structure and function of MCM from archaeal *M. thermoautotrophicum*. *Nat Struct Biol* **10**, 160–7.

Fogg, M. J., Pearl, L. H. & Connolly, B. A. (2002) Structural basis for uracil recognition by archaeal family B DNA polymerases. *Nat Struct Biol* **9**, 922–7.

Forbes, A. J., Patrie, S. M., Taylor, G. K., Kim, Y.-B., Jiang, L. & Kelleher, N. L. (2004) Targeted analysis and discovery of posttranslational modifications in proteins from methanogenic archaea by top-down MS. *Proc Natl Acad Sci USA* **101**, 2678–83.

Forget, D., Langelier, M.-F., Thérien, C., Trinh, V. & Coulombe, B. (2004) Photo-cross-linking of a purified preinitiation complex reveals central roles for the RNA polymerase II mobile clamp and TFIIE in initiation mechanisms. *Mol Cell Biol* **24**, 1122–31.

Forterre, P. (1997) Archaea: what can we learn from their sequence? *Curr Opin Genet Dev* **7**, 764–70.

Forterre, P. (1999) Displacement of cellular proteins by functional analogues from plasmids or viruses could explain puzzling phylogenies of many DNA informational proteins. *Mol Microbiol* **33**, 457–65.

Forterre, P. (2001) Genomics and early cellular evolution. The origin of the DNA world. *C R Acad Sci III* **324**, 1067–76.

Forterre, P. (2002) A hot story from comparative genomics: reverse gyrase is the only hyperthermophile-specific protein. *Trends Genet* **18**, 236–7.

Forterre, P. (2005) The two ages of the RNA world, and the transition to the DNA world: a story of viruses and cells. *Biochimie* **87**, 793–803.

Forterre, P., Bouthier de la Tour, C., Philippe, H. & Duguet, M. (2000) Reverse gyrase from hyperthermophiles: probable transfer of a thermoadaptation trait from archaea to bacteria. *Trends Genet* **16**, 152–4.

Forterre, P. & Philippe, H. (1999) Where is the root of the universal tree of life? *BioEssays* **21**, 871–9.

Fox, G. E., Magrum, L. J., Balch, W. E., Wolfe, R. S. & Woese, C. R. (1977b) Classification of methanogenic bacteria by 16S ribosomal RNA characterization. *Proc Natl Acad Sci USA* **74**, 4537–41.

Fox, G. E., Pechman, K. R. & Woese, C. R. (1977a) Comparative cataloging of 16S ribosomal RNA: molecular approach to procaryotic systematics. *Int J Syst Bacteriol* **27**, 44–57.

Fox, G. E., Stackebrandt, E., Hespell, R. B. et al. (1980) The phylogeny of prokaryotes. *Science* **209**, 457–63.

Fox, J. D., Routzahn, K. M., Bucher, M. H. & Waugh, D. S. (2003) Maltodextrin-binding proteins from diverse bacteria and archaea are potent solubility enhancers. *FEBS Lett* **537**, 53–7.

Frey, G., Thomm, M., Brüdigam, B., Gohl, H. & Hausner, W. (1990) An archaebacterial cell-free transcription system. The expression of tRNA genes from *Methanococcus vannielii* is mediated by a transcription factor. *Nucleic Acids Res* **18**, 1361–7.

Frey, P. A. (2001) Radical mechanisms of enzymatic catalysis. *Annu Rev Biochem* **70**, 121–48.

Frick, D. N. & Richardson, C. C. (2001) DNA primases. *Ann Rev Biochem* **70**, 39–80.

Friedrich, C. G., Bardischewsky, F., Rother, D., Quentmeier, A. & Fischer, J. (2005) Prokaryotic sulfur oxidation. *Curr Opin Microbiol* **8**, 253–9.

Friedrich, C. G., Rother, D., Bardischewsky, F., Quentmeier, A. & Fischer, J. (2001) Oxidation of reduced inorganic sulfur compounds by bacteria: emergence of a common mechanism? *Appl Environ Microbiol* **67**, 2873–82.

Friedrich, M. W., Schmitt-Wagner, D., Lueders, T. & Brune, A. (2001) Axial differences in community structure of *Crenarchaeota* and *Euryarchaeota* in the highly compartmentalized gut of the soil-feeding termite *Cubitermes orthognathus*. *Appl Environ Microbiol* **67**, 4880–90.

Fritz, G., Buchert, T. & Kroneck, P. M. (2002a) The function of the [4Fe-4S] clusters and FAD in bacterial and archaeal adenylylsulfate reductases. Evidence for flavin-catalyzed reduction of adenosine 5′-phosphosulfate. *J Biol Chem* **277**, 26066–73.

Fritz, G., Roth, A., Schiffer, A. et al. (2002b) Structure of adenylylsulfate reductase from the hyperthermophilic *Archaeoglobus fulgidus* at 1.6-Å resolution. *Proc Natl Acad Sci USA* **99**, 1836–41.

Fuhrman, J. A., McCallum, K. & Davis, A. A. (1992) Novel major archaebacterial group from marine plankton. *Nature* **356**, 148–9.

Fukui, T., Atomi, H., Kanai, T., Matsumi, R., Fujiwara, S. & Imanaka, T. (2005) Complete genome sequence of the hyperthermophilic archaeon *Thermococcus kodakaraensis* KOD1 and comparison with *Pyrococcus* genomes. *Genome Res* **15**, 352–63.

Fukui, T., Yamauchi, K., Muroya, T. et al. (2004) Distinct roles of DNA polymerases delta and epsilon at the replication fork in *Xenopus* egg extracts. *Genes Cells* **9**, 179–91.

Fukushima, T., Mizuki, T., Echigo, A., Inoue, A. & Usami, R. (2005) Organic solvent tolerance of halophilic alpha-amylase from a haloarchaeon, *Haloarcula* sp. strain S-1. *Extremophiles* **9**, 85–9.

Furter, R. (1998) Expansion of the genetic code: site-directed p-fluoro-phenylalanine incorporation in *Escherichia coli*. *Protein Sci* **7**, 419–26.

Fusi, P., Grisa, M., Tedeschi, G., Negri, A., Guerritore, A. & Tortora, P. (1995) An 8.5-kDa ribonuclease from the extreme thermophilic archaebacterium *Sulfolobus solfataricus*. *FEBS Lett* **360**, 187–90.

Fütterer, O., Angelov, A., Liesegang, H. et al. (2004) Genome sequence of *Picrophilus torridus* and its implications for life around pH 0. *Proc Natl Acad Sci USA* **101**, 9091–6.

Gadd, G. M. (1993) Microbial formation and transformation of organometallic and organometalloid compounds. *FEMS Microbiol Rev* **11**, 297–316.

Gadelle, D., Filee, J., Buhler, C. & Forterre, P. (2003) Phylogenomics of type II DNA topoisomerases. *Bioessays* **25**, 232–42.

Gajiwala, K. S. & Burley, S. K. (2000) Winged helix proteins. *Curr Opin Struct Biol* **10**, 110–16.

Galagan, J. E., Nusbaum, C., Roy, A. et al. (2002) The genome of *M. acetivorans* reveals extensive metabolic and physiological diversity. *Genome Res* **12**, 532–42.

Galperin, M. Y. & Koonin, E. V. (2000) Who's your neighbor? New computational approaches for functional genomics. *Nat Biotechnol* **18**, 609–13.

Gamow, G. (1954) Possible relations between deoxyribonucleic acid and protein structures. *Nature* **173**, 318.

Ganot, P., Caizergues-Ferrer, M. & Kiss, T. (1997) The family of box H/ACA small nucleolar RNAs is defined by an evolutionarily conserved secondary structure and ubiquitous sequence elements essential for RNA accumulation. *Genes Dev* **11**, 941–56.

Gans, J., Wolinsky, M. & Dunbar, J. (2005) Computational improvements reveal great bacterial diversity and high metal toxicity in soil. *Science* **309**, 1387–90.

Ganther, H. E. (1966) Enzymatic synthesis of dimethyl selenide from sodium selenite in mouse liver extracts. *Biochemistry* **3**, 1089–98.

Gao, J., Bauer, M. W., Shockley, K. R., Pysz, M. A. & Kelly, R. M. (2003) Growth of hyperthermophilic archaeon *Pyrococcus furiosus* on chitin involves two family 18 chitinases. *Appl Environ Microbiol* **69**, 3119–28.

Gardner, W. L. & Whitman, W. B. (1999) Expression vectors for *Methanococcus maripaludis*: overexpression of acetohydroxyacid synthase and β-galactosidase. *Genetics* **152**, 1439–47.

Garrett, R. A., Redder, P., Greve, B., Brügger, K., Chen, L. & She, Q. (2004) Archaeal Plasmids. *In:* B. E. Funnel & G. J. Phillips (eds) *Plasmid Biology.* ASM Press, Washington, DC, pp. 377–92.

Gaspin, C., Cavaille, J., Erauso, G. & Bachellerie, J. P. (2000) Archaeal homologs of eukaryotic methylation guide small nucleolar RNAs: lessons from the *Pyrococcus* genomes. *J Mol Biol* **297**, 895–906.

Geiduschek, E. P. & Ouhammouch, M. (2005) Archaeal transcription and its regulators. *Mol Microbiol* **56**, 1397–407.

Gerlt, J. A. & Babbitt, P. C. (2000) Can sequence determine function? *Genome Biol* **1**, 0005.1–0005.10.

Gernhardt, P., Possot, O., Foglino, M., Sibold, L. & Klein, A. (1990) Construction of an integration vector for use in the archaebacterium *Methanococcus voltae* and expression of a eubacterial resistance gene. *Mol Gen Genet* **221**, 273–9.

Ghane, F. & Grogan, D. W. (1998) Chromosomal marker exchange in the thermophilic archaeon *Sulfolobus acidocaldarius*: physiological and cellular aspects. *Microbiology* **144**, 1649–57.

Ghosh, A., Wondimagegn, T. & Ryeng, H. (2001a) Deconstructing F430: quantum chemical perspectives of biological methanogenesis. *Curr Opin Chem Biol* **5**, 744–50.

Ghosh, P., Ishihama, A. & Chatterji, D. (2001b) *Escherichia coli* RNA polymerase subunit ω and its N-terminal domain bind full-length β' to facilitate incorporation into the $\alpha_2\beta$ subassembly. *Eur J Biochem* **268**, 4621–7.

Giege, R., Puglisi, J. D. & Florentz, C. (1993) tRNA structure and aminoacylation efficiency. *Prog Nucleic Acid Res Mol Biol* **45**, 129–206.

Giovannoni, S. J., Tripp, H. J., Givan, S. et al. (2005) Genome streamlining in a cosmopolitan oceanic bacterium. *Science* **309**, 1242–5.

Girguis, P. R., Cozen, A. E. & DeLong, E. F. (2005) Growth and population dynamics of anaerobic methane-oxidizing archaea and sulfate-reducing bacteria in a continuous-flow bioreactor. *Appl Environ Microbiol* **71**, 3725–33.

Girguis, P. R., Orphan, V. J., Hallam, S. J. & DeLong, E. F. (2003) Growth and methane oxidation rates of anaerobic methanotrophic archaea in a continuous flow bioreactor. *Appl Environ Microbiol* **69**, 5472–82.

Glansdorff, N. (2000) About the last common ancestor, the universal life-tree and lateral gene transfer: a reappraisal. *Mol Microbiol* **38**, 177–85.

Glick, B. R., Chladek, S. & Ganoza, M. C. (1979) Peptide bond formation stimulated by protein synthesis factor EF-P depends on the aminoacyl moiety of the acceptor. *Eur J Biochem* **97**, 23–8.

Gnatt, A. L., Cramer, P., Fu, J., Bushnell, D. A. & Kornberg, R. D. (2001) Structural basis of transcription: an RNA polymerase II elongation complex at 3.3 Å resolution. *Science* **292**, 1876–82.

Goenrich, M., Duin, E. C., Mahlert, F. & Thauer, R. K. (2005) Temperature dependence of methyl-coenzyme M reductase activity and of the formation of the methyl-coenzyme M reductase red2 state induced by coenzyme B. *J Biol Inorg Chem* **10**, 333–42.

Goenrich, M., Mahlert, F., Duin, E. C., Bauer, C., Jaun, B. & Thauer, R. K. (2004) Probing the reactivity of Ni in the active site of methyl-coenzyme M reductase with substrate analogues. *J Biol Inorg Chem* **9**, 691–705.

Goffeau, A., Barrell, B. G., Bussey, H. et al. (1996) Life with 6000 genes. *Science* **274**, 563–7.

Gogarten, J. P., Doolittle, W. F. & Lawrence, J. G. (2002) Prokaryotic evolution in light of gene transfer. *Mol Biol Evol* **19**, 2226–38.

Gogarten-Boekels, M., Hilario, E. & Gogarten, J. P. (1995) The effects of heavy meteorite bombardment on the early evolution – the emergence of the three domains of life. *Orig Life Evol Biosph* **25**, 251–64.

Gohl, H. P., Gröndahl, B. & Thomm, M. (1995) Promoter recognition in archaea is mediated by transcription factors: identification of a TFB from *Methanococcus thermolithotrophicus* as archaeal TATA-binding protein. *Nucleic Acids Res* **23**, 3837–41.

Gomes, J. & Steiner, W. (2004) The biocatalytic potential of extremophiles and extremozymes. *Food Technol Biotechnol* **42**, 223–35.

Goo, Y. A., Roach, J, Glusman, G. et al. (2004) Low-pass sequencing for microbial comparative genomics. *BMC Genomics* **5**, 1–19.

Goodchild, A., Saunders, N. F., Ertan, H. et al. (2004) A proteomic determination of cold adaptation in the

Antarctic archaeon, *Methanococcoides burtonii*. *Mol Microbiol* **53**, 309–21.

Gophna, U., Charlebois, R. L. & Doolittle, W. F. (2004) Have archaeal genes contributed to bacterial virulence? *Trends Microbiol* **12**, 213–19.

Gophna, U., Doolittle, W. F. & Charlebois, R. L. (2005) Weighted genome trees: refinements and applications. *J Bacteriol* **187**, 1305–16.

Gorbalenya, A. E., Koonin, E. V. & Wolf, Y. I. (1990) A new superfamily of putative NTP-binding domains encoded by genomes of small DNA and RNA viruses. *FEBS Lett* **262**, 145–8.

Gosio, B. (1897) Zur Frage, wodurch die Giftigkeit arsenhaltiger Tapeten bedingt wird. *Ber Dtsch Chem Ges* **30**, 1024–7.

Gottesman, S. (2002) Stealth regulation: biological circuits with small RNA switches. *Genes Dev* **16**, 2829–42.

Grabarse, W., Mahlert, F., Duin, E. C. et al. (2001a) On the mechanism of biological methane formation: Structural evidence for conformational changes in methylcoenzyme M reductase upon substrate binding. *J Mol Biol* **309**, 315–30.

Grabarse, W., Mahlert, F., Shima, S., Thauer, R. K. & Ermler, U. (2000) Comparison of three methyl-coenzyme M reductases from phylogenetically distant organisms: Unusual amino acid modification, conservation and adaptation. *J Mol Biol* **303**, 329–44.

Grabarse, W., Shima, S., Mahlert, F., Duin, E. C., Thauer, R. K. & Ermler, U. (2001b) *In:* A. Messerschmidt, R. Huber, T. Poulos & K. Wieghardt (eds) *Handbook of Metalloproteins*. John Wiley & Sons, Chichester, pp. 897–914.

Graham, D. E., Overbeek, R., Olsen, G. J. & Woese, C. R. (2000) An archaeal genomic signature. *Proc Natl Acad Sci USA* **97**, 3304–8.

Graham, D. E., Xu, H. & White, R. H. (2002a) Identification of coenzyme M biosynthetic phosphosulfolactate synthase. A new family of sulfonate-biosynthesizing enzymes. *J Biol Chem* **277**, 13421–9.

Graham, D. E., Xu, H. & White, R. H. (2002b) A divergent archaeal member of the alkaline phosphatase binuclear metalloenzyme superfamily has phosphoglycerate mutase activity. *FEBS Lett* **517**, 190–4.

Grainge, I., Scaife, S. & Wigley, D. (2003) Biochemical analysis of components of the pre-replication complex of *Archaeoglobus fulgidus*. *Nucleic Acids Res* **31**, 4888–98.

Granneman, S. & Baserga, S. J. (2005) Crosstalk in gene expression: coupling and co-regulation of rDNA transcription, pre-ribosome assembly and pre-rRNA processing. *Curr Opin Cell Biol* **17**, 281–6.

Gregor, D. & Pfeifer, F. (2001) The use of a halobacterial *bgaH* reporter gene to analyse the regulation of gene expression in halophilic archaea. *Microbiology* **147**, 1745–54.

Gregor, D. & Pfeifer, F. (2005) *In vivo* analyses of constitutive and regulated promoters in halophilic archaea. *Microbiology* **151**, 25–33.

Greve, B., Jensen, S., Brügger, K., Zillig, W. & Garrett, R. A. (2004) Genomic comparison of archaeal conjugative plasmids from *Sulfolobus*. *Archaea* **1**, 231–9.

Greve, B., Jensen, S., Phan, H. et al. (2005) Novel RepA-MCM proteins encoded in plasmids pTAU4, pORA1 and pTIK4 from *Sulfolobus neozealandicus*. *Archaea* **1**, 319–25.

Gribaldo, S., Lumia, V., Creti, R., de Macario, E. C., Sanangelantoni, A. & Cammarano, P. (1999) Discontinuous occurrence of the hsp70 (dnaK) gene among Archaea and sequence features of HSP70 suggest a novel outlook on phylogenies inferred from this protein. *J Bacteriol* **181**, 434–43.

Gribaldo, S. & Philippe, H. (2002) Ancient phylogenetic relationships. *Theor Popul Biol* **61**, 391–408.

Grill, S., Gualerzi, C. O., Londei, P. & Blasi, U. (2000) Selective stimulation of translation of leaderless mRNA by initiation factor 2: evolutionary implications for translation. *EMBO J* **19**, 4101–10.

Groft, C. M., Beckmann, R., Sali, A. & Burley, S. K. (2000) Crystal structures of ribosome anti-association factor IF6. *Nat Struct Biol* **7**, 1156–64.

Grogan, D. W. (1996) Exchange of genetic markers at extremely high temperatures in the archaeon *Sulfolobus acidocaldarius*. *J Bacteriol* **178**, 3207–11.

Grogan, D. W. (2000) The question of DNA repair in hyperthermophilic archaea. *Trends Microbiol* **8**, 180–5.

Grogan, D. W. (2003) Cytosine methylation by the SuaI restriction-modification system: implications for genetic fidelity in a hyperthermophilic archaeon. *J Bacteriol* **185**, 4657–61.

Grogan, D. W. (2004) Stability and repair of DNA in hyperthermophilic Archaea. *Curr Issues Mol Biol* **6**, 137–44.

Grogan, D. W., Carver, G. T. & Drake, J. W. (2001) Genetic fidelity under harsh conditions: analysis of spontaneous mutation in the thermoacidophilic archaeon *Sulfolobus acidocaldarius*. *Proc Natl Acad Sci USA* **98**, 7928–33.

Grogan, D. W. & Hansen, J. E. (2003) Molecular characteristics of spontaneous deletions in the hyperthermophilic archaeon *Sulfolobus acidocaldarius*. *J Bacteriol* **185**, 1266–72.

Gropp, F. & Betlach, M. (1994) The *bat* gene of *Halobacterim halobium* encodes a transacting oxygen inducible factor. *Proc Natl Acad Sci USA* **91**, 5475–9.

Gropp, F., Grampp, B., Stolt, P., Palm, P. & Zillig, W. (1992) The immunity-conferring plasmid phi HL from the *Halobacterium salinarium* phage phi H: nucleotide sequence and transcription. *Virology* **190**, 45–54.

Gropp, F., Gropp, R. & Betlach, M. (1994) A fourth gene in the *bop* gene cluster of *Halobacterium halobium* is coregulated with the *bop* gene. *Syst Appl Microbiol* **16**, 716–24.

Gropp, F., Gropp, R. & Betlach, M. (1995) Effects of upstream deletions on light- and oxygen-regulated bacterio-opsin gene expression in *Halobacterium halobium*. *Mol Microbiol* **16**, 357–64.

Grosskopf, R., Janssen, P. H. & Liesack, W. (1998) Diversity and structure of the methanogenic community in anoxic rice paddy soil microcosms as examined by cultivation and direct 16S rRNA gene sequence retrieval. *Appl Environ Microbiol* **64**, 960–9.

Guagliardi, A., Cerchia, L. & Rossi, M. (2002) The Sso7d protein of *Sulfolobus solfataricus*: in vitro relationship among different activities. *Archaea* **1**, 87–93.

Guagliardi, A., Mancusi, L. & Rossi, M. (2004) Reversion of protein aggregation mediated by Sso7d in cell extracts of *Sulfolobus solfataricus*. *Biochem J* **381**, 249–55.

Gualerzi, C. & Pon, C. L. (1990) Initiation of mRNA translation in prokaryotes. *Biochemistry* **29**, 5881–9.

Gueguen, Y., Rolland, J. L., Schroeck, S. et al. (2001) Characterization of the maltooligosyl trehalose synthase from the thermophilic archaeon *Sulfolobus acidocaldarius*. *FEMS Microbiol Lett* **194**, 201–6.

Guenneugues, M., Caserta, E., Brandi, L. et al. (2000) Mapping the fMet-tRNA(f)(Met) binding site of initiation factor IF2. *EMBO J* **19**, 5233–40.

Guerrier-Takada, C. & Altaman, S. (1984) Catalytic activity of an RNA molecule prepared by transcription *in vitro*. *Science* **223**, 285–6.

Guindon, S. & Gascuel, O. (2003) A simple, fast, and accurate algorithm to estimate large phylogenies by maximum likelihood. *Syst Biol* **52**, 696–704.

Guo, R., Xue, H. & Huang, L. (2003) Ssh10b, a conserved thermophilic archaeal protein, binds RNA *in vivo*. *Mol Microbiol* **50**, 1605–15.

Gupta, R. S. (1998) Protein phylogenies and signature sequences: A reappraisal of evolutionary relationships among archaebacteria, eubacteria, and eukaryotes. *Microbiol Mol Biol Rev* **62**, 1435–91.

Guy, C. P. & Bolt, E. L. (2005) Archaeal Hel308 helicase targets replication forks in vivo and in vitro and unwinds lagging strands. *Nucleic Acids Res* **33**, 3678–90.

Guy, J. E., Isupov, M. N. & Littlechild, J. A. (2003) The structure of an alcohol dehydrogenase from the hyperthermophilic archaeon *Aeropyrum pernix*. *J Mol Biol* **331**, 1041–51.

Ha, I., Roberts, S., Maldonado, E. et al. (1993) Multiple functional domains of human transcription factor IIB: distinct interactions with two general transcription factors and RNA polymerase II. *Genes Dev* **7**, 1021–32.

Haber, J. E. & Heyer, W. D. (2001) The fuss about Mus81. *Cell* **107**, 551–4.

Hackett, N. R., Krebs, M. P., DasSarma, S. et al. (1990) Nucleotide sequence of a high copy number plasmid from *Halobacterium* strain GRB. *Nucleic Acids Res* **18**, 3408.

Hackstein, J. H. & Stumm, C. K. (1994) Methane production in terrestrial arthropods. *Proc Natl Acad Sci USA* **91**, 5441–5.

Hackstein, J. H. & Vogels, G. D. (1997) Endosymbiotic interactions in anaerobic protozoa. *Antonie Van Leeuwenhoek* **71**, 151–8.

Hafenbradl, D., Keller, M., Thiericke, R. & Stetter, K. O. (1993) A novel unsaturated archaeal ether core lipid from the hyperthermophile *Methanopyrus kandleri*. *Syst Appl Microbiol* **16**, 165–9.

Halic, M., Becker, T., Pool, M. R. et al. (2004) Structure of the signal recognition particle interacting with the elongation-arrested ribosome. *Nature* **427**, 808–14.

Hall, M. J. & Hackett, N. R. (1989) DNA sequence of a small plasmid from *Halobacterium* strain GN101. *Nucleic Acids Res* **17**, 10501.

Hallam, S. J., Girguis, P. R., Preston, C. M., Richardson, P. M. & DeLong, E. F. (2003) Identification of methyl coenzyme M reductase A (*mcrA*) genes associated with methane-oxidizing archaea. *Appl Environ Microbiol* **69**, 5483–91.

Hallam, S. J., Putnam, N., Preston, C. M. et al. (2004) Reverse methanogenesis: testing the hypothesis with environmental genomics. *Science* **305**, 1457–62.

Hamilton-Brehm, S. D., Schut, G. J. & Adams, M. W. W. (2005) Metabolic and evolutionary relationships among *Pyrococcus* species: genetic exchange within a hydrothermal vent environment. *J Bacteriol* **187**, 7492–9.

Hanawa-Suetsugu, K., Sekine, S., Sakai, H. et al. (2004) Crystal structure of elongation factor P from *Thermus thermophilus* HB8. *Proc Natl Acad Sci USA* **101**, 9595–600.

Hannington, M. D., Jonasson, I. R., Herzig, P. M. & Peterson, S. (1995) Physical and chemical processes of seafloor mineralization at mid-ocean ridge. *In*: S. E. Humphris, R. A. Zierenberg, L. S. Mullineaux & R. E. Thomson (eds) *Seafloor Hydrothermal Systems: Physical, Chemical, Biological, and Geological Interactions*. American Geophysical Union, Washington, DC, pp. 115–57.

Hansen, J. E., Dill, A. C. & Grogan, D. W. (2005a) Conjugational genetic exchange in the hyperthermophilic archaeon *Sulfolobus acidocaldarius*: intragenic recombination with minimal dependence on marker separation. *J Bacteriol* **187**, 805–9.

Hansen, T., Schlichting, B., Felgendreher, M. et al. (2005b) Cupin-type phosphoglucose isomerases (Cupin-PGIs) constitute a novel metal-dependent PGI family representing a convergent line of PGI evolution. *J Bacteriol* **187**, 1621–31.

Hansen, T. & Schönheit, P. (2001) Sequence, expression, and characterization of the first archaeal ATP-dependent 6-phosphofructokinase, a non-allosteric enzyme related to the phosphofructokinase-B sugar kinase family, from the hyperthermophilic crenarchaeote *Aeropyrum pernix*. *Arch Microbiol* **177**, 62–9.

Hansen, T., Wendorff, D. & Schönheit, P. (2004) Bifunctional phosphoglucose/phosphomannose isomerases from the Archaea *Aeropyrum pernix* and *Thermoplasma acidophilum* constitute a novel enzyme family within the phosphoglucose isomerase superfamily. *J Biol Chem* **279**, 2262–72.

Hanzelka, B. L., Darcy, T. J. & Reeve, J. N. (2001) TFE, an archaeal transcription factor in *Methanobacterium thermoautotrophicum* related to eucaryal transcription factor TFIIFα. *J Bacteriol* **183**, 1813–18.

Hao, B., Gong, W., Ferguson, T. K., James, C. M., Krzycki, J. A. & Chan, M. K. (2002) A new UAG-encoded residue in the structure of a methanogen methyltransferase. *Science* **296**, 1462–6.

Häring, M., Peng, X., Brugger, K. et al. (2004) Morphology and genome organization of the virus PSV of the hyperthermophilic archaeal genera *Pyrobaculum* and *Thermoproteus*: a novel virus family, the *Globuloviridae*. *Virology* **323**, 233–42.

Häring, M., Rachel, R., Peng, X., Garrett, R. A. & Prangishvili, D. (2005a) Viral diversity in hot springs of Pozzuoli, Italy, and characterization of a unique archaeal virus, *Acidianus* bottle-shaped virus, from a new family, the *Ampullaviridae*. *J Virol* **79**, 9904–11.

Häring, M., Vestergaard, G., Brügger, K., Rachel, R., Garrett, R. A. & Prangishvili, D. (2005b) Structure and genome organization of AFV2, a novel archaeal lipothrixvirus with unusual terminal and core structures. *J Bacteriol* **187**, 3855–8.

Häring, M., Vestergaard, G., Rachel, R., Chen, L., Garrett, R. A. & Prangishvili, D. (2005c) Virology: independent virus development outside a host. *Nature* **436**, 1101–12.

Harmer, J., Finazzo, C., Piskorski, R. et al. (2005) Spin density and coenzyme M coordination geometry of the ox1 form of methyl-coenzyme M reductase: A pulse EPR study. *J Am Chem Soc* **127**, 17744–55.

Hashimoto, H., Nishioka, M., Fujiwara, S. et al. (2001) Crystal structure of DNA polymerase from hyperthermophilic archaeon *Pyrococcus kodakaraensis* KOD1. *J Mol Biol* **306**, 469–77.

Hatfield, D. L. & Gladyshev, V. N. (2002) How selenium has altered our understanding of the Genetic Code. *Mol Cell Biol* **22**, 3565–76.

Hausner, W., Lange, U. & Musfeldt, M. (2000) Transcription factor S, a cleavage induction factor of the archaeal RNA polymerase. *J Biol Chem* **275**, 12393–9.

Hausner, W. & Thomm, M. (2001) Events during initiation of archaeal transcription: Open complex formation and DNA-protein interactions. *J Bacteriol* **183**, 3025–31.

Hausner, W., Wettach, J., Hethke, C. & Thomm, M. (1996) Two transcription factors related with the eucaryal transcription factors TATA-binding protein and transcription factor IIB direct promoter recognition by an archaeal RNA polymerase. *J Biol Chem* **271**, 30144–8.

Hayashi, I., Kawai, G. & Watanabe, K. (1998) Higher-order structure and thermal instability of bovine mitochondrial tRNA$_{UGA}^{Ser}$ investigated by proton NMR spectroscopy. *J Mol Biol* **284**, 57–69.

He, Z., Li, Y., Zhou, P. & Liu, S. (2000) Cloning and heterologous expression of a sulfur oxygenase/reductase gene from the thermoacidophilic archaeon *Acidianus* sp. S5 in *Escherichia coli*. *FEMS Microbiol Lett* **193**, 217–21.

Healy, S. M., Zakharyan, R. A. & Aposhian, H. V. (1997) Enzymatic methylation of arsenic compounds: IV. *In vitro* and *in vivo* deficiency of the methylation of arsenite and monomethylarsonic acid in the guinea pig. *Mutat Res* **386**, 229–39.

Hedderich, R., Klimmek, O., Kröger, A., Dirmeier, R., Keller, M. & Stetter, K. O. (1999) Anaerobic respiration with elemental sulfur and with sulfides. *FEMS Microbiol Rev* **22**, 353–81.

Heinicke, I., Müller, J., Pittelkow, M. & Klein, A. (2004) Mutational analysis of genes encoding chromatin proteins in the archaeon *Methanococcus voltae* indicates their involvement in the regulation of gene expression. *Mol Genet Genomics* **272**, 76–87.

Hemmi, H., Ikejiri, S., Nakayama, T. & Nishino, T. (2003) Fusion-type lycopene β-cyclase from a thermoacidophilic archaeon *Sulfolobus solfataricus*. *Biochem Biophys Res Commun* **305**, 586–91.

Hendrickson, E. L., Kaul, R., Zhou, Y. et al. (2004) Complete genome sequence of the genetically tractable hydrogenotrophic methanogen *Methanococcus maripaludis*. *J Bacteriol* **186**, 6956–69.

Hendrix, R. W., Smith, M. C., Burns, R. N., Ford, M. E. & Hatfull, G. F. (1999) Evolutionary relationships among diverse bacteriophages and prophages: all the world's a phage. *Proc Natl Acad Sci USA* **96**, 2192–7.

Henneke, G., Flament, D., Hubscher, U., Querellou, J. & Raffin, J. P. (2005) The hyperthermophilic euryarchaeota *Pyrococcus abyssi* likely requires the two DNA polymerases D and B for DNA replication. *J Mol Biol* **350**, 53–64.

Hernandez, N. (1993) TBP, a universal eukaryotic transcription factor? *Genes Dev* **7**, 1291–308.

Herrmann, U. & Soppa, J. (2002) Cell cycle-dependent expression of an essential SMC-like protein and dynamic chromosome localization in the archaeon *Halobacterium salinarum*. *Mol Microbiol* **46**, 395–409.

Hezayen, F. F., Rehm, B. H., Eberhardt, R. & Steinbüchel, A. (2000) Polymer production by two newly isolated extremely halophilic archaea: application of a novel corrosion-resistant bioreactor. *Appl Microbiol Biotechnol* **54**, 319–25.

Hezayen, F. F., Rehm, B. H., Tindall, B. J. & Steinbüchel, A. (2001) Transfer of *Natrialba asiatica* B1T to *Natrialba taiwanensis* sp. nov. and description of *Natrialba aegyptiaca* sp. nov., a novel extremely halophilic, aerobic, non-pigmented member of the Archaea from Egypt that produces extracellular poly(glutamic acid). *Int J Syst Evol Microbiol* **51**, 1133–42.

Hickey, A. J., Conway de Macario, E. & Macario, A. J. L. (2002) Transcription in the Archaea: basal factors, regulation, and stress-gene expression. *Crit Rev Biochem Mol Biol* **37**, 537–99.

Higashibata, H., Siddiqui, M. A., Takagi, M., Imanaka, T. & Fujiwara, S. (2003) Surface histidine residue of archaeal histone affects DNA compaction and thermostability. *FEMS Microbiol Lett* **224**, 17–22.

Hinrichs, K. U., Hayes, J. M., Sylva, S. P., Brewer, P. G. & DeLong, E. F. (1999) Methane-consuming archaebacteria in marine sediments. *Nature* **398**, 802–5.

Hjort, K. & Bernander, R. (1999) Changes in cell size and DNA content in *Sulfolobus* cultures during dilution and temperature shift experiments. *J Bacteriol* **181**, 5669–75.

Hjort, K. & Bernander, R. (2001) Cell cycle regulation in the hyperthermophilic crenarchaeon *Sulfolobus acidocaldarius*. *Mol Microbiol* **40**, 225–34.

Hoehler, T., Alperin, M. J., Albert, D. B. & Martens, C. S. (1994) Field and laboratory studies of methane oxidation in an anoxic marine sediment: Evidence for a methanogen-sulfate reducer consortium. *Global Biogeochem Cycles* **8**, 451–63.

Hoehler, T., Alperin, M. J., Albert, D. B. & Martens, C. S. (2001) Apparent minimum free energy requirements for methanogenic Archaea and sulfate-reducing bacteria in an anoxic marine sediment. *FEMS Microbiol Ecol* **38**, 33–41.

Hofacker, A., Schmitz, K. M., Cichonczyk, A., Sartorius-Neef, S. & Pfeifer, F. (2004) GvpE- and GvpD-mediated transcription regulation of the p-*gvp* genes encoding gas vesicles in *Halobacterium salinarum*. *Microbiol* **150**, 1829–38.

Hohn, M. J., Hedlund, B. P. & Huber, H. (2002). Detection of 16S rDNA sequences representing the novel phylum "Nanoarchaeota": indication for a broad distribution in high temperature. *System Appl Microbiol* **25**, 551–4.

Holden, J. F., Poole, F. L. & Tollaksen, S. L. et al. (2001) Identification of membrane proteins in the hyperthermophilic archaeon *Pyrococcus furiosus* using proteomics and prediction programs. *Comp Funct Genomics* **2**, 275–88.

Holmes, M. L. & Dyall-Smith, M. L. (1990) A plasmid vector with a selectable marker for halophilic archaebacteria. *J Bacteriol* **172**, 756–61.

Holmes, M. L. & Dyall-Smith, M. L. (1991) Mutations in DNA gyrase result in novobiocin resistance in halophilic archaebacteria. *J Bacteriol* **173**, 642–8.

Holmes, M. L. & Dyall-Smith, M. L. (2000) Sequence and expression of a halobacterial β-galactosidase gene. *Mol Microbiol* **36**, 114–22.

Holmes, M. L., Nuttall, S. D. & Dyall-Smith, M. L. (1991) Construction and use of halobacterial shuttle vectors and further studies on *Haloferax* DNA gyrase. *J Bacteriol* **173**, 3807–13.

Holmes, M. L., Pfeifer, F. & Dyall-Smith, M. L. (1994) Improved shuttle vectors for *Haloferax volcanii* including a dual-resistance plasmid. *Gene* **146**, 117–21.

Holstege, F. C., Fiedler, U. & Timmers, H. T. (1997) Three transitions in the RNA polymerase II transcription complex during initiation. *EMBO J* **16**, 7468–80.

Hopfner, K. P., Karcher, A., Craig, L., Woo, T. T., Carney, J. P. & Tainer, J. A. (2001) Structural biochemistry and interaction architecture of the DNA double-strand break repair Mre11 nuclease and Rad50-ATPase. *Cell* **105**, 473–85.

Horikoshi, K. (1999) Alkaliphiles: some applications of their products for biotechnology. *Microbiol Mol Biol Rev* **63**, 735–50.

Horne, M., Englert, C. & Pfeifer, F. (1991) A DNA region of 9 kbp contains all genes necessary for gas vesicle synthesis in halophilic archaebacteria. *Mol Microbiol* **5**, 1159–74.

Horng, Y. C., Becker, D. F. & Ragsdale, S. W. (2001) Mechanistic studies of methane biogenesis by methyl-coenzyme M reductase: evidence that coenzyme B participates in cleaving the C–S bond of methyl-coenzyme M. *Biochemistry* **40**, 12875–85.

Hotta, Y., Ezaki, S., Atomi, H. & Imanaka, T. (2002) Extremely stable and versatile carboxylesterase from a hyperthermophilic archaeon. *Appl Environ Microbiol* **68**, 3925–31.

Hsiao, W. W. L., Ung, K., Aeschliman, D., Bryan, J., Finlay, B. B. & Brinkman, F. S. L. (2005) Evidence of a large novel gene pool associated with prokaryotic genomic islands. *PloS Genetics* **1**, e62.

Huang, Y., Krauss, G., Cottaz, S., Driguez, H. & Lipps, G. (2005) A highly acid-stable and thermostable endo-beta-glucanase from the thermoacidophilic archaeon *Sulfolobus solfataricus*. *Biochem J* **385**, 581–8.

Huber, H., Burggraf, S., Mayer, T., Wyschkony, I., Rachel, R. & Stetter, K. O. (2000) *Ignicoccus* gen. nov., a novel genus of hyperthermophilic, chemolithoautotrophic Archaea, represented by two new species, *Ignicoccus islandicus* sp. nov. and *Ignicoccus pacificus*. sp. nov. *Int J Syst Evol Microbiol* **50**, 2093–100.

Huber, H., Hohn, M. J., Rachel, R., Fuchs, T., Wimmer, V. C. & Stetter, K. O. (2002a) A new phylum of Archaea represented by a nano-sized hyperthermophilic symbiont. *Nature* **417**, 63–7.

Huber, H., Hohn, M. J., Rachel, R. & Stetter, K. O. (2003a) Nanoarchaeota. In: M. Dworkin et al. (eds) *The Prokaryotes: An Evolving Electronic Resource for the Microbiological Community*, 3rd edn, release 3.15 (www.prokaryotes.com). Springer-Verlag, New York.

Huber, H., Hohn, M. J., Stetter, K. O. & Rachel, R. (2003b) The phylum *Nanoarchaeota*: present knowledge and future perspectives of a unique form of life. *Res Microbiol* **154**, 165–71.

Huber, H., Huber, R. & Stetter, K. (2002b) The order *Thermoproteales*. In: M. Dworkin (ed.) *The Prokaryotes: An Evolving Electronic Resource for the Microbiological Community*, 3rd edn, release 3.8. Springer-Verlag, New York.

Huber, H. & Prangishvili, D. (2005) The order *Sulfolobales*. In: M. Dworkin (ed.) *The Prokaryotes: An Evolving Electronic Resource for the Microbiological Community*, 3rd edition, release 3.19. Springer-Verlag, New York.

Huber, R. & Stetter, K. O. (2001) Discovery of hyperthermophilic microorganisms. *Methods Enzymol* **330**, 11–24.

Hubscher, U., Maga, G. & Spadari, S. (2002) Eukaryotic DNA polymerases. *Annu Rev Biochem* **71**, 133–63.

Hüdepohl, U., Reiter, W. D. & Zillig, W. (1990) *In vitro* transcription of two rRNA genes of the archaebacterium *Sulfolobus* sp. B 12 indicates a factor requirement for specific initiation. *Proc Natl Acad Sci USA* **87**, 5851–5.

Hügler, M., Huber, H., Stetter, K. O. & Fuchs, G. (2003) Autotrophic CO_2 fixation pathways in archaea (Crenarchaeota). *Arch Microbiol* **179**, 160–73.

Huet, J., Schnabel, R., Sentenac, A. & Zillig, W. (1983) Archaebacteria and eukaryotes possess DNA-dependent RNA polymerases of a common type. *EMBO J* **2**, 1291–4.

Huttenhofer, A., Kiefmann, M., Meier-Ewert, S. et al. (2001) RNomics: an experimental approach that identifies 201 candidates for novel, small, non-messenger RNAs in mouse. *EMBO J* **20**, 2943–53.

Hutton, J. (1788) The theory of the earth. *Trans R Soc Edin* **1**, 214.

Iakhiaeva, E., Yin, J. & Zwieb, C. (2005) Identification of an RNA-binding Domain in Human SRP72. *J Mol Biol* **345**, 659–66.

Ibba, M., Bono, J. L., Rosa, P. A. & Söll, D. (1997a) Archaeal-type lysyl-tRNA synthetase in the Lyme disease spirochete *Borrelia burgdorferi*. *Proc Natl Acad Sci USA* **94**, 14383–8.

Ibba, M., Morgan, S., Curnow, A. W. et al. (1997b) A euryarchaeal lysyl-tRNA synthetase: resemblance to class I synthetases. *Science* **278**, 1119–22.

Ibba, M. & Söll, D. (2000) Aminoacyl-tRNA synthesis. *Annu Rev Biochem* **69**, 617–50.

Ibba, M. & Söll, D. (2001) The renaissance of aminoacyl-tRNA synthesis. *EMBO Rep* **2**, 382–7.

Ibba, M. & Söll, D. (2004) Aminoacyl-tRNAs: setting the limits of the genetic code. *Genes Dev* **18**, 731–8.

Ideno, A., Furutani, M., Iwabuchi, T. et al. (2004) Expression of foreign proteins in *Escherichia coli* by fusing with an archaeal FK506 binding protein. *Appl Microbiol Biotechnol* **64**, 99–105.

Iida, T., Iwabuchi, T., Ideno, A., Suzuki, S. & Maruyama, T. (2000) FK506-binding protein-type peptidyl-prolyl cis-trans isomerase from a halophilic archaeum, *Halobacterium cutirubrum*. *Gene* **256**, 319–26.

Ikeda, M. & Clark, D. S. (1998) Molecular cloning of extremely thermostable esterase gene from hyperthermophilic archaeon *Pyrococcus furiosus* in *Escherichia coli*. *Biotechnol Bioeng* **57**, 624–9.

Ilyina, T. V. & Koonin, E. V. (1992) Conserved sequence motifs in the initiator proteins for rolling circle DNA replication encoded by diverse replicons from eubacteria, eucaryotes and archaebacteria. *Nucleic Acids Res* **20**, 3279–85.

Imamura, H., Fushinobu, S., Jeon, B. S., Wakagi, T. & Matsuzawa, H. (2001) Identification of the catalytic residue of *Thermococcus litoralis* 4-alpha-glucanotransferase through mechanism-based labeling. *Biochemistry* **40**, 12400–6.

Imanaka, T., Fukui, T. & Fujiwara, S. (2001) Chitinase from *Thermococcus kodakaraensis* KOD1. *Methods Enzymol* **330**, 319–29.

Ishikawa, K., Ishida, H., Matsui, I., Kawarabayasi, Y. & Kikuchi, H. (2001) Novel bifunctional hyperthermostable carboxypeptidase/aminoacylase from *Pyrococcus horikoshii* OT3. *Appl Environ Microbiol* **67**, 673–9.

Ishikawa, K. & Mino, K. (2004) Heat resistant cysteine synthase. USA Patent Number 20040002075. Accepted January 1, 2004.

Ishino, Y., Tsurimoto, T., Ishino, S. & Cann, I. K. (2001) Functional interactions of an archaeal sliding clamp with mammalian clamp loader and DNA polymerase delta. *Genes Cells* **6**, 699–706.

Ito, N., Nureki, O., Shirouzu, M., Yokoyama, S. & Hanaoka, F. (2003) Crystal structure of the *Pyrococcus horikoshii* DNA primase–UTP complex: implications for the mechanism of primer synthesis. *Genes Cells* **8**, 913–23.

Iwabe, N., Kuma, K., Hasegawa, M., Osawa, S. & Miyata, T. (1989) Evolutionary relationship of archaebacteria, eubacteria, and eukaryotes inferred from phylogenetic trees of duplicated genes. *Proc Natl Acad Sci USA* **86**, 9355–9.

Iyer, L. M., Aravind, L. & Koonin, E. V. (2001) Common origin of four diverse families of large eukaryotic DNA viruses. *J Virol* **75**, 11720–34.

Iyer, L. M., Leipe, D. D., Koonin, E. V. & Aravind, L. (2004a) Evolutionary history and higher order classification of AAA+ ATPases. *J Struct Biol* **146**, 11–31.

Iyer, L. M., Makarova, K. S., Koonin, E. V. & Aravind, L. (2004b) Comparative genomics of the FtsK-HerA superfamily of pumping ATPases: implications for the origins of chromosome segregation, cell division and viral capsid packaging. *Nucleic Acids Res* **32**, 5260–79.

Jackson, B. E. & McInerney, M. J. (2002) Anaerobic microbial metabolism can proceed close to thermodynamic limits. *Nature* **415**, 454–6.

Jacob, F. (1997) Evolution and tinkering. *Science* **196**, 1161–6.

Jacob, F., Brenner, S. & Kuzin, F. (1963) On the regulation of DNA replication in bacteria. *Cold Spring Harbor Symp Quant Biol* **28**, 329–48.

Jacob, F. & Monod, J. (1961) Genetic regulatory mechanisms in the synthesis of proteins. *J Mol Biol* **3**, 318–56.

Jäger, A., Samorski, R., Pfeifer, F. & Klug, G. (2002) Individual *gvp* transcript segments in *Haloferax mediterranei* exhibit varying half-lives, which are differentially affected by salt concentration and growth phase. *Nucleic Acids Res* **30**, 5436–43.

Jahn, U., Summons, R., Sturt, H., Grosjean, E. & Huber, H. (2004) Composition of the lipids of *Nanoarchaeum equitans* and their origin from its host *Ignicoccus* sp. strain KIN4/1. *Arch Microbiol* **182**, 404–13.

Jain, R., Rivera, M. C. & Lake, J. A. (1999) Horizontal gene transfer among genomes: the complexity hypothesis. *Proc Natl Acad Sci USA* **96**, 3801–6.

Janekovic, D., Wunderl, S., Holz, I., Zillig, W., Gierl, A. & Neumann, H. (1983) TTV1, TTV2 and TTV3, a family of viruses of the extremely thermophilic, anaerobic sulfur reducing archaebacterium *Thermoproteus tenax*. *Mol Gen Genet* **192**, 39–45.

Jansen, R., Embden, J. D., Gaastra, W. & Schouls, L. M. (2002) Identification of genes that are associated with DNA repeats in prokaryotes. *Mol Microbiol* **43**, 1565–75.

Jaun, B. (1993) Properties of metal alkyl derivatives. *In:* H. Sigel & A. Sigel (eds) *Metal Ions in Biological Systems*, Vol 29. Marcel Dekker, New York, pp. 287–337.

Jeffery, C. J. (1999) Moonlighting proteins. *Trends Biochem Sci* **24**, 8–11.

Jelinska, C., Conroy, M. J. & Craven, C. J. et al. (2005) Obligate heterodimerisation of the archaeal Alba2 protein with Alba1 provides a mechanism for control of DNA packaging. *Structure* **13**, 963–71.

Jeon, C. & Agarwal, K. (1996) Fidelity of RNA polymerase II transcription controlled by elongation factor TFIIS. *Proc Natl Acad Sci USA* **93**, 13677–82.

Jeruzalmi, D., Yurieva, O., Zhao, Y. X. et al. (2001) Mechanism of processivity clamp opening by the delta subunit wrench of the clamp loader complex of *E. coli* DNA polymerase III. *Cell* **106**, 417–28.

Johnson, E. F. & Mukhopadhyay, B. (2005) A new type of sulfite reductase – a novel coenzyme F420-dependent enzyme from the a methanarchaeon *Methanocaldococcus jannaschii*. *J Biol Chem* **280**, 38776–86.

Jolivet, E., Corre, E., L'Haridon, S., Forterre, P. & Prieur, D. (2004) *Thermococcus marinus* sp. nov. and *Thermococcus radiotolerans* sp. nov., two hyperthermophilic archaea from deep-sea hydrothermal vents that resist ionizing radiation. *Extremophiles* **8**, 219–27.

Jolivet, E., L'Haridon, S., Corre, E., Forterre, P. & Prieur, D. (2003a) *Thermococcus gammatolerans* sp. nov., a hyperthermophilic archaeon from a deep-sea hydrothermal vent that resists ionizing radiation. *Int J Syst Evol Microbiol* **53**, 847–51.

Jolivet, E., Matsunaga, F., Ishino, Y., Forterre, P., Prieur, D. & Myllykallio, H. (2003b) Physiological responses of the hyperthermophilic archaeon *Pyrococcus abyssi* to DNA damage caused by ionizing radiation. *J Bacteriol* **185**, 3958–61.

Jolley, K. A., Rapaport, E., Hough, D. W., Danson, M. J., Woods, W. G. & Dyall-Smith, M. L. (1996) Dihydrolipoamide dehydrogenase from the halophilic archaeon *Haloferax volcanii*: homologous overexpression of the cloned gene. *J Bacteriol* **178**, 3044–8.

Jonuscheit, M., Martusewitsch, E., Stedman, K. M. & Schleper, C. (2003) A reporter gene system for the hyperthermophilic archaeon *Sulfolobus solfataricus* based on a selectable and integrative shuttle vector. *Mol Microbiol* **48**, 1241–52.

Joye, S. B., Boetius, A., Orcutt, B. N. et al. (2004) The anaerobic oxidation of methane and sulfate reduction in sediments from Gulf of Mexico cold seeps. *Chem Geol* **205**, 219–38.

Judson, H. F. (1996) *The Eighth Day of Creation: Makers of the Revolution in Biology*. Cold Spring Harbor Laboratory Press, Plainview, NY.

Jurgens, G., Glöckner, F., Amann, R. et al. (2000) Identification of novel Archaea in bacterioplankton of a boreal forest lake by phylogenetic analysis and fluorescent in situ hybridization. *FEMS Microbiol Ecol* **34**, 45–56.

Kaczowka, S. J., Reuter, C. J., Talarico, L. A. & Maupin-Furlow, J. A. (2005) Recombinant production of *Zymomonas mobilis* pyruvate decarboxylase in the haloarchaeon *Haloferax volcanii*. *Archaea* **1**, 327–34.

Kahl, B. F., Li, H. & Paule, M. R. (2000) DNA melting and promoter clearance by eukaryotic RNA polymerase I. *J Mol Biol* **299**, 75–89.

Kahler, M. & Antranikian, G. (2000) Cloning and characterization of a family B DNA polymerase from the hyperthermophilic crenarchaeon *Pyrobaculum islandicum*. *J Bacteriol* **182**, 655.

Kahsai, M. A., Vogler, B., Clark, A. T., Edmondson, S. P. & Shriver, J. W. (2005) Solution structure, stability, and flexibility of Sso10a: a hyperthermophile coiled-coil DNA-binding protein. *Biochemistry* **44**, 2822–32.

Kaine, B. P., Gupta, R. & Woese, C. R. (1983) Putative introns in tRNA genes of prokaryotes. *Proc Natl Acad Sci USA* **80**, 3309–12.

Kaine, B. P., Mehr, I. J. & Woese, C. R. (1994) The sequence, and its evolutionary implications, of a *Thermococcus celer* protein associated with transcription. *Proc Natl Acad Sci USA* **91**, 3854–6.

Kaiser, J. T., Gromadski, K., Rother, M., Engelhardt, H., Rodnina, M. V. & Wahl, M. C. (2005) Structural and functional investigation of a putative archaeal selenocysteine synthase. *Biochemistry* **44**, 13315–27.

Kallmeyer, J. & Boetius, A. (2004) Effects of temperature and pressure on sulfate reduction and anaerobic oxidation of methane in hydrothermal sediments of Guaymas Basin. *Appl Environ Microbiol* **70**, 1231–3.

Kamp, A. F. Jr, La Riviere, J. W. M. & Verhoeven, W. (1959) *Albert Jan Kluyver: His Life and Work*. North Holland, Amsterdam.

Kanai, T., Ito, S. & Imanaka, T. (2003) Characterization of a cytosolic NiFe-hydrogenase from the hyperthermophilic archaeon *Thermococcus kodakaraensis* KOD1. *J Bacteriol* **185**, 1705–11.

Kandler, O. & Hippe, H. (1977) Lack of peptidoglycan in the cell walls of *Methanosarcina barkeri*. *Arch Microbiol* **113**, 57–60.

Kandler, O. & König, H. (1993) Cell envelopes of Archaea: structure and chemistry. *In:* M. Kates, D. J. Kushner & A. T. Matheson (eds) *The Biochemistry of Archaea (Archaebacteria)*. Elsevier Science, Amsterdam, pp. 223–59.

Kanemaki, M., Makino, Y., Yoshida, T. et al. (1997) Molecular cloning of a rat 49-kDa TBP-interacting protein (TIP49) that is highly homologous to the bacterial RuvB. *Biochem Biophys Res Comm* **235**, 64–8.

Kang, S., Vieille, C. & Zeikus, J. G. (2004) Identification of *Pyrococcus furiosus* amylopullulanase catalytic residues. *Appl Microbiol Biotechnol* **66**, 408–13.

Kannan, Y., Koga, Y., Inoue, Y. et al. (2001) Active subtilisin-like protease from a hyperthermophilic archaeon in a form with a putative prosequence. *Appl Environ Microbiol* **67**, 2445–52.

Kanugula, S., Pauly, G. T., Moschel, R. C. & Pegg, A. E. (2005) A bifunctional DNA repair protein from *Ferroplasma acidarmanus* exhibits O^6-alkylguanine-DNA alkyltransferase and endonuclease V activities. *Proc Natl Acad Sci USA* **102**, 3617–22.

Kaper, T., Talik, B., Ettema, T. J., Bos, H., van der Maarel, M. J. & Dijkhuizen, L. (2005) Amylomaltase of *Pyrobaculum aerophilum* IM2 produces thermoreversible starch gels. *Appl Environ Microbiol* **71**, 5098–106.

Kaper, T., van der Maarel, M. J., Euverink, G. J. & Dijkhuizen, L. (2004) Exploring and exploiting starch-modifying amylomaltases from thermophiles. *Biochem Soc Trans* **32**, 279–82.

Kappler, U. & Dahl, C. (2001) Enzymology and molecular biology of prokaryotic sulfite oxidation. *FEMS Microbiol Lett* **203**, 1–9.

Karner, M. B., DeLong, E. F. & Karl, D. M. (2001) Archaeal dominance in the mesopelagic zone of the Pacific Ocean. *Nature* **409**, 507–10.

Kashima, Y., Mori, K., Fukada, H. & Ishikawa, K. (2005) Analysis of the function of a hyperthermophilic endoglucanase from *Pyrococcus horikoshii* that hydrolyzes crystalline cellulose. *Extremophiles* **9**, 37–43.

Kates, M. (1972) Ether-linked lipids in extremely halophilic bacteria. *In:* F. Snyder (ed.) *Ether Lipids, Chemistry and Biology*. Academic Press, New York, pp. 351–98.

Kates, M., Yengoyan, L. S. & Sastry, P. S. (1965). A diether analog of phosphatidyl glycerophosphate in *Halobacterium cutirubrum*. *Biochim Biophys Acta* **98**, 252–68.

Kaufmann, S. (1995) *At Home in the Universe*. Oxford University Press, New York.

Kawarabayasi, Y., Hino, Y., Horikawa, H. et al. (1999) Complete genome sequence of an aerobic hyper-thermophilic crenarchaeon, *Aeropyrum pernix*. *DNA Res* **6**, 83–101.

Kawarabayasi, Y., Hino, Y., Horikawa, H. et al. (2001) Complete genome sequence of an aerobic thermoacidophilic crenarchaeon, *Sulfolobus tokodaii* strain 7. *DNA Res* **8**, 123–40.

Kawarabayasi, Y., Sawada, M., Horikawa, H. et al. (1998) Complete sequence and gene organization of the genome of a hyper-thermophilic archaebacterium, *Pyrococcus horikoshii* OT3. *DNA Res* **5**, 55–76.

Kawashima, T., Amano, N., Koike, H. et al. (2000) Archaeal adaptation to higher temperatures revealed by genomic sequence of *Thermoplasma volcanium*. *Proc Natl Acad Sci USA* **97**, 14257–62.

Keeling, P. J., Charlebois, R. L. & Doolittle, W. F. (1994) Archaebacterial genomes: eubacterial form and eukaryotic content. *Curr Opin Genet Dev* **4**, 816–22.

Keeling, P. J., Klenk, H.-P., Singh, R. K. et al. (1996) Complete nucleotide sequence of the *Sulfolobus islandicus* multicopy plasmid pRN1. *Plasmid* **35**, 141–4.

Keeling, P. J., Klenk, H.-P., Singh, R. K. et al. (1998) *Sulfolobus islandicus* plasmids pRN1 and pRN2 share distant but common evolutionary ancestry. *Extremophiles* **2**, 391–3.

Keenan, R. J., Freymann, D. M., Stroud, R. M. & Walter, P. (2001) The signal recognition particle. *Annu Rev Biochem* **70**, 755–75.

Kelly, D. P., Shergill, J. K., Lu, W. P. & Wood, A. P. (1997) Oxidative metabolism of inorganic sulfur compounds by bacteria. *Antonie Van Leeuwenhoek* **71**, 95–107.

Kelly, R. M. & Shockley, K. R. (2004) Enzyme discovery and microbial genomics. *In:* C. M. Fraser, T. D. Read & K. E. Nelson (eds) *Microbial Genomics*. Humana Press, Totowa, NJ, pp. 461–83.

Keltjens, J. T. & van der Drift, C. (1986) Electron-transfer reactions in methanogens. *FEMS Microbiol Rev* **39**, 259–303.

Kengen, S. W. M. & Stams, A. J. M. (1994) Growth and energy conversation in batch cultures of *Pyrococcus furiosus*. *FEMS Microbiol Lett* **117**, 305–10.

Kennedy, S. P., Ng, W. V., Salzberg, S. L., Hood, L. & DasSarma, S. (2001) Understanding the adaptation of *Halobacterium* species NRC-1 to its extreme environment through computational analysis of its genome sequence. *Genome Res* **11**, 1641–50.

Keough, B. P., Schmidt, T. M. & Hicks, R. E. (2003) Archaeal nucleic acids in picoplankton from great lakes on three continents. *Microb Ecol* **46**, 238–48.

Kessler, A., Brinkman, A. B., van der Oost, J. & Prangishvili, D. (2004) Transcription of the rod-shaped viruses SIRV1 and SIRV2 of the hyperthermophilic archaeon *Sulfolobus*. *J Bacteriol* **186**, 7745–53.

Kettenberger, H., Armache, K. J. & Cramer, P. (2003) Architecture of the RNA polymerase II-TFIIS complex and implications for mRNA cleavage. *Cell* **114**, 347–57.

Kettenberger, H., Armache, K. J. & Cramer, P. (2004) Complete RNA polymerase II elongation complex structure and its interactions with NTP and TFIIS. *Mol Cell* **16**, 955–65.

Khan, S. A. (1997) Rolling-circle replication of bacterial plasmids. *Microbiol Mol Biol Rev* **61**, 442–55.

Kim, J. W., Flowers, L. O., Whiteley, M. & Peeples, T. L. (2001) Biochemical confirmation and characterization of the family-57-like α-amylase of *Methanococcus jannaschii*. *Folia Microbiol (Prague)* **46**, 467–73.

Kim, K. K., Hung, L. W., Yokota, H., Kim, R. & Kim, S. H. (1998) Crystal structures of eukaryotic translation initiation factor 5A from *Methanococcus jannaschii* at 1.8 Å resolution. *Proc Natl Acad Sci USA* **95**, 10419–24.

Kim, M. S., Park, J. T., Kim, Y. W. et al. (2004) Properties of a novel thermostable glucoamylase from the hyperthermophilic archaeon *Sulfolobus solfataricus* in relation to starch processing. *Appl Environ Microbiol* **70**, 3933–40.

Kim, S. & Lee, S. B. (2004) Thermostable esterase from a thermoacidophilic archaeon: purification and characterization for enzymatic resolution of a chiral compound. *Biosci Biotechnol Biochem* **68**, 2289–98.

Kim, S., Sussman, J. L. & Suddath, F. L. et al. (1974) The general structure of transfer RNA molecules. *Proc Natl Acad Sci USA* **71**, 4970–4.

Kim, W. & Whitman, W. B. (1999) Isolation of acetate auxotrophs of the methane-producing archaeon *Methanococcus maripaludis* by random insertional mutagenesis. *Genetics* **152**, 1429–37.

Kimball, S. R. (1999) Eukaryotic initiation factor eIF2. *Int J Biochem Cell Biol* **31**, 25–9.

Kimura, M. & Ishihama, A. (2000) Involvement of multiple subunit–subunit contacts in the assembly of RNA polymerase II. *Nucleic Acids Res* **28**, 952–9.

Kireeva, M. L., Hancock, B., Cremona, G. H., Walter, W., Studitsky, V. M. & Kashlev, M. (2005) Nature of the nucleosomal barrier to RNA polymerase II. *Mol Cell* **18**, 97–108.

Kireeva, M. L., Komissarova, N., Waugh, D. S. & Kashlev, M. (2000) The 8-nucleotide-long RNA: DNA hybrid is a primary stability determinant of the RNA polymerase II elongation complex. *J Biol Chem* **275**, 6530–6.

Kiss, T. (2001) Small nucleolar RNA-guided post-transcriptional modification of cellular RNAs. *EMBO J* **20**, 3617–22.

Kiss-Laszlo, Z., Henry, Y., Bachellerie, J.-P., Caizergues-Ferrer, M. & Kiss, T. (1996) Site-specific ribose methylation of preribosomal RNA: a novel function for small nucleolar RNAs. *Cell* **85**, 1077–88.

Kisselev, L. L. & Buckingham, R. H. (2000) Translational termination comes of age. *Trends Biochem Sci* **25**, 561–6.

Kjems, J. & Garrett, R. A. (1988) Novel splicing mechanism for the rRNA intron in the archaebacterium *Desulfurococcus mobilis*. *Cell* **54**, 693–703.

Kjems, J., Jensen, J., Olesen, T. & Garrett, R. A. (1989) Comparison of transfer RNA and ribosomal RNA intron splicing in the extreme thermophile and archaebacterium *Desulfurococcus mobilis*. *Can J Microbiol* **35**, 210–14.

Klein, D. J., Schmeing, T. M., Moore, P. B. & Steitz, T. A. (2001) The kink-turn: a new RNA secondary structure motif. *EMBO J* **20**, 4214–21.

Klein, R., Baranyi, U. & Rossler, N. et al. (2002) *Natrialba magadii* virus phiCh1: first complete nucleotide sequence and functional organization of a virus infecting a haloalkaliphilic archaeon. *Mol Microbiol* **45**, 851–63.

Kleman-Leyer, K., Armbruster, D. W. & Daniels C. J. (1997) Properties of *H. volcanii* intron endonuclease reveal a relationship between the archaeal and eucaryal tRNA intron processing systems. *Cell* **89**, 839–47.

Klenk, H.-P., Clayton, R. A. & Tomb, J.-F. et al. (1997a) The complete genome sequence of the hyperthermophilic, sulphate-reducing archaeon *Archaeoglobus fulgidus*. *Nature* **390**, 364–70.

Klenk, H.-P., Palm, P. & Zillig, W. (1994) DNA-dependent RNA polymerases as phylogenetic marker molecules. *System Appl Microbiol* **16**, 638–47.

Klenk, H.-P., Zhou, L. & Venter, J. C. (1997b) Understanding life on this planet in the age of genomics. *In:* R. B. Hoover (ed.) *Instruments, Methods, and Missions for the Investigation of Extraterrestrial Microorganisms. Proc Intl Soc Opt Engin* **3111**, 306–17.

Klenk, H.-P. & Zillig, W. (1994) DNA-dependent RNA polymerase subunit B as a tool for phylogenetic reconstructions: branching topology of the archaeal domain. *J Mol Evol* **38**, 420–32.

Kletzin, A. (1989) Coupled enzymatic production of sulfite, thiosulfate, and hydrogen sulfide from sulfur: purification and properties of a sulfur oxygenase reductase from the facultatively anaerobic archaebacterium *Desulfurolobus ambivalens*. *J Bacteriol* **171**, 1638–43.

Kletzin, A. (1992) Molecular characterization of the *sor* gene, which encodes the sulfur oxygenase/reductase of the thermoacidophilic archaeum *Desulfurolobus ambivalens*. *J Bacteriol* **174**, 5854–9.

Kletzin, A., Lieke, A., Urich, T., Charlebois, R. L. & Sensen, C. W. (1999) Molecular analysis of pDL10 from *Acidianus ambivalens* reveals a family of related plasmids from extremely thermophilic and acidophilic archaea. *Genetics* **152**, 1307–14.

Kletzin, A., Urich, T., Müller, F., Bandeiras, T. M. & Gomes, C. M. (2004) Dissimilatory oxidation and reduction of elemental sulfur in thermophilic archaea. *J Bioenerg Biomembr* **36**, 77–91.

Kluyver, A. J. (1931) *The Chemical Activities of Microorganisms*. University of London Press, London.

Kluyver, A. J. & Donker, H. J. L. (1926) Die Einheit in der Biochemie. *Chem Zelle Gewebe* **13**, 134–90.

Knappe, J. & Wagner, A. F. (2001) Stable glycyl radical from pyruvate formate-lyase and ribonucleotide reductase (III). *Adv Protein Chem* **58**, 277–315.

Knittel, K., Boetius, A. & Lemke, A. et al. (2003) Activity, distribution, and diversity of sulfate reducers and other bacteria in sediments above gas hydrate (Cascadia margin, Oregon). *Geomicrobiol J* **20**, 269–94.

Knittel, K., Losekann, T., Boetius, A., Kort, R. & Amann, R. (2005) Diversity and distribution of methanotrophic archaea at cold seeps. *Appl Environ Microbiol* **71**, 467–79.

Ko, T. P., Chu, H. M., Chen, C. Y., Chou, C. C. & Wang, A. H. (2004) Structures of the hyperthermophilic chromosomal protein Sac7d in complex with DNA decamers. *Acta Crystallogr D Biol Crystallogr* **60**, 1381–417.

Kobayashi, T., Kanai, H., Aono, R., Horikoshi, K. & Kudo, T. (1994) Cloning, expression, and nucleotide sequence of the α-amylase gene from the haloalkaliphilic archaeon *Natronococcus* sp. strain Ah-36. *J Bacteriol* **176**, 5131–4.

Köhrer, C., Sullivan, E. L. & RajBhandary, U. L. (2004) Complete set of orthogonal 21st aminoacyl-tRNA synthetase-amber, ochre and opal suppressor tRNA pairs: concomitant suppression of three different termination codons in an mRNA in mammalian cells. *Nucleic Acids Res* **32**, 6200–11.

Kokoska, R. J., Bebenek, K., Boudsocq, F., Woodgate, R. & Kunkel, T. A. (2002) Low fidelity DNA synthesis by a Y family DNA polymerase due to misalignment in the active site. *J Biol Chem* **277**, 19633–8.

Komori, K., Hidaka, M., Horiuchi, T., Fujikane, R., Shinagawa, H. & Ishino, Y. (2004) Cooperation of the N-terminal Helicase and C-terminal endonuclease activities of archaeal Hef protein in processing stalled replication forks. *J Biol Chem* **279**, 53175–85.

Komori, K., Miyata, T., DiRuggiero, J. et al. (2000a) Both RadA and RadB are involved in homologous recombination in *Pyrococcus furiosus*. *J Biol Chem* **275**, 33782–90.

Komori, K., Sakae, S., Fujikane, R., Morikawa, K., Shinagawa, H. & Ishino, Y. (2000b) Biochemical characterisation of the Hjc Holliday junction resolvase of *Pyrococcus furiosus*. *Nucleic Acids Res* **28**, 4544–51.

Komori, K., Sakae, S., Shinagawa, H., Morikawa, K. & Ishino, Y. (1999) A Holliday junction resolvase from *Pyrococcus furiosus*: functional similarity to *Escherichia coli* RuvC provides evidence for conserved mechanism of homologous recombination in bacteria, eukarya, and archaea. *Proc Natl Acad Sci USA* **96**, 8873–8.

Könneke, M., Bernhard, A., de la Torre, J., Walker, C. B., Waterbury, J. & Stahl, D. A. (2005) Isolation of an autotrophic ammonia oxidizing archaeon. *Nature* **437**, 543–6.

Koonin, E. V. & Galperin, M. Y. (2003) *Sequence, Evolution, Function: Computational Approaches in Comparative Genomics*. Kluwer Academic, Dordrecht.

Koonin, E. V., Makarova, K. S. & Aravind, L. (2001) Horizontal gene transfer in prokaryotes: quantification and classification. *Annu Rev Microbiol* **55**, 709–42.

Koonin, E. V., Mushegian, A. R. & Bork, P. (1996) Non-orthologous gene displacement. *Trends Genet* **12**, 334–6.

Koonin, E. V., Mushegian, A. R., Galperin, M. Y. & Walker, D. R. (1997) Comparison of archaeal and bacterial genomes: computer analysis of protein sequences predicts novel functions and suggests a chimeric origin for the Archaea. *Mol Microbiol* **25**, 619–37.

Korbel, J. O., Jensen, L. J. & von Mehring, C. et al. (2004) Analysis of genomic context: prediction of functional associations from conserved bidirectionally transcribed gene pairs. *Nat Biotechnol* **22**, 911–17.

Kosa, P. F., Ghosh, G., DeDecker, B. S. & Sigler, P. B. (1997) The 2.1-Å crystal structure of an archaeal preinitiation complex: TATA-box-binding protein/transcription factor (II) B core/TATA-box. *Proc Natl Acad Sci USA* **94**, 6042–7.

Koski, L. B. & Golding, G. B. (2001) The closest BLAST hit is often not the nearest neighbor. *J Mol Evol* **52**, 540–2.

Kowalczykowski, S., Chedin, F. & Seitz, E. (2005) Single stranded DNA binding proteins from archaea. US Patent Number 6,852,832. Accepted February 8, 2005.

Kraft, A., Lutz, C., Lingenhel, A., Grobner, P. & Piendl, W. (1999) Control of ribosomal protein L1 synthesis in mesophilic and thermophilic archaea. *Genetics* **152**, 1363–72.

Krah, R., Kozyavkin, S. A., Slesarev, A. I. & Gellert, M. (1996) A two-subunit type I DNA topoisomerase (reverse gyrase) from an extreme hyperthermophile. *Proc Natl Acad Sci USA* **93**, 106–10.

Krieger, C. J., Roseboom, W., Albracht, S. P. J. & Spormann, A. M. (2001) A stable organic free radical in anaerobic benzylsuccinate synthase of *Azoarcus* sp. strain T. *J Biol Chem* **276**, 12924–7.

Krüger, K., Hermann, T., Armbruster, V. & Pfeifer, F. (1998) The transcriptional activator GvpE for the halobacterial gas vesicle genes resembles a basic region leucine-zipper regulatory protein. *J Mol Biol* **279**, 761–71.

Krüger, K. & Pfeifer, F. (1996) Transcript analysis of the c-vac region, and differential synthesis of the two regulatory gas-vesicle proteins GvpD and GvpE in *Halobacterium salinarium* PHH4. *J Bacteriol* **178**, 4012–19.

Krüger, M., Meyerdierks, A. & Glöckner, F. O. et al. (2003) A conspicuous nickel protein in microbial mats that oxidize methane anaerobically. *Nature* **426**, 878–81.

Kuhn, J. F., Tran, E. J. & Maxwell, E. S. (2002) Archaeal ribosomal protein L7 is a functional homolog of the eukaryotic 15.5kD/Snu13p snoRNP core protein. *Nucleic Acids Res* **30**, 931–41.

Kuldell, N. H. & Buratowski, S. (1997) Genetic analysis of the large subunit of yeast transcription factor IIE reveals two regions with distinct functions. *Mol Cell Biol* **17**, 5288–98.

Kulms, D., Schäfer, G. & Hahn, U. (1995) SaRD, a new protein isolated from the extremophile archaeon *Sulfolobus acidocaldarius*, is a thermostable ribonuclease with DNA-binding properties. *Biochem Biophys Res Comm* **214**, 646–52.

Kunkel, T. A. & Erie, D. A. (2004) DNA mismatch repair. *Annu Rev Biochem* **74**, 681–710.

Kurland, C. G. (2000) Something for everyone. Horizontal gene transfer in evolution. *EMBO Rep* **1**, 92–5.

Kurokawa, Y., Kanemaki, M., Makino, Y. & Tamura, T. A. (1999) A notable example of an evolutionary conserved gene: studies on a putative DNA helicase TIP49. *DNA Seq* **10**, 37–42.

Kurosawa, N. & Grogan, D. W. (2005) Homologous recombination of exogenous DNA with the *Sulfolobus acidocaldarius* genome: properties and uses. *FEMS Microb Lett* **253**, 141–9.

Kutach, A. K. & Kadonaga, J. T. (2000) The downstream promoter element DPE appears to be as widely used as the TATA box in *Drosophila* core promoters. *Mol Cell Biol* **20**, 4754–64.

Kuypers, M. M., Blokker, P., Erbacher, J. et al. (2001) Massive expansion of marine archaea during a mid-Cretaceous oceanic anoxic event. *Science* **293**, 92–5.

Kvaratskhelia, M., Wardleworth, B. N., Bond, C. S., Fogg, J. M., Lilley, D. M. & White, M. F. (2002) Holliday junction resolution is modulated by archaeal chromatin components *in vitro*. *J Biol Chem* **277**, 2992–6.

Kvaratskhelia, M. & White, M. F. (2000) Two Holliday junction resolving enzymes in *Sulfolobus solfataricus*. *J Mol Biol* **297**, 923–32.

Kvint, K., Nachin, L., Diez, A. & Nystrom, T. (2003) The bacterial universal stress protein: function and regulation. *Curr Opin Microbiol* **6**, 140–5.

Kyrpides, N. C. & Woese, C. R. (1998a) Archaeal translation initiation revisited: the initiation factor 2 and eukaryotic initiation factor 2B α-β-δ subunit families. *Proc Natl Acad Sci USA* **95**, 3726–30.

Kyrpides, N. C. & Woese, C. R. (1998b) Universally conserved translation initiation factors. *Proc Natl Acad Sci USA* **95**, 224–8.

Labib, K. & Diffley, J. F. (2001) Is the MCM2-7 complex the eukaryotic DNA replication fork helicase? *Curr Opin Genet Dev* **11**, 64–70.

Ladapo, J. & Whitman, W. B. (1990) Method for isolation of auxotrophs in the methanogenic archaebacteria: role of the acetyl-CoA pathway of autotrophic CO_2 fixation in *Methanococcus maripaludis*. *Proc Natl Acad Sci USA* **87**, 5598–602.

Lai, X., Shao, H., Hao, F. & Huang, L. (2002) Biochemical characterization of an ATP-dependent DNA ligase from the hyperthermophilic crenarchaeon *Sulfolobus shibatae*. *Extremophiles* **6**, 469–77.

Lake, J. A. (1985) Evolving ribosome structure: domains in archaebacteria, eubacteria, eocytes and eukaryotes. *Annu Rev Biochem* **54**, 507–30.

Laksanalamai, P., Jiemjit, A., Bu, Z., Maeder, D. L. & Robb, F. T. (2003) Multi-subunit assembly of the *Pyrococcus furiosus* small heat shock protein is essential for cellular protection at high temperature. *Extremophiles* **7**, 79–83.

Laksanalamai, P., Maeder, D. L. & Robb, F. T. (2001) Regulation and mechanism of action of the small heat shock protein from the hyperthermophilic archaeon *Pyrococcus furiosus*. *J Bacteriol* **183**, 5198–202.

Lam, W. L. & Doolittle, W. F. (1989) Shuttle vectors for the archaebacterium *Halobacterium volcanii*. *Proc Natl Acad Sci USA* **86**, 5478–82.

Lam, W. L. & Doolittle, W. F. (1992) Mevinolin-resistant mutations identify a promoter and the gene for a eukaryote-like 3-hydroxy-3-methylglutaryl-coenzyme A reductase in the archaebacterium *Haloferax volcanii*. *J Biol Chem* **267**, 5829–34.

Lamble, H. J., Danson, M. J., Hough, D. W. & Bull, S. D. (2005) Engineering stereocontrol into an aldolase-catalysed reaction. *Chem Commun (Camb)* **1**, 124–6.

Lamble, H. J., Heyer, N. I. & Bull, S. D. et al. (2003) Metabolic pathway promiscuity in the archaeon *Sulfolobus solfataricus* revealed by studies on glucose dehydrogenase and 2-keto-3-deoxygluconate aldolase. *J Biol Chem* **278**, 34066–72.

Lamosa, P., Burke, A. & Peist, R. et al. (2000) Thermostabilization of proteins by diglycerol phosphate, a new compatible solute from the hyperthermophile *Archaeoglobus fulgidus*. *Appl Environ Microbiol* **66**, 1974–9.

Landegren, U., Schallmeiner, E. & Nilsson, M. et al. (2004) Molecular tools for a molecular medicine: analyzing genes, transcripts and proteins using padlock and proximity probes. *J Mol Recognit* **17**, 194–7.

Langaraju, G. M., Sartori, A. A., Kostrewa, D., Prota, A. E., Jiricny, J. & Winkler, F. K. (2005) A DNA glycosylase from *Pyrobaculum aerophilum* with an 8-oxoguanine binding mode and a noncanonical helix-hairpin-helix structure. *Structure (Camb)* **13**, 87–98.

Lange, M., Westermann, P. & Ahring, B. K. (2005) Archaea in protozoa and metazoa. *Appl Microbiol Biotechnol* **66**, 465–74.

Lange, U. & Hausner, W. (2004) Transcriptional fidelity and proofreading in Archaea and implications for the mechanism of TFS-induced RNA cleavage. *Mol Microbiol* **52**, 1133–43.

Langer, D., Hain, J., Thuriaux, P. & Zillig, W. (1995) Transcription in Archaea: similarity to that in Eukarya. *Proc Natl Acad Sci USA* **92**, 5768–72.

Langer, D. & Zillig, W. (1993) Putative tfIIs gene of *Sulfolobus acidocaldarius* encoding an archaeal transcription elongation factor is situated directly downstream of the gene for a small subunit of DNA-dependent RNA polymerase. *Nucleic Acids Res* **21**, 2251.

Langworthy, T. A., Smith, M. E. & Mayberry, W. R. (1972) Long-chain glycerol diether and polyol dialkyl glycerol triether lipids of *Sulfolobus acidocaldarius*. *J Bacteriol* **112**, 1193–200.

Langworthy, T. A., Smith, M. E. & Mayberry, W. R. (1974) A new class of lipopolysaccharide from *Thermoplasma acidophilum*. *J Bacteriol* **119**, 106–16.

Lao-Sirieix, S. H. & Bell, S. D. (2004) The heterodimeric primase of the hyperthermophilic archaeon *Sulfolobus solfataricus* possesses DNA and RNA primase, polymerase and 3′-terminal nucleotidyl transferase activities. *J Mol Biol* **344**, 1251–63.

Lao-Sirieix, S. H., Nookala, R. K., Roversi, P., Bell, S. D. & Pellegrini, L. (2005a) Structure of the heterodimeric core primase. *Nat Struct Mol Biol* **12**, 1137–44.

Lao-Sirieix, S. H., Pellegrini, L. & Bell, S. D. (2005b) The promiscuous primase. *Trends Genet* **21**, 568–72.

Laska, S., Lottspeich, F. & Kletzin, A. (2003) Membrane-bound hydrogenase and sulfur reductase of the hyperthermophilic and acidophilic archaeon *Acidianus ambivalens*. *Microbiology* **149**, 2357–71.

Lathe, W. C. III, Snel, B. & Bork, P. (2000) Gene context conservation of a higher order than operons. *Trends Biochem Sci* **25**, 474–9.

Lauhon, C. T. (2002) Requirement for IscS in biosynthesis of all thionucleosides in *Escherichia coli*. *J Bacteriol* **184**, 6820–9.

Lawrence, J. G. (2003) Gene organization: selection, selfishness, and serenity. *Annu Rev Microbiol* **57**, 419–40.

Lawrence, J. G. & Hendrickson, H. (2003) Lateral gene transfer: when will adolescence end? *Mol Microbiology* **50**, 739–49.

Lebbink, J. H., Kaper, T., Kengen, S. W., van der Oost, J. & de Vos, W. M. (2001) β-Glucosidase CelB from *Pyrococcus furiosus*: production by *Escherichia coli*, purification, and *in vitro* evolution. *Methods Enzymol* **330**, 364–79.

Leclere, M. M., Nishioka, M., Yuasa, T., Fujiwara, S., Takagi, M. & Imanaka, T. (1998) The O^6-methylguanine-DNA methyltransferase from the hyperthermophilic archaeon *Pyrococcus* sp. KOD1: a thermostable repair enzyme. *Mol Gen Genet* **258**, 69–77.

Lecomte, O., Ripp, R., Puzos-Barbe, V. et al. (2002a) Genome evolution at the genus level: comparison of three complete genomes of hyperthermophilic archaea. *Genome Res* **11**, 981–93.

Lecompte, O., Ripp, R., Thierry, J. C., Moras, D. & Poch, O. (2002b) Comparative analysis of ribosomal proteins in complete genomes: an example of reductive evolution at the domain scale. *Nucleic Acids Res* **30**, 5382–90.

Lee, H.-S., Shockley, K. R. & Schut, G. J. et al. (2006) Transcriptional and biochemical analysis of starch metabolism in the hyperthermophilic archaeon *Pyrococcus furiosus*. *J Bacteriol* **188**, 2115–25.

Lee, J. H., Choi, S. K., Roll-Mecak, A., Burley, S. K. & Dever, T. E. (1999) Universal conservation in translation initiation revealed by human and archaeal homologs of bacterial translation initiation factor IF2. *Proc Natl Acad Sci USA* **96**, 4342–7.

Lee, S. J., Engelmann, A. & Horlacher, R. et al. (2003) TrmB, a sugar-specific transcriptional regulator of the trehalose/maltose ABC transporter from the hyperthermophilic archaeon *Thermococcus litoralis*. *J Biol Chem* **278**, 983–90.

Lee, S. J., Moulakakis, C., Hausner, W., Koning, S. M., Thomm, M. & Boos, W. (2005) TrmB, a sugar sensing regulator of ABC transporter genes in *Pyrococcus furiosus* exhibits dual promoter specificity and is controlled by different inducers. *Mol Microbiol* **57**, 1797–807.

Leipe, D. D., Aravind, L. & Koonin, E. V. (1999) Did DNA replication evolve twice independently? *Nucleic Acids Res* **27**, 3389–401.

Leong, D., Boyer, H. W. & Betlach, M. (1988a) Transcription of genes involved in bacterioopsin gene expression in mutants of a halophilic archaebacterium. *J Bacteriol* **170**, 4910–15.

Leong, D., Pfeifer, F., Boyer, H. W. & Betlach, M. (1988b) Characterization of a second gene involved in bacterioopsin gene expression in a halophilic archaebacterium. *J Bacteriol* **170**, 4903–9.

Lerat, E., Daubin, V. & Moran, N. A. (2003) From gene trees to organismal phylogeny in prokaryotes: the case of the gamma-Proteobacteria. *PLoS Biol* **1**, e19.

Letzelter, C., Duguet, M. & Serre, M. C. (2004) Mutational analysis of the archaeal tyrosine recombinase SSV1 integrase suggests a mechanism of DNA cleavage in trans. *J Biol Chem* **279**, 28936–44.

Leveque, E., Janecek, S., Haye, B. & Belarbi, A. (2000) Thermophilic archaeal amylolytic enzymes. *Enzyme Microb Technol* **26**, 3–14.

Levin, I., Giladi, M., Altman-Price, N., Ortenberg, R. & Mevarech, M. (2004) An alternative pathway for reduced folate biosynthesis in bacteria and halophilic archaea. *Mol Microbiol* **54**, 1307–18.

Li, D. & Stevenson, K. J. (2001) Alcohol dehydrogenase from *Thermococcus* strain AN1. *Methods Enzymol* **331**, 201–7.

Li, L. & Wang, C. C. (2004) Capped mRNA with a single nucleotide leader is optimally translated in a primitive eukaryote, *Giardia lamblia*. *J Biol Chem* **279**, 14656–64.

Li, T., Sun, F., Ji, X., Feng, Y. & Rao, Z. (2003) Structure based hyperthermostability of archaeal histone HPhA from *Pyrococcus horikoshii*. *J Mol Biol* **325**, 1031–7.

Li, W. & Hoffman, D. W. (2001) Structure and dynamics of translation initiation factor aIF-1A from the archaeon *Methanococcus jannaschii* determined by NMR spectroscopy. *Protein Sci* **10**, 2426–38.

Li, W.-T., Sandman, K., Pereira, S. L. & Reeve, J. N. (2000) MJ1647, an open reading frame in the genome of the hyperthermophile *Methanococcus jannaschii*, encodes a very thermostable archaeal histone with a C-terminal extension. *Extremophiles* **4**, 43–51.

Li, Y., Flanagan, P. M., Tschochner, H. & Kornberg, R. D. (1994) RNA polymerase II initiation factor interactions and transcription start site selection. *Science* **263**, 805–7.

Liberi, G. & Foiani, M. (2004) Initiation of DNA replication: a new hint from archaea. *Cell* **116**, 3–4.

Lie, T. J., Wood, G. E. & Leigh, J. A. (2005) Regulation of *nif* expression in *Methanococcus maripaludis*. *J Biol Chem* **200**, 5236–41.

Lim, H., Eng, J. & Yates, J. R. III. et al. (2003) Identification of 2D-gel proteins: a comparison of MALDI/TOF peptide mass mapping to mu LC-ESI tandem mass spectrometry. *J Am Soc Mass Spectrom* **14**, 957–70.

Limauro, D., Cannio, R., Fiorentino, G., Rossi, M. & Bartolucci, S. (2001) Identification and molecular characterization of an endoglucanase gene, celS, from the extremely thermophilic archaeon *Sulfolobus solfataricus*. *Extremophiles* **5**, 213–19.

Lindahl, T. (1993) Instability and decay of the primary structure of DNA. *Nature* **362**, 709–15.

Ling, H., Boudsocq, F., Plosky, B. S., Woodgate, R. & Yang, W. (2003) Replication of a cis-syn thymine dimer at atomic resolution. *Nature* **424**, 1083–7.

Ling, H., Boudsocq, F., Woodgate, R. & Yang, W. (2004) Snapshots of replication through an abasic lesion: structural basis for base substitutions and frameshifts. *Mol Cell* **13**, 751–62.

Lingaraju, G. M., Sartori, A. A., Kostrewa, D., Prota, A. E., Jiricny, J. & Winkler, F. K. (2005) A DNA glycosylase from *Pyrobaculum aerophilum* with an 8-oxyguanine binding mode and a noncanonical helix–hairpin–helix structure. *Structure (Camb)* **13**, 87–98.

Linn, S. C. & Luse, D. S. (1991) RNA polymerase II elongation complexes paused after the synthesis of 15- or 35-base transcripts have different structures. *Mol Cell Biol* **11**, 1508–22.

Lipps, G., Ibanez, P., Strössenreuther, T., Hekimian, K. & Krauss, G. (2001a) The protein ORF80 from the acidophilic and thermophilic archaeon *Sulfolobus islandicus* binds highly site-specifically to double-stranded DNA and represents a novel type of basic leucine zipper protein. *Nucleic Acids Res* **29**, 4973–82.

Lipps, G., Rother, S., Hart, C. & Krauss, G. (2003) A novel type of replicative enzyme harbouring ATPase, primase and DNA polymerase activity. *EMBO J* **22**, 2516–25.

Lipps, G., Stegert, M. & Krauss, G. (2001b) Thermostable and site-specific DNA binding of the gene product ORF56 from the *Sulfolobus islandicus* plasmid pRN1, a putative archael plasmid copy control protein. *Nucleic Acids Res* **29**, 904–13.

Lipps, G., Weinzierl, A. O., von Scheven, G., Buchen, C. & Cramer, P. (2004) Structure of a bifunctional DNA primase-polymerase. *Nat Struct Mol Biol* **11**, 157–62.

Liston, D. R. & Johnson, P. J. (1999) Analysis of a ubiquitous promoter element in a primitive eukaryote: early evolution of the initiator element. *Mol Cell Biol* **19**, 2380–8.

Littlefield, O., Korkhin, Y. & Sigler, P. B. (1999) The structural basis for the oriented assembly of a TBP/TFB/promoter complex. *Proc Natl Acad Sci USA* **96**, 13668–73.

Liu, J., Smith, C. L., DeRyckere, D., DeAngelis, K., Martin, G. S. & Berger, J. M. (2000) Structure and function of Cdc6/Cdc18: implications for origin recognition and checkpoint control. *Mol Cell* **6**, 637–48.

Liu, X., Fan, K. & Wang, W. (2004) The number of protein folds and their distribution over families in nature. *Proteins* **54**, 491–9.

Liu, Y. & Schepartz, A. (2001) Kinetic preference for oriented DNA binding by the yeast TATA-binding protein TBP. *Biochemistry* **40**, 6257–66.

Liu, Y. & West, S. C. (2004) Happy Holidays: 40th anniversary of the Holliday junction. *Nat Rev Mol Cell Biol* **5**, 937–44.

Lloyd, J., Loveley, D. R. & Macaskie, L. E. (2003) Biotechnological application of metal-reducing microorganisms. *Adv Appl Microbio* **53**, 85–128.

Logan, D. T., Mulliez, E. & Larsson, K. M. et al. (2003) A metal-binding site in the catalytic subunit of anaerobic ribonucleotide reductase. *Proc Natl Acad Sci USA* **100**, 3826–31.

Londei, P. (2005) Evolution of translational initiation: new insights from the Archaea. *FEMS Microbiol Rev* **29**, 185–200.

Lopez, P., Forterre, P. & Philippe, H. (1999) The root of the tree of life in the light of the covarion model. *J Mol Evol* **49**, 496–508.

Lopez-Garcia, P., Brochier, C., Moreira, D. & Rodriguez-Valera, F. (2004) Comparative analysis of a genome fragment of an uncultivated mesopelagic crenarchaeote reveals multiple horizontal gene transfers. *Environ Microbiol* **6**, 19–34.

Lopez-Garcia, P., Forterre, P., van der Oost, J. & Erauso, G. (2000) Plasmid pGS5 from the hyperthermophilic archaeon *Archaeoglobus profundus* is negatively supercoiled. *J Bacteriol* **182**, 4998–5000.

Lopez-Garcia, P. & Moreira, D. (1999) Metabolic symbiosis at the origin of eukaryotes. *Trends Biochem Sci* **24**, 88–93.

Lorentzen, E., Hensel, R., Knura, T. et al. (2004) Structural Basis of allosteric regulation and substrate specificity of the non-phosphorylating glyceraldehyde 3-phosphate dehydrogenase from *Thermoproteus tenax*. *J Mol Biol* **341**, 815–28.

Lu, Y. & Conrad, R. (2005) In situ stable isotope probing of methanogenic archaea in the rice rhizosphere. *Science* **309**, 1088–90.

Lucas, S., Toffin, L. & Zivanovic, Y. et al. (2002) Construction of a shuttle vector for, and spheroplast transformation of the hyperthermophilic archaeon *Pyrococcus abyssi*. *Appl Environ Microbiol* **68**, 5528–36.

Ludwig, W. & Strunk, O. (2001) ARB: a software environment for sequence data (www.arb-home.de/arb/documentation.html).

Lueders, T., Chin, K. J., Conrad, R. & Friedrich, M. (2001) Molecular analyses of methyl-coenzyme M reductase alpha-subunit (mcrA) genes in rice field soil and enrichment cultures reveal the methanogenic phenotype of a novel archaeal lineage. *Environ Microbiol* **3**, 194–204.

Luger, K., Mader, A. W., Richmond, R. K., Sargent, D. F. & Richmond, T. J. (1997) Crystal structure of the nucleosome core particle at 2.8 Å resolution. *Nature* **389**, 251–60.

Luke, G. T. & Tedeschi, M. D. (1982) The minamata disease. *Am J For Med Pat* **3**, 335–8.

Lund, P. A., Large, A. T. & Kapatai, G. (2003) The chaperonins: perspectives from the Archaea. *Biochem Soc Trans* **31**, 681–5.

Lundgren, M., Andersson, A., Chen, L., Nilsson, P. & Bernander, R. (2004) Three replication origins in *Sulfolobus* species: synchronous initiation of chromosome replication and asynchronous termination. *Proc Natl Acad Sci USA* **101**, 7046–51.

Luo, H. W., Zhang, H., Suzuki, T., Hattori, S. & Kamagata, Y. (2002) Differential expression of methanogenesis genes of *Methanothermobacter thermoautotrophicus* (formerly *Methanobacterium thermoautotrophicum*) in pure culture and in cocultures with fatty acid-oxidizing syntrophs. *Appl Environ Microbiol* **68**, 1173–9.

Luo, Y., Leisinger, T. & Wasserfallen, A. (2001) Comparative sequence analysis of plasmids pME2001 and pME2200 of *Methanothermobacter marburgensis* strains Marburg and ZH3. *Plasmid* **45**, 18–30.

Luo, Y., Pfister, P., Leisinger, T. & Wasserfallen, A. (2001) The genome of archaeal prophage PsiM100 encodes the lytic enzyme responsible for autolysis of *Methanothermobacter wolfeii*. *J Bacteriol* **183**, 5788–92.

Luo, Y. & Wasserfallen, A. (2001) Gene transfer systems and their applications in Archaea. *Syst Appl Microbiol* **24**, 15–25.

Lurz, R., Grote, M., Dijk, J., Reinhardt, R. & Dobrinski, B. (1986) Electron microscopic study of DNA complexes with proteins from the archaebacterium *Sulfolobus acidocaldarius*. *EMBO J* **5**, 3715–21.

Luton, P. E., Wayne, J. M., Sharp, R. J. & Riley, P. W. (2002) The mcrA gene as an alternative to 16S rRNA in the phylogenetic analysis of methanogen populations in landfill. *Microbiology* **148**, 3521–30.

Lykke-Andersen, J., Aagaard, C., Semionenkov, M. & Garrett, R. A. (1997) Archaeal introns: splicing, intercellular mobility and evolution. *Trends Biochem Sci* **22**, 326–31.

Lykke-Andersen, J. & Garrett, R. A. (1994) Structural characteristics of the stable RNA introns of archaeal hyperthermophiles and their splicing junctions. *J Mol Biol* **243**, 846–55.

Lykke-Andersen, J. & Garrett, R. A. (1997) RNA-protein interactions of an archaeal homotetrameric splicing endonuclease with an exceptional evolutionary history. *EMBO J* **16**, 6290–300.

Ma, H. W. & Zeng, A. P. (2004) Phylogenetic comparison of metabolic capacities of organisms at genome level. *Mol Phylogenet Evol* **31**, 204–13.

Ma, K. & Adams, M. W. (1994) Sulfide dehydrogenase from the hyperthermophilic archaeon *Pyrococcus furiosus*: a new multifunctional enzyme involved in the reduction of elemental sulfur. *J Bacteriol* **176**, 6509–17.

Ma, K. & Adams, M. W. (2001a) Alcohol dehydrogenases from *Thermococcus litoralis* and *Thermococcus* strain ES-1. *Methods Enzymol* **331**, 195–201.

Ma, K. & Adams, M. W. (2001b) Hydrogenases I and I I from *Pyrococcus furiosus*. *Methods Enzymol* **331**, 208–16.

McBride, B. C. & Wolfe, R. S. (1971) Biosynthesis of dimethylarsine by a methanobacterium. *Biochemistry* **10**, 4312–17.

McCloskey, J. A., Graham, D. E. & Zhou, S. et al. (2001) Post-transcriptional modification in archaeal tRNAs: identities and phylogenetic relations of nucleotides from mesophilic and hyperthermophilic *Methanococcales*. *Nucleic Acids Res* **29**, 4699–706.

McCready, S., Müller, J. A., Boubriak, I., Berquistr, B. R., Ng, W. L. & DasSarma, S. (2005) UV irradiation induces homologous recombination genes in the model archaeon, *Halobacterium* sp. NRC-1. *Saline Systems* **1**, 1–9.

McGeoch, A. T., Trakselis, M. A., Laskey, R. A. & Bell, S. D. (2005) Organization of the archaeal MCM complex on DNA and implications for helicase mechanism. *Nat Struct Mol Biol* **12**, 756–62.

McGlynn, P. (2004) Links between DNA replication and recombination in prokaryotes. *Curr Opin Genet Dev* **14**, 107–12.

MacGregor, B. J., Moser, D. P., Alm, E. W., Nealson, K. H. & Stahl, D. A. (1997) Crenarchaeota in Lake Michigan sediment. *Appl Environ Microbiol* **63**, 1178–81.

McInerney, J. O., Wilkinson, M., Patching, J. W., Embley, T. M. & Powell, R. (1995) Recovery and phylogenetic analysis of novel archaeal rRNA sequences from a deep-sea deposit feeder. *Appl Environ Microbiol* **61**, 1646–8.

Maenpaa, P. H. & Bernfield, M. R. (1970) A specific hepatic transfer RNA for phosphoserine. *Proc Natl Acad Sci USA* **67**, 688–95.

Maga, G., Villani, G. & Tillement, V. et al. (2001) Okazaki fragment processing: Modulation of the strand displacement activity of DNA polymerase delta by the concerted action of replication protein A, proliferating cell nuclear antigen, and flap endonuclease-1. *Proc Natl Acad Sci USA* **98**, 14298–303.

Magill, C. P., Jackson, S. P. & Bell, S. D. (2001) Identification of a conserved archaeal RNA polymerase subunit contacted by the basal transcription factor TFB. *J Biol Chem* **276**, 46693–6.

Magrum, L. J., Luehrsen, K. R. & Woese, C. R. (1978) Are extreme halophiles actually "bacteria"? *J Mol Evol* **111**, 1–8.

Mahdi, A. A., Briggs, G. S., Sharples, G. J., Wen, Q. & Lloyd, R. G. (2003) A model for dsDNA translocation revealed by a structural motif common to RecG and Mfd proteins. *EMBO J* **22**, 724–34.

Mahlert, F., Bauer, C., Jaun, B., Thauer, R. K. & Duin, E. C. (2002a) The nickel enzyme methyl-coenzyme M reductase from methanogenic archaea: *in vitro* induction of the nickel-based MCR-ox EPR signals from MCR-red2. *J Biol Inorg Chem* **7**, 500–13.

Mahlert, F., Grabarse, W., Kahnt, J., Thauer, R. K. & Duin, E. C. (2002b) The nickel enzyme methyl-coenzyme M reductase from methanogenic archaea: *in vitro* interconversions among the EPR detectable MCR-red1 and MCR-red2 states. *J Biol Inorg Chem* **7**, 101–12.

Majernik, A. I., Lundgren, M., McDermott, P., Bernander, R. & Chong, J. P. (2005) DNA content and nucleoid distribution in *Methanothermobacter thermautotrophicus*. *J Bacteriol* **187**, 1856–8.

Makarova, K. S., Aravind, L. & Galperin, M. Y. et al. (1999) Comparative genomics of the Archaea (Euryarchaeota): evolution of conserved protein families, the stable core, and the variable shell. *Genome Res* **9**, 608–28.

Makarova, K. S., Aravind, L., Grishin, N. V., Rogozin, I. B. & Koonin, E. V. (2002) A DNA repair system specific for thermophilic archaea and bacteria predicted by genomic context analysis. *Nucleic Acids Res* **30**, 482–96.

Makarova, K. S., Grishin, N. V., Shabalina, S. A. & Koonin, E. V. (2006) A putative RNA-inference-based immune system in prokaryotes: computationalanalysis of the predicted enzymatic machinery, functional analogies with eukaryotic RNAi, and hypothetical mechanism of action. *Biol Direct* **16**, 1–7.

Makarova, K. S. & Koonin, E. V. (2003a) Comparative genomics of Archaea: how much have we learned in six years, and what's next? *Genome Biol* **4**, 115–16.

Makarova, K. S. & Koonin, E. V. (2003b) Filling a gap in the central metabolism of archaea: prediction of a novel aconitase by comparative-genomic analysis. *FEMS Microbiol Lett* **227**, 17–23.

Makarova, K. S. & Koonin, E. V. (2005) Evolutionary and functional genomics of the Archaea. *Curr Opin Microbiol* **8**, 586–94.

Makarova, K. S., Wolf, Y. I. & Koonin, E. V. (2003) Potential genomic determinants of hyperthermophily. *Trends Genet* **19**, 172–6.

Makino, S., Amano, N., Koike, H. & Suzuki, M. (1999) Prophages inserted in archaebacterial genomes. *Proc Japan Acad Ser B* **75**, 166–71.

Malik, H. S. & Henikoff, S. (2003) Phylogenomics of the nucleosome. *Nat Struct Biol* **10**, 882–91.

Mallik, P., Boutz, D. R., Eisenberg, D. & Yeates, T. D. (2002) Genomic evidence that the intracellular proteins of archaeal microbes contain disulfide bonds. *Proc Natl Acad Sci USA* **99**, 9679–84.

Manco, G., Carrea, G., Giosue, E., Ottolina, G., Adamo, G. & Rossi, M. (2002) Modification of the enantioselectivity of two homologous thermophilic carboxylesterases from *Alicyclobacillus acidocaldarius* and *Archaeoglobus fulgidus* by random mutagenesis and screening. *Extremophiles* **6**, 325–31.

Manco, G., Giosue, E. D., Auria, S., Herman, P., Carrea, G. & Rossi, M. (2000) Cloning, overexpression, and properties of a new thermophilic and thermostable esterase with sequence similarity to hormone-sensitive lipase subfamily from the archaeon *Archaeoglobus fulgidus*. *Arch Biochem Biophys* **373**, 182–92.

Mander, G. J., Weiss, M. S., Hedderich, R., Kahnt, J., Ermler, U. & Warkentin, E. (2005) X-ray structure of the gamma-subunit of a dissimilatory sulfite reductase: fixed and flexible C-terminal arms. *FEBS Lett* **589**, 4600–4.

Manzan, A., Pfeiffer, G., Hefferin, M. L., Lang, C. E., Carney, J. P. & Hopfner, K. P. (2004) MlaA, a hexameric ATPase linked to the Mre11 complex in archaeal genomes. *EMBO Rep* **5**, 54–9.

Manzur, K. L. & Zhou, M.-M. (2005) An archaeal SET domain protein exhibits distinct lysine methyltransferase activity towards DNA-associated protein MC1-α. *FEBS Lett* **579**, 3859–65.

Mao, H., White, S. A. & Williamson, J. R. (1999) A novel loop-loop recognition motif in the yeast ribosomal protein L30 autoregulatory RNA complex. *Nat Struct Biol* **6**, 1139–47.

Marc, F., Sandman, K., Lurz, R. & Reeve, J. N. (2002) Archaeal histone tetramerization determines DNA affinity and the direction of DNA supercoiling. *J Biol Chem* **277**, 30879–86.

Marck, C. & Grosjean, H. (2003) Identification of BHB splicing motifs in intron-containing tRNAs from 18 archaea: evolutionary implications. *RNA* **9**, 1516–31.

Margesin, R. & Schinner, F. (2001) Potential of halotolerant and halophilic microorganisms for biotechnology. *Extremophiles* **5**, 73–83.

Marhuenda-Egea, F. C., Piera-Velazquez, S., Cadenas, C. & Cadenas, E. (2002) Mechanism of adaptation of an atypical alkaline p-nitrophenyl phosphatase from the archaeon *Halobacterium salinarum* at low-water environments. *Biotechnol Bioeng* **78**, 497–502.

Marintchev, A., Kolupaeva, V. G., Pestova, T. V. & Wagner, G. (2003) Mapping the binding interface between human eukaryotic initiation factors 1A and 5B: a new interaction between old partners. *Proc Natl Acad Sci USA* **100**, 1535–40.

Markine-Goriaynoff, N., Gillet, L., van Etten, J. L., Korres, H., Verma, N. & Vanderplasschen, A. (2004) Glycosyltransferases encoded by viruses. *J Gen Virol* **85**, 2741–54.

Marsh, V. L., Peak-Chew, S. Y. & Bell, S. D. (2005) Sir2 and the acetyltransferase, Pat, regulate the archaeal chromatin protein, Alba. *J Biol Chem* **280**, 21122–8.

Marsin, S. & Forterre, P. (1998) A rolling circle replication initiator protein with a nucleotidyl-transferase activity encoded by the plasmid pGT5 from the hyperthermophilic archaeon *Pyrococcus abyssi*. *Mol Microbiol* **27**, 1183–92.

Marsin, S. & Forterre, P. (1999) The active site of the rolling circle replication protein Rep75 is involved in site-specific nuclease, ligase and nucleotidyl transferase activities. *Mol Microbiol* **33**, 537–45.

Martens, J. A. & Winston, F. (2003) Recent advances in understanding chromatin remodeling by Swi/Snf complexes. *Curr Opin Genet Dev* **13**, 136–42.

Martin, A., Yeats, S., Janekovic, D., Reiter, W.-D., Aicher, W. & Zillig, W. (1984) SAV1, a temperate UV-inducible DNA virus-like particle from the archaebacterium *Sulfolobus acidocaldarius* isolate B12. *EMBO J* **3**, 2165–8.

Martusewitsch, E., Sensen, C. W. & Schleper, C. (2000) High spontaneous mutation rate in the hyperthermophilic archaeon *Sulfolobus solfataricus* is mediated by transposable elements. *J Bacteriol* **182**, 2574–81.

Maruyama, T., Suzuki, R. & Furutani, M. (2004) Archaeal peptidyl prolyl cis-trans isomerases (PPIases) update 2004. *Front Biosci* **9**, 1680–720.

Massana, R., Murray, A. E., Preston, C. M. & DeLong, E. F. (1997) Vertical distribution and phylogenetic characterization of marine planktonic Archaea in the Santa Barbara Channel. *Appl Environ Microbiol* **63**, 50–6.

Matsuda, T., Fujikawa, M., Ezaki, S. & Imanaka, T., Morikawa, M. & Kanaya, S. (2001) Interaction of TIP26 from a hyperthermophilic archaeon with TFB/TBP/DNA ternary complex. *Extremophiles* **5**, 177–82.

Matsui, E., Kawasaki, S., Ishida, H. et al. (1999) Thermostable flap endonuclease from the archaeon, *Pyrococcus horikoshii*, cleaves the replication fork-like structure endo/exonucleolytically. *J Biol Chem* **274**, 18297–309.

Matsui, E., Nishio, M., Yokoyama, H., Harata, K., Darnis, S. & Matsui, I. (2003) Distinct domain functions regulating de novo DNA synthesis of thermostable DNA primase from hyperthermophile *Pyrococcus horikoshii*. *Biochemistry* **42**, 14968–76.

Matsui, I., Ishikawa, K., Ishida, H., Kosugi, Y. & Tahara, Y. (2002) Agency of industrial science and technology (Tokyo), Patent 222866/1998.

Matsumiya, S., Ishino, Y. & Morikawa, K. (2001) Crystal structure of an archaeal DNA sliding clamp: proliferating cell nuclear antigen from *Pyrococcus furiosus*. *Protein Sci* **10**, 17–23.

Matsunaga, F., Forterre, P., Ishino, Y. & Myllykallio, H. (2001) In vivo interactions of archaeal Cdc6/Orc1 and minichromosome maintenance proteins with the replication origin. *Proc Natl Acad Sci USA* **98**, 11152–7.

Matsunaga, F., Norais, C., Forterre, P. & Myllykallio, H. (2003) Identification of short "eukaryotic" Okazaki fragments synthesized from a prokaryotic replication origin. *EMBO Rep* **4**, 154–8.

Matte-Tailliez, O., Brochier, C., Forterre, P. & Philippe, H. (2002) Archaeal phylogeny based on ribosomal proteins. *Mol Biol Evol* **19**, 631–9.

Mattick, J. S. (2004) RNA regulation: a new genetics? *Nat Rev Genet* **5**, 316–23.

Maxon, M. E., Goodrich, J. A. & Tjian, R. (1994) Transcription factor IIE binds preferentially to RNA polymerase IIa and recruits TFIIH: a model for promoter clearance. *Genes Dev* **8**, 515–24.

Maxwell, E. S. & Fournier, M. J. (1995) The small nucleolar RNAs. *Annu Rev Biochem* **64**, 897–934.

Mayr, A. & Pfeifer, F. (1997) The characterization of the *nv-gvpACNOFGH* gene cluster involved in gas vesicle formation in *Natronobacterium vacuolatum*. *Arch Microbiol* **168**, 24–32.

Mayr, J., Lupas, A., Kellermann, J., Eckerskorn, C., Baumeister, W. & Peters, J. (1996) A hyperthermostable protease of the subtilisin family bound to the surface layer of the archaeon *Staphylothermus marinus*. *Curr Biol* **6**, 739–49.

Meinhart, A., Blobel, J. & Cramer, P. (2003) An extended winged helix domain in general transcription factor E/IIE alpha. *J Biol Chem* **278**, 48267–74.

Melendez-Hevia, E., Waddell, T. G., Heinrich, R. et al. (1997) Theoretical approaches to the evolutionary optimization of glycolysis–chemical analysis. *Eur J Biochem* **244**, 527–43.

Mellor, J. (2005) The dynamics of chromatin remodeling at promoters. *Mol Cell* **19**, 147–57.

Mescher, M. F., Hansen, U. & Strominger, J. L. (1976) Formation of lipid-linked sugar compounds in *Halobacterium salinarium*. Presumed intermediates in glycoprotein synthesis. *J Biol Chem* **251**, 7289–94.

Messer, W. (2002) The bacterial replication initiator DnaA. DnaA and oriC, the bacterial mode to initiate DNA replication. *FEMS Microbiol Rev* **26**, 355–74.

Metcalf, W. W., Zhang, J. K., Apolinario, E., Sowers, K. R. & Wolfe, R. S. (1997) A genetic system for Archaea of the genus *Methanosarcina*: liposome-mediated transformation and construction of shuttle vectors. *Proc Natl Acad Sci USA* **94**, 2626–31.

Metzger, W., Schickor, P. & Heumann, H. (1989) A cinematographic view of *Escherichia coli* RNA polymerase translocation. *EMBO J* **8**, 2745–54.

Mevarech, M. & Werczberger, R. (1985) Genetic transfer in *Halobacterium volcanii*. *J Bacteriol* **162**, 461–2.

Michaelis, W., Seifert, R., Nauhaus, K. et al. (2002) Microbial reefs in the Black Sea fueled by anaerobic oxidation of methane. *Science* **297**, 1013–15.

Michalke, K. & Hensel, R. (2003) Microbial biotransformations of metal(loid)s. *In:* A. V. Hirner & H. Emons (eds) *Organic Metal and Metalloid Species in the Environment: Analysis, Distribution, Processes and Toxicological Evaluation*. Springer-Verlag, Heidelberg, pp. 137–50.

Michalke, K., Meyer, J., Hirner, A. V. & Hensel, R. (2002) Biomethylation of bismuth by the methanogen *Methanobacterium formicicum*. *Appl Organometal Chem* **16**, 221–7.

Michalke, K., Wickenheiser, E. B. & Mehring, M. et al. (2000) Production of volatile derivatives of metal(loid)s by microflora involved in anaerobic digestion of sewage sludge. *Appl Environ Microbiol* **66**, 2791–6.

Michelitsch, M. D. & Weissman, J. S. (2000) A census of glutamine/asparagine-rich regions: implications for their conserved function and the prediction of novel prions. *Proc Natl Acad Sci USA* **97**, 11910–15.

Middelburg, J. J. (2000) The geochemical sulfur cycle. *In:* P. N. L. Lens & P. Hulshoff (eds) *Environmental Technologies to Treat Sulfur Pollution*. IWA Publishing, London, pp. 33–46.

Middleton, C. L., Parker, J. L., Richard, D. J., White, M. F. & Bond, C. S. (2004) Substrate recognition and catalysis by the Holliday junction resolving enzyme Hje. *Nucleic Acids Res* **32**, 5442–51.

Mijts, B. N. & Patel, B. K. (2002) Cloning, sequencing and expression of an α-amylase gene, *amyA*, from the thermophilic halophile *Halothermothrix orenii* and purification and biochemical characterization of the recombinant enzyme. *Microbiol* **148**, 2343–9.

Miller, T. L. & Wolin, M. J. (1982) Isolation of *Methanobrevibacter smithii* from human feces. *Appl Environ Microbiol* **43**, 227–32.

Miller, T.L. & Wolin, M. J. (1985) *Methanosphaera stadtmaniae* gen. nov., sp. nov.: a species that forms methane by reducing methanol with hydrogen. *Arch Microbiol* **141**, 116–22.

Min, B., Pelaschier, J. T., Graham, D. E., Tumbula-Hansen, D. & Söll, D. (2002) Transfer RNA-dependent amino acid biosynthesis: an essential route to asparagine formation. *Proc Natl Acad Sci USA* **99**, 2678–83.

Minakhin, L., Bhagat, S. & Brunning, A. et al. (2001) Bacterial RNA polymerase subunit ω and eukaryotic RNA polymerase subunit RPB6 are sequence, structural, and functional homologs and promote RNA polymerase assembly. *Proc Natl Acad Sci USA* **98**, 892–7.

Miyao, T. & Woychik, N. A. (1998) RNA polymerase subunit RPB5 plays a role in transcriptional activation. *Proc Natl Acad Sci USA* **95**, 15281–6.

Miyata, T., Oyama, T., Mayanagi, K., Ishino, S., Ishino, Y. & Morikawa, K. (2004) The clamp-loading complex for processive DNA replication. *Nat Struct Mol Biol* **11**, 632–6.

Moissl, C., Rachel, R., Briegel, A., Engelhardt, H. & Huber, R. (2005) The unique structure of archaeal "hami" highly complex cell appendages with nano-grappling hooks. *Mol Microbiol* **56**, 361–70.

Moissl, C., Rudolph, C., Rachel, R., Koch, M. & Huber, R. (2003) In situ growth of the novel SM1 euryarchaeon from a string-of-pearls-like microbial community in its cold biotope, its physical separation and insights into its structure and physiology. *Arch Microbiol* **180**, 211–17.

Mojica, F. J. M., Diez-Villasenor, C., Garcia-Martinez, J. & Soria. E. (2005) Intervening sequences of regularly spaced prokaryotic repeats derive from foreign genetic elements. *J Mol Evol* **60**, 174–82.

Mojica, F. J. M., Ferrer, C., Juez, G. & Rodriguez-Valera, F. (1995) Long stretches of short tandem repeats are present in the largest replicons of the archaea *Haloferax mediterranei* and *Haloferax volcanii* and could be involved in replicon partitioning. *Mol Microbiol* **17**, 85–93.

Moore, B. C. & Leigh, J. A. (2005) Markerless mutagenesis in *Methanococcus maripaludis* demonstrates roles for alanine dehydrogenase, alanine racemase, and alanine permease. *J Bacteriol* **187**, 972–9.

Moore, T., Zhang, Y., Fenley, M. O. & Li, H. (2004) Molecular basis of box C/D RNA–protein interactions; cocrystal structure of archaeal L7Ae and a box C/D RNA. *Structure* **12**, 807–18.

Moran, J. J., House, C. H., Freeman, K. H. & Ferry, J. G. (2004) Trace methane oxidation studied in several Euryarchaeota under diverse conditions. *Archaea* **1**, 303–9.

Moran, N. A. (1996) Accelerated evolution and Muller's rachet in endosymbiotic bacteria. *Proc Natl Acad Sci USA* **93**, 2873–8.

Moreira, D. & Lopez-Garcia, P. (2005) Comment on The 1.2-megabase genome sequence of Mimivirus. *Science* **308**, 1114; author reply 1114.

Moreira, D., Rodriguez-Valera, F. & Lopez-Garcia, P. (2004) Analysis of a genome fragment of a deep-sea uncultivated Group II euryarchaeote containing 16S rDNA, a spectinomycin-like operon and several energy metabolism genes. *Environ Microbiol* **6**, 959–69.

Morikawa, M. & Imanaka, T. (2001) Thiol protease from *Thermococcus kodakaraensis* KOD1. *Methods Enzymol* **330**, 424–33.

Morimoto, R. I. (2002) Dynamic remodelling of transcription complexes by molecular chaperones. *Cell* **110**, 281–4.

Mozier, N. M., McConnell, K. P. & Hoffman, J. L. (1988) S-adenosyl-L-methionine thioether S-methyltransferase, a new enzyme in sulfur and selenium metabolism. *J Biol Chem* **10**, 4527–31.

Müller, F. H., Bandeiras, T. M., Urich, T., Teixeira, M., Gomes, C. M. & Kletzin, A. (2004) Coupling of the pathway of sulphur oxidation to dioxygen reduction: characterization of a novel membrane bound thiosulphate:quinone oxidoreductase. *Mol Microbiol* 1147–60.

Müller, P., Egorova, K., Vorgias, C. E. et al. (2006) Cloning, overexpression, and characterization of a thermoactive nitrilase from the hyperthermophilic archaeon *Pyrococcus abyssi*. *Prot Expr Purif* **47**, 672–81.

Murray, A. E., Preston, C. M., Massana, R. et al. (1998) Seasonal and spatial variability of bacterial and archaeal assemblages in the coastal waters near Anvers Island, Antarctica. *Appl Environ Microbiol* **64**, 2585–95.

Murray, R. G. E. (1962) Fine structure and taxonomy of bacteria. *In:* G. C. Ainsworth & P. H. A. Sneath (eds) *Microbial Classification*. The Society for General Microbiology Symposium 12, pp. 119–44.

Murray, R. G. E. (1974) A place for bacteria in the living world. *In:* R. E. Buchanan & N. E. Gibbons (eds) *Bergey's Manual of Determinative Bacteriology*, 8th edn, Williams & Wilkins, Baltimore, pp. 4–9.

Musgrave, D., Forterre, P. & Slesarev, A. (2000) Negative constrained DNA supercoiling in archaeal nucleosomes. *Mol Microbiol* **34**, 341–9.

Mushegian, A. R. & Koonin, E. V. (1996a) A minimal gene set for cellular life derived by comparison of complete bacterial genomes. *Proc Natl Acad Sci USA* **93**, 10268–73.

Mushegian, A. R. & Koonin, E. V. (1996b) Gene order is not conserved in bacterial evolution. *Trends Genet* **12**, 289–90.

Muskhelishvili, G., Palm, P. & Zillig, W. (1993) SSV1-encoded site-specific recombination system in *Sulfolobus shibatae*. *Mol Gen Genet* **237**, 334–42.

Myllykallio, H., Lopez, P., Lopez-Garcia, P. et al. (2000) Bacterial mode of replication with eukaryotic-like machinery in a hyperthermophilic archaeon. *Science* **288**, 2212–15.

Nadal, M., Mirambeau, G., Forterre, P., Reiter, W. D. & Duget, M. (1986) Positively supercoiled DNA in a virus-like particle of an archaebacterium. *Nature* **321**, 256–8.

Nagatomo, H., Matsushita, Y., Sugamoto, K. & Matsui, T. (2005) Preparation and properties of gelatin-immobilized beta-glucosidase from *Pyrococcus furiosus*. *Biosci Biotechnol Biochem* **69**, 128–36.

Nakatani, M., Ezaki, S., Atomi, H. & Imanaka, T. (2002) Substrate recognition and fidelity of strand joining by an archaeal DNA ligase. *Eur J Biochem* **269**, 650–6.

Namy, O., Rousset, J.-P., Napthine, S. & Brierley, I. (2004) Reprogrammed genetic decoding in cellular gene expression. *Mol Cell* **13**, 157–68.

Napoli, A., Valenti, A., Salerno, V. et al. (2004) Reverse gyrase recruitment to DNA after UV light irradiation in *Sulfolobus solfataricus*. *J Biol Chem* **279**, 33192–8.

Napoli, A., van der Oost, J., Sensen, C. W., Charlebois, R. L., Rossi, M. & Ciaramella, M. (1999) An Lrp-like protein of the hyperthermophilic archaeon *Sulfolobus solfataricus* which binds to its own promoter. *J Bacteriol* **181**, 1474–80.

Napoli, A., Zivanovic, Y. & Bocs, C. et al. (2002) DNA bending, compaction and negative supercoiling by the architectural protein Sso7d of *Sulfolobus solfataricus*. *Nucleic Acids Res* **30**, 2656–62.

Naryshkin, N., Revyakin, A., Kim, Y., Mekler, V. & Ebright, R. H. (2000) Structural organization of the RNA polymerase-promoter open complex. *Cell* **101**, 601–11.

Nastopoulos, V., Pisani, F. M., Savino, C., Federici, L., Rossi, M. & Tsernoglou, D. (1998) Crystallization and preliminary X-ray diffraction studies of DNA polymerase from the thermophilic archaeon *Sulfolobus solfataricus*. *Acta Crystallogr D Biol Crystallogr* **54**, 1002–4.

Nauhaus, K., Boetius, A., Krüger, M. & Widdel, F. (2002) *In vitro* demonstration of anaerobic oxidation of methane coupled to sulphate reduction in sediment from a marine gas hydrate area. *Environ Microbiol* **4**, 296–305.

Nauhaus, K., Treude, T., Boetius, A. & Krüger, M. (2005) Environmental regulation of the anaerobic oxidation of methane: a comparison of ANME-I and ANME-II communities. *Environ Microbiol* **7**, 98–106.

Nelson, P., Kiriakidou, M., Sharma, A., Maniataki, E. & Mourelatos, Z. (2003) The microRNA world: small is mighty. *Trends Biochem Sci* **28**, 534–40.

Nercessian, O., Fouquet, Y., Pierre, C., Prieur, D. & Jeanthon, C. (2005) Diversity of Bacteria and Archaea associated with a carbonate-rich metalliferous sediment sample from the Rainbow vent field on the Mid-Atlantic Ridge. *Environ Microbiol* **7**, 698–714.

Nesbo, C. L., Boucher, Y. & Doolittle, W. F. (2001) Defining the core of non-transferable prokaryotic genes: The euryarchaeal core. *J Mol Evol* **53**, 340–50.

Neumann, H. (1988) Struktur, Funktion und Variabilität eines archaebacterialen Genomes. PhD dissertation, Ludwig-Maximillians-Universität München, Munich.

Neumann, H. & Zillig, W. (1990) Structure variability in the genome of the *Thermoproteus tenax* virus TTV1. *Mol Gen Genet* **222**, 435–7.

Newman, M., Murray-Rust, J. & Lally, J. et al. (2005) Structure of an XPF endonuclease with and without DNA suggests a model for substrate recognition. *EMBO J* **24**, 895–905.

Ng, W. L. & DasSarma, S. (1993) Minimal replication origin of the 200-kilobase *Halobacterium* plasmid pNRC100. *J Bacteriol* **175**, 4584–96.

Ng, W. L., Kenney, S. & Mahairas, G., et al. (2000) Genome sequence of *Halobacterium* species NRC-1. *Proc Natl Acad Sci USA* **97**, 12176–81.

Ng, W. L., Kothakota, S. & DasSarma, S. (1991) Structure of the gas vesicle plasmid in *Halobacterium halobium*: inversion isomers, inverted repeats, and insertion sequences. *J Bacteriol* **173**, 1958–64.

Ng, W. V., Ciufo, S. A., Smith, T. M. et al. (1998) Snapshot of a large dynamic replicon in a halophilic archaeon: megaplasmid or minichromosome? *Genome Res* **8**, 1131–41.

Nicol, G. W., Glover, L. A. & Prosser, J. I. (2003a) Spatial analysis of archaeal community structure in grassland soil. *Appl Environ Microbiol* **69**, 7420–9.

Nicol, G. W., Glover, L. A. & Prosser, J. I. (2003b) The impact of grassland management on archaeal community structure in upland pasture rhizosphere soil. *Environ Microbiol* **5**, 152–62.

Nicol, G. W., Tscherko, D., Embley, T. M. & Prosser, J. I. (2005) Primary succession of soil *Crenarchaeota* across a receding glacier foreland. *Environ Microbiol* **7**, 337–47.

Niehaus, F., Bertoldo, C., Kahler, M. & Antranikian, G. (1999) Extremophiles as a source of novel enzymes for industrial application. *Appl Microbiol Biotechnol* **51**, 711–29.

Niehaus, F., Peters, A., Groudieva, T. & Antranikian, G. (2000) Cloning, expression and biochemical characterisation of a unique thermostable pullulan-hydrolysing enzyme from the hyperthermophilic archaeon *Thermococcus aggregans*. *FEMS Microbiol Lett* **190**, 223–9.

Nikolov, D., Hu, S.-H., Lin, J. et al. (1992) Crystal structure of TFIID TATA-box binding protein. *Nature* **360**, 40–6.

Nishino, T., Komori, K., Ishino, Y. & Morikawa, K. (2003) X-ray and biochemical anatomy of an archaeal XPF/Rad1/Mus81 family nuclease: similarity between its endonuclease domain and restriction enzymes. *Structure (Camb)* **11**, 445–57.

Nishino, T., Komori, K., Tsuchiya, D., Ishino, Y. & Morikawa, K. (2001) Crystal structure of the archaeal Holliday junction resolvase Hjc and implications for DNA recognition. *Structure* **9**, 197–204.

Nolling, J., van Eeden, F. J., Eggen, R. I. & de Vos, W. M. (1992) Modular organization of related Archaeal plasmids encoding different restriction-modification systems in *Methanobacterium thermoformicicum*. *Nucleic Acids Res* **20**, 6501–7.

Noon, K. R., Bruenger, E. & McCloskey, J. A. (1998) Post-transcriptional modifications in 16S and 23S rRNAs of the archaeal hyperthermophile *Sulfolobus solfataricus*. *J Bacteriol* **180**, 2883–8.

Norris, P. R., Burton, N. & Foulis, N. (2000) Acidophiles in bioreactor mineral processing. *Extremophiles* **4**, 71–6.

Norris, P. R. & Johnson, D. B. (1998) Acidophilic Microorganisms. In: K. Horikoshi & W. D. Grant (eds) *Extremophiles: Microbial Life in Extreme Environments*. John Wiley, New York, pp. 133–54.

Nunes-Duby, S. E., Kwon, H. J., Tirumalai, R. S., Ellenberger, T. & Landy A. (1998) Similarities and differences among 105 members of the Int family of site-specific recombinases, *Nucleic Acids Res* **26**, 391–406.

Nuttall, S. D., Deutschel, S. E., Irving, R. A., Serrano-Gomicia, J. A. & Dyall-Smith, M. L. (2000) The ShBle resistance determinant from *Streptoalloteichus hindustanus* is expressed in *Haloferax volcanii* and confers resistance to bleomycin. *Biochem J* **346**, 251–4.

Ochsenreiter, T., Selezi, D., Quaiser, A., Bonch-Osmolovskaya, L. & Schleper, C. (2003) Diversity and abundance of Crenarchaeota in terrestrial habitats studied by 16S RNA surveys and real time PCR. *Environ Microbiol* **5**, 787–97.

O'Connor, E. & Shand, R. (2002) Halocins and sulfolobicins: The emerging story of archaeal protein and peptide antibiotics. *J Indust Microbiol Biotechnol* **28**, 23–31.

O'Donnell, M., Jeruzalmi, D. & Kuriyan, J. (2001) Clamp loader structure predicts the architecture of DNA polymerase III holoenzyme and RFC. *Curr Biol* **11**, R935–46.

Ofengand, J. & Fournier, M. J. (1998) The pseudouridine residues in rRNA: number, location, biosynthesis, and function. In: H. Grosjean & R. Benne (eds). *Modification and Editing of RNA*. American Society for Microbiology, Washington, DC, pp. 229–53.

Offner, S., Hofacker, A., Wanner, G. & Pfeifer, F. (2000) Eight of fourteen *gvp* genes are sufficient for formation of gas vesicles in halophilic archaea. *J Bacteriol* **182**, 4328–36.

Offner, S. & Pfeifer, F. (1995) Complementation studies with the gas vesicle-encoding pvac region of *Halobacterium salinarium* PHH1 reveal a regulatory role for the p-*gvpDE* genes. *Mol Microbiol* **16**, 9–19.

Ohkuma, Y. (1997) Multiple functions of general transcription factors TFIIE and TFIIH in transcription:

possible points of regulation by *trans*-acting factors. *J Biochem (Tokyo)* **122**, 481–9.

Ohkuma, Y., Hashimoto, S., Wang, C. K., Horikoshi, M. & Roeder, R. G. (1995) Analysis of the role of TFIIE in basal transcription and TFIIH-mediated carboxy-terminal domain phosphorylation through structure-function studies of TFIIEα. *Mol Cell Biol* **15**, 4856–66.

Ohkuma, Y., Sumimoto, H., Hoffmann, A., Shimasaki, S., Horikoshi, M. & Roeder, R. G. (1991) Structural motifs and potential β homologies in the large subunit of human general transcription factor TFIIE. *Nature* **354**, 398–401.

Olsen, G. J., Lane, D. J., Giovannoni, S. J., Pace, N. R. & Stahl, D. A. (1986) Microbial ecology and evolution: a ribosomal RNA approach. *Annu Rev Microbiol* **40**, 337–65.

Olsen, G. J. & Woese, C. R. (1996) Lessons from an archaeal genome: what are we learning from *Methanococcus jannaschii*? *Trends Genet* **12**, 377–9.

Omer, A. D., Lowe, T. M., Russell, A. G., Ebhardt, H., Eddy, S. R. & Dennis, P. P. (2000) Homologs of small nucleolar RNAs in Archaea. *Science* **288**, 517–22.

Omer, A. D., Ziesche, S., Ebhardt, H. & Dennis, P. P. (2002) In vitro reconstitution and activity of a C/D box methylation guide ribonucleoprotein complex. *Proc Natl Acad Sci* **99**, 5289–94.

Opalka, N., Chlenov, M., Chacon, P., Rice, W. J., Wriggers, W. & Darst, S. A. (2003) Structure and function of the transcription elongation factor GreB bound to bacterial RNA polymerase. *Cell* **114**, 335–45.

Orcutt, B. N., Boetius, A., Lugo, S. K., MacDonal, I. R., Samarkin, V. A. & Joye, S. B. (2004) Life at the edge of methane ice: microbial cycling of carbon and sulfur in Gulf of Mexico gas hydrates. *Chem Geol* **205**, 239–51.

Oren, A. (2002) Diversity of halophilic microorganisms: environments, phylogeny, physiology, and applications. *J Ind Microbiol Biotechnol* **28**, 56–63.

Orita, I., Yurimoto, H., Hirai, R. et al. (2005) The archaeon *Pyrococcus horikoshii* possesses a bifunctional enzyme for formaldehyde fixation via the ribulose monophosphate pathway. *J Bacteriol* **187**, 3636–42.

Orlova, M., Newlands, J., Das, A., Goldfarb, A. & Borukhov, S. (1995) Intrinsic transcript cleavage activity of RNA polymerase. *Proc Natl Acad Sci USA* **92**, 4596–600.

Orphan, V. J., Hinrichs, K. U., Ussler, W. et al. (2001a) Comparative analysis of methane oxidizing archaea and sulfate-reducing bacteria in anoxic marine sediments. *Appl Environ Microbiol* **67**, 1922–34.

Orphan, V. J., House, C. H., Hinrichs, K. U., McKeegan, K. D. & DeLong, E. F. (2001b) Methane-consuming archaea revealed by directly coupled isotopic and phylogenetic analysis. *Science* **293**, 484–7.

Orphan, V. J., House, C. H., Hinrichs, K. U., McKeegan, K. D. & DeLong, E. F. (2002) Multiple archaeal groups mediate methane oxidation in anoxic cold seep sediments. *Proc Natl Acad Sci USA* **99**, 7663–8.

Ortenberg, R., Rozenblatt-Rosen, O. & Mevarech, M. (2000) The extremely halophilic archaeon *Haloferax volcanii* has two very different dihydrofolate reductases. *Mol Microbiol* **35**, 1493–505.

Ortenberg, R., Tchelet, R. & Mevarech, M. (1998) A model for the genetic exchange system of the extremely halophilic archaeon *Haloferax volcanii*. *In:* A. Oren (ed.) *Microbiology and Biogeochemistry of Hypersaline Environments*. CRC Press, Boca Raton, FL, pp. 331–8.

Oshikane, H., Sheppard, K., Fukai, S. et al. (2006) Structural basis of RNA-dependent recruitment of glutamine to the genetic code. *Science* in press.

Ouhammouch, M. (2004) Transcriptional regulation in Archaea. *Curr Opin Genet Dev* **14**, 133–8.

Ouhammouch, M., Dewhurst, R. E., Hausner, W., Thomm, M. & Geiduschek, E. P. (2003) Activation of archaeal transcription by recruitment of the TATA-binding protein. *Proc Natl Acad Sci USA* **100**, 5097–102.

Ouhammouch, M., Langham, G. E., Hausner, W., Simpson, A. J., El-Sayed, N. M. & Geiduschek, E. P. (2005) Promoter architecture and response to a positive regulator of archaeal transcription. *Mol Microbiol* **56**, 625–37.

Ouhammouch, M., Werner, F., Weinzierl, R. O. & Geiduschek, E. P. (2004) A fully recombinant system for activator-dependent archaeal transcription. *J Biol Chem* **279**, 51719–21.

Ouzounis, C. & Sander, C. (1992) TFIIB, an evolutionary link between the transcription machineries of archaebacteria and eukaryotes. *Cell* **71**, 189–90.

Oyama, T., Ishino, Y., Cann, I. K., Ishino, S. & Morikawa, K. (2001) Atomic structure of the clamp loader small subunit from *Pyrococcus furiosus*. *Mol Cell* **8**, 455–63.

Pal, M. & Luse, D. S. (2003) The initiation-elongation transition: lateral mobility of RNA in RNA polymerase II complexes is greatly reduced at +8/+9 and absent by +23. *Proc Natl Acad Sci USA* **100**, 5700–5.

Pal, M., Ponticelli, A. S. & Luse, D. S. (2005) The role of the transcription bubble and TFIIB in promoter clearance by RNA polymerase II. *Mol Cell* **19**, 101–10.

Palm, P., Schleper, C., Grampp, B. et al. (1991) Complete nucleotide sequence of the virus SSV1 of the archaebacterium *Sulfolobus shibatae*. *Virology* **185**, 242–50.

Palmer, J. R. & Daniels, C. J. (1995) *In vivo* definition of an archaeal promoter. *J Bacteriol* **177**, 1844–9.

Pancost, R. D., Hopmans, E. C. & Damste, J. S. S. (2001) Archaeal lipids in Mediterranean sold seeps: molecular proxies for anaerobic oxidation. *Geochim Cosmochim Acta* **65**, 1611–27.

Papke, R. T., König, J. E., Rodriguez-Valera, F. Doolittle, W. F. (2004) Frequent recombination in a saltern population of *Halorubrum*. *Science* **306**, 1928–9.

Paquet, F., Culard, F., Barbault, F., Maurizot, J.-C. & Lancelot, G. (2004) NMR solution structure of the archaebacterial chromosomal protein MC1 reveals a new protein fold. *Biochemistry* **43**, 14971–8.

Paradinas, C., Gervais, A., Maurizot, J.-C. & Culard, F. (1998) Structure specific binding recognition of a methanogen chromosomal protein. *Eur J Biochem* **257**, 372–9.

Parker, J. L. & White, M. F. (2005) The endonuclease Hje catalyses rapid, multiple turnover resolution of Holliday junctions. *J Mol Biol* **350**, 1–6.

Patel, G. B., Nash, J. H. E., Agnew, B. J. & Sprott, G. D. (1994) Natural and electroporation-mediated transformation of *Methanococcus voltae* protoplasts. *Appl Environ Microbiol* **60**, 903–7.

Patel, G. B. & Sprott, G. D. (1999) Archaeobacterial ether lipid liposomes (archaeosomes) as novel vaccine and drug delivery systems. *Crit Rev Biotechnol* **19**, 317–57.

Patenge, N., Haase, A., Bolhuis, H. & Oesterhelt, D. (2000) The gene for a halophilic β-galactosidase (*bgaH*) of *Haloferax alicantei* as a reporter gene for promoter analyses in *Halobacterium salinarum*. *Mol Microbiol* **36**, 105–13.

Patikoglou, G. A., Kim, J. L., Sun, L., Yang, S. H., Kodadek, T. & Burley, S. K. (1999) TATA element recognition by the TATA box-binding protein has been conserved throughout evolution. *Genes Dev* **13**, 3217–30.

Pavlov, N. A., Cherny, D. I., Jovin, T. M. & Slesarev, A. I. (2002a) Nucleosome-like complex of the histone from the hyperthermophile *Methanopyrus kandleri* (MkaH) with linear DNA. *J Biomol Struct Dyn* **20**, 207–14.

Pavlov, N. A., Cherny, D. I., Nazimov, I. V., Slesarev, A. I. & Subramaniam, V. (2002b) Identification, cloning and characterization of a new DNA-binding protein from the hyperthermophilic methanogen *Methanopyrus kandleri*. *Nucleic Acids Res* **30**, 685–94.

Peck, R. F., DasSarma, S. & Krebs, M. P. (2000) Homologous gene knockout in the archaeon *Halobacterium salinarum* with *ura3* as a counterselectable marker. *Mol Microbiol* **35**, 667–76.

Peck, R. F., Echavarri-Erasun C., Johnson, E. et al. (2001) *brp* and *blh* are required for synthesis of the retinal cofactor of bacteriorhodopsin in *Halobacterium salinarum*. *J Biol Chem* **276**, 5739–44.

Pedulla, N., Palermo, R., Hasenohrl, D., Blasi, U., Cammarano, P. & Londei, P. (2005) The archaeal eIF2 homologue: functional properties of an ancient translation initiation factor. *Nucleic Acids Res* **33**, 1804–12.

Pelmenschikov, V., Blomberg, M. R., Siegbahn, P. E. & Crabtree, R. H. (2002) A mechanism from quantum chemical studies for methane formation in methanogenesis. *J Am Chem Soc* **124**, 4039–49.

Pelmenschikov, V. & Siegbahn, P. E. (2003) Catalysis by methyl-coenzyme M reductase: a theoretical study for heterodisulfide product formation. *J Biol Inorg Chem* **8**, 653–62.

Peng, X., Blum, H., She, Q. et al. (2001) Sequences and replication of genomes of the archaeal rudiviruses SIRV1 and SIRV2: relationships to the archaeal lipothrixvirus SIFV and some eukaryal viruses. *Virology* **291**, 226–34.

Peng, X., Brügger, K., Shen, B., Chen, L. She, Q. & Garrett, R. A. (2003) Genus-specific protein binding to the large clusters of DNA repeats (short regularly spaced repeats) present in *Sulfolobus* genomes. *J Bacteriol* **185**, 2410–17.

Peng, X., Holz, I., Zillig, W., Garrett, R. A. & She, Q. (2000) Evolution of the family of pRN plasmids and their integrase-mediated insertion into the chromosome of the crenarchaeon *Sulfolobus solfataricus*. *J Mol Biol* **303**, 449–54.

Peng, X., Kessler, A., Phan, H., Garrett, R. A. & Prangishvili, D. (2004) Multiple variants of the archaeal DNA rudivirus SIRV1 in a single host and a novel mechanism of genomic variation. *Mol Microbiol* **54**, 366–75.

Penny, D. & Poole, A. (1999) The nature of the last universal common ancestor. *Curr Opin Genet Dev* **9**, 672–7.

Pereira, M. M., Bandeiras, T. M., Fernandes, A. S., Lemos, R. S., Melo, A. M. & Teixeira, M. (2004) Respiratory chains from aerobic thermophilic prokaryotes. *J Bioenerg Biomembr* **36**, 93–105.

Pereira, S. L., Grayling, R. A., Lurz, R. & Reeve, J. N. (1997) Archaeal nucleosomes. *Proc Natl Acad Sci USA* **94**, 12633–7.

Pereira, S. L. & Reeve, J. N. (1999) Archaeal nucleosome positioning sequence from *Methanothermus fervidus*. *J Mol Biol* **289**, 675–81.

Pereiro, I. R., Wasik, A. & Lobinski, R. (1999) Speciation of organotin in sediments by multicapillary gas chromatography with atomic emission detection after microwave assisted leaching and solvent extraction-derivatization. *Fresen J Anal Chem* **363**, 460–5.

Perez-Pomares, F., Bautista, V., Ferrer, J., Pire, C., Marhuenda-Egea, F. C. & Bonete, M. J. (2003) α-Amylase activity from the halophilic archaeon *Haloferax mediterranei*. *Extremophiles* **7**, 299–306.

Perez-Rueda, E., Collado-Vides, J. & Segovia, L. (2004) Phylogenetic distribution of DNA-binding transcription factors in bacteria and archaea. *Comput Biol Chem* **28**, 341–50.

Perler, F. (2002) InBase, the intein database. *Nucleic Acids Res* **30**, 383–4.

Persson, B. C., Jager, G. & Gustafsson, C. (1997) The *spoU* gene of *Escherichia coli*, the fourth gene of the *spoT* operon, is essential for tRNA (Gm18) 2′-O-methyltransferase activity. *Nucleic Acids Res* **25**, 4093–7.

Pestova, T. V. & Kolupaeva, V. G. (2002) The roles of individual eukaryotic translation initiation factors in ribosomal scanning and initiation codon selection. *Genes Dev* **16**, 2906–22.

Pestova, T. V., Lomakin, I. B., Lee, J. H., Choi, S. K., Dever, T. E. & Hellen, C. U. (2000) The joining of ribosomal subunits in eukaryotes requires eIF5B. *Nature* **403**, 332–5.

Peters, W. B., Edmondson, S. P. & Shriver, J. W. (2005) Effects of mutation of the Sac7d intercalating residues on the temperature dependence of DNA distortion and binding thermodynamics. *Biochemistry* **44**, 4794–804.

Pfeifer, F. (1986) Insertion elements and genome organization of *Halobacterium halobium*. *System Appl Microbiol* **7**, 36–40.

Pfeifer, F. (2004) Gas vesicle genes in halophlic archaea and bacteria. *In:* A. Ventosa (ed.) *Halophilic Microorganisms*. Springer, Heidelberg, pp. 229–39.

Pfeifer, F. & Betlach, M. (1985) Genome organization of *Halobacterium halobium*: a 70 kb island of more (AT) rich DNA in the chromosome. *Mol Gen Genet* **198**, 449–55.

Pfeifer, F. & Blaseio, U. (1989) Insertion elements and deletion formation in a halophilic archaebacterium. *J Bacteriol* **171**, 5135–40.

Pfeifer, F., Friedman, J., Boyer, H. W. & Betlach, M. (1984) Characterization of insertions affecting the expression of the bacterio-opsin gene in *Halobacterium halobium*. *Nucleic Acids Res* **12**, 2489–97.

Pfeifer, F. & Ghahraman, P. (1993) Plasmid pHH1 of *Halobacterium salinarium*: characterization of the replicon region, the gas vesicle gene cluster and insertion elements. *Mol Gen Genet* **238**, 193–200.

Pfeifer, F., Gregor, D., Hofacker, A., Plößer, P., Zimmermann, P. (2002) Regulation of gas vesicle formation in halophilic archaea. *J Mol Microbiol Biotechnol* **4**, 175–81.

Pfeifer, F., Offner, S., Krüger, K., Gahraman, P. & Englert, C. (1994) Transformation of halophilic archaea and investigation of gas vesicle synthesis. *System Appl Microbiol* **16**, 569–77.

Pfeifer, F., Weidinger, G. & Goebel, W. (1981) Genetic variability in *Halobacterium halobium*. *J Bacteriol* **145**, 375–81.

Pfeifer, F., Zotzel, J., Kurenbach, B., Röder, R. & Zimmermann, P. (2001) A p-loop motif and two basic regions in the regulatory protein GvpD are important for the repression of gas vesicle formation in the archaeon *Haloferax mediterranei*. *Microbiol* **147**, 63–73.

Pfister, P., Wasserfallen, A., Stettler, R. & Leisinger T. (1998) Molecular analysis of *Methanobacterium* phage psiM. *Mol Microbiol* **30**, 233–44.

Pfluger, K., Baumann, S., Gottschalk, G., Lin, W., Santos, H. & Müller, V. (2003) Lysine-2,3-aminomutase and β-lysine acetyltransferase genes of methanogenic archaea are salt induced and are essential for the biosynthesis of Nepsilon-acetyl-beta-lysine and growth at high salinity. *Appl Environ Microbiol* **69**, 6047–55.

Phelan, M. L., Sif, S., Narlikar, G. J. & Kingston, R. E. (1999) Reconstitution of a core chromatin remodeling complex from SWI/SNF subunits. *Mol Cell* **3**, 247–53.

Pihl, T. D., Black, L. K., Schulman, B. A. & Maier, R. J. (1992) Hydrogen-oxidizing electron transport components in the hyperthermophilic archaebacterium *Pyrodictium brockii*. *J Bacteriol* **174**, 137–43.

Pihl, T. D. & Maier, R. J. (1991) Purification and characterization of the hydrogen uptake hydrogenase from the hyperthermophilic archaebacterium *Pyrodictium brockii*. *J Bacteriol* **173**, 1839–44.

Pimenov, N. V. & Ivanova, A. E. (2005) Anaerobic methane oxidation and sulfate reduction in bacterial mats on coral-like carbonate structures in the Black Sea. *Microbiology* **74**, 362–70.

Pimenov, N. V., Rusanov, I. I., Poglazova, M. N. et al. (1997) Bacterial mats on coral-like structures at methane seeps in the Black Sea. *Microbiology* **66**, 354–60.

Pinto, I., Ware, D. E. & Hampsey, M. (1992) The yeast SUA7 gene encodes a homolog of human transcription factor TFIIB and is required for normal start site selection *in vivo*. *Cell* **68**, 977–88.

Piskorski, R. & Jaun, B. (2003) Direct determination of the number of electrons needed to reduce coenzyme F430 pentamethyl ester to the Ni(I) species exhibiting the electron paramagnetic resonance and ultraviolet-visible spectra characteristic for the MCR(red1) state of methyl-coenzyme M reductase. *J Am Chem Soc* **125**, 13120–5.

Plößer, P. & Pfeifer, F. (2002) A bZIP protein from halophilic archaea: structural features and dimer formation of cGvpE from *Halobacterium salinarum*. *Mol Microbiol* **45**, 511–20.

Polycarpo, C., Ambrogelly, A., Bérubé, A. et al. (2004) An aminoacyl-tRNA synthetase that specifically activates pyrrolysine. *Proc Natl Acad Sci USA* **101**, 12450–4.

Polycarpo, C., Ambrogelly, A., Ruan, B. et al. (2003) Activation of the pyrrolysine suppressor tRNA requires formation of a ternary complex with class I and class II lysyl-tRNA synthetases. *Mol Cell* **12**, 287–94.

Poole, A., Jeffares, D. & Penny, D. (1999) Early evolution: prokaryotes, the new kids on the block. *Bioessays* **21**, 880–9.

Poole, F. L. II, Gerwe, B. A., Hopkins, R. C. et al. (2005) Defining genes in the genome of the hyperthermophilic archaeon *Pyrococcus furiosus*: implications for all microbial genomes. *J Bacteriol* **187**, 7325–32.

Possot, O., Gernhardt, P., Klein, A. & Sibold, L. (1988) Analysis of drug resistance in the archaebacterium *Methanococcus voltae* with respect to potential use in genetic engineering. *Appl Environ Microbiol* **54**, 734–40.

Prangishvili, D. (2003) Evolutionary insights from studies on viruses of hyperthermophilic archaea. *Res Microbiol* **154**, 289–94.

Prangishvili, D., Albers, S. V., Holz, I. et al. (1998a) Conjugation in archaea: frequent occurrence of conjugative plasmids in *Sulfolobus*. *Plasmid* **40**, 190–202.

Prangishvili, D., Arnold, H. P., Götz, D. et al. (1999) A novel virus family, the *Rudiviridae*: Structure, virus-host interactions and genome variability of the *Sulfolobus* viruses SIRV1 and SIRV2. *Genetics* **152**, 1387–96.

Prangishvili, D. & Garrett, R. A. (2005) Viruses of hyperthermophilic crenachaea. *Trends Microbiol* **13**, 535–42.

Prangishvili, D., Garrett, R. A. & Koonin, E. V. (2006) Evolutionary genomics of archaeal viruses: unique viral genomes in the third domain of life. *Virus Res* **117**, 52–67.

Prangishvili, D., Klenk, H.-P., Jakobs, G. et al. (1998b) Biochemical and physiological characterization of the dUTPase from the archaeal virus SIRV. *J Biol Chem* **273**, 6024–9.

Prangishvili, D., Stedman, K. & Zillig, W. (2001) Viruses of the extremely thermophilic archaeon *Sulfolobus*. *Trends Microbiol* **9**, 39–43.

Prangishvili, D., Vashakidze, R. P., Chelidze, M. G. & Gabriadze, I. Y. (1985) A restriction endonuclease SuaI from the thermoacidophilic archaebacterium *Sulfolobus acidocaldarius*. *FEBS Lett* **192**, 57–60.

Prato, S., Cannio, R., Klenk, H.-P., Contursi, P., Rossi, M. & Bartolucci, S. (2006) pIT3, a cryptic plasmid isolated from the hyperthermophilic crenarchaeon *Sulfolobus solfactaricus* IT3. *Plasmid* **56**, 35–45.

Preston, C. M., Wu, K. Y., Molinski, T. F. & De Long, E. F. (1996) A psychrophilic crenarchaeon inhabits a marine sponge: *Cenarchaeum symbiosum* gen. nov., sp. nov. *Proc Natl Acad Sci USA* **93**, 6241–6.

Prieur, D., Erauso, G., Geslin, C. et al. (2004) Genetic elements of *Thermococcales*. *Biochem Soc Trans* **32**, 184–7.

Pritchett, M. A., Zhang, J. K. & Metcalf, W. W. (2004) Development of a markerless genetic exchange method for *Methanosarcina acetivorans* C2A and its use in construction of new genetic tools for methanogenic archaea. *Appl Environ Microbiol* **70**, 1425–33.

Protein Data Bank, PDB (2005) http://targetdb.pdb.org/statistics/TargetStatistics.html

Protein Structure Initiative, NIGMS/NIH (2005) posting date (www.nigms.nih.gov/psi/).

Pühler, G., Leffers, H., Gropp, F. et al. (1989) Archaebacterial DNA-dependent RNA polymerases testify to the evolution of the eukaryotic nuclear genome. *Proc Natl Acad Sci USA* **86**, 4569–73.

Quaiser, A., Ochsenreiter, T., Klenk, H.-P. et al. (2002) First insight into the genome of an uncultivated crenarchaeote from soil. *Environ Microbiol* **4**, 603–11.

Qiu Y., Tereshko, V., Kim, Y. et al. (2005) The crystal structure of AQ_328 from the hyperthermophilic bacteria *Aquifex aeolicus* shows an ancestral histone fold. *Proteins* **62**, 8–16.

Qureshi, S. A., Bell, S. D. & Jackson, S. P. (1997) Factor requirements for transcription in the archaeon *Sulfolobus shibatae*. *EMBO J* **16**, 2927–36.

Qureshi, S. A. & Jackson, S. P. (1998) Sequence-specific DNA binding by the *S. shibatae* TFIIB homolog, TFB, and its effect on promoter strength. *Mol Cell* **1**, 389–400.

Rabus, R., Wilkes, H., Behrends, A. et al. (2001) Anaerobic initial reaction of n-alkanes in a denitrifying bacterium: Evidence for (1-methylpentyl)succinate as initial product and for involvement of an organic radical in n-hexane metabolism. *J Bacteriol* **183**, 1707–15.

Rachel, R., Bettstetter, M., Hedlund, B. P. et al. (2002a) Remarkable morphological diversity of viruses and virus-like particles in hot terrestrial environments. *Arch Virol* 147, 2419–29.

Rachel, R., Wyschkony, I., Riehl, S. & Huber, H. (2002b) The ultrastructure of *Ignicoccus*: Evidence for a novel outer membrane and for intracellular vesicle budding in an archaeon. *Archaea* **1**, 9–18.

Radianingtyas, H. & Wright, P. C. (2003) Alcohol dehydrogenases from thermophilic and hyperthermophilic archaea and bacteria. *FEMS Microbiol Rev* **27**, 593–616.

Ragan, M. A. & Charlebois, R. L. (2002) Distributional profiles of homologous open reading frames among bacterial phyla: implications for vertical and lateral transmission. *Int J Syst Evol Microbiol* **52**, 777–87.

Raghoebarsing, A. A., Pol, A., van de Pas-Schoonen, K. T. et al. (2006) A microbial consortium couples anaerobic methan oxidation to denitrification. *Nature* **440**, 918–21.

Randau, L., Münch, R., Hohn, M. J., Jahn, D. & Söll, D. (2005a) *Nanoarchaeum equitans* creates functional tRNAs from separate genes for their 5′- and 3′-halves. *Nature* **433**, 537–41.

Randau, L., Pearson, M. & Söll, D. (2005b) The complete set of tRNA species in *Nanoarchaeum equitans*. *FEBS Lett* **579**, 2945–7.

Raoult, D., Audic, S., Robert, C. et al. (2004) The 1.2-megabase genome sequence of Mimivirus. *Science* **306**, 1344–50.

Rappe, M. S. & Giovannoni, S. J. (2003) The uncultured microbial majority. *Annu Rev Microbiol* **57**, 369–94.

Rashid, N., Cornista, J., Ezaki, S., Fukui, T., Atomi, H. & Imanaka, T. (2002) Characterization of an archaeal cyclodextrin glucanotransferase with a novel C-terminal domain. *J Bacteriol* **184**, 777–84.

Rashid, N., Imanaka, H., Fukui, T. et al. (2004) Presence of a novel phosphopentomutase and a 2-deoxyribose 5-phosphate aldolase reveals a metabolic link between pentoses and central carbon metabolism in the hyperthermophilic archaeon *Thermococcus kodakaraensis*. *J Bacteriol* **186**, 4185–91.

Rashid, R., Aittaleb, M., Chen, Q., Spiegel, K., Demeler, B. & Li, H. (2003) Functional requirement for symmetric assembly of archaeal box C/D small ribonucleoprotein particles. *J Mol Biol* **333**, 295–306.

Redder, P. & Garrett, R. A. (2006) Mutations and rearrangements in the genome of *Sulfolobus solfataricus* P2. *J Bact* **188**, 4198–206.

Redder, P., She, Q. & Garrett. R. A. (2001) Non-autonomous elements in the crenarchaeon *Sulfolobus solfataricus*. *J Mol Biol* **306**, 1–6.

Reeve, J. N. (2003) Archaeal chromatin and transcription. *Mol. Microbiol.* **48**, 587–98.

Reguera, G., McCarthy, K. D., Mehta, T., Nicoll, J. S., Tuominen, M. T. & Lovley, D. R. (2005) Extracellular electron transfer via microbial nanowires. *Nature* **435**, 1098–101.

Reilly, M. S. & Grogan, D. W. (2001) Characterization of intragenic recombination in a hyperthermophilic archaeon via conjugational DNA exchange. *J Bacteriol* **183**, 2943–6.

Reiter, W.-D., Palm P. & Yeats S (1989) Transfer RNA genes frequently serve as integration sites for prokaryotic genetic elements. *Nucleic Acids Res* **17**, 1907–14.

Renalier, M. H., Joseph, N., Gaspin, C., Thebault, P. & Mougin, A. (2005) The Cm56 tRNA modification in archaea is catalyzed either by a specific 2′-O-methylase, or a C/D sRNP. *RNA* **11**, 1051–63.

Renfrow, M. B., Naryshkin, N., Lewis, L. M., Chen, H. T., Ebright, R. M. & Scott, R. A. (2004) Transcription factor B contacts promoter DNA near the transcription start site of the archaeal transcription initiation complex. *J Biol Chem* **279**, 2825–31.

Rest, J. S. & Mindell, D. P. (2003) Retroids in Archaea: phylogeny and lateral origins. *Mol Biol Evol* **20**, 1134–42.

Rettenberger, M. (1990) Das Virus TTV1 das extrem thermophilen Schwefel-Archaebakteriums *Thermoproteus tenax*: Zusammensetzung und Struktur. PhD dissertation, Ludwig-Maximillians-Universität München, Munich.

Reuter, C. J. & Maupin-Furlow, J. A. (2004) Analysis of proteasome-dependent proteolysis in *Haloferax volcanii* cells, using short-lived green fluorescent proteins. *Appl Environ Microbiol* **70**, 7530–8.

Reysenbach, A. L. & Cady, S. L. (2001) Microbiology of ancient and modern hydrothermal systems. *Trends Microbiol* **9**, 79–86.

Ribas de Pouplana, L. & Schimmel, P. (2001) Two classes of tRNA synthetases suggested by sterically compatible dockings on tRNA acceptor stem. *Cell* **104**, 191–3.

Rice, G., Stedman, K., Snyder, J., et al. (2001) Viruses from extreme thermal environments. *Proc Natl Acad Sci USA* **98**, 13341–5.

Rice, G., Tang, L., Stedman, K. et al. (2004) The structure of a thermophilic archaeal virus shows a double-stranded DNA viral capsid type that spans all domains of life. *Proc Natl Acad Sci USA* **101**, 7716–20.

Robb, F. T. (2004) Genomics of thermophiles *In:* C. M. Fraser, T. D. Read & K. E. Nelson (eds) *Microbial Genomics*. Humana Press, Totowa, NJ, pp. 245–67.

Robb, F. T., Maeder, D. L., Brown, J. R. et al. (2001) Genomic sequence of hyperthermophile, *Pyrococcus furiosus*: implications for physiology and enzymology. *Methods Enzymol* **330**, 134–57.

Roberts, J. & Park, J. S. (2004) Mfd, the bacterial transcription repair coupling factor: translocation, repair and termination. *Curr Opin Microbiol* **7**, 120–5.

Roberts, J. A., Bell, S. D. & White, M. F. (2003) An archaeal XPF repair endonuclease dependent on a heterotrimeric PCNA. *Mol Micro* **48**, 361–71.

Roberts, J. A. & White, M. F. (2004) An archaeal endonuclease displays key properties of both eukaryal XPF-ERCC1 and Mus81. *J Biol Chem* **280**, 5924–8.

Robertus, J. D., Ladner, J. E., Finch, J. T. et al. (1974) Structure of yeast phenylalanine tRNA at 3 Å resolution. *Nature* **250**, 546–51.

Robinson, H., Gao, Y. G., McCrary, B. S., Edmondson, S. P., Shriver, J. W. & Wang, A. H. (1998) The hyperthermophile chromosomal protein Sac7d sharply kinks DNA. *Nature* **392**, 202–5.

Robinson, K. A. & Schreier, H. J. (1994) Isolation, sequence and characterization of the maltose-regulated mlrA gene from the hyperthermophilic archaeum *Pyrococcus furiosus*. *Gene* **151**, 173–6.

Robinson, N. P. & Bell, S. D. (2005) Origins of DNA replication in the three domains of life. *FEBS J* **272**, 3757–66.

Robinson, N. P., Dionne, I., Lundgren, M., Marsh, V. L., Bernander, R. & Bell, S. D. (2004) Identification of two origins of replication in the single chromosome of the archaeon *Sulfolobus solfataricus*. *Cell* **116**, 25–38.

Rockel, B., Jakana, J., Chiu, W. & Baumeister, W. (2002) Electron cryo-microscopy of VAT, the archaeal p97/CDC48 homologue from *Thermoplasma acidophilum*. *J Mol Biol* **317**, 673–81.

Röder, R. & Pfeifer, F. (1996) Influence of salt on the transcription of the gas vesicle genes of *Haloferax mediterranei* and identification of the endogenous transcriptional activator gene. *Microbiol* **142**, 1715–23.

Rodionov, D. A., Vitreschak, A. G., Mironov, A. A. & Gelfand, M. S. (2002) Comparative genomics of thiamin biosynthesis in procaryotes. New genes and regulatory mechanisms. *J Biol Chem* **277**, 48949–59.

Rodriguez-Valera, F., Juez, G. & Kushner, D. J. (1983) *Halobacterium mediterranei* spec. nov., a new carbohydrate-utilizing extreme halophile. *System Appl Microbiol* **4**, 369–81.

Rogozin, I. B., Makarova, K. S., Murvai, J. et al. (2002) Connected gene neighborhoods in prokaryotic genomes. *Nucleic Acids Res* **30**, 2212–23.

Rogozin, I. B., Makarova, K. S., Wolf, Y. I. & Koonin, E. V. (2004) Computational approaches for the analysis of gene neighborhoods in prokaryotic genomes. *Brief Bioinform* **5**, 131–49.

Rohwerder, T. & Sand, W. (2003) The sulfane sulfur of persulfides is the actual substrate of the sulfur-oxidizing enzymes from *Acidithiobacillus* and *Acidiphilium* spp. *Microbiology* **149**, 1699–710.

Rolfsmeier, M., Haseltine, C., Bini, E., Clark, A. & Blum, P. (1998) Molecular characterization of the alpha-glucosidase gene (*malA*) from the hyperthermophilic archaeon *Sulfolobus solfataricus*. *J Bacteriol* **180**, 1287–95.

Roll-Mecak, A., Alone, P., Cao, C., Dever, T. E. & Burley, S. K. (2004) X-ray structure of translation initiation factor eIF2γ: implications for tRNA and eIF2α binding. *J Biol Chem* **279**, 10634–42.

Roll-Mecak, A., Cao, C., Dever, T. E. & Burley, S. K. (2000) X-ray structures of the universal translation initiation factor IF2/eIF5B: conformational changes on GDP and GTP binding. *Cell* **103**, 781–92.

Rolland, J. L., Gueguen, Y., Persillon, C., Masson, J. M. & Dietrich, J. (2004) Characterization of a thermophilic DNA ligase from the archaeon *Thermococcus fumicolans*. *FEMS Microbiol Lett* **236**, 267–73.

Ronimus, R. S. & Morgan, H. W. (2003) Distribution and phylogenies of enzymes of the Embden–Meyerhof–Parnas pathway from archaea and hyperthermophilic bacteria support a gluconeogenic origin of metabolism. *Archaea* **1**, 199–221.

Ronimus, R. S. & Morgan, H. W. (2004) Cloning and biochemical characterization of a novel mouse ADP-dependent glucokinase. *Biochem Biophys Res Commun* **315**, 652–8.

Rosenshine, I. & Mevarech, M. (1991) The kinetics of the genetic exchange process in *Halobacterium volcanii* mating. In: F. Rodriguez-Valera (ed.) *General and Applied Aspects of Halophilic Microorganisms*. Plenum Press, New York, pp. 265–70.

Rosenshine, I., Tchelet, R. & Mevarech, M. (1989) The mechanism of DNA transfer in the mating system of an archaebacterium. *Science* **245**, 1387–9.

Rother, M., Mathes, I., Lottspeich, F. & Böck, A. (2003) Inactivation of the selB gene in *Methanococcus maripaludis*: effect on synthesis of selenoproteins and their sulfur-containing homologs. *J Bacteriol* **185**, 107–14.

Rowlands, T., Baumann, P. & Jackson, S. P. (1994) The TATA-binding protein: a general transcription factor in Eukaryotes and Archaebacteria. *Science* **264**, 1326–9.

Roy, A. B. & Trudinger, P. A. (1970) The chemistry of some sulphur compounds. In: *The Biochemistry of Inorganic Compounds of Sulphur*. Cambridge University Press, Cambridge.

Rozhdestvensky, T. S., Tang, T. H., Tchirkova, I. V., Brosius, J., Bachellerie, J. P. & Huttenhofer, A. (2003) Binding of L7Ae protein to the K-turn of archaeal snoRNAs: a shared RNA binding motif for C/D and H/ACA box snoRNAs in Archaea. *Nucleic Acids Res* **31**, 869–77.

Ruan, B. & Söll, D. (2005) The bacterial YbaK protein is a Cys-tRNA^{Pro} and Cys-tRNA^{Cys} deacylase. *J Biol Chem* **280**, 25887–91.

Ruepp, A., Graml, W., Santos-Martinez, M. L. et al. (2000) The genome sequence of the thermoacidophilic scavenger *Thermoplasma acidophilum*. *Nature* **407**, 508–13.

Rueter, P., Rabus, R., Wilkes, H. et al. (1994) Anaerobic oxidation of hydrocarbons in crude-oil by new types of sulfate-reducing bacteria. *Nature* **372**, 455–8.

Ryu, K., Kim, J. & Dordick, J. S. (1994) Catalytic properties and potential of an extracellular protease from an extreme halophile. *Enzyme Microb Technol* **16**, 266–75.

Saha, A., Wittmeyer, J. & Cairns, B. R. (2002) Chromatin remodeling by RSC involves ATP-dependent DNA translocation. *Genes Dev* **16**, 2120–34.

St Jean, A., Trieselmann, B. A. & Charlebois, R. L. (1994) Physical map and set of overlapping cosmid clones representing the genome of the archaeon *Halobacterium* sp. GRB. *Nucleic Acids Res* **22**, 1476–83.

Sakuraba, H., Goda, S. & Ohshima, T. (2004) Unique sugar metabolism and novel enzymes of hyperthermophilic archaea. *Chem Rec* **3**, 281–7.

Sakuraba, H., Utsumi, E., Kujo, C. et al. (1999) An AMP-dependent (ATP-forming) kinase in the hyperthermophilic archaeon *Pyrococcus furiosus*: characterization and novel physiological role. *Arch Biochem Biophys* **364**, 125–8.

Sakuraba, H., Yoshioka, I., Koga, S. et al. (2002) ADP-dependent glucokinase/phosphofructokinase, a novel bifunctional enzyme from the hyperthermophilic archaeon *Methanococcus jannaschii*. *J Biol Chem* **277**, 12495–8.

Salerno, V., Napoli, A., White, M. F., Rossi, M. & Ciaramella, M. (2003) Transcriptional response to DNA damage in the archaeon *Sulfolobus solfataricus*. *Nucleic Acids Res* **31**, 6127–38.

Samkurashvili, I. & Luse, D. S. (1998) Structural changes in the RNA polymerase II transcription complex during transition from initiation to elongation. *Mol Cell Biol* **18**, 5343–54.

Sandaa, R. A., Enger, O. & Torsvik, V. (1999) Abundance and diversity of Archaea in heavy-metal-contaminated soils. *Appl Environ Microbiol* **65**, 3293–7.

Sandler, S. J., Hugenholtz, P., Schleper, C., DeLong, E. F., Pace, N. R. & Clark, A. J. (1999) Diversity of radA genes from cultured and uncultured archaea: comparative analysis of putative RadA proteins and their use as a phylogenetic marker. *J Bacteriol* **181**, 907–15.

Sanger, F., Brownlee, G. G. & Barrell, B. G. (1965). A two-dimensional fractionation procedure for radioactive nucleotides. *J Mol Biol* **13**, 373–98.

Sanger, F. & Thompson, E. O. P. (1953) The amino-acid sequence in the glycyl chain of insulin. *Biochem J* **53**, 353–74.

Sanger, F. & Tuppy, H. (1951). The amino-acid sequence in the phenylalanyl chain of insulin. *Biochem J* **49**, 481–90.

Santoro, N. & Konisky, J. (1987) Characterization of bromoethanesulfonate-resistant mutants of *Methanococcus voltae* – evidence of a coenzyme-M transport system. *J Bacteriol* **169**, 660–5.

Santos, H. & Da Costa, M. (2001) Organic solutes from thermophiles and hyperthermophiles. *Methods Enzymol* **334**, 302–15.

Santos, H. & Da Costa, M. S. (2002) Compatible solutes of organisms that live in hot saline environments. *Environ Microbiol* **4**, 501–9.

Sapp, J. (2005) The prokaryote–eukaryote dichotomy: meanings and mythology. *Microbiol Mol Biol Rev* **69**, 292–305.

Sapra, R., Bagramyan, K. & Adams, M. W. (2003) A simple energy-conserving system: proton reduction coupled to proton translocation. *Proc Natl Acad Sci USA* **100**, 7545–50.

Sartori, A. A. & Jiricny, J. (2003) Enzymology of base excision repair in the hyperthermophilic archaeon *Pyrobaculum aerophilum*. *J Biol Chem* **278**, 24563–76.

Sartori, A. A., Lingaraju, G. M., Hunziker, P., Winkler, F. K. & Jiricny, J. (2004) Pa-AGOG, the founding member of a new family of archaeal 8-oxoguanine DNA-glycosylases. *Nucleic Acids Res* **32**, 6531–9.

Sato, T., Fukui, T., Atomi, H. & Imanaka, T. (2003) Targeted gene disruption by homologous recombination in the hyperthermophilic archaeon *Thermococcus kodakaraensis* KOD1. *J Bacteriol* **185**, 210–20.

Sato, T., Fukui, T., Atomi, H. & Imanaka, T. (2005) Improved and versatile transformation system allowing multiple genetic manipulations of the hyperthermophilic archaeon *Thermococcus kodakaraensis*. *Appl Environ Microbiol* **71**, 3889–99.

Sauerwald, A., Zhu, W., Major, T. A. et al. (2005) RNA-dependent cysteine biosynthesis in archaea. *Science* **307**, 1969–72.

Savino, C., Federici, L., Johnson, K. A. et al. (2004) Insights into DNA replication: The crystal structure of DNA polymerase B1 from the archaeon *Sulfolobus solfataricus*. *Structure* **12**, 2001–8.

Schäfer, T. & Schönheit, P. (1993) Glyconeogenesis from pyruvate in the hyperthermophilic archaeon *Pyrococcus furiosus*: involvement of reactions of the Embden–Meyerhof pathway. *Arch Microbiol* **159**, 354–63.

Schauder, R. & Kröger, A. (1993) Bacterial sulphur respiration. *Arch Microbiol* **159**, 491–7.

Scherer, P. A., Vollmer, G. R., Fakhouri, T. & Martensen, S. (2000) Development of a methanogenic process to degrade exhaustively the organic fraction of municipal "grey waste" under thermophilic and hyperthermophilic conditions. *Water Sci Technol* **41**, 83–91.

Schieg, P. & Herzel, H. (2004) Periodicities of 10–11 bp as indicators of the supercoiled state of genomic DNA. *J Mol Biol* **343**, 891–901.

Schiraldi, C., Giuliano, M. & De Rosa, M. (2002) Perspectives on biotechnological applications of archaea. *Archaea* **1**, 75–86.

Schlach, T., Duda, S., Sargent, D. F. & Richmond, T. J. (2005) X-ray structure of a tetranucleosome and its implications for the chromatin fibre. *Nature* **436**, 138–41.

Schleper, C., DeLong, E. F., Preston, C. M., Feldman, R. A., Wu, K. Y. & Swanson, R. V. (1998) Genomic analysis reveals chromosomal variation in natural populations of the uncultured psychrophilic archaeon *Cenarchaeum symbiosum*. *J Bacteriol* **180**, 5003–9.

Schleper, C., Holben, W. & Klenk, H.-P. (1997a) Recovery of crenarchaeotal ribosomal DNA sequences from freshwater-lake sediments. *Appl Environ Microbiol* **63**, 321–3.

Schleper, C., Holz, I., Janekovic, D., Murphy, J. & Zillig, W. (1995) A multicopy plasmid of the extremely thermophilic archaeon *Sulfolobus* effects its transfer to recipients by mating. *J Bacteriol* **177**, 4417–26.

Schleper, C., Jurgens, G. & Jonuscheit, M. (2005) Genomic studies of uncultivated archaea. *Nat Rev Microbiol* **3**, 479–88.

Schleper, C., Kubo, K. & Zillig, W. (1992) The particle SSV1 from the extremely thermophilic archaeon *Sulfolobus* is a virus: demonstration of infectivity and of transfection with viral DNA. *Proc Natl Acad Sci USA* **89**, 7645–9.

Schleper, C., Roder, R., Singer, T. & Zillig, W. (1994) An insertion element of the extremely thermophilic archaeon *Sulfolobus solfataricus* transposes into the endogenous β-galactosidase gene. *Mol Gen Genet* **243**, 91–6.

Schleper, C., Swanson, R. V., Mathur, E. J. & DeLong, E. F. (1997b) Characterization of a DNA polymerase from the uncultivated psychrophilic archaeon *Cenarchaeum symbiosum*. *J Bacteriol* **179**, 7803–11.

Schmidt, I., van Spanning, R. J. & Jetten, M. S. (2004) Denitrification and ammonia oxidation by N*itrosomonas europaea* wild-type, and NirK- and NorB-deficient mutants. *Microbiology* **150**, 4107–14.

Schmidt, K. J., Beck, K. E. & Grogan, D. W. (1999) UV stimulation of chromosomal marker exchange in *Sulfolobus acidocaldarius*: implications for DNA repair, conjugation and homologous recombination at extremely high temperatures. *Genetics* **152**, 1407–15.

Schmitt, E., Blanquet, S. & Mechulam, Y. (2002) The large subunit of initiation factor aIF2 is a close structural homologue of elongation factors. *EMBO J* **21**, 1821–32.

Schmitt, E., Panvert, M., Blanquet, S. & Mechulam, Y. (2005) Structural basis for tRNA-dependent amidotransferase function. *Structure* **13**, 1421–33.

Schnell, R., Sandalova, T., Hellman, U., Lindqvist, Y. & Schneider, G. (2005) Siroheme- and [Fe4-S4]-dependent NirA from *Mycobacterium tuberculosis* is a sulfite reductase with a covalent Cys-Tyr bond in the active site. *J Biol Chem* **280**, 27319–28.

Scholz, S., Wolff, S. & Hensel, R. (1998) The biosynthesis pathway of di-myo-inositol-1,1'-phosphate in *Pyrococcus woesei*. *FEMS Microbiol. Lett.* **168**, 37–42.

Schönheit, P. & Schäfer, T. (1995) Metabolism of hyperthermophiles. *World J Microbiol Biotechnol* **11**, 26–57.

Schouten, S., Hopmans, E. C., Pancost, R. D. & Damste, J. S. (2000) Widespread occurrence of structurally diverse tetraether membrane lipids: evidence for the ubiquitous presence of low-temperature relatives of hyperthermophiles. *Proc Natl Acad Sci USA* **97**, 14421–6.

Schouten, S., Wakeham, S. G., Hopmans, E. C. & Damste, J. S. S. (2003) Biogeochemical evidence that thermophilic archaea mediate the anaerobic oxidation of methane. *Appl Environ Microbiol* **69**, 1680–6.

Schramm, A., Siebers, B., Tjaden, B. et al. (2000) Pyruvate kinase of the hyperthermophilic crenarchaeote *Thermoproteus tenax*: physiological role and phylogenetic aspects. *J Bacteriol* **182**, 2001–9.

Schrödinger, E. (1954) *Nature and the Greeks*. Cambridge University Press, Cambridge.

Schulein, M. (2000) Protein engineering of cellulases. *Biochim Biophys Acta* **1543**, 239–52.

Schut, G. J., Brehm, S. D., Datta, S. & Adams, M. W. (2003) Whole-genome DNA microarray analysis of a hyperthermophile and an archaeon: *Pyrococcus furiosus* grown on carbohydrates or peptides. *J Bacteriol* **185**, 3935–47.

Schut, G. J., Zhou, J. & Adams, M. W. (2001) DNA microarray analysis of the hyperthermophilic archaeon *Pyrococcus furiosus*: evidence for an new type of sulfur-reducing enzyme complex. *J Bacteriol* **183**, 7027–36.

Segerer, A., Stetter, K. O. & Klink, F. (1985) Two contrary modes of chemolithotrophy in the same archaebacterium. *Nature* **313**, 787–9.

Sehgal, A. C. & Kelly, R. M. (2002) Enantiomeric resolution of 2-aryl propionic esters with hyperthermophilic and mesophilic esterases: contrasting thermodynamic mechanisms. *J Am Chem Soc* **124**, 8190–1.

Sehgal, A. C. & Kelly, R. M. (2003) Strategic selection of hyperthermophilic esterases for resolution of 2-arylpropionic esters. *Biotechnol Prog* **19**, 1410–16.

Sehgal, A. C., Tompson, R., Cavanagh, J. & Kelly, R. M. (2002) Structural and catalytic response to temperature and cosolvents of carboxylesterase EST1 from the extremely thermoacidophilic archaeon *Sulfolobus solfataricus* P1. *Biotechnol Bioeng* **80**, 784–93.

Selmer, T., Kahnt, J., Goubeaud, M. et al. (2000) The biosynthesis of methylated amino acids in the active site region of methyl-coenzyme M reductase. *J Biol Chem* **275**, 3755–60.

Sensen, C. W., Klenk, H.-P., Singh, R. K. et al. (1996) Organizational characteristics and information content of an archaeal genome: 156 kb of sequence from *Sulfolobus solfataricus* P2. *Mol Microbiol* **22**, 175–91.

Serour, E. & Antranikian, G. (2002) Novel thermoactive glucoamylases from the thermoacidophilic Archaea *Thermoplasma acidophilum*, *Picrophilus torridus* and *Picrophilus oshimae*. *Antonie van Leeuwenhoek* **81**, 73–83.

Serre, M.-C., Letzelter, C., Garel, J. R. & Duguet, M. (2002) Cleavage properties of an archaeal site-specific recombinase, the SSV1 integrase. *J Biol Chem* **277**, 16758–67.

Seybert, A., Scott, D. J., Scaife, S., Singleton, M. R. & Wigley, D. B. (2002) Biochemical characterisation of the clamp/clamp loader proteins from the euryarchaeon *Archaeoglobus fulgidus*. *Nucleic Acids Res* **30**, 4329–38.

Seybert, A. & Wigley, D. B. (2004) Distinct roles for ATP binding and hydrolysis at individual subunits of an archaeal clamp loader. *EMBO J* **23**, 1360–71.

Shand, R. & Betlach, M. (1991) Expression of the *bop* gene cluster of *Halobacterium halobium* is induced by low oxygen tension and by light. *J Bacteriol* **173**, 4692–9.

She, Q., Brügger, K. & Chen, L. (2002) Archaeal integrative genetic elements and their impact on evolution. *Res Microbiol* **153**, 325–32.

She, Q., Peng, X., Zillig, W. & Garrett, R. A. (2001a) Gene capture in archaeal chromosomes. *Nature* **409**, 478.

She, Q., Phan, H., Garrett, R. A. et al. (1998) Genetic profile of pNOB8 from *Sulfolobus*: the first conjugative plasmid from an archaeon. *Extremophiles* **2**, 417–25.

She, Q., Shen, B. & Chen, L. (2004) Archaeal integrases and mechanisms of gene capture. *Biochem Soc Trans* **22**, 222–6.

She, Q., Singh, R. K., Confalonieri, F. et al. (2001b) The complete genome of the crenarchaeon *Sulfolobus solfataricus* P2. *Proc Natl Acad Sci USA* **98**, 7835–40.

Shehi, E., Granata, V., Del Vecchio, P. et al. (2003) Thermal stability and DNA binding activity of a variant form of the Sso7d protein from the archaeon *Sulfolobus solfataricus* truncated at leucine 54. *Biochemistry* **42**, 8362–8.

Shehi, E., Serina, S., Fumagalli, G. et al. (2001) The Sso7d DNA-binding protein from *Sulfolobus solfataricus* has ribonuclease activity. *FEBS Lett* **497**, 131–6.

Shen, P. & Huang, H. V. (1986) Homologous recombination in *Escherichia coli*: dependence on substrate length and homology. *Genetics* **112**, 441–57.

Shen, X., Mizuguchi, G., Hamiche, A. & Wu, C. (2000) A chromatin remodelling complex involved in transcription and DNA processing. *Nature* **406**, 541–4.

Sherratt, D. J. (2003) Bacterial chromosome dynamics. *Science* **301**, 780–5.

Shilov, A. E., Koldasheva, E. M., Kovalenko, S. V. et al. (1999) Methanogenesis is reversible: the formation of acetate in methane carboxilation by bacteria of methanigenic biocenose. *Doklady Akademii Nauk* **367**, 557–9.

Shilov, A. E. & Shul'pin, G. B. (1997) Activation of C–H bonds by metal complexes. *Chem Rev* **97**, 2879–932.

Shima, S. & Thauer, R. K. (2005) Methyl-coenzyme M reductase (MCR) and the anaerobic oxidation of methane (AOM) in methanotrophic archaea. *Curr Opin Microbiol* **8**, 643–8.

Shimmin, L. C., Newton, C. H., Ramirez, C. et al. (1989) Organization of genes encoding the L11, L1, L10, and L12 equivalent ribosomal proteins in eubacteria, archaebacteria, and eucaryotes. *Can J Microbiol* **35**, 164–70.

Shimodaira, H. (2002) An approximately unbiased test of phylogenetic tree selection. *Syst Biol* **51**, 492–508.

Shimodaira, H. & Hasegawa, M. (1999) Multiple comparisons of log-likelihoods with applications to phylogenetic inference. *Mol Biol Evol* **16**, 1114–16.

Shin, D. S., Chahwan, C., Huffman, J. L. & Tainer, J. A. (2004) Structure and function of the double-strand break repair machinery. *DNA Repair (Amst)* **3**, 863–73.

Shin, J. H., Grabowski, B., Kasiviswanathan, R., Bell, S. D. & Kelman, Z. (2003) Regulation of minichromosome maintenance helicase activity by Cdc6. *J Biol Chem* **278**, 38059–67.

Shockley, K. R. (2004) Functional genomics investigation of microbial physiology in the hyperthermophilic microorganisms *Pyrococcus furiosus* and *Thermotoga maritima*. PhD thesis, North Carolina State University, Raleigh.

Shockley, K. R., Ward, D. E., Chhabra, S. R., Conners, S. B., Montero, C. I. & Kelly, R. M. (2003) Heat shock response by the hyperthermophilic archaeon *Pyrococcus furiosus*. *Appl Environ Microbiol* **69**, 2365–71.

Shukla, H. D. & DasSarma, S. (2004) Complexity of gas vesicle biogenesis in *Halobacterium* sp. strain NRC-1: identification of five new proteins. *J Bacteriol* **186**, 3182–6.

Siebers, B., Brinkmann, H., Dörr, C. et al. (2001) Archaeal fructose-1,6-bisphosphate aldolases constitute a new family of archaeal type class I aldolase. *J Biol Chem* **276**, 28710–18.

Siebers, B., Klenk, H.-P. & Hensel, R. (1998) PPi-dependent phosphofructokinase from *Thermoproteus tenax*, an archaeal descendant of an ancient line in phosphofructokinase evolution. *J Bacteriol* **180**, 2137–43.

Siebers, B. & Schönheit, P. (2005) Unusual pathways and enzymes of central carbohydrate metabolism in Archaea. *Curr Opin Microbiol* **8**, 287–98.

Siebers, B., Tjaden, B., Michalke, K. et al. (2004) Reconstruction of the central carbohydrate metabolism of *Thermoproteus tenax* by use of genomic and biochemical data. *J Bacteriol* **186**, 2179–94.

Simon, H. M., Jahn, C. E., Bergerud, L. T. et al. (2005) Cultivation of mesophilic soil crenarchaeotes in enrichment cultures from plant roots. *Appl Environ Microbiol* **71**, 4751–60.

Sims, R. J. III, Belotserkovskaya, R. & Reinberg, D. (2004) Elongation by RNA polymerase II: the short and long of it. *Genes Dev* **18**, 2437–68.

Singh, K., Horng, Y. C. & Ragsdale, S. W. (2003) Rapid ligand exchange in the MCRred1 form of methyl-coenzyme M reductase. *J Am Chem Soc* **125**, 2436–43.

Singh, S. K., Gurha, P., Tran, E. J., Maxwell, E. S. & Gupta, R. (2004) Sequential 2′-O-methylation of archaeal pre-tRNATrp nucleotides is guided by the intron-encoded but trans-acting box C/D ribonucleoprotein of pre-tRNA. *J Biol Chem* **279**, 47661–71.

Sioud, M., Baldacci, G., Forterre, P. & de Recondo, A. M. (1987a) Antitumor drugs inhibit the growth of halophilic archaebacteria. *Eur J Biochem* **169**, 231–6.

Sioud, M., Forterre, P. & de Recondo, A. M. (1987b) Effects of the antitumor drug VP16 (etoposide) on the archaebacterial *Halobacterium* GRB 1.7 kb plasmid *in vivo*. *Nucleic Acids Res* **15**, 8217–34.

Slesarev, A. I., Belova, G. I., Kozyavkin, S. A. & Lake, J. A. (1998) Evidence for an early prokaryotic origin of histones H2A and H4 prior to the emergence of eukaryotes. *Nucleic Acids Res* **26**, 427–30.

Slesarev, A. I., Mezhevaya, K. V., Makarova, K. S. et al. (2002) The complete genome of hyperthermophile *Methanopyrus kandleri* AV19 and monophyly of archaeal methanogens. *Proc Natl Acad Sci USA* **99**, 4644–9.

Sleytr, U. B., Bayley, H., Sara, M. et al. (1997) Applications of S-layers. *FEMS Microbiol Rev* **20**, 151–75.

Slikkerveer, A. & de Wolff, F. A. (1989) Pharmacokinetics and toxicity of bismuth compounds. *Med. Toxicol. Adverse Drug Exp* **4**, 303–23.

Sliwinski, M. K. & Goodman, R. M. (2004) Spatial heterogeneity of crenarchaeal assemblages within mesophilic soil ecosystems as revealed by PCR-single-stranded conformation polymorphism profiling. *Appl Environ Microbiol* **70**, 1811–20.

Slupska, M. M., King, A. G., Fitz-Gibbon, S., Besemer, J., Borodovsky, M. & Miller, J. H. (2001) Leaderless transcripts of the crenarchaeal hyperthermophile *Pyrobaculum aerophilum*. *J Mol Biol* **309**, 347–60.

Smale, S. T. & Kadonaga, J. T. (2003) The RNA polymerase II core promoter. *Annu Rev Biochem* **72**, 449–79.

Smith, D. R., Doucette-Stamm, L. A., Deloughery, C. et al. (1997). Complete genome sequence of *Methanobacterium thermoautotrophicum* ΔH: functional analysis and comparative genomics. *J Bacteriol* **179**, 7135–55.

Sniezko, I., Dobson-Stone, C. & Klein, A. (1998) The *treA* gene of *Bacillus subtilis* is a suitable reporter gene for the archaeon *Methanococcus voltae*. *FEMS Microbiol Lett* **164**, 237–42.

Snijders, A., Walther, J., Peter, S. et al. (2006) Reconstruction of central carbohydrate metabolism in

Sulfolobus solfataricus using a two-dimensional gel electrophoresis map, stable isotope labelling and DNA microarray analysis. *Proteomics* **6**, 1518–29.

So, C. M., Phelps, C. D. & Young, L. Y. (2003) Anaerobic transformation of alkanes to fatty acids by a sulfate-reducing bacterium, strain Hxd3. *Appl Environ Microbiol* **69**, 3892–900.

Soares, D., Dahlke, I., Li, W.-T. et al. (1998) Archaeal histone stability, DNA binding, and transcription inhibition above 90 °C. *Extremophiles* **2**, 75–81.

Soares, D. J., Marc, F. & Reeve, J. N. (2003) Conserved eukaryotic histone-fold residues substituted into an archaeal histone increase DNA affinity but reduce complex flexibility. *J Bacteriol* **185**, 3453–7.

Soderberg, T. (2005) Biosynthesis of ribose-5-phosphate and erythrose-4-phosphate in archaea: a phylogenetic analysis of archaeal genomes. *Archaea* **1**, 347–52.

Soderberg, T. & Alver, R. (2004) Transaldolase of *Methanocaldococcus jannaschii*. *Archaea* **1**, 255–62.

Sollmann, T. & Seifter, J. (1939) The pharmacology of trimethyl bismuth. *Pharmacol Exp Ther* **67**, 17–49.

Sonea, S. & Paniset, M. (1976) Vers une nouvelle bactériologie. *Rev Can Biol* **35**, 103–67.

Soppa, J. (1999a) Transcription initiation in Archaea: facts, factors and future aspects. *Mol Microbiol* **31**, 1295–305.

Soppa, J. (1999b) Normalized nucleotide frequencies allow the definition of archaeal promoter elements for different archaeal groups and reveal base-specific TFB contacts upstream of the TATA box. *Mol Microbiol* **31**, 1589–601.

Soppa, J. & Oesterhelt, D. (1989) Bacteriorhodopsin mutants of *Halobacterium* sp. GRB. I. The 5-bromo-2′-deoxyuridine selection as a method to isolate point mutants in halobacteria. *J Biol Chem* **264**, 13043–8.

Sorokin, D. Y., Tourova, T. P. & Muyzer, G. (2005) Oxidation of thiosulfate to tetrathionate by an haloarchaeon isolated from hypersaline habitat. *Extremophiles* **9**, 501–4.

Sosunova, E., Sosunov, V., Kozlov, M., Nikiforov, V., Goldfarb, A. & Mustaev, A. (2003) Donation of catalytic residues to RNA polymerase active center by transcription factor Gre. *Proc Natl Acad Sci USA* **100**, 15469–74.

Southeast Collaboratory for Structural Genomics (2005) www.secsg.org

Sperling, D., Kappler, U., Trüper, H. G. & Dahl, C. (2001) Dissimilatory ATP sulfurylase from *Archaeoglobus fulgidus*. *Methods Enzymol* **331**, 419–27.

Spitalny, P. & Thomm, M. (2003) Analysis of the open region and of DNA-protein contacts of archaeal RNA polymerase transcription complexes during transition from initiation to elongation. *J Biol Chem* **278**, 30497–505.

Spormann, A. M. & Widdel, F. (2000) Metabolism of alkylbenzenes, alkanes, and other hydrocarbons in anaerobic bacteria. *Biodegradation* **11**, 85–105.

Springer, E., Sachs, M. S., Woese, C. R. & Boone, D. R. (1995) Partial gene sequences for the α-subunit of methyl-coenzyme-M reductase (Mcri) as a phylogenetic tool for the family *Methanosarcinaceae*. *Int J Syst Bacteriol* **45**, 554–9.

Sprott, G. D., Patel, G. B. & Krishnan, L. (2003) Archaeobacterial ether lipid liposomes as vaccine adjuvants. *Methods Enzymol* **373**, 155–72.

Srinivasan, G., James, C. M. & Krzycki, J. A. (2002) Pyrrolysine encoded by UAG in Archaea: charging of a UAG-decoding specialized tRNA. *Science* **296**, 1459–62.

Stadnitskaia, A., Muyzer, G., Abbas, B. et al. (2005) Biomarker and 16S rDNA evidence for anaerobic oxidation of methane and related carbonate precipitation in deep-sea mud volcanoes of the Sorokin Trough, Black Sea. *Mar Geol* **217**, 67–96.

Stan-Lotter, H., Doppler, E., Jarosch, M., Radax, C., Gruber, C. & Inatomi, K. I. (1999) Isolation of a chymotrypsinogen B-like enzyme from the archaeon *Natronomonas pharaonis* and other halobacteria. *Extremophiles* **3**, 153–61.

Stanier, R. Y. (1970) In: H. P. Charles & B. C. J. G. Knight (eds) *Organization and Control in Prokaryotic and Eukaryotic Cells*. The Society for General Microbiology: Symposium 20. Cambridge University Press, Cambridge.

Stanier, R. Y., Doudoroff, M. & Adelberg, E. A. (1957) *The Microbial World*. Prentice Hall, Englewood Cliffs, NJ.

Stanier, R. Y., Doudoroff, M. & Adelberg, E. A. (1963) *The Microbial World*, 2nd edn. Prentice Hall, Englewood Cliffs, NJ.

Stanier, R. Y., Doudoroff, M. & Adelberg, E. A. (1970) *The Microbial World*. 3rd edn, Prentice Hall, Englewood Cliffs, NJ.

Stanier, R. Y. & van Niel, C. B. (1941) The main outlines of bacterial classification. *J Bacteriol* **42**, 437–66.

Stanier, R. Y. & van Niel, C. B. (1962) The concept of a bacterium. *Archiv für Mikrobiologie* **42**, 17–35.

Starostina, N. G., Marshburn, S., Johnson, L. S., Eddy, S. R., Terns, R. M. & Terns, M. P. (2004) Circular box C/D RNAs in *Pyrococcus furiosus*. *Proc Natl Acad Sci USA* **101**, 14097–101.

Stathopoulos, C., Kim, W., Li, T. et al. (2001) Cysteinyl-tRNA synthetase is not essential for viability of the archaeon *Methanococcus maripaludis*. *Proc Natl Acad Sci USA* **98**, 14292–7.

Stedman, K. M., She, Q., Phan, H. et al. (2000) pING family of conjugative plasmids from the extremely thermophilic archaeon *Sulfolobus islandicus*: insights into recombination and conjugation in *Crenarchaeota*. *J Bacteriol* **182**, 7014–20.

Stedman, K. M., She, Q., Phan, H. et al. (2003) Relationships between fuselloviruses infecting the extremely thermophilic archaeon *Sulfolobus*: SSV1 and SSV2. *Res Microbiol* **154**, 295–302.

Stein, J. L., Marsh, T. L., Wu, K. Y., Shizuya, H. & DeLong, E. F. (1996) Characterization of uncultivated prokaryotes: isolation and analysis of a 40-kilobase-pair genome fragment from a planktonic marine archaeon. *J Bacteriol* **178**, 591–9.

Steitz, J. A. & Tycowski, K. T. (1995) Small RNA chaperones for ribosome biogenesis. *Science* **270**, 1626–7.

Stetter, K. O. (1982) Ultrathin mycelia-forming organisms from submarine volcanic areas having an optimum growth temperature of 105 °C. *Nature* **300**, 258–60.

Stetter, K. O. (1992) Life at the upper temperature border. *In:* J. Tran Thanh Van, K. Tran Thanh Van, J. C. Mounolou, J. Schneider & C. McKay (eds) *Frontiers of Life*. Editions Frontières, Gif-sur-Yvette, pp. 195–219.

Stetter, K. O. (1999) Extremophiles and their adaptation to hot environments. *FEBS Lett* **452**, 22–5.

Stetter, K. O. & Gaag, G. (1983) Reduction of molecular sulfur by methanogenic bacteria. *Nature* **305**, 309–11.

Stetter, K. O., Hohn, M. J., Huber, H. et al. (2005) A novel kingdom of parasitic Archaea. Proceedings of the first biannual workshop on geothermal biology and geochemistry in Yellowstone National Park.

Steudel, R. (2000) The chemical sulfur cycle. *In:* P. Lens & L. Hulshoff Pol (eds) *Environmental Technologies to Treat Sulfur Pollution*. IWA Publishing, London, pp. 1–32.

Stoiber, R. E. (1995) Volcanic gases from subaerial volcanoes on earth. *In:* T. J. Ahrens (ed.) *Global Earth Physics: A Handbook of Physical Constants*. American Geophysical Union, Washington, DC, pp. 308–19.

Story, S. V., Grunden, A. M. & Adams, M. W. W. (2001) Characterization of an aminoacylase from the hyperthermophilic archaeon *Pyrococcus furiosus*. *Journal of Bacteriology* **183**, 4259–68.

Storz, G., Opdyke, J. A. & Zhang, A. (2004) Controlling mRNA stability and translation with small, noncoding RNAs. *Curr Opin Microbiol* **7**, 140–4.

Strous, M. & Jetten, M. S. M. (2004) Anaerobic oxidation of methane and ammonium. *Annu Rev Microbiol* **58**, 99–117.

Studitsky, V. M., Kassavetis, G. A., Geiduschek, E. P. & Felsenfeld, G. (1997) Mechanism of transcription through the nucleosome by eukaryotic RNA polymerase. *Science* **278**, 1899–901.

Su, S., Gao, Y.-G., Robinson, H. et al. (2000) Crystal structures of the chromosomal proteins Sso7d/Sac7d bound to DNA containing T-G mismatched base pairs. *J Mol Biol* **303**, 395–403.

Sugai, A., Sakuma, R., Fukuda, I. et al. (1995) The structure of the core polyol of the ether lipids from *Sulfolobus acidocaldarius*. *Lipids* **30**, 339–44.

Sugar, F. J., Jenney, F. E. J., Poole, F. L. I., Brereton, P. S., Izumi, M., Shah, C. & Adams, M. W. W. (2005) Comparison of small- and large-scale expression of selected *Pyrococcus furiosus* genes as an aid to high-throughput protein production. *J Struct Funct Genomics* **6**, 149–58.

Suhre, K. & Claverie, J. M. (2003) Genomic correlates of hyperthermostability, an update. *J Biol Chem* **278**, 17198–202.

Sukhodolets, M. V., Cabrera, J. E., Zhi, H. & Jin, D. J. (2001) RapA, a bacterial homolog of SWI2/SNF2, stimulates RNA polymerase recycling in transcription. *Genes Dev* **15**, 3330–41.

Sun, C. W., Chen, Z. W., He, Z. G., Zhou, P. J. & Liu, S. J. (2003) Purification and properties of the sulfur oxygenase/reductase from the acidothermophilic archaeon, *Acidianus* strain S5. *Extremophiles* **7**, 131–4.

Suryadi, J., Tran, E. J., Maxwell, E. S. & Brown, B. A. II (2005) The crystal structure of the *Methanocaldococcus jannaschii* multifunctional L7Ae RNA-binding protein reveals an induced-fit interaction with the box C/D RNAs. *Biochemistry* **44**, 9657–72.

Suyama, M., Latte, W. C. & Bork. P. (2005) Palindromic repetitive DNA elements with coding potenital in *M. jannaschii*. *FEBS Lett* **579**, 5281–6.

Suzuki, R., Nagata, K., Yumoto, F. et al. (2003) Three-dimensional solution structure of an archaeal FKBP with a dual function of peptidyl prolyl cis-trans isomerase and chaperone-like activities. *J Mol Biol* **328**, 1149–60.

Suzuki, T., Iwasaki, T., Uzawa, T. et al. (2002) *Sulfolobus tokodaii* sp. nov. (f. *Sulfolobus* sp. strain 7), a new member of the genus *Sulfolobus* isolated from Beppu Hot Springs, Japan. *Extremophiles* **6**, 39–44.

Suzuki, Y., Miyamoto, K. & Ohta, H. (2004) A novel thermostable esterase from the thermoacidophilic archaeon *Sulfolobus tokodaii* strain 7. *FEMS Microbiol Lett* **236**, 97–102.

Suzuki, Y., Tsunoda, T., Sese, J. et al. (2001) Identification and characterization of the potential promoter regions of 1031 kinds of human genes. *Genome Res* **11**, 677–84.

Tachibana, Y., Kuramura, A., Shirasaka, N. et al. (1999) Purification and characterization of an extremely thermostable cyclomaltodextrin glucanotransferase from a newly isolated hyperthermophilic archaeon, a *Thermococcus* sp. *Appl Environ Microbiol* **65**, 1991–7.

Taguchi, Y., Sugishima, M. & Fukuyama, K. (2004) Crystal structure of a novel zinc-binding ATP sulfurylase from *Thermus thermophilus* HB8. *Biochemistry* **43**, 4111–18.

Tahara, M., Ohsawa, A., Saito, S. & Kimura, M. (2004) *In vitro* phosphorylation of initiation factor 2 α (aIFα) from hyperthermophilic archaeon *Pyrococcus horikoshii* OT3. *J Biochem (Tokyo)* **135**, 479–85.

Takai, K. & Horikoshi, K. (1999) Genetic diversity of archaea in deep-sea hydrothermal vent environments. *Genetics* **152**,1285–97.

Takai, K., Moser, D. P., DeFlaun, M., Onstott, T. C. & Fredrickson, J. K. (2001) Archaeal diversity in waters from deep South African gold mines. *Appl Environ Microbiol* **67**, 5750–60.

Takakuwa, S. (1992) Biochemical aspects of microbial oxidation of sulfur compounds. *In:* S. Oae (ed.) *Organic Sulfur Chemistry: Biochemical Aspects*. CRC Press, Boca Raton, FL, pp. 1–43.

Tamames, J. (2001) Evolution of gene order conservation in prokaryotes. *Genome Biol* **2**, RESEARCH0020 e-pub.

Tamas, I., Klasson, L., Canback, B. et al. (2002) 50 million years of genomic stasis in endosymbiotic bacteria. *Science* **296**, 2376–9.

Tanaka, T., Fujiwara, S., Nishikori, S., Fukui, T., Takagi, M. & Imanaka, T. (1999) A unique chitinase with dual active sites and triple substrate binding sites from the hyperthermophilic archaeon *Pyrococcus kodakaraensis* KOD1. *Appl Environ Microbiol* **65**, 5338–44.

Tanaka, T., Fukui, T., Atomi, H. & Imanaka, T. (2003) Characterization of an exo-β-D-glucosaminidase involved in a novel chitinolytic pathway from the hyperthermophilic archaeon *Thermococcus kodakaraensis* KOD1. *J Bacteriol* **185**, 5175–81.

Tanaka, T., Fukui, T., Fujiwara, S., Atomi, H. & Imanaka, T. (2004) Concerted action of diacetylchitobiose deacetylase and exo-β-D-glucosaminidase in a novel chitinolytic pathway in the hyperthermophilic archaeon *Thermococcus kodakaraensis* KOD1. *J Biol Chem* **279**, 30021–7.

Tanaka, T., Fukui, T. & Imanaka, T. (2001) Different cleavage specificities of the dual catalytic domains in chitinase from the hyperthermophilic archaeon *Thermococcus kodakaraensis* KOD1. *J Biol Chem* **276**, 35629–35.

Tang, S. L., Nuttall, S. & Dyall-Smith, M. (2004) Haloviruses HF1 and HF2: evidence for a recent and large recombination event. *J Bacteriol* **186**, 2810–17.

Tang, T. H., Bachellerie. J. P., Rozhdestvensky, T. et al. (2002a) Identification of 86 candidates for small non-messenger RNAs from the archaeon *Archaeoglobus fulgidus*. *Proc Natl Acad Sci USA* **99**, 7536–41.

Tang, T. H., Polacek, N., Zywicki, M. et al. (2005) Identification of novel non-coding RNAs as potential antisense regulators in the archaeon *Sulfolobus solfataricus*. *Mol Microbiol* **55**, 469–81.

Tang, T. H., Rozhdestvensky, T. S., d'Orval, B. C. et al. (2002b) RNomics in Archaea reveals a further link between splicing of archaeal introns and rRNA processing *Nucleic Acids Res* **30**, 921–30.

Tang, Q., Carrington, P. E., Horng, Y. C., Maroney, M. J., Ragsdale, S. W. & Bocian, D. F. (2002) X-ray absorption and resonance Raman studies of methylcoenzyme M reductase indicating that ligand exchange and macrocycle reduction accompany reductive activation. *J Am Chem Soc* **124**, 13242–56.

Tanner, R. S., McInerney, M. J. & Nagle, D. P. Jr (1989) Formate auxotroph of *Methanobacterium thermoautotrophicum* Marburg. *J Bacteriol* **171**, 6534–8.

Tatusov, R. L., Fedorova, N. D. & Jackson, J. D. (2003) The COG database: an updated version includes eukaryotes. *BMC Bioinformatics* **4**, 41.

Tatusov, R. L., Natale, D. A., Garkavtsev, I. V. et al. (2001) The COG database: new developments in phylogenetic classification of proteins from complete genomes. *Nucleic Acids Res* **29**, 22–8.

Taylor, I., Brown, R., Bycroft, M. et al. (2004) Application of thermophilic enzymes in commercial biotransformation processes. *Biochem Soc Trans* **32**, 290–2.

Tchelet, R. & Mevarech, M. (1994) Interspecies genetic transfer in halophilic archaebacteria. *Syst Appl Microbiol* **16**, 578–81.

Tchong, S. I., Xu, H. & White, R. H. (2005) L-cysteine desulfidase: an [4Fe-4S] enzyme isolated from *Methanocaldococcus jannaschii* that catalyzes the breakdown of L-cysteine into pyruvate, ammonia, and sulfide. *Biochemistry* **44**, 1659–70.

Teichmann, S. A. & Mitchison, G. (1999) Is there a phylogenetic signal in prokaryote proteins? *J Mol Evol* **49**, 98–107.

Telser, J., Davydov, R., Horng, Y. C., Ragsdale, S. W. & Hoffman, B. M. (2001) Cryoreduction of methyl-coenzyme M reductase: EPR characterization of forms, MCR(ox1) and MCR (red1). *J Am Chem Soc* **123**, 5853–60.

Telser, J., Horng, Y. C., Becker, D. F., Hoffman, B. M. & Ragsdale, S. W. (2000) On the assignment of nickel oxidation states of the Ox1, Ox2 forms of methylcoenzyme M reductase. *J Am Chem Soc* **122**, 182–3.

Terada, T., Nureki, O., Ishitani, R., et al. (2002) Functional convergence of two lysyl-tRNA synthetases with unrelated topologies. *Nat Struct Biol* **9**, 257–62.

Terry, N., Zayed, A. M., De Souza, M. P. et al. (2000) Selenium in higher plants. *Annu Rev Plant Physiol Plant Mol Biol* **51**, 401–32.

Thauer, R. K. (1998) Biochemistry of methanogenesis: a tribute to Marjory Stephenson. *Microbiology* **144**, 2377–406.

Thauer, R. K. & Shima, S. (2006) Biogeochemistry, methane and microbes. *Nature* **440**, 878–9.

Thayer, J. S. (2002) Biological methylation of less studied elements. *Appl Organometallic Chem* **16**, 677–91.

Theissen, U., Hoffmeister, M., Grieshaber, M. & Martin, W. (2003) Single eubacterial origin of eukaryotic sulfide:quinone oxidoreductase, a mitochondrial enzyme conserved from the early evolution of eukaryotes during anoxic and sulfidic times. *Mol Biol Evol* **20**, 1564–74.

Theobald-Dietrich, A., Giegé, R. & Rudinger-Thirion, J. (2005) Evidence for the existence in mRNAs of a hairpin element responsible for ribosome dependent pyrrolysine insertion into proteins. *Biochimie* **87**, 813–17.

Thiel, V., Peckmann, J., Richnow, H. H., Luth, U., Reitner, J. & Michaelis, W. (2001) Molecular signals for anaerobic methane oxidation in Black Sea seep carbonates and a microbial mat. *Marine Chemistry* **73**, 97–112.

Thomas, D. J., Waters, S. B. & Styblo, M. (2004) Elucidating the pathway for arsenic methylation. *Toxicol Appl Pharmacol* **198**, 319–26.

Thomas, M. J., Platas, A. A. & Hawley, D. K. (1998) Transcriptional fidelity and proofreading by RNA polymerase II. *Cell* **93**, 627–37.

Thomm, M. (1996) Archaeal transcription factors and their role in transcription initiation. *FEMS Microbiol Rev* **18**, 159–71.

Thomm, M., Hausner, W. & Hethke, C. (1994) Transcription factors and termination of transcription in *Methanococcus*. *System Appl Microbiol* **16**, 148–55.

Thompson, D. K. & Daniels, C. J. (1998) Heat shock inducibility of an archaeal TATA-like promoter is controlled by adjacent sequence elements. *Mol Microbiol* **27**, 541–51.

Thompson, L. D. & Daniels, C. J. (1988) A tRNA(Trp) intron endonuclease from *Halobacterium volcanii*. Unique substrate recognition properties. *J Biol Chem* **263**, 17951–9.

Thompson, L. D. & Daniels, C. J. (1990) Recognition of exon–intron boundaries by the *Halobacterium volcanii* tRNA intron endonuclease. *J Biol Chem* **265**, 18104–11.

Tietze, M., Beuchle, A., Lamla, I. et al. (2003) Redox potentials of methanophenazine and CoB-S-S-CoM, factors involved in electron transport in methanogenic archaea. *Chembiochem* **4**, 333–5.

Tjaden, B., Plagens, A., Dörr, C., Siebers, B. & Hensel, R. (2006) Phosphoenolpyruvate synthetase and pyruvate, phosphate dikinase of *Thermoproteus tenax*: key pieces of the puzzle of archaeal carbohydrate metabolism. *Mol Microbiol* **60**, 287–98.

Tocchini-Valentini, G. D., Fruscoloni, P. & Tocchini-Valentini, G. P. (2005) Structure, function and evolution of the tRNA endonucleases of archaea: an example of subfunctionalization. *Proc Natl Acad Sci USA* **102**, 8933–8.

Tollervey, D. (1996) Small nucleolar RNAs guide ribosomal RNA methylation. *Science* **273**, 1056–7.

Tollervey, D. & Kiss, T. (1997) Function and synthesis of small nucleolar RNAs. *Curr Opin Cell Biol* **9**, 337–43.

Tolstrup, N., Sensen, C. W., Garrett, R. A. & Clausen, I. G. (2000) Two different and highly organized mechanisms of translation initiation in the archaeon *Sulfolobus solfataricus*. *Extremophiles* **4**, 175–9.

Tomschik, M., Karymov, M. A., Zlatanova, J. & Leuba, S. H. (2001) The archaeal histone-fold protein HMf organizes DNA in *bona fide* chromatin fibres. *Structure* **9**, 1201–11.

Toogood, H. S., Hollingsworth, E. J., Brown, R. C. et al. (2002) A thermostable L-aminoacylase from *Thermococcus litoralis*: cloning, overexpression, characterization, and applications in biotransformations. *Extremophiles* **6**, 111–22.

Topping, T. & Gloss, L. M. (2004) Stability and folding mechanism of mesophilic, thermophilic and hyperthermophilic archaeal histones: the importance of folding intermediates. *J Mol Biol* **342**, 247–60.

Torarinsson, E., Klenk, H.-P. & Garrett, R. A. (2005) Divergent transcriptional and translational signals in Archaea. *Environ Microbiol* **7**, 47–54.

Tornabene, T. G. & Langworthy, T. A. (1979) Diphytanyl and dibiphytanyl glycerol ether lipids of methanogenic archaebacteria. *Science* **203**, 51–3.

Torsvik, T. & Dundas, I. D. (1974) Bacteriophage of *Halobacterium salinarium*. *Nature* **248**, 680–1.

Tourova, T. P., Kolganova, T. V., Kuznetsov, B. B. & Pimenov, N. V. (2002) Phylogenetic diversity of the archaeal component in microbial mats on coral like structures associated with methane seeps in the Black Sea. *Microbiology* **71**, 196–201.

Tran, E. J., Zhang, X., Lackey, L. & Maxwell, E. S. (2005) Conserved spacing between the box C/D and C'/D' RNPs of the archaeal box C/D sRNP complex is required for efficient 2'-O-methylation of target RNAs. *RNA* **11**, 285–93.

Tran, E. J., Zhang, X. & Maxwell, E. S. (2003) Efficient RNA 2'-O-methylation requires juxtaposed and symmetrically assembled archaeal box C/D and C'/D' RNPs. *EMBO J* **22**, 3930–40.

Treude, T., Boetius, A., Knittel, K., Wallmann, K. & Jorgensen, B. B. (2003) Anaerobic oxidation of methane above gas hydrates at hydrate ridge, NE Pacific Ocean. *Marine Ecology-Progress Series* **264**, 1–14.

Treude, T., Niggemann, J., Kallmeyer, J. et al. (2005) Anaerobic oxidation of methane and sulfate reduction along the Chilean continental margin. *Geochim Cosmochim Acta* **69**, 2767–79.

Treusch, A. H., Kletzin, A., Raddatz, G. et al. (2004) Characterization of large-insert DNA libraries from soil for environmental genomic studies of Archaea. *Environ Microbiol* **6**, 970–80.

Treusch, A. H., Leininger, S., Kletzin, A., Schuster, S., Klenk, H.-P. & Schleper, C. (2005) Novel genes for nitrite reductase and Amo-related proteins indicate a role of uncultivated mesophilic crenarchaeota in nitrogen cycling. *Environm Microbiol* **7**, 1985–95.

Treusch, A. & Schleper, C. (2005) The microbial soil flora: novel approaches for accessing the phylogenetic and physiological diversity of prokaroytes. *In:* H. König & A. Varma (eds) *Intestinal Microorganisms of Termites and Other Soil Invertebrates.* Springer-Verlag, Heidelberg, pp. 407–24.

Tringe, S. G., von Mering, C., Kobayashi, A. et al. (2005) Comparative metagenomics of microbial communities. *Science* **308**, 554–7.

Tumbula, D. L., Becker, H. D., Chang, W.-Z. & Söll, D. (2000) Domain-specific recruitment of amide amino acids for protein synthesis. *Nature* **407**, 106–10.

Tumbula, D. L., Bowen, T. L. & Whitman, W. B. (1997a) Characterization of pURB500 from the archaeon *Methanococcus maripaludis* and construction of a shuttle vector. *J Bacteriol* **179**, 2976–86.

Tumbula, D. L., Makula, R. A. & Whitman, W. B. (1994) Transformation of *Methanococcus maripaludis* and identification of a Pst I-like restriction system. *FEMS Microbiol Lett* **121**, 309–14.

Tumbula, D. L., Teng, Q. & Bartlett, M. G. (1997b) Ribose biosynthesis and evidence for an alternative first step in the common aromatic amino acid pathway in *Methanococcus maripaludis.* *J Bacteriol* **179**, 6010–13.

Tyson, G. W., Chapman, J., Hugenholtz, P. et al. (2004) Community structure and metabolism through reconstruction of microbial genomes from the environment. *Nature* **428**, 37–43.

Uchida, T., Bonen, L., Schaup, H. W., Lewis, B. J., Zablen, L. & Woese C. (1974) The use of ribonuclease U2 in RNA sequence determination. Some corrections in the catalog of oligomers produced by ribonuclease T1 digestion of *Escherichia coli* 16S ribosomal RNA. *J Mol Evol* **3**, 63–77.

Udagawa, T., Shimizu, Y. & Ueda, T. (2004) Evidence for the translation initiation of leaderless mRNAs by the intact 70 S ribosome without its dissociation into subunits in eubacteria. *J Biol Chem* **279**, 8539–46.

Uotsu-Tomita, R., Tonozuka, T., Sakai, H. & Sakano, Y. (2001) Novel glucoamylase-type enzymes from *Thermoactinomyces vulgaris* and *Methanococcus jannaschii* whose genes are found in the flanking region of the alpha-amylase genes. *Appl Microbiol Biotechnol* **56**, 465–73.

Upreti, R. K., Kumar, M. & Shankar, V. (2003) Bacterial glycoproteins: functions, biosynthesis and applications. *Proteomics* **3**, 363–79.

Urich, T., Bandeiras, T. M., Leal, S. S. et al. (2004) The sulphur oxygenase reductase from *Acidianus ambivalens* is a multimeric protein containing a low-potential mononuclear non-haem iron centre. *Biochem J* **381**, 137–46.

Urich, T., Coelho, R., Kletzin, A. & Frazao, C. (2005a) The sulfur oxygenase reductase from *Acidianus ambivalens* is an icosatetramer as shown by crystallization and Patterson analysis. *Biochim Biophys Acta* **1747**, 267–70.

Urich, T., Gomes, C. M., Kletzin, A. & Frazao, C. (2006) X-ray structure of a self-compartmentalizing sulfur cycle metalloenzyme. *Science* **311**, 996–1000.

Urich, T., Kroke, A., Bauer, C., Seyfarth, K., Reuff, M. & Kletzin, A. (2005b) Identification of core active site residues of the sulfur oxygenase reductase from *Acidianus ambivalens* by site-directed mutagenesis. *FEMS Microbiol Lett* **248**, 171–6.

Valentine, D. L. (2002) Biogeochemistry and microbial ecology of methane oxidation in anoxic environments: a review. *Antonie Van Leeuwenhoek* **81**, 271–82.

Valentine, D. L. & Reeburgh, W. S. (2000) New perspectives on anaerobic methane oxidation. *Environ Microbiol* **2**, 477–84.

Van Bruggen, J. J. A., Zwart, K. B., Hermans, J. G. F. et al. (1986) Isolation and characterization of *Methanoplanus endosymbiosus* sp. nov., an endosymbiont of the marine sapropelic ciliate *Metopus contortus* quennerstedt. *Arch Microbiol* **144**, 367–74.

Van den Ent, F., Amos, L. & Löwe, J. (2001) Bacterial ancestry of actin and tubulin. *Curr Opin Microbiol* **4**, 634–8.

Van der Horn, P. B., Davis, M. C., Cunniff, J. J. et al. (1997) Thermo Sequenase DNA polymerase and *T. acidophilum* pyrophosphatase: new thermostable enzymes for DNA sequencing. *Biotechniques* **22**, 758–62, 764–5.

Van der Oost, J., De Boer, A. P. N., De Gier, J. W. L. et al. (1994) The heme-copper oxidase family consists of three distinct types of terminal oxidases and is related to nitric oxide reductase. *FEMS Microbiol Lett* **121**, 1–10.

Van der Oost, J., Huynen, M. A. & Verhees, C. H. (2002) Molecular characterization of phosphoglycerate mutase in archaea. *FEMS Microbiol Lett* **212**, 111–20.

Van der Oost, J., Schut, G., Kengen, S. W. et al. (1998) The ferredoxin-dependent conversion of glyceraldehyde-3-phosphate in the hyperthermophilic archaeon *Pyrococcus furiosus* represents a novel site of glycolytic regulation. *J Biol Chem* **273**, 28149–54.

Van der Oost, J., Voorhorst, W. G., Kengen, S. W. et al. (2001) Genetic and biochemical characterization of a short-chain alcohol dehydrogenase from the hyperthermophilic archaeon *Pyrococcus furiosus*. *Eur J Biochem* **268**, 3062–8.

Van der Oost, J., Walther, J., Brouns S. J. J. et al. (2006) Functional genomics of the thermo-acidophilic archaeon *Sulfolobus solfataricus*. *In:* F. Rainey & A. Oren (eds) *Methods in Microbiology: Extremophiles* in the press.

Van de Werken, Verhees, C. H., Akerboom, J., De Vos, W. M. & Van der Oost, J. (2006) Identification of a glycolytic regulon in the archaea *Pyrococcus* and *Thermococcus*. *FEMS Micr Lett* in the press.

Van der Wielen, P. W., Bolhuis, H., Borin, S. et al. (2005) The enigma of prokaryotic life in deep hypersaline anoxic basins. *Science* **307**, 121–3.

Van Duyne, G. D. (2001) A structural view of cre-loxp site-specific recombination. *Annu Rev Biophys Biomol Struct* **30**, 87–104.

Van Houten, B., Croteau, D. L., Dellavecchia, M. J., Wang, H. & Kisker, C. (2005) "Close-fitting sleeves": DNA damage recognition by the UvrABC nuclease system. *Mutat Res* **577**, 92–117.

Van Iterson, G. Jr, den Dooren de Jong, L. E. & Kluyver, A. J. (1983) *Martinus Beijerinck: His Life and Work*. Science Tech Inc., Madison, WI.

Van Lieshout, J., Faijes, M., Nieto, J., van der Oost, J. & Planas, A. (2004) Hydrolase and glycosynthase activity of endo-1,3-β-glucanase from the thermophile *Pyrococcus furiosus*. *Archaea* **1**, 285–92.

Van Niel, C. B. (1949) The Delft school and the rise of general microbiology. *Bacteriol Revs* **13**, 161–74.

Venter, J. C., Remington, K., Heidelberg, J. F. et al. (2004) Environmental genome shotgun sequencing of the Sargasso Sea. *Science* **304**, 66–74.

Verfurth, K., Pierik, A. J., Leutwein, C., Zorn, S. & Heider, J. (2004) Substrate specificities and electron paramagnetic resonance properties of benzylsuccinate synthases in anaerobic toluene and m-xylene metabolism. *Arch Microbiol* **181**, 155–62.

Verhagen, M. F., Menon, A. L., Schut, G. J. & Adams, M. W. (2001) *Pyrococcus furiosus*: large-scale cultivation and enzyme purification *Methods Enzymol* **330**, 25–30.

Verhees, C. H., Kengen, S. W. M, Tuininga, J. E. et al. (2003) The unique features of glycolytic pathways in Archaea. *Biochem J* **375**, 231–46.

Vestergaard, G., Häring, M., Peng, X., Rachel, R., Garrett, R. A. & Prangishvili, D. (2005) A novel rudivirus, ARV1, of the hyperthermophilic archaeal genus *Acidianus*. *Virology* **336**, 83–92.

Vidovic, I., Nottrott, S., Hartmuth, K., Luhrmann, R. & Ficner, R. (2000) Crystal structure of the spliceosomal 15.5 kDa protein bound to a U4 snRNA fragment. *Mol Cell* **6**, 1331–42.

Vieille, C. & Zeikus, G. J. (2001) Hyperthermophilic enzymes: sources, uses, and molecular mechanisms for thermostability. *Microbiol Mol Biol Rev* **65**, 1–43.

Vierke, G., Engelmann, A., Hebbeln, C. & Thomm, M. (2003) A novel archaeal transcriptional regulator of heat shock response. *Biol Chem* **278**, 18–26.

Visser, J. M., de Jong, G. A. H., Robertson, L. A. & Kuenen, J. G. (1997) Purification and characterization of a periplasmic thiosulfate dehydrogenase from the obligately autotrophic *Thiobacillus* sp. W5. *Arch Microbiol* **166**, 372–8.

Vossbrinck, C. R., Maddox, J. V., Friedman, S., Debrunner-Vossbrinck, B. A. & Woese, C. R. (1987) Ribosomal RNA sequence suggests microsporidia are extremely ancient eukaryotes. *Nature* **326**, 411–14.

Waino, M. & Ingvorsen, K. (2003) Production of β-xylanase and β-xylosidase by the extremely halophilic archaeon *Halorhabdus utahensis*. *Extremophiles* **7**, 87–93.

Wais, A. C., Kon, M., MacDonald, R. E. & Stollar, B. D. (1975) Salt-dependent bacteriophage infecting *Halobacterium cutirubrum* and *H. halobium*. *Nature* **256**, 314–15.

Wakagi, T., Lee, C. H. & Oshima, T. (1992) An extremely stable inorganic pyrophosphatase purified from the cytosol of a thermoacidophilic archaebacterium, *Sulfolobus acidocaldarius* strain 7. *Biochim Biophys Acta* **1120**, 289–96.

Wakeham, S. G., Hopmans, E. C., Schouten, S. & Damste, J. S. S. (2004) Archaeal lipids and anaerobic oxidation of methane in euxinic water columns: a comparative study of the Black Sea and Cariaco Basin. *Chem Geol* **205**, 427–42.

Wakeham, S. G., Lewis, C. M., Hopmans, E. C., Schouten, S. & Damste, J. S. S. (2003) Archaea mediate anaerobic oxidation of methane in deep euxinic waters of the Black Sea. *Geochim Cosmochim Acta* **67**, 1359–74.

Walter, W., Kireeva, M. L., Studitsky, V. M. & Kashlev, M. (2003) Bacterial polymerase and yeast polymerase II use similar mechanisms for transcription through nucleosomes. *J Biol Chem* **278**, 36148–56.

Walter, W. & Studitsky, V. M. (2001) Facilitated transcription through the nucleosome at high ionic strength occurs via a histone octamer transfer mechanism. *J Biol Chem* **276**, 29104–10.

Wang, B., Lu, D., Gao, R., Yang, Z., Cao, S. & Feng, Y. (2004a) A novel phospholipase A2/esterase from hyperthermophilic archaeon *Aeropyrum pernix* K1. *Protein Expr Purif* **35**, 199–205.

Wang, G., Guo, R., Bartlam, M. et al. (2003) Crystal structure of a DNA binding protein from the hyperthermophilic euryarchaeon *Methanococcus jannaschii*. *Prot Sci* **12**, 2815–22.

Wang, G., Kennedy, S. P., Fasiludeen, S., Rensing, C. & DasSarma, S. (2004b) Arsenic resistance in *Halobacterium* sp. strain NRC-1 examined by using an improved gene knockout system. *J Bacteriol* **186**, 3187–94.

Wang, L. & Schultz, P. G. (2004) Expanding the genetic code. *Angew Chem Int Ed Engl* **44**, 34–66.

Wang, Y., Prosen, D. E., Mei, L., Sullivan, J. C., Finney, M. & van der Horn, P. B. (2004c) A novel strategy to engineer DNA polymerase for enhanced processivity and improved performance *in vitro*. *Nucleic Acids Res* **32**, 1197–207.

Warbrick, E. (2000) The puzzle of PCNA's many partners. *Bioessays* **22**, 997–1006.

Ward, D. E., Revet, I. M., Nandakumar, R. et al. (2002) Characterization of plasmid pRT1 from *Pyrococcus* sp. strain JT1. *J Bacteriol* **184**, 2561–6.

Ward, D. E., Shockley, K. R., Chang, L. S. et al. (2002) Proteolysis in hyperthermophilic microorganisms. *Archaea* **1**, 63–74.

Wardleworth, B. N, Russell, R. J, Bell, S. D., Taylor, G. L. & White, M. F. (2002) Structure of Alba: an archaeal chromatin protein modulated by acetylation. *EMBO J* **21**, 4654–62.

Wassarman, K. M., Repoila, F., Rosenow, C., Storz, G. & Gottesman, S. (2001) Identification of novel small RNAs using comparative genomics and microarrays. *Genes Dev* **15**, 1637–51.

Watanabe, T., Hayashi, K., Tanaka, A., Furumoto, T., Hanaoka, F. & Ohkuma, Y. (2003) The carboxy terminus of the small subunit of TFIIE regulates the transition from transcription initiation to elongation by RNA polymerase II. *Mol Cell Biol* **23**, 2914–26.

Watanabe, Y., Yokobori, S., Inabe, T. et al. (2002) Introns in protein-coding genes in Archaea. *FEBS Letts* **510**, 27–30.

Waters, E., Hohn, M. J., Ahel, I. et al. (2003) The genome of *Nanoarchaeum equitans*: insights into early archaeal evolution and derived parasitism. *Proc Natl Acad Sci USA* **100**, 12984–8.

Watrin, L., Lucas, S., Purcarea, C., Legrain, C. & Prieur, D. (1999) Isolation and characterization of pyrimidine auxotrophs, and molecular cloning of the *pyrE* gene from the hyperthermophilic archaeon *Pyrococcus abyssi*. *Mol Gen Genet* **262**, 378–81.

Watrin, L. & Prieur, D. (1996) UV and ethyl methanesulfonate effects in hyperthermophilic archaea and isolation of auxotrophic mutants of *Pyrococcus* strains. *Curr Microbiol* **33**, 377–82.

Watson, J. D. & Crick, F. H. C. (1953) Genetical implications of the structure of deoxyribonucleic acid. *Nature* **171**, 964–6.

Weber, S. (2005) Light-driven enzymatic catalysis of DNA repair: a review of recent biophysical studies on photolyase. *Biochim Biophys Acta* **1707**, 1–23.

Webster, N. S., Negri, A. P., Munro, M. M. & Battershill, C. N. (2004) Diverse microbial communities inhabit Antarctic sponges. *Environ Microbiol* **6**, 288–300.

Weinberg, M. V., Schut, G. J., Brehm, S., Datta, S. & Adams, M. W. (2005) Cold shock of a hyperthermophilic archaeon: *Pyrococcus furiosus* exhibits multiple responses to a suboptimal growth temperature with a key role for membrane-bound glycoproteins *J Bacteriol* **187**, 336–48.

Wendoloski, D., Ferrer, C. & Dyall-Smith, M. L. (2001) A new simvastatin (mevinolin)-resistance marker from *Haloarcula hispanica* and a new *Haloferax volcanii* strain cured of plasmid pHV2. *Microbiology* **147**, 959–64.

Werne, J. P., Haese, R. R., Zitter, T. et al. (2004) Life at cold seeps: a synthesis of biogeochemical and ecological data from Kazan mud volcano, eastern Mediterranean Sea. *Chem Geol* **205**, 367–90.

Werner, F. & Weinzierl, R. O. (2002) A recombinant RNA polymerase II-like enzyme capable of promoter-specific transcription. *Mol Cell* **10**, 635–46.

Werner, F. & Weinzierl, R. O. (2005) Direct modulation of RNA polymerase core functions by basal transcription factors. *Mol Cell Biol* **25**, 8344–55.

Westoo, G. (1966) Determination of methylmercury compounds in foodstuff. 1-methylmercury in fish, identification and determination. *Acta Chem Scand* **20**, 2131–7.

Westover, K. D., Bushnell, D. A. & Kornberg, R. D. (2004) Structural basis of transcription: separation of RNA from DNA by RNA polymerase II. *Science* **303**, 1014–16.

White, M. F. (2003a) Archaeal DNA repair: paradigms and puzzles. *Biochem Soc Trans* **31**, 690–3.

White, R. H. (2003b) The biosynthesis of cysteine and homocysteine in *Methanococcus jannaschii*. *Biochim Biophys Acta* **1624**, 46–53.

White, R. H. (2004) L-Aspartate semialdehyde and a 6-deoxy-5-ketohexose 1-phosphate are the precursors to the aromatic amino acids in *Methanocaldococcus jannaschii*. *Biochemistry* **43**, 7618–27.

Whitman, W. B., Coleman, D. C. & Wiebe, W. J. (1998) Prokaryotes: the unseen majority. *Proc Natl Acad Sci USA* **95**, 6578–83.

Wickenheiser, E. B., Michalke, K., Drescher, C. et al. (1998a) Development and application of liquid and gas-chromatographic speciation techniques with element specific (ICP-MS) detection to the study of anaerobic arsenic metabolism. *Fresenius J Anal Chem* **362**, 498–501.

Wickenheiser, E. B., Michalke, K., Hensel, R. et al. (1998b) Volatile compounds in gases emitted from the wetland bogs near Lake Cadagno. In: R. Peduzzi, R. Bachofen & M. Tonolla (eds) *Lake Cadagno: A Meromictic Alpine Lake*. Documenta dell'Instituto Italiano Idrobiologia, Verbania Pallanza, 63, pp. 137–40.

Wickenheiser, E. B., Michalke, K., Hirner, A. V. et al. (2000) The biological methylation of bismuth; evidence for the involvement of polydimetylsiloxanes in the biologically-mediated methylation of metals. In: J. A. Centeno, P. Collery, G. Vernet, R. B. Finkelman, H. Gibb. & J. C. Etienne (eds) *Metal Ions in Biology and Medicine*, Vol. 6. John Libbey Eurotext, Paris, pp. 120–3.

Widdel, F. & Rabus, R. (2001) Anaerobic biodegradation of saturated and aromatic hydrocarbons. *Curr Opin Biotech* **12**, 259–76.

Wiedenheft, B., Stedman, K., Roberto, F. et al. (2004) Comparative genomic analysis of hyperthermophilic archaeal *Fuselloviridae* viruses. *J Virol* **78**, 1954–61.

Wiezer, A. & Merkl, R. (2005) A comparative categorization of gene flux in diverse microbial species. *Genomics* **86**, 462–75.

Wilcox, M. (1969) Gamma-phosphoryl ester of Glu-tRNA^{Gln}n as an intermediate in *Bacillus subtilis* glutaminyl-tRNA synthesis. *Cold Spring Harb Symp Quant Biol* **34**, 521–8.

Wilkes, H., Kuhner, S., Bolm, C. et al. (2003) Formation of n-alkane- and cycloalkane-derived organic acids during anaerobic growth of a denitrifying bacterium with crude oil. *Org Geochem* **34**, 1313–23.

Williams, K. P. (2002) Integration sites for genetic elements in prokaryotic tRNA and tmRNA genes: sublocation preference of integrase subfamilies. *Nucleic Acids Res* **30**, 866–75.

Winkler, G. S., Araujo, S. J., Fiedler, U. et al. (2000) TFIIH with inactive XPD helicase functions in transcription initiation but is defective in DNA repair. *J Biol Chem* **275**, 4258–66.

Winkler, H. (1920) *Verbreitung und Ursache der Parthenogenesis im Pflanzen- und Tierreiche*. Gustav Fischer Verlag, Jena.

Witkowski, J. A. (1988) The discovery of "split" genes: a scientific revolution. *Trends Biochem Sci* **13**, 110–13.

Woese, C. R. (1982) Archaebacteria and cellular origins: an overview. *Zbl Bakt Hyg I. Abt Orig C* **3**, 1–17.

Woese, C. R. (1987) Bacterial evolution. *Microbiol Rev* **51**, 221–71.

Woese, C. R. (1993) The Archaea: their history and significance. *In:* M. Kates, D. Kushner & A. Matheson (eds) *The Biochemistry of Archaea (Archaebacteria)*. Elsevier Science Publishers, Amsterdam, pp. vii–xxix.

Woese, C. R. (1998) A manifesto for microbial genomics. *Curr Biol* **8**, R781–3.

Woese, C. R. (2000) Interpreting the universal phylogenetic tree. *Proc Natl Acad Sci USA* **97**, 8392–6.

Woese, C. R. (2002) On the evolution of cells. *Proc Natl Acad Sci USA* **99**, 8742–7.

Woese, C. R. (2004) A new biology for a new century. *Microbiol Mol Biol Rev* **68**, 173–86.

Woese, C. R., Achenbach, L., Rouviere, P. & Mandelco, L. (1991) Archaeal phylogeny: reexamination of the phylogenetic position of *Archaeoglobus fulgidus* in light of certain composition-induced artifacts. *Syst Appl Microbiol* **14**, 364–71.

Woese, C. R. & Fox, G. E. (1977a) The concept of cellular evolution. *J Mol Evol* **10**, 1–6.

Woese, C. R. & Fox, G. E. (1977b) The phylogenetic structure of the procaryotic domain: The primary kingdoms. *Proc Natl Acad Sci USA* **74**, 5088–90.

Woese, C. R, Gupta, R., Hahn, C. M., Zillig, W. & Tu, J. (1984) The phylogenetic relationships of three sulfur dependent archaebacteria. *Syst Appl Microbiol* **5**, 97–105.

Woese, C. R., Kandler, O. & Wheelis M. L. (1990) Towards a natural system of organisms: Proposal for the domains Archaea, Bacteria and Eukarya. *Proc Natl Acad Sci USA* **87**, 4576–9.

Woese, C. R., Magrum, L. J. & Fox, G. E. (1978) Archaebacteria. *J Mol Evol* **11**, 245–51.

Woese, C. R. & Olsen, G. J. (1986) Archaebacterial phylogeny: perspectives on the urkingdoms. *Syst Appl Microbiol* **7**, 161–77.

Woese, C. R., Olsen, G. J., Ibba, M. & Söll, D. (2000) Aminoacyl-tRNA synthetases, the genetic code, and the evolutionary process. *Microbiol Mol Biol Rev* **64**, 202–36.

Wolf, Y. I., Rogozin, I. B., Grishin, N. V. & Koonin, E. V. (2002) Genome trees and the tree of life. *Trends Genet* **18**, 472–9.

Wolf, Y. I., Rogozin, I. B., Grishin, N. V., Tatusov, R. L. & Koonin, E. V. (2001) Genome trees constructed using five different approaches suggest new major bacterial clades. *BMC Evol Biol* **1**, 8.

Wolfe, R. S (2001) The Archaea: a personal overview of the formative years. *In:* M. Dworkin et al. (eds) *The Prokaryotes: An Evolving Electronic Resource for the Microbiological Community*, 3rd edn, release 3.7. Springer-Verlag, New York.

Wolin, M. J. (1981) Fermentation in the rumen and human large intestine. *Science* **213**, 1463–8.

Wong, J. T. Y., New, D. C., Wong, J. C. W. & Hung, V. K. L. (2003) Histone-like proteins of the dinoflagellate *Crypthecodinium cohnii* have homologies to bacterial DNA-binding proteins. *Euk Cell* **2**, 646–50.

Wood, E. R., Ghane, F. & Grogan, D. W. (1997) Genetic responses of the thermophilic archaeon *Sulfolobus acidocaldarius* to short-wavelength UV light. *J Bacteriol* **179**, 5693–8.

Wood, J. M., Kennedy, F. S. & Rosen, C. G. (1968) Synthesis of methyl-mercury compounds by extracts of a methanogenic bacterium. *Nature* **220**, 173–4.

Woods, W. G. & Dyall-Smith, M. L. (1997) Construction and analysis of a recombination-deficient (*radA*) mutant of *Haloferax volcanii*. *Mol Microbiol* **23**, 791–7.

Worthington, P., Hoang, V., Perez-Pomares, F. & Blum, P. (2003) Targeted disruption of the alpha-amylase gene in the hyperthermophilic archaeon *Sulfolobus solfataricus*. *J Bacteriol* **185**, 482–8.

Wuchter, C., Schouten, S., Boschker, H. T. & Sinninghe Damste, J. S. (2003) Bicarbonate uptake by marine Crenarchaeota. *FEMS Microbiol Lett* **219**, 203–7.

Xiang, X., Chen, L., Huang, X., Luo, Y., She, Q. & Huang, L. (2005) *Sulfolobus tengchongensis* spindle-shaped virus STSV1: virus-host interactions and genomic features. *J Virol* **79**, 8677–86.

Xie, Y. (2005) Nucleosomes, transcription and transcription regulation in Archaea. PhD thesis, Ohio State University, Columbus.

Xie, Y. & Reeve, J. N. (2004a) Transcription by an archaeal RNA polymerase is slowed but not blocked by an archaeal nucleosome. *J Bacteriol* **186**, 3492–8.

Xie, Y. & Reeve, J. N. (2004b) Transcription by *Methanothermobacter thermoautotrophicus* RNA polymerase in vitro releases archaeal transcription factor B but not TATA-box binding protein from the template DNA. *J Bacteriol* **186**, 6306–10.

Xue, H., Guo, R., Wen, Y., Liu, D. & Huang, L. (2000) An abundant DNA binding protein from the hyperthermophilic archaeon *Sulfolobus shibatae* affects DNA supercoiling in a temperature-dependent fashion. *J Bacteriol* **182**, 3929–33.

Yan, M. & Gralla, J. D. (1997) Multiple ATP-dependent steps in RNA polymerase II promoter melting and initiation. *EMBO J* **16**, 7457–67.

Yang, J. M. & Wang, A. H. (2004) Engineering a thermostable protein with two DNA-binding domains using the hyperthermophilic protein Sac7d. *J Biomol Struct Dyn* **21**, 513–26.

Yang, S., Doolittle, R. F. & Bourne, P. E. (2005) Phylogeny determined by protein domain content. *Proc Natl Acad Sci USA* **102**, 373–8.

Yang, S. J., Lee, H. S., Park, C. S., Kim, Y. R., Moon, T. W. & Park, K. H. (2004) Enzymatic analysis of an amylolytic enzyme from the hyperthermophilic archaeon *Pyrococcus furiosus* reveals its novel catalytic properties as both an α-amylase and a cyclodextrin-hydrolyzing enzyme. *Appl Environ Microbiol* **70**, 5988–95.

Yankofsky, S. A. & Spiegelman, S. (1962). The identification of the ribosomal RNA cistron by sequence complementarity. II. Saturation of and competitive interaction at the RNA cistron. *Proc Natl Acad Sci USA* **48**, 1466–72.

Yatime, L., Schmitt, E., Blanquet, S. & Mechulam, Y. (2004) Functional molecular mapping of archaeal translation initiation factor 2. *J Biol Chem* **279**, 15984–93.

Yatime, L., Schmitt, E., Blanquet, S. & Mechulam, Y. (2005) Structure-function relationships of the intact aIF2α subunit from the archaeon *Pyrococcus abyssi*. *Biochemistry* **44**, 8749–56.

Ye, X., Ou, J., Ni, L., Shi, W. & Shen, P. (2003) Characterization of a novel plasmid from extremely halophilic Archaea: nucleotide sequence and function analysis. *FEMS Microbiol Lett* **221**, 53–7.

Yokomori, K., Verrijzer, C. P. & Tjian, R. (1998) An interplay between TATA box-binding protein and transcription factors IIE and IIA modulates DNA binding and transcription. *Proc Natl Acad Sci USA* **95**, 6722–7.

Yura, T. & Nakahigashi, K. (1999) Regulation of the heat-shock response. *Curr Opin Microbiol* **2**, 153–8.

Zago, M. A., Dennis, P. P. & Omer, A. D. (2005) The expanding world of small RNAs in the hyperthermophilic archaeon *S. solfataricus*. *Mol Microbiol* **54**, 980–93.

Zakharyan, R. A., Tsaprailis, G., Chowdhury, U. K. et al. (2005) Interactions of sodium selenite, glutathione, arsenic species, and omega class human glutathione transferase. *Chem Res Toxicol* **18**, 1287–95.

Zalenskaya, K., Lee, J., Gujuluva, C. N., Shin, Y. K., Slutsky, M. & Goldfarb, A. (1990) Recombinant RNA polymerase: inducible overexpression, purification and assembly of *Escherichia coli rpo* gene products. *Gene* **89**, 7–12.

Zappa, S., Rolland, J. L., Flament, D., Gueguen, Y., Boudrant, J. & Dietrich, J. (2001) Characterization of a highly thermostable alkaline phosphatase from the euryarchaeon *Pyrococcus abyssi*. *Appl Environ Microbiol* **67**, 4504–11.

Zavialov, A. V., Mora, L., Buckingham, R. H. & Ehrenberg, M. (2002) Release of peptide promoted by the GGQ motif of class 1 release factors regulates the GTPase activity of RF3. *Mol Cell* **10**, 789–98.

Zaychikov, E., Denissova, L. & Heumann, H. (1995) Translocation of the *Escherichia coli* transcription complex observed in the registers 11 to 20: "jumping" of RNA polymerase and asymmetric expansion and contraction of the "transcription bubble." *Proc Natl Acad Sci USA* **92**, 1739–43.

Zehnder, A. J. B. & Brock, T. D. (1979) Methane formation and methane oxidation by methanogenic bacteria. *J Bacteriol* **137**, 420–32.

Zehnder, A. J. B. & Brock, T. D. (1980) Anaerobic methane oxidation – occurrence and ecology. *Appl Environ Microbiol* **39**, 194–204.

Zeikus, J. G. & Wolfe, R. S. (1972) *Methanobacterium thermoautotrophicus* sp. n., an anaerobic, autotrophic, extreme thermophile. *J Bacteriol* **109**, 707–15.

Zhang, C. L., Pancost, R. D., Sassen, R., Qian, Y. & Macko, S. A. (2003) Archaeal lipid biomarkers and isotopic evidence of anaerobic methane oxidation associated with gas hydrates in the Gulf of Mexico. *Org Geochem* **34**, 827–36.

Zhang, D., Vaidehi, N., Goddard, W. A. III et al. (2002) Structure-based design of mutant *Methanococcus jannaschii* tyrosyl-tRNA synthetase for incorporation of O-methyl-L-tyrosine. *Proc Natl Acad Sci USA* **99**, 6579–84.

Zhang, D. Y., Carson, D. J. & Ma, J. (2002) The role of TFIIB-RNA polymerase II interaction in start site selection in yeast cells. *Nucleic Acids Res* **30**, 3078–85.

Zhang, J. K., Pritchett, M. A., Lampe, D. J., Robertson, H. M. & Metcalf, W. W. (2000) *In vivo* transposon mutagenesis of the methanogenic archaeon *Methanosarcina acetivorans* C2A using a modified version of the insect mariner-family transposable element Himar1. *Proc Natl Acad Sci USA* **97**, 9665–70.

Zhang, R. & Zhang, C. T. (2003) Multiple replication origins of the archaeon *Halobacterium* species NRC-1. *Biochem Biophys Res Commun* **302**, 728–34.

Zhang, Y., Baranov, P. V., Atkins, J. F. & Gladyshev, V. N. (2005) Pyrrolysine and selenocysteine use dissimilar decoding strategies. *J Biol Chem* **280**, 20740–51.

Zhao, K., Chai, X. & Marmorstein, R. (2003) Structure of a Sir2 substrate, Alba, reveals a mechanism for deacetylation-induced enhancement of DNA binding. *J Biol Chem* **278**, 26071–7.

Zhao, S. & Williams, K. P. (2002) Integrative genetic element that reverses the usual target gene orientation. *J Bacteriol* **184**, 859–60.

Zheng, L., Roeder, R. G. & Luo, Y. (2003) S phase activation of the histone H2B promoter by OCA-S, a coactivator complex that contains GAPDH as a key component. *Cell* **114**, 255–66.

Zhou, M., Xiang, H., Sun, C. & Tan, H. (2004) Construction of a novel shuttle vector based on an RCR-plasmid from a haloalkaliphilic archaeon and transformation into other haloarchaea. *Biotechnol Lett* **26**, 1107–13.

Zhu, W., Zeng, Q., Colangelo, C. M., Lewis, M., Summers, M. F., Scott, R. A. 1996. The N-terminal domain of TFIIB from *Pyrococcus furiosus* forms a zinc ribbon. *Nat Struct Biol* **3**, 122–4.

Ziesche, S., Omer, A. D. & Dennis P. P. (2004) RNA guided nucleotide modification of ribosomal and non-ribosomal RNAs in Archaea. *Mol Microbiol* **55**, 1812–28.

Zillig, W., Arnold, H. P., Holz, I. et al. (1998) Genetic elements in the extremely thermophilic archaeon *Sulfolobus*. *Extremophiles* **2**, 131–40.

Zillig, W., Holz, I., Klenk, H.-P. et al. (1987) *Pyrococcus woesei*, sp. nov., an ultra-thermophilic marine archaebacterium, representing a novel order, Thermococcales. *System Appl Microbiol* **9**, 62–70.

Zillig, W., Kletzin, A., Schleper, C. et al. (1994) Screening for *Sulfolobales*, their plasmids and their viruses in Icelandic solfataras. *Syst Appl Microbiol* **16**, 609–28.

Zillig, W., Palm, P., Klenk, H.-P. et al. (1993) Transcription in Archaea. *In:* M. Kates, D. J. Kushner & A. T. Matheson (eds) *Biochemistry of Archaea*. Elsevier, Amsterdam, pp. 367–91.

Zillig, W., Prangishvilli, D., Schleper, C. et al. (1996) Viruses, plasmids and other genetic elements of thermophilic and hyperthermophilic Archaea. *FEMS Microbiol Rev* **18**, 225–36.

Zillig, W., Stetter, K. O. & Janekovic, D. (1979) DNA-dependent RNA polymerase from the archaebacterium *Sulfolobus acidocaldarius*. *Eur J Biochem* **96**, 597–604.

Zillig, W., Stetter, K. O. & Tobien, M. (1978) DNA-dependent RNA polymerase from *Halobacterium halobium*. *Eur J Biochem* **91**, 193–9.

Zillig, W., Stetter, K. O., Wunderl, S., Schulz, W., Priess, H. & Scholz, I. (1980) The *Sulfolobus-"Caldariella"* group: taxonomy on the basis of the structure of DNA-dependent RNA polymerases. *Arch Microbiol* **125**, 259–69.

Zillig, W., Yeats, S., Holz, I., Böck, A., Gropp, F., Rettenberger, M. & Lutz, S. (1985) Plasmid-related anaerobic autotrophy of the novel archaebacterium *Sulfolobus ambivalens*. *Nature* **313**, 789–91.

Zimmermann, P., Laska, S. & Kletzin, A. (1999) Two modes of sulfite oxidation in the extremely thermophilic and acidophilic archaeon *Acidianus ambivalens*. *Arch Microbiol* **172**, 76–82.

Zimmermann, P. & Pfeifer, F. (2003) Regulation of the expression of gas vesicle genes in *Haloferax mediterranei*: the two regulatory proteins GvpD and GvpE interact. *Mol Microbiol* **49**, 783–94.

Zivanovic, Y., Lopez, P., Philippe, H. & Forterre, P. (2002) *Pyrococcus* genome comparison evidences chromosome shuffling-driven evolution. *Nucleic Acids Res* **30**, 1902–10.

Zona, R., Chang-Pi-Hin, F., O'Donohue, M. J. & Janecek, S. (2004) Bioinformatics of the glycoside hydrolase family 57 and identification of catalytic residues in amylopullulanase from *Thermococcus hydrothermalis*. *Eur J Biochem* **271**, 2863–72.

Zuckerkandl, E. & Pauling, L. (1965) Molecules as documents of evolutionary history. *J Theor Biol* **8**, 357–66.

Index

Page numbers in *italics* refer to figures and those in **bold** to tables; but note that figures and tables are only indicated when they are separated from their text references.

ABC transporters 195
ABV (*Acidianus* bottle-shaped virus) 69, *70*
acetylation, Alba proteins 151, 154
N-acetyl-β-lysine **318**, 320
O-acetylserine sulfhydrolases *263*, 314
Acidianus
  plasmids **108**, 109–10
  sulfur metabolism 267, 268–74
  viruses 60, 62–3, 68–9
Acidianus ambivalens
  sulfur oxidation 270–4
  sulfur reductase and hydrogenase 268–70
Acidianus brierleyi **91**
Acidithiobacillus tengchongensis, sulfur oxidation 270–3
acyltransferase 72
S-adenosyl methionine (SAM) 231, 289
adenylate kinase 274
adenylylsulfate:phosphate adenylyltransferase (APAT) *271*, 274
adenylylsulfate (APS) reductases 86, *263*, 265–6, 274, Plate 23.2
ADP-dependent phosphofructokinase/glucokinase (ADP-PFK/GLK) 252–5
aEF1A/aEF2 227
a/eIF2 223, 224–5, Plate 19.1, Plate 19.2
Aeropyrum pernix
  genome **77**, *79*, 85, 86
  MITEs 96–7

translation and protein synthesis 218
AFV1 (*Acidianus* filamentous virus 1) *61*, 62–3, 72, 73
AFV2 (*Acidianus* filamentous virus 2) *61*, 63, 73
Alba *see* DBNP-B/SSo10b/Alba proteins
alcohol dehydrogenases (ADH) 130, 309–10, **311**
aldehyde dehydrogenase (Ald DH), NAD-dependent *253*, 256
aldehyde-ferredoxin oxidoreductase 256
aldolases 255, **311**, 313
  see also specific aldolases
alkaline phosphatases 309
alkanes, anaerobic oxidation 282–3
O⁶-alkylguanine-DNA alkyltransferase (AGT)TendoV 173
alkylsuccinate 282–3
Alphalipothrixvirus 60–1
amidases **311**, 313
amino acids
  aromatic, biosynthesis 258
  extreme and hyperthermophiles 85
  unusual *79*, 81, 212–15
aminoacylases **311**, 313
aminoacyl-tRNAs (aatRNA) 207–15
  noncanonical amino acids 212–15
  synthesis via two pathways 207–8
aminoacyl-tRNA synthetase (aaRS) 207
aminopeptidases 306

ammonia
  in Gln-tRNA formation 209
  oxidisers 42, 44–5, Plate 4.4
ammonia monooxygenases (*amo* gene products) 42, 44–5, Plate 4.4
Ampullaviridae 59, 69, *70*
Ampullavirus 69, *70*
amylases 296–9
amylomaltases **298**, 300–1
amylopullulanases 299–300
anaerobic oxidation of alkanes 282–3
anaerobic oxidation of methane (AOM) 43–4, 275, 279–82
  alternative mechanisms 282–3
ANME lineages 43–4, Plate 4.1
  anaerobic oxidation of methane 43–4, 275, 279–80
  metagenomics 43–4, **49**
antibiotics
  archaeal 316, **319**
  for genetic selection 130–1
antimony (Sb), biomethylation 287, **288**, 289
antisense RNAs 98, 237–8
appendages, pili-like 46
arcABC operon 205
Archaeoglobus, dissimilatory sulfate and sulfite reduction 265–6
Archaeoglobus fulgidus 19–20
archaeol 55
archaeosomes 315
aRF1 227
arsenic compounds, biomethylation 285, 286, 287–9
arthropods, symbiotic associations 47

ARV1 (*Acidianus* rod-shaped virus 1) 63–5, 72
ARV2 (*Acidianus* rod-shaped virus 2) 103
asparaginases 208, 209
asparaginyl (Asn)-tRNA 207, 208
aSUI1 **224**, 226
ATP, formation 256, 257
ATPases 53, 72, 109
ATP-dependent glucokinase (ATP-GLK) **253**, 255
ATP-dependent phosphofructokinase (ATP-PFK) **253**, 255
ATP sulfurylases (ATP: sulfate adenylyltransferases) *263*, 265
ATV (*Acidianus* two-tailed virus) 68–9, 72
auxotrophic selectable markers 131

bacitracin 126
bacteria
  chromatin proteins 147
  evolutionary relationships 25–7
  genome sequencing 75, **76**
  genome size 76
  horizontal gene transfer 24, 30–2
  methylation of metal(loid)s 287–90
  MMF proteins 21
  reverse gyrases 22
bacterioopsin 204–6
bacteriorhodopsin (BR) 199
  commercial uses 316–17, **318**
  regulation of synthesis 204, 205
base excision repair (BER) 172–4
bat gene 204, 205
Betalipothrixvirus 61–2, 73
bgaH gene 135
Bicaudaviridae 59, 68–9
Bicaudavirus 68–9
"Big Tree" project 9
biocatalysis, whole-cell 314–15, **319**
biocatalysts 296–314
biomethylation of metal(loid)s 285–94
  analytical techniques 286–7
  bismuth 290–3
  ecotoxicology 290–3
  health concerns 292–3
  in three domains of life 287–90
  versatility of Methanoarchaea 287
biomining 314–15
biopolymers 317, **319**

biotechnology 295–321
  white (industrial) 295
biphytane 55
bismuth (Bi), biomethylation **288**, 290–4
bleomycin 130
blp gene 204, 205, *206*
bop gene cluster 199
  proteins encoded 204–5
  regulation 204–6
BRE 185
5-bromo-2′-deoxyuridine (BrdU) 126
2-bromoethanesulfonate 281
brp gene 204, 205, *206*
bulge–helix–bulge (BHB) motif 99, 100, *101*, 234, 237

caldarchaeol 55
caldariella quinone *271*, 273–4
canthaxanthin 317, **318**
carbohydrate metabolism
  enzymes of industrial interest 296–301
  regulation 258–9
carbon dioxide (CO_2)
  fixation/assimilation 42, 52, 53
  reduction, methanogens 46, 87
carbon monoxide (CO) dehydrogenase 88
carbon source, transcriptional response to changes 242
carotenoids (including β-carotene) 205, 314, 317, **318**
Cbf5 protein
  exon–intron splicing 100, *101*
  RNA binding 230, 235–6, Plate 20.1
C–C bond-forming enzymes 313
C/D box ribonucleoproteins (RNPs) 229, 230–5, Plate 20.1
  circular 234–5
  methylation of tRNA 233–4
  structure–function constraints 231–3
  tRNA gene introns 234
C/D box RNAs 230
cdc6 genes 161
Cdc6 proteins 111–12, 162–3, 164
cell cycle 159–60
cell division 53–4, 159–60
cell envelope/wall
  bacterial 5
  extreme halophiles 90

  Igniococcus 53
  Nanoarchaeum equitans 53
  Sulfolobus 137, *138*
cell growth inhibitors 130–1
cellulose-degrading enzymes 301–4
Cenarchaeum symbiosum 42–3, 47, **49**, 91
chaperones/chaperonins 156, 316, **318**
chemolithoautotrophs 95
  sulfur metabolism 267–70
chitin-degrading enzymes **302**, 305–6
chromosomes 76–9
  DNA exchange 98–9, 138–9
  halophilic archaea 105, **106**
  segregation 159–60
  variation in *Sulfolobus* 96–7, Plate 8.1, Plate 8.2
CipA/CipB 243
coenzyme B 275, *276*, 277–8
coenzyme M 211, 277, 278
cofactor F_{430} 275, 276, 277–8
  methanotrophic Archaea 280–1
  redox state 277
cold shock/cold adaptation 242–3
compatible solutes 317–20
conjugation 127–8, 138–9
conjugative plasmids 98–9, 110, 138–9
convergent evolution 6–7
  metabolic enzymes 249, Plate 22.2
copper mining 314
Crenarchaeota (Crenarchaea) 8, 17
  associated with eukaryotes 46–7
  diversity of uncultivated 40–1, Plate 4.1
  genomes 83, 84, 85–6, Plate 7.2
  marine planktonic 41–3
  metagenomic studies **49**
  phylogenetic analysis 19–20
crtB1 gene 204, 205, *206*
cyclic-2, 3-bisphosphoglycerate **318**, 320
cyclic polydimethylsiloxanes (cyclic PDMS) 291–3
cyclodextrin glucanotransferase (CGTase) **298**, 300, 301
cyclodextrins 301
cyclopyrimidine dimers 175
cysteine desulfidase 211–12
cysteine desulfurase 211, *263*
cysteine synthase **311**, 314

cysteinyl-tRNA synthetase (CysRS) 209–10, 211, 212
DBNP-B/SSo10b/Alba proteins 137, 151–4
  DNA/RNA binding and complex formation 153–4
  regulation of gene expression 154
Delft School 12, 13
Deltalipothrixvirus 61, 63, 73
2-deoxyribose-5-phosphate aldolase (DERA) 258
diglycerol phosphate **318**, 320
dimethylarsine 286, 287, 289
dimethylmercury 285, 290
di-*myo*-inositol-1,1'-phosphate 317–20
dissimilatory siroheme-containing sulfite reductases (DSRs) *263*, 265, 266
diversity, uncultivated Archaea 39–50
DNA
  histone binding 149–50
  MC1 binding 156
  processing enzymes 307–9
  Sac10a/Sso10a binding 155
  Sso7d/Sac7d/Ssh7d/HSPN-C' binding 155
DNA binding nucleoid protein B *see* DBNP-B/SSo10b/Alba
DNA-binding proteins 137, 147–57, 194
DNA damage 171, *172*
  response 181–2
  reversal and bypass 175–6
  see also DNA repair
DNA gyrase, horizontal gene transfer 24–5
DNA ligases 172, **308**, 309
DNA polymerases 166
  DNA repair 172–3
  error bypass Dpo4 **175**, 176
  PCNA interaction 167
  plasmid replication 106
  uracil scanning 174
DNA repair 171–83
  control 181–2
  distribution in Archaea 85, 86, 89
  non-reciprocal recombination and 142
  pathways 172–81
  Sulfolobus 140, 145
DNA replication 159, 161–9
  accuracy in *Sulfolobus* 139–40

domain-specific differences 26–7, 37
double strand break repair and 178
origins 161–3
rolling circle 106, *107*
Sulfolobus chromosomes 137
termination mechanism 162
DNA topoisomerases 309, **318**
double strand break repair (DSBR) 177–81
Dpo4-type polymerase **175**, 176
dUTPase 72

ecology, microbial 39–50
ecotoxicology, biomethylated bismuth 290–3
electron transport chains 271, 273–4
elongation complex 190–1
elongation factors, translation 227
Embden–Meyerhof–Parnas (EMP) pathway 249–56
  regulation 258–9
endoglucanases 301–4
endonucleases
  intron excision 234, 237
  IV (AP) 172–3
  RecB-family 72
  V **173**
  XPF 177–8
enolase (ENO) 248, 249, **253**, 256
Entner–Doudoroff (ED) pathway 250–1, **253**, 256–7
  regulation 258–9
environmental genome tags (EGTs) 48
enzymes, archaeal 295–314
  advantages and limitations of using 320–1
  alcohol dehydrogenases 309–10, **311**
  aminoacylases 313
  C–C bond-forming 313
  cellulose-degrading 301–4
  chitin-degrading **302**, 305–6
  DNA-processing 307–9
  esterases 310–13
  heterologous expression 321
  nitrile-degrading 313
  proteolytic **303**, 306
  starch-processing 296–301
  thermostability 320–1
  xylan-degrading **303**, 304–5
episomal integrative plasmids and viruses 113

eRF1 227
error bypass polymerase Dpo4 **175**, 176
erythrose-4-phosphate 257, 258
esterases 310–13
ethylmethanesulfonate (EMS)-induced mutagenesis 126
eukaryotes 10
  Archaea associated with 46–7
  evolutionary relationships 25–7
  histones 147
  methylation of metal(loid)s 287–90
  rRNA phylogeny 18
  translation initiation 223
Euryarchaeota (Euryarchaea) 8, 17
  diversity of uncultivated 40, 41, Plate 4.1
  freshwater 46
  genomes 83, 84, Plate 7.2
  metagenomic studies **49**
  phylogenetic analysis 19–20
  plasmids 105
evolution
  cell 2, 10–11
  convergent 6–7, 249
  metabolic pathways 247–8
  role of integrated elements 120–2
exoglucanases 301
exopolysaccharides 317, **319**
extremophilic Archaea 295–6
  advantages and limitations of using 320–1
  enzymes/other compounds derived from 296–320
extremozymes 295–314, 320

Fen1 167–8, **175**, 177
ferredoxin:NADP oxidoreductase 267
Ferroplasma acidarmanus 47, **91**, 173
Fib (fibrillarin) 230, 231, Plate 20.1
fluorescent *in situ* hybridization (FISH) 40, 41, Plate 4.2
5-fluoroorotic acid (5-FOA) selection 131, 134, 139–40
formaldehyde 21, 257–8
fosmids 42, 43, **49**
fructose bisphosphate aldolase (FBA) **253**, 255
*fts*Z gene 53, 86, 87
fumarate, anaerobic oxidation 282–3
Fuselloviridae 59, 65–6
Fusellovirus 65–6, 73

β-galactosidase **311**, 314
Gammalipothrixvirus *61*, 62–3, 73
GAP *see* glyceraldehyde-3-phosphate
Gar1 230, 235–6, Plate 20.1
gas vesicles 199, 200–1
   biotechnological uses 317
   genes *see gvp* genes
   regulation of synthesis 201–4
GatCAB 208
GatDE 208–9, Plate 18.1
G+C content **77–8**, 79–80
   extreme and hyperthermophiles 85
   marine planktonic crenarchaeotes 42, 43
gene(s)
   collective signals 32–3
   conserved archaeal core 32–4, 83–4, 247, Plate 3.1
   conversion 142
   co-regulation 81
   duplication 247–8
   knockout methods 133–5
   non-orthologous displacements 248, 252, Plate 22.1
   split 53
   defining core of genes 33–4
   organismal 30, 33–5
gene expression, regulation
   DBNP-B/SSo10b/Alba 154
   histones 150–1, *152*
   post-transcriptional 237–8
   Sso7d/Sac7d/Ssh7d/HSPN-C′ 156
   see also transcription, regulation
genealogy
genetic exchange 98–9, 127–8, 138–9
genetics 125–36
   additional recombinant DNA tools 135–6
   reverse 133–5
   selectable markers *see* selectable markers
   Sulfolobus acidocaldarius 137–45
genome(s) 75–94
   annotation 92, 94
   bacterial gene content 30–2
   engineering by recombination 144
   mechanisms of plasticity 80–1, 95–104
   mutation rates 140
   variation in archaeal viruses 102–4, Plate 8.3

genome sequencing 75, **76**
   future 93–4
   in progress **91**, 92–3
   published sequences **77–8**
genome size 76, **77–8**
   lateral gene transfer and 30, *32*
   methanogens 88
genomics 75
   comparative 84, 248
   crenarchaeal viruses 71–3, Plate 6.1
   functional 92, 94
   future prospects 92–4
   structural 75, 245–6
genomic trees, whole 18–19
Gln-tRNA *see* glutaminyl-tRNA
Globuloviridae 59, 67–8, 73
Globulovirus 67–8
glucoamylases 296–9
glucokinases (GLK) 252–5
gluconate dehydratase (G-hydr) **253**
gluconate kinase (GlcA-kin) **253**
gluconeogenesis 249, 257
gluconolactonase (Glc-lac) **253**
glucose dehydrogenase (GlcDH) **253**
glucose-6-phosphate dehydrogenase (G6PDH) **253**, **254**
α-glucosidases 296–9
β-glucosidases 301, 304
glutaminyl-tRNA (Gln-tRNA) 208
   synthesis 207, 208–9, Plate 18.1
glyceraldehyde-3-phosphate (GAP) dehydrogenase
   classical (GAPDH) 248, **253**, 255–7
   non-phosphorylating (GAPN) **253**, 256, 257, 258
glyceraldehyde-3-phosphate (GAP) oxidoreductases, non-phosphorylating (GAPOR) **253**, 256
glycerate kinase (Gly-kin) **253**, 257
glycogen 249
glycolytic pathways 249–59
   Embden–Meyerhof–Parnas pathway 249–56
   Entner–Doudoroff pathway *250–1*, **253**, 256–7
   multiple-function proteins 248
   pentose phosphate pathway *250–1*, **254**, 257–8
   regulation 258–9
glycosylases, DNA repair 172–4
glycosyltransferases 72

gold extraction 314
GreA/GreB 192
GroES/L systems 88
growth phase/rate, transcriptional responses 240–1, Plate 21.1
Guttaviridae 59, 66–7
Guttavirus 66–7
gvp (gas vesicle) genes 111, 199, 200–1
   regulation of expression 201–4, 206
gyrB gene 130

H/ACA box ribonucleoproteins (RNPs) 229–30, 235–6, Plate 20.1
Haloarchaea *see* halophilic Archaea
Haloarchaeales 17–18, 19–20
Haloarcula marismortui 199
   genome 76, **77**, **79**, 90
   plasmids **106**, 111
   replicons 105, **106**
   ribosomes 220
Haloarcula sp. AS7094, plasmid **112**
Halobacteriaceae 199
Halobacterium
   cell cycle 160
   DNA replication 161, 163, 165
   lipids 6
   phylogenetic studies 17
   plasmids **106**, 111
   replicons 105, **106**
   selectable genetic markers 131
   transcription 187
   viruses 59
Halobacterium salinarum 199
   gas vesicle formation 200–4
   genome **91**, 94
   øH virus 59
   plasmids 106, **108**, **112**
   purple membrane synthesis 204–6
Halobacterium salinarum strain NRC-1 199
   DNA damage response 181
   genome **77**, **79**, 90
   histones 149
   horizontal gene transfer 31
   megaplasmid pNRC100 111
   phylogenetic analysis 19
   replicons **106**
halocins 316, **319**
Haloferax mediterranei 200
   biopolymer formation 317
   gas vesicle formation 200–4

Haloferax volcanii **91**
  genetic exchange system   127–8, 138
  plasmids   **106**, **108**, 111
  replicons   105, **106**
  reverse genetics   133–4
  selectable markers   130, 131
  small noncoding RNAs   234
  transformation   128–9
  vectors   132
halophilic Archaea   4–5
  biotechnological products   296, 306, 314, 316, 317
  waste decontamination   315
Halorubrum vacuolatum, gas vesicles   200, 201, 203
hdrA gene   131
health concerns, biomethylation of metal(loid)s   292–3
heat map analysis, phylogenetic congruence   33, Plate 3.1
heat shock response   195–7, 243–4, Plate 21.2
helicase(s)   164, 180
  superfamily 3 proteins, plasmid   109–10
  XPB and XPD   **176**, 177
helix–hairpin–helix (HhH) DNA glycosylases   **173**, 174
helix-stabilizing nucleoid proteins (HSNP)   151
helix–turn–helix proteins   194
HerA   **179**, 180
heterotrophs, sulfur metabolism   267
3-hexulose-6-phosphate isomerase (PHI)   **254**, 257–8
3-hexulose-6-phosphate synthase (HPS)   **254**, 257–8
high-throughput (HTP) technologies   245–6
histone(s)   147–51, 194
  DNA binding and complex formation   149–50
  regulation of gene expression   150–1, *152*
  sequences and structures   148–9, Plate 13.1
histone folds (HF)   147, 148–9, 150, Plate 13.1
Holliday junction-resolving enzymes   65, 72, 180
  Hjc   **179**, 180
  Hje   180

homologous recombination (HR)   141–4
  double strand break repair and   177, *179*, 180
  non-reciprocal   142, *143*
  reciprocal   142–4
horizontal (or lateral) gene transfer (HGT or LGT)   14, 29–30, 248
  within Archaea   25, 47, 86, 89
  between Archaea and bacteria   24, 30–2, 90
  detection methods   30
  genes not involved in   32–3, Plate 3.1
  informational genes   24–5
  mechanisms   127–8
  methanogens   21, 24
  numbers of genes involved   30–5
  organismal genealogy and   33–5
  role in archaeal history   24–5
  role of integration   113, 120–2
host–parasite association
  Nanoarchaeum equitans   51–2, 53, 57
hpt gene   135
HSNP proteins   151
Hsp20   196
Hsp70/DnaK system   195–6, 197
HSPN-C′   *see* Sso7d/Sac7d/Ssh7d/HSPN-C′ proteins
HTa protein   148
HU proteins   147, 148
hydrogen (H_2)
  oxidation   262, 265, 266
  production   315
hydrogenases   315
  Acidianus ambivalens   268–70
  NiFe   268–70, **319**
  Pyrodictium   267–8
hydrogen sulfide (H_2S)   261, 263, 266–7
  oxidation   273–4
hyperthermophiles
  DNA repair   140, 145
  industrial applications   295–6
  last archaeal common ancestor   22
  phylogenetic studies   17–18
  viruses   59–73

Hyperthermus butylicus   **91**
hypusine   226

Igniococcus   23, 51–2, 57, Plate 5.1
  genome   **91**
  lipids   55
  physiology   52
  ultrastructure   54, 55
initiation codons   222
initiation complex formation   189–90, Plate 16.2
initiation factors (IFs), translation   221, 223–6
initiator element (INR)   186
initiator proteins, plasmid replication   106, *107*
inorganic sulfur compounds   *see* sulfur compounds, inorganic
insertion sequence (IS) elements   **77–8**
  extreme halophiles   90
  mobility and regulation   97–8
  Sulfolobus   96, 139, 141
integrases (Ints)   113–16, 122–3
  pNOB8-type   99, 115–16
  SSV1-type   113–16
  XerC/D type   *see* XerC/D integrases/recombinases
integrated elements (IEs)   113, 116–20
  pNOB8-type   118, **119**, 122
  role in archaeal fitness   120–1
  SSV1-type   118, **119**, 121
  stability and evolution   121–2
integration mechanisms   113–23
inteins   **79**, 81–2, 86
  Sulfolobus   99–100
intestinal flora, biomethylation of metal(loid)s   293
introns
  Sulfolobus   99–100, *101*
  tRNA genes   **79**, 81, 86, 99–100, 234
IS elements   *see* insertion sequence (IS) elements
isoleucyl tRNA synthase   130
Isp1 protein   268

2-keto-3-deoxy(-6-phospho)gluconate (KD(P)G) aldolase   **253**, 256–7
2-keto-3-deoxygluconate (KDG) kinase   **253**, 256–7
Korarchaeota   18, 28, 40

K-turn motif 230, 236–8, Plate 20.1, Plate 20.2

L1 ribosomal protein operon 220
L7Ae protein 220
   C/D box RNPs 230–1, 233, Plate 20.1
   H/ACA box RNPs 235–6
   sense and antisense RNAs 237–8, Plate 20.3
   small RNAs binding 236–8, Plate 20.2
lacS gene 135, 144
lacZ gene 135
laminarinase 301–4
last archaeal common ancestor (LACA) 22
last universal common ancestor (LUCA) 22, 26–7, 29–30, 35–7
lateral gene transfer *see* horizontal gene transfer
Ligamenvirales 73
light repair pathway 175–6
lipids 6–7
   environmental studies 40
   industrial applications 315–16, **319**
liposomes 315, **319**
Lipothrixviridae 59, 60–3, 73
long clusters of tandem repeats (LCTRs) 86
Lrp proteins 195
Lrs14 195
LXa ribosomal protein 218, 220
lycopene β-cyclase **312**, 314

maltodextrin-binding proteins **318**
maltose binding protein 248
maltose-grown cells, transcriptional responses 242
mannosylglyceramide **318**
mannosylglycerate **318**, 320
marine planktonic Archaea 40, 41–3, **49**, Plate 4.2
MC1 156, Plate 13.4
MCM complex 164
MCM proteins 109–10, 164
mcr genes 275, 276–7, 280
Mdr1 195
megaplasmids 105, 111
membrane-bound dehydrogenase (MBH) 267
mercury, biomethylation 285, 286, 287, 290

metabolism, archaeal 248–9
   evolution 247–8
   inorganic sulfur compounds 261–74
   regulation 258–9
   see also glycolytic pathways
metabolomics 75
metagenomics 39, 42, 47–8, **49**, 93
metal(loid)s, biomethylation *see* biomethylation of metal(loid)s
metal sulfides (MeS and MeS$_2$) 263, **264**
metalloprotease **303**
meta-proteomics 48, 94
meta-transcriptomics 48, 94
methane
   anaerobic oxidation *see* anaerobic oxidation of methane
   formation *see* methanogenesis
methane monooxygenase, particulate 45
Methanobacteriales 18, 20, 21, 87
Methanobacterium bryantii 287, 289, 290
Methanocaldococcus jannaschii
   aminoacyl-tRNAs 207–8, 209–12
   DNA replication 165
   genome 75, **77**, **79**, 80, 87, Plate 7.1
   histones 150
   methyl-coenzyme M reductase 276–7
   MITEs 97
   plasmids **112**
Methanococcus maripaludis
   aminoacyl-tRNAs 211
   genome **77**, **79**, 87
   mutagenesis 126
   pURB500 plasmid 111, **112**
   transformation 129
   vectors 132
Methanococcus voltae **91**
   pMip1 plasmid 133
   transformation 128, 129
methanogenesis 275, *276*, 277–8
   genes 87
   origin 21–2, 41
   reverse 44, 89, 278–9
methanogens 3
   eukaryote-associated 46–7
   group I 87, 88
   group II 87, 88
   horizontal gene transfer 21, 24
   lipid analysis 7

   methylation of metal(loids) 285–94
   methyl-coenzyme M reductase 275, 276–9
   pyrrolysine 213–15
   reporter genes 135
   vectors 132
   viruses 59
Methanopyrus kandleri
   genome **77**, **79**, 80, 88
   histones 149
   methyl-coenzyme M reductase 276–7
   phylogenetic position 17–18, 20–1
Methanosarcinales 21, 87, 88
Methanosarcina acetivorans
   genome 76, **77**, **79**, 88
   mutagenesis 127
   pC2A plasmid 111, **112**
Methanosarcina barkeri
   genome **77**, **79**, 88
   methylation of metal(loid)s **288**, 291–2
   pyrrolysine 213–15
Methanosarcina mazei
   genome **77**, **79**, 88
   methylation of metal(loid)s **288**
Methanosarcina thermophila
   genome **91**
   MC1 protein 156, Plate 13.4
Methanosphaera stadtmanae
   genome **91**
   methylation of metal(loid)s **288**, 293
Methanothermobacter marburgensis, methyl-coenzyme M reductase 276–7, 279, Plate 24.1
Methanothermobacter thermautotrophicus (formerly *Methanobacterium thermoautotrophicum*)
   aminoacyl-tRNA synthesis 208–9, Plate 18.1
   biocatalysts 309
   cell cycle 160
   DNA repair 180
   DNA replication 163, 164, 165
   genome **77**, **79**, 87
   histones 148
   methylation of metal(loid)s 287, **288**
Methanothermus fervidus
   genome **91**
   histones 148, 150–1

methanotrophic Archaea 21, 22
    association with sulfate reducers
        279–80
    in marine sediments 43–4
    methyl-coenzyme M reductase
        275, 279–82, Plate 24.1
methylation
    C/D box RNP-guided 231–4
    chromatin proteins 155, 156
    intron excision and 234
    metal(loid)s see biomethylation of
        metal(loid)s
methylcobalamin 289, 291
methyl-coenzyme M reductase
    (MCR) 275–83
    methanogenic Archaea 276–9,
        Plate 24.1
    methanotrophic Archaea 44,
        279–82, Plate 24.1
O6-methylguanine-DNA
    methyltransferase 309
methylmercury 285, 286, 287, 290
methylmetal(loid)s 285
    analytical techniques 286–7
    see also biomethylation
2-methylsuccinate 283
methyltransferases 72, 233, 234,
    312
met-tRNAi 222, 223–4, 225
mevinolin 131
microarray analysis
    to assess species relationships 244
    DNA damage response 181
    glycolytic pathway 259
    Pyrococcus furiosus 240–4
microbial mats 43, Plate 4.3
    anaerobic oxidation of methane
        275, 280, 281
    metagenomics studies 44, **49**
microhalocins 316
Microsporidia 18
mineral biomining 314–15
miniature inverted repeat
    transposable elements (MITEs)
    96–7, 98
minimal efficient processing segment
    (MEPS) 142
mismatch repair (MMR) 140, 145,
    174–5
MMF proteins 21–2
monomethylmercury 285, 290
moonlighting proteins 92, 248
Mre11 178–80
mRNA 221–2

K-turn motifs 237–8, Plate 20.3
    leaderless 221–3
    ribosome interaction 222–3
MtbB 213, *214*, 215
mth810 180
MtmB 213, *214*, 215
MttB 213, *214*, 215
mutagenesis
    insertional 126
    random 126–7
    site-specific chromosomal 133–5
    transposon 126–7
mutational analysis, complex
    processes 144–5
MutS/MutL homologs 174, **175**
Myoviridae 59

Nanoarchaeota 51–7, 87
    diversity 40, 55
    phylogeny 23, 28, 55–7
    see also *Nanoarchaeum equitans*
Nanoarchaeum equitans 51–7, Plate
    5.1
    discovery and cultivation 51–2
    DNA repair 176, 181
    genome and genes 52–3, 76, **77**,
        79, 86–7
    histones 149
    lipids 55
    phylogeny 18, 20, 22–4, 55–7, 87
New York Air metagenomics project
    93
Nhp2 230, 235
nickel cofactor, methyl-coenzyme M
    reductase 44, 275, 276, 277–8
    methanotrophic Archaea 280–1
    redox state 277
NiFe hydrogenases 268–70, **319**
nitrilase **312**, 313
nitrile-degrading enzymes 313
nitrite reductase (nirK) 45, Plate
    4.4
nitrogenases 88
Nitrosopulminus 42
non-orthologous gene displacements
    248, 252, Plate 22.1
Nop10 230, 235–6, Plate 20.1
Nop56/58 (Nop5) 230, 231, Plate
    20.1
novobiocin 130
NrpR 195
nucleoside-diphosphate-sugar
    epimerase 72
nucleosomes 147, 149–50

regulation of gene expression
    150–1, *152*
transcription through 151, *152*,
    194
nucleotide excision repair
    (NER) 145, 176–8
    proteins **175**, 176–7
NurA **179**, 180

octomethylcyclotetrasiloxane
    (OMCTS) 291
Okazaki fragments 165
Orcl/Cdc6 162–3, 164
ORFTPX 104, Plate 8.3
organometal(loid)s see
    methylmetal(loid)s
origin recognition boxes (ORBs)
    163
origin recognition complex (ORC)
    162–3
origins of replication 161–3
orphan repair proteins (ORPs) 181
orthologous genes
    crenarchaeal viruses 72–3, Plate
        6.1
    resistant to lateral transfer 32–3
8-oxoguanine DNA glycosylase
    (AGOG) 173–4

parasitic association 51–2, 53, 57
pARN3 98, **110**
pARN4 98, 99, **110**
pathogens 47, 293
PCNA 164, 166–9
    DNA repair and 177, 180
pentose phosphate pathway (PPP)
    250–1, **254**, 257–8
peptide(s)
    biotechnological uses 316–17,
        319
    grown cells, transcriptional
        responses 242
peptidyl prolyl *cis-trans* isomerase
    (PPIase) 316, **318**
pGRB1 106, **108**
pGT5 106, *107*, **108**
pHGN1 106
øH (phiH) virus 59
phosphatases **308**, 309
phosphatidylethanolamine
    N-methyltransferase 314
phosphoenolpyruvate synthetase
    (PEPS) 252, **253**, 256, 259
phosphofructokinases (PFK) 252–5

6-phosphogluconate dehydrogenase (PG-DHDC) **254**, 257
6-phosphogluconate kinase (PG-hydr) **253**
6-phosphogluconolactonase (6PG-lac) **253**, **254**
phosphoglucose isomerase (PGI) **253**, 255, Plate 22.2
phosphoglycerate kinase (PGK) **253**, 255–6
phosphoglycerate mutase (PGM) **253**, 255
phosphopentomutase 258
phosphoserine *210*
O-phosphoserine sulfhydrolases *263*
phosphoseryl-tRNA:Cys-tRNA synthase (SepCysS) 210–11, 212
phosphoseryl-tRNA$^{Ser(Sec)}$ kinase (PSTK) *210*, 212
phosphoseryl-tRNA synthetase (SepRS) 210–11, 212
photolyase 175–6
photoproducts, DNA 175–6
pHSB2 106, **108**
pHVE14 98, **110**
phytane 55
phytoene synthase homolog 204, 205
Picrophilus oshimae, horizontal gene transfer 25
Picrophilus torridus **78**, **79**, 89
pING1 98, 102, **110**
pKEF9 98, **101**, **110**
plasmids **49**, 105–12
 conjugative 98–9, 110, 138–9
 genomic variation 102
 integrases 118
 integration mechanisms 115–16
 poorly characterized replication proteins 111–12
 rolling circle replication 106, *107*, **108**
 Sulfolobus see under Sulfolobus
 vectors 131–3, 144
plasmid–virus hybrid, pSSVx 108, 109
plrA gene 110
pNOB8 98–9, 102, **110**
 integrase 99, 115–16
 SRSRs **101**, 102
pNRC100 111
poly(β-hydroxy butyric acid) 317, **319**

poly(γ-D-glutamic acid) 317, **319**
polysulfide reductase (PSR) *263*, 267–8
polysulfides 263, **264**
polythionates 263, **264**, 265
pop-in/pop-out method, gene knockout 133–4
 counter-selectable 134–5
PP$_i$-dependent phosphofructokinase (PP$_i$PFK) **253**, 255
pps gene 256, 259
pre-rRNA processing complex 236–7
prim/pol domain, plasmid proteins 109, 110
primase 164–5
pRN type 108–10
pRN1 105, 108–9, 110
Prochlorococcus 42
progenote 10, 26–7, 35
programmed frameshifting 81
proline-tRNA synthetase (ProRS) 211
promoters
 AT content 80
 bop cluster genes 205–6
 controllable 135–6
 escape 190
 gvp genes 201–3
 structure 185–6
proteasome 243
proteins
 biotechnological uses 316–17, **319**
 cysteine incorporation 209–12
 high expression systems 135–6
 "low-hanging fruit" 245, 246
 structure determination 245–6
 synthesis 217–28
proteolytic enzymes **303**, 306
proteomics 75, 94, 244
proviruses 116–17
pRT1 *107*, **108**
pseudomonic acid 130
pseudouridine modification 235–6
pSOG 98
PSV (*Pyrobaculum* spherical virus) 67–8
pTAU4 **108**, 109–10
Ptr2 188, 195, 204
pullulan hydrolases **298**, 300
pullulanases
 type I 299
 type II **298**, 299–300

purple membrane 199, 204
 gene cluster *see bop* gene cluster
 regulation of synthesis 204–6
pylT 213
pyrE/pyrF genes 98, 129, 131, 134, 139–40
Pyrobaculum
 dissimilatory sulfate and sulfite reduction 265–6
 viruses 60, 67–8
Pyrobaculum aerophilum
 DNA repair 172–3
 genome **77**, **79**, 85, 86
Pyrobaculum spherical virus (PSV) 67–8
Pyrococcus abyssi
 DNA replication 165
 genome **77**, **79**, 80, 86, Plate 7.1
 plasmids **108**
 transformation 129
 vectors 132–3
Pyrococcus furiosus 239–46
 anaerobic sulfur metabolism 267
 biocatalysts 300, 301–4, 305–6, 309
 compatible solutes 317–20
 DNA damage response 181
 genome **77**, **79**, 86, 92, 94, 239, 240
 glycolytic pathway 249–52, 255–6, 259
 heat shock response 195–7, 243–4
 horizontal gene transfer 25
 PCNA 167, 168
 proteomics 244
 RNA polymerase 188
 small noncoding RNAs 234–5
 structural genomics 245–6
 transcriptomics 240–4, Plate 21.1, Plate 21.2
Pyrococcus horikoshii
 biocatalysts 304
 genome **78**, **79**, 80, 86, Plate 7.1
 primase 165
 small noncoding RNAs *232*, 233
Pyrococcus woesei 187, 244
Pyrodictium, sulfur reductase and hydrogenase 267–8
Pyrolobus fumarii **78**
pyrophosphatase, inorganic 307–9
pyrrolysine 213–15
 containing genes **79**, 81, 88
pyrrolysyl-tRNA (Pyl-tRNA) 213

pyrrolysyl-tRNA synthetase (PylRS) 213–15
pyruvate kinase (PYK) 249–52, **253**, 255
pyruvate phosphate dikinase (PPDK) 252, **253**

Rad50 178–80
RadA/Rad51 43, 178–80
radiation, ionizing 181
RapA 194
recombinases
  phylogeny 115, *116*
  tyrosine 113–16
  XerC/D type *see* XerC/D integrases/recombinases
  see also integrases
recombination
  genome engineering by 144
  homologous *see* homologous recombination
Rep75 protein 106
repair associated mysterious proteins (RAMPs) 181
repH replication proteins 111
replication proteins **160**, 161–9
  pRN-type plasmids 108–10
replicons
  halophilic archaea 105, **106**
reporter genes 135
restriction-modification (R-M) system 138
reverse genetics 133–5
reverse gyrase
  Alba proteins and 153
  extreme and hyperthermophiles 85
  Nanoarchaeum equitans 53
RF1/RF2 227
RF3 227
RFC 167, 168–9
ribbon–helix–helix (RHH) domain proteins 72
ribonucleoproteins (RNPs) 229–30
  C/D box 229, 230–5, Plate 20.1
  H/ACA box 229–30, 235–6, Plate 20.1
  K-turn motif and L7Ae binding 236–8, Plate 20.2
  small nucleolar (snoRNPs) 229–30, Plate 20.1

ribose-5-phosphate 249, 257
ribose-5-phosphate isomerase (R5P iso) **254**
ribosomal proteins (r-proteins) 218–20
  gene clusters *36*, 37, 80, 219–20
  horizontal gene transfer 25
  phylogenetic analysis 19–20, 23, Plate 2.1
ribosomal RNA *see* rRNA
ribosomes 218–20
  mRNA interaction 222–3
  structure 220
ribulose-5-phosphate *252*
ribulose-5-phosphate 3-epimerase (RP-epi) **254**
ribulose monophosphate pathway (RuMP) **254**, 257–8
  reverse 257, 258
Rice Cluster I (RCI) Archaea 45–6, **49**
RNA
  7S 237
  Alba protein binding 153–4
  antisense 98, 237–8
  C/D box 230
  K-turn motif and L7Ae binding 236–8, Plate 20.2
  modification, crenarchaeal viruses 72
  processing 99–100
  sense 237–8
  small noncoding 229–38
  see also mRNA; rRNA; tRNA
RNA cleavage enzyme 99
RNA genomes 26–7, 37
RNA polymerases (RNAP) 8, 188–9
  elongation complex binding 190–1
  initiation of transcription 189–90, Plate 16.2
  phylogenetic analysis 18, 20, 23, Plate 2.1, Plate 2.2
  regulation by MC1 156
  RpoM subunit 181
  TFS-induced cleavage and 192, *193*
  transcription through nucleosomes 151, *152*, 194
RNAses 151, 156
RNPs *see* ribonucleoproteins
rolling circle replication 106, *107*

r-proteins *see* ribosomal proteins
RRF 227
rRNA 2, 218
  environmental surveys 40, 41, 42
  genes **79**, 84, 218
  introns 81, 99
  oligonucleotide cataloging 2, 3–4, 7–8, 17
  phylogenetic analysis 17–18, 20, 40, Plate 4.1
rubber recycling 314–15
Rudiviridae 59, 63–5, 73
Rudivirus 63–5, 73

S10 operon *36*, 37
Sac7d *see* Sso7d/Sac7d/Ssh7d/HSPN-C′ proteins
Sac8/Sso8 proteins 151
Sac10a/Sso10a proteins 151, 154–5
Sac10b 137, 151
  see also DBNP-B/SSo10b/Alba proteins
Sanger patterns 1, 3–4
Sargasso Sea 42, 43, 45, 48, **49**, 93, Plate 4.4
SEED system 92
SelB 212, 213
selectable markers 128, 130–1
  gene knockouts 133, *134*, 134–5
selenium (Se), biomethylation **288**, 289–90
selenocysteine (Sec) *210*, 212–13
  containing genes **79**, 81, 87, 88
selenocysteine insertion sequence (SECIS) *210*, 212, 213
selenocysteine synthetase (SelA) 212
selenoproteins 212
Sep *see* phosphoserine
serine proteases **303**, 306
seryl-tRNA (Ser-tRNA) *210*
seryl-tRNA synthetase (SerRS) 212
Shine–Dalgarno (SD) motifs 221, 222
short non-messenger RNAs (snos) **79**, 85
short regularly spaced direct repeats (SRSRs) **77–8**, 85–6
Sulfolobus 100–2
shuttle vectors 106, 132–3
SIFV 61–2, 72, 73
signal recognition particle 237

single-stranded DNA binding (SSB) protein   175
SipAB   242, 267
Siphoviridae   59
siroheme   266
SIRV1   63–5, 105
   genes   72
   genome variation   102, 103–4, Plate 8.3
SIRV2   63–5, 72, 103
S-layer   137, *138*
S-layer proteins   316, **319**
SM1   46
small noncoding RNAs   229–38
small nucleolar ribonucleoproteins (snoRNPs)   229, 230–6
SNDV   66–7
soils   40, 41, **49**
   ammonia-oxidizing Crenarchaea   44–5, Plate 4.4
   rice fields   45–6
solfataras   262
solutes, compatible   317–20
Sox complex   *263*, 270
spc operon   36, 37
spheroplast transformation   128, 129
sponges   42–3, 47
*sre*ABCDE gene cluster   268–9
SRSRs   *see* short regularly spaced direct repeats
Sso7d/Sac7d/Ssh7d/HSPN-C′ (Sul7d) proteins   137, 151, 155–6, Plate 13.3
Sso8/Sac8 proteins   151
Sso10a/Sac10a proteins   151, 154–5
Sso10b proteins   *see* DBNP-B/SSo10b/Alba proteins
SSV1 (*Sulfolobus* spindle-shaped virus 1)   65–6, 72, 102
   helper virus function   108
   integrase   113–16, *117*
   vector construction   133, 144
SSV2 (*Sulfolobus* spindle-shaped virus 2)   65–6, 108
SSV-K1 (*Sulfolobus* spindle-shaped virus, Kamchatka 1)   65–6
SSV-Y1 (*Sulfolobus* spindle-shaped virus, Yellowstone 1)   65–6
starch-processing enzymes   296–301
STIV (*Sulfolobus* turreted icosahedral virus)   69–70, *71*
structural genomics (SG)   75, 245–6
STSV1 (*Sulfolobus tengchongensis* spindle-shaped virus)   71, 72

Stygioglobus   8
SUI1   **224**, 226
Sul7d proteins   *see* Sso7d/Sac7d/Ssh7d/HSPN-C′ proteins
sulfate   261
   anaerobic oxidation of methane and   279, 281–2
   dissimilatory reduction   265–6
   inorganic chemistry   263, **264**
   reducers   89, 279–80, 282–3
sulfide:cytochrome *c* oxidoreductase   *263*
sulfide:quinone oxidoreductase (SQR)   *263*, 273
sulfides (HS⁻)   263
sulfite
   dissimilatory reduction   265–6
   inorganic chemistry   263, **264**, 265
   methyl-coenzyme M reductase inhibition   282
   oxidation   270, 273–4
sulfite:acceptor oxidoreductases (SAOR)   *263*, *271*, 274
sulfite:cytochrome *c* oxidoreductase   270
sulfite oxidase   *263*
sulfite reductases, dissimilatory siroheme-containing (DSRs)   *263*, 265, 266
Sulfolobales
   genomes   79–80, 85–6
   horizontal gene transfer   25
   plasmids   105
sulfolobicin   **319**
Sulfolobus acidocaldarius   95
   chromosome variation   96, Plate 8.1, Plate 8.2
   DNA repair   180
   DNA replication   165
   genetic exchange system   98–9, 128, 138–9
   genetics   137–45
   genome   **78**, **79**, 85–6, 144
   introns   100, *101*
   IS elements/MITEs   96, 97, 139
   lipids   6
   RNA polymerase   8, 95
   small noncoding RNAs   231, *232*, 233–4
   SRSRs   **101**, 102
Sulfolobus islandicus   95
   genome   **91**
   IS elements   **96**

   MITEs   97, 98
   plasmids   105, 108, **110**
   SRSRs   **101**, 102
   viruses   62, 63–4, 66
Sulfolobus islandicus filamentous virus   *see* SIFV
Sulfolobus islandicus rod-shaped viruses   *see* SIRV1; SIRV2
Sulfolobus neozealandicus, plasmids   **108**, 109–10
Sulfolobus neozealandicus droplet-shaped virus (SNDV)   66–7
Sulfolobus solfataricus   95, 137
   biocatalysts   299, 304, 310–13
   chromosome variation   96, Plate 8.1, Plate 8.2
   DNA repair   176, 177, 180, 181
   DNA replication   161–2, 164, 165, 167
   genome   **78**, **79**, 85
   glycolytic pathway   257, 258–9
   high expression systems   136
   horizontal gene transfer   25
   integrated elements   **119**, 121–2
   introns   100, *101*
   IS elements   96, 97–8
   mechanisms of genome plasticity   97–8
   MITEs   97
   plasmids   **108**, 109
   small noncoding RNAs   230, 234, 236–8, Plate 20.2
   SRSRs   101, 102
   vectors   133
   viruses   66, 70
Sulfolobus spindle-shaped viruses   *see* SSV1; SSV2
Sulfolobus tengchongensis spindle-shaped virus (STSV1)   71, 72
Sulfolobus tokodaii   95, 137
   chromosome variation   96, Plate 8.1, Plate 8.2
   genome   **78**, **79**, 85
   integrated elements   118, **119**, 122
   introns   100, *101*
   IS elements   **96**
   MITEs   96–7, 97
   SRSRs   100–1, 102
Sulfolobus turreted icosahedral virus (STIV)   69–70, *71*
2-sulfotrehalose   **318**, 320
sulfur, elemental (S°)
   anaerobic reduction   266–70

chemistry 263–5
geochemistry 261–3, Plate 23.1
inorganic chemistry 263–4
oxidation 270–3
transcriptional response 241–2
sulfur compounds, inorganic (ISCs) 261–74
 biologically relevant **264**
 chemistry 263–5
 geochemistry and physiology 261–3
 oxidation 270–4
sulfur cycle, biological *263*
sulfur-dependent Archaea 261
 anaerobic sulfur reduction 266–70
 dissimilatory sulfate and sulfite reduction 265–6
 electron donors, acceptors and products **262**
 non-thermophilic 262–3
 physiology 261–3
 sulfur oxidation 270–4
sulfur dioxide (SO_2) 261
sulfur oxygenase *263*, 270, **272**
sulfur oxygenase reductase (SOR) *263*, 270–3, Plate 23.3
sulfur reductases (SR) *263*, 267–70
sulfur trioxide (SO_3) 261
supermatrix approach 19–20
symbiotic associations 47
syntheny 80

TATA box 185, 186, 189
TATA-box binding protein (TBP) 90, 186–7, Plate 16.1
 initiation complex formation 189, Plate 16.2
 multiple copies 187–8
 TFE interaction 188
TBP-associated factors (TAFs) 186
TBP-interaction proteins (TIPs) 186–7
tellurium (Te), biomethylation **288**, 289–90
TenA 241
termination factors, translation 227
tetrathionate 274
tetrathionate hydrolase *263*, 270
tetrathionate reductase *263*
tetrathionate synthase 274
TFB 90, 187–8
 initiation complex binding 189, Plate 16.2

promoter escape 190
TFE 187, 188
TFS 192, *193*
thermoacidophiles 8
 horizontal gene transfer 25
 industrial applications 295–6
 see also thermophiles
Thermococcales 23, 86, 105
Thermococcus
 anaerobic sulfur metabolism 267
 biocatalysts 300, 305, 306, 307, 310
 plasmids **108**
Thermococcus kodakarensis
 genome **78**, **79**, 86
 glycolytic pathway 252–5, 256, 259
 integrated elements **119**
 reverse gyrase 22
 transformation 129
Thermococcus litoralis 25, 86
thermophiles 41, 295–6
 biocatalysts 296–301
 genomes 85
 lipids 6–7
 transformation 129
 see also hyperthermophiles; thermoacidophiles
Thermoplasma acidophilum
 genome **78**, **79**, 89
 horizontal gene transfer 25
 inorganic pyrophosphatase (TAP) 309
 lipids 6
Thermoplasma volcanium **78**, **79**, 89
Thermoproteus, viruses 60–1, 67–8
Thermoproteus tenax
 genome **91**
 glycolytic pathway 252, 256, 257, 258–9
 viruses 60–1, 67–8
thermosome 196
thiamine pyrophosphate biosynthesis 241
thiol protease **303**, 306
thiols, organic **264**
thiostrepton 130–1
thiosulfate 263, **264**, 265
 oxidation 270, 273–4
 production 270
thiosulfate: acceptor oxidoreductase *263*, 274
thiosulfate oxidoreductase 270

thiosulfate: quinone oxidoreductase (TQO) *271*, 273–4
thiosulfate reductase *263*
Thiotrix 46
three domains (of life)
 concept 19, 25–6, 29
 lateral gene transfer and 29–37
 origin 26–7
three viruses three domains theory 27
thymidylate synthase, flavin-dependent (ThyX) 72
TKV4 118
topoisomerases 309, **318**
transcription 185–97
 activators 195
 cleavage induction factor TFS 192, *193*
 elongation complex 190–1
 initiation complex 189–90, Plate 16.2
 promoter escape 190
 regulation 151, *152*, 156, 194–7, 199–206, 259
 repressors 195
 termination 192–4
transcription factor IIB homolog *see* TFB
transcription factor IIE homolog (TFE) 187, 188
transcription factor IIH (TFIIH) 190
transcription factors 186–8, 192
transcriptomics 75
 Pyrococcus furiosus 240–4
transketolase **254**, 258
translation 217–28
 elongation and termination 227
 initiation 222–6
 initiation factors (IFs) 221, 223–6
 mechanism 222–7
 regulation 259
transposases 96
transposon related proteins 238, Plate 20.3
tree of life 2
 Darwin's concept 29
 hypothesis 37
 lateral gene transfer and 29–37
 position of Archaea 25–6
 root 27, 29, 35–7; *see also* last universal common ancestor
 three-domain concept 19, 25–6, 29
 whole genome based approaches 18–19

trehalose 301, **318**
trehalose synthase **298**
trehalosyl dextrin-forming enzyme **298**
trimethoprim 131
trimethylarsine 285, 287
trimethylbismuth ((CH$_3$)$_3$Bi) 285, 290–4
triose-phosphate isomerase (TIM) 249, **253**
trithionate hydrolase 270
TrmB 195, 259
tRNA
   dependent formation of amide aminoacyl-tRNAs 208
   initiator (met-tRNAi) 222, 223–4, 225
   methylation 233–4
tRNA genes
   introns **79**, 81, 86, 99–100, 234
   mechanisms of integration into 118–20
tRNA^{Pyl} 215
tRNA-ribosyltransferase 72
tRNA^{Trp} gene intron 234
TTSV1 (*Thermoproteus tenax* spherical virus 1) 67–8, 72

TTV1 (*Thermoproteus tenax* virus 1) 60–1
   genome variation 102, 104, Plate 8.3
tyrosine recombinases 113–16

über-operon 81
uncultivated Archaea
   diversity 39–50
   habitats occupied 40–7
   metagenomic studies 47–8, **49**
   phylogenetic analysis 40–1, Plate 4.1
uracil auxotrophs 131, 134, 139–40
uracil DNA glycosylase (UDG) 172–3, 174
uracil scanning polymerases 174
urkingdom concept 4–5, 6–7
UV radiation 126, 141–2, 175–6, 181
UvrABC homologs 174–5, 176

vac regions 200–1
vectors 131–3, 144
   expression 135–6
   shuttle 106, 132–3
viruses 27
   genome variation 102–4, Plate 8.3

hyperthermophilic Crenarchaea 59–73, Plate 6.1
integrated *see* proviruses
integration mechanisms 113–16
integrative elements 118
virus–plasmid hybrid (pSSVx) 108, 109

waste treatment 315
whole-cell biocatalysis 314–15, **319**
winged helix–turn–helix proteins 195

XerC/D integrases/recombinases 72
   DNA replication 162
   integration mechanisms 111, **112**, 115
XPB **175**, 177
XPB1 gene 181
XPD **175**, 177
XPF endonuclease **175**, 177–8
X-ray crystallography 245
xylan-degrading enzymes **303**, 304–5

YddF 72

zinc (Zn) finger proteins 42, 72
zinc ribbon motif 187